T0180702

# Geophysical Fluid Dynamics

**Springer**
*New York*
*Berlin*
*Heidelberg*
*Barcelona*
*Hong Kong*
*London*
*Milan*
*Paris*
*Singapore*
*Tokyo*

Joseph Pedlosky

# Geophysical Fluid Dynamics

Second Edition

With 198 Illustrations

 Springer

Joseph Pedlosky
Woods Hole Oceanographic Institution
Woods Hole, Massachusetts 02543
U.S.A.

Springer study edition based on second edition of original hard cover (Springer-Verlag New York, 1979, 1987).

Library of Congress Cataloging in Publication Data
Pedlosky, Joseph
  Geophysical fluid dynamics.
  Bibliography: p.
  Includes index.
  1.  Fluid dynamics     2.  Geophysics.     I.  Title.
QC809.F5P43   1986      551      86-13941

© 1979, 1982, 1987 by Springer-Verlag New York Inc.
All rights reserved. This work may not be translated or copied in whole or in part without the written permission of the publisher (Springer-Verlag., 175 Fifth Avenue, New York, New York 10010, U.S.A.), except for brief excerpts in connection with reviews or scholarly analysis. Use in connection with any form of information storage and retrieval, electronic adaptation, computer software, or by similar or dissimilar methodology now known or hereafter developed is forbidden.

Printed and bound by Sheridan Books, Inc., Ann Arbor, MI.
Printed in the United States of America.

9 8 7

ISBN 0-387-96387-1
ISBN 3-540-96387-1          SPIN 10836364

Springer-Verlag  New York Berlin Heidelberg
*A member of BertelsmannSpringer Science+Business Media GmbH*

# Preface to the Second Edition

I have taken the opportunity afforded by the appearance of a second edition of *Geophysical Fluid Dynamics* to make a number of revisions and additions to the text. Since the purpose of the book is largely pedagogic, the changes are intended to strengthen the treatment of the fundamentals of the subject rather than bringing up to the current moment the results of ongoing research in the field as would be appropriate for a research monograph. Chapters 3 and 4 now contain a discussion of the elements of geostrophic turbulence whose treatment was lacking in the first edition. In Chapter 5, I have substantially replaced Section 5.13 with a new discussion of the role of topography in the wind-driven circulation with an example of its influence much more striking than earlier. Chapter 6 is the most extensively revised section of the book. Sections on wave-mean flow interaction and the theory of the thermocline have been completely rewritten reflecting the increase in understanding of these fundamental subjects since the preparation of the first edition. Chapter 6 also contains a new derivation of the planetary and synoptic scale geostrophic potential vorticity equations. The new derivation supplements the more traditional derivation of the first edition and illuminates the relationship between the two regimes. In Chapter 7, I have reworked the section on the classical Charney problem in a way which should considerably simplify and illuminate the essential aspects of the problem. The discussion of the finite amplitude problem has been completely reworked to offer the reader an introduction to the role of dissipation in the nonlinear dynamics and the appearance of limit cycle and chaotic behavior. Finally, Chapter 7 has a new section describing some useful theorems on the instability of nonparallel flows.

I am indebted to several correspondents for pointing out errors in the first edition which are corrected herein. The preparation of this second edition was done during a sabbatical year in Venice, Italy.

I gratefully acknowledge the continuing support of the Woods Hole Oceanographic Institution during the period of the sabbatical. I wish also to express my gratitude to the staff of the Istituto per lo Studio della Dinamica delle Grandi Masse for their warm hospitality during the period of the preparation of the revisions for this book. To paraphrase Henry James on the subject, to be able to work and live in Venice led to perhaps the greatest state of happiness consistent with the preservation of reason.

*Woods Hole*                                                         Joseph Pedlosky
*September 1986*

# Preface to the First Edition

The content of this book is based, largely, on the core curriculum in geophysical fluid dynamics which I and my colleagues in the Department of Geophysical Sciences at The University of Chicago have taught for the past decade. Our purpose in developing a core curriculum was to provide to advanced undergraduates and entering graduate students a coherent and systematic introduction to the theory of geophysical fluid dynamics. The curriculum and the outline of this book were devised to form a sequence of courses of roughly one and a half academic years (five academic quarters) in length. The goal of the sequence is to help the student rapidly advance to the point where independent study and research are practical expectations. It quickly became apparent that several topics (e.g., some aspects of potential theory) usually thought of as forming the foundations of a fluid-dynamics curriculum were merely classical rather than essential and could be, however sadly, dispensed with for our purposes. At the same time, the diversity of interests of our students is so great that no curriculum can truly be exhaustive in such a curriculum period. It seems to me that the best that can be achieved as a compromise is a systematic introduction to some important segment of the total scope of geophysical fluid dynamics which is illustrative of its most fruitful methods. The focus of this book is thus the application of fluid mechanics to the dynamics of large-scale flows in the oceans and the atmosphere. The overall viewpoint taken is a theoretical, unified approach to the study of both the atmosphere and the oceans.

One of the key features of geophysical fluid dynamics is the need to combine approximate forms of the basic fluid-dynamical equations of motion with careful and precise analysis. The approximations are required to make any progress possible, while precision is demanded to make the progress meaningful. This combination is often the most elusive feature for the beginning student to appreciate. Therefore, much of the discussion of this book is directed towards the development of the basic notions of scaling and the subsequent derivation of systematic approximations to the equations of motion. The union of physical and intuitive reasoning with mathematical analysis forms the central theme. The ideas of geostrophic scaling, for example, are repeated several times, in various contexts, to illustrate the ideas by example.

The development of physical intuition is always a slow process for the beginner, and the book has a structure which aims to ease that important process. Chapters 1 and 2 discuss certain elementary but fundamental ideas in general terms before the complexities of scaling are required. In Chapter 3 the inviscid dynamics of a homogeneous fluid is discussed in order to expose, in the simplest context, the nature of quasigeostrophic motion. It has been my experience that the absence of the complexities necessarily associated with density stratification is a great help in penetrating quickly to rather basic concepts of potential vorticity dynamics. Rossby waves, inertial boundary currents, the $\beta$-plane, energy propagation, and wave interaction, etc. are all topics whose first treatment is clearer and simpler for the fluid of constant density. Similarly, Chapter 4 describes some of the simple ideas of the influence of friction on large-scale flows in the context of a homogeneous fluid. The vexing problem of turbulence receives short shrift here. Only the simplest model of turbulent mixing is formulated. It is my view that, unsatisfactory as such a model is as a theory of turbulence, it is sufficient for the purposes to which it is generally applied in the theory of large-scale flows. Chapter 5 serves to exemplify the use of the homogeneous model in the discussion of a problem of major geophysical interest, i.e., the wind-driven ocean circulation.

Chapter 6 has two main purposes. First is the systematic development of the quasigeostrophic dynamics of a stratified fluid for flow on a sphere. Careful attention is given to the development of the $\beta$-plane model on logical and straightforward lines. I believe many of the elements of the derivation have been hitherto unfortunately obscure. The second major goal is the application of quasigeostrophic dynamics to a few problems which I feel are central to both meteorology and oceanography and whose outlines, at least, should be familiar to the serious student.

Chapter 7 is reserved for instability theory. Since the publication of the pioneering papers of Charney and Eady, instability theory has held a central position in the conceptual foundation of dynamic meteorology. Recent advances in oceanography suggest a significant role for instability theory also in oceanic dynamics. Baroclinic and barotropic instability are both discussed in Chapter 7, not exhaustively, but to the degree I feel is necessary to provide a clear picture of the basic issues. The final chapter discusses certain topics, not easily grouped into the broad categories of earlier chapters, and chosen primarily to illustrate the way in which the ideas previously developed can be extended by similar methods.

The task of writing a text is made especially difficult by the evident impossibility of being truly comprehensive. The limitations of size make it necessary to omit topics of interest. To begin with, certain introductory aspects of fluid mechanics, such as the derivation of the Navier–Stokes equations (which is essential to a core curriculum) are deleted. Such topics may be found already in such excellent texts as Batchelor's *Fluid Dynamics* or Sommerfeld's *Mechanics of Deformable Bodies*. In other cases, when confronted by difficult choices, I have tried to include material which illustrates principles of general utility in fluid mechanics, e.g., boundary-layer concepts and the application of multiple-time-scale ideas to nonlinear problems. In this way I believe that the problems of geophysical fluid dynamics serve additionally as an excellent vehicle for the teaching of broader dynamical concepts. For example, the relationship between group velocity and phase speed in the Rossby wave is discussed at length in Chapter 3. There is, perhaps, no more dramatic example of the distinction between the two concepts in all fluid dynamics, and it can serve as a useful example of such a distinction for students of varying fluid-dynamical interests.

Naturally, in many cases I have chosen topics for discussion on the basis of my own interest and judgement. To that extent the text is a personal expression of my view of the subject.

It was my happy good fortune as a student to have had a series of marvelous teachers of fluid dynamics. Each in their own way made the subject vivid and beautiful to me. By now, no doubt, many of their ideas and attitudes are so intimately mixed into my own view that they appear implicitly here to the benefit of the text.

It is a pleasure, however, to explicitly acknowledge the singular influence of my teacher and colleague, Professor Jule Charney. His prodigious contributions to the study of the dynamics of the atmosphere and oceans as well as his example of scholarly integrity have been a continuing source of inspiration.

This book was largely written during a sabbatical year made possible by a fellowship from the John Simon Guggenheim Foundation, as well as by the continued support of the University of Chicago. The Woods Hole Oceanographic Institution kindly provided an office for me for the year and their warm hospitality considerably eased the task of writing and preparing the original manuscript. Special thanks are due to the students of the M.I.T.–Woods Hole joint program in physical oceanography, who read the evolving manuscript and made numerous helpful corrections and suggestions. Doris Haight typed the manuscript with skill, patience, and good humor.

*Woods Hole*                                                     Joseph Pedlosky
*September 1979*

# Contents

# CHAPTER 1

# Preliminaries

## 1.1 Geophysical Fluid Dynamics

The atmosphere and the ocean have so many fluid-dynamical properties in common that the study of one often enriches our understanding of the other. Experience has also shown that the recognition of the underlying dynamical concepts applicable to both the atmosphere and the oceans is an excellent starting point for the study of either. Geophysical fluid dynamics is the subject whose concerns are the fundamental dynamical concepts essential to an understanding of the atmosphere and the oceans. In principle, though, geophysical fluid dynamics deals with all naturally occurring fluid motions. Such motions are present on an enormous range of spatial and temporal scales, from the ephemeral flutter of the softest breeze to the massive and persistent oceanic and atmospheric current systems. Indeed, even the "solid" earth itself undergoes a fluidlike internal circulation on time scales of millions of years, the surface expression of which is sea-floor spreading and continental drift. All these phenomena can properly be included within the domain of geophysical fluid dynamics. Partly for historical reasons, however, the subject has tended to focus on the dynamics of large-scale phenomena in the atmosphere and the oceans. It is on large scales that the common character of atmospheric and oceanic dynamics is most evident, while at the same time the majestic nature of currents like the Gulf Stream in the ocean and the atmospheric jet stream makes such a focus of attention emotionally compelling and satisfying. This limitation will be observed in the following discussion, which consequently provides an introductory rather than exhaustive treatment of the subject. In particular the present text does not discuss the observational and descriptive features of meteorology and oceanography, although a familiarity with such evidence is a necessity for the proper formulation of new fluid-dynamical theories. Reference will be made from time to time in the text to the description of particular phenomena for the purpose of clarifying the motive for particular lines of study.

The principles to be derived are largely theoretical concepts which can be applied to an understanding of the natural phenomena. Such principles spring most naturally from the study of model problems whose goal is the development of conceptual comprehension rather than detailed simulation of the complete geophysical phenomenon. Geophysical fluid dynamics has historically progressed by the consideration of a study sequence within a hierarchy of increasingly complex models where each stage builds on the intuition developed by the precise analysis of simpler models.

## 1.2 The Rossby Number

The attribute "large scale" requires a more precise definition. A phenomenon whose characteristic length scale is fifty kilometers might be considered small scale in the atmosphere, while motions of just that scale in the oceans could be considered accurately as large scale. Whether a phenomenon is to be considered a large-scale one *dynamically* depends on more than its size.

For the purpose of this text large-scale motions are those which are significantly influenced by the earth's rotation. An important measure of the significance of rotation for a particular phenomenon is the Rossby number, which we define as follows. Let $L$ be a characteristic length scale of the *motion*. Figure 1.2.1, for example, shows a typical wave pattern observed in the pressure field of the troposphere. A typical and appropriate length scale of the motion, i.e., one that characterizes the horizontal spatial variations of the dynamical fields, could be the distance between a pressure peak and a succeeding trough. Similarly let $U$ be a horizontal velocity scale characteristic of the motion. In Figure 1.2.1 $L$ would be $O(1,000 \text{ km})$, while $U$ would be $O(20 \text{ m s}^{-1})$.*

The time it takes a fluid element moving with speed $U$ to traverse the distance $L$ is $L/U$. If that period of time is much less than the period of rotation of the earth, the fluid can scarcely sense the earth's rotation over the time scale of the motion. For rotation to be important, then, we anticipate

---

* The symbol $O(\ )$ is used in two quite separate ways in this text. The statement that the functions $f(x)$ and $g(x)$ are in the relation $f(x) = O(g(x))$ (in some limit) implies that $f(x)/g(x) \to$ constant in that limit in a formal asymptotic sense. The symbol will also be used to mean that a variable quantity, in this case $U$, has a size exemplified by the value following the ordering symbol. No limit or approximation criterion is implied in the latter case. The two usages are distinct and the particular context will show clearly which is meant.

that

$$\frac{L}{U} \geq \Omega^{-1}, \tag{1.2.1}$$

or, equivalently,

$$\varepsilon = \frac{U}{2\Omega L} \leq 1. \tag{1.2.2}$$

The nondimensional parameter $\varepsilon$ is the Rossby number. Large-scale flows are defined as those with sufficiently large $L$ for $\varepsilon$ to be order one or less. For the earth $\Omega = 7.3 \times 10^{-5}$ s$^{-1}$. For the $L$ and $U$ given above, $\varepsilon = 0.137$ and we can expect the earth's rotation to be important.

Such estimates must often be more refined. For planetary motions we shall see that it is really only the component of the planetary rotation perpendicular to the earth's surface which naturally enters the estimate of $\varepsilon$. Hence (1.2.2) could seriously underestimate the Rossby number for phenomena in low latitudes. Such elaborations and qualifications will be taken up later.

Note that the smaller the characteristic velocity is, the smaller $L$ can be and yet still qualify for a large-scale flow. The Gulf Stream has velocities of order 100 cm s$^{-1}$. Although its characteristic horizontal scale as shown in Figure 1.2.2 is only O(100 km), the associated Rossby number is 0.07. Although the use of the local normal component of the earth's rotation would double this value at a latitude of 30°, it is still clear that such currents meet the criterion of large-scale motion.

Now these considerations have been essentially kinematic. However, the important dynamical consequence of even a moderately small Rossby number follows from the fact that small $\varepsilon$ implies that large-scale motions are slow compared to the velocity imposed by the solid-body rotation of the earth. To a first approximation—i.e., to O($\varepsilon$)—the atmosphere and oceans rotate with the planet with small but significant deviations which we, also rotating with the earth, identify as winds and currents. It is useful to recognize explicitly that the interesting motions are small departures from solid-body rotation by describing the motions in a rotating coordinate frame which kinematically eliminates the rigid rotation. In a frame rotating at a rate $\Omega$ only the deviations from solid-body rotation will be seen. Since such a rotating frame is an accelerating rather than an inertial frame, certain well-known "inertial forces" will be sensed, i.e., the centrifugal force and the subtle and important Coriolis force. We shall see that whenever the Rossby number is small, the Coriolis force is a dominant participant in the balance of forces. The study of the dynamics of large scale oceanic or atmospheric motions must include the Coriolis force to be geophysically relevant, and once the Coriolis force is included a host of subtle and fascinating dynamical phenomena are possible.

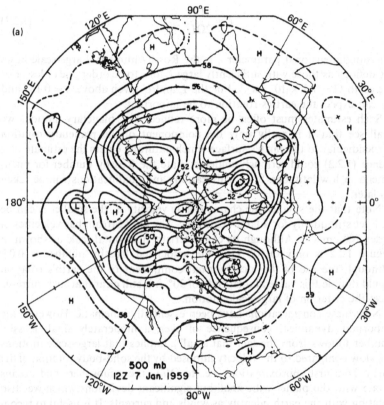

**Figure 1.2.1(a)**   Isolines of constant pressure (isobars) at a level which is above roughly one-half the atmosphere's mass. The isobars very nearly mark the streamlines of the flow (Palmén and Newton, 1969).

(b)

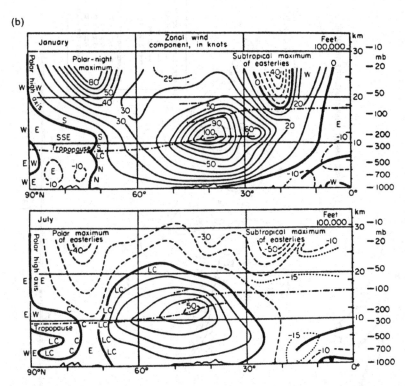

**Figure 1.2.1(b)** Cross section of the zonal wind (i.e., along latitude circles) showing the distribution of wind speed. (One knot $\sim 50\,\mathrm{cm\,s^{-1}}$) (Palmén and Newton, 1969, after Kochanski, 1955).

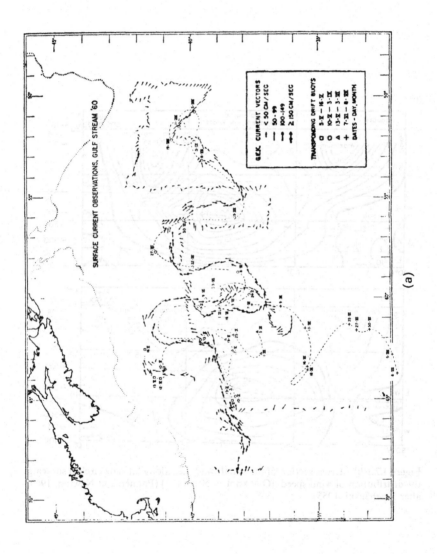

(a)

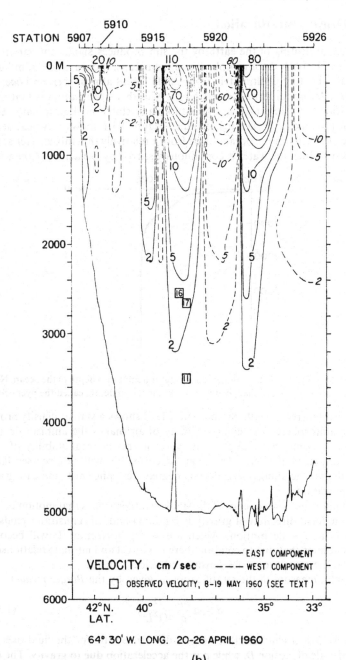

**Figure 1.2.2 (a)**   (*Facing page*) The path of the Gulf Stream as revealed by surface observations, and **(b)** a cross section through the Stream which displays the structure of the current velocity (Fuglister 1963).

## 1.3 Density Stratification

Differential heating of the earth by the sun produces significant variations of density in both the atmosphere and the oceans. Indeed, aside from lunar tides, this heating is ultimately responsible for all atmospheric and oceanic motions. An important observed property of this stratification is that on a large scale it is almost always gravitationally stable in the ordinary sense that heavy fluid underlies lighter fluid. Figure 1.3.1 shows a typical depth profile of density in the ocean. Aside from a relatively thin mixed layer at the surface in which the water is continuously stirred by the action of the wind,

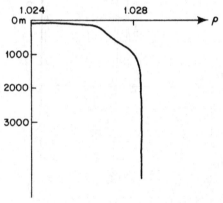

**Figure 1.3.1**   Typical distribution of density (gram/cm$^3$) with depth in the ocean. Note the sharp zone of density change at about 1 kilometer in depth, called the pycnocline.

the density increases with depth. Figure 1.3.2 shows a similar density profile for the atmosphere. The compressibility of air makes the ambient density decrease a somewhat deceptive measure of gravitational stability of vertically displaced elements. The required correction will be discussed later. Nevertheless the atmospheric density decrease is sufficient to indicate gravitational stability.

An important consequence of the stable stratification is that motion parallel to the local direction of gravity is inhibited and this constraint tends to produce large-scale motions which are *nearly* horizontal. It will become apparent later that due to rotation there is a direct and intimate relationship between the structure of the density and velocity fields.

A useful measure of the stratification is given by the *Burger number*,

$$S = g \frac{\Delta\rho}{\rho} \frac{D}{4\Omega^2 L^2}, \tag{1.3.1}$$

where $\Delta\rho/\rho$ is a characteristic density-difference ratio for the fluid over its *vertical* scale of motion $D$, while $g$ is the acceleration due to gravity. The parameter $S$ may be written in terms of the ratio of length scales,

$$S = \left(\frac{L_D}{L}\right)^2, \tag{1.3.2}$$

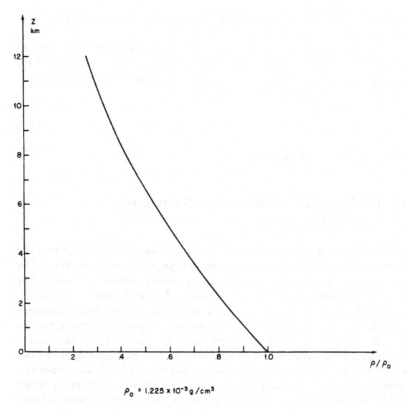

$$\rho_0 = 1.225 \times 10^{-3} \, g/cm^3$$

**Figure 1.3.2** The distribution of density with height in the atmosphere (after the NASA 1962 Standard Atmosphere).

where the length

$$L_D = \frac{1}{2\Omega} \left( g \frac{\Delta\rho}{\rho} D \right)^{1/2} \tag{1.3.3}$$

is called the *Rossby deformation radius*. $S$ depends on the ratio of the deformation radius to the geometric scale $L$, but does *not* depend on the velocity scale $U$.

As in the case of the Rossby number, the Burger number must in practice be written in terms of the local normal component of the earth's rotation, which means that $S$ will be larger for low-latitude motions if all other factors are unchanged.

Another important property of large-scale geophysical motions related to the stable density field is the disparity between horizontal and vertical scales of motion. The depth of the ocean rarely exceeds six kilometers, and the vertical extent of major current systems is usually much less than that. Yet the horizontal scale is hundreds or even thousands of kilometers. Similarly, major atmospheric phenomena like the large-scale waves shown in Figure

1.2.1 have a characteristic vertical scale of ten kilometers, while $L$ is O(1,000 km).

In general large-scale geophysical motions occur within an extraordinarily thin sheet of fluid, and given the grand horizontal scale of the motion, this geometrical constraint produces fluid trajectories which are very flat. That is, the *aspect ratio* for the motion,

$$\delta = \frac{D}{L},$$   (1.3.4)

is usually a very small number.

## 1.4 The Equations of Motion in a Nonrotating Coordinate Frame

The starting point of geophysical fluid dynamics is the premise that the dynamics of meteorological and oceanographic motions are determined by the systematic application of the fluid continuum equations of motion. The dynamic variables generally required to describe the motion are the density $\rho$, the pressure $p$, the vector velocity $\mathbf{u}$, and certain further thermodynamic variables like the temperature $T$, the internal energy per unit mass $e$, and the specific entropy $s$. In specific situations depending on the physical nature of the fluid, additional variables such as salinity may be required, or in cases where the thermodynamic state relations may be simplified, some state variables can be dispensed with. In the Eulerian kinematic description, which will be used without exception, the dynamic variables are functions of time $t$ and the vector position coordinate $\mathbf{r}$.

In this section we shall state the equations of motions in an inertial or nonrotating frame of reference. Their derivation may be found in any one of a number of elementary fluid-dynamics texts.

In the absence of sources or sinks of mass within the fluid, the condition of mass conservation is expressed by the *continuity equation*,

$$\frac{\partial \rho}{\partial t} + \nabla \cdot \rho \mathbf{u} = 0.$$   (1.4.1)

This equation states that the local increase of density with time must be balanced by a divergence of the mass flux $\rho \mathbf{u}$. This may also be written

$$\frac{d\rho}{dt} + \rho \nabla \cdot \mathbf{u} = 0,$$   (1.4.2)

where

$$\frac{d}{dt} \equiv \frac{\partial}{\partial t} + \mathbf{u} \cdot \nabla$$   (1.4.3)

is the total derivative (often called the substantial derivative) with respect to time of any property following individual fluid elements.

Newton's law of motion written for a fluid continuum takes the form

$$\rho \frac{d\mathbf{u}}{dt} = -\nabla p + \rho \nabla \phi + \mathscr{F}(\mathbf{u}), \tag{1.4.4}$$

or that the mass per unit volume times the acceleration is equal to the sum of the pressure gradient force, the body force $\rho \nabla \phi$, and the force $\mathscr{F}$, where $\phi$ is the potential by which conservative body forces such as gravity can be represented. $\mathscr{F}$ is, in principle, any nonconservative force, but in fact, in all cases to be discussed, $\mathscr{F}$ is the frictional force in the fluid. For Newtonian fluids like air or water

$$\mathscr{F} = \mu \nabla^2 \mathbf{u} + \frac{\mu}{3} \nabla(\nabla \cdot \mathbf{u}), \tag{1.4.5}$$

where $\mu$ is the molecular viscosity. This representation of $\mathscr{F}$ is exact when $\mu$, in principle a function of the thermodynamic state variables, is considered constant over the field of motion. This approximation will be adequate for our purposes.

A particularly vexing question is whether this representation for $\mathscr{F}$ is adequate if the state variables are to describe *only* the large-scale motions. It is already clear from the very form of (1.4.1) and (1.4.4) that the equations of motion are quadratically nonlinear, i.e., contain products of the dynamic variables. This implies that in principle it is not possible to simply superpose solutions of the equations. In physical terms, motions on one spatial scale interact with motions on other scales. There is therefore an *a priori* possibility that small-scale motions, which are not the focus of our interest, may yet influence the large-scale motions. One common but not very precise notion is that the small-scale motions, which appear sporadic on longer time scales, act to smooth and mix properties on the larger scales by processes analogous to molecular, diffusive transports. Further discussion of this point is deferred to Chapter 4, but in view of possible consequent alterations of the representation of $\mathscr{F}$, we consider (1.4.4) as our fundamental equation of motion, while (1.4.5) can be thought of as one model of the frictional force (which of course is precise for laboratory experiments).

Unless the density is considered a constant (and there are many interesting phenomena for which this is a useful idealization), the momentum and continuity equations are insufficient to close the dynamical system. The first law of thermodynamics must be considered; it can be written as

$$\rho \frac{de}{dt} = -p\rho \frac{d}{dt} \rho^{-1} + k\nabla^2 T + \chi + \rho Q, \tag{1.4.6}$$

where $e$ is the internal energy per unit mass, $T$ is the temperature, $k$ is the thermal conductivity, $Q$ is the rate of heat addition per unit mass by internal heat sources, and $\chi$ is the addition of heat due to viscous dissipation. For all situations to be discussed, the change of internal energy due to viscous dissipation is negligible and will henceforth be deleted. There are some phenomena, such as the slow circulation of the earth's mantle, where $\chi$ may be important, but such problems are not the concern of this book.

It is convenient to introduce the thermodynamic state property $s$, the specific entropy. The entropy is related to the other state properties by the relation

$$T \, \Delta s = \Delta e + p \, \Delta\left(\frac{1}{\rho}\right), \tag{1.4.7}$$

where $\Delta s$, $\Delta e$, and $\Delta(1/\rho)$ are arbitrary increments in $s$, $e$, and $1/\rho$. In particular

$$T\frac{ds}{dt} = \frac{de}{dt} + p\frac{d}{dt}\frac{1}{\rho}, \tag{1.4.8}$$

so that (1.4.6) can be rewritten as

$$T\frac{ds}{dt} = \frac{k}{\rho}\nabla^2 T + Q. \tag{1.4.9}$$

To complete the system, further thermodynamic state relations expressing the physical nature of the fluid are required. In the simplest case, such as dry air or pure water, they take the general form

$$\rho = \rho(p, T) \tag{1.4.10}$$

and

$$s = s(p, T). \tag{1.4.11}$$

Using (1.4.11), we may rewrite (1.4.9) in the form

$$C_p\frac{dT}{dt} + T\left(\frac{\partial s}{\partial p}\right)_T \frac{dp}{dt} = \frac{k}{\rho}\nabla^2 T + Q, \tag{1.4.12}$$

where the definition of the specific heat at constant pressure,

$$C_p = T\left(\frac{\partial s}{\partial T}\right)_p,$$

has been used. With the further use of the thermodynamic relation*

$$\left(\frac{\partial s}{\partial p}\right)_T = \frac{1}{\rho^2}\left(\frac{\partial \rho}{\partial T}\right)_p, \tag{1.4.13}$$

(1.14.12) takes the form

$$C_p\frac{dT}{dt} - \frac{T}{\rho}\alpha\frac{dp}{dt} = \frac{k}{\rho}\nabla^2 T + Q, \tag{1.4.14}$$

where $\alpha$ is the coefficient of thermal expansion defined by the relation

$$\alpha = -\frac{1}{\rho}\left(\frac{\partial \rho}{\partial T}\right)_p. \tag{1.4.15}$$

---

* See, for example, Batchelor (1967), Chapter 1.

For example, in the atmosphere the state relation for dry air is well represented by the ideal-gas law

$$\rho = \frac{p}{RT}, \tag{1.4.16}$$

where $R$ is the gas constant for dry air. The specific entropy for such an ideal gas can be represented by

$$s = C_p \ln T - R \ln p, \tag{1.4.17}$$

in which case $\alpha = 1/T$, and (1.4.14) becomes

$$\frac{d\theta}{dt} = \frac{\theta}{C_p T} \left\{ \frac{k}{\rho} \nabla^2 T + Q \right\}. \tag{1.4.18}$$

The quantity

$$\theta = T \left( \frac{p_0}{p} \right)^{R/C_p}, \tag{1.4.19}$$

where $p_0$ is a constant reference pressure, is called the potential temperature. Note that in the absence of conductive and internal heating $\theta$ is a conserved property for each fluid element. For a pure liquid where effects of compressibility are minor, the simple state relation

$$\rho = \rho_0 (1 - \alpha(T - T_0)) \tag{1.4.20}$$

is often adequate, in which case the second term on the left-hand side of (1.4.14) is also neglected. In this case the *heat equation*, or first thermodynamic law, takes the form

$$\frac{dT}{dt} = \kappa \nabla^2 T + \frac{Q}{C_p}, \tag{1.4.21}$$

where $\kappa = k/\rho C_p$ is the coefficient of thermal diffusivity. Equation (1.4.21) can be written entirely in terms of the density with the aid of (1.4.20), i.e.,

$$\frac{d\rho}{dt} = \kappa \nabla^2 \rho - \frac{\alpha \rho_0}{C_p} Q. \tag{1.4.22}$$

It is important to note that (1.4.2) and (1.4.22) are two quite different physical principles, the former expressing mass conservation, while the latter, for a liquid, approximately expresses energy conservation.

For an incompressible liquid, or one which is very nearly so, density differences are so slight that they have a negligible effect on the mass balance, so that (1.4.2) can be approximated for an incompressible liquid by

$$\nabla \cdot \mathbf{u} = 0. \tag{1.4.23}$$

This does *not* imply that $d\rho/dt$ vanishes; indeed, $d\rho/dt$ is given by (1.4.22) and vanishes only if conductive and internal heating can be neglected, i.e., only if the motion is adiabatic. The validity of (1.4.23) must be examined in each situation by systematic scaling arguments, and we shall return to this question in Chapter 6. For the time being we can take (1.4.23) as our definition of incompressibility.

For more complicated liquids, as seawater, where the salt content or salinity contributes to the determination of the density, the simple form (1.4.20) must be replaced by a more complicated relation between density, salinity, temperature, and pressure. To a fair approximation (Bryan and Cox 1972)

$$\rho = \rho_0(1 - \alpha_T(T - T_0) + \alpha_S(S - S_0)), \qquad (1.4.24)$$

where $S$ is the salinity (grams/kilogram) of the water, and $\alpha_T$ and $\alpha_S$ are empirically determined functions of depth. In such cases we need an additional equation expressing the budget of salinity, of the form

$$\frac{dS}{dt} = F(S), \qquad (1.4.25)$$

where $F(S)$ must be specified to represent sources, sinks, or diffusive redistribution of salinity.

## 1.5 Rotating Coordinate Frames

We noted earlier that the most natural frame from which to describe atmospheric and oceanic motions is one which rotates with the planetary angular velocity $\Omega$. Naturally the phenomena themselves are unaffected by the choice of frame of reference, but we must be prepared to accept the fact that the description of the phenomena depends on our choice. To an observer in a rotating frame objects fixed in inertial space will appear to rotate and, because of the curvature of their apparent trajectory, to be accelerating. This ambiguity of viewpoint is resolved by recalling that Newton's law of motion and its derived form (1.4.4) are valid in that form only in an inertial frame. We must now find the proper but altered form of the equations of motion when written entirely in terms of quantities directly observed from the rotating frame.

Consider first a vector $\mathbf{A}$ which has constant magnitude but which rotates with angular velocity $\Omega$ (Figure 1.5.1). Let the angle between $\mathbf{A}$ and $\Omega$ be $\gamma$. In a small time $\Delta t$, $\mathbf{A}$ is rotated through the angle $\Delta\theta = |\Omega|\,\Delta t$, where $|\Omega|$ is the magnitude of $\Omega$. It is apparent from Figure 1.5.2 that the small change in $\mathbf{A}$ is given by

$$\mathbf{A}(t + \Delta t) - \mathbf{A}(t) \equiv \Delta\mathbf{A} = \mathbf{n}|\mathbf{A}|\sin\gamma\,\Delta\theta + O((\Delta\theta)^2), \qquad (1.5.1)$$

where $\mathbf{n}$ is the unit vector in the direction of change of $\mathbf{A}$, which must be perpendicular to $\mathbf{A}$ (since its length is fixed) and perpendicular to $\Omega$ (by the definition of the rotation). Thus

$$\mathbf{n} = \frac{\Omega \times \mathbf{A}}{|\Omega \times \mathbf{A}|}. \qquad (1.5.2)$$

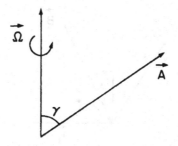

**Figure 1.5.1**  A is a vector of constant length oriented at an angle $\gamma$ with respect to the axis of its rotation.

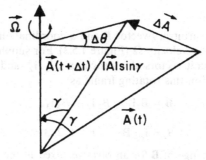

**Figure 1.5.2**  A at times $t$ and $t + \Delta t$, showing the infinitesimal change $\Delta \mathbf{A}$.

In the limit $\Delta t \to 0$,

$$\lim_{\Delta t \to 0} \frac{\Delta \mathbf{A}}{\Delta t} = \frac{d\mathbf{A}}{dt} = |\mathbf{A}| \sin \gamma \, \frac{d\theta}{dt} \frac{\mathbf{\Omega} \times \mathbf{A}}{|\mathbf{\Omega} \times \mathbf{A}|}, \tag{1.5.3}$$

but since

$$|\mathbf{\Omega} \times \mathbf{A}| = |\mathbf{\Omega}| \, |\mathbf{A}| \sin \gamma, \tag{1.5.4}$$

we have finally

$$\boxed{\frac{d\mathbf{A}}{dt} = \mathbf{\Omega} \times \mathbf{A}} \tag{1.5.5}$$

for a vector of *fixed* magnitude.

An observer who is fixed in the rotating frame of reference would see *no* change in A, while an observer in a nonrotating frame would see A change as described by (1.5.5). Both observers would see the same vector A, since by definition a vector A is independent of the coordinate frame used to describe it. However, their perceptions of the rate of change of A differ markedly. Note though that since

$$\frac{d|\mathbf{A}|^2}{dt} = 2\mathbf{A} \cdot \frac{d\mathbf{A}}{dt} = 2\mathbf{A} \cdot (\mathbf{\Omega} \times \mathbf{A}) = 0, \tag{1.5.6}$$

both observers agree that the magnitude of A is unaltered.

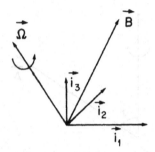

**Figure 1.5.3** The orthogonal coordinate frame whose base vectors are $i_1$, $i_2$, and $i_3$, and the vector **B**. The coordinate frame rotates with angular velocity $\Omega$ about the axis as shown.

Consider now an arbitrary vector **B** and a coordinate frame of reference rotating with angular velocity $\Omega$ (Figure 1.5.3). For simplicity let the frame be Cartesian with unit vectors along each axis, $i_1$, $i_2$, and $i_3$. The vector **B** may be represented in this rotating frame as

$$\mathbf{B} = B_1 \mathbf{i}_1 + B_2 \mathbf{i}_2 + B_3 \mathbf{i}_3, \tag{1.5.7}$$

where

$$B_j = \mathbf{i}_j \cdot \mathbf{B}, \qquad j = 1, 2, 3. \tag{1.5.8}$$

The time rate of change of **B** for an observer fixed in the rotating frame is simply

$$\left(\frac{d\mathbf{B}}{dt}\right)_R = \frac{dB_1}{dt}\mathbf{i}_1 + \frac{dB_2}{dt}\mathbf{i}_2 + \frac{dB_3}{dt}\mathbf{i}_3, \tag{1.5.9}$$

for in this frame the unit vectors are fixed in length *and* direction. (The subscript $R$ in (1.5.9) will remind us that (1.5.9) is valid for the rotating observer.) On the other hand, for a nonrotating observer both the components of **B** and the unit vectors change with time. The rate of change of the *scalar* components $B_1$, $B_2$, $B_3$ are common to both observers, so

$$\left(\frac{d\mathbf{B}}{dt}\right)_I = \frac{dB_1}{dt}\mathbf{i}_1 + \frac{dB_2}{dt}\mathbf{i}_2 + \frac{dB_3}{dt}\mathbf{i}_3$$
$$+ B_1 \frac{d\mathbf{i}_1}{dt} + B_2 \frac{d\mathbf{i}_2}{dt} + B_3 \frac{d\mathbf{i}_3}{dt}, \tag{1.5.10}$$

where the subscript $I$ denotes rates of change as seen by the observer in the nonrotating, inertial frame. Application of (1.5.5) to each of the three unit vectors yields

$$B_1 \frac{d\mathbf{i}_1}{dt} + B_2 \frac{d\mathbf{i}_2}{dt} + B_3 \frac{d\mathbf{i}_3}{dt}$$
$$= B_1 \Omega \times \mathbf{i}_1 + B_2 \Omega \times \mathbf{i}_2 + B_3 \Omega \times \mathbf{i}_3 \tag{1.5.11}$$
$$= \Omega \times (B_1 \mathbf{i}_1 + B_2 \mathbf{i}_2 + B_3 \mathbf{i}_3)$$
$$= \Omega \times \mathbf{B},$$

so that (1.5.10) becomes

$$\boxed{\left(\frac{d\mathbf{B}}{dt}\right)_I = \left(\frac{d\mathbf{B}}{dt}\right)_R + \boldsymbol{\Omega} \times \mathbf{B}}$$

(1.5.12)

Thus the rates of change with time of the same vector $\mathbf{B}$ are perceived differently in the rotating and nonrotating frames. The term $\boldsymbol{\Omega} \times \mathbf{B}$ must be added to the time rate of change as seen in the rotating frame to accurately describe the rate of change of $\mathbf{B}$ as seen in the inertial, nonrotating system. Note that observers in both frames *agree* on the time rate of change of $\boldsymbol{\Omega}$, since $\boldsymbol{\Omega} \times \boldsymbol{\Omega}$ identically vanishes, so that (1.5.12) applies without alteration to cases where $\boldsymbol{\Omega}$ is changing in magnitude and direction.

## 1.6 Equations of Motion in a Rotating Coordinate Frame

Let $\mathbf{r}$ be the position vector of an arbitrary fluid element. According to (1.5.12)

$$\left(\frac{d\mathbf{r}}{dt}\right)_I = \left(\frac{d\mathbf{r}}{dt}\right)_R + \boldsymbol{\Omega} \times \mathbf{r},$$

(1.6.1)

so that the velocity seen in the nonrotating frame, $\mathbf{u}_I$, is equal to the velocity observed in the rotating frame augmented by the velocity imparted to the fluid element by the solid-body rotation $\boldsymbol{\Omega} \times \mathbf{r}$. We may write this as

$$\mathbf{u}_I = \mathbf{u}_R + \boldsymbol{\Omega} \times \mathbf{r},$$

(1.6.2)

where $\mathbf{u}_R$ is called the *relative velocity*. Now Newton's law of motion equates the applied forces per unit mass to the acceleration in inertial space, i.e., to the rate of change of $\mathbf{u}_I$ as seen in an inertial frame. Applying (1.5.12) to $\mathbf{u}_I$ yields

$$\left(\frac{d\mathbf{u}_I}{dt}\right)_I = \left(\frac{d\mathbf{u}_I}{dt}\right)_R + \boldsymbol{\Omega} \times \mathbf{u}_I.$$

(1.6.3)

Our goal, however, is to obtain a description of the motion entirely in terms of quantities which are directly observed in the rotating frame. With that goal in mind (1.6.2) is used to eliminate $\mathbf{u}_I$ from the right-hand side of (1.6.3) to obtain

$$
\begin{aligned}
\left(\frac{d\mathbf{u}_I}{dt}\right)_I &= \left(\frac{d\mathbf{u}_R}{dt}\right)_R + \frac{d\boldsymbol{\Omega}}{dt} \times \mathbf{r} + \boldsymbol{\Omega} \times \left(\frac{d\mathbf{r}}{dt}\right)_R \\
&\quad + \boldsymbol{\Omega} \times (\mathbf{u}_R + \boldsymbol{\Omega} \times \mathbf{r}) \\
&= \left(\frac{d\mathbf{u}_R}{dt}\right)_R + 2\boldsymbol{\Omega} \times \mathbf{u}_R + \boldsymbol{\Omega} \times (\boldsymbol{\Omega} \times \mathbf{r}) \\
&\quad + \frac{d\boldsymbol{\Omega}}{dt} \times \mathbf{r}.
\end{aligned}
$$

(1.6.4)

The discrepancy between the accelerations perceived in the different frames is equal to the three additional terms on the right-hand side of (1.6.4). They are the *Coriolis acceleration* $2\mathbf{\Omega} \times \mathbf{u}_R$, the centripetal acceleration $\mathbf{\Omega} \times (\mathbf{\Omega} \times \mathbf{r})$, and the acceleration due to variations in the rotation rate itself.

The last of these is unimportant for *most* oceanographic or atmospheric phenomena except for those whose time scales are unusually long. For our

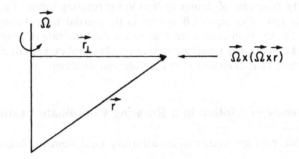

**Figure 1.6.1**   The centripetal acceleration, $\mathbf{\Omega} \times (\mathbf{\Omega} \times \mathbf{r})$.

purposes $\mathbf{\Omega}$ can be considered a constant. The centripetal acceleration term can be rewritten in terms of $\mathbf{r}_\perp$, the perpendicular distance vector from the rotation axis to the position of the fluid element at $\mathbf{r}$ (Figure 1.6.1), i.e., since

$$\mathbf{\Omega} \times \mathbf{r} = \mathbf{\Omega} \times \mathbf{r}_\perp,$$

we have

$$\mathbf{\Omega} \times (\mathbf{\Omega} \times \mathbf{r}) = -|\mathbf{\Omega}|^2 \mathbf{r}_\perp, \qquad (1.6.5)$$

with the aid of the formula for the triple vector product

$$\mathbf{A} \times (\mathbf{B} \times \mathbf{C}) = (\mathbf{A} \cdot \mathbf{C})\mathbf{B} - (\mathbf{A} \cdot \mathbf{B})\mathbf{C}.$$

It follows that the centripetal acceleration may be written in terms of a potential function $\phi_c$,

$$\mathbf{\Omega} \times (\mathbf{\Omega} \times \mathbf{r}) = -\nabla\phi_c,$$

where

$$\phi_c = \frac{|\mathbf{\Omega}|^2|\mathbf{r}_\perp|^2}{2} = \frac{|\mathbf{\Omega} \times \mathbf{r}|^2}{2}. \qquad (1.6.6)$$

This allows us to consider the centripetal acceleration instead as an additional force per unit mass with a sign opposite to (1.6.5) by D'Alembert's principle. Since this apparent force, the centrifugal force, can be written as a potential, it can be included with the force potential of (1.4.4) to yield a total force potential

$$\Phi = \phi + \phi_c.$$

The remaining acceleration is the Coriolis acceleration $2\mathbf{\Omega} \times \mathbf{u}_R$. It is the only new term which explicitly involves the fluid velocity, and it is responsible for the only structural change of the momentum equation when written for a uniformly rotating frame, since the centrifugal force merely introduces an altered force potential.

If we note that spatial gradients are perceived identically in rotating and in nonrotating coordinate frames, the momentum equation (1.4.4) becomes, *for an observer in a uniformly rotating coordinate frame,*

$$\rho \left[ \frac{d\mathbf{u}}{dt} + 2\mathbf{\Omega} \times \mathbf{u} \right] = -\nabla p + \rho \nabla \Phi + \mathscr{F}, \qquad (1.6.7)$$

where $\mathbf{u}$ is the velocity seen in the rotating frame. Although $p$, $\Phi$, and their spatial gradients will obviously be independent of the frame of reference, the invariance of the form of $\mathscr{F}$ from one frame to another depends on the way $\mathscr{F}$ depends on the velocity field. For a Newtonian fluid for which (1.4.5) is valid, it is a simple matter to show that

$$\mathscr{F}(\mathbf{u}_I) = \mathscr{F}(\mathbf{u}_R). \qquad (1.6.8)$$

Since without exception we shall always describe fluid motions from a rotating coordinate frame, the convention is adopted that all dynamical variables unless otherwise stated are measured in a rotating coordinate frame.

If the Coriolis acceleration is moved to the right-hand side of (1.6.7), it appears to an observer in the rotating frame as an additional force. This *Coriolis force* is always perpendicular to the velocity and hence *does no work.* For an observer aligned with $\mathbf{\Omega}$, the Coriolis force $-2\mathbf{\Omega} \times \mathbf{u}$ appears (Figure 1.6.2) as a force tending to deflect moving fluid elements to their right.

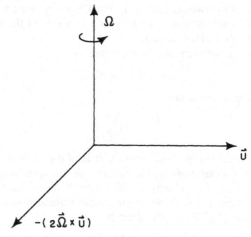

**Figure 1.6.2**   The relationship between $\mathbf{u}$, $\mathbf{\Omega}$, and the Coriolis force per unit mass $(-2\mathbf{\Omega} \times \mathbf{u})$.

It is important to note that the *total* time rate of change of any *scalar* such as the temperature is the same in rotating as in nonrotating frames. Thus the equations of conservation of mass (1.4.2) and the various thermodynamic equations (1.4.6) through (1.4.25) are unaffected by the choice of coordinate frame and therefore can be applied directly to the description of motion in a rotating coordinate system.

It should be stressed that although the *total* derivative of a scalar is the same in both frames, the individual components of the substantial derivative are not invariant. In particular for any scalar property $P$, it can be simply shown that

$$\left(\frac{\partial P}{\partial t}\right)_I = \left(\frac{\partial P}{\partial t}\right)_R - (\mathbf{\Omega} \times \mathbf{r}) \cdot \nabla P, \tag{1.6.9}$$

while

$$\mathbf{u}_I \cdot \nabla P = (\mathbf{u}_R + \mathbf{\Omega} \times \mathbf{r}) \cdot \nabla P, \tag{1.6.10}$$

from which it may be verified that

$$\left(\frac{dP}{dt}\right)_I = \left(\frac{dP}{dt}\right)_R. \tag{1.6.11}$$

## 1.7 Coriolis Acceleration and the Rossby Number

The sum of the forces on the right-hand side of the momentum equation (1.6.7) must be balanced by the sum of the acceleration of the relative velocity observed in the rotating frame and the Coriolis acceleration, each multiplied by the density. The nature of the perceived acceleration depends on the relative size of the two. We can make a preliminary estimate of the importance of the Coriolis acceleration by examining the order of magnitude of the ratio of the two accelerations. For this purpose we write the order of magnitude of each of these accelerations in terms of the characteristic velocity and length scales $U$ and $L$.

The Coriolis acceleration can be estimated as

$$2\mathbf{\Omega} \times \mathbf{u} = O(2\Omega U) \tag{1.7.1}$$

while the relative acceleration

$$\frac{d\mathbf{u}}{dt} = O\left(\frac{U^2}{L}\right), \tag{1.7.2}$$

where we have assumed in (1.7.2) that $\partial \mathbf{u}/\partial t$ is of the same order as $\mathbf{u} \cdot \nabla \mathbf{u}$.

The Coriolis acceleration is *independent* of the geometric scale and depends linearly on the velocity $U$. The size of the relative acceleration *decreases* with increasing scale and depends quadratically on $U$. The ratio of the relative to the Coriolis acceleration is

$$\varepsilon = \frac{U}{2\Omega L}, \tag{1.7.3}$$

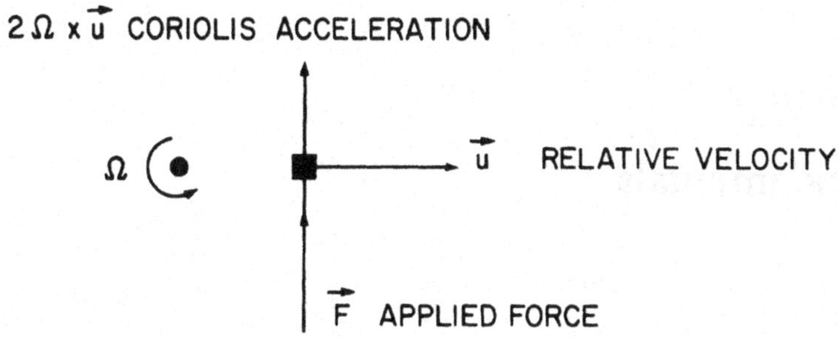

$2\,\Omega \times \vec{u}$  CORIOLIS ACCELERATION

$\Omega \ (\bullet$  ⟲   ■————→ $\vec{u}$   RELATIVE VELOCITY

$\vec{F}$  APPLIED FORCE

**Figure 1.7.1**  The relationship between force and velocity when the Coriolis acceleration dominates.

or the Rossby number. We noted in (1.3) that $\varepsilon$ was small for large-scale flows. It is now apparent that for large-scale flows the applied forces produce primarily a *Coriolis acceleration*. The acceleration, of course, must be in the direction of the resultant of the force, so that the relative velocity required to produce the Coriolis acceleration must be directed *to the right of the applied force*, as shown in Figure 1.7.1. The Coriolis acceleration becomes dominant for large-scale flows, not because the Coriolis acceleration becomes large, but because the relative accelerations become so feeble when the scale of the motion is large. Recalling the kinematic discussion of Section 1.3, we conclude that when the time scale of the *motion* exceeds the rotation period, the Coriolis acceleration exceeds the relative acceleration in importance. The gyroscopic tendency of fluid elements to move at right angles to the applied force is in fact responsible for the unique character of geophysical fluid dynamics.

# CHAPTER 2

# Fundamentals

## 2.1 Vorticity

In principle the accurate solution of the governing equations with appropriate boundary and initial conditions will reveal all required information in any particular problem. However, the equations of motion are so complex that only rarely can exact solutions be found, and any method of approximation of the equations requires first an understanding of the broad, general, physical principles with which any approximation must be consistent.

There are a few basic facts about fluid motion that are at the same time sufficiently general and relevant to geophysical situations. They are of great importance because they alert us to certain fundamental dynamic mechanisms which occur often in various guises in particular problems of interest. It is important to keep in mind that the facts or theorems to be described are simply *derived* relations from the fundamental differential equations. In deriving general theorems or relations, a number of conditions are prescribed on the nature of the motion, and certain consequences follow from the equations of motion. Naturally, the more stringent the *a priori* hypothesis, the stronger the consequences will be.

A derived dynamic variable of preeminent importance in geophysical fluid dynamics is the vorticity vector $\boldsymbol{\omega}$, defined as the curl of the velocity field, i.e.,

$$\boldsymbol{\omega} = \nabla \times \mathbf{u}. \tag{2.1.1}$$

In a Cartesian coordinate frame with coordinates $(x, y, z)$ and corresponding

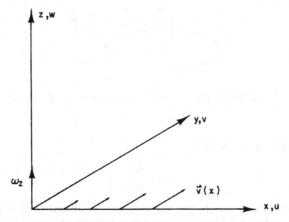

**Figure 2.1.1** Vorticity in a shear flow.

velocity components $u$, $v$, $w$ (Figure 2.1.1) the relationship between the components of vorticity and velocity which follows from (2.1.1) is

$$\omega_x = \frac{\partial w}{\partial y} - \frac{\partial v}{\partial z}, \tag{2.1.2a}$$

$$\omega_y = \frac{\partial u}{\partial z} - \frac{\partial w}{\partial x}, \tag{2.1.2b}$$

$$\omega_z = \frac{\partial v}{\partial x} - \frac{\partial u}{\partial y}. \tag{2.1.2c}$$

In Figure 2.1.1, as an example, only $v(x)$ is different from zero, giving rise to a vorticity vector aligned with the $z$-axis with magnitude $\partial v/\partial x$. For a fluid which is undergoing a uniform rotation, i.e., rotating as a solid body with uniform angular velocity $\mathbf{\Omega}_0$,

$$\mathbf{u} = \mathbf{\Omega}_0 \times \mathbf{r} \tag{2.1.3}$$

and the vorticity is

$$\mathbf{\omega} = 2\mathbf{\Omega}_0. \tag{2.1.4}$$

In this simple case the vorticity is simply *twice* the angular velocity of the fluid. The generalization of this notion to the general case follows from Stokes theorem, viz.:

$$\iint_A \mathbf{\omega} \cdot \mathbf{n} \, dA = \oint_C \mathbf{u} \cdot d\mathbf{r}, \tag{2.1.5}$$

where $C$ is a contour enclosing the surface $A$ whose normal at each point is $\mathbf{n}$ (Figure 2.1.2). In the limit of small surface elements $\delta A$, the mean-value

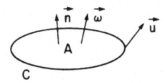

**Figure 2.1.2**   The circuit used to calculate the relation between angular velocity and vorticity.

theorem allows us to write

$$\bar{\omega}_n = \frac{\bar{u}_c l}{\delta A},$$   (2.1.6)

where $\bar{\omega}_n$ is the average of the vorticity vector normal to the elementary surface $\delta A$, and $\bar{u}_c$ is the average circumferential velocity around the contour $C$, whose length is $l$. If $\delta A$ is a small circle of radius $r_0$,

$$\bar{\omega}_n = \bar{u}_c \frac{2\pi r_0}{\pi r_0^2} = 2\frac{\bar{u}_c}{r_0},$$   (2.1.7)

so that the average vorticity $\bar{\omega}_n$ is twice the average angular velocity $\bar{u}_c/r_0$. It is important to note that curved trajectories are not required for a fluid motion to have vorticity; the unidirectional shear flow in Figure 2.1.1 provides an illustration. The vorticity of the fluid as observed from an inertial, nonrotating frame is called the *absolute vorticity*, $\omega_a$, and it is simply defined as the curl of the velocity observed in the nonrotating frame, i.e.,

$$\omega_a = \nabla \times \{\mathbf{u} + \mathbf{\Omega} \times \mathbf{r}\} = \omega + 2\mathbf{\Omega},$$   (2.1.8)

where $\omega$ is the *relative vorticity*, i.e., the curl of the relative velocity. Thus the absolute vorticity of each fluid element is the sum of the *planetary vorticity* $2\mathbf{\Omega}$ and the relative vorticity $\omega$. The relative magnitudes of these two can be estimated in a preliminary manner as follows. The component of the planetary vorticity normal to the earth's surface is the *Coriolis parameter*

$$f = 2\Omega \sin \theta,$$   (2.1.9)

where $\theta$ is the latitude of the fluid element in question. The corresponding estimate of the vertical component of the relative vorticity is given by the characteristic value of the velocity tangent to the earth's surface (the horizontal velocity) divided by $L$, or

$$\omega_n = O\left(\frac{U}{L}\right),$$   (2.1.10)

whose ratio to $f$ is

$$\frac{\omega_n}{f} = \frac{U}{fL} = \frac{U}{2\Omega L \sin \theta} = \frac{\varepsilon}{\sin \theta} \equiv R_0,$$   (2.1.11)

where $R_0$ is defined as the *local Rossby number*. In regions excluding the equator, $\sin \theta$ is $O(1)$, so that low-Rossby-number flows have the property that their *relative vorticity is small compared to the planetary vorticity*. One immediate consequence of this fact is that large-scale flows are hardly ever free of vorticity and their vorticity is primarily the planetary vorticity.

Since the vorticity is defined as the curl of the velocity field, it follows trivially that

$$\nabla \cdot \boldsymbol{\omega} = \nabla \cdot (\nabla \times \mathbf{u}) = 0, \tag{2.1.12a}$$

$$\nabla \cdot \boldsymbol{\omega}_a = \nabla \cdot (\nabla \times \mathbf{u}_I) = 0, \tag{2.1.12b}$$

so that the vorticity vector is nondivergent. A *vortex line* or *vortex filament* is a line in the fluid which at each point is parallel to the vorticity vector at that point, as shown in Figure 2.1.3. The vortex filaments associated with the planetary vorticity are straight lines everywhere parallel to the rotation axis.

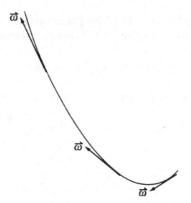

**Figure 2.1.3**   A vortex filament.

The vortex filament associated with the complete flow distorts this uniform field of filaments to a degree which depends on the strength of the relative motion.

A *vortex tube* (Figure 2.1.4) is formed by the surface consisting of the vortex filaments which pass through a closed curve $C$. Since the filaments change their orientation from one position to another, the bounding curve $C'$ of the tube at another position along the vortex tube will differ from $C$ in both length and orientation. However, by definition, at all points along the surface of the tube there can be no component of vorticity which penetrates the tube surface. Further, since $\boldsymbol{\omega}_a$ is divergence free, it follows that for arbitrary volumes $V$

$$\iiint_V dV \, \nabla \cdot \boldsymbol{\omega}_a = 0. \tag{2.1.13}$$

Now consider the volume of integration consisting of any finite segment of

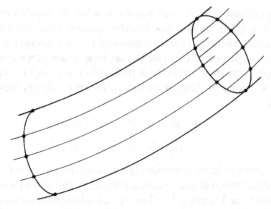

**Figure 2.1.4**   A vortex tube.

the vortex tube sliced at $C$ and $C'$ by plane surfaces with areas $A$ and $A'$ respectively, as shown in Figure 2.1.5. From the divergence theorem,

$$\iiint_V dV \, \nabla \cdot \boldsymbol{\omega}_a = \iint_A \boldsymbol{\omega}_a \cdot \mathbf{n} \, dA, \qquad (2.1.14)$$

where $\mathbf{n}$ is the outward normal at each point on the volume's bounding

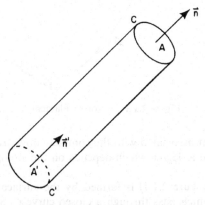

**Figure 2.1.5**   The vortex tube sliced at two sections with areas $A$ and $A'$, used to calculate the flux of vorticity at each section.

surface. Since $\boldsymbol{\omega}_a \cdot \mathbf{n}$ is identically zero on the cylindrical surface formed by the vortex filaments, it follows from (2.1.13) and (2.1.14) that

$$0 = \iint_A \boldsymbol{\omega}_a \cdot \mathbf{n}_A \, dA + \iint_{A'} \boldsymbol{\omega}_a \cdot (-\mathbf{n}_{A'}) \, dA, \qquad (2.1.15)$$

where $-\mathbf{n}_{A'}$ is the *outward* normal on the surface $A'$ or

$$\iint_A \boldsymbol{\omega}_a \cdot \mathbf{n}_A \, dA = \iint_{A'} \boldsymbol{\omega}_a \cdot \mathbf{n}_{A'} \, dA. \qquad (2.1.16)$$

Note that $n_A$ and $n_{A'}$ have the same orientation with respect to the tube axis. We define the *absolute strength* or *flux* of a vortex tube as

$$\Gamma_a = \iint \omega_a \cdot n \, dA, \tag{2.1.17}$$

and from (2.1.16) it follows, from these purely *kinematic* considerations, that the strength of a vortex tube is *constant along the length of the tube*. Consequently absolute vortex tubes and the filaments of which they are composed cannot end in the interior of the fluid. Filaments of relative vorticity also define *relative* vorticity tubes, and the preceding arguments, which apply equally well to the nondivergent vector $\omega$, demonstrate that relative-vorticity tubes are also constant along *their* lengths and that the filaments of the relative vorticity must also thread through the fluid and end only on the surface which bounds the fluid or else close on themselves.

The flux of *absolute* vorticity, $\Gamma_a$, through a surface $A$ is related to the *relative* vorticity flux $\Gamma$ by

$$\Gamma_a = \iint_A \omega_a \cdot n \, dA = \iint_A \omega \cdot n \, dA + \iint_A 2\Omega \cdot n \, dA$$

$$= \Gamma + \iint_A 2\Omega \cdot n \, dA. \tag{2.1.18}$$

Since $\Omega$ is a spatially constant vector,

$$\Gamma_a = \Gamma + 2\Omega A_n, \tag{2.1.19}$$

where $A_n$ is the area of $A$ projected on a surface perpendicular to $\Omega$ as shown in Figure 2.1.6. If we think of the planetary vorticity field as a uniform field

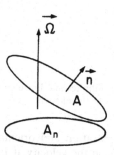

**Figure 2.1.6**   $A_n$ is the projection of the area $A$ on the plane which is perpendicular to $\Omega$.

of straight vortex filaments each parallel to $\Omega$, the difference between $\Gamma_a$ and $\Gamma$ is then proportional to the *number* of planetary vorticity filaments which penetrate the area $A$, or equivalently the number which are girdled by the contour $C$ which bounds $A$.

## 2.2 The Circulation

By Stokes's theorem, the absolute flux of vorticity through an open surface $A$ bounded by a completely reducible curve $C$ is

$$\Gamma_a = \iint_A \omega_a \cdot \mathbf{n} \, dA = \oint_C \mathbf{u}_I \cdot d\mathbf{r} \tag{2.2.1}$$

The line integral of $\mathbf{u}_I$ around $C$ is called the *circulation* of the velocity $\mathbf{u}_I$. Similarly the circulation of the relative velocity is

$$\oint_C \mathbf{u} \cdot d\mathbf{r} = \iint_A \omega \cdot \mathbf{n} \, dA = \Gamma. \tag{2.2.2}$$

The circulation is therefore another way to measure the vortex-tube strength. Consider now the rate of change of the circulation $\Gamma$ when $C$ is a *material curve*, i.e., a curve which always consists of the same fluid elements and which therefore is a curve moving with the fluid:

$$\frac{d\Gamma}{dt} = \frac{d}{dt} \oint_C \mathbf{u} \cdot d\mathbf{r} = \oint_C \frac{d\mathbf{u}}{dt} \cdot d\mathbf{r} + \oint_C \mathbf{u} \cdot \frac{d}{dt} (d\mathbf{r}). \tag{2.2.3}$$

The final term in (2.2.3) involves the rate of change of the differential vector line segments which connect adjacent fluid elements on $C$. Consider two fluid elements on $C$ whose positions differ by $\Delta \mathbf{r}$ as shown in Figure 2.2.1.

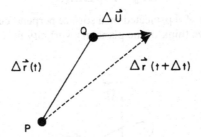

**Figure 2.2.1**   The change of an infinitesimal line element is determined by the velocity difference at the end points of the element.

Clearly, in a time $dt$ the separation and orientation of $\Delta \mathbf{r}$ will alter only if the velocity at $Q$ differs from the velocity at $P$, i.e.,

$$\Delta \mathbf{r}(t + \Delta t) - \Delta \mathbf{r}(t) = (\Delta \mathbf{u}) \, \Delta t, \tag{2.2.4}$$

or in the limit $\Delta t \to 0$

$$\frac{d}{dt} d\mathbf{r} = d\mathbf{u}. \tag{2.2.5}$$

This step depends entirely on $C$ being a *material* curve, so that the rate of change of $d\mathbf{r}$ is uniquely determined by the fluid velocity. We can then write (2.2.3) as

$$\frac{d\Gamma}{dt} = \oint_C \frac{d\mathbf{u}}{dt} \cdot d\mathbf{r} + \oint_C \mathbf{u} \cdot d\mathbf{u}$$

$$= \oint_C \frac{d\mathbf{u}}{dt} \cdot d\mathbf{r} + \frac{1}{2} \oint_C d|u|^2 \qquad (2.2.6)$$

$$= \oint_C \frac{d\mathbf{u}}{dt} \cdot d\mathbf{r},$$

since the second integral is the integral around a closed curve of a perfect differential. Thus, the rate of change of the relative circulation is the line integral around $C$ of the relative acceleration. From (1.6.7) it follows that

$$\frac{d\Gamma}{dt} = -\oint_C (2\boldsymbol{\Omega} \times \mathbf{u}) \cdot d\mathbf{r} - \oint_C \frac{\nabla p}{\rho} \cdot d\mathbf{r}$$

$$+ \oint_C \frac{\mathscr{F}}{\rho} \cdot d\mathbf{r}, \qquad (2.2.7)$$

since the line integral of $\nabla\Phi$ around the closed curve $C$ must vanish if the force potential $\Phi$ is single valued.

There are three mechanisms, according to (2.2.7) which are responsible for changes in $\Gamma$, i.e., which can alter the strength of a relative-vorticity tube during the motion. They are (1) the circulation of the Coriolis *force* per unit mass, (2) the circulation of the pressure force per unit mass, and (3) the circulation of $\mathscr{F}$.

(1) The effect of the Coriolis force can easily be illustrated. Consider a curve $C$, as in Figure 2.2.2, with $\boldsymbol{\Omega}$ directed out of the paper. Suppose, *on* $C$, the relative velocity vector is directed everywhere outward. Then aside from all other influences, the effect of the Coriolis force will deflect each fluid element to the right. This will tend to produce a

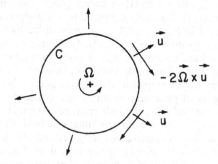

**Figure 2.2.2** A generally radial flow $\mathbf{u}$ gives rise to a clockwise Coriolis force and therefore a clockwise circulation.

relative circulation in the *clockwise* sense around $C$, i.e., $d\Gamma/dt < 0$, which is the content of (2.2.7).

Another way to represent the Coriolis-force contribution to the change of the relative circulation follows from the identity

$$-(2\mathbf{\Omega} \times \mathbf{u}) \cdot d\mathbf{r} = -2\mathbf{\Omega} \cdot (\mathbf{u} \times d\mathbf{r})$$
$$= -2\mathbf{\Omega} \cdot \mathbf{n}_A u_\perp \, d\mathbf{r}, \tag{2.2.8}$$

where $u_\perp$ is the magnitude of the velocity component perpendicular to $d\mathbf{r}$, and $\mathbf{n}_A$ is a unit vector parallel to $\mathbf{u} \times d\mathbf{r}$. In a time $dt$, the effect of $u_\perp$ on $d\mathbf{r}$ is to sweep out an area adjacent to $d\mathbf{r}$ whose size is

$$\delta A = dr \, dl, \tag{2.2.9}$$

where

$$dl = u_\perp \, dt. \tag{2.2.10}$$

The unit vector normal to this differential surface is $\mathbf{n}_A$; hence the rate of increase at the point $P$ on $C$, of the area encompassed by $C$, is

$$\frac{d}{dt} \delta A = u_\perp \, dr, \tag{2.2.11}$$

so that (2.2.8) becomes

$$-(2\mathbf{\Omega} \times \mathbf{u}) \cdot d\mathbf{r} = -2\Omega \frac{d}{dt} \delta A_n, \tag{2.2.8'}$$

where $\delta A_n$ is the increment of the surface enclosed by $C$ perpendicular to $\mathbf{\Omega}$. Integrating over all line elements $d\mathbf{r}$ around $C$ yields

$$-\oint_C 2\mathbf{\Omega} \times \mathbf{u} \cdot d\mathbf{r} = -2\Omega \frac{dA_n}{dt}, \tag{2.2.12}$$

where $A_n$ is the total area enclosed by $C$ projected on a plane perpendicular to $\mathbf{\Omega}$. Thus in the presence of the planetary vorticity $\Omega$, an increase of the area $A_n$ leads to a decrease of the relative circulation. The flux of relative vorticity through the loop $C$ will be decreased (in the absence of other effects) in direct proportion as the number of planetary-vorticity filaments captured by the loop is increased. In the example shown in Figure 2.2.3 the area bounded by $C$ expands in size, encompassing more planetary-vorticity filaments, and consequently tends to produce a negative (clockwise) circulation of the relative velocity around $C$. The same result is achieved if a surface does not expand but is merely tipped into the plane perpendicular to $\mathbf{\Omega}$, so that more filaments of planetary vorticity are collected by $C$. This phenomenon is precisely like the induction of electrical current in a wire loop due to its motion in a uniform magnetic field. In our case we have instead induction of relative vorticity around the loop $C$ due to its motion in the uniform planetary-vorticity field. The ability of relative motion to gain vorticity by induction of planetary vorticity is of particular importance for large-scale motions and explains why large-scale

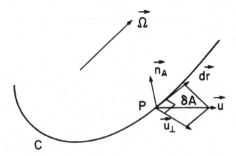

**Figure 2.2.3** The component of **u** perpendicular to $c$, $\bar{u}_\perp$, will, in an interval of time $dt$, increase the area bounded by $c$ by an amount $u_\perp\,dr\,dt$ for each line element $dr$ of $c$.

motions whose natural boundaries encompass many planetary-vorticity filaments are hardly ever free of relative vorticity.

(2) The meaning of the second term on the right-hand side of (2.2.7) which contributes to the rate of change of the relative circulation can be better understood by rewriting it as

$$-\oint_c \frac{\nabla p}{\rho}\cdot d\mathbf{r} = -\iint_A \nabla\times\left(\frac{\nabla p}{\rho}\right)\cdot\mathbf{n}\,dA$$

$$= \iint_A \frac{\nabla\rho\times\nabla p}{\rho^2}\cdot\mathbf{n}\,dA \qquad (2.2.13)$$

with the aid of Stokes's theorem. If the surfaces of constant pressure and constant density do *not* coincide, the state of the fluid is termed *baroclinic*. If the fluid is baroclinic, then the *baroclinic vector* $(\nabla\rho\times\nabla p)/\rho^2 \neq 0$, and the relative circulation will change with time if the average normal component of this vector on the surface $A$ is different from zero. To illustrate this mechanism consider the simple but archetypal situation depicted in Figure 2.2.4, wherein $\nabla\rho$ and $\nabla p$ are depicted as constant vectors which lie in the surface $A$. In the situation shown, the lighter fluid on the right feels the same upward pressure force, $-\nabla p$, as the heavier fluid on the left. The lighter fluid will therefore (in the absence of other effects) tend to rise more rapidly than the heavier fluid, resulting in a net counterclockwise circulation around $C$ in accordance with (2.2.7) and (2.2.13). If the density field has even the least tendency to be convected with the velocity field, the resulting circulation will tend to align the density and pressure surfaces.

Note that if the surfaces of constant $p$ and the surfaces of constant $\rho$ coincide (i.e., if the fluid is barotropic) then it is possible to find a relation between $p$ and $\rho$ such that we can write

$$\rho = \rho(p). \qquad (2.2.14)$$

If this is true on $C$, then

$$\oint_C \frac{\nabla p\cdot d\mathbf{r}}{\rho} = \oint_C \frac{dp}{\rho(p)} = \int_{p(A)}^{p(A)} \frac{dp}{\rho(p)} = 0, \qquad (2.2.15)$$

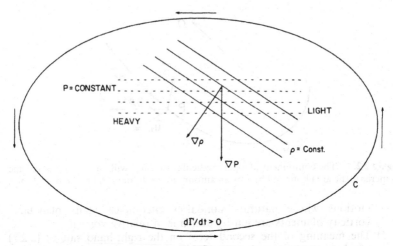

**Figure 2.2.4** The baroclinic production of vorticity. Surfaces of constant $p$ and $\rho$ which are not coincident will tend to increase the circulation on $c$ as shown.

for then (2.2.15) is an integral of a function of a single variable over an interval whose end points coincide.

(3) The final term in the budget for $\Gamma$ is the line integral of $\mathscr{F}$ around $C$. In general it is difficult to give a clear-cut interpretation of the effect of friction. In the case where $\mathscr{F}$ is given by (1.4.5),

$$\oint_C \frac{\mathscr{F}}{\rho} \cdot d\mathbf{r} = v \oint_C (\nabla^2 \mathbf{u}) \cdot d\mathbf{r} \qquad (2.2.16)$$

if the kinematic viscosity $v = \mu/\rho$ is sensibly constant on $C$. Note that

$$v \oint_C \nabla^2 \mathbf{u} \cdot d\mathbf{r} = -v \oint_C (\nabla \times \boldsymbol{\omega}) \cdot d\mathbf{r}, \qquad (2.2.17)$$

since

$$\nabla \times (\nabla \times \mathbf{u}) = \nabla(\nabla \cdot \mathbf{u}) - \nabla^2 \mathbf{u} \qquad (2.2.18)$$

has been used. Consider, for the purposes of illustration, the situation depicted in Figure 2.2.5. In this case the vorticity in the vicinity of the point $P$ on $C$ has a component, $\omega_z$, in the $z$-direction only, while the $x$-axis is oriented tangent to $C$ at $P$. For this case

$$-v(\nabla \times \boldsymbol{\omega}) \cdot d\mathbf{r} = -v \frac{\partial \omega_z}{\partial y} dx, \qquad (2.2.19)$$

so that the effect of the viscosity is to reduce the vortex-tube strength encompassed by $C$ by an amount proportional to the vorticity gradients on $C$. If the filaments encircled by $C$ are stronger inside rather than outside the loop, the vortex-tube strength will suffer a decrease in direct analogy with the flow of heat from a region as a result of

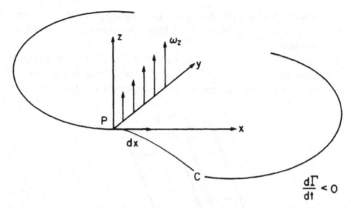

**Figure 2.2.5** At the point $P$, vorticity will diffuse away from the fluid bounded by $c$ if, as shown, the vorticity increases inwards from $c$.

temperature gradients on the boundary. Similarly the vorticity within $C$ will tend to diffuse by viscous action down the vorticity gradient and pass through the material elements which compose $C$.

## 2.3 Kelvin's Theorem

Since from (2.1.19)

$$\Gamma_a = \Gamma + 2\Omega A_n, \tag{2.3.1}$$

it follows from (2.2.7) and (2.2.12) that

$$\frac{d\Gamma_a}{dt} = - \oint_C \frac{\nabla p}{\rho} \cdot d\mathbf{r} + \oint_C \frac{\mathscr{F}}{\rho} \cdot d\mathbf{r}. \tag{2.3.2}$$

If

(1) the fluid is barotropic on $C$, and
(2) the frictional force vanishes on $C$

then the absolute circulation is conserved following the motion, i.e.,

$$\frac{d\Gamma_a}{dt} = 0. \tag{2.3.3}$$

This is *Kelvin's theorem*, and we will have many occasions to observe its application. Since $\Gamma_a$ is the sum of the planetary and relative vortex-tube strengths, a decrease of one produces a corresponding increase of the other. Kelvin's theorem therefore shows that the mechanism of vorticity induction is responsible for a transfer between relative and planetary vorticity so as to conserve the absolute vortex-tube strength. Thus as $C$ moves and deforms with the fluid, it always forms the boundary of a tube of fixed absolute vortex strength.

If the cross section of a tube decreases, the intensity of the tube strength, i.e., the vorticity $\omega_a$ in the tube, must increase in inverse proportion to the tube cross-sectional area.

Consider now the absolute-vorticity tube in Figure 2.3.1, which has been wrapped initially with a cylindrical "blanket" whose four corners are $A$, $A'$, $B$, and $B'$. Since at this initial instant this surface wraps a vortex tube, there is

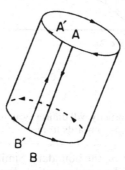

**Figure 2.3.1** The surface $BAA'B'$ about whose edge a contour is chosen to prove that vortex filaments move with the fluid.

no flux of vorticity through the surface, and the circulation about the circuit $A \rightarrow B \rightarrow B' \rightarrow A' \rightarrow A$ must also vanish. Let this blanket now move with the fluid. If Kelvin's theorem (2.3.3) applies, the circulation about the circuit must remain zero. That is, no absolute-vorticity filaments can penetrate the blanket. In the limit where $A'$ coalesces with $A$ while $B'$ coalesces with $B$, it follows that the surface of the absolute-vorticity tube remains coincident with the "blanket" of material elements. In other words, *absolute vortex tubes move with the fluid when Kelvin's theorem applies.* The limit of a vortex tube of vanishingly small cross section demonstrates the further result: When Kelvin's theorem is valid, individual filaments of absolute vorticity remain material lines, i.e., a line which is once a filament of absolute vorticity is always a filament, so that *absolute-vorticity filaments move with the fluid.*

Although, as we have seen, it is possible to construct tubes and filaments of relative vorticity, it is only the strength of the absolute-vorticity tube that is conserved as a consequence of Kelvin's theorem, and only the filaments of absolute vorticity that remain coincident with material lines.

The presence of viscosity allows vorticity filaments to diffuse away from their original fluid elements, while baroclinic effects can actually produce new vorticity filaments.

## 2.4  The Vorticity Equation

The circulation is a scalar measure of vorticity. Since vorticity is a vector, a discussion of the circulation yields, at best, an incomplete picture of the vector dynamics of the vorticity field. To further examine the vector nature of the vorticity, we will now derive directly an equation for the relative vorticity vector.

The vector identity

$$\boldsymbol{\omega} \times \mathbf{u} = (\mathbf{u} \cdot \nabla)\mathbf{u} - \nabla \frac{(|\mathbf{u}|^2)}{2} \tag{2.4.1}$$

allows the momentum equation to be rewritten as

$$\frac{\partial \mathbf{u}}{\partial t} + (2\boldsymbol{\Omega} + \boldsymbol{\omega}) \times \mathbf{u} = -\frac{\nabla p}{\rho} + \nabla\left\{\Phi - \frac{|\mathbf{u}|^2}{2}\right\} + \frac{\mathscr{F}}{\rho}. \tag{2.4.2}$$

To find an equation for $\boldsymbol{\omega}$, we now take the curl of (2.4.2) and obtain

$$\frac{\partial \boldsymbol{\omega}}{\partial t} + \nabla \times \{(2\boldsymbol{\Omega} + \boldsymbol{\omega}) \times \mathbf{u}\} = \frac{\nabla\rho \times \nabla p}{\rho^2} + \nabla \times \left(\frac{\mathscr{F}}{\rho}\right). \tag{2.4.3}$$

Now for any two vectors $\mathbf{A}$ and $\mathbf{B}$,

$$\nabla \times (\mathbf{A} \times \mathbf{B}) = \mathbf{A}\nabla \cdot \mathbf{B} + (\mathbf{B} \cdot \nabla)\mathbf{A} - \mathbf{B}\nabla \cdot \mathbf{A} - (\mathbf{A} \cdot \nabla)\mathbf{B}, \tag{2.4.4}$$

so that

$$\nabla \times \{(2\boldsymbol{\Omega} + \boldsymbol{\omega}) \times \mathbf{u}\}$$
$$= (2\boldsymbol{\Omega} + \boldsymbol{\omega})\nabla \cdot \mathbf{u} + (\mathbf{u} \cdot \nabla)(2\boldsymbol{\Omega} + \boldsymbol{\omega}) - (2\boldsymbol{\Omega} + \boldsymbol{\omega}) \cdot \nabla\mathbf{u}, \tag{2.4.5}$$

since $\boldsymbol{\omega} + 2\boldsymbol{\Omega}$ has zero divergence. Hence (2.4.3) becomes

$$\frac{d\boldsymbol{\omega}}{dt} = \boldsymbol{\omega}_a \cdot \nabla\mathbf{u} - \boldsymbol{\omega}_a\nabla \cdot \mathbf{u} + \frac{\nabla\rho \times \nabla p}{\rho^2} + \nabla \times \frac{\mathscr{F}}{\rho}, \tag{2.4.6}$$

where, recall,

$$\boldsymbol{\omega}_a = \boldsymbol{\omega} + 2\boldsymbol{\Omega}. \tag{2.4.7}$$

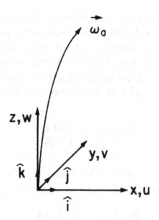

**Figure 2.4.1**   A local coordinate frame for the interpretation of (2.4.6).

The rate of change of the absolute vorticity following a fluid element is due to the four terms on the right-hand side of (2.4.6). The last two terms are the baroclinic vector and the curl of the friction force; the interpretation of these vorticity sources has already been given in our earlier discussion of the rate of change of the circulation. The first two terms on the right-hand side of (2.4.6) warrant further discussion.

Consider an arbitrary point $P$ in the fluid, and choose a local Cartesian coordinate frame such that at $P$ the $z$-axis of the frame is tangent to the absolute-vorticity filament passing through $P$, as shown in Figure 2.4.1.

In this coordinate frame $\boldsymbol{\omega}_a$ at the point $P$ is simply the vector $\omega_a \hat{\mathbf{k}}$, where $\hat{\mathbf{k}}$ is a unit vector along the $z$-axis. It then follows that at $P$, the first two terms on the right-hand side of (2.4.6) are

$$\boldsymbol{\omega}_a \cdot \nabla \mathbf{u} - \boldsymbol{\omega}_a \nabla \cdot \mathbf{u}$$

$$= \omega_a \frac{\partial}{\partial z} \{ u\hat{\mathbf{i}} + v\hat{\mathbf{j}} + w\hat{\mathbf{k}} \} - \omega_a \hat{\mathbf{k}} \left| \frac{\partial u}{\partial x} + \frac{\partial v}{\partial y} + \frac{\partial w}{\partial z} \right| \qquad (2.4.8)$$

$$= \hat{\mathbf{i}} \omega_a \frac{\partial u}{\partial z} + \hat{\mathbf{j}} \omega_a \frac{\partial v}{\partial z} - \hat{\mathbf{k}} \omega_a \left| \frac{\partial u}{\partial x} + \frac{\partial v}{\partial y} \right|,$$

where $\hat{\mathbf{i}}$ and $\hat{\mathbf{j}}$ are unit vectors along the $x$- and $y$-axes respectively. The relative velocity components parallel to $\hat{\mathbf{i}}$, $\hat{\mathbf{j}}$, and $\hat{\mathbf{k}}$ are $u$, $v$, and $w$.

Consider now the rate of change of the vorticity component parallel to the $z$-axis, i.e., in the direction of the preexisting vorticity. According to (2.4.8) the contribution made by the terms under discussion to the rate of change of the $z$-component is given (in the absence of baroclinicity and friction) by

$$\frac{d\omega_z}{dt} = -(\omega_a)_z \left( \frac{\partial u}{\partial x} + \frac{\partial v}{\partial y} \right) \qquad (2.4.9)$$

and is simply the product of the *total* vorticity originally present multiplied by the *convergence* of the velocity component in the plane perpendicular to $\boldsymbol{\omega}_a$. If the convergence is positive (i.e., if $\partial u/\partial x + \partial v/\partial y < 0$), absolute-vorticity filaments will be gathered closer together, increasing the magnitude of the vorticity vector by decreasing the cross-sectional area of the local vortex tube in accordance with Kelvin's theorem. Indeed, it can be shown that

$$\frac{\partial u}{\partial x} + \frac{\partial v}{\partial y} = \frac{1}{A_\perp} \frac{dA_\perp}{dt} \qquad (2.4.10)$$

where $A_\perp$ is a differential area element in the $x$, $y$ plane; thus (2.4.9) can be written, since $\boldsymbol{\Omega}$ is a constant, as

$$\frac{d(\omega_a)_z}{dt} = -\frac{(\omega_a)_z}{A_\perp} \frac{dA_\perp}{dt},$$

or

$$\frac{d}{dt}(\omega_a)_z A_\perp = 0, \tag{2.4.11}$$

which is clearly the same information Kelvin's theorem would yield. This mechanism for the increase of the component of vorticity in the direction of the filament is referred to as *vortex-tube stretching*, since the decrease of the cross section of the tube in a nearly incompressible fluid can be accomplished only by extension of the material tube along its length so as to preserve its volume.

Examination of (2.4.8) shows, however, that this is by no means the whole story. Consider the rate of increase of the component of vorticity in the x-direction, i.e., perpendicular to the direction of the vorticity filament. According to (2.4.8) the rate of increase of this component is $\omega_a \, \partial u/\partial z$, so that after an interval $\Delta t$ (see Figure 2.4.2)

$$\frac{(\Delta\omega_a)_x}{\omega_a} = \frac{\partial u}{\partial z} \Delta t = \frac{\Delta x}{\Delta z} = \tan \gamma, \tag{2.4.12}$$

where $\Delta x$ is the projection on the x-axis of the material line element $\Delta l$ which initially was parallel to $\omega_a$ and had length $\Delta z$. This material line element is tilted by the shear of the velocity field through an angle $\gamma$, and (2.4.12) reveals that the new *vorticity* parallel to the x-axis is the x-component of the vorticity vector at $t + \Delta t$, which itself has been tilted by the same angle as the material line. This is no more than the manifestation of the fact that

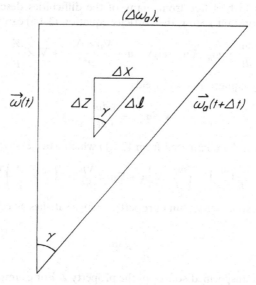

**Figure 2.4.2** The vorticity vector at $t$ and $t + \Delta t$, showing the tendency for $\omega_a$ to tip the same angle $\gamma$ as a material element initially aligned parallel to $\Omega_a$.

absolute-vorticity filaments move with the fluid in the absence of baroclinicity and friction. The same argument applies to the $y$ direction.

Therefore, the rate of change of the relative vorticity is equal to the sum of

(i) the production of vorticity by baroclinicity,
(ii) the diffusive effects of friction,
(iii) the vortex-tube stretching, which alters the vorticity parallel to the filament by convergence of the filaments, and
(iv) the vortex tilting by the variation, along the direction of the filament, of the velocity component perpendicular to the filaments of $\omega_a$.

It is important to note that if $\omega$ is initially zero, the twisting and convergence of vortex lines in the fluid by the relative velocity in the presence of the planetary vorticity will, if allowed to proceed, introduce relative vorticity.

## 2.5 Potential Vorticity

The vorticity equation is illuminating because it deals directly with the vector character of vorticity. However, it is more a description of how vorticity is changed than a useful constraint on that change. Kelvin's theorem is more powerful, but is an integral theorem dealing with a scalar and requires a knowledge of the detailed evolution of material surfaces in the fluid. Moreover, it is valid only in the absence of baroclinic effects, which substantially restricts its geophysical application. A beautiful and unusually useful theorem due to Ertel (1942) can be derived, which provides a constraint on the vorticity which is free from many of the difficulties described above. Since $\Omega$ is a constant vector, the vorticity equation (2.4.6) can be written as

$$\frac{d\omega_a}{dt} = \omega_a \cdot \nabla u - \omega_a \nabla \cdot u + \frac{\nabla \rho \times \nabla p}{\rho^2} + \nabla \times \frac{\mathscr{F}}{\rho}. \tag{2.5.1}$$

The continuity equation (1.4.2) yields

$$\nabla \cdot u = -\frac{1}{\rho} \frac{d\rho}{dt}, \tag{2.5.2}$$

so that $\nabla \cdot u$ can be eliminated from (2.5.1) which can be rewritten

$$\frac{d}{dt}\left(\frac{\omega_a}{\rho}\right) = \left(\frac{\omega_a}{\rho} \cdot \nabla\right) u + \nabla\rho \times \frac{\nabla p}{\rho^3} + \left(\nabla \times \frac{\mathscr{F}}{\rho}\right)\frac{1}{\rho}. \tag{2.5.3}$$

Consider now some scalar fluid property $\lambda$ which satisfies an equation of the form

$$\frac{d\lambda}{dt} = \Psi, \tag{2.5.4}$$

where $\Psi$ is an unspecified source of the property $\lambda$. For example, for atmospheric motions $\lambda$ could be the potential temperature $\theta$, and $\Psi$ would then be the collection of terms on the right-hand side of (1.4.18). Or, for a liquid whose state equation can be approximated by (1.4.20), $\lambda$ might be chosen as the density, in which case $\Psi$ would be given by the right-hand side of (1.4.22). There are many other possibilities that will suggest themselves in particular situations.

Now

$$\frac{\omega_a}{\rho} \cdot \frac{d}{dt} \nabla \lambda = \left( \frac{\omega_a}{\rho} \cdot \nabla \right) \frac{d\lambda}{dt} - \left[ \left( \frac{\omega_a}{\rho} \cdot \nabla \mathbf{u} \right) \right] \cdot \nabla \lambda, \tag{2.5.5}$$

which can be verified by expanding (2.5.5) into component form, an exercise which is left to the reader. If the dot product of $\nabla \lambda$ and (2.5.3) is taken, we obtain

$$\nabla \lambda \cdot \frac{d}{dt} \left( \frac{\omega_a}{\rho} \right) = \left[ \left( \frac{\omega_a}{\rho} \cdot \nabla \right) \mathbf{u} \right] \cdot \nabla \lambda + \nabla \lambda \cdot \left[ \frac{\nabla \rho \times \nabla p}{\rho^3} \right] + \frac{\nabla \lambda}{\rho} \cdot \left\{ \nabla \times \frac{\mathscr{F}}{\rho} \right\} \tag{2.5.6}$$

and the sum of (2.5.5) and (2.5.6) then yields, with (2.5.4),

$$\frac{d}{dt} \left\{ \frac{\omega_a}{\rho} \cdot \nabla \lambda \right\} = \frac{\omega_a}{\rho} \cdot \nabla \Psi + \nabla \lambda \cdot \left[ \frac{\nabla \rho \times \nabla p}{\rho^3} \right] + \frac{\nabla \lambda}{\rho} \cdot \left\{ \nabla \times \frac{\mathscr{F}}{\rho} \right\}. \tag{2.5.7}$$

If

(1) $\lambda$ is a conserved quantity for each fluid element, i.e., $\Psi = 0$,
(2) the frictional force is negligible, i.e., $\mathscr{F} = 0$,
    and *either*
(3a) the fluid is barotropic, i.e., $\nabla \rho \times \nabla p = 0$,
    *or*
(3b) $\lambda$ can be considered a function only of $p$ and $\rho$,

then the *potential vorticity*

$$\Pi = \frac{(\omega + 2\Omega)}{\rho} \cdot \nabla \lambda \tag{2.5.8}$$

is conserved by each fluid element, i.e.,

$$\boxed{\frac{d\Pi}{dt} = 0} . \tag{2.5.9}$$

Note that the scalar $\Pi$ involves the component of $\omega_a$ parallel to the gradient of $\lambda$.

In the examples listed above for the choice of $\lambda$, the condition that $\lambda$ is conserved is equivalent to the condition that the motion is adiabatic, but this is not a general requirement. It is important to emphasize that the theorem does not require the fluid to be barotropic, for if condition (3b) obtains [i.e., if $\lambda = \lambda(p, \rho)$, i.e., if $\lambda$ is a thermodynamic function], then

$$\nabla \lambda = \frac{\partial \lambda}{\partial p} \nabla p + \frac{\partial \lambda}{\partial \rho} \nabla \rho, \tag{2.5.10}$$

whose dot product with the baroclinic vector $\nabla\rho \times \nabla p$ trivially vanishes. However, if $\nabla\rho \times \nabla p$ should itself vanish, the possible candidates for the choice of $\lambda$ need not be restricted to a thermodynamic function, and a wider class of functions $\lambda$ are allowable, only the condition that $\lambda$ is conserved being required.

The concept of potential vorticity is so important that it is useful to consider an alternative derivation of the theorem which is more physically revealing.

It follows from the considerations of Section 2.3 that in the absence of friction

$$\frac{d}{dt} \iint_A \boldsymbol{\omega}_a \cdot \mathbf{n} \, dA = \iint_A \left( \frac{\nabla\rho \times \nabla p}{\rho^2} \right) \cdot \mathbf{n} \, dA, \qquad (2.5.11)$$

where $A$ is an area in *any* material surface and $\mathbf{n}$ is the unit normal to that surface. Suppose that the contour $C$ which encloses $A$ is *chosen* to lie initially in a surface of constant $\lambda$, as shown in Figure 2.5.1. Since $\lambda$ is conserved by

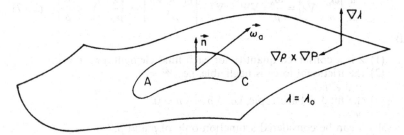

**Figure 2.5.1** The contour for the application of the circulation theorem in a surface of constant $\lambda$.

each fluid element, the surface $\lambda = \lambda_0$ remains composed of the same fluid elements, i.e., $\lambda = \lambda_0$ is a material surface, so that $C$, which is a material line, remains in the same surface. If $\lambda$ is a function of $p$ and $\rho$ only, then $\nabla\rho \times \nabla p$ must lie in the surface of constant $\lambda$, since by (2.5.10) it is perpendicular to $\nabla\lambda$. Therefore, for the circuit $C$ as chosen, Kelvin's theorem applies, since the right-hand side of (2.5.11) vanishes to yield

$$\frac{d}{dt} \iint_A \boldsymbol{\omega}_a \cdot \mathbf{n} \, dA = 0, \qquad (2.5.12)$$

which remains true for all $t$, since $C$ remains in a surface of constant $\lambda$. Consider now a tiny circuit $C$ which is sufficiently small that (2.5.12) can be written

$$\frac{d}{dt} (\boldsymbol{\omega}_a \cdot \mathbf{n} \, \delta A) = 0, \qquad (2.5.13)$$

where $\delta A$ is the small surface element enclosed by $C$, and $\boldsymbol{\omega}_a$ is the mean value of $\boldsymbol{\omega}_a$ over the differential surface, as shown in Figure 2.5.2.

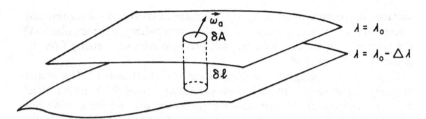

**Figure 2.5.2** The element of mass bounded by the two $\lambda$-surfaces. The mass in the elementary pillbox is $\rho \, \delta l \, \delta A$.

To find a useful expression for $\delta A$ it is only necessary to consider the mass contained in a tiny cylindrical pillbox between two surfaces of constant $\lambda$. The separation of the surfaces is given by $\delta l$, which is related to the value of $\lambda$ on the neighboring surface by

$$\delta l |\nabla \lambda| = \Delta \lambda \qquad (2.5.14)$$

by the definition of $\nabla \lambda$. The mass in the pillbox is

$$\delta m = \rho \, \delta A \, \delta l$$

$$= \rho \, \delta A \, \frac{\Delta \lambda}{|\nabla \lambda|} \qquad (2.5.15)$$

and is conserved, since the boundary surfaces of the box are material surfaces. It follows that (2.5.13) can be written

$$\frac{d}{dt} \left\{ \frac{\omega_a \cdot \mathbf{n}}{\rho} |\nabla \lambda| \frac{\delta m}{\Delta \lambda} \right\} = 0. \qquad (2.5.16)$$

Since $\delta m$ and $\Delta \lambda$ are constant following the motion and

$$\nabla \lambda = \mathbf{n} |\nabla \lambda|, \qquad (2.5.17)$$

this becomes

$$\frac{d}{dt} \left( \omega_a \cdot \frac{\nabla \lambda}{\rho} \right) = 0, \qquad (2.5.18)$$

which is identical to (2.5.8) and (2.5.9). The conservation of potential vorticity is therefore really Kelvin's theorem for a very special but useful contour.

In what sense is $\Pi$ "potential" vorticity? The name seems poorly chosen, for $\Pi$ does not even have the dimensions of vorticity. The term is by now traditional, however, and the motive for it rests in the following considerations. If $\Pi$ is conserved following a fluid element, when the distance between two adjacent $\lambda$-surfaces increases, $\nabla \lambda$ must decrease, and the component of the vector $\omega_a/\rho$ parallel to $\nabla \lambda$ must increase proportionally to keep $\Pi$ constant. If $\rho$ doesn't vary very much, this will be manifested as an increase of $\omega_a$ and we may consider that there is a reservoir of vorticity associated with the packing together of $\lambda$-surfaces which can be released as the $\lambda$-

surfaces are stretched apart by the mechanism of vortex-tube stretching. Except in the pathological case where $\nabla \lambda$ is everywhere perpendicular to $\Omega$, the stretching apart of $\lambda$-surfaces will produce relative vorticity from the planetary vorticity field.

It is hard to exaggerate the importance of the theorem of potential-vorticity conservation. Indeed, a proper statement of the theorem which explicitly recognizes the constraints rotation imposes on the motion provides, as we shall see, the governing equation for large-scale nondissipative motions in the atmosphere and ocean.

## 2.6 The Thermal Wind

In the discussion so far, we have not exploited the fact that for the motions we will be concerned with, the Rossby number is small, i.e., the relative vorticity is small compared to the planetary vorticity. Suppose we now ask if any conditions are required for the persistence of motions whose relative vorticity remains small with respect to $2\Omega$ and whose time scales for change are long compared to a rotation period. Clearly some constraint is necessary, else the shear in the relative velocity field alone will rapidly produce vorticity, by the tilting mechanism, at a rate of $O(2\Omega\, U/D)$, which yields a time scale for relative-vorticity change of the order of the rotation period, in contradiction with our hypotheses (and observation) about the time scale of large-scale motions.

The desired constraint can be derived directly from (2.4.6). In the absence of friction, and if $\omega \ll 2\Omega$, (2.4.6) can be approximated by

$$\frac{d\omega}{dt} = 2\Omega \cdot \nabla u - 2\Omega \nabla \cdot u + \frac{\nabla\rho \times \nabla p}{\rho^2}. \tag{2.6.1}$$

Now

$$\frac{d\omega}{dt} = \frac{\partial\omega}{\partial t} + u \cdot \nabla\omega, \tag{2.6.2}$$

so that the rate of change of vorticity is the sum of the local rate of change plus the convective rate of change. The magnitude of the first is $O(\omega/\tau)$, where $\tau$ is a characteristic time scale for the motion, while the order of magnitude of the second term $u \cdot \nabla\omega$ is $\omega U/L$. Since $\omega$ is of order* $U/L$, the ratio of the left hand side of (2.6.1) to either $(2\Omega \cdot \nabla)u$ or $2\Omega\nabla \cdot u$, which are both $O(2\Omega U/L)$, is

$$\frac{d\omega/dt}{2\Omega U/L} = \left\{ (2\Omega\tau)^{-1}, \frac{U}{2\Omega L} \right\}. \tag{2.6.3}$$

* In all of these estimates we are assuming there is a single scale which characterizes spatial variations. For motions of small aspect ratio this is not true, but the arguments proceed with only minor variation.

The order of the ratio is the larger of two terms in curly brackets in (2.6.3). We recognize the second term as the Rossby number $\varepsilon$. The other is also a Rossby number, but based on the local rather than the convective time scale. If both time scales are long compared with the rotation period, i.e., if the Rossby number is small, the approximate vorticity equation becomes

$$(2\mathbf{\Omega} \cdot \nabla)\mathbf{u} - 2\mathbf{\Omega}\nabla \cdot \mathbf{u} = -\frac{\nabla\rho \times \nabla p}{\rho^2}, \qquad (2.6.4)$$

which is the condition for the persistence of motions with time scales long compared to the rotation period. That is, the baroclinic production of vorticity must *cancel* the production of relative vorticity by the stretching and twisting terms.

Let us write (2.6.4) in component form in a frame whose $z$ axis is parallel to $\mathbf{\Omega}$. Then

$$2\Omega\frac{\partial u}{\partial z} = -\frac{1}{\rho^2}\left(\frac{\partial p}{\partial z}\frac{\partial \rho}{\partial y} - \frac{\partial p}{\partial y}\frac{\partial \rho}{\partial z}\right) \qquad (2.6.5a)$$

$$2\Omega\frac{\partial v}{\partial z} = \frac{1}{\rho^2}\left(\frac{\partial p}{\partial z}\frac{\partial \rho}{\partial x} - \frac{\partial p}{\partial x}\frac{\partial \rho}{\partial z}\right) \qquad (2.6.5b)$$

$$2\Omega\left(\frac{\partial u}{\partial x} + \frac{\partial v}{\partial y}\right) = -\frac{1}{\rho^2}\left(\frac{\partial p}{\partial x}\frac{\partial \rho}{\partial y} - \frac{\partial p}{\partial y}\frac{\partial \rho}{\partial x}\right). \qquad (2.6.5c)$$

The first two of these component equations relate the variation along the rotation axis of the velocity component in the plane perpendicular to that axis to the existence of density variations. Since the density variations are commonly connected with temperature variations, the winds or currents implied by (2.6.5a,b) are called the *thermal wind*. In Section 2.8 we shall return to a more detailed discussion of (2.6.5a,b). The final component equation (2.6.5c) places a limit on the allowable horizontal divergence of velocity. We shall show in (2.8) that the terms on the right-hand side of (2.6.5c) are, in reality, no larger than terms in the vorticity equation we have already ignored as small, so that for consistency we must write (2.6.5c) as

$$2\Omega\left(\frac{\partial u}{\partial x} + \frac{\partial v}{\partial y}\right) = 0, \qquad (2.6.6)$$

so that the velocity component in the plane perpendicular to $\mathbf{\Omega}$ must be nondivergent in order for the Rossby number to remain small.

## 2.7 The Taylor–Proudman Theorem

If the fluid satisfies the conditions required for the validity of (2.6.5)—i.e., if the Rossby number is small and friction can be ignored—and if *in addition* the baroclinic vector is identically zero, then it follows from (2.6.5) that

$$\frac{\partial u}{\partial z} = \frac{\partial v}{\partial z} = \frac{\partial u}{\partial x} + \frac{\partial v}{\partial y} = 0. \tag{2.7.1}$$

The velocity in planes perpendicular to $\Omega$ must be independent of the coordinate parallel to $\Omega$. Furthermore, that velocity must be nondivergent in the plane perpendicular to $\Omega$. This is a very powerful constraint, for it implies that a material line once parallel to $\Omega$ must always remain so. This constraint is a direct consequence of the fact, previously shown, that in the absence of baroclinicity and friction, absolute-vorticity filaments must be material lines. For small Rossby number the absolute vorticity is primarily the planetary vorticity $2\Omega$. With an error of $O(\varepsilon)$ the absolute vorticity filaments are always parallel to $\Omega$, and therefore material lines must also remain parallel to $\Omega$. Otherwise, unbalanced tilting of the panetary vorticity filaments would occur which would violate the conditions of small $\varepsilon$.

If the fluid is essentially incompressible, then from (1.4.23)

$$\frac{\partial u}{\partial x} + \frac{\partial v}{\partial y} + \frac{\partial w}{\partial z} = 0, \tag{2.7.2}$$

which with (2.7.1) implies that

$$\frac{\partial w}{\partial z} = 0. \tag{2.7.3}$$

In this case all three components of the relative velocity are independent of the direction parallel to the rotation axis. In vector form, this follows immediately from (2.6.4) if $\nabla \cdot \mathbf{u} = 0$, i.e.,

$$(2\Omega \cdot \nabla)\mathbf{u} = 0. \tag{2.7.4}$$

This constraint is called the Taylor–Proudman theorem. It follows from (2.7.3) that if $w$ is zero at some level, for example at a rigid, horizontal surface, it remains zero for all $z$. The motion is then completely two dimensional and can be pictured as moving in columns, each column oriented parallel to the rotation axis and moving so as to maintain this orientation. The columns themselves are most often referred to as Taylor columns and more rarely (but more alliteratively) as Proudman pillars.

The simplest situation in which such motions can occur is in the slow, relative motion of a homogeneous fluid (i.e., a fluid of uniform density). Motions of this type, extraordinary as it may seem, can easily be observed in laboratory experiments. If a body, such as a sphere or cylinder, is towed in a homogeneous fluid on a path perpendicular to the rotation axis, fluid must stream around the object as it passes through the fluid. If (2.7.4) is correct, the motion is strictly two dimensional. Fluid above and below the body must imitate the fluid parted by the body and allow a phantom body, consisting of the fluid contained in the Taylor column formed by the projection of the body along the rotation axis, to pass through the fluid as if it too were solid. "The idea," to quote from the seminal paper of Taylor (1923), "appears fantastic, but the experiments ... show that the true motion does, in fact, approximate to this curious type," as indeed he demonstrated in a beautiful set of simple and elegant experiments described in his paper.

It is important to recall that this theorem is based on a series of approximations to the complete vorticity equation. The neglected terms never exactly vanish, so the Taylor–Proudman constraint is only as strong as the neglected terms are weak, and departures from columnar motion are to be expected. Nevertheless, the tendency for strong coupling of the motion in the direction along the rotation axis is well demonstrated by (2.7.4).

## 2.8 Geostrophic Motion

The same conditions that led to the thermal-wind approximation to the vorticity equation and the Taylor–Proudman theorem have similarly important consequences at the more fundamental level of the momentum equation,

$$\frac{d\mathbf{u}}{dt} + 2\mathbf{\Omega} \times \mathbf{u} = -\frac{\nabla p}{\rho} + \nabla \Phi + \frac{\mathscr{F}}{\rho}. \tag{2.8.1}$$

The order of magnitude of the relative acceleration is given by the estimate

$$\frac{d\mathbf{u}}{dt} = \frac{\partial \mathbf{u}}{\partial t} + (\mathbf{u} \cdot \nabla)\mathbf{u} = O\left(\frac{U}{\tau}, \frac{U^2}{L}\right), \tag{2.8.2}$$

whose ratio to the estimate of the Coriolis acceleration

$$2\mathbf{\Omega} \times \mathbf{u} = O(2\Omega U)$$

is

$$\frac{|d\mathbf{u}/dt|}{|2\mathbf{\Omega} \times \mathbf{u}|} = O\left[(2\Omega\tau)^{-1}, \frac{U}{2\Omega L}\right], \tag{2.8.3}$$

which is identical to (2.6.3). Hence the relative accelerations are negligible to the lowest order.

To estimate the frictional force a representation of $\mathscr{F}$ must be specified. If $\mathscr{F}$ is given by (1.4.5), then

$$\frac{\mathscr{F}}{\rho} = O\left(\frac{\nu U}{L^2}\right) \tag{2.8.4}$$

where $\nu$ is the kinematic viscosity. Again we have assumed that a single length scale characterizes the variation of $U$. If not, the smallest relevant length scale must be used for (2.8.4). For now we retain (2.8.4) as suitable to our purpose. The ratio of the frictional force per unit mass to the Coriolis acceleration is a nondimensional parameter, called the Ekman number, $E$:

$$E = \frac{\nu U/L^2}{2\Omega U} = \frac{\nu}{2\Omega L^2}. \tag{2.8.5}$$

If $\nu$ is the molecular, kinematic viscosity of water, for example, a straight-forward estimate for $E$ for oceanic motions would be, for $L = 10^3$ km, $\nu = 10^{-2}$ cm$^2$ s$^{-1}$,

$$E = \frac{10^{-2} \text{ cm}^2 \text{ s}^{-1}}{(10^{-4} \text{ s}^{-1})(10^{16} \text{ cm}^2)} = 10^{-14}. \tag{2.8.6}$$

This is a terribly small number, and such frictional forces are clearly negligible for large scale motions. The important issue is whether (1.4.5) is an adequate representation of the dissipation of large-scale motions, and therefore whether (2.8.5) is an adequate measure of the importance of friction. By focusing our attention on large-scale motions alone, that is, by characterizing the motion by single scales for velocity and length, the details of the interaction of the large-scale motion with motions of smaller scales and *different* dynamical characteristics has been ignored. This is done of necessity, since the complexity involved in determining the interaction of motions of widely varying scales is overwhelming. It is simply out of the question for us to attempt to deal exactly with all interacting scales. Instead, some sort of *ad hoc* assumption is sought which attempts to describe at least qualitatively the transfer of energy and momentum between the scales of interest and the much smaller-scale, usually turbulent motions we do not wish to deal with explicitly. This question is discussed at greater length in Chapter 4. For the present purposes it is only necessary to note that one way to estimate the dissipative influence of smaller-scale motions is to retain (2.8.5) but replace $\nu$ by a *turbulent viscosity*, $A$, of much larger magnitude than the molecular value, supposedly because of the greater efficiency of momentum transport by macroscopic chunks of fluid. This is, at best, an empirical concept—hard to justify, even harder to quantify, and impossible to derive rigorously. In some cases it is clearly wrong to imagine that small-scale turbulence acts on the large-scale flow as massive molecules. Nevertheless, *a posteriori* estimates of $A$, difficult and ambiguous as they are to make, still give us a more realistic measure of the importance of friction and the accompanying estimate of the Ekman number. We defer a more detailed discussion of this vexing issue and remark here that observations indicate that friction is small enough that the appropriately defined Ekman number is small.

If both the Ekman number and the Rossby number are small, the first approximation to the momentum equation is

$$\rho \, 2\boldsymbol{\Omega} \times \mathbf{u} = -\nabla p + \rho\nabla\Phi. \tag{2.8.7}$$

Consider the application of (2.8.7) to the earth's atmosphere or ocean, that is, to a thin layer of fluid on a sphere. The surface of the earth is essentially a geopotential surface (i.e., a surface of constant $\Phi$), and $\nabla\Phi$ yields the effective

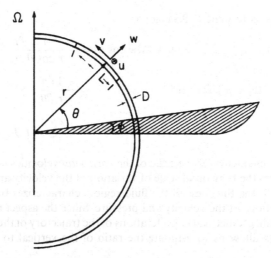

**Figure 2.8.1**   The spherical coordinate system $r$, $\theta$, $\phi$ used to describe motion with a characteristic length scale $L$ and depth $D$ in the vicinity of latitude $\theta$.

gravitational acceleration, g, normal to the earth's surface. Let $u$, $v$, and $w$ refer to velocities which are eastward, northward, and vertical as shown in Figure 2.8.1. The appropriate spherical coordinates are $r$, $\theta$, and $\phi$ as shown in the figure.

In component form (2.8.7) is

$$\rho[-2\Omega v \sin\theta + 2\Omega w \cos\theta] = -\frac{1}{r\cos\theta}\frac{\partial p}{\partial \phi}, \qquad (2.8.8a)$$

$$\rho 2\Omega u \sin\theta = -\frac{1}{r}\frac{\partial p}{\partial \theta}, \qquad (2.8.8b)$$

$$-\rho 2\Omega u \cos\theta = -\frac{\partial p}{\partial r} - \rho g. \qquad (2.8.8c)$$

It is helpful at this point to partition the pressure and density fields into two parts. In the absence of relative motion, $u = v = w = 0$, (2.8.8) implies that $p$ must be independent of $\phi$ and $\theta$ and therefore be a function only of $r$. The density must also, by (2.8.8c), be a function only of $r$. We therefore write

$$p = p_s(r) + p'(r, \theta, \phi), \qquad (2.8.9a)$$

$$\rho = \rho_s(r) + \rho'(r, \theta, \phi), \qquad (2.8.9b)$$

where $p_s(r)$ and $\rho_s(r)$ are the fields that would be present in the absence of motion, while $p'$ and $\rho'$ are the departures from this basic state due to the existence of winds and currents. It follows from (2.8.8c) that

$$-\frac{\partial p_s}{\partial r} = \rho_s g. \qquad (2.8.10)$$

This allows us to write (2.8.8a,b,c) as

$$(\rho_s + \rho')[-2\Omega v \sin \theta + 2\Omega w \cos \theta] = -\frac{1}{r \cos \theta} \frac{\partial p'}{\partial \phi}, \quad (2.8.11a)$$

$$(\rho_s + \rho')2\Omega u \sin \theta = -\frac{1}{r} \frac{\partial p'}{\partial \theta}, \quad (2.8.11b)$$

$$-(\rho_s + \rho')2\Omega u \cos \theta = -\frac{\partial p'}{\partial r} - \rho'g. \quad (2.8.11c)$$

Suppose $U$ characterizes the scale of the *horizontal* velocities $u$ and $v$, and $L$ characterizes the horizontal scale of variation of the velocity and pressure fields, while $D$, the thickness of the fluid region, characterizes the scale for vertical variations of the velocity and pressure. Since the aspect ratio of the motion is small, geometrical considerations of the trajectory of the motion of fluid elements allow us to estimate the ratio of the vertical to horizontal velocities as

$$\frac{w}{u} = O\left(\frac{w}{v}\right) = O\left(\frac{D}{L}\right) \ll 1, \quad (2.8.12)$$

so that the second term in the square bracket can be ignored* as long as $\delta = D/L$ is small. Since the horizontal scales of variation of $p'$ are $O(L)$, the pressure-gradient terms on the right-hand sides of (2.8.11a,b) are of order $p'/L$. In order for the horizontal pressure gradient to balance the Coriolis acceleration, we must have

$$p' = O(\rho 2\Omega U L), \quad (2.8.13)$$

which determines the magnitude of the pressure change due to the motion. In the vertical equation of motion the vertical pressure gradient can now be estimated, i.e.,

$$\frac{\partial p'}{\partial r} = O\left(\frac{p'}{D}\right) = O\left(\rho \frac{2\Omega U L}{D}\right). \quad (2.8.14)$$

The vertical component of the Coriolis acceleration is

$$\rho 2\Omega u \cos \theta = O(\rho 2\Omega U), \quad (2.8.15)$$

and the ratio of the two is

$$\frac{\rho 2\Omega u \cos \theta}{\partial p'/\partial r} = O\left(\frac{D}{L}\right) = \delta \ll 1, \quad (2.8.16)$$

so that to the same order, $\delta \ll 1$, the Coriolis term proportional to $2\Omega \cos \theta$ in both the horizontal and vertical equations can be neglected. That is, the

---

* Near the equator the terms proportional to $2\Omega \sin \theta$ are small. This does not imply that the Coriolis terms proportional to $2\Omega \cos \theta$ are significant. They still remain unimportant, but other dynamical terms, ignored in the approximation required to obtain (2.8.11a,b,c), must be retained near the equator.

horizontal Coriolis acceleration due to the vertical motion and the vertical Coriolis acceleration due to the horizontal motion are both small terms of $O(\delta)$ when compared to the pressure gradients in their *respective* equations.

This implies that only the local normal component of the earth's rotation, $\Omega \sin \theta$, is dynamically significant.

From (2.8.11c) it is possible to determine an upper bound on the magnitude of the density perturbation $\rho'$, for it is clear that $g\rho'$ can be no larger than $\partial p'/\partial r$, so that

$$\rho' \leq O\left(\frac{p'}{gD}\right) = O\left(\frac{\rho 2\Omega U L}{gD}\right), \tag{2.8.17}$$

or

$$\frac{\rho'}{\rho} = O\left(\frac{U}{2\Omega L}\right)\frac{4\Omega^2 L^2}{gD}$$

$$= \varepsilon\frac{4\Omega^2 L^2}{gD}. \tag{2.8.18}$$

The parameter $4\Omega^2 L^2/gD$ is independent of the intensity of the motion and depends only on its geometric scale. For example, for large-scale atmospheric motions for which $L$ is $10^3$ km and $D$ is 10 km, $4\Omega^2 L^2/gD$ is 0.196. Thus, as long as the Rossby number is small,

$$\frac{\rho'}{\rho} \leq O(\varepsilon) \ll 1, \tag{2.8.19}$$

so that

$$\rho' \ll \rho_s(r). \tag{2.8.20}$$

If we collect the consequences of these order-of-magnitude estimates, we find that (2.8.11a,b,c) can be legitimately approximated by

$$fv = +\frac{1}{\rho_s r \cos \theta}\frac{\partial p}{\partial \phi}, \tag{2.8.21a}$$

$$fu = -\frac{1}{\rho_s r}\frac{\partial p}{\partial \theta}, \tag{2.8.21b}$$

$$\rho g = -\frac{\partial p}{\partial r}, \tag{2.8.21c}$$

where the fact that $\rho' \ll \rho_s$ has been explicitly used in (2.8.21a,b). The fact that (2.8.21a,b) can be written in terms of $p$ follows trivially from the fact that $p_s$ is independent of $\phi$ and $\theta$, while (2.8.21c) follows from (2.8.10) and the approximation (2.8.16).

The notation

$$f = 2\Omega \sin \theta \tag{2.8.22}$$

has been introduced. The parameter $f$ is the local component of the planetary vorticity normal to the earth's surface and is called the *Coriolis parameter*.

The fact that both the atmosphere and oceans are thin layers of fluid suggests that we introduce as the radial coordinate

$$z = r - r_0, \tag{2.8.23}$$

where $r_0$ is the distance from the earth's center to its surface. Since $z \leq O(D)$, it follows that $z \ll r_0$ or that to $O(D/r_0)$ (2.8.21a,b,c) can be written

$$fv = \frac{1}{\rho_s r_0 \cos \theta} \frac{\partial p}{\partial \phi}, \tag{2.8.24a}$$

$$fu = -\frac{1}{\rho_s r_0} \frac{\partial p}{\partial \theta}, \tag{2.8.24b}$$

$$\rho g = -\frac{\partial p}{\partial z}. \tag{2.8.24c}$$

The balance of terms indicated by (2.8.21a,b) or (2.8.24a,b) is the *geostrophic approximation* to the full momentum equation. In this limiting form the momentum balance for the *horizontal velocity* reduces to a balance between the horizontal pressure gradient and the horizontal component of the Coriolis acceleration. The velocity derived from this relation is called the *geo strophic velocity*. The geostrophic approximation can be written in vector form as

$$\mathbf{u}_H = \frac{1}{f\rho_s} \mathbf{k} \times \nabla p, \tag{2.8.25}$$

where $\mathbf{k}$ is the unit vector perpendicular to the surface of the sphere and $\mathbf{u}_H$ is the horizontal velocity vector. Note that the geostrophic approximation gives no direct information about the vertical velocity.

The third equation (2.8.24c) does not involve the velocity at all, but describes a balance in the vertical direction between the vertical pressure gradient and the buoyancy force. This is the *hydrostatic approximation*. Even though the fluid is in motion, the nearly horizontal character of the fluid trajectories makes the vertical accelerations so small that the Archimedian principle for a static fluid is applicable. That is, the pressure difference between any two points on the same vertical line depends only on the weight of the fluid between those points, *as if the fluid were at rest*, though in fact it is in motion.

## 2.9 Consequences of the Geostrophic and Hydrostatic Approximations

The precise derivation of the geostrophic and hydrostatic approximations is given in Chapter 6 as the necessary first step in an asymptotic expansion of the equations of motion for small Rossby and Ekman numbers. Assuming for now the validity of the heuristic argument given in the previous section, let us examine some of the consequences which can be deduced from geostrophy and the hydrostatic balance.

### 2.9a The Geostrophic Stream Function

The horizontal, geostrophic velocities given by (2.8.24) are *perpendicular* to the horizontal pressure gradient. This is a remarkable feature of geostrophic flow. The fluid flows along and not across the lines of constant pressure (isobars). It is this feature that enables the isobars on a weather map to be representative of the pattern of atmospheric flow. In a nonrotating fluid the flow would normally be in the direction of the pressure force. In the geostrophic approximation each fluid element acts like a tiny spinning top; pushed in one direction by the pressure force, it responds by moving at right angles to the force and its rotation, as shown in Figure 2.9.1. Note again that

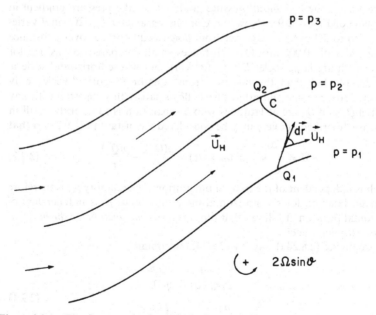

**Figure 2.9.1**   The flow, $\mathbf{u}_H$, is along the isobars, and the total flow of mass between points $Q_1$ and $Q_2$ depends only on $P_2 - P_1$ and not on the contour $c$.

because of the slimness of the fluid region ($\delta \ll 1$) the *effective* rotation in the balance is $\Omega \sin \theta = f/2$. In the northern hemisphere, where $f > 0$, the geostrophic velocity is such that an observer facing downstream will have higher pressure on his right. In the southern hemisphere ($f < 0$) the higher pressure is on his left.

Consider now the mass flow in a unit vertical distance between the two points $Q_1$ and $Q_2$ on the isobars $p = p_1$ and $p = p_2$. As seen in Figure 2.9.1, this flux through the curve $C$ is

$$
M = \int_{Q_1}^{Q_2} \mathbf{k} \cdot \rho_s(\mathbf{u}_H \times d\mathbf{r})
$$

$$
= \int_{Q_1}^{Q_2} (\mathbf{k} \times \rho_s \mathbf{u}_H) \cdot d\mathbf{r}, \tag{2.9.1}
$$

which by (2.8.24) can be written

$$
M = \int_{Q_1}^{Q_2} \frac{[\mathbf{k} \times (\mathbf{k} \times \nabla p)]}{f} \cdot d\mathbf{r}
$$

$$
= -\int_{Q_1}^{Q_2} \{\nabla p - \mathbf{k}(\mathbf{k} \cdot \nabla p)\} \cdot \frac{d\mathbf{r}}{f} \tag{2.9.2}
$$

$$
= -\int_{Q_1}^{Q_2} \frac{\nabla_H p}{f} \cdot d\mathbf{r},
$$

where $\nabla_H p$ is the *horizontal* pressure gradient, i.e., the pressure gradient in the plane of Figure 2.9.1. Now the Coriolis parameter $f = 2\Omega \sin \theta$ varies from zero to $2\Omega$ over the distance from pole to equator, i.e., over a distance of the order of 10,000 km. Over distances small compared to this, i.e., for flows which are large scale ($L = O(10^3 \text{ km})$) but whose horizontal scale is considerably less than the distance from equator to pole, $f$ varies only slightly. Most oceanic and atmospheric flows satisfy this constraint. In any event, if $Q_1$ and $Q_2$ are at latitudes whose separation is sufficiently small in the sense described above, $f$ may be considered a constant in (2.9.2), so that

$$
M = \int_{Q_1}^{Q_2} \mathbf{k} \cdot (\rho_s \mathbf{u}_H \times d\mathbf{r}) = \frac{p(Q_1)}{f} - \frac{p(Q_2)}{f}, \tag{2.9.3}
$$

which is independent of the path of integration. The quantity $p/f$ is therefore a stream function for the horizontal mass flux. Since $\rho_s$ is independent of horizontal position, it follows that $p/\rho_s f$ is a stream function for the horizontal, geostrophic velocity.

If we divide (2.8.24a), say, by (2.8.24c) we obtain

$$
fv = -\frac{\rho g}{\rho_s r_0 \cos \theta} \frac{\partial p/\partial \phi}{\partial p/\partial z}
$$

$$
= +\frac{\rho}{\rho_s r_0 \cos \theta} \frac{1}{g}\left(\frac{\partial z}{\partial \phi}\right)_p, \tag{2.9.4}
$$

or

$$fv = + \frac{1}{r_0 \cos \theta} g \left( \frac{\partial z}{\partial \phi} \right)_p, \tag{2.9.5}$$

where the fact that $\rho' = \rho - \rho_s$ is $O(\varepsilon)$ has been used to go from (2.9.4) to (2.9.5). Similarly

$$fu = - \frac{1}{r_0} g \left( \frac{\partial z}{\partial \theta} \right)_p, \tag{2.9.6}$$

where the subscript $p$ refers to differentiation at constant pressure and $z$ is the height of a constant-pressure surface.

Thus the geostrophic velocity is given equally well in terms of the slope of surfaces of constant pressure, viz.

$$\mathbf{u}_H = \frac{g\mathbf{k}}{f} \times (\nabla z)_p. \tag{2.9.7}$$

## 2.9b The Thermal Wind

Differentiating (2.8.24a) with respect to $z$ yields, with (2.8.24c),

$$f \frac{\partial v}{\partial z} = - \frac{g}{\rho_s r_0 \cos \theta} \frac{\partial \rho}{\partial \phi} - \frac{1}{\rho_s^2} \frac{\partial \rho_s}{\partial z} \frac{1}{r_0 \cos \theta} \frac{\partial p}{\partial \phi}$$

$$= - \frac{g}{\rho_s r_0 \cos \theta} \frac{\partial \rho}{\partial \phi} - \frac{1}{\rho_s} \frac{\partial \rho_s}{\partial z} fv, \tag{2.9.8}$$

which, with (2.9.5), yields

$$f \frac{\partial v}{\partial z} = - \frac{g}{\rho_s r_0 \cos \theta} \frac{\partial \rho}{\partial \phi} - \frac{1}{\rho_s} \frac{\partial \rho_s}{\partial z} \frac{g}{r_0 \cos \theta} \left( \frac{\partial z}{\partial \phi} \right)_p, \tag{2.9.9}$$

or

$$f \frac{\partial v}{\partial z} = - \frac{g}{\rho_s r_0 \cos \theta} \left[ \frac{\partial \rho}{\partial \phi} + \left( \frac{\partial z}{\partial \phi} \right)_p \frac{\partial \rho_s}{\partial z} \right], \tag{2.9.10}$$

which to $O(\varepsilon)$ is

$$f \frac{\partial v}{\partial z} = - \frac{g}{\rho r_0 \cos \theta} \left[ \frac{\partial \rho}{\partial \phi} + \frac{\partial \rho}{\partial z} \left( \frac{\partial z}{\partial \phi} \right)_p \right], \tag{2.9.11}$$

or

$$\frac{\partial v}{\partial z} = - \frac{g}{f \rho r_0 \cos \theta} \left( \frac{\partial \rho}{\partial \phi} \right)_p. \tag{2.9.12}$$

Similarly,

$$\frac{\partial u}{\partial z} = \frac{g}{f \rho r_0} \left( \frac{\partial \rho}{\partial \theta} \right)_p. \tag{2.9.13}$$

In vector form

$$\frac{\partial \mathbf{u}_H}{\partial z} = -\frac{g}{\rho f} \mathbf{k} \times (\nabla \rho)_p, \qquad (2.9.14)$$

so that, in accordance with (2.6.5a,b) the increase of the horizontal velocity with height depends on the horizontal gradient of the density within surfaces of constant pressure. If the density and pressure surfaces are coincident, the geostrophic velocity must be independent of height. The geostrophic approximation therefore implies the Taylor–Proudman theorem.

For the atmosphere, the perfect-gas law applies, and therefore

$$\frac{1}{\rho}(\nabla \rho)_p = -\frac{1}{T}(\nabla T)_p = -\frac{1}{\theta}(\nabla \theta)_p, \qquad (2.9.15)$$

so that the increase of the wind with height can be naturally written directly in terms of the horizontal gradient of temperature or potential temperature in constant-pressure surfaces.

In the case of the oceans, the ratio of the terms on the right hand side of (2.9.8) is

$$\frac{fv\, \partial \rho_s/\partial z}{(g/r_0 \cos \theta)\, \partial \rho/\partial \phi} = O\left(\frac{fU/H}{fU/D}\right) = \frac{D}{H}, \qquad (2.9.16)$$

where $H$ is the density scale height

$$H = \left(\frac{-1}{\rho_s}\frac{\partial \rho_s}{\partial z}\right)^{-1} \qquad (2.9.17)$$

and $D$ is the depth scale of the motion. The ratio $D/H$ is therefore of the order $\Delta \rho_s/\rho_s$, i.e., the order of the proportional change of the density over the vertical scale of motion. This never exceeds an amount of $O(10^{-3})$, so that to an excellent approximation for the oceans the density gradients in (2.9.12) and (2.9.13) can be evaluated in surfaces of constant height.

## 2.9c The Divergence and Vorticity of the Geostrophic Wind

The divergence of the geostrophic velocity is, by (2.8.24a,b),

$$\frac{1}{r_0 \cos \theta}\frac{\partial}{\partial \theta}(v \cos \theta) + \frac{1}{r_0 \cos \theta}\frac{\partial u}{\partial \phi} = \frac{-v}{fr_0}\frac{\partial f}{\partial \theta} = \frac{-v \cos \theta}{r_0 \sin \theta}. \quad (2.9.18)$$

The characteristic magnitude of *each* of the terms on the left-hand side of (2.9.18) is $O(U/L)$, while the term on the right is $O(U/r_0)$.

For motions whose scales $L$ satisfy

$$\frac{L}{r_0} < 1, \qquad (2.9.19)$$

it follows that to lowest order, the horizontal geostrophic velocity is divergence free. For future reference we now derive an important relation for the relative vorticity. The same considerations show that to the same order the vertical component of relative vorticity is directly proportional to the horizontal Laplacian of the pressure:

$$
\zeta = \frac{1}{r_0 \cos \theta} \frac{\partial v}{\partial \phi} - \frac{1}{r_0} \frac{\partial u}{\partial \theta} = \frac{1}{\rho_s f} \nabla_H^2 p
$$

$$
= \frac{1}{\rho_s f} \left| \frac{1}{\cos \theta \, r_0} \frac{\partial}{\partial \theta} \left( \frac{\cos \theta}{r_0} \frac{\partial p}{\partial \theta} \right) + \frac{1}{r_0^2 \cos^2 \theta} \frac{\partial^2 p}{\partial \phi^2} \right|.
$$

(2.9.20)

## 2.10 Geostrophic Degeneracy

The geostrophic approximation is extraordinarily useful and esthetically appealing. Once the pressure field is known, the horizontal velocities, their vertical shear, and the vertical component of vorticity are immediately determined. Naturally, as is evident from the results of the preceding section, the approximation fails in the vicinity of the equator, and we expect a more complicated dynamical framework to be required in equatorial regions. Even at higher latitudes, though, a fundamental difficulty remains for there is no way with only the geostrophic relations to calculate the pressure field or predict its evolution with time. The geostrophic approximation is a diagnostic relationship and all it tells us is that as the velocity field slowly alters in time its Coriolis acceleration will continue to balance the evolving pressure gradient. Even if we are interested in the dynamics of steady motions, the geostrophic approximation would not suffice, for then, typically, only information about the motion on the boundaries of the domain of interest is given, but the structure in the interior must be calculated.

The geostrophic approximation alone simply does not contain enough information to complete the dynamical determination of the motion. Any pressure field can yield a consistent geostrophic velocity as long as its Rossby number is small, and there is no way, at the level of the geostrophic approximation, to distinguish which of the proposed pressure fields is the correct one. The resolution of this difficulty, it is clear, will require the consideration of higher-order dynamics, i.e., the effects of small *departures* from complete geostrophy. These small departures involve either the relative acceleration terms, of the order of the Rossby number, and/or the frictional forces, of the order of the Ekman number. The dependence of the *determination* of the field of motion on small forces, each one negligible compared to the dominant Coriolis force, sets a problem of great delicacy and sensitivity. Very roughly, this complication arises because of the very near balance of the pressure gradient and Coriolis acceleration. However, the curl of the pressure gradient is identically zero, while the curl of the Coriolis acceleration yields the divergence of the geostrophic velocity, which is very nearly zero; so the process of taking the curl of the equations of motion annihilates each of them. That is, neither the pressure gradient nor the Coriolis acceleration is as effective a contributor to the vorticity balance as to the momentum balance. This implies that the sensitive issue of the structure of the higher-order dynamics is best approached through considerations of vorticity dynamics—or, as we shall observe, more precisely through the consideration of the potential vorticity.

CHAPTER 3

# Inviscid Shallow-Water Theory

## 3.1 Introduction

In this chapter we take up the study of the dynamics of a shallow, rotating layer of homogeneous incompressible and inviscid fluid. There are two purposes to our consideration of this physical system. It is first of all simple enough so that the issues raised by the problem of geostrophic degeneracy can be dealt with directly without the need to simultaneously treat the complexities of the thermodynamics of a density-stratified fluid. The first goal of the present chapter is to illustrate how the geostrophic approximation can be systematically exploited to produce a deterministic dynamical framework adequate for the calculation of motions of large time and space scales. Furthermore, the method of analysis to be presented also can be generalized to the study of thermodynamically active fluids. The key technique of the analysis is the formulation of a *systematic* approximation scheme in which the geostrophic approximation is merely the first step.

The second purpose of this chapter is to then *use* the model to study motions of atmospheric and oceanic relevance. Of course it is not clear ahead of time that a model which completely ignores the presence of stratification will be at all useful for this purpose. At this point, such a suggestion simply presents for consideration a model of the real atmosphere or ocean, whose validity must be examined after the fact. This step of modeling is always, to some degree or other, a leap of faith. Whether the step is a useful one depends on the intuitive skill of the modeler in anticipating the way the physical model shares the essential dynamical character of the

57

geophysical system. The step of choosing the model is informed by experience, personal intuition, and a degree of explicit dynamical reasoning. It is important to emphasize that once the model has been chosen the proper test of its usefulness can only come from the systematic, logical, and precise working out of the dynamical predictions of the laws of motion as they apply to the model.

Experience has shown that the shallow-water model is capable of describing important aspects of atmospheric and oceanic motions so that we can now directly take to our advantage the profound intuitive insights of earlier workers* in the field. Necessarily, the motions described by such a model can only be expected to apply to phenomena which do not depend in a crucial way on stratification.

## 3.2 The Shallow-Water Model

We start by considering a sheet of fluid with constant and uniform density as shown in Figure 3.2.1. The height of the surface of the fluid above the reference level $z = 0$ is $h(x, y, t)$. With application to the earth's atmosphere

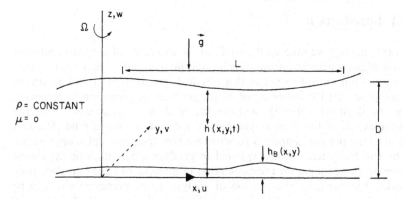

**Figure 3.2.1**   The shallow-water model.

or ocean in mind, we model the body force arising from the potential $\Phi$ as a vector, $\mathbf{g}$, directed perpendicular to the $z = 0$ surface, i.e., antiparallel to the vertical axis. The rotation axis of the fluid coincides with the $z$-axis in the model, i.e., $\mathbf{\Omega} = \mathbf{k}\Omega$, so that in this case the Coriolis parameter $f$ is simply $2\Omega$. The rigid bottom is defined by the surface $z = h_B(x, y)$. The velocity has components $u$, $v$, and $w$ parallel to the $x$-, $y$-, and $z$-axes respectively. The pressure of the fluid surface can be arbitrarily imposed, but for our purposes we may take it to be constant. Finally, the fluid is assumed inviscid, that is,

---

* This intuitive leap is astoundingly clear in the work of Rossby (1939) and Stommel (1948).

only motions for which viscosity is unimportant are considered in this chapter.

Although the depth of the fluid, $h - h_B$, varies in space and time, we suppose that a characteristic value for the depth can be sensibly chosen. Call that number $D$. $D$ could be chosen, for example, to be the average depth of the layer. We also suppose that $D$ characterizes the vertical scale of the motion as well. Similarly we suppose there exists a characteristic horizontal length scale for the *motion*, which we call $L$. The fundamental parametric condition which characterizes shallow-water* theory is

$$\delta = \frac{D}{L} \ll 1. \tag{3.2.1}$$

The shallow-water model therefore contains some of the important dynamical features of the atmosphere and ocean as described in the first chapter. The fluid is rotating, so that Coriolis accelerations can be important and the aspect ratio of the motion is small. The fluid layer is, however, flat rather than forming a spherical shell, but we shall see in Section 3.17 that this does not pose a fundamental problem for application of the model. Its major physical deficiency as noted above is the absence of the density stratification present in the real atmosphere and oceans.

## 3.3 The Shallow-Water Equations

In this section we now trace the consequences of the model for the dynamical equations of motion. The *specification* of incompressibility and constant density immediately decouples the dynamics from the thermodynamics and reduces the equation of mass conservation to the condition of incompressibility:

$$\frac{\partial u}{\partial x} + \frac{\partial v}{\partial y} + \frac{\partial w}{\partial z} = 0. \tag{3.3.1}$$

Each of the first two terms of (3.3.1) is $O(U/L)$, where $U$ is a characteristic scale for the horizontal velocity. If $W$ is the scale for the vertical velocity, it follows that $W/D$ can be no larger than $O(U/L)$, i.e., that

$$W \leq O(\delta U). \tag{3.3.2}$$

The vertical velocity can be *smaller* than the estimate of (3.3.2) if there is cancellation between $\partial u/\partial x$ and $\partial v/\partial y$ so that (3.3.2) in fact represents an upper bound on $W$.

---

* It is traditional to think in terms of a layer of water, but obviously any incompressible fluid such as air flowing at low speeds will be just as apt.

Now let us estimate the terms in the momentum equation. In component form, they are the inviscid form of (1.6.7), viz.

$$\frac{\partial u}{\partial t} + \left[ u\frac{\partial u}{\partial x} + v\frac{\partial u}{\partial y} + w\frac{\partial u}{\partial z} \right] - fv = -\frac{1}{\rho}\frac{\partial \tilde{p}}{\partial x},$$

$$\frac{U}{T} \qquad \frac{U^2}{L} \quad \frac{U^2}{L} \quad \frac{UW}{D} \qquad fU = \frac{P}{\rho L}$$

(3.3.3)

$$\frac{\partial v}{\partial t} + u\frac{\partial v}{\partial x} + v\frac{\partial v}{\partial y} + w\frac{\partial v}{\partial z} + fu = -\frac{1}{\rho}\frac{\partial \tilde{p}}{\partial y},$$

$$\frac{U}{T} \qquad \frac{U^2}{L} \quad \frac{U^2}{L} \quad \frac{UW}{D} \qquad fU = \frac{P}{\rho L}$$

(3.3.4)

$$\frac{\partial w}{\partial t} + u\frac{\partial w}{\partial x} + v\frac{\partial w}{\partial y} + w\frac{\partial w}{\partial z} = -\frac{1}{\rho}\frac{\partial \tilde{p}}{\partial z},$$

$$\frac{W}{T} \qquad \frac{UW}{L} \quad \frac{UW}{L} \quad \frac{WW}{D} = \frac{P}{\rho D}$$

(3.3.5)

where the order of magnitude of each term is written below it in terms of characteristic scales. Note that $T$ is a characteristic scale for time, i.e., the proper scale for time change, while $P$ is the scale for the variable pressure field—i.e., the total pressure $p$ has already been written

$$p(x, y, z, t) = -\rho g z + \tilde{p}(x, y, z, t),$$

(3.3.6)

the first part of which cancels the constant gravitational force per unit mass in the fluid. Since, by (3.3.2),

$$\frac{UW}{D} = O\left(\frac{U^2}{L}\right),$$

(3.3.7)

it follows from (3.3.3) and (3.3.4) that the pressure scale is given by

$$P = \rho U \left[\frac{L}{T}, U, fL\right]_{\max},$$

(3.3.8)

(i.e., by the largest of the three entries on the right-hand side) in order that the horizontal pressure gradient may enter as a forcing term in the horizontal momentum balance, for otherwise the flow would be unaccelerated. This in turn implies that the ratio of the terms on the left-hand side of (3.3.5) to the vertical pressure gradient is bounded by the larger of

$$\frac{\rho[W/T, WU/L]}{P/D} = O\left(\rho \frac{dw/dt}{\partial \tilde{p}/\partial z}\right),$$

(3.3.9)

or from (3.3.8)

$$\rho \frac{dw/dt}{\partial \tilde{p}/\partial z} = \frac{\delta^2 (1/T, U/L)_{\max}}{(1/T, U/L, f)_{\max}}.$$

(3.3.10)

There are two cases of interest. If the Rossby number, $U/fL$, is $O(1)$ or greater, the estimate in (3.3.10) yields a ratio of $O(\delta^2)$. If the Rossby number is small, the ratio is $O(\delta^2[1/fT, U/fL]_{\max})$, which for small Rossby number is even smaller than $O(\delta^2)$. Since by hypothesis $\delta^2 \ll 1$, it follows that *at least* to $O(\delta^2)$ $\partial\tilde{p}/\partial z$ is negligible, or more succinctly, in terms of the total pressure,

$$\frac{\partial p}{\partial z} = -\rho g + O(\delta^2), \tag{3.3.11}$$

which is the hydrostatic approximation. Note again that the correct scaling argument does *not* compare the vertical acceleration with $g$. Rather the vertical acceleration, which is small because of (3.3.2), must be compared with the vertical pressure gradient *due to the motion*. From another point of view (3.3.11) can be taken as the definition of the shallow-fluid model.

It is possible to immediately integrate (3.3.11) and obtain

$$p = -\rho g z + A(x, y, t). \tag{3.3.12}$$

The boundary condition

$$p(x, y, h) = p_0,$$

where $p_0$ is a constant, implies that

$$p = \rho g(h - z) + p_0, \tag{3.3.13}$$

so that the pressure in excess of $p_0$ at any point is simply equal to the weight of the unit column of fluid above the point *at that instant*.

Next note that the horizontal pressure gradient is independent of $z$, i.e.,

$$\frac{\partial p}{\partial x} = \rho g \frac{\partial h}{\partial x}, \tag{3.3.14a}$$

$$\frac{\partial p}{\partial y} = \rho g \frac{\partial h}{\partial y}, \tag{3.3.14b}$$

so that the horizontal *accelerations* must be independent of $z$. It is therefore *consistent* to assume that the horizontal velocities themselves remain $z$-independent if they are so initially. We shall in fact do just this in the anticipation that for low Rossby number the Taylor–Proudman theorem, applied to this homogeneous fluid, will *require* the velocities to be independent of $z$.

The horizontal momentum equations then become

$$\frac{\partial u}{\partial t} + u\frac{\partial u}{\partial x} + v\frac{\partial u}{\partial y} - fv = -g\frac{\partial h}{\partial x}, \tag{3.3.15a}$$

$$\frac{\partial v}{\partial t} + u\frac{\partial v}{\partial x} + v\frac{\partial v}{\partial y} + fu = -g\frac{\partial h}{\partial y}. \tag{3.3.15b}$$

The condition that $u$ and $v$ are independent of $z$ now allows (3.3.1) to be integrated in $z$ to yield

$$w(x, y, z, t) = -z\left(\frac{\partial u}{\partial x} + \frac{\partial v}{\partial y}\right) + \tilde{\omega}(x, y, t). \qquad (3.3.16)$$

Now the condition of no normal flow at the rigid surface $z = h_B$ requires that

$$w(x, y, h_B, t) = u\frac{\partial h_B}{\partial x} + v\frac{\partial h_B}{\partial y}. \qquad (3.3.17)$$

Therefore

$$\tilde{\omega}(x, y, t) = u\frac{\partial h_B}{\partial x} + v\frac{\partial h_B}{\partial y} + h_B\left(\frac{\partial u}{\partial x} + \frac{\partial v}{\partial y}\right), \qquad (3.3.18)$$

so that

$$w(x, y, z, t) = (h_B - z)\left(\frac{\partial u}{\partial x} + \frac{\partial v}{\partial y}\right) + u\frac{\partial h_B}{\partial x} + v\frac{\partial h_B}{\partial y}. \qquad (3.3.19)$$

The corresponding kinematic condition at the surface $z = h$ is

$$w = \frac{\partial h}{\partial t} + u\frac{\partial h}{\partial x} + v\frac{\partial h}{\partial y}, \qquad z = h(x, y, t), \qquad (3.3.20)$$

which, when combined with (3.3.19), yields

$$\frac{\partial h}{\partial t} + \frac{\partial}{\partial x}\{(h - h_B)u\} + \frac{\partial}{\partial y}\{(h - h_B)v\} = 0. \qquad (3.3.21)$$

Define the total depth

$$H = h - h_B,$$

in terms of which the equation of mass conservation (3.3.21) becomes

$$\frac{\partial H}{\partial t} + \frac{\partial}{\partial x}(uH) + \frac{\partial}{\partial y}(vH) = 0, \qquad (3.3.22a)$$

or equivalently

$$\frac{dH}{dt} + H\left(\frac{\partial u}{\partial x} + \frac{\partial v}{\partial y}\right) = 0. \qquad (3.3.22b)$$

The first of these, (3.3.22a), states that if the local horizontal divergence of volume, $\nabla \cdot (\mathbf{u}_H H)$, is positive, it must be balanced by a local decrease of the layer thickness due to a drop in the free surface. The second statement, (3.3.22b), which is equivalent to the first, is that following the fluid, as the cross section $A$ of a fluid column increases at a rate

$$\frac{1}{A}\frac{dA}{dt} = \frac{\partial u}{\partial x} + \frac{\partial v}{\partial y} \qquad (3.3.23)$$

the total thickness must decrease so that

$$\frac{1}{H}\frac{dH}{dt} + \frac{1}{A}\frac{dA}{dt} = 0 \tag{3.3.24}$$

—i.e., so that the volume $HA$ remains constant.

The shallow-water equations are (3.3.15a,b) and (3.3.22a) or its equivalent (3.3.22b). Note that the consequences of the condition $\delta \ll 1$ have reduced the number of dynamical equations by one, have reduced the number of dependent variables by one (by eliminating $w$ explicitly from the dynamics), and have reduced the number of independent variables by one (since $z$ no longer explicitly appears in the dynamical equations). The remaining variables are $u$, $v$, and $h$, and they are functions of $x$, $y$, and $t$ only.

The fact that $w$ is a simple linear function of $z$ has an important further implication. If we use (3.3.22b) to eliminate $\partial u/\partial x + \partial v/\partial y$ from (3.3.19), we obtain

$$w \equiv \frac{dz}{dt} = \frac{z - h_B}{H}\frac{dH}{dt} + u\frac{\partial h_B}{\partial x} + v\frac{\partial h_B}{\partial y}, \tag{3.3.25}$$

which implies that

$$\frac{d}{dt}\left(\frac{|z - h_B|}{H}\right) = 0, \tag{3.3.26}$$

so that the function $(z - h_B)/H$ is conserved following the motion of each fluid element. Now $(z - h_B)/H$ is the relative height from the bottom of each fluid element, i.e., its status, which ranges from zero at the bottom to unity at the free surface. During the motion of the fluid the fact that $u$ and $v$ are independent of $z$ implies that the fluid moves as a set of columns oriented parallel to the $z$-axis. The condition (3.3.26) simply states that during the stretching or contraction of each column the *relative* position of a fluid element in the column is unchanged.

## 3.4 Potential-Vorticity Conservation: Shallow-Water Theory

The three components of relative vorticity are, in a Cartesian frame,

$$\omega_x = \frac{\partial w}{\partial y} - \frac{\partial v}{\partial z}, \tag{3.4.1a}$$

$$\omega_y = \frac{\partial u}{\partial z} - \frac{\partial w}{\partial x}, \tag{3.4.1b}$$

$$\omega_z = \frac{\partial v}{\partial x} - \frac{\partial u}{\partial y}. \tag{3.4.1c}$$

In the present case, where $u$ and $v$ are independent of $z$,

$$\omega_x = \frac{\partial w}{\partial y} = O\left(\frac{W}{L}\right) = O\left(\delta \frac{U}{L}\right), \tag{3.4.2a}$$

$$\omega_y = -\frac{\partial w}{\partial x} = O\left(\frac{W}{L}\right) = O\left(\delta \frac{U}{L}\right), \tag{3.4.2b}$$

$$\omega_z = \frac{\partial v}{\partial x} - \frac{\partial u}{\partial y} = O\left(\frac{U}{L}\right), \tag{3.4.2c}$$

so that the horizontal components of relative vorticity are $O(\delta)$ smaller than the vertical component of vorticity. If $h$ is eliminated from (3.3.15a,b) by cross differentiation of the first equation with respect to $y$ and the second with respect to $x$, we obtain

$$\frac{d\zeta}{dt} \equiv \frac{\partial \zeta}{\partial t} + u\frac{\partial \zeta}{\partial x} + v\frac{\partial \zeta}{\partial y} = -(\zeta + f)\left(\frac{\partial u}{\partial x} + \frac{\partial v}{\partial y}\right), \tag{3.4.3}$$

where the notation

$$\zeta \equiv \omega_z \tag{3.4.4}$$

has been introduced. Following fluid columns, the only mechanism which can change the relative vorticity is the convergence of absolute-vorticity filaments. Using (3.3.22b), we can rewrite (3.4.3) as

$$\frac{d\zeta}{dt} = \frac{\zeta + f}{H}\frac{dH}{dt}, \tag{3.4.5}$$

which expresses the same result instead in terms of vortex-tube stretching—namely, for $dH/dt > 0$ (stretching) the vorticity $\zeta$ is intensified by an amount proportional to the product of the column stretching and the absolute vorticity, $\zeta + f$, which is already present. Note that $\zeta$ cannot change by vortex tilting, which mechanism is absent in the columnar motion of shallow-water theory.

Since $f$ is a constant, (3.4.5) can be put in the form

$$\boxed{\frac{d}{dt}\left(\frac{\zeta + f}{H}\right) = 0}. \tag{3.4.6}$$

Following the motion of each fluid column,

$$\Pi_s = \frac{\zeta + f}{H} \tag{3.4.7}$$

is conserved. If $H$ increases, the absolute (and hence relative) vorticity must increase to keep $\Pi_s$ constant for the column. Note that if $\zeta$ is originally zero, it will remain so *only* if $H$ remains constant; more typically, relative vorticity is produced by column stretching in the field of planetary vorticity, $f$. It should be clear from the physical implications of (3.4.6) that $\Pi_s$ is directly related to the general potential vorticity defined by (2.5.8). That this is so can be seen from the fact that since the fluid is trivially barotropic, any conservative fluid property $\lambda$ can be used to define $\Pi$. In particular the status function (3.3.26), i.e.,

$$\lambda = \frac{z - h_B}{H}, \tag{3.4.8}$$

is a possible candidate. Since the only component of vorticity, to $O(\delta)$, is the vertical component, we have

$$\Pi = \frac{\boldsymbol{\omega}_a}{\rho} \cdot \nabla \lambda = \frac{(\zeta + f)\mathbf{k} \cdot \nabla \left| z - h_B \right|}{\rho \left| H \right|} + O(\delta)$$

$$= \frac{\zeta + f}{\rho} \frac{\partial}{\partial z} \left( \frac{z - h_B}{H} \right) \tag{3.4.9}$$

$$= \frac{\zeta + f}{\rho H}.$$

Since $\rho$ is constant, $\Pi_s$ is, aside from a constant factor, truly the *potential vorticity*. Largely for historical reasons, it is conventional usage to call $\Pi_s$ the potential vorticity in the context of shallow-water theory, and this sensible terminology is used throughout this chapter.

## 3.5 Integral Constraints

Before proceeding to the consideration of particular problems, it will be helpful to first derive certain general integral constraints which apply to the shallow-water model for later use.

Consider the motion governed by (3.3.15a,b) and (3.3.22a) of a fluid layer in a *closed* region $R$ bounded by a curve $C_0$. Further, let $R$ contain $J$ internal "islands" each bounded by the curve $C_j, j = 1, 2, \ldots, J$, as shown in Figure 3.5.1. Each boundary $C_0, C_1, \ldots, C_J$ is rigid and impermeable to fluid motion, and therefore on each boundary

$$\mathbf{u}_H \cdot \mathbf{n}_j = 0, \tag{3.5.1}$$

where $\mathbf{n}_j$ is the outward unit normal vector for each contour and $\mathbf{u}_H$ is the horizontal velocity, i.e.,

$$\mathbf{u}_H = \mathbf{i}u + \mathbf{j}v. \tag{3.5.2}$$

The horizontal momentum equations can be rewritten in vector form as

$$\frac{\partial}{\partial t} \mathbf{u}_H + \nabla \frac{\mathbf{u}_H \cdot \mathbf{u}_H}{2} + (\zeta + f)\mathbf{k} \times \mathbf{u}_H = -g\nabla h. \tag{3.5.3}$$

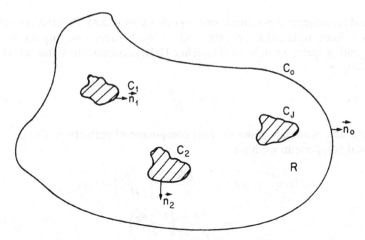

**Figure 3.5.1**    The contour $c$ encircles the fluid region $R$, which contains islands, each of which is bounded by the curve $c_j$, $j = 1, 2, \ldots, J$.

If (3.5.3) is integrated around any closed contour $c_j$, $j = 0, 1, \ldots, J$, then

$$\frac{\partial}{\partial t} \oint_{C_j} \mathbf{u}_H \cdot d\mathbf{r} = -\oint_{C_j} (\zeta + f)(\mathbf{k} \times \mathbf{u}_H) \cdot d\mathbf{r}, \qquad (3.5.4)$$

where $d\mathbf{r}$ is a vector line element tangent to $C_j$. The other terms in (3.5.3) vanish when integrated about a closed loop. Since

$$(\mathbf{k} \times \mathbf{u}_H) \cdot d\mathbf{r} = \mathbf{k} \cdot (\mathbf{u}_H \times d\mathbf{r}),$$

the *integrand* on the right-hand side of (3.5.4) identically vanishes, since by (3.5.1) $\mathbf{u}_H$ is parallel to $d\mathbf{r}$ on $C_j$. Thus the circulation of the relative velocity about each curve $C_j$ is independent of time, i.e.,

$$\boxed{\frac{\partial}{\partial t} \oint_{C_j} \mathbf{u}_H \cdot d\mathbf{r} = 0}, \qquad j = 0, 1, 2, \ldots, J. \qquad (3.5.5)$$

The vorticity equation, (3.4.3), can be written in the more compact form

$$\frac{\partial \zeta}{\partial t} + \nabla \cdot [(\zeta + f)\mathbf{u}_H] = 0. \qquad (3.5.6)$$

If (3.5.6) is integrated over the area $A$ of the *fluid* region of $R$, i.e., the area between $C_0$ and all the islands $C_j$, use of the divergence theorem yields, with (3.5.1),

$$\frac{\partial}{\partial t} \iint_A \zeta \, dx \, dy = -\oint_{C_0} (\zeta + f)\mathbf{u}_H \cdot \mathbf{n}_0 \, dr + \sum_{j=1}^{J} \oint_{C_j} (\zeta + f)\mathbf{u}_H \cdot \mathbf{n}_j \, dr = 0, \ (3.5.7)$$

so that the *area*-averaged relative vorticity must be constant in time. It is left to the reader to show that (3.5.7) also follows directly from (3.5.5) and Stokes's theorem.

A similar area average of (3.3.22a) yields

$$\frac{\partial}{\partial t} \iint_A h \, dx \, dy = 0, \tag{3.5.8}$$

which, of course, is simply the conservation of mass for the region as a whole.

If (3.5.3) is multiplied by $H\mathbf{u}_H$ and (3.3.22b) is used, it follows that

$$\frac{\partial}{\partial t} \frac{H}{2} |\mathbf{u}_H|^2 + \nabla \cdot \tfrac{1}{2}\{\mathbf{u}_H H |\mathbf{u}_H|^2\} = -gH\mathbf{u}_H \cdot \nabla h$$

$$= -g\nabla \cdot H\mathbf{u}_H h + hg\nabla \cdot \mathbf{u}_H H. \tag{3.5.9}$$

A further use of (3.3.22a) to eliminate the second term on the right-hand side of (3.5.9) yields

$$\frac{\partial}{\partial t} \left\{ \frac{H}{2} |\mathbf{u}_H|^2 + g\frac{h^2}{2} \right\} = -\nabla \cdot \{H\mathbf{u}_H[hg + \tfrac{1}{2}|\mathbf{u}_H|^2]\}, \tag{3.5.10}$$

which when integrated over the domain $A$ yields

$$\frac{\partial}{\partial t} \iint_A dx \, dy \left[ \frac{H}{2} |\mathbf{u}_H|^2 + \frac{gh^2}{2} \right] = 0. \tag{3.5.11}$$

The term $(H/2)|\mathbf{u}_H|^2$ is the kinetic energy per unit volume multiplied by $H$ to yield the kinetic energy per unit area of the columnar motion. The kinetic energy contribution of $w$ is $O(\delta^2)$ and is absent from (3.5.11). The potential energy per unit cross sectional area is

$$\int_0^h gz \, dz = g\frac{h^2}{2}, \tag{3.5.12}$$

where the level $z = 0$ has been arbitrarily chosen as the zero reference level for potential energy. Thus (3.5.11) is simply the statement of conservation of kinetic plus potential energy in shallow-water theory. These integral constraints also apply when the fluid is unbounded if the velocity fields vanish at infinity.

## 3.6 Small-Amplitude Motions

A logical first step in the analysis of a new set of dynamical equations is the study of small-amplitude motions, which allow a linearization of the equations of motion. In particular the presence of solutions representing free oscillations or waves often illustrates fundamental mechanisms which occur in more complicated situations. Furthermore, for our purpose it is important to identify the intrinsic time scales of the natural modes of oscillation of the system. In this section the governing equation for linearized motions will be derived.

Let the thickness of the fluid layer in the absence of motion be $H_0(x, y)$. Then in general

$$H(x, y, t) = H_0(x, y) + \eta(x, y, t). \tag{3.6.1}$$

The condition that the amplitude is small implies that $\eta \ll H_0$. Further we suppose $u$ and $v$ are small enough that

$$\frac{\partial \mathbf{u}_H}{\partial t} \gg \mathbf{u}_H \cdot \nabla \mathbf{u}_H. \tag{3.6.2}$$

Then the linearized forms of (3.3.15a,b) and (3.3.22a), which ignore all quadratic terms in the dynamical variables $u$, $v$, $\eta$ with respect to the linear terms, are

$$\frac{\partial u}{\partial t} - fv = -g\frac{\partial \eta}{\partial x}, \tag{3.6.3a}$$

$$\frac{\partial v}{\partial t} + fu = -g\frac{\partial \eta}{\partial y}, \tag{3.6.3b}$$

$$\frac{\partial \eta}{\partial t} + \frac{\partial}{\partial x}(uH_0) + \frac{\partial}{\partial y}(vH_0) = 0. \tag{3.6.3c}$$

Define the linearized mass flux vector by $\mathbf{U} = \mathbf{i}U + \mathbf{j}V$, where

$$\begin{aligned} U &= uH_0, \\ V &= vH_0, \end{aligned} \tag{3.6.4}$$

in terms of which (3.6.3a,b,c) become

$$\frac{\partial U}{\partial t} - fV = -gH_0\frac{\partial \eta}{\partial x}, \tag{3.6.5a}$$

$$\frac{\partial V}{\partial t} + fU = -gH_0\frac{\partial \eta}{\partial y}, \tag{3.6.5b}$$

$$\frac{\partial \eta}{\partial t} + \frac{\partial U}{\partial x} + \frac{\partial V}{\partial y} = 0. \tag{3.6.5c}$$

Manipulation of the first two of the above yields

$$\frac{\partial}{\partial t}\left|\frac{\partial V}{\partial y} + \frac{\partial U}{\partial x}\right| - f\left|\frac{\partial V}{\partial x} - \frac{\partial U}{\partial y}\right| = -g\nabla \cdot \{H_0\nabla\eta\}, \tag{3.6.6}$$

$$\frac{\partial}{\partial t}\left|\frac{\partial V}{\partial x} - \frac{\partial U}{\partial y}\right| + f\left|\frac{\partial V}{\partial y} + \frac{\partial U}{\partial x}\right| = -g\left|\frac{\partial H_0}{\partial x}\frac{\partial \eta}{\partial y} - \frac{\partial H_0}{\partial y}\frac{\partial \eta}{\partial x}\right|, \tag{3.6.7}$$

from which it follows that

$$\left(\frac{\partial^2}{\partial t^2} + f^2\right)\left(\frac{\partial V}{\partial y} + \frac{\partial U}{\partial x}\right)$$

$$= -g\frac{\partial}{\partial t}\nabla \cdot \{H_0\nabla\eta\} - fg\left|\frac{\partial H_0}{\partial x}\frac{\partial\eta}{\partial y} - \frac{\partial H_0}{\partial y}\frac{\partial\eta}{\partial x}\right|.$$

(3.6.8)

An equation in the single variable $\eta$ can now be obtained with the use of (3.6.5c) and (3.6.8), namely

$$\boxed{\frac{\partial}{\partial t}\left[\left(\frac{\partial^2}{\partial t^2} + f^2\right)\eta - \nabla \cdot (C_0^2\nabla\eta)\right] - gfJ(H_0, \eta) = 0}, \quad (3.6.9)$$

where

$$C_0^2 = gH_0 \tag{3.6.10}$$

and the useful notation for the Jacobian of two functions

$$J(A, B) \equiv \frac{\partial A}{\partial x}\frac{\partial B}{\partial y} - \frac{\partial A}{\partial y}\frac{\partial B}{\partial x} \tag{3.6.11}$$

has been introduced. The velocities $u$ and $v$ can be found in terms of $\eta$ from the solution of the following ordinary differential equations, derived from (3.6.3a,b):

$$\left(\frac{\partial^2}{\partial t^2} + f^2\right)u = -g\left(\frac{\partial^2\eta}{\partial x\,\partial t} + f\frac{\partial\eta}{\partial y}\right), \tag{3.6.12a}$$

$$\left(\frac{\partial^2}{\partial t^2} + f^2\right)v = -g\left(\frac{\partial^2\eta}{\partial y\,\partial t} - f\frac{\partial\eta}{\partial x}\right). \tag{3.6.12b}$$

## 3.7 Linearized Geostrophic Motion

Consider, first, the time-*independent* forms of (3.6.9) and (3.6.12a,b). The latter two imply that

$$u = -\frac{g}{f}\frac{\partial\eta}{\partial y}, \tag{3.7.1a}$$

$$v = \frac{g}{f}\frac{\partial\eta}{\partial x}, \tag{3.7.1b}$$

which, recalling the relationship between $\eta$ and the pressure field, are recognized as the geostrophic relation for the horizontal motions. In particular the isolines of $\eta$ are streamlines for the steady geostrophic flow, since

$$u\frac{\partial\eta}{\partial x} + v\frac{\partial\eta}{\partial y} = 0 \tag{3.7.2}$$

by (3.7.1a,b). Furthermore, the steady form of (3.6.9) is simply

$$J(H_0, \eta) = 0, \tag{3.7.3}$$

or

$$\left(\frac{\partial y}{\partial x}\right)_{H_0} \equiv -\frac{\partial H_0/\partial x}{\partial H_0/\partial y} = -\frac{\partial \eta/\partial x}{\partial \eta/\partial y} \equiv \left(\frac{\partial y}{\partial x}\right)_{\eta}, \tag{3.7.4}$$

so that lines of constant undisturbed depth, $H_0$, must coincide with lines of constant $\eta$ in the $x$, $y$ plane. Since the isolines of $\eta$ are the streamlines of the steady geostrophic flow, it follows that the linearized geostrophic flow must be along lines of constant depth as shown in Figure 3.7.1. This, of course, is a

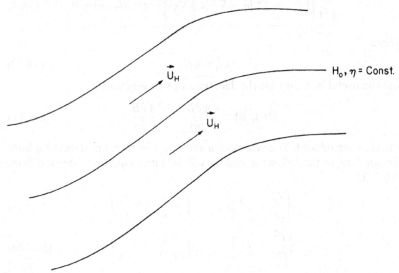

**Figure 3.7.1**   The isobaths and streamlines for steady, purely geostrophic flow must coincide as shown.

direct consequence of potential-vorticity conservation (3.4.6), which for steady, linearized motion reduces to

$$\mathbf{u}_H \cdot \nabla\left(\frac{f}{H_0}\right) = 0. \tag{3.7.5}$$

Therefore, if $\eta$ is given at one point on each $H_0$-contour, it is determined everywhere on that contour. Clearly if contour lines of $H_0$ intersect a rigid boundary, where the normal velocity vanishes, steady *linearized* motion on these blocked contours is impossible. Strictly geostrophic motion is thus possible only if the contour lines of $H_0$ close on themselves or extend to infinity. Note that if $H_0$ is constant, any geostrophic flow is possible.

Of course, real motions are not *precisely* geostrophic, and in the next sections we consider what happens when either the constraints of steadiness or those of linearity are relaxed.

## 3.8 Plane Waves in a Layer of Constant Depth

Consider the free oscillations possible in a layer of uniform depth of sufficient lateral extent to be idealized as an infinite plane—that is, an extent vastly greater than the wavelength of the oscillations to be described. For $H_0$ a constant, the coefficients of (3.6.9) are constant, so that solutions may be sought in the form of a *plane wave*, i.e.,

$$\eta = \text{Re } \eta_0 e^{i(kx + ly - \sigma t)} \qquad (3.8.1)$$

where the symbol Re denotes the real part of the function so labeled. The *amplitude* of the oscillation is $\eta_0$, and its *phase* $\theta$ is given by

$$\theta = kx + ly - \sigma t. \qquad (3.8.2)$$

At a given *instant* the phase (and therefore the surface height) is constant on the lines of constant $kx + ly$ as shown in Figure 3.8.1, where lines of

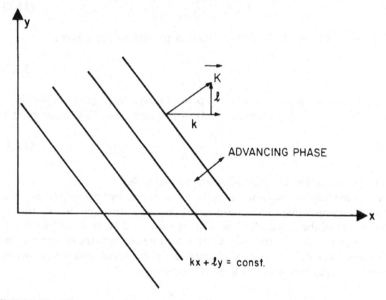

**Figure 3.8.1**   The geometry of a plane wave. The wave vector **k** is perpendicular to the lines of constant wave phase.

constant phase, say crests of the wave, are drawn. The wave properties are constant along the lines of constant phase for the plane wave. The normal to the surfaces of constant phase is given by the wave vector

$$\mathbf{K} = \nabla\theta = \hat{\mathbf{i}}k + \hat{\mathbf{j}}l. \qquad (3.8.3)$$

The perpendicular distance $\lambda$ between two adjacent lines of the same phase is determined by the condition $|\mathbf{K}|\lambda = 2\pi$, so that the *wavelength* $\lambda$ is related to the *wave number*, $K = |\mathbf{K}|$ by

$$\lambda = \frac{2\pi}{K}. \tag{3.8.4}$$

The wave number gives the number of spatial undulations in a unit distance perpendicular to the crest.

At any given *point* the *phase* changes linearly with time at a rate

$$\sigma = -\frac{\partial\theta}{\partial t}, \tag{3.8.5}$$

which gives the number of crests passing the point per unit time. The speed with which the phase advances along the $x$-axis is determined by the condition that $\theta$ is constant for fixed $y$ as $x$ and $t$ vary, i.e., that $\theta$ is constant for fixed $y$ to an observer moving parallel to the $x$-axis at a rate

$$C_x = -\frac{\partial\theta/\partial t}{\partial\theta/\partial x} = \frac{\sigma}{k}. \tag{3.8.6}$$

The speed with which the phase advances parallel to the $y$-axis is

$$C_y = -\frac{\partial\theta/\partial t}{\partial\theta/\partial y} = \frac{\sigma}{l}. \tag{3.8.7}$$

The fundamental *phase speed* is the speed of phase propagation in the direction of $\mathbf{K}$, which is the direction of advance of each crest. This phase speed is

$$C = -\frac{1}{|\nabla\theta|}\frac{\partial\theta}{\partial t} = \frac{\sigma}{K} = \frac{\sigma}{(k^2 + l^2)^{1/2}} \tag{3.8.8}$$

and gives the speed of phase advance parallel to $\mathbf{K}$.

It is important to emphasize that the phase speed *does not satisfy the rules of vector composition*. The phase speed in the $x$-direction is *not* the $x$-component of the speed along $\mathbf{K}$. That speed would be $\sigma k/K^2$, which is not $C_x$. The phase speed in the $x$-direction is the rate at which the intersection of the phase lines with the $x$-axis advance along the $x$-axis, and this *increases* with *decreasing* projection of $\mathbf{K}$ on the $x$-axis, i.e.,

$$C_x = \frac{CK}{k}, \tag{3.8.9a}$$

while similarly

$$C_y = \frac{CK}{l}. \tag{3.8.9b}$$

If (3.8.1) is substituted in (3.6.9) for constant $H_0$, the condition for (3.8.1) to be a solution is

$$\sigma\eta_0\{f^2 - \sigma^2 + C_0^2 K^2\} = 0. \tag{3.8.10}$$

If $\sigma \neq 0$, i.e., if the motion is unsteady (otherwise the results of section (3.7) immediately apply) and if $\eta_0 \neq 0$ (the wave has a nontrivial amplitude), the plane wave (3.8.1) will be a solution only if $\sigma$ is a specific function of $\mathbf{K}$ determined by the zeros of (3.8.10). This relationship is the dispersion relation, and in this simple case is given by

$$\sigma = \sigma(\mathbf{K}) = \pm \{f^2 + C_0^2 K^2\}^{1/2}. \qquad (3.8.11)$$

In the present case the frequency is a function only of the absolute value of $\mathbf{K}$ and not of its orientation. For each $\mathbf{K}$, two free oscillations are present, representing waves with phase speeds

$$C = \pm \left\{ C_0^2 + \frac{f^2}{K^2} \right\}^{1/2}, \qquad (3.8.12)$$

whose crests are moving parallel and antiparallel to $\mathbf{K}$.

In the absence of rotation ($f = 0$), the phase at all wavelengths moves with the same phase speed $(gH_0)^{1/2}$, which is the shallow-water speed of classical linear theory. The presence of rotation increases the wave speed. Indeed, from (3.8.11) it is clear that all these free waves have frequencies which *exceed f*, i.e., have periods less than half a rotation period and consequently are at frequencies considerably in excess of those characteristic of large-scale, slow atmospheric and oceanic flows.

Now (3.8.1) can be written as

$$\eta = |\eta_0| \cos(kx + ly - \sigma t + \phi) \qquad (3.8.13)$$

if $\eta_0 = |\eta_0| e^{i\phi}$. It is now possible to use (3.6.12a,b) to obtain $u$ and $v$, or more revealingly

$$u_{\parallel} = \frac{|\eta_0|}{H_0} C \cos(kx + ly - \sigma t + \phi),$$

$$u_{\perp} = \frac{|\eta_0|}{H_0} \frac{f}{\sigma} C \sin(kx + ly - \sigma t + \phi), \qquad (3.8.14)$$

where

$$u_{\parallel} = \left[ \mathbf{u}_H \cdot \frac{\mathbf{K}}{K} \right],$$

$$u_{\perp} = [\mathbf{u}_H - u_{\parallel}\mathbf{K}], \qquad (3.8.15)$$

i.e., $u_{\parallel}$ is the particle velocity parallel to $\mathbf{K}$ or perpendicular to the wave crests, while $u_{\perp}$ is the particle speed along the crests. It follows from (3.8.14) that the horizontal velocity vector traces an ellipse as time progresses, as shown in Figure 3.8.2, whose equation is

$$u_{\parallel}^2 + u_{\perp}^2 \frac{\sigma^2}{f^2} = \frac{C^2 \eta_0^2}{H_0^2}. \qquad (3.8.16)$$

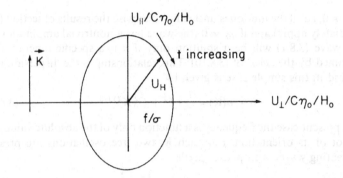

**Figure 3.8.2**   The ellipse traced by the velocity vector in a Poincaré wave.

The tip of the velocity vector proceeds once in the clockwise direction around the ellipse in one period of oscillation. Note that since $\sigma/f > 1$, the maximum velocity parallel to **K** (i.e., parallel to the gradients of $\eta$ and therefore parallel to the pressure gradient) exceeds the maximum velocity along the lines of constant phase of $\eta$, which are also isobars. Consequently, these waves are far from being in geostrophic balance. The fluid flow is primarily in the direction of the pressure gradient.

The vorticity in the wave is simply

$$\zeta = \frac{\mathbf{K}}{K} \cdot \nabla u_\perp$$

$$= \frac{|\eta_0|}{H_0} \frac{f}{\sigma} KC \cos(kx + ly - \sigma t + \phi) \qquad (3.8.17)$$

$$= \frac{f\eta}{H_0},$$

which follows directly from the linearized form of the potential vorticity equation (3.4.6), viz.

$$\frac{\partial}{\partial t}\left[\zeta - \frac{f}{H_0}\eta\right] = 0. \qquad (3.8.18)$$

For a periodic flow, integration immediately yields (3.8.17). As the free surface rises and falls during the passage of the wave, the vortex tubes are stretched in the presence of the background planetary vorticity $f$, and this produces positive relative vorticity at the wave crests and negative relative vorticity at the wave troughs.

The horizontal divergence in the wave field is also given in terms of the wave height, namely

$$\nabla \cdot \mathbf{u}_H = -\frac{1}{H_0}\frac{\partial \eta}{\partial t} = O\left(\frac{\sigma\eta}{H}\right), \qquad (3.8.19)$$

and consequently the ratio of the order of the relative vorticity to the horizontal divergence is

$$\frac{\zeta}{\nabla \cdot \mathbf{u}_H} = O\left(\frac{f}{\sigma}\right) < 1,$$    (3.8.20)

whose smallness is a measure of the departure of the flow from geostrophy.


## 3.9 Poincaré and Kelvin Waves

In this section we continue our study of the free linear modes of oscillation of a shallow, rotating fluid layer. We now turn our attention to the modes which appear in a partially bounded region, a channel of width $L$ oriented parallel to the $x$-axis as shown in Figure 3.9.1. The fact that the region is

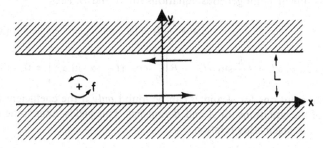

**Figure 3.9.1**    The infinite channel of width $L$, rotating with angular velocity $f/2$.

bounded in $y$ removes the implicit constraint suffered by the flow in the infinite region that the solution for $\eta$ be periodic in $x$ *and* $y$. This is no longer the case. Instead, on each of the two rigid walls of the channel the velocity in the $y$-direction must vanish, which implies, in view of (3.6.12b), that

$$\frac{\partial^2 \eta}{\partial y \, \partial t} - f \frac{\partial \eta}{\partial x} = 0, \qquad y = 0, L.$$    (3.9.1)

The governing equation for $\eta$ remains, for constant $H_0$,

$$\frac{\partial}{\partial t}\left\{\left(\frac{\partial^2}{\partial t^2} + f^2\right)\eta - C_0^2 \nabla^2 \eta\right\} = 0.$$    (3.9.2)

Wave solutions which are periodic in $x$ and $t$ can be sought in the form

$$\eta = \text{Re } \bar{\eta}(y)e^{i(kx - \sigma t)},$$    (3.9.3)

where $\bar{\eta}(y)$ is the (complex) wave amplitude which varies with the cross-channel coordinate $y$. Substitution of (3.9.3) into (3.9.1) yields the *eigenvalue problem* for $\bar{\eta}$, namely

$$\frac{d^2\bar{\eta}}{dy^2} + \left|\frac{\sigma^2 - f^2}{C_0^2} - k^2\right|\bar{\eta} = 0, \tag{3.9.4}$$

$$\frac{d\bar{\eta}}{dy} + f\frac{k}{\sigma}\bar{\eta} = 0, \qquad y = 0, L. \tag{3.9.5}$$

The general solution of (3.9.4) is

$$\bar{\eta} = A \sin \alpha y + B \cos \alpha y, \tag{3.9.6}$$

where

$$\alpha^2 = \frac{\sigma^2 - f^2}{C_0^2} - k^2, \tag{3.9.7}$$

and the application of the boundary condition (3.9.5) at $y = 0$ and $y = L$ yields two linear homogeneous equations for $A$ and $B$, i.e.,

$$\alpha A + \frac{fk}{\sigma}B = 0, \tag{3.9.8a}$$

$$A\left[\alpha \cos \alpha L + f\frac{k}{\sigma}\sin \alpha L\right] + B\left[\frac{fk}{\sigma}\cos \alpha L - \alpha \sin \alpha L\right] = 0. \tag{3.9.8b}$$

Nontrivial solutions for $A$ and $B$ can be found *only* if the coefficient determinant of the equations for $A$ and $B$ vanishes. This yields, after some manipulation, the eigenvalue relation

$$(\sigma^2 - f^2)(\sigma^2 - C_0^2 k^2)\sin \alpha L = 0. \tag{3.9.9}$$

There are three apparent possibilities: either $\sin \alpha L$ vanishes, or $\sigma^2 = f^2$, or finally, $\sigma^2 = C_0^2 k^2$. Let us consider these three possibilities in turn.

(i) The equation

$$\sin \alpha L = 0 \tag{3.9.10}$$

can be satisfied if $\alpha$ satisfies

$$\alpha = \frac{n\pi}{L}, \qquad n = 1, 2, 3, \ldots, \tag{3.9.11}$$

i.e., there are an infinite number of solutions. Note, however, that $\alpha = n = 0$ is not a possible solution. Such a solution represents a plane wave with crests oriented parallel to the $y$-axis, i.e., with no $y$-variation in the wave field. Although such a solution is possible for a non-rotating fluid, it follows from (3.6.12b) that for $\partial\eta/\partial y = 0$,

$$v = \frac{gf}{f^2 - \sigma^2}\frac{\partial\eta}{\partial x}(x, t), \tag{3.9.12}$$

so that $v$ is different from zero and independent of $y$, and consequently cannot satisfy the boundary condition of vanishing $v$ on

$y = 0$ and $y = L$. The eigenvalue relation (3.9.11) implies that

$$\alpha^2 = \frac{\sigma^2 - f^2}{C_0^2} - k^2 = \frac{n^2 \pi^2}{L^2}, \tag{3.9.13}$$

or

$$\sigma = \sigma_n = \pm \left\{ f^2 + C_0^2 \left( k^2 + \frac{n^2 \pi^2}{L^2} \right) \right\}^{1/2}, \qquad n = 1, 2, 3, \ldots, \tag{3.9.14}$$

which is identical to the dispersion relation (3.8.11) for free plane waves with the important exception that the $y$-component of the wave vector is now quantized as an integral multiple of $\pi/L$. These modes are called Poincaré waves and are dynamically similar to the plane-wave oscillations described in the previous section.

The solutions (3.9.14) for $\sigma$ of the same magnitude but opposite sign imply that the Poincaré waves propagate their phase equally well in both the positive and the negative $x$-direction. Note that the frequency always exceeds $f$; indeed, because the cross-stream wave number is quantized,

$$\sigma \geq \left\{ f^2 + \frac{C_0^2 \pi^2}{L^2} \right\}^{1/2}. \tag{3.9.15}$$

The dynamical fields can be found using (3.9.8a) and (3.9.14). For each mode

$$\eta = \eta_0 \left[ \cos \frac{n \pi y}{L} - \frac{L}{n \pi} \frac{f}{C_x} \sin \frac{n \pi y}{L} \right] \cos(kx - \sigma t + \phi),$$

$$u = \frac{\eta_0}{H_0} \left[ \frac{C_0^2}{C_x} \cos \frac{n \pi y}{L} - \frac{fL}{n \pi} \sin \frac{n \pi y}{L} \right] \cos(kx - \sigma t + \phi),$$

$$v = -\frac{\eta_0}{H_0} \frac{L}{\sigma n \pi} \left[ f^2 + \frac{C_0^2 n^2 \pi^2}{L^2} \right] \sin \frac{n \pi y}{L} \sin(kx - \sigma t + \phi),$$

$$\tag{3.9.16}$$

where $C_x = \sigma/k$ is the phase speed in the $x$-direction, and $\eta_0$ is an arbitrary amplitude and $\phi$ an arbitrary wave phase. Note the dependence of the structure of the solution in $y$ on the direction of the phase speed, i.e., on the sign of $C_x$.

(ii) The second solution to (3.9.9) occurs when

$$\sigma = \pm C_0 k. \tag{3.9.17}$$

This is a rather remarkable result, for it is nothing more than the dispersion relation for a plane wave whose crests are parallel to the $y$-axis for a *nonrotating* fluid. It is already evident that this solution supplements the set of Poincaré modes and plays the role of the $n = 0$ mode of that set which was disallowed due to the effects of rotation. This mode is the *Kelvin wave*, and it has a truly remarkable dynamical structure as well.

Consider the solution which propagates in the positive $x$-direction, i.e., $\sigma = C_0 k$. Then

$$\alpha^2 = -\frac{f^2}{C_0^2}, \tag{3.9.18}$$

so that $\alpha$ is purely imaginary, i.e., $\alpha = \pm if/C_0$. Without loss of generality, take $\alpha = if/C_0$. It follows that the dynamical fields can then be written in the form

$$\eta = \eta_0 e^{-fy/C_0} \cos(k[x - C_0 t] + \phi), \tag{3.9.19a}$$

$$u = \frac{\eta_0}{H_0} C_0 e^{-fy/C_0} \cos(k[x - C_0 t] + \phi) \tag{3.9.19b}$$

$$= -\frac{g}{f} \frac{\partial \eta}{\partial y}, \tag{3.9.19c}$$

$$v = 0.$$

There are several extraordinary features to note. First of all, the cross-channel velocity is identically zero. Second, the flow in the $x$-direction is in precise geostrophic balance even though the frequency is not, in general, small with respect to $f$. For *this* motion the detailed dynamical balances are

$$\frac{\partial u}{\partial t} = -g \frac{\partial \eta}{\partial x}, \tag{3.9.20a}$$

$$\frac{\partial \eta}{\partial t} = -H_0 \frac{\partial u}{\partial x}, \tag{3.9.20b}$$

$$u = -\frac{g}{f} \frac{\partial \eta}{\partial y}. \tag{3.9.20c}$$

The first two of these yield the classical wave equation

$$\frac{\partial^2 \eta}{\partial t^2} = C_0^2 \frac{\partial^2 \eta}{\partial x^2}, \tag{3.9.21}$$

from which (3.9.17) is obtained, while the geostrophic balance (3.9.20c) produces a free-surface slope to balance the Coriolis acceleration due to $u$. This cross-channel slope of the wave height is exponential. For an observer facing the direction of propagation, the wave height is highest to his right. The $e$-folding scale for the cross-channel amplitude variation is

$$R = \frac{C_0}{f}, \tag{3.9.22}$$

which is independent of any property of the wave field. This intrinsic length scale is the *Rossby radius of deformation* and is the distance over which the gravitational tendency to render the free surface flat is balanced by the tendency of the Coriolis acceleration to deform the surface. Note that $R \to \infty$ as $f \to 0$, so that the Kelvin wave becomes, in that limit, the $n = 0$ mode of (3.9.14). Hence the Kelvin mode is indeed the missing gravest mode of the Poincaré set. For $\sigma = -C_0 k$, i.e., for a Kelvin wave propagating in the negative $x$-direction, the same considerations apply. The wave amplitude diminishes exponentially from a maximum at the wall of $y = L$ as $y$ *decreases*, so that again the wave height is a maximum on the right hand of an observer looking in the direction of the wave propagation. The Kelvin wave requires at least one internal boundary for its existence. The exponentially increasing character of the wave renders it unacceptable in a completely infinite region. Note also that low-frequency Kelvin waves, for which $\sigma \ll f$, require $kR \ll 1$—i.e., the wavelength in the $x$-direction is much larger than the cross-stream scale $R$. This anisotropy is responsible for the nongeostrophy of (3.9.20a), i.e., the down-channel pressure gradient (which is weak for $kR \ll 1$ compared to the cross-channel gradient) is not balanced by a Coriolis acceleration, since $v$ is identically zero. Rather it is balanced by the similarly weak acceleration of $u$. The Kelvin wave illustrates that the simple balance-of-terms considerations of the first two chapters need to be refined when the horizontal scales of the motion are not isotropic.

(iii) The third apparent solution of (3.9.9) is an oscillation whose frequency is the Coriolis parameter, also called the inertial frequency—i.e., an oscillation for which

$$\sigma = \pm f. \tag{3.9.23}$$

However, when $\sigma = f$ the operator $\partial^2/\partial t^2 + f^2$, which in (3.6.12a,b) must be inverted to obtain $u$ and $v$ in terms of $\eta$, is identically zero, and we must exercise care to determine whether this root is spurious. Indeed it is, as can be seen by returning to the original set (3.6.3a,b,c). If (3.9.3) is used with $\sigma = f$, (3.6.3a,b) become

$$f\bar{u} - if\bar{v} = gk\bar{\eta}, \tag{3.9.24a}$$

$$f\bar{u} - if\bar{v} = -g\frac{d\bar{\eta}}{dy}, \tag{3.9.24b}$$

where $\bar{u}(y)$ and $\bar{v}(y)$ are the complex amplitudes of $u$ and $v$, i.e.,

$$u = \operatorname{Re} \bar{u}e^{i(kx - ft)}, \tag{3.9.25a}$$

$$v = \operatorname{Re} \bar{v}e^{i(kx - ft)}. \tag{3.9.25b}$$

Note that (3.9.24a,b) cannot be solved to yield $\bar{u}$ and $\bar{v}$ in terms of $\bar{\eta}$. Instead, by subtracting (3.9.24a,b), the constraint

$$\frac{d\bar{\eta}}{dy} + k\bar{\eta} = 0 \tag{3.9.26}$$

is required, or

$$\bar{\eta} = \eta_0 e^{-ky}. \tag{3.9.27}$$

On the other hand (3.6.3c) requires, for $\sigma = f$ that

$$-if\bar{\eta} + H_0\left(ik\bar{u} + \frac{d\bar{v}}{dy}\right) = 0, \qquad (3.9.28)$$

or from (3.9.24a) and (3.9.27),

$$\frac{d\bar{v}}{dy} - k\bar{v} = \frac{if\bar{\eta}}{H_0}\left(1 - \frac{k^2C_0^2}{f^2}\right)$$

$$= \frac{if\eta_0}{H_0}(1 - k^2R^2)e^{-ky}, \qquad (3.9.29)$$

the general solution of which is

$$\bar{v} = V_0 e^{ky} - \frac{if}{2k}\frac{\eta_0}{H_0}(1 - k^2R^2)e^{-ky}, \qquad (3.9.30)$$

where $V_0$ is arbitrary. To satisfy the condition that $\bar{v}(0)$ vanishes, $V_0$ must be chosen so that

$$\bar{v} = \frac{if\eta_0}{kH_0}(1 - k^2R^2)\sinh ky. \qquad (3.9.31)$$

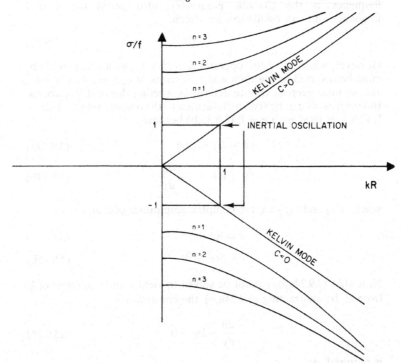

**Figure 3.9.2**   The dispersion diagram for Poincaré and Kelvin waves, showing the coincidence of the inertial oscillation $\sigma/f = \pm 1$ and the Kelvin mode at $kR = 1$.

The final condition, that $\bar{v}$ vanishes on $y = L$, is *impossible* except in the special case that $k = R^{-1}$, in which $v$ vanishes identically, $\sigma = C_0 k = f$, and the wave is indistinguishable from the Kelvin wave at that wave number. The oscillation $\sigma = f$ is therefore a spurious root of the eigenvalue problem, and the complete spectrum of the solution consists of the Kelvin mode, the Poincaré modes, and of course the $\sigma = 0$ mode, which is the geostrophic flow. Figure 3.9.2 shows the dispersion diagram, i.e., a graph of the frequency as a function of wave number.

## 3.10 The Rossby Wave

Consider now the free oscillations that are possible in a channel of width $L$, as in Section 3.9, with the single important difference that $H_0$ varies slightly in the $y$-direction, i.e.,

$$H_0 = D_0\left(1 - \frac{sy}{L}\right), \tag{3.10.1}$$

where the slope

$$s \ll 1. \tag{3.10.2}$$

Lines of constant $H_0$ run parallel to the $x$-axis, and pure geostrophic motion is possible only if $v$ is identically zero. Motion across the isobaths (the isolines of $H_0$) of fluid columns will cause them to stretch or contract, which when coupled to $f$ will produce relative vorticity. We are obliged to anticipate the possibility of a new mode of motion whose existence depends on the joint effect of rotation and bottom slope.

Solutions to (3.6.9) of the form

$$\eta = \text{Re } \bar{\eta}(y)e^{i(kx - \sigma t)}$$

may still be sought. The problem for $\bar{\eta}(y)$ is now

$$\left(1 - s\frac{y}{L}\right)\frac{d^2\bar{\eta}}{dy^2} - \frac{s}{L}\frac{d\bar{\eta}}{dy} + \bar{\eta}\left[\frac{\sigma^2 - f^2}{gD_0} - k^2\left(1 - s\frac{y}{L}\right) - \frac{fs}{L\sigma}k\right] = 0, \tag{3.10.3}$$

where again

$$\frac{d\bar{\eta}}{dy} + \frac{fk}{\sigma}\bar{\eta} = 0 \quad \text{on } y = 0, L. \tag{3.10.4}$$

Now $y/L$ is never larger than unity. Therefore, for small $s$, i.e., small depth variations, a consistent and excellent approximation to (3.10.3) is simply

$$\frac{d^2\bar{\eta}}{dy^2} - \frac{s}{L}\frac{d\bar{\eta}}{dy} + \bar{\eta}\left[\frac{\sigma^2 - f^2}{C_0^2} - k^2 - \frac{f}{L}\frac{s}{\sigma}k\right] = 0. \tag{3.10.5}$$

Note that only in terms where the slope parameter can be compared with quantities known to be of order *unity* has it been neglected. The last term in the square bracket in (3.10.5) depends on the ratio $s/\sigma$ and cannot be estimated *a priori*. Indeed, if $\sigma$ is O($s$), that term will be O(1). The solution to (3.10.5) is

$$\bar{\eta} = e^{sy/2L}[A \sin \alpha y + B \cos \alpha y], \qquad (3.10.6)$$

where now

$$\alpha^2 = \frac{\sigma^2 - f^2}{C_0^2} - \left(k^2 + \frac{s^2}{4L^2}\right) - \frac{fks}{\sigma L}. \qquad (3.10.7)$$

The application of (3.10.4) now yields the eigenvalue relation

$$(\sigma^2 - f^2)(\sigma^2 - k^2 C_0^2)\sin \alpha L = 0. \qquad (3.10.8)$$

It is important to note that the factors multiplying $\sin \alpha L$ are the same as for the case $s = 0$, i.e., for the case of a flat bottom. Hence, to lowest order in $s$ the slope does not alter the Kelvin mode. The roots corresponding to the zeros of $\sin \alpha L$ now yield

$$\sigma^2 - \frac{fksC_0^2}{L\sigma} - C_0^2\left(k^2 + \frac{n^2\pi^2}{L^2} + \frac{f^2}{C_0^2}\right) = 0, \qquad (3.10.9)$$

where a term O($s^2/L^2$) has been ignored with respect to $n^2\pi^2/L^2$, since $s \ll 1$. As before, $n$ must be greater than zero.

There are now two distinctly separate classes of solutions to (3.10.9).

The first class has frequencies each of which exceeds $f$. For this class the term in $s$ is negligible, and to O($s$) we recover the Poincaré modes, i.e.,

$$\sigma^2 = f^2 + C_0^2\left(k^2 + \frac{n^2\pi^2}{L^2}\right) + O(s), \qquad n = 1, 2, \ldots, \qquad (3.10.10)$$

so that the high-frequency Poincaré waves are also essentially unaffected by the presence of the small bottom slope.

The important new solution, i.e., the third root of the cubic, has a frequency $\sigma = $ O($s$), for which the first term in (3.10.9) is negligible, while the second is O(1). This leads to the dispersion relation for the topographic *Rossby wave*, i.e.,

$$\sigma = -s\left(\frac{f}{L}\right)\frac{k}{k^2 + n^2\pi^2/L^2 + f^2/C_0^2}, \qquad n = 1, 2, \ldots, \qquad (3.10.11)$$

A more complete discussion of the Rossby wave will be given in Section 3.15. It is useful at this point to discuss certain salient features of the dynamics of the wave in order to aid our formulation of a consistent dynamical framework for low-frequency motions. The Rossby-wave frequency attains its *maximum* value when

$$k = k_n = \left(\frac{n^2\pi^2}{L^2} + \frac{f^2}{C_0^2}\right)^{1/2}, \qquad (3.10.12)$$

for which

$$\sigma = \sigma_{max} = -\frac{s}{2}\frac{f}{(n^2\pi^2 + f^2 L^2/C_0^2)^{1/2}}, \qquad (3.10.13)$$

so that, for small $s$, the Rossby-wave frequency is always *less than f*. Thus the Rossby wave, whose existence requires both $s$ and $f$ to be nonzero, is a low-frequency wave oscillation in the sense described in the first chapter, i.e., its period is greater than a rotation period.

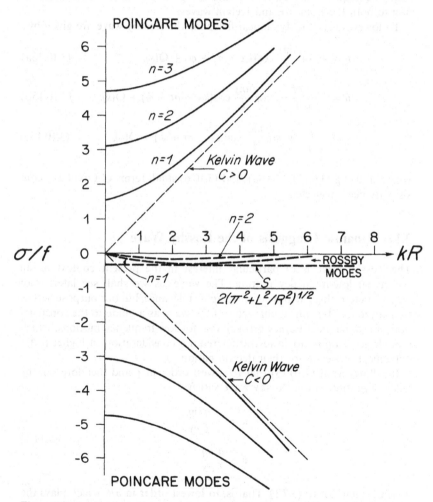

**Figure 3.10.1**  A schematic representation of the dispersion diagram for Poincaré, Kelvin, and topographic Rossby waves in a channel.

A truly remarkable feature of the Rossby wave is that its phase speed in the $x$-direction,

$$C_x = \frac{\sigma}{k} = -\left(\frac{sf}{L}\right)\bigg/\left(k^2 + \frac{n^2\pi^2}{L^2} + \frac{f^2}{C_0^2}\right), \qquad (3.10.14)$$

is *always negative*. The phase of the wave, for all $k$, propagates so that an observer riding on a wave crest sees shallow fluid—that is, smaller $H_0$—on his right (left) hand if $f$ is positive (negative). The complete dispersion diagram is shown in Figure 3.10.1. Note that for high wave number, i.e., small scale, the frequency *decreases* with increasing wave number in contradistinction to both the Poincaré and Kelvin waves.

To lowest order the dynamical fields for the Rossby wave are given by

$$\eta = \eta_0 \sin\frac{n\pi y}{L}\cos(kx - \sigma t + \phi) + O(s), \qquad (3.10.15a)$$

$$u = -\frac{g}{f}\frac{n\pi}{L}\eta_0 \cos\frac{n\pi y}{L}\cos(kx - \sigma t + \phi) + O(s). \qquad (3.10.15b)$$

$$v = -\frac{g}{f}k\eta_0 \sin\frac{n\pi y}{L}\sin(kx - \sigma t + \phi), \qquad (3.10.15c)$$

where in using (3.6.12), (3.9.8a), and (3.10.6) small terms of $O(s)$ have consistently been neglected.

## 3.11 Dynamic Diagnosis of the Rossby Wave

The Rossby wave is of particular interest in the present context as an archetypal low-frequency motion. The wave, as we shall see later, is of central meteorological and oceanographic relevance, but our purpose here is to analyze the dynamical character of the wave as a guide to the construction of a dynamical framework suitable for low-frequency motions, which though geostrophic to lowest order require consideration of higher-order dynamical processes for their determination.

It follows from (3.10.15) that to lowest order in $s$, and therefore $\sigma/f$, the fields of motion in the Rossby wave satisfy

$$u = -\frac{g}{f}\frac{\partial\eta}{\partial y},$$

$$v = \frac{g}{f}\frac{\partial\eta}{\partial x}, \qquad (3.11.1)$$

which is identical to (3.7.1). That is, to lowest order in $\sigma/f$, which plays the role of the Rossby number in the present case, the velocity fields, though changing with time, remain continuously in geostrophic balance with the pressure field. Nonetheless the flow is not exactly geostrophic, for then the flow would be restricted to travel parallel to the isobaths, i.e., $v$ would vanish. For $v$ to vanish, $k$ would have to be zero by (3.10.15c), and if $k$ were zero, then $\sigma$, according to (3.10.11), would itself vanish. Even though the velocity fields are geostrophic to lowest order, it is the very small departures from geostrophy that give rise to the wave. It is the small cross-isobath flow, which is a nongeostrophic effect, which produces the oscillation.

For small $\sigma/f$, it follows that

$$\frac{\partial^2}{\partial t^2} \eta \ll f^2 \eta, \tag{3.11.2}$$

so that the governing equation (3.6.9) reduces, for small $s$, to

$$\frac{\partial}{\partial t}[gH_0 \nabla^2 \eta - f^2 \eta] - gf \frac{\partial H_0}{\partial y} \frac{\partial \eta}{\partial x} = 0, \tag{3.11.3}$$

where the relatively small term $g(\partial \eta/\partial y)(\partial H_0/\partial y)$ has been ignored in the square bracket, as an $O(s)$ term. It follows from (3.11.1) that

$$gH_0 \nabla^2 \eta = H_0 f \left( \frac{\partial v}{\partial x} - \frac{\partial u}{\partial y} \right) = H_0 f \zeta, \tag{3.11.4}$$

which should be compared with (2.9.20). Thus, (3.11.3) can be written with the aid of (3.11.1) as

$$\frac{\partial}{\partial t}\left[ \zeta - \frac{f\eta}{H_0} \right] - v \frac{f}{H_0} \frac{dH_0}{dy} = 0. \tag{3.11.5}$$

Since the motion is linear,

$$\frac{\partial}{\partial t}\left[ \zeta - \frac{f\eta}{H_0} \right] \approx \frac{d}{dt}\left[ \zeta - \frac{f\eta}{H_0} \right], \tag{3.11.6}$$

and since $H_0$ is independent of $t$,

$$v \frac{dH_0}{dy} = \frac{dH_0}{dt}, \tag{3.11.7}$$

so that within the accuracy of the linear theory,

$$\frac{d}{dt}\left[ \zeta/H_0 - \frac{f}{H_0^2}(H_0 + \eta) \right] = 0, \tag{3.11.8}$$

or neglecting quadratic terms,

$$\frac{d}{dt}\frac{\zeta + f}{H} = 0. \tag{3.11.9}$$

Thus for slow, nearly geostrophic motion (i.e., quasigeostrophic motion), the governing equation of motion (3.11.3), which determines the dynamics, is the statement of potential-vorticity conservation (3.11.9) wherein the velocity and vorticity fields are approximated by their geostrophic values. Thus at least for this linear problem, the potential vorticity equation yields a deterministic dynamical system for the geostrophic velocities in a way which *implicitly* takes into account the small departures for geostrophy without the need to calculate the departures directly. We shall see that this result can be generalized so that the geostrophic degeneracy can be resolved by the systematic use of the geostrophic approximation in conjunction with the potential-vorticity theorem.

## 3.12  Quasigeostrophic Scaling in Shallow-Water Theory

We now return to the complete shallow-water equations (3.3.15a, b), (3.3.22) and focus our attention on motions whose time scales are long compared to $f^{-1}$, and attempt to derive a nonlinear analogue to the quasi-geostrophic potential-vorticity equation of the preceding section. There the governing equation was extracted with the aid of the solution. In general we cannot do this, since the solutions of more complex problems are not available to us in their general form. Were they available, the problem associated with approximations such as geostrophy would be moot. In fact, in nonlinear problems and frequently in linear problems, progress can only be made within the framework of an approximate geostrophic theory. What is required now is the systematic use of *qualitative* a priori statements about the motion (e.g., long time scales) to deduce *quantitative* rules for the consistent calculation of that motion while disregarding the unnecessary complications inherent in the original theory, which had to be complex enough to compute the higher-frequency motions also.

   To proceed systematically it is essential to introduce nondimensional variables. First we *choose* scales $L$, $T$, $U$, and $N_0$, which we *demand* characterize the magnitudes of length, time, velocity, and free-surface elevation, respectively. We then use these scales to define nondimensional dependent and independent variables, denoted by primes, as follows:

$$(x, y) = L(x', y'),$$

$$t = Tt',$$

$$(u, v) = U(u', v'), \tag{3.12.1}$$

$$\eta = N_0\eta'.$$

It is useful to introduce the following notation:

$$H = H_0(x, y) + \eta = D + \eta - h_B, \tag{3.12.2a}$$

where

$$H_0 = D - h_B(x, y). \tag{3.12.2b}$$

As shown in Figure 3.2.1, $\eta$ is the departure of the free surface from its level, rest height, and $h_B$ is the measure of the bottom variation which produces a departure of $H_0$, the depth of the resting fluid, from the constant value $D$.

The important presumption is that the scales $L$, $U$, etc. have been accurately chosen so that the magnitude of the nondimensional variables are *of order* unity and that the product of any combination of variables is accurately measured by the product of the scale factors. If the variables introduced in (3.12.1) are used to express (3.3.15a,b) and (3.3.22) in terms of primed variables, we obtain

$$\frac{U}{T}\frac{\partial u'}{\partial t'} + \frac{U^2}{L}\left\{ u'\frac{\partial u'}{\partial x'} + v'\frac{\partial u'}{\partial y'} \right\} - fUv' = -g\frac{N_0}{L}\frac{\partial \eta'}{\partial x'}. \tag{3.12.3a}$$

$$\frac{U}{T}\frac{\partial v'}{\partial t'} + \frac{U^2}{L}\left\{ u'\frac{\partial v'}{\partial x'} + v'\frac{\partial v'}{\partial y'} \right\} + fUu' = -g\frac{N_0}{L}\frac{\partial \eta'}{\partial y'} \tag{3.12.3b}$$

$$\frac{N_0}{T}\frac{\partial \eta'}{\partial t'} + \frac{U}{L}\left\{ u'\frac{\partial}{\partial x'}(N_0\eta' - h_B) + v'\frac{\partial}{\partial y'}(N_0\eta' - h_B) \right\}$$

$$+ \frac{U}{L}[D + N_0\eta' - h_B]\left[ \frac{\partial u'}{\partial x'} + \frac{\partial v'}{\partial y'} \right] = 0. \tag{3.12.3c}$$

We now *insist* that the motions to be described should be such that the time, length, and velocity scales satisfy

$$\varepsilon = \frac{U}{fL} \ll 1,$$

$$\varepsilon_T = \frac{1}{fT} \ll 1. \tag{3.12.4}$$

In this case the relative acceleration terms in (3.12.3) will be small compared to the Coriolis acceleration terms by $O(\varepsilon)$ or $O(\varepsilon_T)$. In order for $u'$ and $v'$ to be different from zero, the pressure-gradient terms must be large enough to balance the Coriolis acceleration. We therefore *choose* the parameter $N_0$ as

$$N_0 = \frac{fUL}{g} = \frac{U}{fL}\frac{f^2L^2}{g}, \tag{3.12.5}$$

which implies that (3.12.2a) may be written

$$H = D\left[ 1 + \varepsilon\frac{L^2}{R^2}\eta' - \frac{h_B}{D} \right] \tag{3.12.6}$$

where $R$ is the Rossby deformation radius for the layer of depth $D$, i.e.,

$$R = (gD)^{1/2}/f. \tag{3.12.7}$$

If $L/R$ is a parameter of order one or less, the departure of the layer thickness from its value in the absence of motion is $O(\varepsilon)$. If the momentum equations are each divided by $fU$, and the mass-conservation statement by $U/L$, then we obtain with the aid of (3.12.5) a set of equations entirely in terms of nondimensional variables and nondimensional parameters, viz.

$$\varepsilon_T \frac{\partial u}{\partial t} + \varepsilon \left\{ u \frac{\partial u}{\partial x} + v \frac{\partial u}{\partial y} \right\} - v = - \frac{\partial \eta}{\partial x}, \tag{3.12.8a}$$

$$\varepsilon_T \frac{\partial v}{\partial t} + \varepsilon \left\{ u \frac{\partial v}{\partial x} + v \frac{\partial v}{\partial y} \right\} + u = - \frac{\partial \eta}{\partial y}, \tag{3.12.8b}$$

$$\varepsilon_T F \frac{\partial \eta}{\partial t} + \varepsilon F \left\{ u \frac{\partial \eta}{\partial x} + v \frac{\partial \eta}{\partial y} \right\} - u \frac{\partial}{\partial x}\left(\frac{h_B}{D}\right) - v \frac{\partial}{\partial y}\left(\frac{h_B}{D}\right)$$

$$+ \left\{ 1 + \varepsilon F \eta - \frac{h_B}{D} \right\}\left\{\frac{\partial u}{\partial x} + \frac{\partial v}{\partial y}\right\} = 0. \tag{3.12.8c}$$

For the sake of neatness the primes have been dropped from the nondimensional variables. For the remainder of the chapter dimensionless variables are unprimed, while the dimensional counterparts will be denoted by an asterisk, e.g., $u_* = Uu$.

The parameter

$$F = \frac{f^2 L^2}{gD} = \left(\frac{L}{R}\right)^2, \tag{3.12.9}$$

which naturally appears in (3.12.8c), is the square of the ratio of the geometric length scale $L$ to the Rossby deformation radius $R$. We shall assume that

$$F = O(1) \tag{3.12.10}$$

throughout our discussion. The Rossby numbers $\varepsilon_T$ and $\varepsilon$ measure the relative importance of the local and convective accelerations to the Coriolis acceleration. We are interested in cases where both $\varepsilon$ and $\varepsilon_T$ are small. The *relative* importance of local to advective accelerations is measured by the ratio

$$\frac{\varepsilon_T}{\varepsilon} = \frac{L}{UT}. \tag{3.12.11}$$

When this ratio is large the equations are essentially linear, i.e., the local time derivative dominates the advective nonlinearity. At this point we wish to allow the fluid velocity $U$ to be sufficiently large that the nonlinear terms can be as important as the linear acceleration term. We therefore choose to set

$$\frac{\varepsilon_T}{\varepsilon} = 1, \tag{3.12.12}$$

i.e., to consider cases where the advective time $L/U$ is as short as the time scale for local change. We can do this without any loss of generality, as we shall see.

We are now in a position to *systematically* examine the orders of magnitude of the various terms in the equations of motion, but more than that, we can find relationships between terms of like order in $\varepsilon$. Consider any solution of (3.12.8a,b,c). It will be a function of $x$, $y$, $t$, *and* the parameter $\varepsilon$. That is, the velocity $u$, for example, is

$$u = u(x, y, t, \varepsilon). \tag{3.12.13}$$

For small $\varepsilon$, we suppose that $u$ can be expanded in an asymptotic series in $\varepsilon$. Since only integral powers of $\varepsilon$ appear in (3.12.8) it seems sensible to suppose the expansion proceeds as

$$u(x, y, t, \varepsilon) = u_0(x, y, t) + \varepsilon u_1(x, y, t) + \varepsilon^2 u_2(x, y, t) + \cdots, \tag{3.12.14}$$

where the functions $u_0$, $u_1$, etc. are independent of $\varepsilon$. Each function $u$, $v$, and $\eta$ is expanded in this way and then inserted into the equations of motion. Since $\varepsilon$, while small, is arbitrary, like powers of $\varepsilon$ must balance if the equations are to be satisfied for arbitrary small $\varepsilon$ for all $x$, $y$, $t$. Since $\varepsilon$ is small, our interest is focused mainly on $u_0$, $v_0$, and $\eta_0$. However, the expansion must in general be carried to higher order than the first to determine the lowest-order fields.

The $O(1)$ terms from (3.12.8a,b) yield

$$v_0 = \frac{\partial \eta_0}{\partial x},$$

$$\tag{3.12.15}$$

$$u_0 = -\frac{\partial \eta_0}{\partial y},$$

so that the lowest-order fields are geostrophic. It follows directly that

$$\frac{\partial u_0}{\partial x} + \frac{\partial v_0}{\partial y} = 0. \tag{3.12.16}$$

We immediately see that (3.12.8c) presents us with a choice. If $h_B/D$ is $O(1)$, then the $O(1)$ terms of (3.12.8c), using (3.12.16), are simply

$$u_0 \frac{\partial}{\partial x}\left(\frac{h_B}{D}\right) + v_0 \frac{\partial}{\partial y}\left(\frac{h_B}{D}\right) = 0, \tag{3.12.17}$$

which we recognize as the strong constraint that purely geostrophic motion must move along the isobaths. The other choice is the more interesting one, namely that the Rossby number and $h_B/D$ are of the same order, as in the case of the Rossby wave where $\sigma/f = \varepsilon_T$ is $O(s)$. That is, while the Rossby number is small, we are interested in cases where it is large enough to break the constraint (3.12.17), i.e., that the motion is sufficiently distant from the exact geostrophic mode. Formally this means we consider

$$\frac{h_B}{D} = \varepsilon \eta_B(x, y) \tag{3.12.18}$$

where $\eta_B$ is $O(1)$. We shall see however that although we proceed along this second path, we shall be able to recover (3.12.17) as a limiting case of the theory to be derived.

With the constraint (3.12.17) removed, we find ourselves in the familiar situation of being unable, with the $O(1)$ portions of (3.12.18), to determine the order-one fields $u_0$, $v_0$, $\eta_0$. We are now able to make progress beyond our earlier crude order-of-magnitude estimates by exploiting the systematic asymptotic expansion of the equations. This is why the nondimensional formulation is so valuable.

The $O(\varepsilon)$ terms in the equations of motion yield

$$\frac{\partial u_0}{\partial t} + u_0 \frac{\partial u_0}{\partial x} + v_0 \frac{\partial u_0}{\partial y} - v_1 = -\frac{\partial \eta_1}{\partial x}, \tag{3.12.19a}$$

$$\frac{\partial v_0}{\partial t} + u_0 \frac{\partial v_0}{\partial x} + v_0 \frac{\partial v_0}{\partial y} + u_1 = -\frac{\partial \eta_1}{\partial y}, \tag{3.12.19b}$$

$$F\left| \frac{\partial \eta_0}{\partial t} + u_0 \frac{\partial \eta_0}{\partial x} + v_0 \frac{\partial \eta_0}{\partial y} \right| - u_0 \frac{\partial \eta_B}{\partial x} - v_0 \frac{\partial \eta_B}{\partial y} + \left( \frac{\partial u_1}{\partial x} + \frac{\partial v_1}{\partial y} \right) = 0. \tag{3.12.19c}$$

The velocities $u_1$ and $v_1$ are not geostrophic. The departures of these velocities from geostrophic balance with the $O(\varepsilon)$ pressure field are produced, according to (3.12.19), entirely by the acceleration of the $O(1)$ velocity fields that are in geostrophic balance. Their horizontal divergence is nonzero and must be balanced by the stretching of fluid columns. To establish a closed dynamical system which explicitly involves *only* the $O(1)$ fields, we eliminate the pressure between (3.12.19a,b) to obtain

$$\frac{d\zeta_0}{dt} = \frac{\partial \zeta_0}{\partial t} + u_0 \frac{\partial \zeta_0}{\partial x} + v_0 \frac{\partial \zeta_0}{\partial y} = -\left( \frac{\partial u_1}{\partial x} + \frac{\partial v_1}{\partial y} \right), \tag{3.12.20}$$

where

$$\zeta_0 = \frac{\partial v_0}{\partial x} - \frac{\partial u_0}{\partial y} = \nabla^2 \eta_0. \tag{3.12.21}$$

Hence to $O(\varepsilon)$ the rate of change of relative vorticity is equal to the convergence present in the $O(\varepsilon)$ nongeostrophic* velocity field. Note that this effect is linear in the velocity field. This is due to the fact that although it is the convergence of total vorticity filaments that originally enters the vorticity equation, e.g., (3.4.3), $\zeta_*$ is so much smaller than $f$ that only the squeezing together of the planetary vorticity filaments enters at the lowest order. If (3.12.19c) is used to evaluate the divergence of the order-$\varepsilon$ velocity field—i.e., if $\partial u_1/\partial x + \partial v_1/\partial y$ is eliminated between (3.12.20) and (3.12.19c)—we obtain

$$\frac{d\zeta_0}{dt} = F \frac{d\eta_0}{dt} - \frac{d\eta_B}{dt}, \tag{3.12.22}$$

---

* Also called the ageostrophic field.

where

$$\frac{d}{dt} = \frac{\partial}{\partial t} + u_0 \frac{\partial}{\partial x} + v_0 \frac{\partial}{\partial y},$$  (3.12.23)

or

$$\boxed{\frac{d}{dt}\{\zeta_0 - F\eta_0 + \eta_B\} = 0}.$$  (3.12.24)

Since $u_0$, $v_0$, and $\zeta_0$ are related to $\eta_0$ by (3.12.15) and (3.12.21), the conservation statement (3.12.24) can be written entirely in terms of $\eta_0$:

$$\boxed{\left[\frac{\partial}{\partial t} + \frac{\partial \eta_0}{\partial x}\frac{\partial}{\partial y} - \frac{\partial \eta_0}{\partial y}\frac{\partial}{\partial x}\right][\nabla^2 \eta_0 - F\eta_0 + \eta_B] = 0},$$  (3.12.25)

which is the nonlinear generalization of (3.11.3).

There are several important points to note.

(i) The problem for quasigeostrophic motion is given entirely in terms of the surface height $\eta_0$, or equivalently, the pressure.
(ii) Once $\eta_0$ is determined as a solution of (3.12.25), $u_0$ and $v_0$ are determined geostrophically from (3.12.15).
(iii) If $h_B/D$ is much greater than $O(\varepsilon)$, $\eta_B$ is then far greater than $O(1)$, so that (3.12.25) reduces to

$$u_0 \frac{\partial \eta_B}{\partial x} + v_0 \frac{\partial \eta_B}{\partial y} = 0,$$

which is the constraint (3.12.17). Therefore that possibility is contained within the theory we have derived.
(iv) The equation for $\eta_0$ is a conservation statement for the quantity

$$\Pi_g = \zeta_0 - F\eta_0 + \eta_B.$$  (3.12.26)

Consider the *dimensional* potential vorticity

$$\Pi_* = \frac{\zeta_* + f}{H_*} = \frac{f}{D}\frac{\varepsilon\zeta + 1}{1 + \varepsilon F\eta - \varepsilon\eta_B}.$$  (3.12.27)

The second equality merely uses the scaling variables and (3.12.6). For small $\varepsilon$

$$\Pi_* \approx \frac{f}{D}(1 + \varepsilon\zeta)(1 - \varepsilon F\eta + \varepsilon\eta_B)$$

$$\approx \frac{f}{D}[1 + \varepsilon(\zeta - F\eta + \eta_B)],$$  (3.12.28)

so that aside from an irrelevant constant term, $\Pi_g$ is the nondimensional potential vorticity and (3.12.25) is the quasigeostrophic potential-vorticity equation. That is, it is the potential-vorticity equation in which all terms are evaluated through the use of their geostrophic values, a result deduced heuristically by our diagnosis of the Rossby wave in the preceding section.

Since $\eta_0$ serves as a stream function for the $O(1)$ velocity field, the notation

$$\eta_0 = \psi(x, y, t) \tag{3.12.29a}$$

is introduced, in terms of which

$$u_0 = -\frac{\partial \psi}{\partial y}, \tag{3.12.29b}$$

$$v_0 = \frac{\partial \psi}{\partial x}, \tag{3.12.29c}$$

while (3.12.25) becomes

$$\left[\frac{\partial}{\partial t} + \frac{\partial \psi}{\partial x}\frac{\partial}{\partial y} - \frac{\partial \psi}{\partial y}\frac{\partial}{\partial x}\right][\nabla^2\psi - F\psi + \eta_B] = 0. \tag{3.12.30}$$

In the limit $\varepsilon/\varepsilon_T \ll 1$ Equation (3.12.30) would still apply, with the single exception that the nonlinear terms would be absent. If, on the other hand, $\varepsilon_T/\varepsilon \ll 1$, it is the steady version of (3.12.30) which obtains. Hence as long as both $\varepsilon$ and $\varepsilon_T$ are small, (3.12.25) is valid for all $\varepsilon_T/\varepsilon$. Indeed, had this ratio been left arbitrary, (3.12.25) would be left unchanged with the single exception that the time derivative

$$\frac{\partial}{\partial t} \to \frac{\varepsilon_T}{\varepsilon}\frac{\partial}{\partial t}.$$

With the above remarks in mind, however, it is sufficient to formally consider $\varepsilon$ and $\varepsilon_T$ equal for the purposes of the scaling and effect any linearization subsequently as desired.

The quasigeostrophic approximation to the potential vorticity $\Pi_g$ is a linear combination of three terms. The first two in (3.12.26) are due entirely to the relative motion. The first is the relative vorticity, while the second is the contribution to the potential vorticity due to variations in the free-surface height. The relative importance of the second term to the first is measured by $F$, i.e., by the ratio of $L$, the scale of motion, to the deformation radius. If $L$ is small compared to $R$, there is, *on the scale of the motion*, a negligible variation of $\eta$ and a consequently negligible contribution to the potential vorticity by vortex-tube stretching. Thus if $R \gg L$, from the point of view of the vorticity balance, the free surface appears no different than a rigid lid. If $L \gg R$, the relative vorticity is unimportant and the fluid velocity appears horizontally uniform. *The Rossby deformation radius is the scale for which the relative vorticity and the surface height (vortex-tube stretching) make equal contributions to the potential vorticity.* The remaining term in (3.12.26) is the *ambient* potential vorticity $\eta_B$, whose presence is independent of the motion. In terms of dimensional quantities

$$\eta_B = \frac{f}{D} \frac{h_B}{U/L},$$ (3.12.31)

so that $\eta_B$ is a relative measure of the variable, ambient potential vorticity measured with respect to the relative vorticity, which is $O(U/L)$. If $\eta_B$ is small, the layer thickness can be considered constant insofar as the balance of potential vorticity is concerned. If $\eta_B$ is very large, the ambient potential-vorticity gradients are so large compared with the relative vorticity that (3.12.17) applies and the quasigeostrophic flow will be along isolines of $\eta_B$.

## 3.13 Steady Quasigeostrophic Motion

If the motion is independent of time, the quasigeostrophic potential-vorticity equation reduces to

$$J(\psi, \Pi_g) = 0,$$ (3.13.1)

i.e., the Jacobian of $\psi$ and the potential vorticity vanishes. This implies that lines of constant $\Pi_g$ and constant $\psi$ coincide in the $x, y$ plane, or that (3.13.1) implies

$$\Pi_g = G(\psi),$$ (3.13.2)

where $G$ is an arbitrary function of $\psi$. In steady flow, where trajectories coincide with streamlines, the potential vorticity must be constant along fluid streamlines. Using the explicit form for $\Pi_g$ in terms of $\psi$, (3.13.2) is

$$\nabla^2\psi - F\psi + \eta_B = G(\psi),$$ (3.13.3)

or

$$\nabla^2\psi + \eta_B = G(\psi) + F\psi = K(\psi),$$ (3.13.4)

where $K(\psi)$ is clearly also an arbitrary function of $\psi$. The sum of the relative vorticity and ambient potential vorticity is conserved along streamlines. For steady flow, vortex stretching due to fluctuations of the free surface is absent. Each fluid element preserves its value of $\psi$ and hence experiences no tube stretching due to the upper surface, which could, as far as the dynamics is concerned, be flat. Note that the vortex tubes will be stretched if fluid crosses isolines of $\eta_B$.

Once $K(\psi)$ has been determined, (3.13.4) determines the steady flow. To find $K(\psi)$ it is necessary to specify $\nabla^2\eta_0 + \eta_B$ at one point on each streamline.

## 3.14 Inertial Boundary Currents

As an application of the ideas of the preceding section, consider the problem of determining the structure of the uniform, steady flow impinging, as shown in Figure 3.14.1, on the wall at $x = 0$ which is perpendicular to the isobaths of a fluid whose undisturbed thickness is given by

$$H_0 = D\left(1 - s\frac{y_*}{L}\right). \tag{3.14.1}$$

Thus

$$\eta_B = \frac{s}{\varepsilon}y = s\left(\frac{fL}{U}\right)y \equiv \beta y, \tag{3.14.2}$$

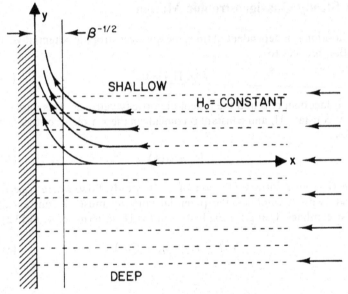

**Figure 3.14.1**   A parallel flow along the isobaths is in purely geostrophic balance until it is turned aside by the boundary at $x = 0$. The width of the turning zone is $\beta^{-1/2}$.

where $U$ is the magnitude of the undisturbed flow at infinity. Note that in the absence of the wall, which blocks the contours of $\eta_B$, the uniform flow along $\eta_B$ would be the appropriate flow everywhere, with each fluid column streaming along its original $\eta_B$ contour—i.e., the motion would be linear and strictly geostrophic. The presence of the barrier introduces a new element. If the flow were *strictly* geostrophic *everywhere*, the results of (3.7) would imply that the velocity must *vanish* everywhere. In the present problem that constraint must be broken, at least near the boundary, where fluid columns forced toward $x = 0$ must leave their isobaths to flow parallel to the wall, and by (3.13.4) they must suffer a change in their relative vorticity to compensate, along the streamline, the change in $\eta_B$. The governing equation for the process is (3.13.4), which with (3.14.2) is

$$\nabla^2\psi + \beta y = K(\psi). \tag{3.14.3}$$

Far from the wall the flow is uniform and directed antiparallel to the $x$-axis with a nondimensional amplitude of unity (since $U$ was chosen as the velocity scale). Thus at infinity

$$\psi = \psi_\infty = (y - y_0), \tag{3.14.4}$$

where $y_0$ is an arbitrary *constant*. Furthermore, at infinity $\nabla^2\psi$ vanishes, so that at infinity

$$\beta y = K(\psi_\infty), \tag{3.14.5}$$

or

$$K(\psi_\infty) = \beta\psi_\infty + \beta y_0. \tag{3.14.6}$$

Since $K(\psi)$ is constant on streamlines, the functional relationship (3.14.6) must hold for all streamlines, at *any* $x$ and $y$, which have their origin at infinity. On those streamlines, then,

$$K(\psi) = \beta\psi + \beta y_0. \tag{3.14.7}$$

$K(\psi)$ will not be a linear function of $y$ except at infinity, but it must remain a linear function of $\psi$. $K(\psi)$ is now fixed, so that (3.14.3) is

$$\nabla^2\psi - \beta\psi = -\beta(y - y_0). \tag{3.14.8}$$

Write $\psi$ as

$$\psi = (y - y_0) + \phi(x, y), \tag{3.14.9}$$

where $\phi$ is the departure of the stream function from the uniform flow caused by the presence of the wall, and $\phi$ satisfies

$$\frac{\partial^2\phi}{\partial x^2} + \frac{\partial^2\phi}{\partial y^2} - \beta\phi = 0. \tag{3.14.10a}$$

The appropriate boundary conditions for $\phi$ are

$$\phi \to 0 \qquad \text{as } x \to \infty \tag{3.14.10b}$$

$$\frac{\partial\phi}{\partial y} = -1 \quad \text{at } x = 0. \tag{3.14.10c}$$

The first of these follows from the requirement that $\psi$ approach $\psi_\infty$ far upstream, while the second is the condition of no normal flow through the wall at $x = 0$, i.e., that $u = -\partial\psi/\partial y$ must vanish at the wall. It seems sensible to try a solution of the form

$$\phi = (y - y_0)X(x), \tag{3.14.11}$$

since $\phi$ must be linear in $y$ on $x = 0$. $X(x)$ must then satisfy

$$\frac{d^2 X}{dx^2} - \beta X = 0, \tag{3.14.12}$$

whose general solution is

$$X = C_1 e^{-\beta^{1/2} x} + C_2 e^{\beta^{1/2} x}. \tag{3.14.13}$$

Since $X$ must vanish at infinity, $C_2$ must vanish. The application of (3.14.10c) yields

$$C_1 = -1, \tag{3.14.14}$$

or

$$\phi = -(y - y_0) e^{-\beta^{1/2} x}, \tag{3.14.15}$$

so that the total stream function is

$$\psi = (y - y_0)[1 - e^{-\beta^{1/2} x}]. \tag{3.14.16}$$

The streamlines of the flow are shown in Figure 3.14.2. Note that the stagnation point $x = 0$, $y = y_0$ where both $u$ *and* $v$ vanish is at an arbitrary value of

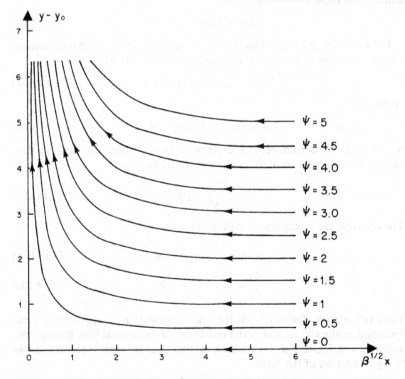

**Figure 3.14.2**   The streamlines of the flow in the inertial boundary current.

$y$. Indeed, for this inviscid flow we can equally well think of the flow as streaming in the corner formed by the intersection of the wall at $x = 0$ with a wall at $y = y_0$. In fact, $y_0$ can be specified by the additional condition, which fixes $y_0$, that

$$v = \frac{\partial \psi}{\partial x} = 0 \quad \text{on } y = y_0.$$

For reasons of symmetry we can choose our origin of coordinates at $x = 0$, $y = y_0$ and take $y_0$ zero in (3.14.16).

For very small $\beta$, i.e., small ambient potential vorticity gradients, (3.14.16) becomes, for $x = O(1)$,

$$\psi = \beta^{1/2} y x, \tag{3.14.17}$$

which is the formula for stagnation-point flow of an irrotational (zero-vorticity) flow in classical potential theory. That is, $\beta$ is so small that the production of relative vorticity is very feeble and the flow remains nearly free of vorticity.

On the other hand for larger $\beta$ the perturbation of the velocity from its constant free-stream value diminishes *exponentially* with increasing distance from the wall. For large $\beta$ the constraint of the ambient potential vorticity holds the columns to their original isobaths (i.e., their original $y$-position) until in the vicinity of the wall they are squeezed into a shallower region. The larger $\beta$ is, the longer the linear constraint (3.12.17) prevails and the nearer the fluid element gets to the wall before it leaves its isobath. The velocity it achieves flowing parallel to the wall is

$$v = \frac{\partial \psi}{\partial x} = y \beta^{1/2} e^{-\beta^{1/2} x}, \tag{3.14.18}$$

which increases with $y$, as more streamlines join those already turned to positive $y$, and also increases with increasing $\beta$. The total deflected transport—which for each $y$ is fixed to be simply $y$, i.e.,

$$\int_0^\infty v \, dx = y, \tag{3.14.19}$$

—is squeezed into a narrower region with increasing $\beta$, intensifying the velocity. The relative vorticity

$$\zeta = \frac{\partial v}{\partial x} - \frac{\partial u}{\partial y} = -\beta y e^{-\beta^{1/2} x}, \tag{3.14.20}$$

so that on $x = 0$, $\zeta$ is equal and opposite to the change in the *ambient* potential vorticity experienced by elements squeezed into shallower water.

The narrow region near the wall where the fluid is channeled into its path along the wall is an *inertial* (i.e., nonviscous) boundary layer whose thickness is $O(\beta^{-1/2})$ and which is narrower for increasing ambient potential-vorticity gradient. This width in dimensional units is, from (3.14.2),

$$\delta_* = L\beta^{-1/2}$$

$$= \left|\frac{U}{-f(dH_0/dy_*)/H_0}\right|^{1/2} \tag{3.14.21}$$

$$= \left|\frac{\text{on-shore velocity}}{\text{ambient potential-vorticity gradient} \times \text{depth}}\right|^{1/2}$$

So although *strictly* geostrophic flow must flow along lines of constant ambient potential vorticity, quasigeostrophic theory shows how the fluid can break this constraint in a narrow region of the inertial boundary current while remaining geostrophic *to the lowest order*. The *local* Rossby number for the boundary-current region is the ratio of the generated vorticity to $f$, i.e.,

$$\frac{v_*}{f\delta_*} = \beta\frac{Uy}{fL}e^{-\beta^{1/2}x} = (sy)e^{-\beta^{1/2}x},$$

and remains small for small $s$ even though the velocity in the inertial boundary current becomes large compared with the flow at infinity.

There is yet one more important fact to be noted. Suppose all aspects of the problem remain unaltered with the single exception that $s$ changes sign, i.e., that the fluid becomes *deeper* with increasing $y$. It follows then that $\beta$ is *negative*, i.e.,

$$\beta = -|\beta|, \tag{3.14.22}$$

so that now (3.14.10a) becomes

$$\frac{\partial^2\phi}{\partial x^2} + \frac{\partial^2\phi}{\partial y^2} + |\beta|\phi = 0. \tag{3.14.23}$$

Thus solutions of the form (3.14.11) become

$$\phi = (y - y_0)[C_1 \cos|\beta|^{1/2}x + C_2 \sin|\beta|^{1/2}x]. \tag{3.14.24}$$

Although $C_1$ and $C_2$ can be chosen to satisfy the boundary condition on $x = 0$, both solutions oscillate in $x$ rather than decay, and no solution can be found which satisfies $\phi \to 0$ as $x \to \infty$. Instead of an inertial boundary current, the presence of a barrier produces a *steady wave* which reacts back on the flow at infinity. The wave is stationary and has a wavelength whose *magnitude* is identical to the inertial boundary-layer thickness for the same magnitude of $\beta$. Indeed, we could have anticipated, from (3.14.21) the general result, that the existence of the inertial boundary current requires

$$\mathbf{k} \cdot \left[\mathbf{U}_\infty \times \nabla\left(\frac{f}{H_0}\right)\right] < 0, \tag{3.14.25}$$

where $\mathbf{U}_\infty$ is the velocity vector far from the boundary and $\mathbf{k}$ is a unit vertical vector. If (3.14.25) does not hold, waves penetrating the region far from the boundary will be produced instead. This clearly raises a number of pressing questions. First, why is there the asymmetry in the nature of the flow under a reversal of $s$ or $U_\infty$? Second, what kind of wave is it that is produced? The two questions are related and lead us naturally to a discussion of quasigeostrophic wave motion.

## 3.15 Quasigeostrophic Rossby Waves

We saw in Section 3.10 that a layer of fluid with varying thickness supported, in general, small-amplitude waves of several types, but that only one wave, the Rossby wave, was in geostrophic balance. We now reexamine the dynamics of that system within the framework of quasigeostrophic theory.

Consider first the case of wave motion in the unbounded fluid. According to quasigeostrophic theory this is described by the potential-vorticity equation, which from (3.12.25) and (3.12.29) can be written

$$\frac{\partial}{\partial t}(\nabla^2\psi - F\psi) + J(\psi, \nabla^2\psi - F\psi) + \frac{\partial\psi}{\partial x}\frac{\partial\eta_B}{\partial y} - \frac{\partial\psi}{\partial y}\frac{\partial\eta_B}{\partial x} = 0. \quad (3.15.1)$$

For simplicity let $\nabla\eta_B$ be a constant vector, which is equivalent to the condition that the bottom slope is sensibly constant over a distance of the order of a wavelength of the motion.

Consider the plane wave as a putative solution to (3.15.1), i.e.,

$$\psi = A\cos(kx + ly - \sigma t + \phi), \quad (3.15.2)$$

where the amplitude $A$ and phase angle $\phi$ are real constants. In this case

$$\nabla^2\psi - F\psi = -(k^2 + l^2 + F)\psi, \quad (3.15.3)$$

so that the potential vorticity of the wave field is a constant multiple of $\psi$. The Jacobian of $\psi$ and this wave potential vorticity therefore identically vanishes. The plane wave is therefore an exact solution of the nonlinear equation (3.15.1) if the dispersion relation

$$\sigma = -\frac{k\,\partial\eta_B/\partial y - l\,\partial\eta_B/\partial x}{k^2 + l^2 + F} \quad (3.15.4)$$

is satisfied.

In the case where $H_0 = D(1 - sy)$, we have $\eta_B = sy/\varepsilon$, and it follows from (3.15.4) that the *dimensional frequency*

$$\sigma_* = \frac{U}{L}\sigma = -\frac{sf}{L}\frac{k_*}{k_*^2 + l_*^2 + (f^2/gH_0)}. \quad (3.15.5)$$

This is precisely the same as the dispersion relation (3.10.11), except that there $l_*$ is quantized in integral units of $\pi/L$ because of the boundedness of the region. The quasigeostrophic theory reproduces *only* the low-frequency Rossby wave. The higher-frequency waves, such as the Poincaré modes, are filtered out of the system by the *a priori*, systematic limitation of the dynamics to quasigeostrophy. In recompense it is now clear that the Rossby wave is a truly finite-amplitude solution of the low-frequency vorticity equation. The formal limitation to small amplitude, required by the general dynamics of the shallow-water system of equations, can be relaxed at least for the plane-wave solution.

The dispersion relation (3.15.4) may be written

$$\sigma = \mathbf{K} \cdot \frac{(\hat{Z} \times \nabla \eta_B)}{(K^2 + F)}, \tag{3.15.6}$$

where $\hat{Z}$ is the vertical unit vector, a notation introduced here to avoid confusion with the wave vector

$$\mathbf{K} = k\hat{\mathbf{i}} + l\hat{\mathbf{j}} \tag{3.15.7}$$

whose magnitude is $K$. Since

$$\mathbf{u} = \hat{Z} \times \nabla \psi = -(\hat{Z} \times \mathbf{K})A \sin(kx + ly - \sigma t + \phi), \tag{3.15.8}$$

it follows that the particle velocity is *perpendicular* to the wave vector and oriented parallel to the crests of the wave, as shown in Figure 3.15.1. When

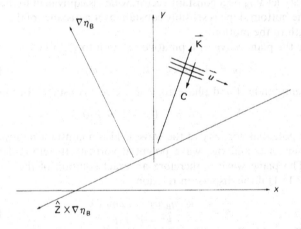

**Figure 3.15.1** The direction of the potential-vorticity gradient is parallel to $\nabla \eta_B$. The advance of the phase of the plane wave must be such that $\mathbf{c} = (\sigma/K)(\mathbf{K}/K)$ always makes an acute angle with $\hat{Z} \times \nabla \eta_B$, where $\hat{Z}$ is a unit vector pointing out of the $x, y$ plane.

the wave vector is perpendicular to $\hat{Z} \times \nabla \eta_B$, i.e., perpendicular to the isobaths, the frequency, by (3.15.6), vanishes. In this case the fluid columns are flowing along the isobaths, which, as we have seen, is the condition for *steady* motion. Only when the fluid columns cross the isobaths and undergo vortex-tube stretching will relative vorticity be produced. When $\mathbf{K}$ is parallel to $\hat{Z} \times \nabla \eta_B$ the frequency takes on its maximum value, for then the fluid columns in the wave move entirely across the isobaths and experience the maximum vortex-tube stretching.

The *velocity* of the advancing crests in the direction of **K** is given by

$$C = \frac{\sigma}{K}\left(\frac{K}{K}\right), \tag{3.15.9}$$

i.e., by the phase speed multiplied by the unit vector $K/K$ in the direction of the wave vector. The inner product of **C** and $\hat{Z} \times \nabla\eta_B$ is therefore, from (3.15.6),

$$C \cdot (\hat{Z} \times \nabla\eta_B) = \frac{[(\hat{Z} \times \nabla\eta_B) \cdot K/K]^2}{K^2 + F}, \tag{3.15.10}$$

which is always *positive*. Regardless of the orientation of **K**, the phase speed always propagates so that **C** makes an acute angle with $\hat{Z} \times \nabla\eta_B$ although it may be parallel or antiparallel to **K**. More simply, the wave must propagate so that an observer travelling with a wave crest always sees larger values of $\eta_B$ on his right. That is, the observer always sees higher *ambient* potential vorticity on his right. In distinction to the Poincaré wave, the phase speed is *unidirectional*. For a given **K** there is *no* solution corresponding to a wave traveling both parallel and antiparallel to **K**. The introduction of an ambient potential-vorticity gradient picks out a unique direction in space, so that dynamically the space is no longer isotropic and symmetry under reflection of direction is lost.

Without loss of generality we can align the $y$-axis in the direction of increasing ambient potential vorticity, so that (3.15.4) becomes

$$\sigma = -\frac{\beta k}{k^2 + l^2 + F}, \tag{3.15.11}$$

where

$$\beta = \frac{\partial\eta_B}{\partial y} = \frac{-1}{\varepsilon D}\frac{\partial H_0}{\partial y}. \tag{3.15.12}$$

Lines of constant phase *always* move in the negative $x$-direction at the rate

$$C_x = -\frac{\beta}{k^2 + l^2 + F}, \tag{3.15.13a}$$

while the phase speed in the $y$-direction,

$$C_y = -\frac{\beta k/l}{k^2 + l^2 + F} \tag{3.15.13b}$$

can be positive or negative depending on the orientation of the wave vector, i.e., on the sign of $k/l$.

For a given $l$, the maximum frequency (numerically) of the Rossby wave is

$$\sigma_m = -\frac{\beta}{2(l^2 + F)^{1/2}}, \tag{3.15.14}$$

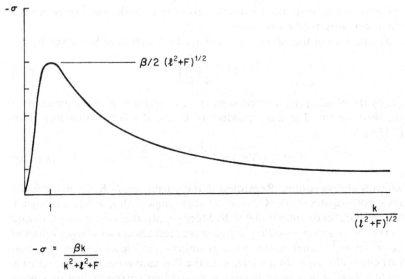

**Figure 3.15.2**    The frequency of a Rossby wave as a function of the component of the wave vector parallel to the isobaths.

which is achieved for

$$k = (l^2 + F)^{1/2}, \tag{3.15.15}$$

as shown in Figure 3.15.2. The absolute maximum occurs for $l = 0$ and is

$$\sigma_M = -\frac{\beta}{2F^{1/2}}, \tag{3.15.16}$$

for which the dimensional wavelength is

$$\lambda_* = \lambda L = 2\pi \frac{L}{F^{1/2}} = (2\pi)R, \tag{3.15.17}$$

or just $2\pi$ times the Rossby deformation radius.

## 3.16  The Mechanism for the Rossby Wave

It is clear from the dispersion relation that the Rossby wave can exist only in the presence of an ambient potential-vorticity gradient, and this provides a clue for the physical explanation of the oscillation. Consider three fluid columns labeled A, B, and C, initially at rest on an isoline of ambient potential vorticity, i.e., $y = $ constant, as shown in Figure 3.16.1. Suppose column B is initially displaced in the positive $y$-direction, which is also the direction of increasing ambient potential vorticity. To preserve its total potential vorticity, the wave potential vorticity $\zeta_0 - F\eta_0$ must decrease to

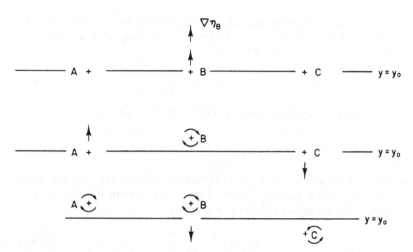

**Figure 3.16.1**    The position of the three-point vortices $A$, $B$, and $C$ at three successive times. Initially collinear and positioned along an isobath, $B$ is displaced upwards, producing velocities at $A$ and $C$ which move them as shown. The vorticity induced on $A$ and $C$ produces a velocity at $B$ tending to restore it to its original position.

offset the increase in ambient potential vorticity B obtains in its new position. This happens in two ways. First the column moving into shallower fluid will tend to be squeezed, and the vortex-tube compression in the presence of the planetary vorticity field will induce a negative relative vorticity on B. Second, since the upper surface is not rigid, the squeezing of the column into the new ambient depth is not complete, and the column will tend to ride partially up the slope so that in its new position it will have a greater value of $\eta_0$ than its neighbors. Both effects, $\zeta_0 < 0$ and $\eta_0 > 0$, reinforce to reduce $\zeta_0 - F\eta_0$. Furthermore, both effects will produce a clockwise circulation in the fluid around B—the first by the circulation induced by any concentrated negative vortex, and the second by the local increase of pressure at B due to the local rise in $\eta_0$, which will geostrophically produce a clockwise flow. The clockwise circulation in the fluid will then move column C into deeper fluid, and column A will be similarly squeezed into shallower fluid as shown. By similar arguments, C will become the center of a counterclockwise circulation, while A will become the center of a clockwise circulation. Both will contribute to a velocity at B that will return it toward its original position, which it will overshoot due to its inertia, and the oscillation will continue. This is obviously a highly oversimplified view of the phenomenon, for an infinite number of fluid columns simultaneously participate in this vorticity dance, whose proper description is given by the formulae of section (3.15). It does reveal a key point. The strength of the restoring mechanism depends on the vigor of the circulation induced on neighboring fluid columns by the displaced column. Consider again the column which is displaced a distance $Y$, and suppose the increment of potential vorticity is concentrated at a point for a column of infinitesimal cross section. Then the increment of ambient potential vorticity of the displaced column, $\beta Y$, induces a stream field $\psi$ which satisfies

$$\nabla^2 \psi - F\psi = -\beta Y \frac{\delta(r)}{2\pi r}, \qquad (3.16.1)$$

where $r$ measures distance from the column center. Note that

$$-\int_0^\infty \int_0^{2\pi} \frac{\beta Y}{2\pi r} \frac{\delta(r)}{} r \, dr \, d\phi = -\beta Y, \qquad (3.16.2)$$

so that the total increment of ambient potential vorticity over the area of the column is $\beta Y$ and is balanced for the fluid by the increment in the value of $\zeta_0 - F\eta_0$. The solution of (3.16.1) for $\psi$ is*

$$\psi = \beta Y \frac{K_0(F^{1/2}r)}{2\pi} \qquad (3.16.3)$$

where $K_0$ is the modified Bessel function of the second kind of order zero.

The azimuthal velocity $v_\theta$ induced on the fluid column is

$$v_\theta = \frac{\partial \psi}{\partial r} = -\frac{\beta Y}{2\pi} F^{1/2} K_1(F^{1/2}r) \qquad (3.16.4)$$

and is shown in Figure 3.16.2. For small values of $F^{1/2}r$, i.e., for fluid elements well within a deformation radius of the displaced column, the asymptotic form of $K_1$ yields

$$v_\theta \approx -\frac{\beta Y}{2\pi r}, \qquad F^{1/2}r \ll 1, \qquad (3.16.5)$$

which is the circulatory speed around a line potential vortex. On the other hand, for large $F^{1/2}r$

$$v_\theta \approx -\frac{\beta Y F^{1/4}}{2^{3/2}\pi^{1/2}r^{1/2}} e^{-F^{1/2}r}. \qquad (3.16.6)$$

If the distance between columns B and C greatly exceeds a deformation radius, (3.16.6) shows that the induced circulation and therefore the restoring mechanism will be very feeble. This explains the dependence of $\sigma$ in (3.15.11) on $F$, i.e., that for fixed wavelength a decrease in the deformation radius (an increase in $F$) lowers the frequency of oscillation.

Note also that for very small wavelengths, the cross-sectional area of each displaced region is small and the vorticity induction effect itself becomes weak, so that for very short wavelengths (high wave number) the frequency of oscillation also becomes small.

---

* Derived in Appendix A at the end of this chapter.

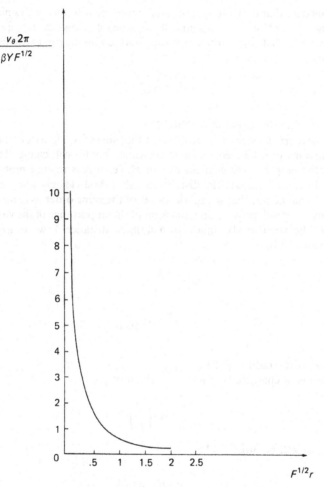

**Figure 3.16.2**   The circulatory velocity produced about a concentrated vortex as a function of radial distance. The velocity decays exponentially on a scale of the Rossby deformation radius.

## 3.17   The Beta-Plane

The wave solution (3.15.2) and the dispersion relation (3.15.11) were studied by Rossby (1939) in quite a different context. In a remarkable paper characterized by brilliant intuitive reasoning, Rossby recommended the study of a homogeneous sheet of fluid on a *sphere* as the simplest relevant model for the dynamics of the observed large-scale waves in the earth's atmosphere. He pointed out a fact we have already noted in Chapter 2, namely that the thinness of the atmospheric shell makes large-scale motions predominantly horizontal so that only the local normal component of the earth's planetary vorticity, $f = 2\Omega \sin \theta$, is dynamically significant. Now let us suppose, as Rossby did, that the motion of the fluid columns preserves potential vorticity*,

$$\Pi = \frac{\zeta + f}{H}, \tag{3.17.1}$$

where $H$ is the thickness of the fluid layer.

The next crucial step is the realization that since $f$ is a function of latitude $\theta$, excursions of fluid elements across latitude circles will change $\Pi$ unless there are compensating changes in $\zeta$ or $H$, i.e., unless relative motions are induced. Let us estimate the distance which a fluid element must move so that the change in $f$ that it experiences is of the same order as $\zeta$, its relative vorticity. This will provide an indication of the importance of the variations of $f$ on the vorticity dynamics. In a latitude distance $Y$ we estimate the increment of $f$ by

$$\Delta f = \frac{1}{r_0} \frac{\partial f}{\partial \theta} Y$$
$$= \frac{Y}{r_0} 2\Omega \cos \theta, \tag{3.17.2}$$

where $r_0$ is the radius of the earth.

In terms of characteristic scales for the motion,

$$\zeta = O\left(\frac{U}{L}\right). \tag{3.17.3}$$

Hence, $\Delta f$ will be order $\zeta$ when

$$\frac{Y}{r_0} = O(\varepsilon \tan \theta). \tag{3.17.4}$$

Hence, in mid latitudes where $\varepsilon = U/fL$ is small, relatively small latitude changes will still give rise to sufficiently large changes in the planetary vorticity to be dynamically significant. Rossby carried this argument even further. Let the scale of the motion be sufficiently small in north–south extent, as in (3.17.4), that *geometrically* a locally flat, Cartesian coordinate system can be used in which the *only* effect of the earth's sphericity is the variation of the Coriolis parameter $f$ with latitude. For such motions $f$ can be linearized about a mean latitude $\theta_0$, i.e., for small $Y/r_0$

$$f \sim f_0 + \beta_0 y, \qquad \beta_0 y \ll f_0, \tag{3.17.5}$$

* In this section all variables are dimensional.

where

$$f_0 = 2\Omega \sin \theta_0, \tag{3.17.6a}$$

$$\beta_0 = \frac{2\Omega}{r_0} \cos \theta_0. \tag{3.17.6b}$$

If

$$H = D + \eta - h_B, \tag{3.17.7}$$

where $\eta$ is the free-surface deviation due to the motion and $h_B$ measures the variation of the layer thickness in the absence of motion, then small variations of $\eta$ and $h_B$ allow $\Pi$ to be written as

$$\Pi = \frac{f_0 + \left(\beta_0 y + f_0 \dfrac{h_B}{D}\right) + \zeta - f_0 \dfrac{\eta}{D}}{D}. \tag{3.17.8}$$

Note that the variable part of the *ambient* potential vorticity is simply $\beta_0 y + f_0 h_B/D$. Within the context of the homogeneous-layer model *there is an exact dynamical equivalence between the variation of the Coriolis parameter with latitude, the β-effect, and variations of topography in the presence of constant f*, that is, $\beta_0 y$ plays the same role in potential-vorticity dynamics as $f_0 h_B/D$. Therefore all the results of the preceding sections in which we have used a linear topographic gradient can be immediately translated in terms of the dynamics of a layer of uniform ambient thickness on the *sphere* if $\beta$ in (3.14.3) and (3.15.11) is interpreted by the relation

$$\beta = \beta_0 \frac{L^2}{U} \tag{3.17.9}$$

and if we identify $x$ and $y$ with eastward and northward coordinates respectively. Such a model, in which the effect of the earth's sphericity is modeled by a linear variation of $f$ in an otherwise planar geometry, is called the β-plane model (as opposed to the $f$-plane model we *explicitly* introduced in this chapter).

The discussion in this section has been intuitive and heuristic. The detailed derivation and justification of the β-plane approximation is deferred to Chapter 6. Nevertheless, the arguments of the present section are essentially correct and immediately convert the topographic Rossby wave of the "flat"-earth model to at least a rudimentary barotropic model of large-scale atmospheric waves. Furthermore the inertial boundary layer of Section 3.14 will be shown in Chapter 5 to have, for the same reason, direct application to the study of strong oceanic currents like the Gulf Stream. Note that the inertial boundary-layer thickness, by (3.14.21), is $L\beta^{-1/2}$, which with (3.17.9) yields a current of width

$$\delta_I = \left(\frac{U}{\beta_0}\right)^{1/2}. \tag{3.17.10}$$

At 30° N, $\beta_0$ is $1.9 \times 10^{-13}$ cm$^{-1}$ s$^{-1}$, so that for an on-shore flow of 5 cm/s the predicted width is 50 kilometers, in fair agreement with the observed horizontal extent of the current. Hence the model discussed in Section 3.14 can be considered as a simple model for the formation region of the Gulf Stream where oceanic flow driven by the wind from the east impinges on the barrier formed by the North American continent in the region between Florida and Cape Hatteras.

## 3.18 Rossby Waves in a Zonal Current

Consider the effect on the Rossby wave of a uniform flow streaming along the isolines of ambient potential vorticity. With

$$\eta_B = \beta y, \tag{3.18.1}$$

a flow whose streamlines are rectilinear and parallel to the $x$-axis will be an exact solution of the potential-vorticity equation. With the $\beta$-plane concept in mind, the coordinates $x$ and $y$ can be thought of as eastward and northward coordinates. A flow along latitude circles, i.e., due eastward, is called a *zonal flow*. The simplest example is

$$\psi = -\hat{U}y. \tag{3.18.2}$$

If $\hat{U}$ is $+1$, the dimensional flow is to the east (a westerly flow in meteorological lingo) with a magnitude $U$, while if $\hat{U}$ is $-1$ the velocity is westward (an easterly flow) with the same magnitude $U$. The dimensionless constant $\hat{U}$ is introduced so we can consider the cases of easterly and westerly flow together.

We now inquire whether a wave pattern superimposed on the mean current is possible; i.e., now let

$$\psi = -\hat{U}y + \phi(x, y, t), \tag{3.18.3}$$

which when substituted in (3.15.1) yields, with (3.18.1),

$$\left[\frac{\partial}{\partial t} + \hat{U}\frac{\partial}{\partial x}\right][\nabla^2\phi - F\phi] + \frac{\partial \phi}{\partial x}[\beta + F\hat{U}] + J(\phi, \nabla^2\phi - F\phi) = 0. \tag{3.18.4}$$

Plane-wave solutions for $\phi$ of the form

$$\phi = A \cos(kx + ly - \sigma t) \tag{3.18.5}$$

have, as before, the property that

$$\nabla^2\phi - F\phi = -(k^2 + l^2 + F)\phi, \tag{3.18.6}$$

so that the nonlinear term in (3.18.4) identically vanishes. Consequently, with no limitation on the size of the amplitude, the plane wave will be an exact solution if

$$\sigma = \frac{k}{K^2 + F}[\hat{U}K^2 - \beta], \tag{3.18.7}$$

where

$$K^2 = k^2 + l^2. \tag{3.18.8}$$

The phase speed in the $x$-direction is

$$\begin{aligned} C_x = \frac{\sigma}{k} &= \frac{\hat{U}K^2 - \beta}{K^2 + F} \\ &= \hat{U} - \frac{\beta + F\hat{U}}{K^2 + F}. \end{aligned} \tag{3.18.9}$$

If $\hat{U} = +1$ (i.e., for westerly flow), $C_x$ will be positive if $K^2 > \beta$ and negative if $K^2 < \beta$. The wave will be *stationary* if

$$K = K_s = \beta^{1/2}. \tag{3.18.10}$$

If $\hat{U} = -1$ (i.e., for easterly flow), $C_x$ will be negative for all $K$, and steady, stationary waves are not possible. Clearly steady Rossby waves are possible only if $\hat{U}\beta > 0$. If $\hat{U}\beta < 0$, stationary solutions exist only if $K^2 < 0$, which would imply exponential growth or decay in at least one spatial direction with an $e$-folding scale $\beta^{-1/2}$. This is clearly ruled out in an unbounded region.

We are now in a better position to understand the results of Section 3.13 on the question of the existence of inertial steady currents. There we saw that inertial boundary currents require $\hat{U}\beta < 0$, while for $\hat{U}\beta > 0$ the response to the boundary was wavelike. It is clear now that in the latter case a stationary Rossby wave is produced by the boundary, while in the former case no stationary Rossby wave is possible, so an exponential decay occurs. Thus whether the steady motion is wavelike or of boundary-current character really is a reflection of the criterion for the existence of a stationary Rossby wave.

In dimensional units the wavelength of the stationary Rossby wave is

$$\begin{aligned} \lambda_* = L\lambda \\ = L\frac{2\pi}{K_s} &= L\, 2\pi\beta^{-1/2} \\ = 2\pi\left(\frac{U}{\beta_0}\right)^{1/2}, \end{aligned} \tag{3.18.11}$$

where $\beta_0$ is given by (3.17.6b), i.e., we have identified $\beta$ as the nondimensional variation of the Coriolis parameter according to the formula (3.17.9). For the atmosphere, where a typical value of $U$ is $O(10 \text{ m/s})$, $\lambda_*$ is 5,400 km if $\beta_0$ is evaluated at a latitude of 45°. It is the similarity of this scale to the characteristic scale of the large waves in the westerly winds that excited so much interest in the Rossby wave. It will become clear in subsequent chapters that the model discussed here is far too simple to adequately predict the structure and motion of large-scale atmospheric waves. It will also become evident, though, that these considerations are a good first step toward that question, and the sensible prediction of (3.18.11) is a manifestation of that fact. For the ocean, the smaller values of $U$ yield much smaller length scales at the same latitude. For example if $U$ is 10 cm/s (i.e., a velocity a hundred times smaller than the atmospheric counterpart), $\lambda_*$ is on the order of 540 km, or ten times smaller.

An intriguing question is presented by (3.18.9). The phase speed of a Rossby wave in the absence of a current is

$$C_x = -\frac{\beta}{K^2 + F} \tag{3.18.12}$$

and retrogresses toward negative $x$ (i.e., the west), and yet the current speed in the eastward direction required to hold the wave fixed is the larger value

$$\hat{U} = \frac{\beta}{K^2}. \tag{3.18.13}$$

It is clear from (3.18.9) that the presence of the mean flow has *two* effects on the wave. The first is a simple Doppler shift, i.e., a simple augmentation of the phase speed by the speed of the carrying medium for the wave. The second effect, according to (3.18.9), is an alteration of the phase speed *relative to the flow* by the replacement of $\beta$ with $\beta + F\hat{U}$. The effective ambient potential-vorticity gradient is altered by the presence of the mean flow in addition to the simple wave-speed shift. The dimensionless thickness, $H_0$, in the absence of wave motion but in the presence of the mean flow is

$$
\begin{aligned}
H_0 &= 1 + \varepsilon F \eta_0 - \varepsilon \eta_B \\
&= 1 + \varepsilon F \psi - \varepsilon \eta_B \\
&= 1 - \varepsilon \{ F\hat{U} + \beta \} y.
\end{aligned}
\tag{3.18.14}
$$

The mean flow tilts the free surface (as shown in Figure 3.18.1), since it is in geostrophic balance. This tilt increases the gradient of the layer thickness in

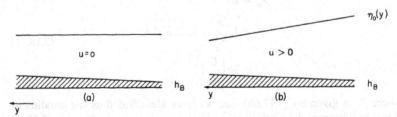

**Figure 3.18.1**   The free surface in the fluid at rest (a) and in uniform motion into the paper (b).

the $y$-direction as indicated in (3.18.14) if $\hat{U} > 0$, and therefore increases the ambient potential vorticity seen by the wave. This increase will increase the frequency of the wave oscillation by increasing the effective value of $\beta$. This increases the intrinsic phase speed of the waves in the negative $x$-direction so that (3.18.13) obtains. If the Rossby radius of deformation is large over a wavelength, then $F \ll 1$, and the tilt of the free surface caused by the flow is imperceptible on the scale of the wave. Note that for $\hat{U} < 0$ the effect of $\beta$ is weakened and may even disappear if $\hat{U} = -\beta/F$.

In the more general case where the current is not uniform in $y$, the effective $\beta$ will be itself a function of $y$ and more complex behavior may be anticipated. Such questions are discussed in Chapter 7. Nevertheless it is still true that the basic idea remains, namely that the dynamics of the wave propagation depends on the *total* potential-vorticity gradient of the flow in which the wave is embedded.

## 3.19  Group Velocity

One of the most fundamental properties of waves is their ability to cause disturbances, i.e., to transmit energy, over great distances compared to the characteristic displacement of the fluid elements in the wave during its passage. The particle displacement over a wave period is

$$l_* = O\left(u_* \frac{2\pi}{\sigma_*}\right), \tag{3.19.1}$$

where $u_*$ is the characteristic fluid velocity and is directly proportional to the wave amplitude. $\sigma_*$ is the frequency. The ratio of $l_*$ to the wavelength is

$$\frac{l_*}{\lambda_*} = O\left(u_* \frac{k_*}{\sigma_*}\right) = \frac{u_*}{C_*}, \tag{3.19.2}$$

where $C_*$ is the phase speed. Waves of small amplitude are characterized by the smallness of $u_*/C_*$, and we see this is equivalent to the relative smallness of particle displacements in the wave. The fluid wave is a cooperative activity of many fluid elements moving slightly but coherently over large scales to effect the propagation of information over distances which are far greater than the excursion of each fluid element—much as a line of men in a bucket brigade can pass pails of water over distances great compared to their individual reach.

The plane wave is a useful idealization for some purposes, especially for revealing the fundamental mechanism which gives rise to the wave oscillation. Yet the restriction that the wave field is strictly periodic with a constant amplitude is highly unrealistic. It may serve as a useful approximation for the wave pattern if the pattern changes slowly in space. However, any real train of waves has a beginning and an end, both in time and in space. In fact we usually observe waves traveling in groups or packets. The plane wave with its uniform amplitude *already* occupies all space, and therefore the question of the transmission of a wave disturbance requires a consideration of more complex wave fields in which the wave amplitude itself is a function of space and time. Of course, to the degree that the more complex waves look locally like a plane wave, the dynamics of the plane wave should apply as a *local* approximation to the true dynamics.

Perhaps the simplest example of a spatially limited wave disturbance is the *wave packet*, e.g.,

$$\psi = A(x, y, t)\cos(kx + ly - \sigma t), \tag{3.19.3}$$

where $A$ is a slowly varying function of $x$, $y$, and $t$. That is,

$$\frac{1}{A}\frac{\partial A}{\partial x} \ll K, \qquad \frac{1}{A}\frac{\partial A}{\partial y} \ll K, \qquad \frac{1}{A}\frac{\partial A}{\partial t} \ll \sigma, \tag{3.19.4}$$

where again $K^2 = k^2 + l^2$. The condition (3.19.4) ensures that over one wavelength, the wave amplitude is very nearly constant, while a similar consideration applies to the amplitude variation over the wave period. A schematic picture of a one-dimensional wave packet is shown in Figure 3.19.1. The chief question is what determines the motion of the envelope of

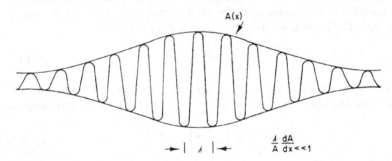

**Figure 3.19.1** A wave packet. $A(x)$ is the envelope of the disturbance, whose wavelength is $\lambda$.

the packet, for it is the amplitude of the envelope which determines where there is wave energy. Where the packet envelope is small, far from the envelope center, there is clearly little wave activity. To determine the laws of propagation of the wave envelope we must see what conditions are required for (3.19.3) to be a solution of the potential-vorticity equation. Now although a pure plane wave is an exact solution of the full equation, a wave packet will not be, and (as usual in wave problems) we must linearize the wave equation.

In order to neglect the nonlinear terms in (3.15.1) we require that

$$J(\psi, \nabla^2\psi) \ll \beta\frac{\partial\psi}{\partial x} \tag{3.19.5}$$

for the archetypal case $\eta_B = \beta y$. Since all scaled variables are by presumption $O(1)$, this requires

$$\beta = \frac{s f L}{U} \gg 1 \tag{3.19.6}$$

for topographic waves, or

$$\beta = \frac{\beta_0 L^2}{U} \gg 1$$

for Rossby waves on a sphere. The dimensional phase speed of the Rossby wave is $O(sfL)$ in the former case and $O(\beta_0 L^2)$ in the latter case, if $L$ is identified with the wavelength. Hence in both cases the condition for linearization is

$$\frac{C_*}{U} \gg 1, \tag{3.19.7}$$

which as we have seen is the condition for small displacements of fluid elements. Thus $\beta \gg 1$ is in fact the ordinary condition for linear wave motion in the context of our scaling scheme. Under these conditions the linearized potential vorticity equation is

$$\frac{\partial}{\partial t}[\nabla^2 \psi - F\psi] + \beta \frac{\partial \psi}{\partial x} = 0. \tag{3.19.8}$$

Substitution of (3.19.3) into (3.19.8) yields

$$\sin(kx + ly - \sigma t)$$

$$\times \left[ -\{\sigma(K^2 + F) + \beta k\}A - 2\left(k\frac{\partial^2 A}{\partial x \, \partial t} + l\frac{\partial^2 A}{\partial y \, \partial t}\right) + \sigma \nabla^2 A \right]$$

$$+ \cos(kx + ly - \sigma t)$$

$$\times \left[ -(K^2 + F)\frac{\partial A}{\partial t} + 2\sigma\left(k\frac{\partial A}{\partial x} + l\frac{\partial A}{\partial y}\right) + \beta \frac{\partial A}{\partial x} + \nabla^2 \frac{\partial A}{\partial t} \right] = 0. \tag{3.19.9}$$

Since $A$ is a slowly varying function of $x$, $y$, and $t$, the coefficients of $\sin(kx + ly - \sigma t)$ and $\cos(kx + ly - \sigma t)$ must vanish separately to the lowest order, i.e., if (3.19.9) is multiplied by $\sin(kx + ly - \sigma t)$ and then averaged over a wave period, we obtain, to *lowest order*,

$$\sigma(K^2 + F) + \beta k = 0, \tag{3.19.10}$$

which is simply the dispersion relation for the *pure* plane wave, and appears here in the first approximation for the slowly varying wave packet. Hence the basic notions developed for the plane wave of phase propagation and the frequency–wave-number relation are directly applicable to the slowly varying wave packet. To lowest order the condition that the coefficient of

$\cos(kx + ly - \sigma t)$ vanishes yields

$$\frac{\partial A}{\partial t} - \frac{2\sigma k + \beta}{K^2 + F}\frac{\partial A}{\partial x} - \frac{2\sigma l}{K^2 + F}\frac{\partial A}{\partial y} = 0, \qquad (3.19.11)$$

or with (3.19.10),

$$\frac{\partial A}{\partial t} + C_{gx}\frac{\partial A}{\partial x} + C_{gy}\frac{\partial A}{\partial y} = 0, \qquad (3.19.12)$$

where

$$C_{gx} = \beta\frac{k^2 - l^2 - F}{(K^2 + F)^2}, \qquad (3.19.13a)$$

$$C_{gy} = 2\frac{\beta kl}{(K^2 + F)^2}. \qquad (3.19.13b)$$

The vector form of (3.19.12) is simply

$$\frac{\partial A}{\partial t} + (\mathbf{C}_g \cdot \nabla)A = 0. \qquad (3.19.14)$$

Therefore $A$ is constant for an observer moving with velocity $\mathbf{C}_g$, or

$$A = A(\mathbf{r} - \mathbf{C}_g t), \qquad (3.19.15)$$

where $\mathbf{r}$ is the position vector in the $x, y$ plane. To the first approximation the envelope of the wave packet moves with $\mathbf{C}_g$, which is called the *group velocity*, the $x$ and $y$ components of which are given by (3.19.13a,b). In contrast to the phase velocity, the group velocity *does* satisfy the usual vector rules of projection.

It can be verified directly from (3.19.10) and (3.19.13a,b) that

$$C_{gx} = \frac{\partial \sigma}{\partial k},$$

$$C_{gy} = \frac{\partial \sigma}{\partial l}, \qquad (3.19.16)$$

or

$$\mathbf{C}_g = \nabla_K \sigma \equiv \hat{\mathbf{i}}\frac{\partial \sigma}{\partial k} + \hat{\mathbf{j}}\frac{\partial \sigma}{\partial l},$$

where $\nabla_K$ denotes the vector gradient with respect to wave number. Since $\sigma = KC$, where $C$ is the speed of the advance of the crests in the direction of the wave vector,

$$\mathbf{C}_g = \nabla_K(KC) = \frac{\mathbf{K}}{K}C + K\,\nabla_K C$$

$$= \mathbf{C} + K\,\nabla_K C. \qquad (3.19.17)$$

Hence unless the phase speed is independent of **K**, the group velocity and the phase speed will be different in *both magnitude and direction*. Waves for which $C_g \neq C$ are called *dispersive*, and the Rossby wave is a particularly striking case of a dispersive wave. The magnitude of $C_g$ differs from the phase speed, and its direction, as we shall see, departs considerably from the direction of **K**. The wave envelope of a disturbance of finite extent will move with the group velocity and not the phase speed. Individual crests which are determined by the phase of the wave will still move with the phase speed, parallel (or antiparallel) to **K**. Therefore individual wave crests will move through the packet, appearing at one end of the packet, threading their way through the envelope, taking up the local value of the envelope amplitude, and then disappearing at the other end of the packet with vanishing amplitude. The crests and troughs of the wave are only manifestations of the *pattern* of the wave, and they vanish like the smiles and frowns of actors at play's end.

The differential relations (3.19.16) which yield the group velocity in terms of derivatives of $\sigma$ are general results which transcend the particular calculation given here for the Rossby wave, as can be seen by the following argument. Consider any linear wave equation with constant coefficients written in the form

$$\mathscr{L}\left(\frac{\partial}{\partial t}, \frac{\partial}{\partial x}, \frac{\partial}{\partial y}\right)\psi = 0, \qquad (3.19.18)$$

where $\mathscr{L}$ is a polynomial in the time and space derivatives. For the Rossby wave, for example,

$$\mathscr{L} = \frac{\partial^3}{\partial t\, \partial x^2} + \frac{\partial^3}{\partial t\, \partial y^2} - F\frac{\partial}{\partial t} + \beta\frac{\partial}{\partial x}. \qquad (3.19.19)$$

If solutions of the form

$$\psi = \operatorname{Re} A e^{i(kx + ly - \sigma t)} \qquad (3.19.20)$$

are inserted in (3.19.18), the equation for $A$ becomes

$$\mathscr{L}\left(-i\sigma + \frac{\partial}{\partial t},\ ik + \frac{\partial}{\partial x},\ il + \frac{\partial}{\partial y}\right)A = 0. \qquad (3.19.18a)$$

If $A$ is a slowly varying function of $x$, $y$, and $t$, the polynomial $\mathscr{L}$ can be expanded in a Taylor series to yield

$$\left\{\mathscr{L}(-i\sigma,\ ik,\ il) + \left[\frac{\partial\mathscr{L}}{\partial(-i\sigma)}\frac{\partial}{\partial t} + \frac{\partial\mathscr{L}}{\partial(ik)}\frac{\partial}{\partial x} + \frac{\partial\mathscr{L}}{\partial(il)}\frac{\partial}{\partial y}\right]\right\}A = 0, \qquad (3.19.21)$$

where higher-order terms have been neglected. All terms in this series in the "small parameters" $(1/A)\,\partial A/\partial t$, $(1/A)\,\partial A/\partial x$, etc. must vanish separately. Hence to lowest order

$$\mathscr{L}(-i\sigma,\ ik,\ il) = 0. \qquad (3.19.22)$$

This is, however, the polynomial for $\sigma$ obtained for the plane wave by replacing time and space derivatives in the original differential operator with frequency and wave-number. Indeed, (3.19.22) is merely the dispersion relation, which for the Rossby plane wave is (3.19.10). In order for the next term in the expansion to also vanish,

$$\frac{\partial A}{\partial t} - \frac{\partial \mathcal{L}/\partial k}{\partial \mathcal{L}/\partial \sigma}\frac{\partial A}{\partial x} - \frac{\partial \mathcal{L}/\partial l}{\partial \mathcal{L}/\partial \sigma}\frac{\partial A}{\partial y} = 0. \tag{3.19.23}$$

Now by definition

$$-\frac{\partial \mathcal{L}/\partial k}{\partial \mathcal{L}/\partial \sigma} = \left(\frac{\partial \sigma}{\partial k}\right)_{l,\,\mathcal{L}},$$

$$-\frac{\partial \mathcal{L}/\partial l}{\partial \mathcal{L}/\partial \sigma} = \left(\frac{\partial \sigma}{\partial l}\right)_{k,\,\mathcal{L}}, \tag{3.19.24}$$

i.e., the coefficients in (3.19.23) are derivatives of $\sigma$ with respect to $k$ and $l$ for a fixed value of $\mathcal{L}$ (indeed, zero). That is, they are derivatives of the frequency fixed by the wave number according to the plane-wave dispersion relation, so that (3.19.16) is the *general* formula for the packet velocity for an arbitrary wave type. Therefore, only the dispersion relation is required to determine the motion of a wave packet.

Returning to the particular case of the Rossby wave, let us examine in more detail the character of the group velocity in that case. In particular, note that although the phase speed in the $x$-direction is always negative, the component of the group velocity in the $x$-direction can be positive or negative, as shown in Figure 3.19.2. A packet containing a wave for which

$$k^2 > l^2 + F \tag{3.19.25}$$

will propagate in the positive $x$-direction (towards the *east* if we think of Rossby waves on the $\beta$-plane), while waves which are long in the $x$-direction, i.e., wave packets for which

$$k^2 < l^2 + F, \tag{3.19.26}$$

will propagate in the negative $x$-direction (to the west on the $\beta$-plane). Waves for which the equality $k^2 = l^2 + F$ holds will have stationary envelopes. In each case the speed of individual crests will be different than the envelope speed, and indeed the short waves, for which (3.19.25) holds, will have crests moving *opposite* to the motion of the envelope. The maximum positive value of $C_{gx}$ obtains at $k = 3^{1/2}(l^2 + F)^{1/2}$ and is

$$\text{Max positive } C_{gx} = \frac{\beta}{8(l^2 + F)}, \tag{3.19.27}$$

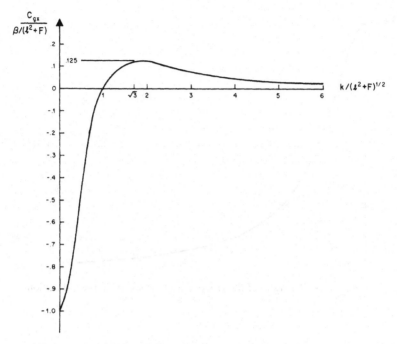

**Figure 3.19.2** The group velocity in the x-direction for a Rossby wave.

while the maximum negative value of $C_{gx}$ is obtained at $k = 0$,

$$\text{Max negative } C_{gx} = -\frac{\beta}{(l^2 + F)}, \qquad (3.19.28)$$

or eight times faster than the maximum positive group velocity in the x-direction.

Figure 3.19.3 shows the ratio of the group to the phase speed in the x-direction. Note that $|C_{gx}| \leq |C_x|$.

Since

$$\frac{C_{gy}}{C_y} = -\frac{2l^2}{K^2 + F} < 0, \qquad (3.19.29)$$

the group velocity in the y-direction (northward on the β-plane) is always *oppositely* directed to the phase speed in the y-direction, e.g., crests will appear to move southward as the region of wave disturbance, defined by the boundaries of the wave envelope, moves northward.

Before proceeding further with the discussion of the transmission of energy by Rossby waves, we discuss in the next section an alternate derivation of (3.19.14) by the method of multiple time scales. The reader content with the derivation of the present section may proceed directly to section (3.21).

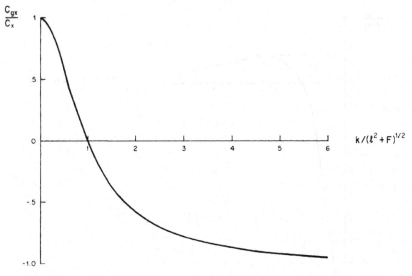

**Figure 3.19.3**    The ratio of the group and phase speeds in the $x$-direction.


## 3.20  The Method of Multiple Time Scales

A more formal, and often a more useful and powerful, approach can be taken to the problem of slowly varying wave fields which explicitly recognizes that the total wave field has two separate scales of oscillation apparent to an observer. There is the so-called "fast time" scale, i.e., the period of local oscillation over which the wave amplitude appears constant, and the "slow time" scale, over which the amplitude of the wave field gradually changes as, for example, a wave packet slowly and smoothly passes the observer. To pass from this intuitive notion to a useful calculation scheme, it is only necessary to consider the $\psi$-field to depend explicitly on the two sets of variables

$$\tilde{t} = t, \qquad \tilde{x} = x, \qquad \tilde{y} = y,$$
$$T = \Delta t, \quad X = \Delta x, \quad Y = \Delta y, \qquad \Delta \ll 1, \tag{3.20.1}$$

where the tilde variables are the fast variables and the capital variables are the slow variables, and $\Delta$ is a small parameter that is a measure of the slowness of the temporal and spatial variations of the field. The function $\psi$ is then written explicitly in terms of both variables, i.e.,

$$\psi = \psi(\tilde{x}, \tilde{y}, \tilde{t}; X, Y, T). \tag{3.20.2}$$

Derivatives, for example, are then given by the usual chain rule

$$\frac{\partial \psi}{\partial t} = \frac{\partial \psi}{\partial \tilde{t}} + \Delta \frac{\partial \psi}{\partial T},$$

$$\frac{\partial \psi}{\partial x} = \frac{\partial \psi}{\partial \tilde{x}} + \Delta \frac{\partial \psi}{\partial X}, \qquad (3.20.3)$$

$$\frac{\partial \psi}{\partial y} = \frac{\partial \psi}{\partial \tilde{y}} + \Delta \frac{\partial \psi}{\partial Y}.$$

In terms of these variables (3.19.8) becomes

$$\frac{\partial}{\partial \tilde{t}} \left\{ \frac{\partial^2 \psi}{\partial \tilde{x}^2} + \frac{\partial^2 \psi}{\partial \tilde{y}^2} - F\psi \right\} + \beta \frac{\partial \psi}{\partial \tilde{x}}$$

$$= -\Delta \left[ \frac{\partial}{\partial T} \left\{ \frac{\partial^2 \psi}{\partial \tilde{x}^2} + \frac{\partial^2 \psi}{\partial \tilde{y}^2} - F\psi \right\} + 2 \left( \frac{\partial}{\partial X} \frac{\partial^2 \psi}{\partial \tilde{x}} \frac{}{\partial \tilde{t}} + \frac{\partial}{\partial Y} \frac{\partial^2 \psi}{\partial \tilde{y} \partial \tilde{t}} \right) + \beta \frac{\partial \psi}{\partial X} \right]$$

$$- \Delta^2 \left[ \frac{\partial}{\partial T} \left( 2 \frac{\partial^2 \psi}{\partial X \partial \tilde{x}} + 2 \frac{\partial^2 \psi}{\partial \tilde{y} \partial Y} \right) + \frac{\partial}{\partial \tilde{t}} \left( \frac{\partial^2 \psi}{\partial X^2} + \frac{\partial^2 \psi}{\partial Y^2} \right) \right]$$

$$- \Delta^3 \left[ \frac{\partial}{\partial T} \left( \frac{\partial^2 \psi}{\partial X^2} + \frac{\partial^2 \psi}{\partial Y^2} \right) \right] \qquad (3.20.4)$$

after collecting common terms in $\Delta$. Since $\Delta$ is a small parameter, $\psi$ can be written

$$\psi = \psi_0 + \Delta \psi_1 + \Delta^2 \psi_2 + \cdots. \qquad (3.20.5)$$

If (3.20.5) is inserted in (3.20.4), terms of like order in $\Delta$ must separately vanish. The order-one terms yield

$$\frac{\partial}{\partial \tilde{t}} \left\{ \frac{\partial^2 \psi_0}{\partial \tilde{x}^2} + \frac{\partial^2 \psi_0}{\partial \tilde{y}^2} - F\psi_0 \right\} + \beta \frac{\partial \psi_0}{\partial \tilde{x}} = 0 \qquad (3.20.6)$$

which *is an equation in the fast variables alone* and consequently admits solutions of the form

$$\psi_0 = \text{Re } A(X, Y, T) e^{i(k\tilde{x} + l\tilde{y} - \sigma \tilde{t})}. \qquad (3.20.7)$$

As far as the operator in (3.20.6) is concerned, $A$ is an unknown constant (i.e., independent of $\tilde{x}$, $\tilde{y}$, and $\tilde{t}$), and the substitution of (3.20.7) into (3.20.6) yields the familiar dispersion relation (3.19.10). The dependence of $A$ on $X$, $Y$, and $T$ is so far unspecified.

The order-$\Delta$ equation is

$$\frac{\partial}{\partial \tilde{t}}\left\{\frac{\partial^2 \psi_1}{\partial \tilde{x}^2} + \frac{\partial^2 \psi_1}{\partial \tilde{y}^2} - F\psi_1\right\} + \beta \frac{\partial \psi_1}{\partial \tilde{x}}$$

$$= -\left[\frac{\partial}{\partial T}\left(\frac{\partial^2 \psi_0}{\partial \tilde{x}^2} + \frac{\partial^2 \psi_0}{\partial \tilde{y}^2} - F\psi_0\right) + 2\frac{\partial}{\partial X}\left(\frac{\partial^2 \psi_0}{\partial \tilde{x}\, \partial \tilde{t}}\right) + 2\frac{\partial}{\partial Y}\left(\frac{\partial^2 \psi_0}{\partial \tilde{y}\, \partial \tilde{t}}\right) + \beta\frac{\partial \psi_0}{\partial X}\right]$$

$$= +(K^2 + F)\left[\frac{\partial A}{\partial T} - \frac{2\sigma k + \beta}{K^2 + F}\frac{\partial A}{\partial X} - \frac{2\sigma l}{K^2 + F}\frac{\partial A}{\partial Y}\right]e^{i(k\tilde{x} + l\tilde{y} - \sigma \tilde{t})} \qquad (3.20.8)$$

after using (3.20.7) to evaluate the right-hand side of (3.20.8). The calculation of $\psi_1$ presents us with an apparent dilemma. The right-hand side of (3.20.8) is, as far as its dependence on $\tilde{x}$, $\tilde{y}$, and $\tilde{t}$ is concerned, oscillating with the frequency of the homogeneous solution of the equation for $\psi_1$. That is, for a fixed $k$ and $l$ the forcing term on the right-hand side of (3.20.8) oscillates at the natural frequency of oscillation of the system. In a manner precisely analogous to the resonant forcing of an undamped oscillator, the solution for $\psi_1$ would then grow linearly with $\tilde{t}$, i.e., would contain a secular growth with $\tilde{t}$, in which case

$$\frac{\Delta \psi_1}{\psi_0} = O(\Delta \tilde{t}) = T \qquad (3.20.9)$$

so that in a time $t = O(\Delta^{-1})$ (for which $T = O(1)$) the second term in the expansion (3.20.5) would become as large as the first. Of course it is precisely for times $t = O(\Delta^{-1})$ that we wish to describe the evolution of the wave amplitude, so we must insist that the expansion (3.20.5) remain valid for this length of time. To accomplish this we must remove the resonant forcing term from the right-hand side of (3.20.8) by insisting that

$$\frac{\partial A}{\partial T} - \frac{2\sigma k + \beta}{K^2 + F}\frac{\partial A}{\partial X} - \frac{2\sigma l}{K^2 + F}\frac{\partial A}{\partial Y} = 0, \qquad (3.20.10)$$

which specifies the dependence of $A$ on the slow variables. Comparison of (3.20.10) with (3.19.11) shows that (3.20.10) is indeed the amplitude equation previously derived by the less systematic methods of the previous section.

The chief virtue of the method of multiple time scales is that it systematically separates the problem for the local dynamics in space and time from the problem of the slow larger-scale variations. In the present problem the method only enriches our understanding of the approximations leading to (3.19.14). It does, however, give a device for the systematic calculation of higher-order corrections if desired. What is much more important is that the method can be efficiently applied to problems for which the heuristic methods are inadequate, for example the problem of nonlinear interaction of Rossby waves discussed in Section 3.26.

## 3.21 Energy and Energy Flux in Rossby Waves

If the linear wave equation (3.19.8) is multiplied by $\psi$, a little manipulation yields

$$\frac{\partial}{\partial t}\left[\frac{(\nabla\psi)^2 + F\psi^2}{2}\right] + \nabla \cdot \left\{-\psi\nabla\frac{\partial\psi}{\partial t} - \hat{\imath}\beta\frac{\psi^2}{2}\right\} = 0, \qquad (3.21.1)$$

which has a direct interpretation in terms of energy. The kinetic energy of the motion in dimensional units is, per unit mass, when integrated over the depth,[*]

$$D\left(\frac{u_*^2 + v_*^2}{2}\right) = \frac{U^2}{2}D\left\{\left(\frac{\partial\psi}{\partial x}\right)^2 + \left(\frac{\partial\psi}{\partial y}\right)^2\right\}, \qquad (3.21.2)$$

while the potential energy, using the static, free surface as zero reference value, is

$$g\int_0^{\eta_*} Z_* \, dZ_* = g\frac{\eta_*^2}{2} = \frac{U^2 D}{2}F\psi^2 \qquad (3.21.3)$$

if the scaling definitions (3.12.1), (3.12.5), and (3.12.29a) are used. The ratio of potential to kinetic energy is therefore $O(F)$, i.e., depends on the ratio of the scale of the motion to the deformation radius. It is apparent then that the term in the square bracket in (3.21.1) is the nondimensional form of the sum of the kinetic plus potential energy, whose rate of increase with time balances the convergence of the flux vector

$$\mathbf{S} = -\psi\nabla\frac{\partial\psi}{\partial t} - \hat{\imath}\frac{\beta\psi^2}{2}, \qquad (3.21.4)$$

so that the energy $E(x, y, t)$ satisfies the conservation law

$$\frac{\partial E}{\partial t} + \nabla \cdot \mathbf{S} = 0. \qquad (3.21.5)$$

For a Rossby wave packet,

$$\psi = A \cos(kx + ly - \sigma t), \qquad (3.21.6)$$

and the energy to the lowest order is

$$E = \frac{K^2 A^2}{2} \sin^2(kx + ly - \sigma t) + \frac{F}{2} A^2 \cos^2(kx + ly - \sigma t), \quad (3.21.7)$$

ignoring the relatively small contribution made by the spatial gradients of $A$. As written, at any fixed point, $E$ varies rapidly with half the period of the

---

[*] The vertical velocity makes no contribution to the kinetic energy, since in shallow-water theory $w = O(\delta u)$.

wave about the average value

$$\langle E \rangle = (K^2 + F)\frac{A^2}{4}, \tag{3.21.8a}$$

so that

$$E = \langle E \rangle + \frac{A^2}{4}(F - K^2)\cos 2(kx + ly - \sigma t). \tag{3.21.8b}$$

It is the average over the period, $\langle E \rangle$, which gives a stable definition of the local wave energy and is the appropriate definition of the wave energy. Note that $\langle E \rangle$ varies slowly over the packet with $A^2$. The energy flux vector, from (3.21.4) and (3.21.6), is

$$\mathbf{S} = -A^2 \sigma \mathbf{K} \cos^2(kx + ly - \sigma t) - \hat{\mathbf{i}}\beta\frac{A^2}{2}\cos^2(kx + ly - \sigma t), \tag{3.21.9}$$

whose average over a period is

$$\langle \mathbf{S} \rangle = \frac{A^2}{2}\left[-\sigma \mathbf{K} - \frac{\beta}{2}\hat{\mathbf{i}}\right], \tag{3.21.10}$$

which, with the aid of the dispersion relation (3.15.11) and (3.19.13a,b) can be written*

$$\langle \mathbf{S} \rangle = \frac{A^2}{4}\left[\hat{\mathbf{i}}\beta\frac{k^2 - l^2 - F}{K^2 + F} + \hat{\mathbf{j}}\frac{\beta 2kl}{K^2 + F}\right] \tag{3.21.11}$$

$$= \mathbf{C}_g \langle E \rangle.$$

Hence *the average energy flux vector is equal to the wave energy multiplied by the group velocity.* Or, since $\mathbf{C}_g$ is independent of space, (3.21.5) becomes, when averaged over a wave period,

$$\frac{\partial \langle E \rangle}{\partial t} + \mathbf{C}_g \cdot \nabla\langle E \rangle = 0. \tag{3.21.12}$$

*The wave energy, defined by (3.21.8), is conserved and propagates with the packet at the group velocity.* The speed of energy transmission is not at all given by the propagation of crests or troughs, i.e., by the phase speed. The fact that energy propagates with the group velocity also follows directly from the earlier result that the wave envelope propagates with the group velocity along with the restriction that the fluid is homogeneous so that $\mathbf{C}_g$, $\mathbf{K}$, and $\sigma$ are independent of space and time.

---

* For a more general discussion of S and its physical meaning see Section (6.9).

## 3.22 The Energy Propagation Diagram

A helpful geometrical representation of the direction of the group velocity, and hence energy flux, in terms of the wave vector was introduced by Longuet-Higgins (1964). Suppose a Rossby wave (or wave packet) has a frequency $\sigma$. For any given $\sigma$ the wave number components $(k, l)$ must satisfy

$$k^2 + l^2 + F - \frac{\beta k}{-\sigma} = 0, \tag{3.22.1}$$

which is merely the dispersion relation rewritten. Without loss of generality $k$ can be considered positive, so that $\sigma < 0$. It follows from (3.22.1) that

$$\left[k - \frac{\beta}{-2\sigma}\right]^2 + l^2 = \frac{\beta^2}{4\sigma^2} - F. \tag{3.22.2}$$

Hence for a given $\sigma$ the allowable wave vector must lie on a circle in the $k, l$ plane, whose center, $O$, is at $(\beta/(-2\sigma), 0)$ and whose radius is $(\beta^2/4\sigma^2 - F)^{1/2}$, as shown in Figure 3.22.1. Since $\sigma \le \beta/2F^{1/2}$ by (3.15.16), the radius is always real. The offset of the circle from the origin is given by the minimum $x$-wave-number consistent with $\sigma$ and is

$$k_m = \frac{\beta}{-2\sigma} - \left(\frac{\beta^2}{4\sigma^2} - F\right)^{1/2} \ge 0, \tag{3.22.3}$$

so that the circle barely touches the $l$-axis in the limiting case $F = 0$, i.e., for very large deformation radius. The direction of $\mathbf{K}$ yields the direction of phase propagation, which for a Rossby wave is antiparallel to $\mathbf{K}$ for $k > 0$.

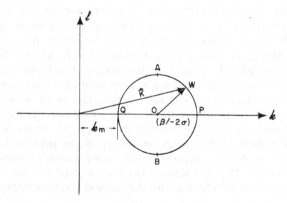

**Figure 3.22.1** The vector **OW** shows the direction of the group velocity of a Rossby wave with frequency $\sigma$ and wave vector **K**.

The energy flux vector is, by (3.21.11),

$$\langle \mathbf{S} \rangle = \frac{A^2}{2} \frac{\beta}{K^2 + F} k \left[ \hat{\mathbf{i}} \frac{k^2 - (l^2 + F)}{2k} + \hat{\mathbf{j}} l \right]$$

$$= \frac{A^2}{2} (-\sigma) \left[ \hat{\mathbf{i}} \left( k - \frac{K^2 + F}{2k} \right) + \hat{\mathbf{j}} l \right],$$

(3.22.4)

or

$$\langle \mathbf{S} \rangle = \frac{A^2}{2} (-\sigma) \left[ \hat{\mathbf{i}} \left( k - \left( \frac{\beta}{-2\sigma} \right) \right) + \hat{\mathbf{j}} l \right]$$

$$= \frac{A^2}{2} (-\sigma) \mathbf{OW},$$

(3.22.5)

where **OW** is the vector from the center of the circle to the tip of the wave vector **K** on the circumference of the circle. Since the magnitude of **OW** is

$$|\mathbf{OW}| = \left( \frac{\beta^2}{4\sigma^2} - F \right)^{1/2},$$

(3.22.6)

the magnitude of the energy flux is the same for all waves with the same amplitude and frequency. Waves whose wave vector **K** lie on the semicircle $APB$ propagate energy to the right (larger $x$), while those whose wave vectors lie on the arc $AQB$ propagate energy to the left. The former are short waves, the latter long waves.

## 3.23 Reflection and the Radiation Condition

An interesting application of the ideas of energy transmission and group velocity occurs in the study of the reflection of Rossby waves at a solid boundary. If the $\beta$-plane model is kept in mind, the reflection problem to be discussed is directly relevant to the reflection of Rossby waves at the boundary of an ocean basin. Consider the situation shown in Figure 3.23.1(a). A Rossby wave (or slowly varying packet) is incident on a rigid boundary at $x = 0$. On the $\beta$-plane this is the western boundary of an ocean. For the packet to be truly approaching the boundary its group velocity must be directed towards the boundary so that its energy flux vector is directed westward. The path of the packet, or its *ray*, is the straight line in the direction of $\mathbf{C}_g$. In the figure the path of the incoming energy is shown by the heavy arrow labeled $\langle \mathbf{S}_i \rangle$. The incident ray is at an angle $\theta_i$ to the normal to the boundary. The stream function for the incident wave can be represented as

$$\psi_i = \text{Re } A_i e^{i(k_i x + l_i y - \sigma_i t)},$$

(3.23.1)

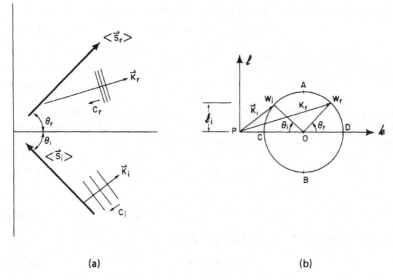

(a)                                              (b)

**Figure 3.23.1**  The incoming and reflected plane waves and their representation in the energy propagation diagram.

i.e., with frequency $\sigma_i$ and wave vector $\mathbf{K}_i$. Of course

$$\sigma_i = \frac{-\beta k_i}{k_i^2 + l_i^2 + F},$$  (3.23.2)

so that $\mathbf{K}_i$ must lie on the circle shown in Figure 3.23.1(b), whose radius is $(\beta^2/4\sigma_i^2 - F)^{1/2}$ and whose center is $\beta/-2\sigma_i$. From the results of the preceding section we see that $\mathbf{K}_i$ is the vector $\mathbf{PW}_i$ in the $k, l$ plane, where $W_i$ is the point on the circle in the $k, l$ plane whose radius vector $\mathbf{OW}_i$ is parallel to $\langle \mathbf{S}_i \rangle$. The vector $\mathbf{K}_i$, so determined, is also shown in Figure 3.23.1(a). The crests of the wave move antiparallel to $\mathbf{K}_i$ in a southwestward direction, while the wave energy is streaming *northwestward*. The incident wave must be a long wave, i.e., its wave vector must lie on the arc $ACB$ in the $k, l$ plane. For a given $\sigma_i$ and $l_i$, $k_i$ from (3.22.2) must be either

$$k = \left( \frac{\beta}{-2\sigma_i} \right) - \left| \frac{\beta^2}{4\sigma_i^2} - (F + l_i^2) \right|^{1/2}$$  (3.23.3a)

or

$$k = \left( \frac{\beta}{-2\sigma_i} \right) + \left| \frac{\beta^2}{4\sigma_i^2} - (F + l_i^2) \right|^{1/2}$$  (3.23.3b)

The former is a long wave whose wave radiation streams toward the boundary; the latter is a short wave whose radiation streams away from the boundary. Consequently it is (3.23.3a) which determines the x-wave-number of the incident wave (recall that $-\sigma_i$ is positive).

The presence of the wall will produce a reflected wave,

$$\psi_r = \text{Re } A_r e^{i(k_r x + l_r y - \sigma_r t)}. \tag{3.23.4}$$

The total stream-function field representing a continuous process of incoming and reflecting energy flux is

$$\psi = \psi_i + \psi_r. \tag{3.23.5}$$

Equivalently, (3.23.5) is the stream-function field during the long "bounce time" required for a large-scale packet to reflect from the boundary. In either case, on $x = 0$, the $x$-velocity must vanish, i.e.,

$$\frac{\partial \psi}{\partial y} = 0, \qquad x = 0, \tag{3.23.6}$$

or

$$l_i A_i e^{i(l_i y - \sigma_i t)} + l_r A_r e^{i(l_r y - \sigma_r t)} = 0. \tag{3.23.7}$$

The only way (3.23.7) can be true for *all t* is for

$$\sigma_i = \sigma_r = \sigma, \tag{3.23.8}$$

so that the frequency is *preserved* in the reflection. One immediate result of this is that the reflected wave must have its wave vector on the same circle in the $k$, $l$ plane as the incident wave, since the radius and center of the circle depend only on frequency. Furthermore, in order for (3.23.7) to be true for *all y*,

$$l_i = l_r = l, \tag{3.23.9}$$

so that the projection of the wave vector parallel to the boundary is preserved under reflection. This completely determines the wave vector of the reflected wave, since if $l_i = l$, the reflected wave vector must have its tip at the point $W_r$ on the wave-number circle in order that the wave flux of the reflected wave may be directed *away* from the boundary. Since $l_i = l_r$, a little geometry shows that the angle of the reflected-ray path, $\theta_r$, is equal to the angle of the incident-ray path, so that the energy flux vector bounces off the wall "elastically," i.e., with angle of reflection equal to angle of incidence for the energy and wave-packet trajectory. The $x$-wave-number of the reflected wave is given by (3.23.3b). So although $\sigma$ and $l$ alone are insufficient to determine which of the two roots for $k$ is appropriate, the *radiation condition* that the reflected wave must have outgoing energy flux determines $k$. Since $k_r > k_i$, the wave vector *increases its length as a consequence of reflection*. Since $l_i = l_r$, this means that the wave vector of the reflected wave will be rotated into a position more nearly aligned with the $x$-axis. For geostrophic motion the velocity is perpendicular to the wave vector, so that the velocity in the reflected wave is rotated in the direction of the $y$-axis (northward) by the reflection. The change in $k$ due to the reflection is

$$k_r - k_i = 2\left(\frac{\beta^2}{4\sigma^2} - (l^2 + F)\right)^{1/2}, \tag{3.23.10}$$

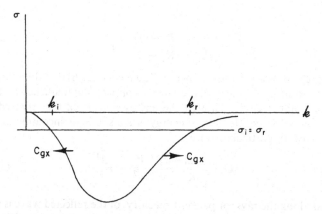

**Figure 3.23.2** The incident and reflected $x$-wave-numbers are $k_i$ and $k_r$, respectively. At low frequencies the difference $k_r - k_i$ becomes large.

so that the change of $k$ increases with decreasing frequency, a result immediately evident from a consideration of the dispersion curve as shown in Figure 3.23.2.

The remaining condition required by (3.23.7) is then

$$A_i = -A_r = A, \tag{3.23.11}$$

so that the amplitude of the pressure field is preserved during reflection while the phase is flipped by 180°. Since the amplitude is preserved, the velocity fields are proportional to the wave number, i.e., for the incident wave

$$u_i = -\operatorname{Re}(il)Ae^{i(k_ix+ly-\sigma t)}$$
$$v_i = \operatorname{Re}(ik_i)Ae^{i(k_ix+ly-\sigma t)}, \tag{3.23.12}$$

while for the reflected wave

$$u_r = \operatorname{Re}(il)Ae^{i(k_rx+ly-\sigma t)},$$
$$v_r = -\operatorname{Re}(ik_r)Ae^{i(k_rx+ly-\sigma t)}. \tag{3.23.13}$$

The magnitude of the $x$-velocity, or zonal component, is preserved under reflection, while the $y$-component of the velocity is increased by the process of reflection. Since $k_r^2 > l^2 + F > l^2$, the reflected velocity must be primarily in the $y$-direction.

Note that the energy density (i.e., the energy per unit area) of the reflected wave is greater than the energy density of the incident wave, since

$$\langle E_i \rangle = (K_i^2 + F)|A_i|^2,$$
$$\langle E_r \rangle = (K_r^2 + F)|A_r|^2 \tag{3.23.14}$$
$$= (K_r^2 + F)|A_i|^2,$$

so that

$$\frac{\langle E_r \rangle}{\langle E_i \rangle} = \frac{K_r^2 + F}{K_i^2 + F} > 1. \tag{3.23.15}$$

However, we have already shown in Equation (3.22.6) that the energy flux $\langle S \rangle$ is the same for all waves of the same amplitude and frequency that lie on the wave-number circle. Though the energy density of the reflected wave goes up by reflection, the flux of energy is the same for incoming and outgoing waves. In particular, since

$$|\mathbf{C}_g| = \frac{2(-\sigma)}{K^2 + F} \left| \frac{\beta^2}{4\sigma^2} - F \right|^{1/2}, \tag{3.23.16}$$

the *speed* along the rays, or packet trajectory, of the reflected wave is slower than the incoming wave in the ratio

$$\frac{|\mathbf{C}_{gr}|}{|\mathbf{C}_{gi}|} = \frac{K_i^2 + F}{K_r^2 + F}, \tag{3.23.17}$$

so that the flux of energy is unchanged by the reflection process. For very low-frequency waves the group velocity of the reflected wave will be very low, since its $x$-wave-number will be high.

Since the group velocity is decreased by reflection, an incoming wave packet will have its length compressed along the rays, as shown in Figure 3.23.3, by the ratio

$$\frac{L_r}{L_i} = \frac{|\mathbf{C}_{gr}|}{|\mathbf{C}_{gi}|} \tag{3.23.18}$$

from purely kinematic considerations. The width of the packet in the direction perpendicular to the ray path is unaffected by the reflection. Thus, since

$$L_i \langle E_i \rangle = L_r \langle E_r \rangle, \tag{3.23.19}$$

from (3.23.18), (3.23.17), and (3.23.15) the *total* energy in the packet is conserved by the reflection, although locally, within the packet, the reflected motion is more energetic as the energy is squeezed into a narrower region traveling more slowly from the boundary.

This remarkable behavior of the reflection process depends on the orientation of the boundary. The singular case of the reflection of a wave from a zonal boundary, i.e., one parallel to the $x$-axis, points this fact out clearly. The method of analysis is the same, and again both the frequency and the component of the wave vector parallel to the boundary are conserved. This means that $k$ is preserved under reflection, and a glance at Figure 3.23.4 shows that $l$ only suffers a change of sign, so that the wavelength, group *speed*, and energy density are unaltered by the reflection in this special case. Even here, though, the remarkable nature of the Rossby wave is evident, for the unwitting eye observing the progression of crests in the two waves would easily confuse the reflected and the incident wave.

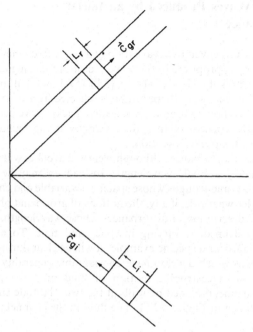

**Figure 3.23.3**   The trajectory of a packet upon reflection from a boundary at $x = 0$. Note the shortening of the packet as a consequence of the reduction in group velocity of the reflected wave.

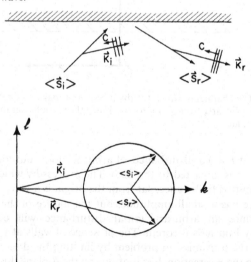

**Figure 3.23.4**   The reflection of a plane wave at a boundary coincident with an isoline of ambient potential vorticity. In this case the wavelength of the wave is preserved by the reflection.

## 3.24 Rossby Waves Produced by an Initial Disturbance

A wave packet which envelops a wave train of a single frequency moves with the group velocity appropriate to the wave vector of that single wave. If the amplitude of the packet is slowly varying, (3.19.15) shows that the amplitude progresses without change of shape until it is reflected from a boundary. The rules governing the propagation of such a wave packet have been discussed at length, and the concept of the group velocity is the foundation of an understanding of the packet's motion.

In many cases a disturbance, although clearly wavelike, will not contain even approximately a single wavelength. Instead we may be aware of a succession of crests and troughs whose spacing is variable and changing. It is natural to wonder what role, if any, the notions of group and phase velocity may play in the description of disturbances whose wavelength, assuming it can be sensibly defined, is varying in space and time. To examine this question let us consider a specific example; it will be clear during the discussion that the ideas which develop have considerable generality.

Imagine the quasigeostrophic motion of fluid on the $\beta$-plane in an infinitely long channel bounded by walls at the two "latitude circles" $y = 0$ and $y = 1$, as shown in Figure 3.24.1. The fluid in the channel is initially at

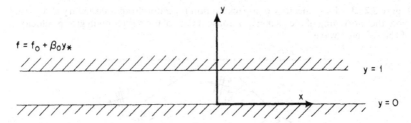

**Figure 3.24.1**   The channel in which the disturbance propagates. Ambient potential vorticity increases linearly from $y = 0$ to $y = 1$ as if the Coriolis parameter increased linearly with latitude.

rest, and then at $t = 0$ a disturbance, of arbitrary form but localized to the vicinity of $x = 0$, is imparted to the fluid. The process by which this disturbance is transmitted to distances far from the origin is a wave mechanism if the disturbance has a small amplitude, but the nature of the propagation is complex, since an arbitrary initial disturbance will contain many wavelengths by Fourier's theorem. The presence of walls at $y = 0, 1$ allows us to simplify the transmission problem by limiting the direction of energy propagation to the $x$-direction. For motions on the $\beta$-plane this is equivalent to a model in which a latitude band is arbitrarily isolated from its surroundings and only the propagation of the energy in longitude is considered.

For small-amplitude disturbances, the governing equation is

$$\frac{\partial}{\partial t}(\nabla^2 \psi - F\psi) + \beta \frac{\partial \psi}{\partial x} = 0 \qquad (3.24.1)$$

with boundary conditions

$$v_0 = \frac{\partial \psi}{\partial x} = 0, \qquad y = 0, 1, \qquad (3.24.2)$$

and initial condition

$$\psi(x, y, 0) = \psi_0(x, y). \qquad (3.24.3)$$

To keep the discussion as simple as possible without sacrificing essentials, let us consider the case where

$$\psi_0(x, y) = \Psi_n(x)\sin n\pi y, \qquad (3.24.4)$$

where $n$ is arbitrary. A general initial disturbance can be obtained by a Fourier sum over $n$ of disturbances of the form (3.24.4), and since our problem is linear, the resulting motion in the general case will simply be the linear superposition of the result for each $n$.

Inspection of (3.24.1) suggests that a solution of the form

$$\psi = \phi_n(x, t)\sin n\pi y,$$

which automatically satisfies (3.24.2), is possible if $\phi_n(x, t)$ satisfies

$$\frac{\partial}{\partial t}\left|\frac{\partial^2 \phi_n}{\partial x^2} - a^2 \phi_n\right| + \beta \frac{\partial \phi_n}{\partial x} = 0, \qquad (3.24.5)$$

where

$$a^2 = n^2 \pi^2 + F.$$

The Fourier transform of $\phi_n$ is defined by the integral

$$\hat{\phi}_n(k, t) = \frac{1}{\sqrt{2\pi}} \int_{-\infty}^{\infty} e^{-ikx} \phi_n(x, t) \, dx, \qquad (3.24.6)$$

and the complementary, return relationship is given by

$$\phi_n(x, t) = \frac{1}{\sqrt{2\pi}} \int_{-\infty}^{\infty} e^{ikx} \hat{\phi}_n(k, t) \, dk. \qquad (3.24.7)$$

Any reasonably well-behaved function on the infinite $x$-interval $(-\infty, \infty)$ can be represented as a Fourier integral (3.24.7) which is the sum (integral) over all wave numbers $k$ of plane waves each of which has the amplitude $\hat{\phi}_n(k, t) \, dk$. At $t = 0$, the Fourier transform of (3.24.4) yields

$$\hat{\phi}_n(k, 0) = A_n(k) = \frac{1}{\sqrt{2\pi}} \int_{-\infty}^{\infty} e^{-ikx} \Psi_n(x) \, dx. \qquad (3.24.8)$$

By integration by parts it follows that

$$\frac{1}{\sqrt{2\pi}}\int_{-\infty}^{\infty} e^{-ikx}\frac{\partial\phi_n}{\partial x}\, dx = \frac{ik}{\sqrt{2\pi}}\int_{-\infty}^{\infty} e^{-ikx}\phi_n\, dx \tag{3.24.9}$$

$$= ik\hat{\phi}_n(k, t)$$

and

$$\frac{1}{\sqrt{2\pi}}\int_{-\infty}^{\infty} e^{-ikx}\frac{\partial^2}{\partial x^2}\frac{\partial\phi_n}{\partial t}\, dx = -\frac{k^2}{\sqrt{2\pi}}\int_{-\infty}^{\infty} e^{-ikx}\frac{\partial\phi_n}{\partial t}\, dx$$

$$= -k^2\frac{\partial\hat{\phi}_n}{\partial t}(k, t),$$

where we have tacitly used the fact that for all finite $t$, $\phi_n$ must still be zero as $|x| \to \infty$, i.e., we have only needed to assume a finite speed of energy propagation. With the aid of (3.24.9), (3.24.5) becomes an ordinary differential equation for $\hat{\phi}_n$ (which fact, indeed, is the very reason for the introduction of the Fourier transform), i.e.,

$$\frac{\partial\hat{\phi}_n}{\partial t} - \frac{i\beta k}{k^2 + a^2}\hat{\phi}_n = 0, \tag{3.24.10}$$

whose general solution is

$$\hat{\phi}_n(k, t) = C_n(k)e^{-i\sigma(k)t}, \tag{3.24.11}$$

where

$$\sigma(k) = -\frac{\beta k}{k^2 + a^2}. \tag{3.24.12}$$

To satisfy the initial condition (3.24.8), $C_n(k)$ must satisfy

$$C_n(k) = A_n(k), \tag{3.24.13}$$

or

$$\hat{\phi}_n = A_n(k)e^{-i\sigma(k)t}, \tag{3.24.14}$$

from which it follows that

$$\boxed{\phi_n(x, t) = \frac{1}{\sqrt{2\pi}}\int_{-\infty}^{\infty} A_n(k)e^{i(kx - \sigma(k)t)}\, dk} \tag{3.24.15}$$

The disturbance consists of an infinite Fourier superposition of plane waves, each with a different wave number $k$, and each oscillating with frequency $\sigma(k)$ equal to the Rossby wave frequency for the plane wave for that wave number, as given by (3.24.12). The amplitude of each Fourier mode, $A_n(k)$, is completely determined by the initial condition (3.24.8). For example, consider the initial disturbance

$$\Psi_n(x) = \frac{1}{2^{1/2}x_0}e^{-(x/x_0)^2}e^{ik_0 x} \tag{3.24.16a}$$

(where the real part is implicitly to be taken), i.e., a disturbance whose characteristic spatial extent is $x_0$ and whose spatial period is $2\pi/k_0$. If $k_0 x_0 \gg 1$ this will look, at least initially, like a wave packet. The Fourier transform of (3.24.16a) is

$$A_n(k) = \frac{e^{-x_0^2(k-k_0)^2/4}}{2}. \qquad (3.24.16b)$$

If $k_0 x_0 \gg 1$, the *spectrum* of $A_n(k)$ (i.e., the distribution of $A_n(k)$ in $k$) is sharply peaked around $k = k_0$. That is, a slowly varying wave packet is a superposition of waves each near the central wave number $k = k_0$, as shown in Figure 3.24.2—i.e., a distribution whose characteristic width in $k$-space

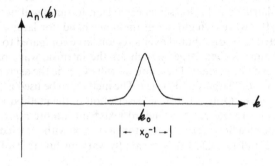

**Figure 3.24.2**   The spectrum of a wave packet. As $x_0$, the packet length increases, the sharpness of the spectrum increases.

goes *inversely* with the spatial spread $x_0$. Thus the broader the packet in $x$, the narrower the spectrum in $k$. For a *very* narrow spectrum, suitable for a slowly varying wave packet, $A_n(k)$ vanishes unless $k$ is near $k_0$ so that it is only necessary for the evaluation of (3.24.15) to consider the form of $\sigma(k)$ near $k_0$, i.e.,

$$\sigma(k) = \sigma(k_0) + \left(\frac{\partial\sigma}{\partial k}\right)_{k=k_0}(k - k_0) + \cdots, \qquad (3.24.17)$$

so that

$$\phi_n(x, t) = \frac{1}{\sqrt{2\pi}}\int_{-\infty}^{\infty} A_n(k)e^{i(k_0 x - \sigma(k_0)t)}$$

$$\times \exp\left(i\left\{x - \left(\frac{\partial\sigma}{\partial k}\right)_{k_0}t\right\}[k - k_0]\right) dk$$

$$= \frac{e^{i(k_0 x - \sigma(k_0)t)}}{\sqrt{2\pi}}\int_{-\infty}^{\infty} A_n(k)$$

$$\times \exp\left(i\left\{x - \left(\frac{\partial\sigma}{\partial k}\right)_{k_0}t\right\}[k - k_0]\right) dk$$

$$= e^{i(k_0 x - \sigma(k_0)t)}\Phi_n(x - C_{gx}(k_0)t), \qquad (3.24.18)$$

where $\Phi_n(x)$ is the envelope shape at the initial instant. This is the formula for the propagation, previously derived, of a slowly varying packet.

In this section we are concerned with the quite different situation, i.e., one for which the initial disturbance is sharply localized in $x$, so that the spectrum of $A_n(k)$ is quite broad. In this case the approximation (3.24.17) and (3.24.18) is utterly inadequate, since contributions of many, widely spread wave numbers must be considered. Without any essential loss of generality let us assume the initial condition is an even function of $x$, i.e., symmetric about $x = 0$.

For a short time after the disturbance is first made, the form of the motion will depend on the particular nature of the initial form of the disturbance This is of course arbitrary, and so the solution (3.24.15) for small $t$ (which can be calculated by a Taylor series expansion in time) will not reveal the essentials of the wave dynamics. On the other hand, for large $t$ the energy can be expected to be distributed over a region large compared to the source. The elapsed time is then large enough for the intrinsic wave-propagation dynamics to become evident. Therefore we will examine the asymptotic form of (3.24.15) for both large $x$ and large $t$. The method to be used is the method of *stationary phase*. Only a heuristic description of the method is presented here. The reader is referred to standard texts* for a more rigorous derivation. The basic idea however is to note that as $t \to \infty$ with $x/t$ fixed, the phase of the exponential in (3.24.15) is a rapidly varying function of $k$, i.e., for large $t$

$$\theta = tq(k) = t\left[\left(\frac{x}{t}\right)k - \sigma(k)\right] \qquad (3.24.19)$$

changes enormously for each small increment of $k$:

$$\frac{\partial \theta}{\partial k} = t\frac{\partial q}{\partial k} = t\left[\frac{x}{t} - \frac{\partial \sigma}{\partial k}\right]. \qquad (3.24.20)$$

Hence both the real and imaginary parts of $e^{itq(k)}$ oscillate wildly between plus and minus one, more and more rapidly as $t$ increases, until finally for large $t$ contributions to the integral from one value of $k$ will be canceled by an immediately neighboring value of $k$ for which the phase has advanced by 180°, since $A_n(k)$, a smooth function, is nearly the same for both values of $k$. This behavior occurs at all intervals in $k$ except near special points $k_s$ where an increment in $k$ does not lead to an increase of $q$, i.e., at those points of *stationary phase* where $\partial q/\partial k$ vanishes, which occur at

$$\frac{x}{t} = \frac{\partial \sigma}{\partial k}(k_s). \qquad (3.24.21)$$

The dominant contributions to the integral in (3.24.15) will occur therefore in the neighborhood of all those values of $k$ which satisfy (3.24.21), which

* A particularly clear discussion is presented in Jeffreys and Jeffreys (1962).

itself has a simple interpretation. At time $t \gg 1$, the dominant contribution to $\phi(x, t)$ at the point $x$ comes from the wave number in the original spectrum whose group velocity has allowed a disturbance to propagate the distance $x$ in the elapsed time $t$. Then (3.24.21) can be solved for such points $k_s$, i.e., (3.24.21) can be inverted to yield

$$k_s = k_s\left(\frac{x}{t}\right). \qquad (3.24.22)$$

By hypothesis $\Psi_n(x)$ is an even function of $x$, so $A_n(k)$ is an even function of $k$, since by (3.24.8)

$$A_n(-k) = \frac{1}{\sqrt{2\pi}} \int_{-\infty}^{\infty} e^{ikx}\Psi_n(x)\, dx. \qquad (3.24.23)$$

Letting $x = -\xi$ as the dummy variable of integration yields

$$A_n(-k) = -\frac{1}{\sqrt{2\pi}} \int_{+\infty}^{-\infty} e^{-ik\xi}\Psi_n(-\xi)\, d\xi$$

$$= \frac{1}{\sqrt{2\pi}} \int_{-\infty}^{\infty} e^{-ik\xi}\Psi_n(\xi)\, d\xi, \qquad (3.24.24)$$

$$= A_n(k)$$

where the evenness of $\Psi_n(\xi)$, i.e.,

$$\Psi_n(\xi) = \Psi_n(-\xi) \qquad (3.24.25)$$

has been used. This in turn implies that (3.24.15) can be written in terms of an integral over the interval $(0, \infty)$ in $k$, i.e., since $A_n(k)$ is even in $k$,

$$\phi_n(x, t) = \frac{1}{\sqrt{2\pi}} \int_{-\infty}^{\infty} A_n(k)\{\cos(kx - \sigma(k)t)\}\, dk$$

$$= \sqrt{\frac{2}{\pi}} \int_{0}^{\infty} A_n(k)\cos(kx - \sigma(k)t)\, dk \qquad (3.24.26)$$

$$= \sqrt{\frac{2}{\pi}} \operatorname{Re} \int_{0}^{\infty} A_n(k)e^{i(kx - \sigma(k)t)}\, dk.$$

Thus in the evaluation of (3.24.26) only positive values of $k$ which are solutions to (3.24.21) need be considered.*

In the small interval around $k = k_s$ which contributes to (3.24.26),

$$\sigma(k) = \sigma(k_s) + \frac{\partial\sigma}{\partial k}(k_s)(k - k_s) + \frac{\partial^2\sigma}{\partial k^2}(k_s)\frac{(k - k_s)^2}{2} + \cdots, \qquad (3.24.27)$$

---

* If $\Psi_n(x)$ is an odd function of $x$, $A_n(k)$ will also be an odd function of $k$, and again the integral can be written in the form (3.24.26), where instead of the real part the imaginary part of the result is retained. Since all functions can be written as the sum of an odd and an even function, the analysis presented here is directly applicable to the general case.

so that

$$kx - \sigma(k)t = [k_s x - \sigma(k_s)t] + \left[ x - \frac{\partial \sigma}{\partial k}(k_s)t \right](k - k_s)$$

$$- t \frac{\partial^2 \sigma}{\partial k^2}(k_s) \frac{(k - k_s)^2}{2} + \cdots,$$

(3.24.28)

the second term of which, linear in $k - k_s$, vanishes by (3.24.21). Thus for large $t$ (3.24.26) can be approximated by

$$\phi_n(x, t) = \text{Re} \sqrt{\frac{2}{\pi}} \int_{k_s - \Delta}^{k_s + \Delta} A_n(k_s) e^{i(k_s x - \sigma(k_s)t)}$$

$$\times \exp\left( -it \frac{\partial^2 \sigma}{\partial k^2}(k_s) \frac{(k - k_s)^2}{2} \right) dk$$

(3.24.29)

$$= \text{Re} \sqrt{\frac{2}{\pi}} A_n(k_s) e^{i(k_s x - \sigma(k_s)t)}$$

$$\times \int_{k_s - \Delta}^{k_s + \Delta} \exp\left( -it \frac{\partial^2 \sigma}{\partial k^2}(k_s) \frac{(k - k_s)^2}{2} \right) dk.$$

We have used the fact that in the tiny interval $2\Delta$ around $k_s$, $A_n(k)$ is a smooth function, so that $A_n(k)$ can be well approximated by its central value $A_n(k_s)$. The result obtained is independent of the precise size of the interval $\Delta$, as long as it is small.

Introduce

$$\alpha = (k - k_s) \left\{ \frac{t}{2} \left| \frac{\partial^2 \sigma}{\partial k^2}(k_s) \right| \right\}^{1/2},$$

(3.24.30)

in terms of which (3.24.29) becomes

$$\phi_n(x, t) \approx \text{Re} \frac{2}{\sqrt{\pi}} \frac{A_n(k_s) e^{i(k_s x - \sigma(k_s)t)}}{\left( t \left| \frac{\partial^2 \sigma}{\partial k^2}(k_s) \right| \right)^{1/2}}$$

(3.24.31)

$$\times \int_{-\alpha_s}^{\alpha_s} \exp\left[ -i\alpha^2 \, \text{sgn} \, \frac{\partial^2 \sigma}{\partial k^2}(k_s) \right] d\alpha,$$

where

$$\text{sgn} \, \frac{\partial^2 \sigma}{\partial k^2}(k_s) = \begin{cases} +1 & \text{if } \frac{\partial^2 \sigma}{\partial k^2}(k_s) > 0, \\ -1 & \text{if } \frac{\partial^2 \sigma}{\partial k^2}(k_s) < 0, \end{cases}$$

(3.24.32)

and

$$\alpha_s = \Delta \left\{ \frac{t}{2} \left| \frac{\partial^2 \sigma}{\partial k^2}(k_s) \right| \right\}^{1/2}.$$

(3.24.33)

As $t \to \infty$, $\alpha_s \to \infty$, so that the integral in (3.24.31) becomes

$$\int_{-\infty}^{\infty} \exp\left\{-i\alpha^2 \operatorname{sgn} \frac{\partial^2 \sigma}{\partial k^2}(k_s)\right\} d\alpha = \sqrt{\frac{\pi}{2}\left(1 - i \operatorname{sgn} \frac{\partial^2 \sigma}{\partial k^2}(k_s)\right)}$$

$$= \sqrt{\pi} \exp\left(-i\frac{\pi}{4} \operatorname{sgn} \frac{\partial^2 \sigma}{\partial k^2}(k_s)\right),$$

(3.24.34)

so that

$$\phi_n(x, t) = 2 \operatorname{Re} A_n(k_s) \frac{e^{i(k_s x - \sigma(k_s)t)}}{\left(t\left|\frac{\partial^2 \sigma}{\partial k^2}(k_s)\right|\right)^{1/2}} \exp\left(-i\frac{\pi}{4} \operatorname{sgn} \frac{\partial^2 \sigma}{\partial k^2}(k_s)\right).$$

An argument similar to that given in the derivation of (3.24.24) shows that since $A_n(k)$ is an even function of $x$, it must be real. Summing over all $k_s > 0$ which are solutions to (3.24.21) then yields the final asymptotic approximation for $\phi_n(x, t)$:

$$\boxed{\begin{aligned} \phi_n(x, t) &\approx \sum_{k_s} \frac{2A_n(k_s)}{\left(t\left|\frac{\partial^2 \sigma}{\partial k^2}(k_s)\right|\right)^{1/2}} \\ &\times \cos\left(k_s x - \sigma(k_s)t - \frac{\pi}{4} \operatorname{sgn} \frac{\partial^2 \sigma}{\partial k^2}(k_s)\right). \end{aligned}}$$

(3.24.35)

It can be shown that the error made by approximating (3.24.15) by (3.24.35) is $O(t^{-1})$. The approximation therefore improves with increasing time.

The solution (3.24.35) appears to have the form of a plane wave with a wave amplitude slowly diminishing like $t^{-1/2}$. This is somewhat misleading, since by (3.24.22) we see that $k_s$ and hence $\sigma(k_s)$ are rather complicated functions of $x/t$. Nevertheless this first impression is basically correct if properly interpreted. The variable part of the wave phase of the oscillating part of $\phi_n$ is

$$\theta(x, t) = k_s x - \sigma(k_s)t. \qquad (3.24.36)$$

The rate of increase of $\theta$ with increasing $x$ is the local wave number $k$, in analogy with (3.8.3). However, now $k$ will be a function of space and time, since by (3.24.21)

$$k = \frac{\partial \theta}{\partial x} = k_s + \left(x - t\frac{\partial \sigma(k_s)}{\partial k_s}\right)\frac{\partial k_s}{\partial x} = k_s(x, t). \qquad (3.24.37)$$

The wave number will appear constant only to an observer moving outward from the origin with the *group velocity appropriate to that wave number*. A stationary observer or one moving at any other speed will see an oscillation whose wavelength is changing with time.

Similarly the local frequency is, from (3.24.36)

$$\sigma = -\frac{\partial \theta}{\partial t} = -\left(x - t\frac{\partial \sigma}{\partial k_s}(k_s)\right)\frac{\partial k_s}{\partial t} + \sigma(k_s) \qquad (3.24.38a)$$

For an observer moving with the group velocity appropriate to the wave number $k_s$, this is simply

$$\sigma = \sigma(k_s), \qquad (3.24.38b)$$

where $\sigma(k_s)$ is the frequency at $k_s$ as determined by the dispersion relation (3.24.12) for plane waves. The local phase speed (i.e., the rate of advance of a particular crest, say) is

$$C = \left(\frac{\partial x}{\partial t}\right)_\theta = \frac{\sigma(k_s)}{k_s} \qquad (3.24.39)$$

Since $k$ is constant only for an observer moving with $C_g(k_s)$ and since $C \neq C_g$ for a dispersive wave such as the Rossby wave, an observer moving so as to follow a particular *wavelength* will observe different crests gliding past. An observer who wishes to travel with a *particular crest* must move at the *local* value of $C$, which differs from $C_g$. Hence an observer moving with a given crest will observe a wave whose wavelength is changing with time, and consequently the speed $C$ will *also* change with time. Thus an observer following the location of a disturbance of particular wavelength will move uniformly with the group velocity, while an observer who wishes to follow a particular crest must accelerate or decelerate as the wavelength of the wave carrying the crest changes with time. From (3.24.37) and (3.24.38)

$$\frac{\partial k_s}{\partial t} + \frac{\partial \sigma_s}{\partial x} = 0. \qquad (3.24.40)$$

Since $k_s$ is the number of complete undulations per unit length and $\sigma(k_s)$ is the rate at which any particular phase (say crests) passes a fixed observation point, (3.24.40) implies that the number of crests in the wave train (3.24.35) is conserved as the disturbance propagates, even though the separation between crests of different wavelengths changes with time (since their phase speed $C$ varies with wavelength). This dispersion of the wave train, however, conserves the number of crests. Since $\sigma_s$ is a function of $x$ only through $k_s$, it follows from (3.24.40) that

$$\frac{\partial k_s}{\partial t} + \left|\frac{\partial \sigma(k_s)}{\partial k}\right|\frac{\partial k_s}{\partial x} = 0 \qquad (3.24.41a)$$

and

$$\frac{\partial \sigma(k_s)}{\partial t} + \frac{\partial \sigma(k_s)}{\partial k}\frac{\partial \sigma}{\partial x} = 0, \qquad (3.24.41b)$$

which is the mathematical statement of the already deduced fact that wavelengths and frequencies propagate with the group velocity.

Consider now the wave energy between two closely spaced planes at $x = x_s$ and $x = x_s + \Delta x_s$. Let each point move with the group velocity of the wave number which at time $t$ is at that point. If $\Delta x_s$ is small, the wavelength of the wave within $\Delta x_s$ will appear sensibly constant, so that locally (3.21.8) can be used to evaluate the energy density. With attention to the fact that the wave amplitude, $A$, is here

$$\frac{2A_n}{\left(t\left|\frac{\partial^2 \sigma}{\partial k^2}(k_s)\right|\right)^{1/2}},$$

we obtain

$$E(x_s, t) = \frac{(k_s^2 + a^2)A_n^2(k_s)\,\Delta x_s}{t\left|\frac{\partial^2 \sigma}{\partial k^2}\right|\,2} \tag{3.24.42}$$

for the energy of the strip of unit length in $y$ and breadth $\Delta x_s$ in $x$. The point $x_s$ at time $t$ corresponds to a wave number $k_s$ given by (3.24.21), so that, at the same time $t$, the interval between neighboring points can be written as

$$\Delta x_s = t\left|\frac{\partial^2 \sigma}{\partial k^2}\right|\Delta k_s, \tag{3.24.43}$$

where the absolute-value sign assures that we are considering a positive interval $\Delta x_s$ for a positive increment in $k_s$. In the limit $\Delta k_s \to 0$ the total energy in the infinitesimal strip $dx_s$ is

$$E(x_s, t)\,dx_s = \tfrac{1}{2}(k_s^2 + a^2)A_n^2(k_s)\,dk_s \tag{3.24.44a}$$

and is *constant* as the strip moves away from the origin with constant $k_s$, i.e., at the group velocity appropriate to $k = k_s$. At the initial instant the total energy in the wave field is

$$\frac{1}{2}\int_{-\infty}^{\infty} dx \int_{0}^{1} dy \left\{\left(\frac{\partial \psi}{\partial x}\right)^2 + \left(\frac{\partial \psi}{\partial y}\right)^2 + F\psi^2\right\}$$

$$= \frac{1}{4}\int_{-\infty}^{\infty} dx \left\{\left(\frac{\partial \phi_n}{\partial x}\right)^2 + a^2(\phi_n)^2\right\}. \tag{3.24.44b}$$

Since

$$\int_{-\infty}^{\infty} \phi_n^2\,dx = \int_{-\infty}^{\infty} |\tilde{\phi}_n(k)|^2\,dk,$$

$$\int_{-\infty}^{\infty} \left(\frac{\partial \phi_n}{\partial x}\right)^2 dx = \int_{-\infty}^{\infty} k^2|\tilde{\phi}_n(k)|^2\,dk$$

by Parseval's theorem, the total energy at time $t = 0$ is

$$\int_{0}^{\infty} \frac{k^2 + a^2}{2} A_n^2(k)\,dk = \int_{-\infty}^{\infty} E(x)\,dx, \tag{3.24.45}$$

which with (3.24.44a) allows the following interpretation. At the initial instant the total energy is localized near the origin and can be considered, by (3.24.45), as an infinite sum of energy bundles, each bundle a wave packet of amplitude $A_n(k)$. As time progresses each wave bundle moves out from the origin at a rate given by the group velocity determined by the packet wave number $k_s$. During its travel from the origin each bundle preserves its frequency and wave number, but the physical length of the bundle gradually stretches linearly with time. Since the energy of each bundle is preserved, the amplitude of the wave must decrease as $t^{-1/2}$ to compensate for the spatial dispersion of the bundles due to the slight variation of $C_g$ from one end of the bundle to the other. Individual crests slip from one bundle to the next, since $C_g(k) \neq C(k)$, so that the shape of the wave train formed by the succession of bundles changes with time.

The bundle with the largest group velocity will outdistance the others. The rest will eventually string themselves out in a line, with the most rapid in the van and the slower bundles behind.

For the Rossby wave, $k_s$ must satisfy

$$\frac{x}{t} = C_{gx}(k_s) = \beta \frac{k_s^2 - a^2}{(k_s^2 + a^2)^2}. \tag{3.24.46}$$

Reference to Figure 3.19.2 above shows that for $x > \beta t/8a^2$ no solutions for $k_s$ are possible, as is also the case for $x < -\beta t/a^2$. Solutions are possible only in the $x$-interval $(-(\beta/a^2)t, \beta t/8a^2)$ which corresponds to the region traversed by the bundles with the *maximum* negative and positive group speeds respectively. Beyond that interval the fluid still remains, to this approximation, undisturbed. Within the interval the wave numbers observed are the solutions of (3.24.46), which are shown in Figure 3.24.3. Note that long waves, for which $k_s < a = (n^2\pi^2 + F)^{1/2}$, are found far to the left (west) of the original disturbance, while shorter waves ($k_s > a$) are found at positive $x$ to the east of the initial disturbance. For $-\beta t/a^2 < x < 0$ one and only one wavelength is found for each $x$, but for $x > 0$ there are two wavelengths at each $x$: a band of medium-scale waves in the wave-number interval $1 \leq k_s/a \leq \sqrt{3}$ and a band of short waves with $k_s/a > \sqrt{3}$. The wave disturbance extends farther, by a factor of 8, to the west of the origin than to the east. From (3.24.46)

$$\frac{k_s}{a} = \left[ \frac{1}{2\xi} - 1 \pm \frac{(1 - 8\xi)^{1/2}}{2\xi} \right]^{1/2}, \tag{3.24.47}$$

where

$$\xi = \frac{x}{t} \frac{a^2}{\beta}. \tag{3.24.48}$$

For *fixed* $x$, $\xi$ becomes small as $t$ increases, so that at a *fixed point* $x$, expansion of the radicand yields two solutions for large $t$:

$$\frac{k_s}{a} = 1$$

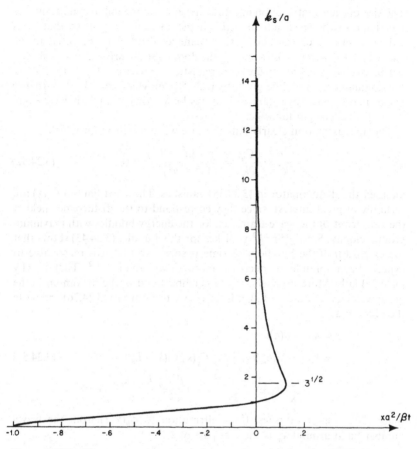

**Figure 3.24.3**  The wave number of stationary phase as a function of distance from the origin of the disturbance. Note that for $x > 0$ there are two values of $k_s$ for each $xa^2/\beta t < 0.125$.

or

$$\frac{k_s}{a} = \xi^{-1/2}, \tag{3.24.49}$$

corresponding to the slowest-moving packets. The latter solution is clearly valid only for $x > 0$. For either positive or negative $x$ the first solution yields a contribution to $\phi_n(x, t)$ of

$$\phi_n^{(1)} = \frac{2A_n(a)}{(t\beta/2a^3)^{1/2}} \cos\left(ax + \frac{\beta t}{2a} - \frac{\pi}{4}\right), \tag{3.24.50}$$

while the contribution which must be added for $x > 0$ is

$$\phi_n^{(2)} = 2^{1/2} A_n\left(\left(\frac{\beta t}{x}\right)^{1/2}\right)\frac{(\beta t)^{1/4}}{x^{3/4}} \cos\left[2(\beta x t)^{1/2} + \frac{\pi}{4}\right]. \tag{3.24.51}$$

The size of each contribution naturally depends on the initial conditions, i.e., the shape of the spectrum of $A_n(k)$. If the spectrum is flat, so that $A_n$ is independent of $k$, corresponding to extreme localization of the initial disturbance in $x$, then eventually as $t \to \infty$ the dominant disturbance at any fixed $x$ will be given by (3.24.51). In the more realistic case where $A_n(k) \to 0$ as $k \to \infty$, the dominant term will be given by (3.24.50), in which case the disturbance left behind by the wave train will always be an oscillation with wavelength $2\pi/a$, the wavelength for which $C_g$ is zero.

The approximation clearly fails at those wave numbers for which

$$\frac{\partial^2 \sigma}{\partial k^2} = \frac{\partial C_{gx}}{\partial k} = \frac{2\beta k(3a^2 - k^2)}{(k^2 + a^2)^3} = 0, \qquad (3.24.52)$$

for then the denominator in (3.24.35) vanishes. These extrema in $C_g(k)$ are certainly of great interest, since they correspond to the disturbance field at the very front of the wave train, i.e., for the energy bundles with maximum group velocity. Since $\partial^2 \sigma/\partial k^2$ vanishes for this bundle, (3.24.43) shows that the extension of the bundle with time is small, so that it is reasonable to expect the amplitude to decrease more slowly than $t^{-1/2}$. That is why (3.24.35) fails. More precisely, in the neighborhood of the maximum in the group velocity (say at $k = k_m$), it is necessary to return to (3.24.26) and write the phase $\theta$ as

$$\theta = kx - \sigma(k)t$$

$$= k_m x - \sigma(k_m)t + [x - C_g(k_m)t](k - k_m) \qquad (3.24.53)$$

$$- \frac{\partial C_g}{\partial k}(k_m)\frac{(k - k_m)^2}{2}t - t\frac{\partial^2 C_g}{\partial k^2}\frac{(k - k_m)^3}{6} + \cdots.$$

Since $(\partial C_g/\partial k)(k_m)$ vanishes, the contribution to $\phi_n(x, t)$ from the wave-number band around $k_m$ is the real part of

$$I = \sqrt{\frac{2}{\pi}} A_n e^{i(k_m x - \sigma(k_m)t)}$$

$$\times \int_{k_m - \Delta}^{k_m + \Delta} dk \exp i\left[x - C_g t(k - k_m) - \frac{C_g''}{6}(k - k_m)^3 t\right], \qquad (3.24.54)$$

where for large $t$ the limits of the integral, as previously, can be extended to plus and minus infinity. At this point we need the integral relation

$$\int_0^\infty \cos(ak^3 + bk)\, dk = \frac{\pi}{(3a)^{1/3}} \text{Ai}((3a)^{-1/3}b). \qquad (3.24.55)$$

$\text{Ai}(x)$ is the tabulated Airy function which satisfies the differential equation

$$\frac{d^2 \text{Ai}}{dx^2} - x\,\text{Ai} = 0. \qquad (3.24.56)$$

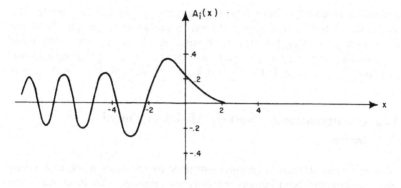

**Figure 3.24.4** The Airy function Ai($x$).

A sketch of Ai($x$) is given in Figure 3.24.4. In terms of Ai, the disturbance near the leading edge of the wave train is

$$\phi_n(x, t) = 2^{1/2}\pi^{1/2} \frac{A_n(k_m)}{|C''_g(k_m(t/2))|^{1/3}} \cos[k_m x - \sigma(k_m)t]$$

$$\times \text{Ai}\left(\frac{x - C_g(k_m)t}{|C''_g(k_m)t/2|^{1/3}}\right),$$

(3.24.57)

where

$$C''_g(k_m) = \frac{\partial^2 C_g}{\partial k^2}(k_m).$$

The Airy function decays exponentially for positive argument and oscillates for negative argument as shown in Figure 3.24.4, as indeed can also be deduced from the character of (3.24.56) for positive and negative $x$. Thus ahead of the point $x = C_g(k_m)t$ the disturbance is evanescent and *decays*, while behind this leading edge the disturbance is wavelike. The transition is therefore smooth rather than abrupt at the leading edge between disturbed

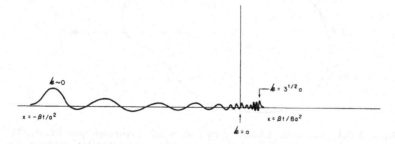

**Figure 3.24.5** A schematic diagram showing the asymptotic distribution of the disturbance for large $t$.

and undisturbed fluid. Since $\phi_n(x, t)$ decays as $t^{-1/3}$ on the leading edge, the amplitude at the leading edge will tend to dominate the amplitude of the wave train unless $A_n(k_m)$ is fortuitously small. A schematic sketch of the wave train which pieces together the qualitative picture we have achieved is shown in Figure 3.24.5. Note that for $x > 0$, $k_m = 3^{+1/2}a$, while for $x < 0$, $k_m = 0$.

## 3.25 Quasigeostrophic Normal Modes in Closed Basins

To this point the wave dynamics sustained by an ambient potential vorticity gradient has been considered only for regions open in at least one direction. This is a natural idealization when the scale of the disturbance is small compared to the dimensions of an enclosed domain and disturbances must therefore travel a long time before meeting a boundary. However, for closed basins like the oceans the scale of the forcing, say by the wind, is sufficiently large that the effect of the boundary is dominant. In such cases the response to a time-dependent forcing can best be discussed in terms of the normal modes of oscillation.

Let us imagine a closed basin on the $\beta$-plane as shown in Figure 3.25.1. Again, the governing equation for small-amplitude oscillations is the linearized form of the potential-vorticity equation, viz.,

$$\frac{\partial}{\partial t}\{\nabla^2\psi - F\psi\} + \beta\frac{\partial\psi}{\partial x} = 0. \tag{3.25.1}$$

The boundary conditions must be applied with care. Since the $O(1)$ velocity field $(u_0, v_0)$ is a vector tangent to the streamlines, the condition that the

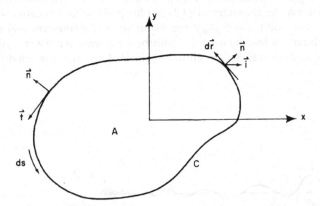

**Figure 3.25.1** The basin $A$ bounded by $c$ for which quasigeostrophic free oscillations are calculated. The vector **n** is normal to $c$, **t** is a tangent vector, and **i** is a unit vector in the $x$-direction.

normal component of velocity vanishes at the basin boundary is satisfied at
$O(1)$ in Rossby number by the specification that the boundary $C$ coincide
with a streamline, i.e., that $\psi$ is a constant on $C$, or

$$\psi = \Gamma_0 \quad \text{on } C \tag{3.25.2}$$

where $\Gamma_0$ may be a function of *time* only.

Although the addition of a spatial constant to $\psi$ does not affect the calculation of the velocities, the fact that $\psi$ is also the $O(1)$ free-surface deviation
$\eta_0$ implies that $\Gamma_0$ is a nontrivial constant related to the fluid depth along the
boundary and is not arbitrary. If $\psi$ is simply sinusoidal along the boundary
(as in the channel model of the preceding section), then the condition that $\psi$
is constant implies $\Gamma_0 = 0$. In the case at hand the motion need not be (and
indeed is not) simply periodic along the boundary, and $\Gamma_0$ must be
determined.

We proved in Section 3.5 that as a consequence of the vanishing of the
normal velocity on a closed boundary,

$$\frac{\partial}{\partial t} \oint_C \mathbf{u}_H \cdot d\mathbf{r} = 0 \quad \text{on } C. \tag{3.25.3}$$

This constraint holds to *all* orders in Rossby number. To *lowest* order in
Rossby number, since

$$\mathbf{u}_0 = \mathbf{i} u_0 + \mathbf{j} v_0 = \mathbf{k} \times \nabla \psi, \tag{3.25.4}$$

this constraint implies that

$$\frac{\partial}{\partial t} \oint (\mathbf{k} \times \nabla \psi) \cdot d\mathbf{r} = \frac{\partial}{\partial t} \oint (\mathbf{k} \times \nabla \psi) \cdot \mathbf{t} \, ds = 0, \tag{3.25.5}$$

where $\mathbf{t}$ is the unit vector tangent to $C$ and $ds$ is the scalar differential arc
length along $C$. Since

$$\mathbf{t} = \mathbf{k} \times \mathbf{n}, \tag{3.25.6}$$

(3.25.3) is simply

$$\frac{\partial}{\partial t} \oint_C (\nabla \psi \cdot \mathbf{n}) \, ds = 0. \tag{3.25.7}$$

A comparison of (3.25.3) and (3.25.5) with (3.12.19a,b) shows that the application of (3.25.3) to the lowest-order velocity field is equivalent to the condition that the $O(\varepsilon)$ correction to the velocity also vanishes on the boundary.
In general (3.25.7) is independent of the condition that $\psi$ is constant on the
boundary. Note, as remarked above, that if $\psi$ is periodic on $C$, (3.25.7) is
trivially satisfied. Thus when $\Gamma_0$ is a significant undetermined constant,
(3.25.7) is nontrivial and could be used to determine $\Gamma_0$. However, Section
3.5 contains another constraint, namely,

$$\frac{\partial}{\partial t} \iint_A dx \, dy \, \psi = 0, \tag{3.25.8}$$

which comes from (3.5.8) at lowest order in $\varepsilon$ and also is implied by the vanishing of the normal component of the $O(\varepsilon)$ velocity on $C$. Are (3.25.8), (3.25.7), and (3.25.2) independent? If so, we have too many conditions for the determination of $\Gamma_0$.

Consider the integral of (3.25.1) over the area $A$ of the basin. Use of the divergence theorem yields

$$\frac{\partial}{\partial t} \oint_C (\nabla \psi \cdot \mathbf{n}) \, ds = F \iint_A \frac{\partial \psi}{\partial t} \, dx \, dy - \beta \iint_A \frac{\partial \psi}{\partial x} \, dx \, dy. \qquad (3.25.9)$$

Now

$$\iint_A \frac{\partial \psi}{\partial x} \, dx \, dy = \int_{y_L}^{y_u} dy \, [\psi(x_R, y) - \psi(x_L, y)] \qquad (3.25.10)$$

as illustrated in Figure 3.25.2. The integrand vanishes if (3.25.2) is applied.

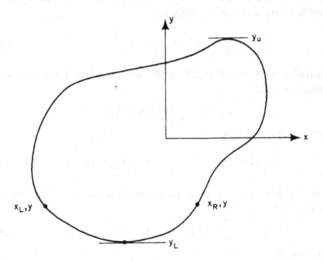

**Figure 3.25.2**  The points $x_R$ and $x_L$ define, for each $y$, the $x$ interval of the basin. The points $y_u$ and $y_L$ define the maximum interval in $y$ covered by the basin.

Hence if $\psi$ is constant on $C$, the satisfaction of (3.25.8) automatically satisfies (3.25.7) and vice versa. Thus either may be used to determine $\Gamma_0$.

Furthermore, (3.25.9) shows that in the limit $F \to 0$, which corresponds to a basin whose length $L$ is very much less than a deformation radius, (3.25.7) will automatically be satisfied for any constant $\Gamma_0$. It makes good physical sense in this limit, which we have already seen is equivalent to a rigid-lid approximation, that the determination of the degree of deformation of the upper surface is irrelevant to the dynamics. Let us first consider this limiting case for the sake of simplicity. The basic ideas are unaffected. Thus in this limit we may arbitrarily set $\Gamma_0$ equal to zero.

Oscillations of the form

$$\psi(x, y, t) = \text{Re } e^{-i\sigma t}\Phi(x, y) \qquad (3.25.11)$$

will be solution of (3.25.1) if, for $F = 0$,

$$\nabla^2\Phi + \frac{i\beta}{\sigma}\frac{\partial\Phi}{\partial x} = 0. \qquad (3.25.12)$$

It is useful to transform $\Phi$ so as to remove the first $x$-derivative from (3.25.12) by writing

$$\Phi = e^{-i\beta x/2\sigma}\phi(x, y). \qquad (3.25.13)$$

Substitution of (3.25.13) into (3.25.12) shows that $\phi$ must satisfy

$$\nabla^2\phi + \lambda^2\phi = 0, \qquad (3.25.14)$$

where $\lambda^2 = \beta^2/4\sigma^2$.

This equation is identical to the equation for the amplitude of a vibrating membrane, and the condition $\phi = \Phi = 0$ on $C$ implies that the solutions $\phi$ are spatially identical to the infinite class of clamped-membrane modes.

For example, consider the rectangular basin $0 \le x \le x_0$, $0 \le y \le y_0$. Then any of the set

$$\phi = \phi_{mn} = \sin\frac{m\pi x}{x_0}\sin\frac{n\pi y}{y_0}, \quad m = 1, 2, 3, \ldots, \qquad n = 1, 2, 3, \ldots, \quad (3.25.15a)$$

will satisfy the boundary conditions and (3.25.14), provided $\sigma$ equals the corresponding eigenvalue

$$\sigma = \sigma_{mn} = -\frac{\beta}{2\pi\{(m^2/x_0^2) + (n^2/y_0^2)\}^{1/2}}. \qquad (3.25.15b)$$

It is important to note that the lowest modes (i.e., those with the smallest $m$ and $n$) have the highest frequencies, while the smaller-scale modes (which do not sense much thickness variation over each separate sector of $\phi$) oscillate more slowly. The form of each mode is fascinating. The stream function $\psi$ is, aside from an arbitrary phase,

$$\psi = \cos\left[\frac{\beta x}{2\sigma_{mn}} + \sigma_{mn}t\right]\sin m\pi\frac{x}{x_0}\sin n\pi\frac{y}{y_0}. \qquad (3.25.16)$$

It consists of a carrier wave $\cos(\beta x/2\sigma_{mn} + \sigma_{mn}t)$, whose phase propagation is always to the left (westward), modulated by an envelope of sine functions which serve to satisfy the boundary conditions. Each mode has fixed and moving nodes separating cells of motion, each cell alternately decreasing and then increasing in size as the moving nodes of the carrier wave approach and then pass a fixed node. The speed of westward propagation of the phase is

$$C = -\frac{2\sigma_{mn}^2}{\beta} = -\frac{\beta}{2\pi^2\{(m^2/x_0^2) + (n^2/y_0^2)\}}. \qquad (3.25.17)$$

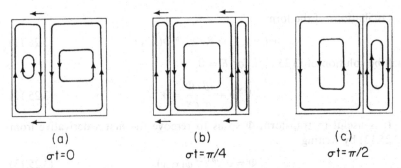

(a)                         (b)                         (c)
$\sigma t = 0$              $\sigma t = \pi/4$          $\sigma t = \pi/2$

**Figure 3.25.3**   The first normal mode in a rectangular basin. Note the propagation of phase toward negative $x$ in the pattern of the oscillation.

This remarkable behaviour is shown in Figure 3.25.3 for the lowest mode $m = n = 1$. Higher modes have more fixed internal nodes with a similar but more complex pattern of contracting and expanding cells of vortical motion.

Each normal mode can be written as the sum of four Rossby plane waves with wave vectors

$$\mathbf{K} = \mathbf{i}\left(\frac{\beta}{-2\sigma} \pm \frac{m\pi}{x_0}\right) + \mathbf{j}\left(\pm \frac{n\pi}{y_0}\right) = \mathbf{i}k + \mathbf{j}l. \qquad (3.25.18)$$

It is easily verified that for each wave vector the dispersion relation for plane Rossby waves,

$$\sigma = -\frac{\beta k}{k^2 + l^2}, \qquad (3.25.19)$$

is satisfied. Figure 3.25.4 shows the four plane waves on the frequency circle

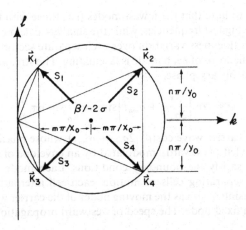

**Figure 3.25.4**   The representation of a Rossby normal mode in the energy propagation diagram as the sum of four plane waves whose energy fluxes add to zero.

(for $F = 0$), from which it is immediately apparent that at each point in the basin the net energy flux vector is identically zero, since the four flux vectors add to zero. Even though there is propagation of phase from east to west in the basin, there is no transmission of energy from one part of the basin to another in each mode.

The normal modes for basins of other shapes have the same character, i.e., a carrier wave propagating to the west modulated by a steady envelope whose nodal lines are fixed and determined by the membrane modes of (3.25.14). For the circular basin of unit radius, for example,

$$\psi_{mn} = \cos(m\theta + \alpha)J_m(k_{nm}r)\cos(\beta x/2\sigma_{mn} + \sigma_{mn}t), \qquad (3.25.20)$$

where $r$ and $\theta$ are the usual polar coordinates and $J_m$ is the Bessel function of order $m$;

$$\sigma_{mn} = -\frac{\beta}{2k_{mn}}; \qquad (3.25.21)$$

$k_{mn}$ is the $n$th zero of $J_m$, i.e.,

$$J_m(k_{nm}) = 0; \qquad (3.25.22)$$

and $\alpha$ is an arbitrary phase angle. The fixed nodes are either circles or radial lines where $\cos(m\theta + \alpha)$ vanishes. The pattern is further dissected by the moving nodes of the carrier wave oriented parallel to the $y$-axis.

The problem for the normal modes with nonzero $F$ is considerably more complex, and we will only briefly outline the analysis which leads to basin modes *qualitatively* similar to those found above. To a large extent the treatment here follows the work of Flierl (1977), who calculated the normal modes for a circular basin. As before, it is useful to consider solutions of the form

$$\psi = \mathrm{Re}\ e^{-i(\beta x/2\sigma + \sigma t)}\phi(x, y), \qquad (3.25.23)$$

where now $\phi$ satisfies

$$\nabla^2\phi + \left(\frac{\beta^2}{4\sigma^2} - F\right)\phi = 0. \qquad (3.25.24)$$

Solutions of (3.25.24) which are symmetric about the $x$-axis can be found in the general form

$$\phi = \sum_{m=0}^{\infty} \phi_m \cos m\theta\, J_m(\lambda r), \qquad (3.25.25)$$

where

$$\lambda^2 = \frac{\beta^2}{4\sigma^2} - F. \qquad (3.25.26)$$

The condition (3.25.2) for the normal modes is

$$\psi(r = 1) = \Gamma_0 = \gamma_0 \cos \sigma t, \qquad (3.25.27a)$$

or

$$\phi = \sum_{m=0}^{\infty} J_m(\lambda)\phi_m \cos m\theta = \gamma_0 e^{i(\beta/2\sigma)\cos \theta}. \qquad (3.25.27b)$$

The $\phi_m$ are determined from the usual Fourier integral, viz.,

$$J_m(\lambda)r_m \pi\phi_m = \gamma_0 \int_0^{2\pi} \cos m\theta \, e^{i(\beta/2\sigma)\cos \theta} \, d\theta, \qquad (3.25.28)$$

where

$$r_m = \begin{cases} 2 & \text{if } m = 0, \\ 1 & \text{if } m \geq 1. \end{cases} \qquad (3.25.29)$$

The standard integral representation for the Bessel function,

$$J_n(z) = \frac{e^{-in\pi/2}}{\pi} \int_0^{\pi} e^{iz \cos \theta} \cos n\theta \, d\theta, \qquad (3.25.30)$$

allows (3.25.28) to be written

$$\phi_m = \gamma_0 \frac{2}{r_m} i^m \frac{J_m(\beta/2\sigma)}{J_m(\lambda)}. \qquad (3.25.31)$$

The remaining condition, (3.25.7), becomes

$$0 = \int_0^{2\pi} \frac{\partial\psi}{\partial r} \, d\theta = \int_0^{2\pi} e^{-i(\beta/2\sigma)\cos \theta} \frac{\partial\phi}{\partial r}(r = 1) \, d\theta$$

$$- \frac{i\beta}{2\sigma} \int_0^{2\pi} \cos \theta \, e^{-i(\beta/2\sigma)\cos \theta} \phi(r = 1) \, d\theta$$

$$= \int_0^{2\pi} e^{-i(\beta/2\sigma)\cos \theta} \frac{\partial\phi}{\partial r}(r = 1) \, d\theta - \frac{i\beta}{2\sigma} \gamma_0 \int_0^{2\pi} \cos \theta \, d\theta \qquad (3.25.32)$$

$$= \int_0^{2\pi} e^{-i(\beta/2\sigma)\cos \theta} \frac{\partial\phi}{\partial r}(r = 1) \, d\theta$$

$$= \int_0^{2\pi} 2\gamma_0 \sum_{m=0}^{\infty} \lambda \frac{i^m}{r_m} \frac{J_m(\beta/2\sigma)}{J_m(\lambda)} J_m'(\lambda)\cos m\theta \, e^{-i(\beta/2\sigma)\cos \theta} \, d\theta.$$

Application of (3.25.30) to (3.25.32) yields the final eigenvalue relation for $\sigma$,

$$\sum_{m=0}^{\infty} \frac{J_m^2(\beta/2\sigma)}{r_m} \frac{J_m'(\lambda)}{J_m(\lambda)} = 0, \qquad (3.25.33)$$

where

$$J_m'(\lambda) = \frac{dJ_m(\lambda)}{d\lambda}. \qquad (3.25.34)$$

Note that if we attempt a solution with $\gamma_0 = 0$, all $\phi_m$ will be zero by

(3.25.31), except for that $m = M$ for which $J_M(\lambda)$ vanishes. In that case (3.25.33) contains the single term $m = M$ and is singular unless $\lambda = \beta/2\sigma$, which is the condition $F = 0$. Thus (3.25.33) will reduce to (3.25.22) when $F = 0$. For $F \neq 0$ the frequencies depart from (3.25.21). However, detailed calculations by Flierl (1977) show that the frequency shift is always less than 5% and the mode structure is similar to (3.25.20), with the important difference that the surface of the fluid at the boundary is alternately raised and lowered uniformly around the basin. Table 3.25.1 shows the frequency shift at $F \neq 0$ due to the application of (3.25.27) and (3.25.32) rather than the incorrect condition $\phi = 0$ for the gravest symmetric mode.

**Table 3.25.1**

| F | $\beta/2\sigma$ | |
|---|---|---|
| | (a) | (b) |
| 0.2 | 2.435 | 2.446 |
| 0.5 | 2.481 | 2.507 |
| 1 | 2.556 | 2.604 |
| 2 | 2.706 | 2.790 |
| 5 | 3.148 | 3.284 |
| 10 | 3.852 | 3.973 |
| 20 | 5.066 | 5.078 |
| 30 | 5.976 | 5.982 |
| 50 | 7.455 | 7.469 |

(a) Calculations using the correct boundary condition (3.25.27) for the lowest symmetric mode.
(b) Calculations for the frequency *incorrectly* applying $\phi = 0$ to the lowest *symmetric* mode. (Courtesy G. R. Flierl.)

Solutions of (3.25.24) which are *antisymmetric* about $y = 0$ can be found in the form

$$\phi = \sum_{m=0}^{\infty} \phi_m \sin m\theta \, J_m(\lambda r). \tag{3.25.35}$$

The analysis proceeds as before, with $\cos m\theta$ replaced by $\sin m\theta$ throughout. It follows however that (3.25.28) is now

$$J_m(\lambda) r_m \pi \phi_m = \gamma_0 \int_0^{2\pi} \sin m\theta \, e^{i(\beta/2\sigma)\cos\theta} \, d\theta \equiv 0, \tag{3.25.36}$$

so that *all* $\phi_m$, $m \neq M$, can be chosen to be zero for arbitrary $\gamma_0$ if $\lambda$ satisfies

$$J_M(\lambda) = 0. \tag{3.25.37a}$$

Thus

$$\lambda = k_{nM},$$

so that for the modes antisymmetric about $y = 0$ the modal shapes are precisely the *same* as in the case $F = 0$, while the frequency is

$$\sigma_{nm} = -\frac{\beta}{2(k_{nm}^2 + F)^{1/2}}, \tag{3.25.37b}$$

and is merely reduced by the deformation of the free surface. It is left to the reader to verify that these antisymmetric modes, of the form

$$\psi = \cos(\beta x/2\sigma + \sigma t)\sin m\theta \, J_m(k_{nm}r), \tag{3.25.38}$$

satisfy (3.25.32) *trivially*. The antisymmetric modes have zero amplitude at the coast. Because of the spatial anisotropy of the dynamics due to $\beta$, it is only the modes *symmetric* about $y = 0$ that have nonzero amplitudes at $r = 1$ and for which the constraint (3.25.7) is nontrivial.

The orthogonality condition for these normal modes, for arbitrary $F$ and arbitrary basin shape, follows directly from the equation for $\Phi = e^{i\sigma t}\psi$. Consider two modes $\Phi_{Kl}$ and $\Phi_{mn}$ corresponding to the frequencies $\sigma_{Kl}$ and $\sigma_{mn}$. Then

$$\sigma_{Kl}[\nabla^2\Phi_{Kl} - F\Phi_{Kl}] + i\beta\frac{\partial\Phi_{Kl}}{\partial x} = 0, \tag{3.25.39a}$$

while

$$\sigma_{mn}^*[\nabla^2\Phi_{mn}^* - F\Phi_{mn}^*] - i\beta\frac{\partial\Phi_{mn}^*}{\partial x} = 0, \tag{3.25.39b}$$

where * denotes the complex conjugate. Multiplying the first equation by $\Phi_{mn}^*$ and the second by $\Phi_{Kl}$ and then integrating the difference over the area of the basin yields

$$(\sigma_{Kl} - \sigma_{mn}^*)\iint_A dx\, dy\, [\nabla\Phi_{Kl}\cdot\nabla\Phi_{mn}^* + F\Phi_{Kl}\Phi_{mn}^*]$$

$$= \sigma_{Kl}\oint_C \Phi_{mn}^*\nabla\Phi_{Kl}\cdot\mathbf{n}\, ds - \sigma_{mn}^*\oint_C \Phi_{Kl}\nabla\Phi_{mn}^*\cdot\mathbf{n}\, ds \tag{3.25.40}$$

$$+ i\beta\oint_C \mathbf{n}\cdot\mathbf{i}\Phi_{Kl}\Phi_{mn}^*\, ds.$$

Since $\Phi_{Kl}$ and $\Phi_{mn}$ are constant on $C$, (3.25.7) when applied to the right-hand side of (3.25.40) shows that the orthogonality condition

$$(\sigma_{Kl} - \sigma_{mn}^*)\iint_A [\nabla\Phi_{Kl}\cdot\nabla\Phi_{mn}^* + F\Phi_{Kl}\Phi_{mn}^*]\, dx\, dy = 0 \tag{3.25.41}$$

must hold. If we choose $\Phi_{mn}^*$ to be equal to $\Phi_{Kl}^*$—i.e., if (3.25.41) is applied to a single mode—it follows that

$$\sigma_{Kl} = \sigma_{Kl}^*, \tag{3.25.42}$$

so that the normal-mode frequencies must be real. This must of course be true, for otherwise the exponential factor $e^{-i\sigma t}$ would yield exponential decay or growth in the absence of an energy source. Indeed, (3.25.41) applied to a single mode is just the statement of energy conservation. For two modes of differing frequencies the integral itself must vanish.

Consider now an arbitrary intial disturbance,

$$\psi(x, y, t) = \Psi(x, y) \quad \text{at } t = 0. \tag{3.25.43}$$

At all future times $\psi$ can be represented as a linear combination of normal modes, viz.

$$\psi(x, y, t) = \sum_{m, n} A_{mn} \Phi_{mn}(x, y) e^{-i\sigma_{mn} t}, \tag{3.25.44}$$

where the real part of the right-hand side is implied. At $t = 0$

$$\psi(x, y, 0) = \Psi(x, y) = \sum_{m, n} A_{mn} \Phi_{mn}(x, y). \tag{3.25.45}$$

It therefore follows that

$$\iint_A dx \, dy \, [\nabla \Phi_{mn}^* \cdot \nabla \Psi + F \Phi_{mn}^* \Psi] = A_{mn} C_{mn}, \tag{3.25.46}$$

where $C_{mn}$ is the mode normalization constant,

$$C_{mn} = \iint_A dx \, dy \, [\,|\nabla \Phi_{mn}|^2 + F \,|\Phi_{mn}|^2]. \tag{3.25.47}$$

Thus the $A_{mn}$'s are completely determined by the projection of the initial conditions on each of the basin normal modes, and the subsequent motion is the simple sum of all of the excited normal modes.

## 3.26 Resonant Interactions

Although a single Rossby plane wave of arbitrary amplitude is an exact solution of the quasigeostrophic potential vorticity equation, a superposition of waves will not be. The nonlinear interaction between the waves, by which the velocity field of one wave advects the vorticity of another, leads to a nonlinear coupling and energy transfer between the waves. This nonlinearity is described by the Jacobian term in (3.15.1). In the archetypal example of Rossby waves on the $\beta$-plane, (3.15.1) is

$$\frac{\partial}{\partial t} (\nabla^2 \psi - F\psi) + \frac{\partial \psi}{\partial x} \frac{\partial \nabla^2 \psi}{\partial y} - \frac{\partial \psi}{\partial y} \frac{\partial \nabla^2 \psi}{\partial x} + \beta \frac{\partial \psi}{\partial x} = 0. \tag{3.26.1}$$

The purpose of this section is to examine the nature of the nonlinear interactions in the case $\beta \gg 1$, i.e., when to *lowest* order a superposition of Rossby waves is possible. When $\beta \gg 1$ the characteristic period of the Rossby wave in dimensional units, $(\beta_0 L)^{-1}$, is much less than the advective time $L/U$, where $U$ is the characteristic velocity of fluid elements in the wave. It is useful at this point to recognize this fact explicitly and rescale the dimensional time $t_*$ in terms of a new dimensionless time $\tilde{t}$:

$$t_* = (\beta_0 L)^{-1}\tilde{t} = \frac{L}{U}t, \tag{3.26.2}$$

or

$$\tilde{t} = \frac{\beta_0 L^2}{U}t = \beta t. \tag{3.26.3}$$

With this redefinition (3.26.1) becomes

$$\frac{\partial}{\partial \tilde{t}}[\nabla^2\psi - F\psi] + \frac{\partial \psi}{\partial x} = \frac{1}{\beta}\left[\frac{\partial \psi}{\partial y}\frac{\partial}{\partial x}\nabla^2\psi - \frac{\partial \psi}{\partial x}\frac{\partial}{\partial y}\nabla^2\psi\right]. \tag{3.26.4}$$

The nonlinear terms in (3.26.4) are now explicitly $O(\beta^{-1})$, which is the ratio of the Rossby period to the advective time, and also the ratio of the relative vorticity gradient $U/L^2$ to the planetary vorticity gradient $\beta_0$. When this ratio is small, a linear-superposition Rossby wave is an apt first description of the motion, but as long as $\beta^{-1}$ is finite, nonlinear interactions among the waves will occur.

For $\beta \gg 1$ it seems logical to look for solutions in the form

$$\psi(x, y, \tilde{t}, \beta) = \psi_0(x, y, \tilde{t}) + \frac{1}{\beta}\psi_1(x, y, \tilde{t})$$

$$+ \frac{1}{\beta^2}\psi_2(x, y, \tilde{t}) + \cdots \tag{3.26.5}$$

whose substitution in (3.26.4) will yield a sequence of problems for $\psi_0$, $\psi_1$, etc. by comparing like orders in $\beta^{-1}$. The lowest order, or $O(1)$, problem for $\psi_0$ is

$$\frac{\partial}{\partial \tilde{t}}(\nabla^2\psi_0 - F\psi_0) + \frac{\partial \psi_0}{\partial x} = 0, \tag{3.26.6}$$

whose solution can be written as a general superposition of Rossby waves, e.g.,

$$\psi_0 = \sum_j a_j \cos \theta_j, \tag{3.26.7}$$

where the range of the sum over the integral index $j$ is formally infinite, but where in fact only a finite number of the amplitudes $a_j$ may be different from zero. The phase of each wave is

$$\theta_j = k_j x + l_j y - \sigma_j \tilde{t} + \phi_j, \tag{3.26.8}$$

where $\phi_j$ is an arbitrary phase angle, and where $\sigma_j$, the frequency of the $j$th wave with wave number $\mathbf{K}_j = \mathbf{i}k_j + \mathbf{j}l_j$, is given by the linear Rossby dispersion relation

$$\sigma_j = -\frac{k_j}{k_j^2 + l_j^2 + F}. \tag{3.26.9}$$

These are the results of linear theory; the nonlinear interactions become evident only in the problem for $\psi_1$, which is

$$\frac{\partial}{\partial t}(\nabla^2 \psi_1 - F\psi_1) + \frac{\partial \psi_1}{\partial x} = \frac{\partial \psi_0}{\partial y}\frac{\partial \nabla^2 \psi_0}{\partial x} - \frac{\partial \psi_0}{\partial x}\frac{\partial}{\partial y}\nabla^2 \psi_0 \tag{3.26.10}$$

$$= \sum_m \sum_n a_m a_n K_m^2(k_n l_m - k_m l_n)\sin \theta_m \sin \theta_n, \tag{3.26.11a}$$

which by the symmetry of the sum in $m$, $n$, is,

$$\frac{\partial}{\partial t}(\nabla^2 \psi_1 - F\psi_1) + \frac{\partial \psi_1}{\partial x}$$

$$= \sum_m \sum_n \frac{a_m a_n}{2}(K_m^2 - K_n^2)(k_n l_m - k_m l_n)\sin \theta_m \sin \theta_n$$

$$= \sum_m \sum_n \frac{a_m a_n}{2} B(K_m, K_n)[\cos(\theta_m + \theta_n) - \cos(\theta_m - \theta_n)], \tag{3.26.11b}$$

where

$$B(K_m, K_n) = \tfrac{1}{2}(K_m^2 - K_n^2)\mathbf{Z} \cdot (\mathbf{K}_m \times \mathbf{K}_n) \tag{3.26.12}$$

and $\mathbf{Z}$ is the unit vertical vector. The interaction of the $m$th and $n$th wave produces a forcing term in the problem for $\psi_1$ which oscillates with the sum and difference of their two phases, i.e., a forcing term with wave vector

$$\mathbf{K}_{mn} \equiv \mathbf{K}_m \pm \mathbf{K}_n \tag{3.26.13a}$$

and frequency

$$\omega_{mn} \equiv \sigma_m \pm \sigma_n. \tag{3.26.13b}$$

Note that two waves of either the same wavelength $(K_m^2 = K_n^2)$ or parallel wave vectors $(\mathbf{K}_m \times \mathbf{K}_n = 0)$ will not interact, for their interaction coefficient $B(K_m, K_n)$ will then be zero. The first case corresponds to a situation where the vorticity of the sum of the pair is a constant multiple of the sum of their stream functions, so that the Jacobian of the two wave fields identically vanishes, while the second case corresponds to parallel motion in a single direction, for which the nonlinear advection term in (3.26.4) is also trivially zero.

The problem for $\psi_1$ is a linear, forced problem, and therefore the response to each forcing term can be considered separately and the results summed. Note that due to the double sum over $m$ and $n$, the contribution of the pair of wave vectors $m$ and $n$ contributes twice to the sum in (3.26.11b). Consider the solution forced by the term $a_m a_n B(K_m, K_n)\cos(\theta_m + \theta_n)$. A periodic forced solution for $\psi_1$ may be sought in the form

$$\psi_1 = A_{1mn} \sin(\theta_m + \theta_n) \tag{3.26.14}$$

where $A_{1mn}$ is determined by (3.26.11b) to be

$$A_{1mn} = \frac{a_m a_n B(K_m, K_n)}{(K_{mn}^2 + F)(\omega_{mn} - \sigma_{mn})}. \tag{3.26.15}$$

Here $\sigma_{mn}$ is the Rossby wave frequency corresponding to a wave with the wave vector $\mathbf{K}_{mn}$, i.e.,

$$\sigma_{mn} = -\frac{k_{mn}}{k_{mn}^2 + l_{mn}^2 + F} = -\frac{(k_m + k_n)}{(k_m + k_n)^2 + (l_m + l_n)^2 + F}. \tag{3.26.16}$$

Clearly the solution given by (3.26.14) and (3.26.15) is valid only if the forcing frequency $\omega_{mn}$ is not equal to the natural frequency of oscillation, $\sigma_{mn}$, of a free Rossby wave with the wave number $K_{mn}$ of the forcing. Otherwise a resonance occurs, that is, two waves then combine to force a third wave with a wave number and frequency appropriate to a free, linear oscillation. Such interactions, for which (3.26.15) predicts an infinite response, are called *resonant interactions*. They are clearly of great interest, for all nonresonant interactions will merely produce, for large $\beta$, a small-amplitude background jangle of forced waves whose amplitudes are small compared to those waves produced by the resonant interactions. When resonance occurs, the solution (3.26.14) must be replaced, as is the usual case for resonant forcing, by the solution

$$\psi_1 = \tilde{t} a_{1mn} \cos(\theta_m + \theta_n), \tag{3.26.17}$$

which leads to $\psi$ of the form

$$\psi = \psi_0 + (\beta^{-1}\tilde{t})a_{1mn} \cos(\theta_m + \theta_n)$$
$$= \psi_0 + t a_{1mn} \cos(\theta_m + \theta_n), \tag{3.26.18}$$

i.e., a slow growth of the $O(\beta^{-1})$ correction on the nonlinear advective time $t$, so that, after a time $t = O(1)$ or $\tilde{t} = O(\beta)$, our initial series (3.26.5) is clearly invalid, for then the second term is as large as the first. We must then answer the following two important questions: are resonant interactions in fact possible, and if so, how do the wave amplitudes evolve over a time $\tilde{t} = O(\beta)$? The answer to the first question is not immediately obvious.

In order for resonance to occur, three Rossby plane waves must clearly satisfy

$$-\theta_j = \theta_m + \theta_n, \tag{3.26.19}$$

where the minus sign (irrelevant, as (3.26.18) shows) is introduced for symmetry. This condition implies that two waves, the $m$th and the $n$th, combine to produce a phase, $\theta_m + \theta_n$, which is equal to the phase angle of a third free mode $\theta_j$. The forcing is then spatially and temporally always in phase with a natural mode of oscillation and resonance, i.e., strong energy interchange between the waves occurs. For (3.26.9) to apply for all $x$, $y$, and $t$, the three conditions required for

$$\theta_j + \theta_m + \theta_n = 0 \tag{3.26.20}$$

are

$$k_j + k_m + k_n = 0, \tag{3.26.21a}$$

$$l_j + l_m + l_n = 0, \tag{3.26.21b}$$

$$\sigma_j(k_j, l_j) + \sigma_m(k_m, l_m) + \sigma_n(k_n, l_n) = 0. \tag{3.26.21c}$$

Resonant forcing at the difference phase $\theta_m - \theta_n$, a possibility according to (3.26.11b), is included in (3.26.21) if we allow the $k$'s, $l$'s, and $\sigma$'s to take on both positive and negative values. The first two conditions can be illustrated graphically, as shown in Figure 3.26.1, i.e., the three wave vectors must add

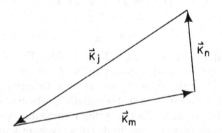

**Figure 3.26.1**   A triad of interacting plane waves.

to zero to form a *resonant triad* in which any two of the waves will, by (3.26.18), force the third resonantly. The difficulty of finding three wave vectors which not only form the geometrical triad of figure (3.26.1) but will *also* satisfy (3.26.21c)—i.e.,

$$\frac{k_m}{k_m^2 + l_m^2 + F} + \frac{k_n}{k_n^2 + l_n^2 + F} + \frac{k_j}{k_j^2 + l_j^2 + F} = 0, \tag{3.26.22}$$

—means that such vectors, if found, form a very restricted set of all possible interactions distinguished by their efficiency of mutual interaction. The analytical search for wave triads can be pictured as follows. Fix in mind an arbitrary single wave vector, say $\mathbf{K}_j$. Then $\mathbf{K}_m$ is found in terms of $\mathbf{K}_j$ and $\mathbf{K}_n$ from (3.26.21a,b). Inserting that result in (3.26.21c) yields a single equation for the wave numbers $(k_n, l_n)$ in terms of $(k_j, l_j)$. This then determines the locus in the $k$, $l$ plane of wave vectors, if any, that can resonantly interact with $\mathbf{K}_j$. Longuet-Higgins and Gill (1967) have investigated this very complicated algebraic problem in great detail and have shown that all wave vectors can participate in a resonant triad with a family of wave vectors. As a concrete but simple example, consider the case where $\mathbf{K}_j$ is directed eastward, i.e., the case of the elemental Rossby wave for which $\mathbf{K}_j = ik_j$, so that $l_j = 0$. Further, choose the scaling length $L$, otherwise free, to be the deformation radius, so that $F = 1$. Finally, for simplicity we study the typical case $k_j = 1$, i.e., a wavelength equal to $2\pi$ times the Rossby deformation radius. Then in this case (3.26.21a,b) becomes

$$l_m = -l_n \equiv l, \qquad (3.26.23)$$

while (3.26.22) is

$$\frac{1}{2} + \frac{k}{k^2 + l^2 + 1} = \frac{k + 1}{(k + 1)^2 + l^2 + 1}. \qquad (3.26.24)$$

This yields an equation for $l$ in terms of $k$, i.e.,

$$l^4 + l^2[k^2 + (k + 1)^2] + k^2(k + 1)^2 + 4k(k + 1) = 0, \qquad (3.26.25)$$

whose solution is

$$l = \pm\{(1 - 3\kappa^2)^{1/2} - (\kappa^2 + \tfrac{1}{4})\}^{1/2}, \qquad (3.26.26)$$

where

$$\kappa = k + \tfrac{1}{2}. \qquad (3.26.27)$$

The locus of solutions $l(k)$ is shown in Figure 3.26.2. The two wave vectors $\mathbf{K}_n$ and $\mathbf{K}_m$ lie on the locus at opposite ends of the line passing through $k = -\tfrac{1}{2}$ and intersecting the locus. Now that we have verified that resonant triads can be found, we must examine the equations which govern the amplitude of the triad. The result, (3.26.18), of the rudimentary perturbation expansion (3.26.5) shows that the amplitudes of a resonantly interacting triad can be expected to grow on the advective time scale. That is, we ought to allow the resonantly interacting waves to be functions of the "slow" variable $t$ as well as the fast variable $\tilde{t}$. More generally, $\psi$ must be written in terms of the multiple time variables $\tilde{t}$ and $t$, i.e.,

$$\psi = \psi_0(x, y, \tilde{t}, t) + \frac{1}{\beta}\psi(x, y, \tilde{t}, t) + \cdots, \qquad (3.26.28)$$

where now the time derivative $\partial/\partial\tilde{t}$ in (3.26.4) must be replaced by

$$\frac{\partial}{\partial\tilde{t}} \rightarrow \frac{\partial}{\partial\tilde{t}} + \frac{1}{\beta}\frac{\partial}{\partial t} \qquad (3.26.29)$$

in accordance with the ideas described in Section 3.20. The equation for $\psi_0$ is identical to (3.26.6), and we consider the solution corresponding to a single triad

$$\psi_0 = a_1 \cos \theta_1 + a_2 \cos \theta_2 + a_3 \cos \theta_3, \qquad (3.26.30)$$

where the conditions (3.26.21a,b,c) are supposed to be satisfied, viz.

$$k_1 + k_2 + k_3 = 0,$$
$$l_1 + l_2 + l_3 = 0, \qquad (3.26.31)$$
$$\sigma_1(k_1, l_1) + \sigma_2(k_2, l_2) + \sigma_3(k_3, l_3) = 0,$$

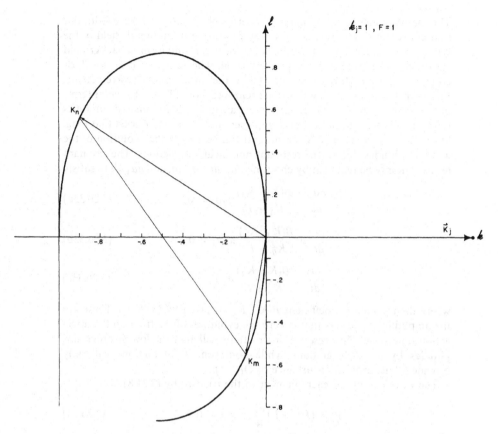

**Figure 3.26.2** The locus of vectors $\mathbf{K}_n$ and $\mathbf{K}_m$ which can resonantly interact with $\mathbf{K}_j$. $\mathbf{K}_n$ and $\mathbf{K}_m$ must, as shown, touch the ellipse-like curve, and their tips must be on either end of the line through $(k, l) = (-0.50, 0)$ and which intersects the locus.

and where each frequency satisfies the linear dispersion relation. The problem for $\psi_1$ is now

$$\frac{\partial}{\partial t}(\nabla^2\psi_1 - F\psi_1) + \frac{\partial\psi_1}{\partial x}$$

$$= + (K_1^2 + F)\frac{da_1}{dt}\cos\theta_1 + a_2a_3B(K_2, K_3)[\cos(\theta_2 + \theta_3) - \cos(\theta_2 - \theta_3)]$$

$$+ (K_2^2 + F)\frac{da_2}{dt}\cos\theta_2 + a_3a_1B(K_3, K_1)[\cos(\theta_3 + \theta_1) - \cos(\theta_3 - \theta_1)]$$

$$+ (K_3^2 + F)\frac{da_3}{dt}\cos\theta_3 + a_1a_2B(K_1, K_2)[\cos(\theta_1 + \theta_2) - \cos(\theta_1 - \theta_2)].$$

$$(3.26.32)$$

The forcing consists of two parts. The terms like $\cos(\theta_2 - \theta_3)$ are (with our sign convention) *nonresonant* and only produce a harmonic $\theta_1$ field of the type given by (3.26.15). This response is a weak one and forms a background wave field in addition to the primary triad. The terms such as $\cos \theta_j$ or $\cos(\theta_m + \theta_n)$ are each resonant for each of the three separate wave vectors. Unless the resonant forcing terms are removed from (3.26.32), secular terms of the form (3.26.18) will render the expansion (3.26.28) invalid for times $t = O(1)$. We wish to preserve our representation correct, at least that long, for as we have seen, that is the natural time scale for the evolution of the wave amplitudes due to the resonant nonlinear interactions. The resonant terms can only be removed by choosing $da_1/dt$, $da_2/dt$, and $da_3/dt$ to satisfy

$$\frac{da_1}{dt} + \frac{B(K_2, K_3)}{K_1^2 + F} a_2 a_3 = 0, \qquad (3.26.33a)$$

$$\frac{da_2}{dt} + \frac{B(K_3, K_1)}{K_2^2 + F} a_3 a_1 = 0, \qquad (3.26.33b)$$

$$\frac{da_3}{dt} + \frac{B(K_1, K_2)}{K_3^2 + F} a_1 a_2 = 0. \qquad (3.26.33c)$$

where the interaction coefficient $B(K_j, K_m)$ is given by (3.26.12). These are the amplitude equations that govern the evolution of the triad on the interaction time scale. Thus *resonant* interactions will alter the lowest-order amplitudes by an $O(1)$ amount, while nonresonant interactions will only provide a background disturbance of $O(\beta^{-1})$.

The wave energy for each member of the triad is, by (3.21.8),

$$E_j = (K_j^2 + F)\frac{a_j^2}{4}, \qquad j = 1, 2, 3. \qquad (3.26.34)$$

If (3.26.33a,b,c) are multiplied by $a_1$, $a_2$, and $a_3$ respectively, it follows that

$$\frac{d}{dt}(E_1 + E_2 + E_3) = -a_1 a_2 a_3 (B(K_2, K_3) + B(K_3, K_1) + B(K_1, K_2))$$

$$= 0 \qquad (3.26.35)$$

if (3.26.12) and (3.26.13) are used. Thus, the energy of the triad is preserved. The waves in this approximation only exchange energy among themselves on the advective time scale. This is a truly remarkable result, for (as we have seen) in addition to providing a resonant energy exchange among triad members, (3.26.32) shows that some energy is also passed to nonresonant waves. However, this transfer is so feeble and slow compared to the resonant exchange, that to the lowest order the resonant interaction is energy preserving.

An additional constraint can be derived from (3.26.33). The potential vorticity of each wave is

$$\Pi_j = -(K_j^2 + F)a_j \cos \theta_j. \qquad (3.26.36)$$

The average of half the square of the potential vorticity is called the *potential enstrophy* and is

$$V_j = (K_j^2 + F)^2 \frac{a_j^2}{2} = (K_j^2 + F)E_j. \tag{3.26.37}$$

and it easily can be verified that

$$\frac{\partial}{\partial t}(V_1 + V_2 + V_3) = 0, \tag{3.26.38}$$

so that the total wave enstrophy is also preserved. These two conservation statements have extraordinary implications if the relation between $E_j$ and $V_j$ given by (3.26.37) is exploited, for then not only must

$$\sum_{j=1}^{3} E_j = \text{constant} = E_0, \tag{3.26.39a}$$

but also

$$\sum_{j=1}^{3} K_j^2 E_j = \text{constant} = K_0^2 E_0. \tag{3.26.39b}$$

Hence, as the amplitudes change they must do so in a way that preserves the radius of gyration of the three waves, i.e.,

$$K_0^2 = \frac{\sum K_j^2 E_j(t)}{\sum E_j(t)} \tag{3.26.40}$$

is a constant of the motion.

Thus it is impossible, for example, for one wave to lose energy to two other waves, *both* of which have larger wave numbers. If some energy is transmitted to larger wave numbers by nonlinear effects, energy must also be transmitted to a *lower* wave number (larger wave length) in order to preserve both energy and enstrophy. For the triad of Figure 3.26.2, for example, $K_3^2 = 0.31$, $K_1^2 = 1$, and $K_2^2 = 1.110$.

The two conservation statements can also be written

$$\frac{1}{K_2^2 - K_3^2} \frac{\partial E_1}{\partial t} = \frac{1}{K_3^2 - K_1^2} \frac{\partial E_2}{\partial t} = \frac{1}{K_1^2 - K_2^2} \frac{\partial E_3}{\partial t}. \tag{3.26.41}$$

Suppose $K_3^2 < K_1^2 < K_2^2$. Then if $\partial E_1/\partial t < 0$, *both* $\partial E_2/\partial t$ and $\partial E_3/\partial t$ must be positive. The direction of energy flow is shown schematically in Figure 3.26.3.

Consider the case where initially $a_2$ is much greater than $a_1$ or $a_3$. We can imagine a Rossby wave of initially constant amplitude $a_2$, which is then perturbed by a small perturbation consisting of the other two waves in the triad. Let, then,

$$a_2 = A_2 + \alpha_2(t), \qquad \alpha_2 \ll A_2,$$

$$a_1 = \phantom{A_2 +} \alpha_1(t), \qquad \alpha_1 \ll A_2, \tag{3.26.42}$$

$$a_3 = \phantom{A_2 +} \alpha_3(t), \qquad \alpha_3 \ll A_2.$$

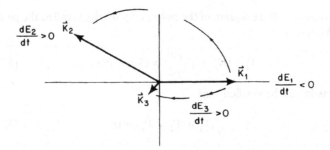

**Figure 3.26.3** A schematic diagram showing the energy transfer from a wave with wave vector $\mathbf{K}_1$ to a larger wavelength $(\mathbf{K}_3)$ and shorter $(\mathbf{K}_2)$ wavelength wave. The direction of the energy transfer is reversible.

Then the linearized forms of (3.26.33), to the lowest order in the $\alpha$'s, are

$$\frac{d\alpha_1}{dt} = -\frac{B(K_2, K_3)}{(K_2^2 + F)} A_2 \alpha_3,$$

$$\frac{d\alpha_3}{dt} = -\frac{B(K_1, K_2)}{(K_3^2 + F)} A_2 \alpha_1, \qquad (3.26.43)$$

$$\frac{d\alpha_2}{dt} = O(\alpha_1 \alpha_3),$$

so that

$$\frac{d^2\alpha_1}{dt^2} = \frac{B(K_2, K_3)B(K_1, K_2)}{(K_3^2 + F)(K_2^2 + F)} A_2^2 \alpha_1. \qquad (3.26.44)$$

Since

$$B(K_2, K_3)B(K_1, K_2)$$
$$= \tfrac{1}{4}(K_2^2 - K_3^2)(K_1^2 - K_2^2)\mathbf{Z} \cdot (\mathbf{K}_2 \times \mathbf{K}_3)\mathbf{Z} \cdot (\mathbf{K}_1 \times \mathbf{K}_2), \quad (3.26.45a)$$

and since

$$\mathbf{Z} \cdot (\mathbf{K}_2 \times \mathbf{K}_3) = -\mathbf{Z} \cdot (\mathbf{K}_2 \times \mathbf{K}_1) \qquad (3.26.45b)$$

by the resonance condition, the coefficient of $\alpha_1$ on the right-hand side of (3.26.44) will be positive if and only if

$$(K_1^2 - K_2^2)(K_2^2 - K_3^2) > 0. \qquad (3.26.46)$$

If this condition is met, both $\alpha_1$ and $\alpha_3$ will grow exponentially at a rate proportional to $A_2$, extracting energy from the original Rossby wave. The condition (3.26.46) again requires that the wave donating energy to the other two members of the triad be of a scale *intermediate* between the other two. The exponential growth predicted by (3.26.44) will continue until $\alpha_1$ and $\alpha_3$ become large enough for the linearization to fail. After all, we know the total energy of the triad must be preserved. The behavior of $a_1$, $a_2$, and $a_3$ for longer times requires the integration of (3.26.33). The constraints on the energy and enstrophy allow any two amplitudes to be eliminated in terms of the third, i.e.,

$$(K_1^2 + F)\frac{a_2^2}{4} + (K_3^2 + F)\frac{a_3^2}{4} = E_0 - (K_1^2 + F)\frac{a_1^2}{4},$$

$$(K_2^2 + F)^2\frac{a_2^2}{4} + (K_3^2 + F)^2\frac{a_3^2}{4} = V_0 - (K_1^2 + F)^2\frac{a_1^2}{4},$$

(3.26.47)

where $E_0$ and $V_0$ are determined by the initial conditions on the wave amplitudes. Solving for $a_2$ and $a_3$ in terms of $a_1$ then yields an equation for $a_1$ of the form

$$\frac{da_1}{dt} = C_3[(C_2 - a_1^2)(C_3 - a_1^2)]^{1/2},$$

(3.26.48)

where $C_1$, $C_2$, and $C_3$ are constants which depend on the initial conditions. Periodic solutions may be found in terms of elliptic functions, and the amplitude behavior is shown schematically in Figure 3.26.4. The details do not

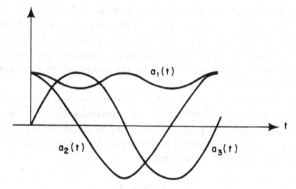

**Figure 3.26.4** A typical oscillation of the wave amplitudes as a consequence of their mutually resonant interaction.

concern us here; they have been described in the paper of Longuet-Higgins and Gill (1967) cited above. The important feature for our purpose is the simple fact that the energy flow among the members of the triad pulsates with time and that the direction of energy flow, shown in Figure 3.26.3, is reversible. The energy flows first, say, to waves 1 and 3, and then reverses direction when the initial instability of wave 2 is halted by nonlinear effects. The pulsation is perpetual, each member of the triad first receiving and then returning energy to the others.

Thus although a plane Rossby wave, alone, is an exact solution, it is clearly an unstable solution. Disturbances which complete a resonant triad will always drain energy from the Rossby wave slowly for large $\beta$, and more rapidly for larger-amplitude Rossby waves, i.e., for smaller $\beta$. The drain will continue until the other members of the triad pair reach an amplitude commensurate with the original wave.

The linearized solution in which other triad members are ignored is a realistic description of the wave field only for times $< O(\beta)$. Eventually nonlinear effects will cause other waves to rise out of the background of imperceptible waves and significantly share the energy of the initial wave. This evolution of the spectrum is inexorable but slow, and the linearized solution *is* an increasingly more accurate description as $\beta$ gets larger. However, unless the initial conditions are unrealistically limited to *precisely* a single wave, the wave spectrum will slowly become broader and more complex by the processes described above. At each stage of this process, though, the dynamics of the *individual* wave, for $\beta \gg 1$, *will* still be governed by linear theory. The nonlinear theory describes the energy exchange between the nearly linear waves.

## 3.27 Energy and Enstrophy

We saw in the preceding section that the nonlinear dynamics of a triad of Rossby waves is severely constrained by the necessity to conserve both energy and enstrophy. As we shall now see, this is a general fact not limited to small-amplitude Rossby waves. In fact, one of the most striking manifestations of these conservation principles arises in the study of strongly nonlinear, turbulent motions to be discussed shortly.

Consider motion of the $\beta$-plane. Suppose the motion is contained, laterally in some region, $A$, on whose boundary, $C$, the normal velocity $\mathbf{u}_0 \cdot \mathbf{n}$ vanishes. An infinite region in which all the velocities vanish at infinity is thus included as a special case in which the lateral boundary of the region recedes to infinity.

If (3.26.1) is multiplied by $\psi$, we obtain

$$\frac{\partial E}{\partial t} + \nabla \cdot \mathbf{S} = 0, \tag{3.27.1}$$

where

$$E = \frac{(\nabla \psi)^2}{2} + F \frac{\psi^2}{2} \tag{3.27.2}$$

is the energy (per unit area) and where

$$\mathbf{S} = -\psi \nabla \frac{\partial \psi}{\partial t} - \beta \frac{\psi^2}{2} \hat{\mathbf{i}} - \mathbf{u}_0 \psi (\nabla^2 \psi - F \psi) \tag{3.27.3}$$

is the nonlinear generalization of the energy flux vector discussed in Section 3.21.

If (3.27.1) is integrated over the domain of the flow, the use of (3.25.7) (which clearly also applies to nonlinear motion), (3.25.10), and the constancy of $\psi$

along the boundary (equivalent to $\mathbf{u}_0 \cdot \mathbf{n} = 0$) implies that

$$\frac{\partial}{\partial t} \iint_A E \, dx \, dy = 0 \qquad (3.27.4)$$

so that total energy is conserved.

If now (3.26.1) is multiplied by $\nabla^2 \psi - F\psi$, we obtain

$$\frac{\partial V}{\partial t} + \nabla \cdot \mathbf{Q} = 0, \qquad (3.27.5)$$

where $V$ is the potential enstrophy, defined by

$$V = \tfrac{1}{2}(\nabla^2 \psi - F\psi)^2 \qquad (3.27.6)$$

in analogy with (3.26.37). The enstrophy flux vector $\mathbf{Q}$ is given by

$$\mathbf{Q} = \mathbf{u}_0 V + \beta\left( \nabla\psi \frac{\partial\psi}{\partial x} - \hat{\mathbf{i}} E \right). \qquad (3.27.7)$$

On the encircling boundary whose outward normal is $\mathbf{n}$

$$\mathbf{Q} \cdot \mathbf{n} = \beta\left[ \frac{\partial\psi}{\partial x} \mathbf{n} \cdot \nabla\psi - \cos\theta E \right], \qquad (3.27.8)$$

where $\theta$ is the angle the outward normal makes with the $x$-axis.

Now on any curve $C$ we have by definition

$$\nabla\psi = \mathbf{n}(\mathbf{n} \cdot \nabla\psi) + \mathbf{t}(\mathbf{t} \cdot \nabla\psi), \qquad (3.27.9)$$

where $\mathbf{n}$ and $\mathbf{t}$ are unit normal and tangent vectors to $C$. However, on the particular curve encircling the domain of the motion, $\mathbf{t} \cdot \nabla\psi$ must vanish in order that $\mathbf{u}_0 \cdot \mathbf{n} = 0$. Thus, (3.27.8) becomes

$$\mathbf{Q} \cdot \mathbf{n} = \frac{\beta}{2}\left[ \left(\frac{\partial\psi}{\partial n}\right)^2 - F\psi^2 \right] \cos\theta, \qquad (3.27.10)$$

where we have used the identity, valid on the boundary,

$$\frac{\partial\psi}{\partial x} = \hat{\mathbf{i}} \cdot \nabla\psi = \cos\theta \frac{\partial\psi}{\partial n},$$

where

$$\frac{\partial\psi}{\partial n} \equiv \mathbf{n} \cdot \nabla\psi.$$

Hence, when (3.27.5) is integrated over the domain of the motion, we obtain

$$\frac{\partial}{\partial t} \iint_A V \, dx \, dy = -\frac{\beta}{2} \oint_C \left[ \left(\frac{\partial\psi}{\partial n}\right)^2 - F\psi^2 \right] \cos\theta \, dl, \qquad (3.27.11)$$

where $dl$ is the infinitesimal line element of integration around the bounding curve $C$. Note that $\tfrac{1}{2}(\partial\psi/\partial n)^2$ is the kinetic energy of the motion at the boundary. Hence the boundary term in (3.27.11) is proportional to the *difference* between the kinetic and potential energies on the boundary.

In general the enstrophy will not be preserved in a bounded region. This has already been seen in Section 3.23 where it was shown that a Rossby wave reflected from a rigid boundary at $x = 0$ preserved its wave energy but increased its total wave number in the reflection process thereby increasing its enstrophy. Thus boundaries may serve as inviscid sources or sinks of enstrophy.

There are three conditions under which the total enstrophy may be conserved. First, there is the rather unlikely possibility that the kinetic energy will exactly balance the potential energy on the boundary. Second, and of more interest, is the case where $\beta$ is zero. This case is of interest when all the significant scales of motion are short (at least initially) compared to a stationary Rossby wavelength, for then $\beta \ll 1$. The third possibility, and the one we shall pursue further in this and following sections is the case where $C$ recedes to infinity and where the motion vanishes at infinity. If $C$ recedes to infinity and $\psi$ becomes spatially constant at infinity, then the boundary term in (3.27.11) vanishes leading to conservation of potential enstrophy, i.e.,

$$\frac{\partial}{\partial t} \iint_A V \, dx \, dy = 0. \tag{3.27.12}$$

On the infinite plane it is convenient to represent $\psi$ in terms of double Fourier integral, i.e.,

$$\psi = \frac{1}{2\pi} \int_{-\infty}^{\infty} \int_{-\infty}^{\infty} dl \, dk \, e^{i(kx+ly)} \tilde{\psi}(k, l). \tag{3.27.13}$$

The total energy can then be expressed in terms of the Fourier amplitude $\tilde{\psi}(k, l)$ as

$$E = \iint_A dx \, dy \left[ \frac{(\nabla \psi)^2 + F\psi^2}{2} \right]$$

$$= \int_{-\infty}^{\infty} \int_{-\infty}^{\infty} dk \, dl \left( \frac{k^2 + l^2 + F}{2} \right) \tilde{\psi}(k, l) \tilde{\psi}^*(k, l) \tag{3.27.14}$$

after repeated use of the fundamental Fourier identity

$$\frac{1}{2\pi} \int_{-\infty}^{\infty} \int_{-\infty}^{\infty} dk \, dx \, e^{i(k+k')x} \tilde{\psi}(k') = \tilde{\psi}(-k). \tag{3.27.15}$$

Note that to ensure that $\psi(x, y)$ is real

$$\tilde{\psi}(-k, -l) = \tilde{\psi}^*(k, l), \tag{3.27.16}$$

where * denotes complex conjugation.

We define the energy density in the wave-number spectrum as

$$\tilde{E}(k, l) = \frac{k^2 + l^2 + F}{2} |\tilde{\psi}(k, l)|^2. \tag{3.27.17}$$

Similarly, the integral of the enstrophy over the plane may be written in terms of $|\tilde{\psi}|^2$ since

$$\iint dx \, dy \frac{[\nabla^2\psi - F\psi]^2}{2} = \int_{-\infty}^{\infty} \int_{-\infty}^{\infty} dk \, dl \, (k^2 + l^2 + F)^2 |\tilde{\psi}(k, l)|^2$$

(3.27.18)

$$= \int_{-\infty}^{\infty} \int_{-\infty}^{\infty} dk \, dl \, (k^2 + l^2 + F)\tilde{E}(k, l).$$

Since both the total energy and total enstrophy are constant in time it follows that the radius of gyration of the energy spectrum

$$K_2^2 \equiv \frac{\int_{-\infty}^{\infty} \int_{-\infty}^{\infty} k^2 \tilde{E}(k, l) \, dk \, dl}{\int_{-\infty}^{\infty} \int_{-\infty}^{\infty} \tilde{E}(k, l) \, dk \, dl}$$

(3.27.19)

is also constant (where $K^2 \equiv k^2 + l^2$). Although the shape of the energy spectrum may change with time, the constancy of $K_2^2$ implies that any transfer of energy to smaller scales (larger $K^2$) by nonlinear processes must be accompanied, in general, by a substantial transfer of energy to larger scales where $K^2$ is smaller, just as in the case of the Rossby wave triad.

To consider the energy and enstrophy spectrum as a function of scale without regard to the orientation of the wave vector, it is useful to introduce the one-dimensional spectra

$$\varepsilon(k) \equiv K \int_0^{2\pi} \tilde{E}(K, \theta) \, d\theta,$$

(3.27.20a)

$$\Omega(k) \equiv K \int_0^{2\pi} K^2 \tilde{E}(K, \theta) \, d\theta,$$

(3.27.20b)

where $\theta$ is the angle each wave vector, $\mathbf{K} = \hat{\imath}k + \hat{\jmath}l$, makes with the x-axis. Note that in (3.27.20a, b) the energy spectrum has been rewritten in terms of polar coordinates in the $\mathbf{K}$-plane. Thus $\varepsilon(K)$ is the energy density in the scalar wave-number interval between $K$ and $K + dK$, i.e., within an annulus in the $\mathbf{K}$-plane.

$$E = \int_0^{\infty} \varepsilon(K) \, dK,$$

(3.27.21a)

$$V = \int_0^{\infty} \Omega(K) \, dK + FE.$$

(3.27.21b)

Since $E$ is conserved, it follows that $\int_0^{\infty} \Omega(k) \, dk$ must *separately* be conserved. Although $\Omega(K)$ is not the enstrophy spectrum, it differs from the true enstrophy spectrum only by a constant times the energy spectrum and is often a more convenient measure of the enstrophy dynamics. We shall call $\Omega(K)$ the *relative* enstrophy spectrum.

Suppose the energy spectrum is initially peaked about the wave number $K_0$ and that due to nonlinear interactions by an unspecified mechanism all that energy is transferred to two other wave-number intervals, sharply peaked about wave numbers $K_1$ and $K_2$ such that $K_1 < K_0 < K_2$. Then, if these peaks are sharp enough, the energy and enstrophy constraints imply that

$$\varepsilon_0 = \varepsilon_1 + \varepsilon_2, \tag{3.27.22a}$$

$$K_0^2 \varepsilon_0 = K_1^2 \varepsilon_1 + K_2^2 \varepsilon_2, \tag{3.27.22b}$$

where $\varepsilon_n \equiv \varepsilon(K_n)$. It follows that

$$\frac{\varepsilon_1}{\varepsilon_2} = \frac{K_2^2 - K_0^2}{K_0^2 - K_1^2} = \left(\frac{K_2 - K_0}{K_0 - K_1}\right)\left(\frac{K_2 + K_0}{K_1 + K_0}\right). \tag{3.27.23}$$

If for the sake of example, the transfer of energy is imagined to go to wave numbers *equidistant* from $K_0$ so that $K_2 - K_0 = K_0 - K_1$, then the ratio $\varepsilon_1/\varepsilon_2$ in (27.33) must clearly be greater than one since $K_2 > K_1$. If $K_2$ and $K_1$ represent neighboring octaves, so that $K_2 = 2K_0$ and $K_1 = K_0/2$, then $\varepsilon_1/\varepsilon_2 = 4$. Thus, because of the enstrophy constraint, more energy is shifted to lower wave numbers (larger scale). In the latter example there is four times as much energy at $K_0/2$ than at $2K_0$.

At the same time, the enstrophy transfer has a different character, namely

$$\frac{\Omega_1}{\Omega_2} = \frac{K_1^2 \varepsilon_1}{K_2^2 \varepsilon_2} = \frac{K_1^2 (K_2^2 - K_0^2)}{K_2^2 (K_0^2 - K_1^2)}. \tag{3.27.24}$$

Thus if $K_2 = 2K_0$ and $K_1 = K_0/2$, $\Omega_1/\Omega_2 = \frac{1}{4}$ and four times the relative enstrophy goes to the *higher* wave number.

Thus while 80 percent of the original energy moves to $K_0/2$, 80 percent of the original enstrophy ends up at $2K_0$. Naturally, the example just described is not only contrived but also rather unrealistic. The processes of energy transfer are generally smoother and less sharply peaked than we have shown. Nevertheless, this example demonstrates the general tendency for energy to

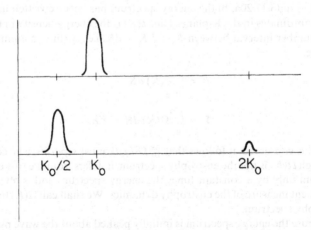

**Figure 3.27.1**   Energy initially peaked at $K = K_0$ is entirely transfered to peaks at $K_0/2$ and $2K_0$. The constraints of energy and enstrophy conservation require that 80 percent of the total energy move to the *smaller* wave number.

mover to larger scales due to nonlinear interactions while the enstrophy moves to smaller scales. A symmetric spread of energy about $K_0$ is simply not allowed by the constraint of enstrophy conservation.

## 3.28 Geostrophic Turbulence

In many cases of interest in oceanography and meteorology, the measure of nonlinearity of the motion field, $\beta^{-1}$, is O(1) rather than being small. We saw in Section 3.26 that when $\beta \gg 1$ the nonlinear interactions were restricted primarily to resonant triads of waves which at lowest order are linear Rossby waves. When $\beta \leq$ O(1), the characterization of the motion as a superposition of Rossby waves is no longer useful. The nonlinear interaction time $L/U$ becomes as rapid as the wave period and interactions between different scales in the spectrum representing the motion become fast and powerful. Indeed, in some cases, the interactions may become so potent that the transfer of energy between different wave numbers proceeds primarily by a nonlinear cascade of energy from one wave number to another rather than by the external forcing at each wave number. In the extreme case, the nonlinear interactions may become so powerful that the internal transfer of energy introduces a random quality to the field of motion. The transfer may occur by a sequence of short period instabilities of one scale to another, rendering the motion time-dependent, chaotic, and only statistically a function of either initial conditions or external forcing. Although for linear waves there is a well-defined relation between frequency and wave number, turbulent motions lack that character and the motion loses its detailed deterministic character.

It must be remembered that such a picture is itself an idealization of meteorological or oceanographic phenomena. It is rare indeed that natural motion systems lack a well-defined deterministic structure in addition to a random field of eddy-like flow. Nevertheless, as is the case in all idealizations, the goal is to produce a conceptual model which, though extreme, provides useful intuition for the more complex natural situation.

Although the conceptual model of a turbulent fluid subject to geostrophic dynamics is already an idealization, the description of the dynamics of geostrophic turbulence remains a subject of great difficulty due to the strongly nonlinear nature of the problem and the random character of the motion. For this reason we shall discuss here only those properties of the motion which we suppose can be well defined statistically, at least in principle, as the energy spectrum. Such quantities can be defined by averaging over a large ensemble of flows characterized by equivalent boundary conditions and external forcings. Furthermore, in this section only the inviscid dynamics of the turbulent field will be discussed while the reader will find additional discussion of the viscous problem in Section 4.14.

For three-dimensional, small-scale turbulent motions not subject to the constraint of geostrophy, common experience suggests that larger eddies are ripped apart by instabilities to form smaller eddies. It is usually imagined that this process continues inexorably to smaller and smaller scales until dissipation transforms the kinetic energy irreversibly to thermal energy. This somewhat grim entropic decay of the energy structure to eventual loss at small scales has become an almost folkloric part of fluid dynamical intuition, embodied, as is most common wisdom, most succinctly in a folk poem

> "Big whirls have little whirls that feed on their velocity,
> and little whirls have littler whirls and so on to viscosity."

However, the results of Section 3.27 suggest that for motions subject to quasigeostrophic dynamics the transformation of energy to smaller scales (higher wave number) is most unlikely.

Following an argument by Rhines (1975) we may consider the matter in the following way. Suppose we define the wave number associated with the "center of gravity" of the energy spectrum as

$$K_1 = \frac{\int_0^\infty K\varepsilon(K)\, dK}{\int_0^\infty \varepsilon(K)\, dK} \tag{3.28.1}$$

so that $K_1$ is the mean wave number of the energy spectrum. To fix ideas, let us suppose that initially the spectrum is smoothly peaked about $K_1$. It is natural to imagine that under the influence of nonlinear interactions of the turbulent fluid the energy spectrum will broaden with time. In principle, this need not happen but our experience with random, nonlinear processes suggests that energy will become increasingly spread over different scales. If this is the case, then we may expect that

$$\frac{\partial}{\partial t}\int_0^\infty (K - K_1)^2 \varepsilon(K)\, dK > 0. \tag{3.28.2}$$

Both $K_1$ and $\varepsilon(K)$ may be functions of time. Since, from (3.28.1)

$$\int_0^\infty (K - K_1)^2 \varepsilon\, dK = \int_0^\infty K^2 \varepsilon\, dK - 2K_1 \int_0^\infty K\varepsilon\, dK + K_1^2 \int_0^\infty \varepsilon\, dK \tag{3.28.3}$$

$$= \int_0^\infty K^2 \varepsilon\, dK - K_1^2 \int_0^\infty \varepsilon\, dK,$$

it follows that if energy and enstrophy are both preserved

$$\frac{\partial}{\partial t} K_1^2 = -\frac{\dfrac{\partial}{\partial t}\displaystyle\int_0^\infty (K - K_1)^2 \varepsilon\, dK}{\displaystyle\int_0^\infty \varepsilon\, dK} < 0. \tag{3.28.4}$$

Thus under the influence of nonlinear interactions which conserve energy and enstrophy, the mean wave number of a spreading spectrum must move to larger scales (smaller wave numbers). The role of dissipation in effecting this result will be examined in Section 4.14. However, it is natural to suppose that energy, in moving to larger scales where dissipation is normally less efficient, will be relatively immune to the effects of dissipation.

On the other hand, the movement of energy to smaller wave numbers will be accompanied by a transfer of enstrophy to larger wave numbers as described in the simple example of Section 3.27. This can alternatively be understood in the following way. Consider the case where the scales of motion, and more especially of enstrophy are small enough so that the $\beta$-effect may be neglected. Then (3.26.1) becomes simply

$$\left(\frac{\partial}{\partial t} + u_0\frac{\partial}{\partial x} + v_0\frac{\partial}{\partial y}\right)(\nabla^2\psi - F\psi) = 0, \tag{3.28.5}$$

so that the potential vorticity associated with the motion field is preserved by each fluid parcel. Thus contours of potential enstrophy become, in the absence of $\beta$, coincident with material lines on the $(x, y)$-plane. Consider two such material lines as shown in Figure 3.28.1. Each contour will be affected by the

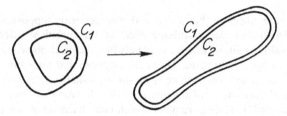

**Figure 3.28.1**  Material contours $C_1$ and $C_2$ tend to be extended by the turbulent velocity field. Since area is preserved to lowest order in quasigeostrophic flow, the contours $C_1$ and $C_2$ must approach one another. Properties, such as potential vorticity, which are constant on each contour must therefore experience an increase in their gradient.

field of turbulent velocity. Again, taking a phenomenological view, experience suggests that the random nature of the turbulent field will tend, on average, to stretch out the contours rather than concentrating the encircled region. Since to order of the Rossby number, area is preserved by quasigeostrophic flows, the distance between the two contours $C_1$ and $C_2$ must decrease. Since the enstrophy remains constant on each such material line, the spatial gradient of the enstrophy field will increase. This implies that in the Fourier representation, the enstrophy will move to higher wave numbers while the energy will move to lower wave numbers. In physical space, this should be manifested by an ever-increasing size for the energy containing eddies while the enstrophy becomes concentrated in narrow shear layers on the eddy boundaries. Naturally, the emergence of these shear layers makes one wonder about the role of viscosity. However, it must be borne in mind that relatively little of the energy of the turbulent field resides in these shear layers. Figure 3.28.2 shows the result of a numerical calculation by Rhines (1977) demonstrating the nonlinear growth in the size of the energy-containing eddies when $\beta = F = 0$. Clearly energy is moving to a smaller wave number.

A helpful solution to the problem for the evolving spectrum may be deduced

**Figure 3.28.2**    The figure on the left shows an initial stream function field which then evolves (Rhines, P. B., *The Sea*, Vol. VI. Copyright © 1977, John Wiley & Sons Inc. Reprinted with permission) according to (3.26.1) for the case $\beta = F = 0$ to the form shown in the right-hand panel. Note the increase in geometric scale of the eddy field.

in a special case. Suppose that $\beta = F = 0$. Furthermore, suppose, following Batchelor (1969) that the scrambling effect of the turbulent velocity field leaves the resulting motion, which completely fills, the $(x, y)$-plane, statistically isotropic in the horizontal plane. That is, we suppose that some time after the inviscid fluid is set in motion there is no distinguishable direction in which the statistics of the motion differ from any other. In that case, there are only two variables and a single parameter which can characterize the turbulent energy spectrum. The single parameter is the total energy, while the energy density depends only on $K$ and $t$.

Since $F = 0$ the energy is entirely kinetic energy. This allows us to identify, for this problem, the scaling velocity $U$ with the root mean square of energy spectrum. Care must be taken in the definition of the energy spectrum when the turbulent motion extends over the infinite plane in a statistically uniform manner, for then the area integrals (3.27.14) and (3.27.18) become formally divergent. It is possible to circumvent this difficulty in a variety of technical ways and the reader may consult Batchelor (1953) for a complete discussion of this point. For our purposes it is sufficient to merely reinterpret $\varepsilon(k)$ so that its integral over all $k$ yields the *average energy density*. Thus, with asterisks denoting dimensional variables, we define $U$ by the relation

$$U^2 = 2 \int_0^\infty \varepsilon_*(K_*) \, dK_*. \tag{3.28.6}$$

Since $K_* = K/L$, it follows that

$$\varepsilon_* = \frac{U^2}{2} L \varepsilon(K, t), \tag{3.28.7}$$

where $\varepsilon$ is the nondimensional energy spectrum. However, $L$ is undefined in terms of any externally imposed scale and the energy density spectrum should not explicitly depend on an arbitrary scale length. However, if $\varepsilon(K, t)$ has the form

$$\varepsilon(K, t) = tg(Kt), \tag{3.28.8}$$

where $g$ is an arbitrary function, it follows that

$$\varepsilon(K_*, t_*) = \frac{U^3}{2} t_* g(K_* U t_*), \qquad (3.28.9)$$

where $t_* = Lt/U$ is the dimensional time variable and thus $\varepsilon_*$ is independent of $L$.

The form (3.28.8) is not unique. The choice $\varepsilon(k, t) = k^{-1}h(kt)$ is equivalent where $h = g(kt)/kt$. The particular choice (3.28.8) is especially convenient, however, since conservation of total energy requires only that

$$1 = \int_0^\infty \varepsilon(K)\, dK = \int_0^\infty tg(Kt)\, dK$$

be independent of time and this constraint is met as long as

$$\int_0^\infty g(\alpha)\, d\alpha = 1. \qquad (3.28.10)$$

The solution

$$\varepsilon(K, t) = tg(Kt), \qquad (3.28.11)$$

due originally to Batchelor (1969), is a *similarity* solution. The shape of the spectrum is a function of $K$ only through the combination $\alpha = Kt$. Thus as time goes on, each particular feature of the spectrum moves to smaller $K$. For example, the mean wave number of the spectrum, as defined by (3.28.1) is

$$K_1 = \frac{\int_0^\infty K\varepsilon\, dK}{\int_0^\infty \varepsilon\, dK}$$

$$= \int_0^\infty Ktg(Kt)\, dK = \frac{C}{t}, \qquad (3.28.12)$$

where

$$C = \int_0^\infty \alpha g(\alpha)\, d\alpha.$$

Thus the mean wave number of the spectrum continuously decreases so that the associated linear scale increases at a constant rate, i.e.,

$$\frac{d}{dt}(K_1^{-1}) = C^{-1}. \qquad (3.28.13)$$

Thus the associated time scale, $\tau$, for the increase in the size of the energy-containing eddies is given by

$$\tau^{-1} = \frac{1}{K_1^{-1}} \frac{d}{dt}(K_1^{-1}) = \frac{K_1}{C}, \qquad (3.28.14)$$

or in dimensional units

$$\tau_* = C/UK_{1*}. \qquad (3.28.15)$$

Now $(UK_*)^{-1}$ is nothing more than the characteristic time for a fluid particle to traverse an eddy circumference at the speed of the energy-containing eddies. Therefore, on the order of one turn over time of the energy-containing eddies, their size has doubled. Small eddies form large eddies at the rate $UK_*$ where $U$ is related to the overall energy level. Rather than "big whirls feeding whirls," geostrophic turbulence more closely resembles big fish eating little fish and thriving on the diet.

It is useful to characterize the rate at which energy is flowing from one part of the spectrum to another. Since energy is conserved, the amount of energy flowing from wave numbers lower than $K$ across $K$ to higher wave numbers must equal

$$\mathscr{F} = -\int_0^K \frac{\partial \varepsilon}{\partial t}\, dk \tag{3.28.16}$$

so that the depletion of energy in the interval $(0, K)$ is balanced by an export to higher wave number. Note that (3.28.16) is equivalent to

$$\frac{\partial \varepsilon}{\partial t} + \frac{\partial \mathscr{F}}{\partial K} = 0. \tag{3.28.17}$$

$\mathscr{F}$ is the spectral energy flux and when it is positive it represents a flow to high wave numbers. If (3.29.8) is used in (3.28.16) it follows that for the similarity solution $\varepsilon = tg(Kt)$

$$\mathscr{F} = -\frac{K\varepsilon(K)}{t}, \tag{3.28.18}$$

so that the energy flux is always negative, i.e., towards lower wave number. As time goes on the energy becomes compressed into an increasingly narrow region in the vicinity of $K = 0$. As a consequence of this compression the energy density associated with $K_1$ increases linearly with time, i.e.,

$$\varepsilon(K_1) = tg(C). \tag{3.28.19}$$

The enstrophy spectrum associated with the similarity solution is

$$\Omega(K) = K^2 \varepsilon(K) = K^2 tg(Kt), \tag{3.28.20}$$

so that the total enstrophy

$$V = \int_0^\infty \Omega\, dk = \frac{A}{t^2}, \tag{3.28.21}$$

where

$$A = \int_0^\infty \alpha^2 g(\alpha)\, d\alpha. \tag{3.28.22}$$

This implies that the similarity solution fails to conserve total enstrophy. Enstrophy in this solution is so efficiently exported to high wave numbers that the enstrophy flux remains finite as $K \to \infty$. If we define the enstrophy flux in analogy with (3.28.18) by the relation

$$\frac{\partial \Omega}{\partial t} + \frac{\partial \mathcal{H}}{\partial K} = 0, \tag{3.28.23}$$

it follows that with (3.28.20)

$$\mathcal{H} = -\varepsilon \frac{K^3}{t} + \frac{2}{t} \int_0^K K^2 \varepsilon \, dK. \tag{3.28.24}$$

If $\varepsilon$ goes to zero as $K \to \infty$ fast enough so that $K^3 \varepsilon$ vanishes there, the enstrophy flux at $K = \infty$ which continuously drains the system of enstrophy is

$$\mathcal{H}(\infty) = \frac{2}{t} V(t). \tag{3.28.25}$$

It is natural to anticipate that viscous dissipation acting at high wave number will provide a sink for the exported enstrophy.

As the energy containing eddies grow in size their characteristic time scale $(UK_*)^{-1}$ increases (since $U$ is preserved). As the eddies become larger and slower, the basic presumption underpinning the idea of the turbulent cascade and the solution (3.28.8) is lost. The integral statements of energy and enstrophy conservation still apply. However, as the time scale of the eddies increases competitive processes which resist the stirring processes of the eddy velocity field begin to be felt and the isotropic character of the eddy field is lost. In particular, in the presence of the $\beta$-effect, wave dynamics will reassert themselves when the scale of the energy containing eddies becomes large enough for $\beta$ to be of order one or greater. When that occurs, the nonlinear transfer begins to be dominated by wave-wave interactions and wave radiation. The eddies metamorphose into Rossby wave packets and their nonlinear interactions become limited by the requirements of the triad resonance conditions and the mutual separation of the interacting fields by the separation due to packet propagation. If we identify $L$ with the scale of the energy containing eddy at $K_{1*}$, then this transformation will occur when $\beta \sim 1$, or, using (3.17.9) when

$$K_{1*} = (\beta_0/U)^{1/2}, \tag{3.28.26}$$

which will occur after a time, $\tau_r$, of $O(K_{1*}U)^{-1}$ or

$$\tau_r = (U\beta_0)^{-1/2}. \tag{3.28.27}$$

After this time the field will evolve according to the resonant interaction process described in Section 3.26. However, instead of a single triad, the field will consist of an infinite number of triads and a statistical description is still required. However, in distinction to the purely turbulent regime there is now a well-defined dispersion relation which links frequency and spatial scale, i.e.,

$$\sigma = -\frac{\beta k}{K^2 + F}. \tag{3.28.28}$$

In Section 3.22, it was seen that a Rossby wave with wave number $K_2$ would become unstable to two other Rossby waves with wave numbers $K_1$ and $K_3$, if the first wave were intermediate in spatial scale, i.e., if from (3.26.46)

$$(K_1^2 - K_2^2)(K_2^2 - K_3^2) > 0. \tag{3.28.29}$$

This condition may be rewritten in terms of wave frequency as

$$\left(\frac{k_1}{\sigma_1} - \frac{k_2}{\sigma_2}\right)\left(\frac{k_2}{\sigma_2} - \frac{k_3}{\sigma_3}\right) > 0, \tag{3.28.30}$$

where

$$\sigma_j \equiv -\frac{\beta k_j}{K_j^2 + F}; \qquad j = 1, 2, 3. \tag{3.28.31}$$

A little algebra allows (3.28.30) to be rewritten as

$$\frac{(\sigma_3 k_2 - \sigma_2 k_3)^2}{\sigma_1 \sigma_3 \sigma_2^2} > 0. \tag{3.28.32}$$

Thus for the wave at $K = K_2$ to be unstable, $\sigma_1$ and $\sigma_3$ must have the same sign. However, the resonance condition requires

$$\sigma_2 = -(\sigma_1 + \sigma_3), \tag{3.28.33}$$

so that the unstable wave must have the numerically largest frequency in the triad. High-frequency Rossby waves are therefore unstable to lower frequency waves. Were the triad to exist alone, the results of Section 3.26 show that the energy fed into waves 1 and 3 would periodically return to wave 2. However, in a random field of many Rossby waves, it is more reasonable to expect that on average the lower frequency recipients of the energy would themselves become unstable to even lower frequency waves before they could return the energy to the higher frequency wave from which it came. If this is the case, energy would then proceed to flow towards lower *frequency* as well as lower wave number. For Rossby waves in a fluid with a rigid upper lid (i.e., $F = 0$), this demands that the spectrum consist of waves with ever-decreasing values of

$$\sigma = -\frac{\beta \cos \theta}{K}, \tag{3.28.34}$$

where $\theta$ is the angle of the wave vector to the x-axis. As we have already noted, $K$ will certainly not increase. Hence we can expect that as time goes on $\cos \theta \to 0$, or equivalently that the wave vector orients itself increasingly parallel to the y-direction. This implies that the velocity in the eddy field should become almost strictly zonal.

Numerical experiments by Rhines (1975, 1977) show precisely this behavior. Figure 3.28.3 shows the eddy field in the case $\beta = 52$ at a time $t = 5.1$. Note that in the case of strong $\beta$, the stream function field is elongated in the zonal direction. Even more significantly the length scale of the eddies are smaller for the strong $\beta$ case. Whereas the eddies continue to increase their size in the absence of $\beta$, the energy cascade to lower wave number is observed to nearly

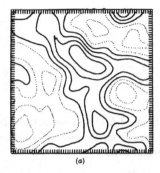 

(a)                              (b)

**Figure 3.28.3**  Streamlines of the turbulent eddy field after several eddy turnover times. The field in the upper panel has evolved with $\beta = 0$. The field in the lower panel has $\beta = 52$. Note the elongation of the eddy field in the $x$-direction and its smaller scale for the case $\beta \neq 0$. (Rhines, P. B., *J. Fluid Mech.*, **69**, Part 3. Copyright © 1975, Cambridge University Press. Reprinted with permission.)

halt in the experiments once (3.28.26) is satisfied. Further interactions serve mainly to alter the *direction* of the wave vectors so as to reduce $\sigma$. Such interactions are statistically preferred because the enstrophy constraint becomes redundant in triads each of whose wave vectors is nearly of equal length. This, in turn, implies a barrier to further energy cascade and the development of a peak in the energy spectrum at

$$K_{1*} = (\beta_*/U)^{1/2},$$

where, however, the ultimate orientation of the eddies leads to jetlike zonal flows.

There are several aspects of the idealized processes here that must be kept in mind. If the turbulence is originally localized in space rather than filling it completely, then as the energy evolves into a wavelike form, it may simply disperse by radiating Rossby waves and the nonlinear interactions will largely cease. The presence of lateral boundaries where Rossby waves may be reflected will also alter the dynamics by providing a region in which energy at low wave numbers may be continuously transformed to higher wave numbers. Each of these processes can be expected to alter the simple conceptual picture described in this section.

Furthermore, the turbulent dynamics we have discussed are completely inviscid. Thus the spectrum freely evolves from an initial state with no dissipation and no external, continuing energy input. A discussion of these issues is found in Section 4.14.

## Appendix to Chapter 3

The solution to (3.16.1),

$$\nabla^2\psi - F\psi = -\frac{\beta Y\,\delta(r)}{2\pi r}, \tag{3.A.1}$$

may be found as follows.

If we recognize the azimuthal symmetry of (3.A.1) and introduce a polar coordinate frame where $r$ measures distance from the origin, solutions of (3.A.1) which are symmetric about the origin satisfy

$$\frac{1}{r}\frac{\partial}{\partial r}r\frac{\partial\psi}{\partial r} - F\psi = -\frac{\beta Y}{2\pi r}\,\delta(r). \tag{3.A.2}$$

For all $r \neq 0$ the right-hand side of (3.A.2) vanishes. The general homogeneous solution of (3.A.2) valid for all nonzero $r$ is

$$\psi = AK_0(F^{1/2}r) + BI_0(F^{1/2}r). \tag{3.A.3}$$

The function $I_0$ exponentially increases with increasing $r$ and must be rejected. Hence $B = 0$. Integration of (3.A.2) over a small interval which includes the origin yields

$$\lim_{r\to 0} r\frac{\partial\psi}{\partial r} = -A = -\frac{\beta Y}{2\pi}, \tag{3.A.4}$$

which determines $A$, viz.

$$A = +\frac{\beta Y}{2\pi}, \tag{3.A.5}$$

so that

$$\psi = \frac{\beta Y}{2\pi}K_0(F^{1/2}r) \tag{3.A.6}$$

# CHAPTER 4

# Friction and Viscous Flow

## 4.1 Introduction

The observed persistence over several days of large-scale waves in the atmosphere and the oceans reinforces the impression that frictional forces are weak, almost everywhere, when compared with the Coriolis acceleration and the pressure gradient. Friction rarely upsets the geostrophic balance to lowest order. Indeed, for many flows it is probably also true that the dissipative time scale is long compared to the advective time scale, i.e., that the frictional forces are weak in comparison with the nonlinear relative acceleration. Nevertheless friction, and the dissipation of mechanical energy it implies, cannot be ignored. The reasons are simple yet fundamental. For the time-averaged flow, i.e., for the general circulation of both the atmosphere and the oceans, the fluid motions respond to a variety of essentially steady external forcing. The atmosphere, for example, is set in motion by the persistent but spatially nonuniform solar heating. This input of energy produces a mechanical response, i.e., kinetic energy of the large-scale motion, and eventually this must be dissipated if a steady state—or at least a statistically stable average state of motion—is to be maintained. This requires frictional dissipation. In addition, the driving force itself may be frictional, as in the case of the wind-driven oceanic circulation. There the wind stress on the ocean surface produces a major component of the oceanic circulation. Finally, even though friction may be weak compared with other forces, its dissipative nature, qualitatively distinct from the conservative nature of the inertial forces, require its consideration if questions of the decay of free motions are to be studied.

The presence of frictional dissipation arises ultimately from the random motion of fluid molecules. The random motion of the fluid molecules is not described by the continuum equations. Rather, only the resulting viscous forces they produce are considered, and empirically, for many fluids, this viscous force can be written in terms of a viscosity coefficient and the macroscopic velocity *alone*. This frictional force is given by (1.4.5). Although this force must ultimately be responsible for the dissipation of kinetic energy and its transformation into the kinetic energy of the disordered molecular motion (i.e., heat), its *direct* effect on the large-scale motion has been shown in Section 2.8 to be utterly negligible. The length scales of large-scale motions are too great for molecular viscosity to be directly significant in the force balance. Yet we have argued above that friction must be important.

The problem, of course, is that the large-scale motion does not exist in isolation. For the very reason that molecular viscosity is so small, both the atmosphere and the oceans contain an enormously broad spectrum of turbulent motions fed to a considerable extent by the energy of the largest-scale flows. These turbulent fluctuations, embedded in the larger flow, tend to drain the large-scale flow of energy by a variety of mechanical processes and in turn pass the energy to finer scales of motion where viscosity can act directly. This notion of the *cascade* of energy from the largest to the smallest scales of motion is far from clear and rigorous. Indeed, we already have seen in the discussion of Section 3.26 that in some cases the cascade must at least partially go from small to large scale. Nevertheless, the gross aspect of the notion of the energy cascade is probably correct and has the following implication. If we wish to formulate a mathematical framework that describes directly the dynamics of only the large-scale motions, the drain of large-scale energy by smaller-scale motions must be represented entirely in terms of the kinematic features of the large scale. The only alternative is the intractable task of describing in detail *all* scales of motion, from the very largest to the very smallest. Although the example of the representation of viscous forces of molecular origin in terms of the macroscopic velocity serves as an encouraging model, the situation with regard to the representation of the turbulent interactions of small- and large-scale motions is considerably less satisfactory. Indeed this problem is one of the most vexing in geophysical fluid dynamics. At this time, there seems to be no tractable theory of turbulence that provides a practical and accurate description of the *effective* frictional force due to the cascade of energy by turbulent fluctuations.

In the following sections of this chapter we instead will describe one very crude representation of the effect of the turbulent cascade on the large-scale flow that quite frankly follows the notions that are appropriate to the representation of the effect of molecular motions on the mean. This is quite clearly a necessary compromise, since considering only the molecular viscosity acting on the large scale severely underestimates the role of friction, while an accurate consideration of the details of turbulent motions is simply impractical to the point of impossibility.

## 4.2 Turbulent Reynolds Stresses

Consider the *total* relative velocity field in the fluid, **u**, to be split into two parts. The first, temporarily denoted by $\langle \mathbf{u} \rangle$, is to represent the large-scale flow we wish to describe in detail. The second part, **u**', is to represent the smaller-scale turbulence, which is not directly of interest except insofar as it affects the large scale. Part of the conceptual problem immediately arises here, for there is no completely logical way to decide how this decomposition is to be effected, i.e., whether a motion is to be considered part of the mean or part of the turbulence. To some degree **u**' is specified by the condition that times short compared to the characteristic evolution time of $\langle \mathbf{u} \rangle$ are still long enough to provide an adequate averaging interval for **u**', so that over those times the average of **u**' vanishes, i.e., that for

$$\mathbf{u} = \langle \mathbf{u} \rangle + \mathbf{u}' \tag{4.2.1}$$

the average of the small-scale velocity

$$\langle \mathbf{u}' \rangle = 0. \tag{4.2.2}$$

Hence the bracket now has an operational meaning in terms of averaging, so that

$$\langle \langle \mathbf{u} \rangle + \mathbf{u}' \rangle = \langle \mathbf{u} \rangle. \tag{4.2.3}$$

Whether it is really possible to find averaging times long enough for (4.2.2) to obtain, while short compared to the natural time scales of the large scale, is problematic. We assume here that it *is* possible.

Consider now the $x$, $y$, and $z$ momentum equations for an incompressible homogeneous fluid rotating about the $z$-axis. Using the decomposition (4.2.1), the $x$-component, for example, is

$$\left[ \frac{\partial}{\partial t} + (\langle u \rangle + u') \frac{\partial}{\partial x} + (\langle v \rangle + v') \frac{\partial}{\partial y} + (\langle w \rangle + w') \frac{\partial}{\partial z} \right] (\langle u \rangle + u')$$

$$- f(\langle v \rangle + v') = -\frac{1}{\rho} \frac{\partial \langle p \rangle}{\partial x} - \frac{1}{\rho} \frac{\partial p'}{\partial x} + \mathscr{F}_x \tag{4.2.4}$$

where $\mathscr{F}_x$ is the $x$-component of the viscous force (1.4.5). The average of (4.2.4) yields an equation for $\langle u \rangle$:

$$\frac{\partial \langle u \rangle}{\partial t} + \langle u \rangle \frac{\partial \langle u \rangle}{\partial x} + \langle v \rangle \frac{\partial \langle u \rangle}{\partial y} + \langle w \rangle \frac{\partial \langle u \rangle}{\partial z} - f \langle v \rangle$$

$$= -\frac{1}{\rho} \frac{\partial \langle p \rangle}{\partial x} + \nu \nabla^2 \langle u \rangle \tag{4.2.5}$$

$$- \frac{\partial}{\partial x} \langle u'u' \rangle - \frac{\partial}{\partial y} \langle v'u' \rangle - \frac{\partial}{\partial z} \langle w'u' \rangle,$$

where the condition of incompressibility, which must hold separately for the average and fluctuation velocity, has been used. The equation for $\langle u \rangle$ is written entirely in terms of the large-scale velocity, with the important exception of the three last terms on the right-hand side of (4.2.5). Although $u'$, $v'$, and $w'$ have zero average, the *momentum flux* of the fluctuations, which is quadratic in the fluctuation velocities, *need not vanish when averaged*. This is precisely analogous to the nonzero momentum flux due to the random thermal motion of molecules in a gas. For example, the term

$$\langle v'u' \rangle \equiv -\frac{\tau_{xy}}{\rho} \qquad (4.2.6)$$

represents an average flux of $x$-momentum due to the small-scale motions across a surface $y =$ constant as shown in Figure 4.2.1. If the flux across $y_1$

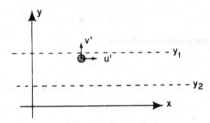

**Figure 4.2.1**  A fluid element with fluctuation velocity $v'$ in the $y$-direction will carry $x$-momentum $\rho u'$ across $y_1$. The net flux is $\rho \langle u'v' \rangle$.

exceeds the flux across $y_2$ and $y_1 > y_2$, then the mean momentum between the two layers must diminish so as to conserve total momentum—i.e., if

$$\frac{\partial}{\partial y} \langle u'v' \rangle > 0, \qquad (4.2.7)$$

the tendency is for

$$\frac{\partial \langle u \rangle}{\partial t} < 0 \qquad (4.2.8)$$

unless balanced by other effects. The momentum flux due to the small scale motion is, as far as the large-scale motion is concerned, equivalent to a *stress* $\tau_{xy}$ on the large-scale flow. In terms of this stress field the averaged equations for the large-scale flow are

$$\frac{\partial u}{\partial t} + u \frac{\partial u}{\partial x} + v \frac{\partial u}{\partial y} + w \frac{\partial u}{\partial z} - fv$$

$$= -\frac{1}{\rho} \frac{\partial p}{\partial x} + \frac{1}{\rho} \left\{ \frac{\partial \tau_{xx}}{\partial x} + \frac{\partial \tau_{xy}}{\partial y} + \frac{\partial \tau_{xz}}{\partial z} \right\} + \nu \nabla^2 u, \qquad (4.2.9a)$$

$$\frac{\partial v}{\partial t} + u\frac{\partial v}{\partial x} + v\frac{\partial v}{\partial y} + w\frac{\partial v}{\partial z} + fu$$

$$= -\frac{1}{\rho}\frac{\partial p}{\partial y} + \frac{1}{\rho}\left|\frac{\partial \tau_{yx}}{\partial x} + \frac{\partial \tau_{yy}}{\partial y} + \frac{\partial \tau_{yz}}{\partial z}\right| + v\nabla^2 v, \qquad (4.2.9b)$$

$$\frac{\partial w}{\partial t} + u\frac{\partial w}{\partial x} + v\frac{\partial w}{\partial y} + w\frac{\partial w}{\partial z}$$

$$= -\frac{1}{\rho}\frac{\partial p}{\partial z} + \frac{1}{\rho}\left|\frac{\partial \tau_{zx}}{\partial x} + \frac{\partial \tau_{zy}}{\partial y} + \frac{\partial \tau_{zz}}{\partial z}\right| + v\nabla^2 w, \qquad (4.2.9c)$$

where the bracket notation for the large-scale flow has been suppressed. It is now to be understood that unprimed velocities refer entirely to the large scale. The stresses that appear in (4.2.9) are given by

$$\tau_{xx} = -\rho\langle u'u'\rangle, \quad \tau_{yy} = -\rho\langle v'v'\rangle, \quad \tau_{zz} = -\rho\langle w'w'\rangle$$

$$\tau_{xy} = \tau_{yx} = -\rho\langle u'v'\rangle,$$

$$\tau_{xz} = \tau_{zx} = -\rho\langle u'w'\rangle, \qquad (4.2.10)$$

$$\tau_{yz} = \tau_{zy} = -\rho\langle v'w'\rangle.$$

These stresses, which appear as a natural result of averaging each momentum equation, are called the Reynolds stresses. The Reynolds stresses appear *only* because of the splitting of the velocity field into large-scale versus turbulent small-scale flow. They represent no new physical mechanism on the fundamental fluid-dynamical level; rather, they are the consequence of our decision to focus entirely on the dynamics of the large scale.

The profoundly difficult question that confronts us now is how to specify the Reynolds stresses in terms of only the large-scale velocities so that (4.2.9a,b,c) represent a closed set of equations in terms of only the large-scale velocities.

One of the most common and also one of the most crude ways to deal with this perplexing question is to imagine that the turbulent motions act on the large-scale flow in a manner that mimics the way in which molecular motions affect the macroscopic flow. That is, the Reynolds stresses are assumed to depend in a linear way on the spatial derivatives of the large-scale flow velocity. A particularly simple model of this type for the stresses— which also preserves the necessary symmetry between $\tau_{xy}$ and $\tau_{yx}$, etc.—is

$$\frac{\tau_{xx}}{\rho} = 2A_H\frac{\partial u}{\partial x}, \qquad \frac{\tau_{yy}}{\rho} = 2A_H\frac{\partial v}{\partial y}, \qquad \frac{\tau_{zz}}{\rho} = 2A_V\frac{\partial w}{\partial z}$$

$$\tau_{xy} = \tau_{yx} = \rho A_H\left(\frac{\partial v}{\partial x} + \frac{\partial u}{\partial y}\right),$$

$$\tau_{xz} = \tau_{zx} = \rho A_V\frac{\partial u}{\partial z} + \rho A_H\frac{\partial w}{\partial x}, \qquad (4.2.11)$$

$$\tau_{yz} = \tau_{zy} = \rho A_V\frac{\partial v}{\partial z} + \rho A_H\frac{\partial w}{\partial y}.$$

The coefficients $A_H$ and $A_V$ are called the horizontal and vertical turbulent viscosity coefficients respectively. There is no *a priori* reason why $A_H$ and $A_V$ should be identical. Indeed the anisotropy of the *large scale* between the horizontal and vertical scales of the flow suggest that the mixing of large-scale momentum in the two directions cannot be expected to be the same. It is important to realize that the model (4.2.11) is a hypothesis of doubtful validity. Although it does provide a simple way of closing the equations for the large-scale flow, there is really no *a priori* justification for the model. Not only is the very form of (4.2.11) open to question, but even if the idea is accepted that the turbulent stresses can be so simply modeled, the *ad hoc* nature of the model absolutely precludes an *a priori* determination of $A_H$ and $A_V$. Since the details of the turbulent flow are ignored, $A_H$ and $A_V$ cannot be calculated. The significant failure of the analogy with molecular friction is further evident in that $A_H$ and $A_V$ cannot be determined even empirically, as the molecular viscosity can be, in a way that is independent of the particular flow configuration. Instead, (4.2.11) must be considered the weakest part of the dynamical structure we are constructing. If (4.2.11) *is* accepted, and $A_V$ and $A_H$ are in addition considered constant, then (4.2.9a,b,c) become

$$\frac{\partial u}{\partial t} + u\frac{\partial u}{\partial x} + v\frac{\partial u}{\partial y} + w\frac{\partial u}{\partial z} - fv$$

$$= -\frac{1}{\rho}\frac{\partial p}{\partial x} + A_H\left(\frac{\partial^2 u}{\partial x^2} + \frac{\partial^2 u}{\partial y^2}\right) + A_V\frac{\partial^2 u}{\partial z^2} + v\nabla^2 u,$$

$$\frac{\partial v}{\partial t} + u\frac{\partial v}{\partial x} + v\frac{\partial v}{\partial y} + w\frac{\partial v}{\partial z} + fu$$

$$= -\frac{1}{\rho}\frac{\partial p}{\partial y} + A_H\left(\frac{\partial^2 v}{\partial x^2} + \frac{\partial^2 v}{\partial y^2}\right) + A_V\frac{\partial^2 v}{\partial z^2} + v\nabla^2 v, \qquad (4.2.12)$$

$$\frac{\partial w}{\partial t} + u\frac{\partial w}{\partial x} + v\frac{\partial w}{\partial y} + w\frac{\partial w}{\partial z}$$

$$= -\frac{1}{\rho}\frac{\partial p}{\partial z} + A_H\left(\frac{\partial^2 w}{\partial x^2} + \frac{\partial^2 w}{\partial y^2}\right) + A_V\frac{\partial^2 w}{\partial z^2} + v\nabla^2 w.$$

If $A_H = A_V = A$, the form of (4.2.12) is identical to that of the ordinary Navier-Stokes equations with an effective viscosity $v + A$. The model (4.2.11) is then equivalent to the notion that the turbulent fluctuations produce stresses which can be "parametrized" by merely increasing the size of the viscosity from its molecular value to account for the greater efficiency of smoothing of momentum on a large scale by the transport of great chunks of fluid, rather than molecules, across the averaged momentum gradient. When $A_H \neq A_V$, the spatial anisotropy makes a detailed identification with the molecular case more difficult. Nevertheless both $A_H$ and $A_V$ can be safely assumed to greatly exceed $v$, so that henceforth the terms proportional to the molecular viscosity will be ignored for the large-scale flow.

Estimates of $A_H$ and $A_V$ in the atmosphere and the ocean vary enormously. Estimates of $A_V$ suggest values of $10^5$ cm²/s near the earth's surface and considerably lower in the free atmosphere. Estimates of $A_H$ are even more difficult to accept with confidence, but some estimates of $O(10^5 \text{ meter}^2/\text{s})$ for $A_H$ have been given. For the ocean estimates of $A_V$ range from 1 cm$^2$/s to $10^3$ cm$^2$/s, while for $A_H$, similarly uncertain estimates of the order of $10^5$ cm$^2$/s to $10^8$ cm$^2$/s have been put forward.

Considering the uncertainty associated with the values of $A_H$ and $A_V$, theories of geophysical phenomena which depend on particular values of these coefficients are suspect, to say the least. On the other hand, unsatisfactory as they may be, their use often gives at least a qualitatively sensible picture of those aspects of large-scale dynamics which demand some frictional forces to be present.

## 4.3  The Ekman Layer

The role of friction in geophysical fluid dynamics is closely related to the structure of the frictional layer which appears on a rigid surface perpendicular to the rotation vector. Such friction layers, in conjunction with the constraints of the Taylor–Proudman theorem (Section 2.7) will be shown to exert a profound influence on the dynamics of the flow remarkably far from the regions which are *directly* affected by viscosity. The fundamental character of the friction region, called the *Ekman layer*, is illustrated by the following example.

Consider the motion of a homogeneous, incompressible fluid rotating with angular velocity $\mathbf{\Omega}$. A rigid wall at $z = 0$ is perpendicular to $\mathbf{\Omega}$ and, as shown in Figure 4.3.1, a horizontally *uniform* flow of velocity $U$ is specified far from the wall. The governing equations of motion are

$$\frac{\partial u}{\partial t} + u\frac{\partial u}{\partial x} + v\frac{\partial u}{\partial y} + w\frac{\partial u}{\partial z} - fv$$

$$= -\frac{1}{\rho}\frac{\partial p}{\partial x} + A_V\frac{\partial^2 u}{\partial z^2} + A_H\left(\frac{\partial^2 u}{\partial x^2} + \frac{\partial^2 u}{\partial y^2}\right), \tag{4.3.1a}$$

$$\frac{\partial v}{\partial t} + u\frac{\partial v}{\partial x} + v\frac{\partial v}{\partial y} + w\frac{\partial v}{\partial z} + fu$$

$$= -\frac{1}{\rho}\frac{\partial p}{\partial y} + A_V\frac{\partial^2 v}{\partial z^2} + A_H\left(\frac{\partial^2 v}{\partial x^2} + \frac{\partial^2 v}{\partial y^2}\right), \tag{4.3.1b}$$

$$\frac{\partial w}{\partial t} + u\frac{\partial w}{\partial x} + v\frac{\partial w}{\partial y} + w\frac{\partial w}{\partial z}$$

$$= -\frac{1}{\rho}\frac{\partial p}{\partial z} - g + A_V\frac{\partial^2 w}{\partial z^2} + A_H\left(\frac{\partial^2 w}{\partial x^2} + \frac{\partial^2 w}{\partial y^2}\right), \tag{4.3.1c}$$

$$\frac{\partial u}{\partial x} + \frac{\partial v}{\partial y} + \frac{\partial w}{\partial z} = 0, \tag{4.3.2}$$

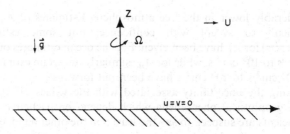

**Figure 4.3.1**    A uniform flow **U**, as seen in a frame rotating with angular velocity $\Omega$, above a plane at $z = 0$ upon which $u$, $v$, and $w$ must vanish.

where $f = 2\Omega$. We may align the $x$-axis in the direction of the velocity at infinity, so that the boundary condition at $z \to \infty$ is simply

$$u = U,$$
$$v = 0, \qquad z \to \infty. \qquad (4.3.3)$$
$$w = 0,$$

At the rigid surface $z = 0$ we suppose that the friction, in direct analogy with the role of molecular viscosity, inhibits the fluid motion to such an extent that both the normal *and* tangential velocities vanish on $z = 0$, i.e.,

$$u = v = w = 0 \qquad (z = 0). \qquad (4.3.4)$$

The crucial condition is the condition of no slip on the tangential veloci-ties $u$ and $v$. In the absence of retarding frictional forces, this boundary condition would be absent. Then, the uniform velocity $U$ would be an exact solution of (4.3.1a,b,c), i.e.,

$$u = U = -\frac{1}{\rho f} \frac{\partial p}{\partial y},$$
$$v = w = 0, \qquad (4.3.5)$$

which is the geostrophic balance. The presence of friction *and* the rigid-wall condition that $u$ and $v$ must vanish at the wall means that, at least in the vicinity of the wall, the flow must significantly depart from geostrophic balance.

An exact solution of (4.3.1a,b,c) which satisfies (4.3.3) and (4.3.4) can be found in the form

$$u = u(z),$$
$$v = v(z), \qquad (4.3.6)$$
$$w = w(z).$$

This form is admissible because neither the boundary condition at $z = 0$ nor the one at infinity requires lateral variations in the velocity fields. Introducing (4.3.6) into (4.3.2) yields

$$\frac{\partial w}{\partial z} = 0, \tag{4.3.7}$$

which, with the condition that $w$ vanishes on $z = 0$, implies that the vertical velocity is zero for *all* $z$, i.e.,

$$w(z) = 0. \tag{4.3.8}$$

This, along with the fact that $u$ and $v$ are functions of $z$ alone, allows (4.3.1a,b,c) to be written without approximation as

$$-fv = -\frac{1}{\rho}\frac{\partial p}{\partial x} + A_V\frac{\partial^2 u}{\partial z^2}, \tag{4.3.9a}$$

$$fu = -\frac{1}{\rho}\frac{\partial p}{\partial y} + A_V\frac{\partial^2 v}{\partial z^2}, \tag{4.3.9b}$$

$$g = -\frac{1}{\rho}\frac{\partial p}{\partial z}. \tag{4.3.9c}$$

Since the fluid is homogeneous, it follows from (4.3.9c) that

$$\frac{\partial}{\partial x}\frac{\partial p}{\partial z} = \frac{\partial}{\partial y}\frac{\partial p}{\partial z} = 0, \tag{4.3.10}$$

or

$$\frac{\partial}{\partial z}\begin{vmatrix} \partial p/\partial x \\ \partial p/\partial y \end{vmatrix} = 0, \tag{4.3.11}$$

so that the *horizontal* pressure gradient must be *independent* of $z$. Now as $z \rightarrow \infty$ both $u$ and $v$ become constant, so that for very large $z$, using (4.3.3),

$$0 = -\frac{1}{\rho}\frac{\partial p}{\partial x},$$

$$fU = -\frac{1}{\rho}\frac{\partial p}{\partial y}. \tag{4.3.12}$$

Since the horizontal pressure gradient is independent of $z$, (4.3.12) must also hold for all $z$. The horizontal pressure gradient is therefore determined entirely by the geostrophic velocity $U$ far from the boundary. This allows (4.3.9a,b) to be written

$$f\tilde{u} = A_V\frac{d^2\tilde{v}}{dz^2},$$

$$-f\tilde{v} = A_V\frac{d^2\tilde{u}}{dz^2}, \tag{4.3.13}$$

where the tilde variables are

$$\tilde{u} = u - U,$$

$$\tilde{v} = v,$$

and are departures from geostrophic flow induced by the presence of the rigid wall and friction. In the absence of friction ($A_V = 0$) both $\tilde{u}$ and $\tilde{v}$ would be zero. It is also important to note that if $A_V$ were replaced by the coefficient of kinematic viscosity $v$, (4.3.13) would apply equally well to the laminar flow of a fluid.

A single equation for $\tilde{u}$ can be derived by eliminating $\tilde{v}$, i.e.,

$$\frac{d^4\tilde{u}}{dz^4} + \frac{f^2}{A_V^2}\tilde{u} = 0, \tag{4.3.14}$$

whose general solution is the sum of the four independent homogeneous solutions, i.e.,

$$\tilde{u} = C_1 e^{(1+i)z/\delta_E} + C_2 e^{(1-i)z/\delta_E} + C_3 e^{-(1+i)z/\delta_E} + C_4 e^{-(1-i)z/\delta_E}, \tag{4.3.15}$$

where $C_1$, $C_2$, $C_3$, and $C_4$ are constants, $i = \sqrt{-1}$, and $\delta_E$ is the Ekman layer thickness given by

$$\delta_E = \left(\frac{A_V}{f/2}\right)^{1/2}. \tag{4.3.16}$$

The first two solutions, proportional to $C_1$ and $C_2$, grow exponentially as $z \to \infty$ and become unbounded for large values of $z/\delta_E$. To keep the solution bounded at infinity we must choose

$$C_1 = C_2 = 0. \tag{4.3.17}$$

This automatically satisfies (4.3.3), which in terms of $\tilde{u}$ and $\tilde{v}$ is simply

$$\tilde{u} = \tilde{v} = 0 \quad \text{as } z \to \infty. \tag{4.3.18}$$

Hence

$$\tilde{u} = C_3 e^{-(1+i)z/\delta_E} + C_4 e^{-(1-i)z/\delta_E}, \tag{4.3.19a}$$

while from (4.3.13)

$$\tilde{v} = -iC_3 e^{-(1+i)z/\delta_E} + iC_4 e^{-(1-i)z/\delta_E}. \tag{4.3.19b}$$

On $z = 0$,

$$\tilde{v} = 0,$$

$$\tilde{u} = -U, \tag{4.3.19c}$$

so that

$$C_3 = C_4 = -\frac{U}{2}. \tag{4.3.19d}$$

Thus

$$\tilde{v} = Ue^{-z/\delta_E} \sin z/\delta_E, \qquad (4.3.20a)$$

$$\tilde{u} = -Ue^{-z/\delta_E} \cos z/\delta_E, \qquad (4.3.20b)$$

or in terms of the total velocity,

$$u = U[1 - e^{-z/\delta_E} \cos z/\delta_E], \qquad (4.3.21a)$$

$$v = Ue^{-z/\delta_E} \sin z/\delta_E. \qquad (4.3.21b)$$

As $z \to \infty$ the velocities approach with exponential rapidity the interior, inviscid geostrophic velocity. The region *directly* affected by friction is of the order of the Ekman layer thickness $\delta_E$, which increases with increasing $A_V$ but *decreases* with increasing rotation. The stronger the rotation, the narrower the region directly affected by viscosity. For very rapidly rotating fluids the Ekman layer will be a thin region pressed against the surface at $z = 0$. It is important to note that $\delta_E$ is independent of $U$.

The physical balances responsible for the Ekman layer are easily understood in terms of the vorticity. If $\xi$ and $\eta$ are the $x$- and $y$-components of vorticity, we have in the present circumstance

$$\xi = -\frac{\partial v}{\partial z},$$
$$\eta = +\frac{\partial u}{\partial z}, \qquad (4.3.22)$$

so that a $z$-derivative of (4.3.9a,b) immediately yields a reduced form of the vorticity equation, i.e.,

$$0 = +f \frac{\partial v}{\partial z} + A_V \frac{\partial^2 \eta}{\partial z^2}, \qquad (4.3.23a)$$

$$0 = +f \frac{\partial u}{\partial z} + A_V \frac{\partial^2 \xi}{\partial z^2}. \qquad (4.3.23b)$$

The last term in each of (4.3.23a,b) represents the diffusion by friction of vorticity, which if unchecked would fill the interior with vorticity continuously generated by the frictional retardation of the flow at the boundary. The terms proportional to $f$ in (4.3.23a,b) represent the production of horizontal vorticity due to the tilting of the planetary vorticity filaments by the shear of the horizontal velocity as described in Section 2.4. The Ekman layer is established as a *balance between the vorticity diffusion from the boundary and the compensating tilting of the planetary vorticity filaments*. The requirement that $u$ should decrease from $U$ to zero as the boundary is approached requires $\eta \neq 0$, and this in turn requires $v \neq 0$, so that the diffusion of $\eta$ can be balanced by the tilting of vortex filaments in the $y$-direction by $\partial v/\partial z$.

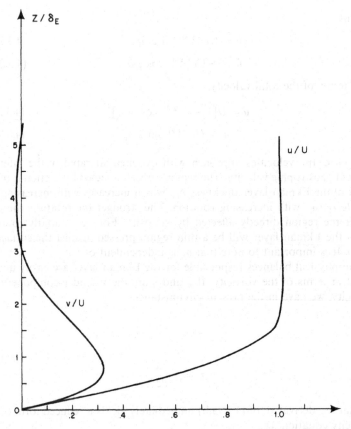

**Figure 4.3.2**   The profile of the velocity component $u$ in the direction of the geostrophic velocity $U$, and of the velocity $v$ at right angles to $U$. The vertical coordinate is the distance from the solid surface measured in units of the Ekman-layer thickness $\delta_E$.

Figure 4.3.2 shows the profiles of the velocity components $u$ and $v$. Far from the wall the velocity is entirely in the $x$-direction, which is the direction of the geostrophic flow. As the wall is approached, the retarding effect of friction decreases $u$. However, the pressure gradient in the $y$-direction is independent of $z$ and is therefore balanced only at infinity by the $y$-component of the Coriolis force. As $u$ decreases, this Coriolis force weakens, and in the presence of the pressure force in the $y$-direction, a velocity $v$ in that direction must be produced, flowing from high pressure to low pressure, retarded only by fluid friction. In the presence of the wall, then, the effect of friction is to break the constraint of exact geostrophic balance and produce a flow across the isobars from high to low pressure, which implies that *work is being done on the fluid in the Ekman layer* by the pressure force. This work by the pressure force supplies the necessary energy to maintain the Ekman layer in the presence of frictional dissipation. The rate, $\dot{W}$, at which work is done by the pressure force in the Ekman layer is given by

$$\dot{W} = \int_0^\infty \left( -v \frac{\partial p}{\partial y} \right) dz = \frac{\rho U^2}{2} f \delta_E$$

$$= \rho \frac{U^2}{2} \{2A_V f\}^{1/2} \tag{4.3.24}$$

and represents the rate energy must be supplied to each unit area of the Ekman layer to maintain it. In the problem we have just solved the energy of the geostrophic flow was held fixed in some unspecified way. We see now, though, that in the presence of the frictional Ekman layer this implicitly requires an energy source to maintain the geostrophic flow. Otherwise the geostrophic flow will decay as the pressure field of the geostrophic flow supplies energy to the Ekman layer. A geostrophic layer of fluid of depth $D$, moving with velocity $U$, possesses a kinetic energy per unit area

$$K = \rho \frac{U^2}{2} D,$$

and supplies energy at a rate $\dot{W}$ to the Ekman layer. Unless externally maintained, the geostrophic flow will therefore decay through viscous dissipation in a time for which a simple estimate is

$$\tau = \frac{K}{\dot{W}} = \left( \frac{D}{\delta_E} \right) f^{-1}$$

$$= \frac{D}{(2A_V f)^{1/2}}. \tag{4.3.25}$$

This characteristic decay time for geostrophic energy will be long compared to a rotation period only if

$$\tau f = \frac{D}{\delta_E}$$

$$= \left( \frac{D^2 f}{2A_V} \right)^{1/2} = E_V^{-1/2} \tag{4.3.26}$$

is *large*. The nondimensional number

$$E_V = 2 \frac{A_V}{f D^2} \tag{4.3.27}$$

is the *Ekman number*. We saw in Section 2.8 that the smallness of this parameter was precisely the condition that viscous forces are small compared to the Coriolis acceleration. We see now that this condition equally well can be interpreted as the condition that frictional forces are sufficiently weak so that the natural decay time $\tau$, due to viscous dissipation in the Ekman layer, is large compared to a rotation period. We shall refer to

$\tau$, which measures the effectiveness of friction, as the "spin-down" time.*
Note that $\tau$ is *independent* of the relative velocity, and depends only on $f$
and $\delta_E$.

Another remarkable feature of the Ekman layer is the graceful turning of
the velocity vector as the wall is approached. Figure 4.3.3 shows the velocity

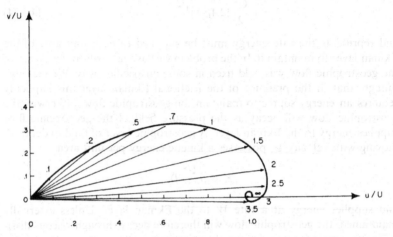

**Figure 4.3.3**   The velocity vector within the Ekman layer. The locus of the tip of the
velocity vector traces the Ekman spiral. The value of $z/\delta_E$ corresponding to each
vector is indicated on the spiral curve.

vector as a function of $z/\delta_E$. Note that the tip of the velocity vector traces a
delicate spiral called the Ekman spiral as $z$ decreases to zero. As each layer of
fluid is frictionally retarded by the layer beneath it, the response of the
velocity of the upper layer, due to the planetary rotation, is to twist to the
right of the applied stress. As the stress is communicated from one lamina to
another, the vector slowly turns. Indeed, as the surface is approached,

$$\lim_{z \to 0} \frac{v}{u} = 1, \qquad (4.3.28)$$

so that the velocity vector has turned 45° to the *left* of the geostrophic velocity.

Two additional important results are derived here for later use. The total
cross-isobar flow is, from (4.3.21b),

$$V_E = \int_0^\infty v \, dz = \frac{U\delta_E}{2} \qquad (4.3.29)$$

and is to the *left* of the geostrophic flow. The stress exerted by the rigid

---

* $\tau$ is also called the spin-*up* time, implying then the time required to achieve an equilibrium
state in the presence of friction.

surface on the fluid is the vector

$$\tau = -\rho A_V \left( \mathbf{i} \frac{\partial u(0)}{\partial z} + \mathbf{j} \frac{\partial v(0)}{\partial z} \right) = -\frac{A_V \rho}{\delta_E} U\{\mathbf{i} + \mathbf{j}\}, \qquad (4.3.30)$$

while the total mass flux per unit area in the Ekman layer *caused by* the *frictional forces* is

$$\mathbf{M}_E = \mathbf{i} \int_0^\infty \tilde{u}\, dz + \hat{j} \int_0^\infty \tilde{v}\, dz = \delta_E \frac{U}{2}\{-\mathbf{i} + \mathbf{j}\}, \qquad (4.3.31)$$

so that

$$\boxed{\mathbf{M}_E = \frac{\tau \times \mathbf{k}}{\rho f}}, \qquad (4.3.32)$$

where **k** is a unit vertical vector.

The total mass flow in excess of the prescribed geostrophic flow is therefore perpendicular and to the right of the frictional stress on the fluid. This important result is necessarily independent of the representation of the turbulent stresses, as illustrated in Figure 4.3.4. The only forces acting on the Ekman layer *as a whole* are the pressure-gradient force and the frictional stress at the lower rigid boundary, since the frictional stress by definition vanishes on the upper edge of the Ekman layer. Now the pressure force is balanced exactly by the Coriolis acceleration of the geostrophic velocity. This requires, then, that the frictional stress produce on average an additional Coriolis acceleration in the direction of the stress. This in turn requires that the average flow departure from the geostrophic velocity, $\mathbf{M}_E$, be perpendicular to $\tau$. $\mathbf{M}_E$ depends only on $\tau$ and *not* on the details of the turbulent stress parametrization, as evidenced by the absence of any reference to the structure of the Ekman layer in (4.3.32) or any dependence on the particular magnitude of $A_V$.

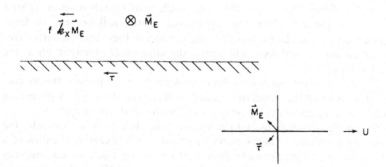

**Figure 4.3.4**  The relationship between the Ekman flux $\mathbf{M}_E$ and the frictional stress $\tau$ on the plate. The drawing at the lower right shows this in plan view with the stress, the flux, and the free-stream geostrophic velocity $U$ shown. Note that $\mathbf{M}_E$ is the total mass flux associated with the deviation of the velocity from its geostrophic value.

For both the atmosphere and the ocean the cross-isobar flow in the Ekman layer represents an important mechanism for the dissipation of kinetic energy. As has been remarked in Section 4.2, we cannot expect the detailed predictions of this simple theory to apply in the real, turbulent ocean and atmosphere, but the basic physical ideas concerning the dissipation, stress, and cross-isobar flow in the friction region are sufficiently general to transcend the limitations of the model and therefore can be expected to give a good qualitative picture of the frictional coupling of geostrophic flow to the earth's surface.

## 4.4 The Nature of Nearly Frictionless Flow

In most cases of interest $A_V$ is sufficiently small (i.e., $E_V \ll 1$) that it might appear that friction could be neglected altogether. However, the condition of no slip on solid surfaces can be expected to still apply for arbitrarily small but nonzero values of $A_V$. Therefore, no matter how small $A_V$ is, friction *must* be important somewhere, and, quite likely, near the boundary where the no-slip condition is applied. This is directly related to the mathematical structure of the equations of motion. It is apparent from (4.3.1a,b,c) that as far as the derivatives are concerned, $A_V$ is the coefficient of the *highest* $z$-derivative. If terms proportional to $A_V$ are neglected, the order of the equations is *reduced* and it is no longer possible to satisfy as many boundary conditions as the original equations demand. That is, if we arbitrarily set $A_V = 0$ we must expect to be unable to satisfy both the kinematic condition on the normal flow and the frictional no-slip condition on the tangential flow. Indeed in the inviscid theory of the previous chapter *only* the kinematic condition on the velocity normal to the fluid surfaces was required. If, though, we insist that the no-slip condition must also be satisfied regardless of the smallness of $A_V$, the only way the necessary friction terms can be retained is if the dynamic variables vary sufficiently rapidly in space, at least near the boundary. Then the highest derivatives will become so large compared to lower derivatives that the product of these highest derivatives with their small coefficients will remain the same order of magnitude as the nonfrictional terms.

This raises several questions. First, to what extent is inviscid theory correct if $A_V$ is small but nonzero? Second, how can the effect of friction, even if small, be included in the dynamical description of the fluid motion?

To give these questions and ideas some concrete expression, consider the following example. As in the previous section, let us examine the flow of a homogeneous, incompressible fluid, again rotating, but now *confined* between two infinite rigid horizontal planes a distance $2D$ apart as shown in Figure 4.4.1. Also, as we did in Section 4.3, let us examine the flow which is horizontally uniform, i.e., independent of $x$ and $y$. For the same reasons as those which led to (4.3.8), we can here set the vertical velocity $w$ equal to

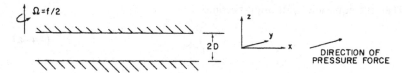

**Figure 4.4.1**   The setting for the discussion of channel flow in the rotating system. The direction of the pressure gradient is parallel to the $y$-axis.

zero. The equations for the horizontal velocity are again

$$-fv = -\frac{1}{\rho}\frac{\partial p}{\partial x} + A_V\frac{\partial^2 u}{\partial z^2}, \qquad (4.4.1a)$$

$$fu = -\frac{1}{\rho}\frac{\partial p}{\partial y} + A_V\frac{\partial^2 v}{\partial z^2}, \qquad (4.4.1b)$$

$$0 = -\frac{1}{\rho}\frac{\partial p}{\partial z} - g. \qquad (4.4.1c)$$

As before, it follows from (4.4.1c) that $\partial p/\partial x$ and $\partial p/\partial y$ must be independent of $z$. Furthermore since $u$ and $v$ are independent of $x$ and $y$, it follows from (4.4.1a,b) that the horizontal pressure gradient must also be independent of $x$ and $y$. Therefore $\partial p/\partial x$ and $\partial p/\partial y$ are constants and can be chosen arbitrarily. In a physical situation they correspond to the externally imposed overall pressure drop across a large but finite region. We choose, without loss of generality, to consider the case where the pressure drop is in the $y$-direction, i.e., let

$$\frac{\partial p}{\partial x} = 0, \qquad (4.4.2a)$$

$$-\frac{\partial p}{\partial y} = \rho f U_0. \qquad (4.4.2b)$$

The second expression merely defines a *number* $U_0$ which has the dimensions of velocity as a measure of the strength of the pressure gradient. It is *not* the geostrophic approximation. It is useful once again to introduce nondimensional variables. Define

$$z' = \frac{z}{D},$$

$$u' = \frac{u}{U_0}, \qquad (4.4.3)$$

$$v' = \frac{v}{U_0}.$$

Then the equations of motion become

$$-v = \frac{E_V}{2} \frac{\partial^2 u}{\partial z^2},$$
(4.4.4a)

$$u = 1 + \frac{E_V}{2} \frac{\partial^2 v}{\partial z^2},$$
(4.4.4b)

where the prime notation has been dropped for neatness. *Unprimed* variables are now *nondimensional*, while the dimensional variables are denoted by an asterisk. Thus the dimensional velocities are

$$u_* = U_0 u,$$
$$v_* = U_0 v,$$
(4.4.5)

etc. The parameter $E_V$ is the Ekman number, i.e.,

$$E_V = 2 \frac{A_V}{fD^2},$$
(4.4.6)

and we are particularly interested in the case where $E_V$ is small.

The appropriate boundary conditions for (4.4.4a,b) are the no-slip conditions

$$u = v = 0 \quad \text{on } z = \pm 1,$$
(4.4.7)

which are independent of the size of $E_V$.

If $E_V$ is small, we might suppose that at least to lowest order the terms proportional to $E_V$ can be neglected. This would yield

$$v = 0,$$
$$u = 1,$$
(4.4.8)

which, in fact satisfies the full equations (4.4.4a,b) for arbitrary $E_V$, but is utterly incapable of satisfying (4.4.7). The approximation (4.4.8) has lowered the order of the differential equation and renders the approximate solution (4.4.8) inadequate although, as we shall see below, not irrelevant. This approximation, which does such violence to the mathematical structure of the equations, is a *singular perturbation* of the equations, and we must seek to understand how the limit $E_V \to 0$ can be taken while still satisfying the full boundary conditions. Fortunately this problem is sufficiently simple that exact solutions for arbitrary $E_V$ may be obtained. Eliminating $u$ between (4.4.4a,b) yields

$$E_V^2 \frac{d^4 v}{dz^4} + 4v = 0.$$
(4.4.9)

Since the boundary conditions are symmetric in $z$, only symmetric solutions need be considered. The general symmetric solution for $v$ is

$$v = A \sinh kz \sin kz + B \cosh kz \cos kz,$$
(4.4.10)

where

$$k = E_V^{-1/2}. \tag{4.4.11}$$

The solution for $u$ is obtained from (4.4.10) and (4.4.4b) and is

$$u = 1 + A \cosh kz \cos kz - B \sinh kz \sin kz. \tag{4.4.12}$$

Application of (4.4.7) determines both $A$ and $B$, so that after a little algebra $u$ and $v$ are determined as

$$v = \frac{\sinh k(1 - z)\sin k(1 + z) + \sinh k(1 + z)\sin k(1 - z)}{\cos 2k + \cosh 2k} \tag{4.4.13}$$

$$u = 1 - \frac{\cos k(1 + z)\cosh k(1 - z) + \cosh k(1 + z)\cos k(1 - z)}{\cos 2k + \cosh 2k}.$$

It is evident $u$ and $v$ satisfy $u = v = 0$ on $z = \pm 1$. For the case of very *large* values of $E_V$, i.e., the *high*-friction limit wherein $k$ is very small, (4.4.13) becomes, after expanding the trigonometric and hyperbolic functions in a Taylor series in $k$,

$$v \approx k^2(1 - z^2) + \cdots,$$

$$u \approx \frac{k^4}{6} (1 - z^2)(5 - z^2) + \cdots, \tag{4.4.14}$$

which in dimensional units is simply

$$v_* = -\left(\frac{\partial p}{\partial y}\right)\frac{D^2 - z_*^2}{2A_V \rho}, \tag{4.4.15}$$

$$u_* = -\left(\frac{\partial p}{\partial y}\right)(D^2 - z_*^2)(5D^2 - z_*^2)\frac{f}{24\rho A_V^2}. \tag{4.4.16}$$

For large friction, i.e., large $E_V$, the flow is predominantly in the $y$-direction and so *down the pressure gradient*. The form of $v$ is the parabolic profile of nonrotating channel flow. There is a weak component of flow $u$, $O(k^2)$ compared to $v$, to the *right* of the down-gradient flow, a consequence of the relatively feeble Coriolis force. However, our interest is in the other extreme, i.e., when $E_V$ is very small ($k$ very large). In this limit, for example,

$$\cosh k(1 + z) \approx \sinh k(1 + z) \approx \frac{e^{k(1 + z)}}{2}, \tag{4.4.17}$$

so that

$$v \approx e^{-k(1 + z)} \sin k(1 + z) + e^{-k(1 - z)} \sin k(1 - z),$$

$$u \approx 1 - e^{-k(1 + z)} \cos k(1 + z) - e^{-k(1 - z)} \cos k(1 - z). \tag{4.4.18}$$

Note that the exponential terms of the form $e^{-k(1 + z)}$ are utterly negligible *unless* $z$ is near $-1$, while similarly the terms of the form $e^{-k(1 - z)}$ are negligible unless $z$ is very near 1. The asymptotic solution (4.4.18) has a simple interpretation. It consists of a uniform geostrophic flow in the $x$-direction (i.e., $u = 1$), supplemented by a thin Ekman layer on $z = 1$ and another Ekman layer on $z = -1$, as reference to (4.3.21a,b) shows. That is, near $z = -1$, for example, the terms like $e^{-k(1-z)}$ are $O(e^{-2k})$ and are so small that there,

$$u = 1 - e^{-k(1+z)} \cos k(1 + z),$$
$$v = e^{-k(1+z)} \sin k(1 + z), \tag{4.4.19}$$

which is the Ekman solution. Similar behavior occurs near $z = 1$. A schematic sketch of the solution is shown in Figure 4.4.2. Note that

$$k = \frac{D}{\delta_E} = \left(\frac{2A_V}{fD^2}\right)^{-1/2}. \tag{4.4.20}$$

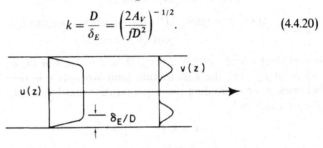

**Figure 4.4.2**    A schematic of $u(z)$ and $v(z)$ in the limit of small $E_V$.

Now consider *any fixed* point $z$ not exactly on the boundary, and examine the nature of the solution (4.4.18) as $E_V$ decreases. No matter how close to the boundary $z$ is, for sufficiently small $E_V$ the point $z$ will be *outside* both Ekman layers. In other words, *one* sensible limit of the solutions for small $E_V$ is

$$\lim{}_1 = \lim_{\substack{E_V \to 0 \\ z \text{ fixed}}} \binom{u}{v} = \binom{1}{0}. \tag{4.4.21}$$

We recognize the result of this limit process as the singular solutions (4.4.8). While inadequate, they are certainly an excellent approximation to the solution, for small $E_V$, in regions *away* from the boundary. Since they are relevant only *outside* the Ekman layers, it is only to be expected that they do not satisfy the frictional boundary conditions on the tangential velocity.

In order to remain within an Ekman layer as $E_V \to 0$ it is necessary to get closer and closer to the boundary as $E_V \to 0$. For example, to remain within the lower Ekman layer at a fixed fraction of the Ekman-layer thickness we must stay at a fixed value of

$$\zeta = (1 + z)k = (1 + z)\frac{D}{\delta_E}, \tag{4.4.22}$$

so that, as $\delta_E/D \to 0$, $z$ approaches the lower boundary at $z = -1$, i.e., $z = -1 + \zeta \, \delta_E/D$. There is therefore a second limit process (call it $\lim_2$),

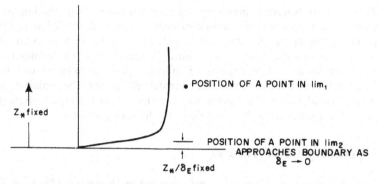

**Figure 4.4.3**  The position of a point in $\lim_1$ is at fixed $z_*$. In $\lim_2$ its position is at fixed $z_*/\delta_E$ and hence approaches the boundary as $E_V \to 0$.

defined as

$$\lim_2 = \lim_{\substack{E_V \to 0 \\ \zeta \text{ fixed}}} \begin{pmatrix} u \\ v \end{pmatrix} = \begin{pmatrix} 1 - e^{-\zeta} \cos \zeta \\ e^{-\zeta} \sin \zeta \end{pmatrix}, \tag{4.4.23}$$

which is illustrated in Figure 4.4.3 and which adequately describes the solution near $z = -1$. A third limit, $\lim_3$, for which $\bar{\zeta} = (1 - z)k$ is held fixed as $E_V \to 0$, gives a similar representation near $z = 1$, viz.

$$\lim_3 = \lim_{\substack{E_V \to 0 \\ \bar{\zeta} \text{ fixed}}} \begin{pmatrix} u \\ v \end{pmatrix} = \begin{pmatrix} 1 - e^{-\bar{\zeta}} \cos \bar{\zeta} \\ e^{-\bar{\zeta}} \sin \bar{\zeta} \end{pmatrix}. \tag{4.4.24}$$

Which limit is correct? The answer is that *all these limits are correct limits of the solution in different regions of the flow.*

For the interior region, the singular limit, $\lim_1$, which corresponds to neglecting viscosity altogether, will yield the appropriate interior inviscid solution. The other limits will yield the solutions for each of the friction regions. In each friction region the correction to the interior solutions is seen to be given by functions of a *stretched variable*, e.g., $\zeta$. That is, in the boundary layer the flow is a function, not of $z_*/D$, but of the vertical coordinate scaled by the very small boundary layer thickness $\delta_E$. This means that as $\delta_E \to 0$, these solutions vary in the physical coordinate $z_*$ so rapidly that the frictional force becomes *locally* important even though $A_V \to 0$. For example, with (4.4.22) we have

$$\frac{\partial v}{\partial z} = \frac{\partial v}{\partial \zeta} \frac{\partial \zeta}{\partial z} = \frac{D}{\delta_E} \frac{\partial v}{\partial \zeta}, \tag{4.4.25}$$

so that

$$E_V \frac{\partial^2 v}{\partial z^2} = \frac{\partial^2 v}{\partial \zeta^2} E_V \frac{D^2}{\delta_E^2} = \frac{\partial^2 v}{\partial \zeta^2} = O(1) \tag{4.4.26}$$

for all $E_V \to 0$.

This singular behavior, involving multiple limits for small friction parameters, is completely characteristic of nearly inviscid flows. What is not apparent in this simple example is how easily we can exploit this behavior to find solutions of more complex problems. The fundamental simplification that clearly arises is that regions of different physical balances can be isolated and treated separately, and then matched together. The exploitation of this physical fact is the fundamental notion of singular perturbation theory, which we describe by example in the following section.

## 4.5 Boundary-Layer Theory

The exact solution described in the previous section showed that for $E_V \ll 1$, the direct effect of friction can be expected to be limited to narrow layers near the fluid surface. In this section we will see how this idea can be exploited in a more general setting, where exact solutions are no longer accessible, to deal with the effects of friction on quasigeostrophic motion.

To fix ideas let us consider again the motion of a fluid between two planes a distance $2D$ apart. The lower plane is fixed in the rotating fluid while the upper plane is moving and imparts a velocity

$$u_{*T} = u_{*T}(x_*, y_*),$$
$$v_{*T} = v_{*T}(x_*, y_*) \tag{4.5.1}$$

to the fluid on the upper surface. Asterisks denote dimensional quantities. The motion of the upper surface will only induce motion due to frictional coupling with the fluid. Since $u_{*T}$ and $v_{*T}$ are functions of horizontal position, we cannot expect to obtain exact solutions. The main questions to be faced here are: (1) How is this frictional forcing communicated to the fluid? (2) If the forcing is due to friction, is the inviscid theory of quasigeostrophic motion relevant? (3) How can the motion be calculated? This relatively simple problem contains all the important elements which must be understood before more realistic oceanographic or meteorological models can be formulated.

For a homogeneous and incompressible fluid, (4.2.12) again are the equations of motion. Let us imagine that $u_{*T}$ and $v_{*T}$ are specified and they have some characteristic magnitude $U$, for example their maximum value over the plane. Similarly, let the typical scale of horizontal variation of $u_{*T}$ and $v_{*T}$ be on a length scale $O(L)$. It seems logical then to introduce the following nondimensional variables (unprimed):

$$(x_*, y_*) = L(x, y), \qquad z_* = Dz,$$

$$(u_*, v_*) = U(u, v), \qquad w_* = U\frac{D}{L}w, \tag{4.5.2}$$

$$t_* = \left(\frac{L}{U}\right)t, \qquad p_* = -\rho g z_* + \rho f U L p.$$

Note that $z_*$ is scaled differently than the horizontal coordinates, since vertical variations can be anticipated to exist on the scale $D$ rather than $L$. The vertical velocity, as in shallow-water theory, is scaled with the presumption that the smallness of the aspect ratio $D/L$ implies that $w_*/u_* = O(D/L)$. This also has the effect of ensuring that, a priori, each of the terms in the continuity equation is of the same order, i.e., in dimensionless units

$$\frac{\partial u}{\partial x} + \frac{\partial v}{\partial y} + \frac{\partial w}{\partial z} = 0. \tag{4.5.3}$$

Even though the vertical mass flux is over a large area $O(L^2)$, its magnitude is $O(w_* L^2)$, i.e., $O(UDL)$. This is precisely the same order as the horizontal mass flux in the depth $D$, across the perimeter, $O(L)$, of the same region. The scaling (4.5.2) ensures that no a priori restriction is placed on the nature of the mass balance.

The scaling for the time that has been chosen is the advective time scale $L/U$, although at this stage it is not clear whether it, rather than the spin-up time, is the better scale. It will be seen below that this depends on the ratio

$$r = \frac{L}{U\tau} \tag{4.5.4}$$

and so for simplicity we will formally assume that

$$r = O(1), \tag{4.5.5}$$

i.e., that a priori the spin-up time and the advective time, both of which are long compared to $f^{-1}$, are of the same order. This allows us to proceed without an a priori restriction that frictional rather than advective processes dominate. The theory, when developed, can then be examined for large or small values of $r$.

The pressure has been scaled in anticipation of the fact that the horizontal pressure gradient will be of the same order as the Coriolis acceleration. This, as we have seen, is certainly true as long as the Rossby number is not *large*.

If (4.5.2) are used to express the momentum equations in nondimensional form, we obtain

$$\varepsilon \left\{ \frac{\partial u}{\partial t} + u \frac{\partial u}{\partial x} + v \frac{\partial u}{\partial y} + w \frac{\partial u}{\partial z} \right\} - v$$

$$= -\frac{\partial p}{\partial x} + \frac{E_V}{2} \frac{\partial^2 u}{\partial z^2} + \frac{E_H}{2} \left( \frac{\partial^2 u}{\partial x^2} + \frac{\partial^2 u}{\partial y^2} \right), \tag{4.5.6a}$$

$$\varepsilon \left\{ \frac{\partial v}{\partial t} + u \frac{\partial v}{\partial x} + v \frac{\partial v}{\partial y} + w \frac{\partial v}{\partial z} \right\} + u$$

$$= -\frac{\partial p}{\partial y} + \frac{E_V}{2} \frac{\partial^2 v}{\partial z^2} + \frac{E_H}{2} \left( \frac{\partial^2 v}{\partial x^2} + \frac{\partial^2 v}{\partial y^2} \right), \tag{4.5.6b}$$

$$\delta^2\varepsilon\left|\frac{\partial w}{\partial t} + u\frac{\partial w}{\partial x} + v\frac{\partial w}{\partial y} + w\frac{\partial w}{\partial z}\right|$$

$$= -\frac{\partial p}{\partial z} + \delta^2\left[\frac{E_V}{2}\frac{\partial^2 w}{\partial z^2} + \frac{E_H}{2}\left(\frac{\partial^2 w}{\partial x^2} + \frac{\partial^2 w}{\partial y^2}\right)\right], \qquad (4.5.6c)$$

where

$$\varepsilon = \frac{U}{fL}, \qquad \text{Rossby number,}$$

$$\delta = \frac{D}{L}, \qquad \text{aspect ratio,}$$

$$\qquad\qquad\qquad\qquad\qquad\qquad\qquad\qquad\qquad (4.5.7)$$

$$E_V = 2\frac{A_V}{fD^2}, \qquad \text{``vertical'' Ekman number,}$$

$$E_H = 2\frac{A_H}{fL^2}, \qquad \text{``horizontal'' Ekman number.}$$

The equation of motion in the vertical direction shows that hydrostatic balance can be expected when $\delta$ is small. Indeed each of the parameters $\varepsilon$, $\delta$, $E_V$, and $E_H$ are small in cases of interest, and one of the essential benefits of systematic scaling is to allow us to see clearly the *relative* importance of different small physical forces, each of which may be a potential contributor to the higher-order dynamics which must be considered to resolve the degeneracy of the geostrophic balance.

The boundary conditions are

$$w = 0, \qquad\qquad z = 0, 1, \qquad\qquad (4.5.8)$$

$$u = v = 0, \qquad\qquad z = 0, \ \Big\}$$

$$u = u_T(x, y), \ \Big\} \qquad\qquad\qquad\qquad (4.5.9)$$
$$\qquad\qquad\qquad\qquad z = 1. \ \Big\}$$
$$v = v_T(x, y), \ \Big\}$$

The first condition, (4.5.8), is the kinematic condition that there be no flow *through* the surfaces on $z = 0, 1$. That condition can be satisfied by an inviscid fluid. The condition (4.5.9) can only be satisfied if small frictional forces exist in the fluid, and in fact in this case the fluid will only be set in motion by those forces produced by the frictional traction of the upper surface. For simplicity we will assume that the imposed velocity $(u_T, v_T)$ is horizontally nondivergent, i.e., that no new plate is produced.

The problem as it stands is still appallingly difficult unless the smallness of $\varepsilon$, $\delta$, $E_V$, and $E_H$ is explicitly exploited.

The solution for each variable, for example $u$, will be a function of $x$, $y$, $z$, and $t$ and of $\varepsilon$, $\delta$, $E_V$, and $E_H$, i.e.,

$$u = u(x, y, z, t, \varepsilon, \delta, E_V, E_H). \qquad\qquad (4.5.10)$$

For small values of the parameters we can expand $u$ in a series in the small parameters, i.e.,

$$u = u_0(x, y, z, t) + \Delta(\varepsilon, \delta, E_V, E_H)u_1(x, y, z, t) + \cdots, \quad (4.5.11)$$

where $\Delta$ is a small parameter and is a function of $\varepsilon$, $\delta$, $E_V$, and $E_H$. It is unknown at this point except that $\Delta$ must go to zero as $\varepsilon$, $\delta$, $E_V$, and $E_H$ go to zero. The particular form of the expansion parameters is not important now. What is important is that (4.5.11) defines $u_0$ by the limiting process

$$u_0 = \lim_{\substack{(x, y, z, t)\,\text{fixed} \\ E_V, \varepsilon, \delta, E_H \to 0}} u. \quad (4.5.12)$$

Insofar as this limit fixes $z$ and lets $E_V \to 0$, the limit must correspond to $\lim_1$ of the preceding section. We therefore can expect that the expansion (4.5.11) will yield only an adequate representation for the *interior* flow. We cannot expect that $u_0$ and $v_0$ will themselves satisfy (4.5.9). Nevertheless, we also saw that the interior limit gave an adequate description of the interior dynamics, so let us examine the consequences of (4.5.11) first. If each variable is expanded as in (4.5.11), the order-one terms of (4.5.6) and (4.5.3) yield for the inviscid interior

$$u_0 = -\frac{\partial p_0}{\partial y}, \quad (4.5.13a)$$

$$v_0 = \frac{\partial p_0}{\partial x}, \quad (4.5.13b)$$

$$0 = \frac{\partial p_0}{\partial z}, \quad (4.5.13c)$$

$$\frac{\partial u_0}{\partial x} + \frac{\partial v_0}{\partial y} + \frac{\partial w_0}{\partial z} = 0. \quad (4.5.14)$$

The interior $O(1)$ horizontal velocity is geostrophic and hydrostatic. It follows from (4.5.13c) that $u_0$ and $v_0$ must also be independent of $z$, so it is clear that $u_0$ and $v_0$ cannot satisfy (4.5.9) unless $u_T = v_T = 0$. If (4.5.13) is used in (4.5.14), it follows that

$$\frac{\partial w_0}{\partial z} = 0, \quad (4.5.15)$$

so that $w_0$ is also independent of $z$. The $O(1)$ interior flow is hydrostatic and geostrophic, satisfies the Taylor-Proudman theorem, and is completely *unknown*. From physical considerations alone it is clear that only by examining the friction regions near $z = 0$ and $z = 1$ will the forcing in the problem enter and aid in the determination of the geostrophic flow.

In order to examine the boundary-layer region near $z = 0$, we need a way of expanding $u$, say, for small $E_V$ which keeps us in the friction layer. That is, we need, as in (4.5.12), to define $u_0$ in the friction layer by a limiting process which retains the importance of friction for small $E_V$. As we saw in the exact solution of the preceding section, this implies that $u$ will be a rapidly varying function of $z$ in the friction layer, a function whose variation is sufficiently rapid so that friction is locally important. This means that in the friction layer, $u$ is really a function of a stretched coordinate in $z$. The stretching ensures that the new boundary-layer coordinate, $\zeta$, remains O(1) as the boundary-layer thickness gets arbitrarily thin.

Another way of expressing this same fact is simply to admit that *in the friction layer*, the presumption of (4.5.2) that the dynamical fields vary in $z_*$ on the scale $D$ is wrong. In the friction layer they vary on a much smaller scale, and the equations should be rescaled with a new vertical coordinate before approximations are made to the full equations of motion. From either point of view we must find a new coordinate

$$\zeta = \frac{z}{l}, \tag{4.5.16}$$

where $l$ is the dimensionless boundary-layer thickness. The dimensional thickness is

$$l_* = Dl. \tag{4.5.17}$$

Now what is $l$? The only way to determine $l$ is to demand that the transformation (4.5.16) retain the frictional terms. If (4.5.6) and (4.5.3) are now rewritten in terms of $\zeta$ (i.e., if each dependent variable is considered a function of $x$, $y$, and $\zeta$), then

$$\varepsilon\left|\frac{\partial u}{\partial t} + u\frac{\partial u}{\partial x} + v\frac{\partial u}{\partial y} + \frac{w}{l}\frac{\partial u}{\partial \zeta}\right| - v$$

$$= -\frac{\partial p}{\partial x} + \frac{E_V}{2l^2}\frac{\partial^2 u}{\partial \zeta^2} + \frac{E_H}{2}\left(\frac{\partial^2 u}{\partial x^2} + \frac{\partial^2 u}{\partial y^2}\right), \tag{4.5.18a}$$

$$\varepsilon\left|\frac{\partial v}{\partial t} + u\frac{\partial v}{\partial x} + v\frac{\partial v}{\partial y} + \frac{w}{l}\frac{\partial v}{\partial \zeta}\right| + u$$

$$= -\frac{\partial p}{\partial y} + \frac{E_V}{2l^2}\frac{\partial^2 v}{\partial \zeta^2} + \frac{E_H}{2}\left(\frac{\partial^2 v}{\partial x^2} + \frac{\partial^2 v}{\partial y^2}\right), \tag{4.5.18b}$$

$$+ \delta^2\varepsilon\left|\frac{\partial w}{\partial t} + u\frac{\partial w}{\partial x} + v\frac{\partial w}{\partial y} + \frac{w}{l}\frac{\partial w}{\partial \zeta}\right|$$

$$= -\frac{1}{l}\frac{\partial p}{\partial \zeta} + \frac{\delta^2}{l^2}\left[\frac{E_V}{2}\frac{\partial^2 w}{\partial \zeta^2}\right] + \delta^2\frac{E_H}{2}\left(\frac{\partial^2 w}{\partial x^2} + \frac{\partial^2 w}{\partial y^2}\right), \tag{4.5.18c}$$

$$\frac{\partial u}{\partial x} + \frac{\partial v}{\partial y} + \frac{1}{l}\frac{\partial w}{\partial \zeta} = 0. \tag{4.5.18d}$$

In order for the friction term proportional to $E_V$ to be as large as the O(1) Coriolis acceleration, we choose

$$l = E_V^{1/2}, \tag{4.5.19}$$

so that

$$l_* = DE_V^{1/2} = \delta_E.$$

The appropriate scaling thickness is therefore the Ekman-layer thickness, even though the Ekman solution is no longer an exact solution. In the boundary layer, then, each variable is written

$$\mathbf{u} = \tilde{\mathbf{u}}(x, y, \zeta, t, E_V, E_H, \varepsilon, \delta) = \tilde{\mathbf{u}}_0(x, y, \zeta, t) + \cdots, \qquad (4.5.20)$$

where the tilde, $\tilde{\ }$, reminds us the representation is valid in the boundary layer near $z = 0$. The expansion in the small parameters in (4.5.20) is equivalent to $\lim_2$, i.e., the limit $E_V \to 0$, for fixed $\zeta = z/E_V^{1/2}$.

Consider now the continuity equation (4.5.18d). It follows from (4.5.19) that

$$\frac{\partial \tilde{w}}{\partial \zeta} = -E_V^{1/2}\left(\frac{\partial \tilde{u}}{\partial x} + \frac{\partial \tilde{v}}{\partial y}\right), \qquad (4.5.21)$$

so that $\partial \tilde{w}/\partial \zeta$ is no larger than $O(E_V^{1/2})$. If $\tilde{w}$ is larger than $O(E_V^{1/2})$ it would require

$$\frac{\partial \tilde{w}}{\partial \zeta} = 0, \qquad (4.5.22)$$

but since the vertical velocity vanishes on the surface $z = \zeta = 0$, it follows that $\tilde{w}$ would then be identically zero. (This argument could not be applied directly to $w_0$ in (4.5.15), since the interior solution is not valid at the boundary where the boundary conditions are applied.) Thus $\tilde{w}$ should be rescaled, i.e., we write, in the boundary layer,

$$\tilde{w}(x, y, \zeta, t, \varepsilon, E_V, E_H, \delta) = E_V^{1/2}\tilde{W}(x, y, \zeta, t, \varepsilon, E_V, E_H, \delta)$$
$$= E_V^{1/2}\{\tilde{W}_0(x, y, \zeta, t) + \cdots\}. \qquad (4.5.23)$$

This is equivalent to recognizing that for the boundary layer, the relevant aspect ratio for the motion is not $D/L$ but $\delta_E/L$, so that

$$w_* = O\left(U\,\frac{\delta_E}{L}\right) = U\,\frac{D}{L}\,\frac{\delta_E}{D} = E_V^{1/2}\,\frac{UD}{L}. \qquad (4.5.24)$$

If $\tilde{u}$, $\tilde{v}$, $\tilde{W}$, and $\tilde{p}$ are expanded in a series for small values of $\varepsilon$, $E_V$, $E_H$, etc., the $O(1)$ variables must satisfy

$$-\tilde{v}_0 = -\frac{\partial \tilde{p}_0}{\partial x} + \frac{1}{2}\frac{\partial^2 \tilde{u}_0}{\partial \zeta^2}, \qquad (4.5.25a)$$

$$\tilde{u}_0 = -\frac{\partial \tilde{p}_0}{\partial y} + \frac{1}{2}\frac{\partial^2 \tilde{v}_0}{\partial \zeta^2}, \qquad (4.5.25b)$$

$$0 = -\frac{\partial \tilde{p}_0}{\partial \zeta}, \qquad (4.5.25c)$$

and

$$\frac{\partial \tilde{W}_0}{\partial \zeta} = -\left(\frac{\partial \tilde{u}_0}{\partial x} + \frac{\partial \tilde{v}_0}{\partial y}\right). \qquad (4.5.26)$$

As long as the Rossby numbers and the Ekman numbers are small with respect to one, the lowest-order equations in the boundary layer are the same equations as for the Ekman layer for a horizontally uniform flow. The scale of variation in $x$ and $y$ is so slow compared to the vertical scale of variation that at each horizontal position the dynamics to lowest order does not recognize that the fields are not constant horizontally. This is one of the great simplifications afforded by the exploitation of the boundary-layer concept.

Now (4.5.25c) implies that $\partial \tilde{p}_0 / \partial x$ and $\partial \tilde{p}_0 / \partial y$ must be independent of $\zeta$, since

$$0 = \frac{\partial}{\partial x} \frac{\partial \tilde{p}_0}{\partial \zeta} = \frac{\partial}{\partial \zeta} \frac{\partial \tilde{p}_0}{\partial x}. \qquad (4.5.27)$$

Consider *any* variable, for example the pressure $p$. We have two representations for $p$: one valid in the interior, $(p_0)$, and one valid in the boundary layer $(\tilde{p}_0)$. As we leave the boundary layer, $\tilde{p}_0$ must merge smoothly with $p_0$ at the edge of the boundary layer. This means that $\tilde{p}_0$ for large $\zeta$ must merge smoothly into $p_0$ for small $z$, i.e.,

$$\lim_{z/l \to \infty} \tilde{p}_0 = \lim_{z \to 0} p_0. \qquad (4.5.28)$$

This *matching principle* is an essential ingredient in ensuring that the approximate solutions we obtain have the same character of smoothness we would anticipate for solutions from the complete equations. The solutions may vary rapidly, but they must merge in an analytically smooth fashion. Since $p_0$ and $\tilde{p}_0$ are independent of $z$ and $\zeta$ respectively, it follows that for *all* $\zeta$ in the boundary layer

$$\frac{\partial \tilde{p}_0}{\partial x} = \frac{\partial p_0}{\partial x} = v_0(x, y),$$

$$\frac{\partial \tilde{p}_0}{\partial y} = \frac{\partial p_0}{\partial y} = -u_0(x, y), \qquad (4.5.29)$$

so that the horizontal pressure gradient in the Ekman layer is given by the *interior* horizontal pressure gradient. The interior pressure is impressed on the boundary layer, and therefore (4.5.25a,b) become

$$\frac{1}{2}\frac{\partial^2 \tilde{u}_0}{\partial \zeta^2} + \tilde{v}_0 = v_0(x, y),$$

$$\frac{1}{2}\frac{\partial^2 \tilde{v}_0}{\partial \zeta^2} - \tilde{u}_0 = -u_0(x, y). \qquad (4.5.30)$$

The solutions of (4.5.30) will be the Ekman-layer solutions of the preceding sections. These can now satisfy the no-slip conditions on $z = 0$, viz.

$$\tilde{u}_0 = u_0(x, y)[1 - e^{-\zeta} \cos \zeta] - v_0(x, y)e^{-\zeta} \sin \zeta,$$
$$\tilde{v}_0 = v_0(x, y)[1 - e^{-\zeta} \cos \zeta] + u_0(x, y)e^{-\zeta} \sin \zeta, \qquad (4.5.31a)$$

or in vector form,

$$\tilde{\mathbf{u}}_0 = \mathbf{u}_0(1 - e^{-\zeta} \cos \zeta) + (\mathbf{k} \times \mathbf{u}_0)e^{-\zeta} \sin \zeta, \qquad (4.5.31b)$$

where $\mathbf{k}$ is a unit vertical vector. Note that the presence of a positive $u_0$ produces a contribution to $\tilde{v}_0$ which is positive (a flow *down* the pressure gradient in $y$), while a positive $v_0$ produces a negative contribution to $\tilde{u}_0$, since then $\partial p_0 / \partial x > 0$ and the flow slipping down the pressure gradient must be in the negative $x$-direction. These Ekman solutions again illustrate the matching principle. Let us write $\tilde{u}_0$, for example, in terms of the interior variable $z$:

$$\tilde{u}_0 = u_0(x, y)[1 - e^{-zE_V^{-1/2}} \cos zE_V^{-1/2}]$$
$$- v_0(x, y)e^{-zE_V^{-1/2}} \sin zE_V^{-1/2}. \qquad (4.5.32)$$

Then for fixed $z$ as $E_V^{1/2} \to 0$, which is the interior limit,

$$\tilde{u}_0 \to u_0(x, y).$$

Note that for *fixed $z$*, the limit $E_V^{1/2} \to 0$ implies $\zeta \to \infty$, since $\zeta = z/E_V^{1/2}$, while for fixed $\zeta$, $E_V^{1/2} \to 0$ implies $z \to 0$. Thus the matching principle (4.5.28) also states that if the boundary-layer solutions are written in terms of the *interior variables*, and the interior solutions are written in terms of the *boundary layer variable*, the subsequent limit $E_V \to 0$ must yield the same result if the solutions merge smoothly,[*] i.e.,

$$\lim_{\substack{\text{fixed } z \\ E_V^{1/2} \to 0}} \tilde{p}_0(z/E_V^{1/2}) = \lim_{\substack{\text{fixed } \zeta \\ E_V^{1/2} \to 0}} p_0(\zeta E_V^{1/2}). \qquad (4.5.33)$$

We now turn our attention to the calculation of the $O(E_V^{1/2})$ vertical velocity given by $\tilde{W}_0$. Even though this velocity is small, the vortex-tube stretching it produces on the planetary vorticity filaments can make it an important factor in the overall dynamics of the layer. Since $\tilde{u}_0$ and $\tilde{v}_0$ are now known, it follows from (4.5.26) that

$$\frac{\partial \tilde{W}_0}{\partial \zeta} = -\left(\frac{\partial \tilde{u}_0}{\partial x} + \frac{\partial \tilde{v}_0}{\partial y}\right)$$
$$= \left(\frac{\partial v_0}{\partial x} - \frac{\partial u_0}{\partial y}\right)e^{-\zeta} \sin \zeta, \qquad (4.5.34)$$

---

[*] Although this form of the matching principle is sufficient for our purposes it must be slightly modified when the solutions contain algebraic powers of $\zeta$. For a discussion of this point, see Cole (1968).

or

$$\tilde{W}_0(x, y, \zeta) = -\frac{1}{2}\left|\frac{\partial v_0}{\partial x} - \frac{\partial u_0}{\partial y}\right|e^{-\zeta}[\cos \zeta + \sin \zeta] + C(x, y), \quad (4.5.35)$$

where $C(x, y)$ is an arbitrary function of horizontal position. However on $\zeta = 0$, $\tilde{W}_0$ must vanish according to (4.5.8). This determines $C(x, y)$, so that

$$\tilde{W}_0(x, y, \zeta) = \frac{1}{2}\left|\frac{\partial v_0}{\partial x} - \frac{\partial u_0}{\partial y}\right|[1 - e^{-\zeta}\{\cos \zeta + \sin \zeta\}], \quad (4.5.36)$$

a graph of which is shown in Figure 4.5.1. The important, even crucial, point to note is that as $\zeta \to \infty$ (i.e., at the upper edge of the boundary layer),

$$\tilde{W}_0(x, y, \infty) = \frac{1}{2}\left|\frac{\partial v_0}{\partial x} - \frac{\partial u_0}{\partial y}\right| \equiv \tfrac{1}{2}\zeta_0. \quad (4.5.37)$$

Thus in the presence of the $O(1)$ interior vorticity, $\zeta_0(x, y)$, a small vertical velocity $O(E_v^{1/2})$ is pumped out of the Ekman layer and into the interior. The physical cause of the pumping is the cross-isobar flow from high to low pressure in the boundary layer. Consider the situation in Figure 4.5.2. A cyclonic region (i.e., a region of positive vorticity, $\zeta_0 > 0$) sits over the plane at $z = 0$. Since the interior flow is geostrophically balanced, the center of the

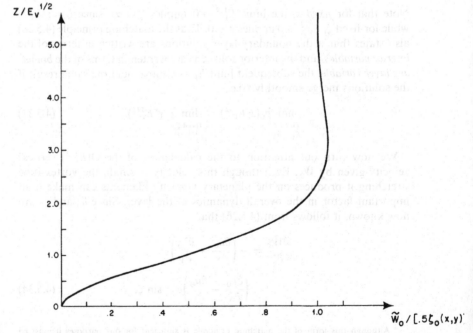

**Figure 4.5.1** The profile of the vertical velocity in the Ekman layer, normalized by the geostrophic vorticity, as a function of distance from the boundary (in units of $\delta_E$).

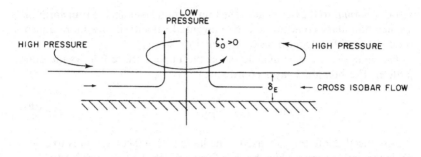

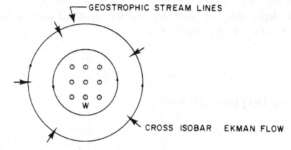

**Figure 4.5.2** A center of low pressure ($\zeta_0 > 0$) produces an influx of mass in the Ekman layer across the geostrophic streamlines and therefore a vertical flux into the interior.

vortex is at low pressure compared to its distant rim. As we noted earlier, a cross-isobar flow of $O(1)$ velocity will occur and will be pushed by the pressure force towards the vortex center. The converging mass flux is order of the velocity times the layer thickness, as given by (4.3.29), which in non-dimensional units is $1 \times E_V^{1/2}$. Since the bottom boundary is impenetrable, the fluid must rise and enter the interior, over the $O(1)$-wide region of the vortex, with a vertical velocity $O(E_V^{1/2})$. An anticyclonic high-pressure center, $\zeta_0 < 0$, will push fluid out radially in the Ekman layer, sucking fluid into the layer from the interior above.

The matching principle when applied to the vertical velocity implies that to lowest order the interior vertical velocity satisfies

$$\lim_{z \to 0} w(x, y, z) = \lim_{\zeta \to \infty} \tilde{w}(x, y, \zeta), \tag{4.5.38}$$

or

$$w(x, y, 0) = \frac{E_V^{1/2}}{2} \left| \frac{\partial v_0}{\partial x} - \frac{\partial u_0}{\partial y} \right|. \tag{4.5.39}$$

Thus, the vertical velocity pumped out of the lower Ekman layer establishes the *lower boundary condition for the interior flow*. Since the interior flow is essentially inviscid, the boundary condition applies to the interior velocity *normal* to the boundary. The thin friction layer acts to transform the applied boundary condition on the tangential velocity into a more appropriate normal flow condition on the interior flow.

The analysis of the Ekman layer on the upper surface follows the same pattern. The boundary-layer coordinate

$$\bar{\zeta} = \frac{1 - z}{E_V^{1/2}} \tag{4.5.40}$$

is introduced. Note that for fixed $\bar{\zeta}$ the limit $E_V^{1/2} \to 0$ corresponds to $z \to 1$, while the limit process which holds $z$ fixed and lets $E_V^{1/2} \to 0$ sends $\bar{\zeta}$ to $+\infty$. If all dependent variables are written as functions of $x$, $y$, and $\bar{\zeta}$, and $z$-derivatives are rewritten using

$$\frac{\partial}{\partial z} = -\frac{1}{E_V^{1/2}} \frac{\partial}{\partial \bar{\zeta}}, \tag{4.5.41}$$

then the $O(1)$ terms of (4.5.6) yield

$$-\hat{v}_0 = -\frac{\partial \hat{p}_0}{\partial x} + \frac{1}{2} \frac{\partial^2 \hat{u}_0}{\partial \bar{\zeta}^2}, \tag{4.5.42a}$$

$$\hat{u}_0 = -\frac{\partial \hat{p}_0}{\partial y} + \frac{1}{2} \frac{\partial^2 \hat{v}_0}{\partial \bar{\zeta}^2}, \tag{4.5.42b}$$

$$0 = -\frac{\partial \hat{p}_0}{\partial \bar{\zeta}}, \tag{4.5.42c}$$

where the caret, $\hat{}$, reminds us that these variables represent the solution in the boundary layer on the *upper* surface.

It follows from (4.5.42c) and the matching principle that the horizontal pressure gradient in the upper boundary is identical to the interior gradient, so that (4.5.42a,b) can be written

$$\frac{1}{2} \frac{\partial^2 \hat{u}_0}{\partial \bar{\zeta}^2} + \hat{v}_0 = v_0(x, y),$$

$$\frac{1}{2} \frac{\partial^2 \hat{v}_0}{\partial \bar{\zeta}^2} - \hat{u}_0 = -u_0(x, y). \tag{4.5.43}$$

The solution of (4.5.43) which satisfies the condition that $\hat{v}_0$ and $\hat{u}_0$ merge smoothly to their interior values as $\bar{\zeta} \to \infty$ is

$$\hat{u}_0 = u_0(x, y) + e^{-\bar{\zeta}}[C_1(x, y)\cos \bar{\zeta} + C_2(x, y)\sin \bar{\zeta}],$$

$$\hat{v}_0 = v_0(x, y) + e^{-\bar{\zeta}}[-C_1(x, y)\sin \bar{\zeta} + C_2(x, y)\cos \bar{\zeta}], \tag{4.5.44}$$

where $C_1$ and $C_2$ are arbitrary functions of $x$ and $y$.

On the upper surface

$$\hat{u}_0 = u_T(x, y),$$
$$\hat{v}_0 = v_T(x, y), \tag{4.5.45}$$

so that

$$C_1 = u_T - u_0,$$
$$C_2 = v_T - v_0, \tag{4.5.46}$$

from which it follows that

$$\hat{u}_0 = u_0(x, y) + e^{-\zeta}[(u_T - u_0)\cos \zeta + (v_T - v_0)\sin \zeta]$$
$$\hat{v}_0 = v_0(x, y) + e^{-\zeta}[(v_T - v_0)\cos \zeta - (u_T - u_0)\sin \zeta]. \tag{4.5.47}$$

Note that if $v_T$ and $u_T$ are zero, (4.5.47) is identical to (4.5.31). That is, the structure of the Ekman layer on $z = 1$ is identical to the structure on $z = 0$. Again, at any horizontal position across isobar, flow will occur in proportion to the difference between the geostrophic velocity and the velocity of the boundary.

The continuity equation becomes

$$\frac{\partial \hat{w}}{\partial \overline{\zeta}} = E_V^{1/2}\left(\frac{\partial \hat{u}_0}{\partial x} + \frac{\partial \hat{v}_0}{\partial y}\right). \tag{4.5.48}$$

Note the change in sign in comparison with (4.5.21). Formally this arises from the minus sign in the coordinate transformation in (4.5.41). Physically it reflects the simple fact that a horizontal convergence of mass in the lower Ekman layer must produce an upward velocity, while an identical convergence in the upper Ekman layer must give rise to a vertical velocity moving downward into the interior.

If (4.5.47) is used to evaluate (4.5.48), $\hat{w}$ is determined by simple integration with the condition that $\hat{w}$ vanishes on $\overline{\zeta} = 0$, i.e.,

$$\hat{w} = \frac{E_V^{1/2}}{2}\left|\frac{\partial}{\partial x}(v_T - v_0) - \frac{\partial}{\partial y}(u_T - u_0)\right|$$
$$\times [1 - e^{-\zeta}(\cos \zeta + \sin \zeta)]. \tag{4.5.49}$$

Again, an $O(E_V^{1/2})$ vertical velocity persists at $\overline{\zeta} \to \infty$, i.e., at the edge of the boundary layer, where it merges into the interior. The matching principle for $w$,

$$\lim_{z \to 1} w(x, y, z) = \lim_{\overline{\zeta} \to \infty} \hat{w}(x, y, \overline{\zeta}), \tag{4.5.50}$$

yields the *upper* boundary condition for the *interior* flow,

$$w(x, y, 1) = \frac{E_V^{1/2}}{2}[\zeta_T(x, y) - \zeta_0(x, y)], \tag{4.5.51}$$

where $\zeta_T$ is the vorticity of the upper boundary and $\zeta_0$ is the $O(1)$ vorticity of the interior flow. If $\zeta_T > \zeta_0$, fluid is flung outward in the upper Ekman layer, sucking fluid vertically from the top of the interior.

Thus boundary-layer theory, which explicitly exploits the thinness of the frictional layer, allows two great simplifications. First, it has allowed us to satisfy the no-slip condition on the tangential velocity on $z = 0$ and $z = 1$ for a *general* geostrophic interior flow. Second, it has translated the frictional or viscous conditions on the tangential flow to conditions on the velocity normal to the boundary *for the interior flow*. For $E_V \ll 1$, the interior region constitutes the bulk of the fluid. The Ekman layers occupy only a small fraction of the fluid thickness. Yet the analysis of the layers is essential in revealing the nature of the coupling of the surface to the fluid interior. For both the upper and lower surfaces any discrepancy between the interior vorticity and the vorticity of the boundary will produce an $O(E_V^{1/2})$ vertical velocity at the edges of the interior flow. The small vortex-tube stretching or compression of the planetary vorticity that results from this is the primary mechanism by which the effects of friction can be communicated into the interior. The tube stretching is an inertial mechanism driven by friction at the boundary. The degree to which this effect is important in the interior vorticity balance is precisely the degree to which friction has an important role to play in the dynamics of the great bulk of the fluid.

## 4.6 Quasigeostrophic Dynamics in the Presence of Friction

In the preceding section the analysis of the interior flow was taken only as far as the lowest-order balance. At the $O(1)$ level (4.5.13a,b,c) demonstrates that the flow is geostrophic and hydrostatic. In order to actually determine the flow in the interior region, the small nongeostrophic forces in the interior must be considered. In particular the asymptotic expansion of the velocity indicated in (4.5.11) must be carried out beyond the first term. The immediate question is what expansion parameter should be used for $\Delta$ in (4.5.11). This parameter measures the departure of the flow from strictly geostrophic balance, and there are several different mechanisms that are responsible for the departure from geostrophy. First, the relative acceleration of the flow will yield nongeostrophic velocities of $O(\varepsilon)$. In the analysis of the quasigeostrophic, inviscid, shallow-water model of Chapter 3, this was the only nongeostrophic mechanism and $\Delta$ was unambiguously equal to $\varepsilon$. The same mechanism exists here, but in addition there are two others. The vortex-tube stretching by the Ekman suction velocity acting on the planetary vorticity filaments will produce an $O(E_V^{1/2})$ variation of the relative vorticity. This is a nongeostrophic effect arising from the influence of friction. The ratio of the stretching by Ekman pumping to advection of relative vorticity is

$$r = \frac{E_V^{1/2}}{\varepsilon} = \left(\frac{2A_V f}{D^2}\right)^{1/2} \frac{L}{U} = \frac{L}{U\tau}, \qquad (4.6.1)$$

where $\tau$ is the spin-up time as derived in Section 4.3. In order to consider the situation where neither the acceleration terms nor the Ekman pumping terms are dominant, we proceed under the assumption that $r$ is $O(1)$. It is possible afterwards to recover the limit of large or small $r$ directly. The third significant contributor to the nongeostrophic velocity field is the presence of the horizontal diffusion of momentum in the interior. It is clear from (4.5.6) that this term is $O(E_H)$. The ratio of this frictional term to the inertial acceleration is

$$\frac{E_H}{\varepsilon} = \frac{2A_H}{UL} = \frac{2}{\text{Re}} \tag{4.6.2}$$

where

$$\text{Re} = \frac{UL}{A_H}$$

is the *Reynolds number* of the interior flow. In most cases of geophysical interest the Reynolds number is quite large. For example, even the fairly large value of $10^5$ m$^2$/s for $A_H$ in the atmosphere gives a value of $\text{Re} = 10^2$ if $U$ is 10 m/s and $L$ is $10^3$ km. Hence $E_H/\varepsilon = \text{Re}^{-1}$ is generally a small parameter. Nevertheless, to ignore this momentum diffusion term altogether would be a singular perturbation of (4.5.6a,b) with regard to the variables $x$ and $y$. That is, if the fluid is enclosed laterally in a basin where no-slip conditions are to be applied on the perimeter, the friction terms must be important near the perimeter. To include that possibility the terms of $O(E_H)$ must be retained in the expectation that locally large $x$- and $y$-derivatives will increase their size.

Thus for the interior velocity we write

$$u = u(x, y, z, t, \varepsilon, E_V, E_H)$$
$$= u_0(x, y, z, t, r, \text{Re}) + \varepsilon u_1(x, y, z, t, r, \text{Re}) + \cdots,$$
$$v = v(x, y, z, t, \varepsilon, E_V, E_H)$$
$$= v_0(x, y, z, t, r, \text{Re}) + \varepsilon v_1(x, y, z, t, r, \text{Re}) + \cdots, \tag{4.6.3}$$
$$w = w(x, y, z, t, \varepsilon, E_V, E_H)$$
$$= w_0(x, y, z, t, r, \text{Re}) + \varepsilon w_1(x, y, z, t, r, \text{Re}) + \cdots,$$
$$p = p(x, y, z, t, \varepsilon, E_V, E_H)$$
$$= p_0(x, y, z, t, r, \text{Re}) + \varepsilon p_1(x, y, z, t, r, \text{Re}) + \cdots,$$

If this expansion is inserted in (4.5.6a,b,c), the $O(1)$ terms of course yield (4.5.13a,b,c) and (4.5.14). It follows from this that $u_0$ and $v_0$ are independent of $z$, while (4.5.15) shows that $w_0$ is also independent of height. Now on $z = 0$, to lowest order,

$$w(x, y, t) = \frac{E_V^{1/2}}{2}\left(\frac{\partial v_0}{\partial x} - \frac{\partial u_0}{\partial y}\right) \tag{4.6.4}$$

by (4.5.39). Since $E_V^{1/2} \ll 1$, it follows that $w_0$ in (4.6.3) must be identically zero for all $z$.

The $O(\varepsilon)$ terms which result from (4.6.3) and (4.5.6a,b,c) are

$$\frac{\partial u_0}{\partial t} + u_0 \frac{\partial u_0}{\partial x} + v_0 \frac{\partial u_0}{\partial y} - v_1 = -\frac{\partial p_1}{\partial x} + \frac{1}{\text{Re}} \left| \frac{\partial^2 u_0}{\partial x^2} + \frac{\partial^2 u_0}{\partial y^2} \right|, \tag{4.6.5a}$$

$$\frac{\partial v_0}{\partial t} + u_0 \frac{\partial v_0}{\partial x} + v_0 \frac{\partial v_0}{\partial y} + u_1 = -\frac{\partial p_1}{\partial y} + \frac{1}{\text{Re}} \left| \frac{\partial^2 v_0}{\partial x^2} + \frac{\partial^2 v_0}{\partial y^2} \right|, \tag{4.6.5b}$$

$$0 = \frac{\partial p_1}{\partial z}, \tag{4.6.5c}$$

$$\frac{\partial u_1}{\partial x} + \frac{\partial v_1}{\partial y} + \frac{\partial w_1}{\partial z} = 0. \tag{4.6.6}$$

Note that the terms $w_0 \, \partial u_0 / \partial z$ and $w_0 \, \partial v_0 / \partial z$ are missing from (4.6.5a,b), not only because $u_0$ and $v_0$ are independent of $z$, but more significantly because $w_0$ is zero. If the pressure gradient is eliminated by cross differentiation with respect to $x$ and $y$, the vorticity equation results:

$$\frac{d\zeta_0}{dt} = \frac{\partial \zeta_0}{\partial t} + u_0 \frac{\partial \zeta_0}{\partial x} + v_0 \frac{\partial \zeta_0}{\partial y}$$

$$= -\left( \frac{\partial u_1}{\partial x} + \frac{\partial v_1}{\partial y} \right) + \frac{1}{\text{Re}} \left( \frac{\partial^2}{\partial x^2} + \frac{\partial^2}{\partial y^2} \right) \zeta_0 \tag{4.6.7}$$

$$= \frac{\partial w_1}{\partial z} + \frac{1}{\text{Re}} \nabla^2 \zeta_0.$$

The total rate of change of the $O(1)$ relative vorticity is due to the planetary vortex tube stretching, $\partial w_1 / \partial z$, and the horizontal diffusion of vorticity, $\text{Re}^{-1} \nabla^2 \zeta_0$. Since $u_0$, $v_0$, and hence $\zeta_0$ are independent of $z$, (4.6.7) may be trivially integrated from $z = 0$ to $z = 1$:

$$\frac{d\zeta_0}{dt} = w_1(x, y, 1) - w_1(x, y, 0) + \frac{1}{\text{Re}} \nabla^2 \zeta_0. \tag{4.6.8}$$

It is at this point that the presence of the viscous layers on $z = 0$ and $z = 1$ becomes important. It follows from (4.5.39) and (4.6.3) that

$$w(x, y, 0) \equiv \varepsilon w_1(x, y, 0) + \cdots = \frac{E_V^{1/2}}{2} \zeta_0(x, y, t) \tag{4.6.9}$$

so that

$$\boxed{w_1(x, y, 0) = \frac{E_V^{1/2}}{2\varepsilon} \zeta_0(x, y, t)}, \tag{4.6.10}$$

and similarly

$$w_1(x, y, 1) = \frac{E_V^{1/2}}{2\varepsilon} \left( \zeta_T(x, y, t) - \zeta_0(x, y, t) \right), \qquad (4.6.11)$$

so that the vorticity equation becomes

$$\frac{d\zeta_0}{dt} = -r \left| \zeta_0 - \frac{\zeta_T}{2} \right| + \frac{1}{Re} \nabla^2 \zeta_0 . \qquad (4.6.12)$$

Since $u_0$ and $v_0$ are geostrophic, (4.6.12) can be written entirely in terms of the pressure $p_0$:

$$\left[ \frac{\partial}{\partial t} + \frac{\partial p_0}{\partial x} \frac{\partial}{\partial y} - \frac{\partial p_0}{\partial y} \frac{\partial}{\partial x} \right] \nabla^2 p_0 = -r \left[ \nabla^2 p_0 - \frac{\zeta_T}{2} \right] + \frac{1}{Re} \nabla^4 p_0 . \quad (4.6.13)$$

In regions removed from lateral boundaries the term proportional to $Re^{-1}$ can be consistently neglected, so that outside friction layers on side boundaries (4.6.12) becomes simply

$$\left[ \frac{\partial}{\partial t} + \frac{\partial p_0}{\partial x} \frac{\partial}{\partial y} - \frac{\partial p_0}{\partial y} \frac{\partial}{\partial x} \right] \nabla^2 p_0 = -r \left[ \nabla^2 p_0 - \frac{\zeta_T}{2} \right] . \qquad (4.6.14)$$

It is useful to compare (4.6.14) with the quasigeostrophic potential vorticity equation (3.12.25) derived in Chapter 3 for an inviscid fluid layer with a free surface. The terms $-F\eta_0 + \eta_B$ in (3.12.25)—the contributions of the free-surface variation and the bottom slope to the potential vorticity—are absent in (4.6.13). This is as it should be, since the model here has two flat and rigid horizontal bounding surfaces. (Note that $F \to 0$ yields the rigid-lid model.) We shall see in Section 4.11 how to include both effects in the presence of viscosity. The essential dynamical difference between (4.6.14) and (3.12.25) is the presence of the damping term for vorticity on the right-hand side of (4.6.14). This represents, *for the interior*, the sole direct effect of the frictional Ekman layers. The magnitude of the effect of friction on the interior flow is measured by the parameter $r$. As is evident in (4.6.1), $r$ is the ratio of the inviscid time scale divided by the spin-up time scale. If the characteristic advective time scale is short compared to the spin-down time, the effect of viscosity can be neglected for periods of time of the order of the advective time. In *linearized* wave problems where the advective time is long compared to a wave period the appropriate Rossby number is really $\varepsilon_T = (fT)^{-1}$, as discussed in Section 3.12, where $T$ is the wave period. In such cases, the parameter $r$ is replaced by

$$r_T = \frac{T}{\tau} . \qquad (4.6.15)$$

Hence if the wave period is short compared to the spin-up time, the effects of viscosity can be neglected for many wave periods, i.e., for intervals of time short compared to $\tau$ but long compared to $T$. The analysis of free inviscid waves in Chapter 3 is therefore valid in the presence of friction if the friction is sufficiently small that $\tau \gg T$. The fluid therefore behaves as an inviscid fluid as long as $\tau \gg (T, L/U)_{\max}$.

## 4.7 Spin-Down

On the basis of very simple energy considerations it was suggested in Section 4.3 that

$$\tau = f^{-1}E_V^{-1/2} \tag{4.7.1}$$

is the appropriate characteristic time for the decay of geostrophic motion under the influence of Ekman-layer friction. With (4.6.14) this notion can be made more precise and the conclusion established more rigorously.

Consider the quasigeostrophic motion in the $x$, $y$ plane in the absence of lateral boundaries and forcing, i.e., let $\zeta_T = 0$. Then for $\text{Re} \gg 1$ the interior vorticity equation is simply

$$\frac{d\zeta_0}{dt} = -r\zeta_0. \tag{4.7.2}$$

Now consider the circulation $\Gamma$ around the closed material curve $C$ in the $x$, $y$ plane. If $C$ is a reducible curve, i.e., if it encompasses no holes or "islands" in the fluid, then

$$\Gamma = \iint_A \zeta_0 \, dA = \oint_C \mathbf{u}_0 \cdot d\mathbf{r}, \tag{4.7.3}$$

where $A$ is the area encompassed by $C$. The rate of change of $\Gamma$ *moving with the fluid* is

$$\frac{d\Gamma}{dt} = \iint_A \frac{d\zeta_0}{dt} \, dA + \iint_A \zeta_0 \frac{d}{dt}(dA). \tag{4.7.4}$$

Since, to lowest order, the geostrophic velocity is nondivergent, $d(dA)/dt$, which is evaluated following the geostrophic velocity, is identically zero, since

$$\frac{d}{dt}(dA) = dA\left|\frac{\partial u_0}{\partial x} + \frac{\partial v_0}{\partial y}\right| = 0. \tag{4.7.5}$$

It follows therefore from (4.7.2) and (4.7.4) that

$$\frac{d\Gamma}{dt} = -r\Gamma, \tag{4.7.6}$$

or

$$\boxed{\Gamma = \Gamma(0)e^{-rt}}, \tag{4.7.7}$$

so that the circulation around any closed contour moving with the fluid exponentially decays with the characteristic time $r^{-1}$. The corresponding dimensional time is simply

$$\frac{L}{U}r^{-1} = \frac{L}{U}\frac{\varepsilon}{E_V^{1/2}} = f^{-1}E_V^{-1/2} = \tau, \tag{4.7.8}$$

i.e., the spin-down time.

The mechanism for the spin-down can be described in several ways. From the viewpoint of the vorticity balance, consider the cyclonic vortex with vorticity $\zeta_0$ as shown in Figure 4.7.1. In the presence of positive relative

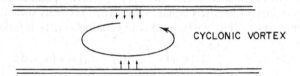

**Figure 4.7.1** A cyclonic vortex sucks fluid out of both upper and lower Ekman layers, producing vortex-tube squashing and an accompanying decrease in the geostrophic relative vorticity.

vorticity the lower Ekman layer will pump fluid into the low-pressure center of the vortex from below while the upper Ekman layer will pump fluid downwards. The combined effect is to compress the planetary vorticity tube at a rate $-r\zeta_0$, and this *inertial* effect reduces the interior vorticity exponentially. From an alternate point of view, the fluid sucked out of the two Ekman layers in the center of the vortex must flow outward in the interior, from the vortex center to its rim. This interior flow must flow from low pressure at the center to high pressure at the vortex edge (being an *interior* cross-isobar flow, it must clearly comprise the small nongeostrophic components of the velocity $u_1$ and $v_1$). This outward mass flux must balance the inward mass flux in the Ekman layers. Since the pressure gradient is the same in the interior as in the Ekman layers, it follows that the rate of energy loss of the interior flow is given precisely by $W$ in (4.3.24) *with the opposite sign*. The small energy loss by the interior flow pushing against the pressure gradient uses up the kinetic energy of the interior at the same rate as it is supplied to the fluid in the Ekman layer, where it is depleted by frictional forces. This yields, as did the argument in Section 4.3, a time scale $\tau$ identical to the decay time.

## 4.8 Steady Motion

If the motion is steady and driven by the motion of the upper lid, (4.6.14) may be written

$$\left(\frac{\partial p_0}{\partial x}\frac{\partial}{\partial y} - \frac{\partial p_0}{\partial y}\frac{\partial}{\partial x}\right)\nabla^2 p_0 = -r\left[\nabla^2 p_0 - \frac{\zeta_T}{2}\right]. \tag{4.8.1}$$

First consider the case where $r \gg 1$, i.e., where the advective time is long compared to the spin-up time. In this case (4.8.1) reduces to the linear equation

$$\zeta_0 = \nabla^2 p_0 = \frac{\zeta_T}{2}.\qquad(4.8.2)$$

In this limit the vorticity of the interior is precisely half of the relative vorticity of the upper-lid motion. This simple solution follows directly from the fact that in the fluid interior, by (4.6.7), the vortex-tube stretching $\partial w/\partial z$ can be no larger than $O(\varepsilon)$, which is the order of $d\zeta/dt$. Thus if $E_V^{1/2} \gg \varepsilon$, this constraint will be violated unless the $O(E_V^{1/2})$ velocity pumped out of the lower Ekman layer is precisely equal to the $O(E_V^{1/2})$ velocity sucked into the upper Ekman layer. Thus if $w(x, y, 0)$ is to equal $w(x, y, 1)$, it follows that

$$w(x, y, 0) = \frac{E_V^{1/2}}{2}\,\zeta_0(x, y) = \frac{E_V^{1/2}}{2}\,(\zeta_T - \zeta_0) = w(x, y, 1),\qquad(4.8.3)$$

or

$$\zeta_0 = \tfrac{1}{2}\zeta_T\qquad(4.8.4)$$

as given by (4.8.2).

This relationship is more general than might at first be imagined. That is, for all $r$ a similar result must also hold. Consider the steady-state form of (4.6.14) written in the form

$$\frac{\partial}{\partial x}(u_0\zeta_0) + \frac{\partial}{\partial y}(v_0\zeta_0) = -r\!\left(\zeta_0 - \frac{\zeta_T}{2}\right),\qquad(4.8.5)$$

and integrate (4.8.5) over an area $A$ in the $x, y$ plane girdled by any closed streamline. Since there is no $O(1)$ motion normal to a streamline, the integral of the left-hand side of (4.8.5) identically vanishes. Thus

$$\iint_A \zeta_0\, dA = \frac{1}{2} \iint \zeta_T\, dA,\qquad(4.8.6)$$

or

$$\oint_C \mathbf{u}_0 \cdot d\mathbf{r} = \frac{1}{2}\oint \mathbf{u}_T \cdot dr,\qquad(4.8.7)$$

so that the geostrophic circulation, in the steady state, on any closed streamline is one-half the circulation of the boundary driving the flow. For the special case of axially symmetric flow where the velocity of the boundary is purely circumferential, the resulting geostrophic velocity will be strictly circumferential and equal to half the boundary velocity. Thus if

$$\mathbf{u}_T = V(r)\hat{\Theta}\qquad(4.8.8)$$

where $\hat{\Theta}$ is a unit vector in the azimuthal direction and $V$, the azimuthal speed, is an arbitrary function of radius, the geostrophic flow is purely azimuthal and equal to $\frac{1}{2}V(r)$.

## 4.9 Ekman Layer on a Sloping Surface

The sensitivity of geostrophic motion to depth variations is strikingly evident in the inviscid theory of Chapter 3. The purpose of this section is to examine to what extent the results of the inviscid theory remain true in the presence of viscosity and, at the same time, investigate how the Ekman-layer contribution to the interior vertical velocity is manifested when the boundary slopes.

The basic idea is revealed in the circumstance illustrated in Figure 4.9.1. A rigid surface slopes upward at an angle $\alpha$ such that its surface is described by

$$z_* = x_* \tan \alpha. \tag{4.9.1}$$

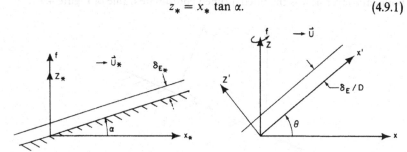

**Figure 4.9.1**  The sloping Ekman layer in the dimensional and nondimensional coordinate frames.

Although the slope considered here is constant, the results to be derived will be valid as long as the radius of curvature of the surface is large compared to the boundary layer thickness.

In dimensionless units

$$z = x \frac{L}{D} \tan \alpha = x \tan \theta, \tag{4.9.2}$$

where $\tan \theta$ is the slope of the surface in the *nondimensional* system. The results of the preceding sections suggest that the viscous boundary layer on the surface results from a balance between the Coriolis acceleration and the viscous forces. The nonlinear terms remain unimportant as long as the Rossby number is much less than unity. The algebra is simplified if that fact is used from the beginning, so we consider, *as appropriate for the boundary layer*, the system

$$-v = -\frac{\partial p}{\partial x} + \frac{E_V}{2} \frac{\partial^2 u}{\partial z^2} + \frac{E_H}{2} \left| \frac{\partial^2 u}{\partial x^2} + \frac{\partial^2 u}{\partial y^2} \right|, \tag{4.9.3a}$$

$$u = -\frac{\partial p}{\partial y} + \frac{E_V}{2} \frac{\partial^2 v}{\partial z^2} + \frac{E_H}{2} \left| \frac{\partial^2 v}{\partial x^2} + \frac{\partial^2 v}{\partial y^2} \right|, \tag{4.9.3b}$$

$$0 = -\frac{\partial p}{\partial z} + \delta^2 \left[ \frac{E_V}{2} \frac{\partial^2 w}{\partial z^2} + \frac{E_H}{2} \left| \frac{\partial^2 w}{\partial x^2} + \frac{\partial^2 w}{\partial y^2} \right| \right], \tag{4.9.3c}$$

$$\frac{\partial u}{\partial x} + \frac{\partial v}{\partial y} + \frac{\partial w}{\partial z} = 0, \tag{4.9.4}$$

with the boundary conditions

$$u = v = w = 0 \quad \text{on } z = x \tan \theta. \tag{4.9.5}$$

Friction is required only to bring the velocity parallel to the boundary to zero. The condition on the velocity normal to the boundary is easily satisfied by inviscid theory. To bring friction into the physical balance, the velocity fields must vary rapidly in the direction perpendicular to the boundary, as shown schematically in Figure 4.9.2. The natural system to describe the boundary-layer flow is the tilted, primed coordinate frame of Figure 4.9.1.

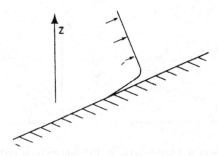

**Figure 4.9.2** The velocity tangential to the boundary changes rapidly in the direction *normal to the boundary.*

The independent and dependent variables in the two frames are related by simple trigonometry as

$$\begin{aligned}
x' &= x \cos \theta + z \sin \theta, & x &= x' \cos \theta - z' \sin \theta, \\
z' &= z \cos \theta - x \sin \theta, & z &= z' \cos \theta + x' \sin \theta, \\
u' &= u \cos \theta + w \sin \theta, & u &= u' \cos \theta - w' \sin \theta, \\
w' &= w \cos \theta - u \sin \theta, & w &= w' \cos \theta + u' \sin \theta,
\end{aligned} \tag{4.9.6}$$

where $v$ and $y$ are the same in both systems. Derivatives in one frame are easily transformed into the other; for example

$$\frac{\partial u}{\partial z} = \frac{\partial u}{\partial z'} \cos \theta + \frac{\partial u}{\partial x'} \sin \theta \tag{4.9.7a}$$

and

$$\frac{\partial^2 u}{\partial z^2} = \frac{\partial^2 u}{\partial z'^2} \cos^2 \theta + 2 \frac{\partial^2 u}{\partial x' \, \partial z'} \sin \theta \cos \theta + \frac{\partial^2 u}{\partial x'^2} \sin^2 \theta, \tag{4.9.7b}$$

etc.

Equations (4.9.3a,b,c) then become, when written in terms of the primed variables,

$$-v = -\frac{\partial p}{\partial x'}\cos\theta + \frac{\partial p}{\partial z'}\sin\theta$$

$$+\frac{E_V}{2}\left[\cos^2\theta\frac{\partial^2}{\partial z'^2} + \sin 2\theta\frac{\partial^2}{\partial x'\,\partial z'} + \sin^2\theta\frac{\partial^2}{\partial x'^2}\right][u'\cos\theta - w'\sin\theta]$$

$$+\frac{E_H}{2}\left[\sin^2\theta\frac{\partial^2}{\partial z'^2} - \sin 2\theta\frac{\partial^2}{\partial x'\,\partial z'} + \cos^2\theta\frac{\partial^2}{\partial x'^2}\right][u'\cos\theta - w'\sin\theta]$$

$$+\frac{E_H}{2}\frac{\partial^2}{\partial y^2}[u'\cos\theta - w'\sin\theta], \tag{4.9.8a}$$

$$u'\cos\theta - w'\sin\theta = -\frac{\partial p}{\partial y} + \frac{E_V}{2}\left[\cos^2\theta\frac{\partial^2}{\partial z'^2} + \sin 2\theta\frac{\partial^2}{\partial x'\,\partial z'} + \sin^2\theta\frac{\partial^2}{\partial x'^2}\right]v$$

$$+\frac{E_H}{2}\left[\sin^2\theta\frac{\partial^2}{\partial z'^2} - \sin 2\theta\frac{\partial^2}{\partial x'\,\partial z'} + \cos^2\theta\frac{\partial^2}{\partial x'^2}\right]v + \frac{E_H}{2}\frac{\partial^2}{\partial y^2}v, \tag{4.9.8b}$$

and

$$0 = -\frac{\partial p}{\partial z'}\cos\theta - \frac{\partial p}{\partial x'}\sin\theta$$

$$+\delta^2\frac{E_V}{2}\left[\cos^2\theta\frac{\partial^2}{\partial z'^2} + \sin 2\theta\frac{\partial^2}{\partial x'\,\partial z'} + \sin^2\theta\frac{\partial^2}{\partial x'^2}\right][w'\cos\theta + u'\sin\theta]$$

$$+\delta^2\frac{E_H}{2}\left[\sin^2\theta\frac{\partial^2}{\partial z'^2} - \sin 2\theta\frac{\partial^2}{\partial x'\,\partial z'} + \cos^2\theta\frac{\partial^2}{\partial x'^2}\right][w'\cos\theta + u'\sin\theta]$$

$$+\delta^2\frac{E_H}{2}\frac{\partial^2}{\partial y^2}[w'\cos\theta + u'\sin\theta]. \tag{4.9.8c}$$

If (4.9.8a) is multiplied by $\cos\theta$ and added to the product of $\sin\theta$ and (4.9.8c), the result is

$$-v\cos\theta = -\frac{\partial p}{\partial x'} + \frac{E_V}{2}\cos^4\theta\{1 + \delta^2\tan^2\theta\}\left\{1 + \frac{E_H}{E_V}\tan^2\theta\right\}\frac{\partial^2 u'}{\partial z'^2}. \tag{4.9.9}$$

In deriving (4.9.9) the fact that derivatives with respect to $z'$ are much greater than derivatives *of the same function* with respect to $x'$ has been used. Similarly, the fact has been used that in the boundary layer the ratio $w'/u'$ must be of the order of $l = l_*/D$, where $l_*$ is the small boundary-layer thickness (as yet unknown) and $D$ the vertical scale of the inviscid flow. If (4.9.8a) is multiplied by $\sin\theta$ and (4.9.8c) by $\cos\theta$, their difference yields

$$-v\sin\theta = \frac{\partial p}{\partial z'} + (1 - \delta^2)\frac{E_V}{2}\cos^3\theta\sin\theta\left\{1 + \frac{E_H}{E_V}\tan^2\theta\right\}\frac{\partial^2 u'}{\partial z'^2}, \tag{4.9.10}$$

while the approximations outlined above allow (4.9.8b) to be written

$$u' \cos \theta = - \frac{\partial p}{\partial y} + \frac{E_V}{2} \cos^2 \theta \left\{ 1 + \frac{E_H}{E_V} \tan^2 \theta \right\} \frac{\partial^2 v}{\partial z'^2}. \qquad (4.9.11)$$

The continuity equation is unchanged in form, i.e.,

$$\frac{\partial u'}{\partial x'} + \frac{\partial v}{\partial y} + \frac{\partial w'}{\partial z'} = 0. \qquad (4.9.12)$$

If the stretched boundary-layer variable

$$\zeta = \frac{z'}{l}$$

is introduced, where $l$ is naturally small, it follows from (4.9.12) that if $u'$ and $v$ are O(1), $w'$ must be O($l$) as anticipated. Further, (4.9.10) becomes

$$-lv \sin \theta = \frac{\partial p}{\partial \zeta} + l \frac{1 - \delta^2}{2} \frac{E_V}{l^2} \cos^3 \theta \sin \theta$$

$$\times \left\{ 1 + \frac{E_H}{E_V} \tan^2 \theta \right\} \frac{\partial^2 u'}{\partial \zeta'^2}. \qquad (4.9.13)$$

Hence, as long as

$$\frac{E_V}{l^2} \le O(1)$$

(assuming $(E_H/E_V)\tan^2 \theta$ is not large), to lowest order in the small parameter $l$, we have

$$\frac{\partial p}{\partial \zeta} = 0, \qquad (4.9.14)$$

i.e., as usual, the boundary layer is so thin that the pressure variation through the layer is negligible. Thus $\partial p/\partial x'$ and $\partial p/\partial y$ in (4.9.9) and (4.9.11) are independent of $\zeta$ to the lowest order and equal to their interior values. Eliminating $u'$ between the two equations yields

$$-v \cos^2 \theta = - \cos \theta \frac{\partial p}{\partial x'} + \frac{\cos^6 \theta}{l^4} \frac{E_V^2}{4} \{1 + \delta^2 \tan^2 \theta\}$$

$$\times \left\{ 1 + \frac{E_H}{E_V} \tan^2 \theta \right\}^2 \frac{\partial^4 v}{\partial \zeta^4}. \qquad (4.9.15)$$

It is evident that the natural choice for $l$ is

$$\boxed{l = (\cos \theta)\left( 1 + \frac{E_H}{E_V} \tan^2 \theta \right)^{1/2} (1 + \delta^2 \tan^2 \theta)^{1/4} E_V^{1/2}, \qquad (4.9.16)}$$

for then (4.9.15) becomes

$$\frac{1}{4}\frac{\partial^4 v}{\partial \zeta^4} + v = \frac{1}{\cos\theta}\frac{\partial p}{\partial x'}. \tag{4.9.17}$$

The general solution of (4.9.17) subject only to the condition of finiteness as $\zeta \to \infty$ is

$$v = \frac{1}{\cos\theta}\frac{\partial p}{\partial x'} + e^{-\zeta}[C_1 \cos\zeta + C_2 \sin\zeta], \tag{4.9.18}$$

while from (4.9.11) it now follows that

$$u' = -\frac{1}{\cos\theta}\frac{\partial p}{\partial y} + \frac{(1 + \delta^2 \tan^2\theta)^{-1/2}}{\cos\theta}e^{-\zeta}[C_1 \sin\zeta - C_2 \cos\zeta], \tag{4.9.19}$$

where $C_1$ and $C_2$ are arbitrary functions of $x'$ and $y$. Note that the boundary-layer solution is written in terms of the variable

$$\frac{z'}{l} = \frac{z\cos\theta - x\sin\theta}{\cos\theta \, E_V^{1/2}(1 + \delta^2 \tan^2\theta)^{1/4}(1 + (E_H/E_V)\tan^2\theta)^{1/2}} \tag{4.9.20}$$

$$= \frac{z - x\tan\theta}{E_V^{1/2}(1 + \delta^2 \tan^2\theta)^{1/4}(1 + (E_H/E_V)\tan^2\theta)^{1/2}}$$

$$= \frac{z_* - x_* \tan\alpha}{(\delta_E/(\cos\alpha)^{1/2})(1 + (E_H/E_V)\tan^2\theta)^{1/2}}, \tag{4.9.21}$$

since

$$1 + \delta^2 \tan^2\theta = 1 + \tan^2\alpha = \frac{1}{\cos^2\alpha}. \tag{4.9.22}$$

In the *laminar* case, where $A_H = A_V = \nu$,

$$1 + \frac{E_H}{E_V}\tan^2\theta = 1 + \delta^2 \tan^2\theta = \frac{1}{\cos^2\alpha}, \tag{4.9.23}$$

so that

$$\frac{z'}{l} = \frac{z_* \cos\alpha - x_* \sin\alpha}{\delta_E(\cos\alpha)^{-1/2}}$$

$$= \frac{z'_*}{\delta_E(\cos\alpha)^{-1/2}}. \tag{4.9.24}$$

In that case the boundary-layer thickness, measured perpendicular to the boundary, is simply

$$\frac{\delta_E}{(\cos\alpha)^{1/2}} = \left(\frac{2\nu}{f\cos\alpha}\right)^{1/2}, \tag{4.9.25}$$

i.e., it is the ordinary Ekman-layer thickness based now on the dynamically relevant component of the Coriolis parameter normal to the boundary, $f \cos \alpha$. When the frictional forces are anisotropic, (4.9.16) shows that $l$ is still $O(E_V^{1/2})$ but is a more complex function of $E_H/E_V$ and the slope.

Returning to (4.9.18) and (4.9.19), the constants $C_1$ and $C_2$ are determined by the condition that $u'$ and $v$ vanish on $\zeta = 0$, or

$$C_1 = -\frac{1}{\cos \theta} \frac{\partial p}{\partial x'}$$

$$C_2 = -\frac{\partial p}{\partial y} (1 + \delta^2 \tan^2 \theta)^{1/2},$$

(4.9.26)

so that

$$u' = -\frac{1}{\cos \theta} \frac{\partial p}{\partial y} [1 - e^{-\zeta} \cos \zeta]$$

$$- \frac{1}{\cos^2 \theta} \frac{\partial p}{\partial x'} (1 + \delta^2 \tan^2 \theta)^{-1/2} e^{-\zeta} \sin \zeta,$$

(4.9.27a)

$$v = \frac{1}{\cos \theta} \frac{\partial p}{\partial x'} [1 - e^{-\zeta} \cos \zeta]$$

$$- \frac{\partial p}{\partial y} (1 + \delta^2 \tan^2 \theta)^{1/2} e^{-\zeta} \sin \zeta.$$

(4.9.27b)

The continuity equation

$$\frac{\partial w'}{\partial \zeta} = -\left( \frac{\partial u'}{\partial x'} + \frac{\partial v}{\partial y} \right) l$$

(4.9.28)

can be integrated to yield

$$w'(x', y, \infty) - w'(x', y, 0) = -l \int_0^\infty \left( \frac{\partial u'}{\partial x'} + \frac{\partial v}{\partial y} \right) d\zeta,$$

or

$$w'(x, y, \infty) = +\frac{l}{2} \left[ \frac{1}{\cos^2 \theta} \frac{\partial^2 p}{\partial x'^2} (1 + \delta^2 \tan^2 \theta)^{-1/2} \right.$$

$$\left. + \frac{\partial^2 p}{\partial y^2} (1 + \delta^2 \tan^2 \theta)^{1/2} \right].$$

(4.9.29)

Now, using the coordinate transformation, we have

$$\frac{\partial^2 p}{\partial x'^2} = \cos^2 \theta \frac{\partial^2 p}{\partial x^2} - \sin 2\theta \frac{\partial^2 p}{\partial x \, \partial z} + \sin^2 \theta \frac{\partial^2 p}{\partial z^2}$$

$$= \cos^2 \theta \frac{\partial^2 p}{\partial x^2},$$

since $\partial p/\partial z$ is zero in the interior to $O(\delta^2)$. Thus the velocity, $w'$, at the edge of the boundary layer can be written entirely in terms of the derivatives of the velocity in the *original* $x$, $y$, $z$ frame, i.e.,

$$w'(x,\, y,\, \infty) = \frac{l}{2}\left[\cos\alpha\,\frac{\partial v}{\partial x} - \frac{1}{\cos\alpha}\frac{\partial u}{\partial y}\right], \tag{4.9.30}$$

where $l$ is given by (4.9.16).

The velocity $w'$ is the velocity perpendicular to the boundary. The velocity which produces the vortex-tube stretching is the velocity $w$ parallel to the rotation vector, i.e., the vertical velocity in the original frame. From (4.9.6)

$$w' = w\cos\theta - u\sin\theta, \tag{4.9.31}$$

so that the *vertical* velocity $w$ at the edge of the boundary layer is given by

$$w = u\tan\theta + \frac{l}{2\cos\theta}\left[\cos\alpha\,\frac{\partial v}{\partial x} - \frac{\partial u}{\partial y}\frac{1}{\cos\alpha}\right], \tag{4.9.32}$$

where $u$ is the *interior* velocity on the edge of the boundary layer.

The first term on the right-hand side of (4.9.32) is the expression for the vertical velocity produced in an inviscid fluid by the lifting of fluid elements flowing parallel to the slope. Reference to Figure 4.9.2 shows that this still occurs, since the fluid layers just external to the friction layer still glide, unimpeded by friction, along the direction of the sloping boundary. The second term in (4.9.32) is the additional *vertical* velocity produced by the pumping out of the Ekman layer.

Since

$$d = L\tan\alpha \tag{4.9.33}$$

represents the thickness variation of the layer over the characteristic horizontal scale $L$, it follows that

$$\tan\theta = \frac{L}{D}\tan\alpha = \frac{d}{D}. \tag{4.9.34}$$

Hence if the thickness variations are $O(1)$ or less, $\tan\alpha$ must be very small, i.e., $O(D/L)$. For such small $\alpha$, $\cos\alpha$ is very nearly unity, in which case (4.9.32) becomes

$$w = u\tan\theta + \frac{E_V^{1/2}}{2}\left(1 + \frac{E_H}{E_V}\tan^2\theta\right)^{1/2}\left(\frac{\partial v}{\partial x} - \frac{\partial u}{\partial y}\right), \tag{4.9.35}$$

so that once again the frictionally induced vertical velocity is directly proportional to the vorticity of the interior flow.

Reference to (4.9.33) and (4.9.34) shows that

$$\tan\theta = \frac{1}{D}\frac{\partial h_B}{\partial x},$$

where $h_B$ is the (dimensional) height of the lower boundary. It was shown in

Chapter 3 that when the variations of $h_B/D$ are O(1), the topographic constraints are so large that the flow is forced to flow entirely along an isobath
and that the more dynamically interesting case occurs when $\tan \theta < 1$. This
suggests that unless $E_H/E_V \gg 1$, the factor $1 + (E_H/E_V)\tan^2 \theta$ might well be
replaced by unity. We shall make this approximation, not only for that
reason, but also to emphasize that any apparent detailed dependence on
topography which arises ultimately from the uncertain nature of the frictional parametrization (i.e., on the ratio $E_H/E_V$) cannot be taken seriously. It
is preferable in such matters—indeed, shows better judgement—to consider
(4.9.35) in its most elementary form, neglecting any elaborate implications
from the very crude frictional modeling. We therefore write (4.9.35) in the
simple form

$$w\left(x, y, \frac{h_B}{D}\right) = \mathbf{u} \cdot \nabla\left(\frac{h_B}{D}\right) + \frac{E_V^{1/2}}{2} \zeta \qquad (4.9.36)$$

In this vector form, which emphasizes that the orientation of the surface was
in fact arbitrary, the vertical velocity is seen to be the simple sum of the
inviscid velocity produced by a slope and the Ekman velocity appropriate
for a level surface.

The ratio of the two contributions to the interior $w$ at $z = h_B/D$ is

$$\frac{h_B}{D} E_V^{-1/2} = \frac{h_B}{\delta_E}, \qquad (4.9.37)$$

i.e., the ratio of the rise of the bottom over the length $L$ to the Ekman-layer
thickness $\delta_E$. If $h_B/\delta_E$ is small, the bottom variations of the fluid are buried
in the Ekman layer and the dynamical effect of the slope is unimportant. If
$h_B/\delta_E$ is large, the vertical velocity produced by topographic variations over-
whelms the weak Ekman pumping.

## 4.10  Ekman Layer on a Free Surface

If the upper surface of the fluid layer is free rather than in contact with a
solid surface, the appropriate boundary condition is continuity of pressure
and continuity of frictional stress across the fluid surface. It is clear that the
continuity condition on the pressure will not be affected by the presence of a
thin frictional layer, since over the width of the layer the pressure is unchanged. The pressure condition is met, as in Chapter 3, by small deviations
of the free surface from its rest position given by (3.3.13). The condition on
continuity of the frictional stress, or traction, at the surface is another
matter. A friction layer is generally required to satisfy this condition.

Consider the situation shown in Figure 4.10.1. The free surface, at a height
$z_* = h$ experiences a stress $\tau_*$. For the moment consider the surface to be
flat and stationary. In order for the tangential stress to be continuously

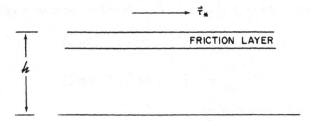

**Figure 4.10.1**  The friction layer at a free surface under an applied stress $\tau_*$.

communicated to the fluid at the free surface, (4.2.11) implies that

$$\left.\begin{array}{l} \tau_*^{(x)} = \rho A_V \dfrac{\partial u_*}{\partial z_*} + \rho A_H \dfrac{\partial w_*}{\partial x_*} \\[2mm] \tau_*^{(y)} = \rho A_V \dfrac{\partial v_*}{\partial z_*} + \rho A_H \dfrac{\partial w_*}{\partial y_*} \end{array}\right\} \quad \text{on } z_* = h, \tag{4.10.1}$$

or in dimensionless units

$$\left.\begin{array}{l} \left(\dfrac{D\tau_0}{\rho A_V U}\right)\tau^{(x)} = \dfrac{\partial u}{\partial z} + E_H/E_V \dfrac{\partial w}{\partial x} \\[2mm] \left(\dfrac{D\tau_0}{\rho A_V U}\right)\tau^{(y)} = \dfrac{\partial v}{\partial z} + E_H/E_V \dfrac{\partial w}{\partial y} \end{array}\right\} \quad \text{on } z = \dfrac{h}{D}, \tag{4.10.2}$$

where $\tau_0$ is the characteristic value of the stress, while $U$ and $D$ are the characteristic values of the horizontal velocity and depth. Since we expect that $\partial u/\partial z = \mathrm{O}(E_V^{-1/2}u)$ while $w = \mathrm{O}(E_V^{1/2}u)$ on the basis of the results of Section 4.5, it follows that the terms in $w$ in (4.10.2) will be $\mathrm{O}(E_H)$ with respect to the terms in $u$ and $v$. Hence they may definitely be neglected as long as $E_H \ll 1$. As in Section 4.5, the boundary-layer coordinate

$$\zeta = \frac{h/D - z}{E_V^{1/2}} \tag{4.10.3}$$

is introduced. The resulting lowest-order equations in the boundary layer are again

$$-\tilde{v} = -\frac{\partial \tilde{p}}{\partial x} + \frac{1}{2}\frac{\partial^2 \tilde{u}}{\partial \bar{\zeta}^2}, \tag{4.10.4a}$$

$$\tilde{u} = -\frac{\partial \tilde{p}}{\partial y} + \frac{1}{2}\frac{\partial^2 \tilde{v}}{\partial \bar{\zeta}^2}, \tag{4.10.4b}$$

$$0 = -\frac{\partial \tilde{p}}{\partial \bar{\zeta}}, \tag{4.10.4c}$$

where the $\tilde{\ }$ reminds us the variables refer to the region of the upper Ekman layer. Again the horizontal pressure gradient in the boundary layer is given by its value in the inviscid interior. Thus the general solution is again

$$\tilde{u} = -\frac{\partial p}{\partial y} + e^{-\zeta}[C_1 \cos \zeta + C_2 \sin \zeta],$$

$$\tilde{v} = \frac{\partial p}{\partial x} + e^{-\zeta}[-C_1 \sin \zeta + C_2 \cos \zeta]. \tag{4.10.5}$$

The boundary conditions are

$$-\frac{\partial \tilde{u}}{\partial \zeta} = \left(\frac{E_V^{1/2} D}{U A_V}\frac{\tau_0}{\rho}\right)\tau^{(x)} = \alpha\tau^{(x)},$$

$$-\frac{\partial \tilde{v}}{\partial \zeta} = \left(\frac{E_V^{1/2} D}{U A_V}\frac{\tau_0}{\rho}\right)\tau^{(y)} = \alpha\tau^{(y)}, \tag{4.10.6}$$

where

$$\alpha = \frac{2\tau_0/\rho}{f \, \delta_E \, U} \tag{4.10.7}$$

and $\delta_E$ is the Ekman thickness, $DE_V^{1/2}$. The application of (4.10.6) to (4.10.5) yields

$$C_1 = \frac{\alpha}{2}(\tau^{(y)} + \tau^{(x)}),$$

$$C_2 = \frac{\alpha}{2}(\tau^{(y)} - \tau^{(x)}), \tag{4.10.8}$$

so that

$$\tilde{u} = u_0 + \frac{e^{-\zeta}}{2}\alpha[(\tau^{(y)} - \tau^{(x)})\sin \zeta$$

$$+ (\tau^{(y)} + \tau^{(x)})\cos \zeta],$$

$$\tilde{v} = v_0 + \frac{e^{-\zeta}}{2}\alpha[(\tau^{(y)} - \tau^{(x)})\cos \zeta$$

$$- (\tau^{(y)} + \tau^{(x)})\sin \zeta], \tag{4.10.9}$$

where $u_0$ and $v_0$ are the interior, geostrophic velocities. The velocity vector

$$\tilde{\mathbf{u}} = \mathbf{i}\tilde{u} + \mathbf{j}\tilde{v} \tag{4.10.10}$$

is given by

$$\tilde{\mathbf{u}} = \mathbf{u}_0 + \frac{\alpha e^{-\zeta}}{2}[\boldsymbol{\tau}(\cos \zeta - \sin \zeta) - (\mathbf{k} \times \boldsymbol{\tau})(\cos \zeta + \sin \zeta)], \tag{4.10.11}$$

where $\mathbf{k}$ is a unit vector in the $z$-direction.

The velocity of the friction layer, i.e., the departure of the velocity from the geostrophic velocity $\mathbf{u}_0$, is simply resolved into components parallel and perpendicular to $\tau$. At the surface $(\bar{\zeta} = 0)$ this velocity $\mathbf{u}_E$ is simply

$$\mathbf{u}_E = \tilde{\mathbf{u}} - \mathbf{u}_0 = \frac{\alpha}{2}[\tau - (\mathbf{k} \times \tau)],   \tag{4.10.12}$$

and, as shown in Figure 4.10.2, is 45° to the *right* of the applied stress.[*]
Figure 4.10.3 shows the velocity components parallel and perpendicular to
the stress, while Figure 4.10.4 shows the position of the vector $\mathbf{u}_E$ in the $x$, $y$
plane as a function of $\bar{\zeta}$, again illustrating the Ekman spiral.

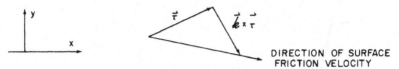

**Figure 4.10.2**    The relation between $\tau$ and the surface velocity.

The horizontal mass flux associated with the Ekman-layer friction velocity is

$$\mathbf{M}_E = \int_0^{h/D} \mathbf{u}_E \, dz = \int_0^\infty \mathbf{u}_E \, d\zeta \, E_V^{1/2} = +\frac{\alpha}{2}(\tau \times \mathbf{k})E_V^{1/2}   \tag{4.10.13}$$

i.e., completely *perpendicular* to the applied stress. In dimensional units

$$\mathbf{M}_{E.} = UDM_E = \frac{\tau_0}{\rho f \delta_E} DE_V^{1/2}(\tau \times \mathbf{k}),$$

or

$$\boxed{\mathbf{M}_{E.} = \frac{\tau_*}{\rho f} \times \mathbf{k}},   \tag{4.10.14}$$

which is identical to (4.3.32). It is vitally important to note that the magnitude of the mass flux in the surface friction layer does not depend on the
turbulent viscosity. It depends only on the applied stress and the Coriolis
parameter. As explained in Section 4.3, this simple result is a direct consequence of the balance between the applied stress and the average Coriolis
acceleration of the friction-layer velocity. Thus (4.10.14) is completely
independent of the details of the parametrization of turbulence.

The vertical velocity flowing into the surface Ekman layer from the interior follows from an integration of the continuity equation.

$$\frac{\partial \tilde{w}}{\partial \bar{\zeta}} = E_V^{1/2}\left(\frac{\partial \tilde{u}}{\partial x} + \frac{\partial \tilde{v}}{\partial y}\right)$$

$$= \frac{\alpha}{2} E_V^{1/2} e^{-\bar{\zeta}}[(\nabla \cdot \tau)(\cos \bar{\zeta} - \sin \bar{\zeta})$$

$$- \nabla \cdot (\mathbf{k} \times \tau)(\cos \bar{\zeta} + \sin \bar{\zeta})].   \tag{4.10.15}$$

---

[*] If $f$ is negative instead of positive, the signs of the Coriolis terms in (4.10.4) are each
changed and the symmetries are reversed, i.e., the rotation produces deviations to the left
instead of to the right.

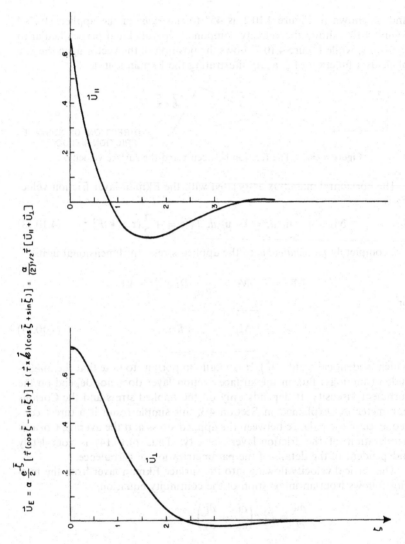

**Figure 4.10.3**   The profile of the velocity components perpendicular and parallel to the direction of the applied stress.

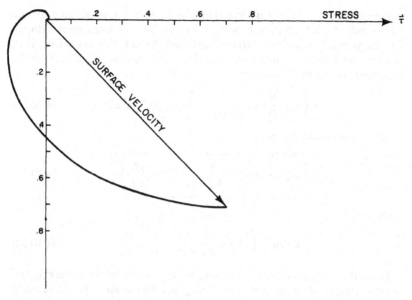

**Figure 4.10.4** The heavy solid curve indicates the spiral of the Ekman velocity vector as a function of depth.

Carrying out the integration yields

$$\tilde{w}(x, y, \infty) - \tilde{w}(x, y, 0) = -\alpha \frac{E_V^{1/2}}{2} \nabla \cdot (\mathbf{k} \times \boldsymbol{\tau})$$

$$= \frac{\tau_0}{\rho f \, UD} \mathbf{k} \cdot \text{curl } \boldsymbol{\tau}, \qquad (4.10.16)$$

where

$$\mathbf{k} \cdot \text{curl } \boldsymbol{\tau} = \frac{\partial \tau^{(y)}}{\partial x} - \frac{\partial \tau^{(x)}}{\partial y}. \qquad (4.10.17)$$

The velocity $\tilde{w}(x, y, 0)$ refers to the change in the position of the upper boundary. If the free surface is horizontal but moving up or down,

$$\tilde{w}(x, y, 0) = \frac{\partial}{\partial t}\left(\frac{h}{D}\right). \qquad (4.10.18)$$

The *interior vertical velocity* $w$ at the lower edge of the boundary layer (i.e., the velocity sucked into the Ekman layer) is given by the matching principle, i.e.,

$$w(x, y, h/D) = \tilde{w}(x, y, \infty) = \frac{\partial}{\partial t}\left(\frac{h}{D}\right) + \frac{\tau_0}{\rho f \, UD} \mathbf{k} \cdot \text{curl } \boldsymbol{\tau}. \quad (4.10.19)$$

If the surface $h$ is sloping, then the right-hand side of (4.10.19) must be augmented by an amount $\mathbf{u} \cdot \nabla(h/D)$, as in Section 4.9, to account for the additional vertical velocity produced by the *interior* horizontal velocity flowing parallel to the free surface—although (recall) this extra term is in fact zero for quasigeostrophic motion. When this effect is added to (4.10.19), we obtain as our final result

$$w\left(x, y, \frac{h}{D}\right) = \frac{d}{dt}\left(\frac{h}{D}\right) + \left[\frac{\tau_0}{\rho f\, UD}\right](\mathbf{k} \cdot \text{curl } \tau), \qquad (4.10.20)$$

which in dimensional units is

$$w_*(\text{surface}) = \frac{dh}{dt_*} + \frac{\mathbf{k}}{\rho} \cdot \text{curl}_*\left(\frac{\tau_*}{f}\right), \qquad (4.10.21)$$

where

$$\mathbf{k} \cdot \text{curl}_*\left(\frac{\tau_*}{f}\right) = \frac{\partial}{\partial x_*}\frac{\tau_*^{(y)}}{f} - \frac{\partial}{\partial y_*}\frac{\tau_*^{(x)}}{f}. \qquad (4.10.22)$$

Thus, the vertical velocity the interior sees at its upper surface is the vertical velocity of an inviscid fluid, $dh/dt_*$, at a free surface plus the velocity sucked into the upper Ekman layer under the action of the applied stress. It is again important to note that the velocity pumped into the upper Ekman layer is independent of our treatment of the small-scale turbulence. Consider the stress field in Figure 4.10.5. The stress is in the $y$-direction, and its magnitude increases in the $x$-direction. The total Ekman flux in the friction layer is in the $x$-direction and is increasing with increasing $x$. To conserve mass, fluid is sucked into the Ekman layer at a rate proportional to the divergence of the Ekman flux. Since the Ekman flux is independent of the details of the turbulence, so also is its divergence and consequently the Ekman suction velocity.

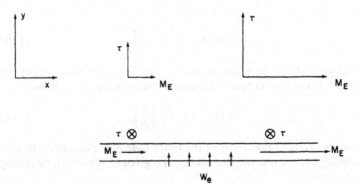

**Figure 4.10.5** In the $x$, $y$ plane, the stress is in the $y$-direction and increases in magnitude with $x$. In the lower panel the consequent increase of $M_E$ to the right is indicated, along with the velocity sucked into the surface layer to compensate for the divergence of $M_E$.

Note that in the absence of an applied surface stress, the upper Ekman layer, though it may exist, has zero intensity, i.e., the correction to the geostrophic flow is zero. The geostrophic flow, independent of $z$, then satisfies the condition of zero shear stress itself.

## 4.11 Quasigeostrophic Potential Vorticity Equation with Friction and Topography

If the fluid layer has a free surface and a sloping bottom, the vorticity balance will reflect these two mechanisms of vorticity production by vortex-tube stretching. The analysis of Section 4.6 still applies, however, insofar as the interior *vorticity* equation is concerned—i.e., (4.6.7), namely

$$\frac{d\zeta_0}{dt} = \frac{\partial w_1}{\partial z} + \frac{1}{Re} \nabla^2 \zeta_0, \tag{4.11.1}$$

is still valid. $\zeta_0$ is the $O(1)$ vorticity of the geostrophic velocity, while in the interior

$$w(x, y, z) = \varepsilon w_1 + \cdots . \tag{4.11.2}$$

For a layer with a sloping lower boundary (4.9.36) applies, so that on $z = h_B/D$, the lower surface,

$$w = \mathbf{u}_0 \cdot \nabla \frac{h_B}{D} + \frac{E_V^{1/2}}{2} \zeta_0 \tag{4.11.3}$$

to lowest order in $\varepsilon$. With (4.11.2),

$$w_1 = \mathbf{u}_0 \cdot \nabla \eta_B + \frac{r}{2} \zeta_0, \tag{4.11.4}$$

where, as in Chapter 3,

$$\frac{h_B}{D} = \varepsilon \eta_B(x, y), \tag{4.11.5}$$

while again

$$r = \frac{E_V^{1/2}}{\varepsilon} . \tag{4.11.6}$$

The upper surface is at

$$z = \frac{h(x, y, t)}{D}, \tag{4.11.7}$$

which, by the results of Section 3.12, is

$$z = 1 + \varepsilon F \eta_0 . \tag{4.11.8}$$

Here $\eta_0$ is O(1) and

$$F = \frac{f^2 L^2}{gD}.$$  (4.11.9)

At the upper surface (4.10.20) applies, so that

$$w_1\left(x, y, \frac{h}{D}\right) = F\frac{d\eta_0}{dt} + \left|\frac{\tau_0}{\rho f\, U D\varepsilon}\right| \mathbf{k} \cdot \text{curl } \tau.$$  (4.11.10)

Since the O(1) geostrophic flow is independent of $z$, (4.11.1) can be integrated from $z = \varepsilon\eta_B$ to $z = 1 + \varepsilon F\eta_0$ to yield (to lowest order in $\varepsilon$) with (4.11.4) and (4.11.10)

$$\frac{d}{dt}\{\zeta_0 - F\eta_0 + \eta_B\} = \left[\frac{\tau_0}{\rho f\, U D\varepsilon}\right]\mathbf{k} \cdot \text{curl } \tau - \frac{r}{2}\zeta_0 + \frac{1}{\text{Re}}\nabla^2\zeta_0,$$  (4.11.11)

or in terms of the geostrophic stream function $\psi$,

$$\left[\frac{\partial}{\partial t} + \frac{\partial\psi}{\partial x}\frac{\partial}{\partial y} - \frac{\partial\psi}{\partial y}\frac{\partial}{\partial x}\right][\nabla^2\psi - F\psi + \eta_B]$$

$$= \left[\frac{\tau_0}{\rho f\, U D\varepsilon}\right]\mathbf{k} \cdot \text{curl } \tau - \frac{r}{2}\nabla^2\psi + \frac{1}{\text{Re}}\nabla^4\psi,$$  (4.11.12)

where

$$\eta_0 = \psi,$$

$$u_0 = -\frac{\partial\psi}{\partial y}$$  (4.11.13)

$$v_0 = \frac{\partial\psi}{\partial x},$$

$$\zeta_0 = \nabla^2\psi.$$

The presence of friction has produced in (4.11.12) three important changes in this generalization of the potential-vorticity equation. In the absence of friction the right-hand side of (4.11.12) is zero and (4.11.12) reduces to the statement of potential-vorticity conservation, i.e., (3.12.25). Friction now allows the potential vorticity of each fluid column to change with time. The curl of the applied stress acts as a source of potential vorticity, while the frictional dissipation in the lower Ekman layer acts as a sink of potential vorticity. The presence of a small amount ($O(\text{Re}^{-1})$) of friction in the interior, although generally negligible, acts to diffuse vorticity laterally.

If the concept of the $\beta$-plane (Section 3.17) is used, so that $\eta_B$ is interpreted as the variable part of the ambient potential vorticity caused by the variation of the Coriolis parameter with latitude $\theta$, then (3.14.2) and (3.17.9) yield

$$\eta_B = \frac{\beta_0 L^2}{U} y = \beta y$$

$$\beta_0 = \frac{2\Omega \cos \theta}{r_0},$$
(4.11.14)

where $r_0$ is the earth's radius.

In the absence of friction, geostrophic flow will tend to be along the lines of constant $\eta_B$, as shown in Section 3.14—that is, for steady, *linear* flow (very small $\varepsilon$ or large $\eta_B$),

$$\frac{d\eta_B}{dt} = 0.$$
(4.11.15)

The input of vorticity by an applied stress will alter (4.11.15) for small $\varepsilon$ or large $\eta_B$ to

$$\frac{d\eta_B}{dt} = \frac{\tau_0}{\rho f U D \varepsilon} \mathbf{k} \cdot \text{curl } \tau,$$
(4.11.16)

i.e., the curl of the stress will drive fluid across the lines of ambient potential vorticity, so that the vorticity input is balanced by the increase of the column's potential vorticity experienced when it moves to an area of larger ambient potential vorticity. This suggests that when the fluid is driven by an applied stress in the presence of an $O(1)$ $\eta_B$, the scaling velocity should be determined by assuring that $d\eta_B/dt$ and the stress term be of the same order, i.e., we *choose* $U$ such that, when (4.11.14) applies,

$$\frac{\beta_0 L^2}{U} = \frac{\tau_0}{\rho f U D \varepsilon} = \frac{\tau_0 L}{\rho U^2 D},$$

or

$$U = \frac{\tau_0}{\rho \beta_0 D L}.$$
(4.11.17)

With this choice of $\eta_B$ and $U$, (4.11.12) becomes

$$\left[ \frac{\partial}{\partial t} + \frac{\partial \psi}{\partial x} \frac{\partial}{\partial y} - \frac{\partial \psi}{\partial y} \frac{\partial}{\partial x} \right] [\nabla^2 \psi - F\psi + \beta y] = \beta \mathbf{k} \cdot \text{curl } \tau - \frac{r}{2} \nabla^2 \psi + \text{Re}^{-1} \nabla^4 \psi.$$
(4.11.18)

For large $\beta$ (i.e., $U \ll \beta_0 L^2$), (4.11.18) reduces to

$$\boxed{v_0 = \mathbf{k} \cdot \text{curl } \tau},$$
(4.11.19)

which yields immediately the northward flow (i.e., the flow across the isolines of ambient potential vorticity) directly in terms of the curl of the applied stress. This relation, called the *Sverdrup balance*, is the keystone of the theory of the wind-driven oceanic circulation which will be discussed in more detail in Chapter 5.

## 4.12 The Decay of a Rossby Wave

If

$$\eta_B = \beta y, \tag{4.12.1}$$

the homogeneous form ($\tau = 0$) of (4.11.11) has plane-wave solutions

$$\psi = A e^{i(kx + ly - \sigma t)} \tag{4.12.2}$$

if $\sigma$ satisfies the dispersion relation

$$\sigma = -\frac{\beta k}{K^2 + F} - \frac{i K^2}{K^2 + F}\left[\frac{r}{2} + \frac{K^2}{Re}\right], \tag{4.12.3}$$

where

$$K^2 = k^2 + l^2. \tag{4.12.4}$$

The frequency is now complex, i.e.,

$$\sigma = \sigma_r + i\sigma_i. \tag{4.12.5}$$

The real part of the frequency is identical to the inviscid result of Chapter 3. The imaginary part of the frequency,

$$\sigma_i = -\frac{K^2}{K^2 + F}\left[\frac{r}{2} + \frac{K^2}{Re}\right], \tag{4.12.6}$$

is always negative. Since

$$e^{-i\sigma t} = e^{-i\sigma_r t} e^{\sigma_i t}, \tag{4.12.7}$$

a $\sigma_i < 0$ yields an exponential decay of the wave. Since Re is generally very large, the decay rate $\sigma_i$ can be approximated by

$$\sigma_i(K) = -\frac{K^2}{K^2 + F}\frac{r}{2}. \tag{4.12.8}$$

The decay rate is very small for waves with wavelengths long compared to a deformation radius, while shorter waves have a larger decay rate as shown in Figure 4.12.1. When $K^2$ is considerably greater than $F$ the decay rate becomes insensitive to scale. The ratio of the decay rate to the frequency is

$$\frac{\sigma_i}{\sigma_r} = \frac{r}{2}\frac{K^2}{\beta k}. \tag{4.12.9}$$

For Rossby waves on the $\beta$-plane, for which $\beta = \beta_0 L^2 / U$,

$$\frac{\sigma_i}{\sigma_r} = \frac{E_V^{1/2}}{2\varepsilon}\frac{K^2 U}{\beta_0 L^2 k} = \frac{E_V^{1/2}}{2}\frac{f}{\beta_0}\frac{K_*^2}{k_*}$$

$$= O\left(\frac{T_{\text{Rossby}}}{\tau}\right), \tag{4.12.10}$$

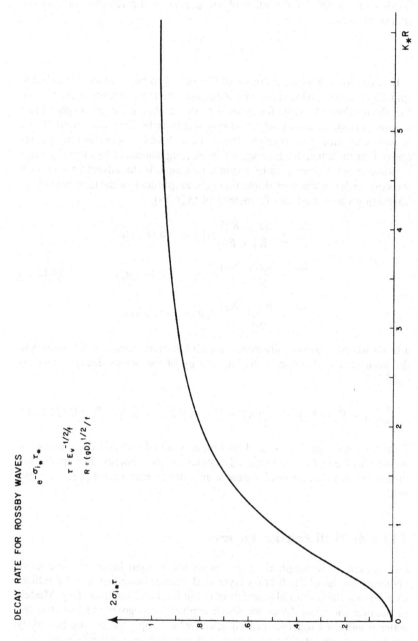

**Figure 4.12.1** The decay rate for a Rossby wave as a function of wave number $k_*$. $R$ is the deformation radius, $\sigma_{1*}$ is the decay rate, and $\tau$ is the spin-up time defined in Section 4.3.

i.e., is of the order of the ratio of the period of the Rossby wave to the spin-down time

$$\tau = \frac{E_V^{-1/2}}{f}.$$

In order for the inviscid theory of Chapter 3 to be valid as it stands, the spin-down time must therefore be long compared to a Rossby-wave period. For the results to be valid for inviscid processes that are slow compared to a Rossby period requires that the spin-down time be long compared to the slower time scale. For example, the inviscid theory of wave interaction requires that the advective time scale $L/U$ be long compared to a Rossby-wave period but short compared to a spin-down time. If the advective time is of the same order as the spin-down time, the amplitudes of the members of the interacting wave triad satisfy, instead of (3.26.33),

$$\frac{da_1}{dt} + \frac{B(K_2, K_3)}{K_1^2 + F} a_2 a_3 = +\sigma_i(K_1)a_1,$$

$$\frac{da_2}{dt} + \frac{B(K_3, K_1)}{K_2^2 + F} a_3 a_1 = +\sigma_i(K_2)a_2, \qquad (4.12.11)$$

$$\frac{da_3}{dt} + \frac{B(K_1, K_2)}{K_3^2 + F} a_1 a_2 = +\sigma_i(K_3)a_3,$$

a result which follows easily from the addition of frictional considerations to the argument of Section 3.26. The energy of the waves decays, since by (3.26.35),

$$\frac{d}{dt}(E_1 + E_2 + E_3) = -\frac{r}{2}[K_1^2|a_1|^2 + K_2^2|a_2|^2 + K_3^2|a_3|^2] < 0. \quad (4.12.12)$$

Thus the waves pass an ever-depleting amount of energy from member to member in the triad until friction dissipates the wave energy. The interaction process is then halted until a new source of vorticity reenergizes the wave field.

## 4.13  Side-Wall Friction Layers

For models of atmospheric phenomena the Ekman layer or some more elaborate model of the friction layer at the lower boundary usually suffices to represent the frictional interaction of the fluid and the boundary. Models of oceanic (or lake) dynamics which explicitly recognize the fact that the water is gathered together in basins, and which apply the no-slip boundary condition at the lateral boundaries, will generally introduce side-wall friction layers whose structure differs considerably from that of the Ekman layer.

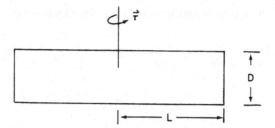

**Figure 4.13.1**   A cylindrical region of depth $D$ experiences a circumferential stress $\tau_* = \tau_0 \tau$.

As an example, consider the situation pictured in Figure 4.13.1. Water in a circular basin of radius $L$ is driven by a surface stress $\tau_* = \tau_0 \tau$. The basin has vertical side walls and has constant depth $D$. Horizontal lengths are then scaled by $L$ and vertical lengths by $D$. In the case at hand it seems natural to introduce polar coordinates. Let $r$ be the nondimensional radial distance (scaled by $L$) from the basin center, and $\theta$ the angle to the $x$-axis, as shown in Figure 4.13.2. The velocity components in the radial and circumferential direction are $u_r$ and $u_\theta$ respectively. The geostrophic relation in vector form,

$$\mathbf{u}_0 = \mathbf{k} \times \nabla\psi, \tag{4.13.1}$$

which the O(1) velocity must satisfy, implies that $u_r$ and $u_\theta$ are given by

$$u_\theta = \frac{\partial\psi}{\partial r},$$

$$u_r = -\frac{1}{r}\frac{\partial\psi}{\partial\theta} \tag{4.13.2}$$

to lowest order in $\varepsilon$.

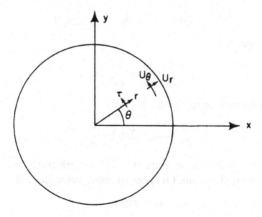

**Figure 4.13.2**   The polar coordinate frame.

For steady flow $\psi$ must satisfy (4.11.11), which may be written for steady flow as

$$\frac{u_\theta}{r}\frac{\partial \zeta_0}{\partial \theta} + u_r \frac{\partial \zeta_0}{\partial r} = \frac{\tau_0}{\rho f U D \varepsilon} \mathbf{k} \cdot \text{curl } \tau - \frac{E_V^{1/2}}{2\varepsilon}\zeta_0 + \frac{1}{\text{Re}}\nabla^2 \zeta_0 \quad (4.13.3)$$

where

$$\zeta_0 = \nabla^2 \psi = \frac{1}{r}\frac{\partial}{\partial r}r\frac{\partial \psi}{\partial r} + \frac{1}{r^2}\frac{\partial^2 \psi}{\partial \theta^2}. \quad (4.13.4)$$

Consider the case where the applied stress is circumferentially directed and independent of $\Theta$, i.e.,

$$\tau = \tau(r)\hat{\Theta} \quad (4.13.5)$$

where $\hat{\Theta}$ is the unit vector in the azimuthal direction. Since the applied stress is independent of azimuth, and since the dynamics is spatially isotropic in the horizontal ($\eta_B = 0$), it is sensible to seek solutions for steady flows which are independent of $\theta$. If $\psi$ is independent of $\theta$, then $u_r$ is zero and (4.13.3) reduces to

$$0 = \frac{\tau_0}{\rho f \, UD\varepsilon} \mathbf{k} \cdot \text{curl } \tau - \frac{E_V^{1/2}}{2\varepsilon}\nabla^2 \psi + \frac{1}{\text{Re}}\nabla^4 \psi. \quad (4.13.6)$$

Although $r$ may be O(1), Re is generally very large for geophysically realistic models. Certainly, away from the lateral boundaries the last term on the right-hand side of (4.13.6) can be neglected. For axially symmetric motion (4.13.6) becomes

$$\frac{E_V^{1/2}}{2\varepsilon}\frac{1}{r}\frac{\partial}{\partial r}r\frac{\partial \psi}{\partial r} = \frac{\tau_0}{\rho f U D\varepsilon}\mathbf{k} \cdot \text{curl } \tau = \frac{\tau_0}{\rho f U D\varepsilon}\frac{1}{r}\frac{\partial}{\partial r}r\tau. \quad (4.13.7)$$

The balance described by (4.13.7) suggests that the proper choice for the scaling velocity in the present case is

$$U = \frac{\tau_0}{\rho f DE_V^{1/2}} = \frac{\tau_0}{\rho f \delta_E} = \frac{\tau_0}{\rho\sqrt{2A_V f}}, \quad (4.13.8)$$

so that (4.13.7) becomes

$$\frac{1}{r}\frac{\partial}{\partial r}r\frac{\partial \psi}{\partial r} = \frac{2}{r}\frac{\partial}{\partial r}r\tau. \quad (4.13.9)$$

This can be integrated immediately to yield

$$\frac{\partial \psi}{\partial r} = u_\theta = 2\tau(r) + \frac{C}{r}. \quad (4.13.10)$$

Since the velocity must remain finite at $r = 0$, the arbitrary constant $C$ must be taken to be zero. Thus the O(1) geostrophic velocities are

$$u_\theta = 2\tau(r),$$
$$u_r = 0. \quad (4.13.11)$$

In the interior of the fluid the $O(1)$ flow is strictly azimuthal, i.e., in the direction of the stress. The dimensional azimuthal velocity is

$$u_{*\theta} = \frac{2}{\rho\sqrt{2A_V f}}\,\tau_*(r). \tag{4.13.12}$$

This of course only represents the solution in the geostrophic interior. In the *upper* Ekman layer the velocity is given by (4.10.11). In the present case

$$\mathbf{u}_0 = u_\theta\,\hat{\Theta} = 2\tau \tag{4.13.13}$$

and

$$\alpha = \frac{2\tau_0}{\rho f\,\delta_E\,U} = 2, \tag{4.13.14}$$

so that the velocity field in the region of the upper Ekman layer is

$$\tilde{\mathbf{u}} = 2\tau + e^{-\zeta}[\tau(\cos\zeta - \sin\zeta) - (\mathbf{k} \times \tau)(\cos\zeta + \sin\zeta)]. \tag{4.13.15}$$

The net mass flux in the upper Ekman layer due to the nongeostrophic velocity is given by (4.10.13),

$$\mathbf{M}_E = (\tau \times \mathbf{k})E_V^{1/2}, \tag{4.13.16}$$

and is directed radially outward. Note that the radial *velocity* in the upper Ekman layer is the same order as the azimuthal geostrophic velocity. Since it is limited, however, to a layer whose nondimensional thickness is $O(E_V^{1/2})$ the radial transport is small compared to the size of the mass transport which circulates azimuthally about the center. This transport is

$$\int_0^1 \mathbf{u}_0\,dz = 2\tau = O(1). \tag{4.13.17}$$

In the lower Ekman layer (4.5.31b) applies, so that the velocity in the vicinity of the lower boundary is

$$\tilde{\mathbf{u}}_0 = 2\tau(1 - e^{-\zeta}\cos\zeta) + 2\mathbf{k} \times \tau e^{-\zeta}\sin\zeta. \tag{4.13.18}$$

The interior azimuthal velocity produces an $O(1)$ radial flow in the lower Ekman layer, flowing inward (for $\tau > 0$) across the isobars. This radial mass flow is

$$E_V^{1/2}\int_0^\infty d\zeta\,2(\mathbf{k} \times \tau)e^{-\zeta}\sin\zeta = -E_V^{1/2}(\tau \times \mathbf{k}) \tag{4.13.19}$$

and exactly balances the outward radial flow in the upper Ekman layer.

The vertical velocity sucked out of the lower Ekman layer is, by (4.5.39) and (4.13.13),

$$w(r, \theta, 0) = \frac{E_V^{1/2}}{2}\zeta_0 = \frac{E_V^{1/2}}{r}\frac{\partial}{\partial r}\,r\tau(r), \tag{4.13.20}$$

while the velocity sucked into the upper Ekman layer is given by (4.10.20), which in the present case yields

$$w(r, \theta, 1) = \frac{\tau_0}{\rho f U D} \frac{1}{r} \frac{\partial}{\partial r} r\tau(r) = \frac{E_V^{1/2}}{r} \frac{\partial}{\partial r} r\tau(r), \qquad (4.13.21)$$

so that the velocity pumped out of the lower Ekman layer is absorbed directly into the upper Ekman layer at the same horizontal position, i.e., $w$ is independent of $z$.

A schematic picture of the circulation is shown in Figure 4.13.3. The surface stress drives an Ekman flux of $O(E_V^{1/2})$ radially outward, sucking a

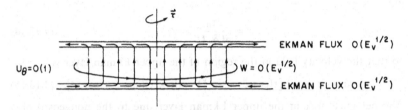

**Figure 4.13.3** A schematic of the motion in the fluid outside the region influenced by the side walls. Fluid circulates azimuthally with $O(1)$ velocity while a small vertical motion arises from Ekman pumping out of the lower Ekman layer where fluid converges, and suction into the upper Ekman layer where fluid elements are flung radially outward.

vertical velocity of $O(E_V^{1/2})$ into the upper Ekman layer while an $O(1)$ azimuthal velocity is produced, which in turn sucks just the right amount of fluid out of the lower Ekman layer to balance the mass flux of the upper Ekman layer. It is important to note that this entire picture has been completed *without* consideration of the dynamics in the region of the side wall of the basin at $r = 1$. It is also important to realize that the solution described above will be deficient in the vicinity of the side wall. Unless $\tau(r)$ is fortuitously zero at the edge of the basin, the order-one azimuthal geostrophic flow will not satisfy the no-slip boundary condition at $r = 1$. Furthermore it is clear that the fluid which rises in the interior of the region under the action of the curl of the applied stress must somewhere descend from the upper to lower Ekman layer to close the radial mass flow.

If we return to (4.13.6), we see that the neglect of the horizontal diffusion of vorticity of $O(\text{Re}^{-1})$ is a singular perturbation of the vorticity equation. We can expect the neglect of this term to yield useful results away from the lateral boundary only. Near the basin perimeter the effect of the horizontal diffusion of momentum must become important so as to satisfy the no-slip condition. Let us refer to the stream function $\psi$ which satisfies (4.13.7) as the *interior* geostrophic stream function. That is, away from the perimeter of the basin

$$\psi = \psi_I(r) = 2 \int_0^r \tau(r') \, dr', \qquad (4.13.22)$$

from (4.13.10). We anticipate that in the vicinity of $r = 1$ the stream function will vary sufficiently rapidly in the radial direction that the horizontal diffusion term in (4.13.6) enters the vorticity balance. Let

$$\eta = \frac{1 - r}{l} \tag{4.13.23}$$

be the stretched radial boundary-layer coordinate, and let $\psi$ in the boundary layer be given by

$$\psi = \psi_I(r) + \hat{\psi}(\eta), \tag{4.13.24}$$

where $\hat{\psi}(\eta)$ is the *correction* to the interior stream function produced in the friction layer. Since $\psi_I$ satisfies (4.13.7), the substitution of (4.13.24) into (4.13.6) yields a homogeneous equation for $\hat{\psi}(\eta)$, viz.

$$0 = -\frac{E_V^{1/2}}{2\varepsilon} \left| \frac{\partial^2 \hat{\psi}}{\partial \eta^2} + \frac{l}{r} \frac{\partial \hat{\psi}}{\partial \eta} \right|$$

$$+ \frac{1}{\operatorname{Re} l^2} \left| \frac{\partial^2}{\partial \eta^2} + \frac{l}{r} \frac{\partial}{\partial \eta} \right| \left| \frac{\partial^2}{\partial \eta^2} + \frac{l}{r} \frac{\partial}{\partial \eta} \right| \hat{\psi}. \tag{4.13.25}$$

The correction function $\hat{\psi}$ must vanish as $\eta \to \infty$ so that (4.13.24) merges smoothly into the interior solution. Thus since $l \ll 1$ by hypothesis, $\hat{\psi}$ will be zero for all $r$ which differ by more than $O(l)$ from the value $r = 1$, i.e., $r = 1 - l\eta$ by (4.13.23). Hence in (4.13.25) $r$ may be replaced by unity. Since $l \ll 1$, (4.13.25) becomes to lowest order

$$\frac{d^4}{d\eta^4} \hat{\psi} - l^2 \frac{E_V^{1/2}}{\varepsilon} \frac{\operatorname{Re}}{2} \frac{d^2 \hat{\psi}}{d\eta^2} = 0. \tag{4.13.26}$$

Since by (4.6.2)

$$\frac{E_V^{1/2}}{\varepsilon} \frac{\operatorname{Re}}{2} = \frac{E_V^{1/2}}{2\varepsilon} \frac{2\varepsilon}{E_H} = \frac{E_V^{1/2}}{E_H}, \tag{4.13.27}$$

$l$ must be given by

$$l = \frac{E_H^{1/2}}{E_V^{1/4}} \tag{4.13.28}$$

in order that the two terms in (4.13.26) may balance. This yields a *dimensional* boundary-layer thickness

$$\delta_S = Ll = \left( \frac{2A_H}{f} \right)^{1/2} \frac{D^{1/2}}{\delta_E^{1/2}}, \tag{4.13.29}$$

where $\delta_E$ is the Ekman-layer thickness on the horizontal boundary.

The length scale $\delta_S$ reflects the physical balance that gives rise to the layer. Vorticity, generated at the boundary by the frictional retardation of the azimuthal flow, diffuses outward. If no other mechanism were present, this diffusion would, in a time $t_*$, reach a distance of the order

$$l_* = (A_H t_*)^{1/2} \tag{4.13.30}$$

by the simple diffusion law. In a time of the order of the spin-down time, though, this vorticity will be eliminated by the pumping effects of Ekman suction as in Section 4.7. This time is, by (4.7.8),

$$\tau = \frac{D}{\delta_E} f^{-1},$$

so that the diffusion length $l_*$ to which vorticity penetrates in (4.13.30) is

$$l_* = (A_H \tau)^{1/2} = \left(\frac{A_H}{f} \frac{D}{\delta_E}\right)^{1/2}, \tag{4.13.31}$$

which aside from a constant factor is (4.13.29).

The general solution to (4.13.26) is then

$$\hat{\psi} = C_1 + C_2 \eta + C_3 e^{\eta} + C_4 e^{-\eta}. \tag{4.13.32}$$

In order that $\hat{\psi}$ may vanish as $\eta \to \infty$, $C_1, C_2$, and $C_3$ must be zero. Thus

$$\psi = \psi_I(r) + C_4 e^{-\eta}. \tag{4.13.33}$$

On $r = 1$ (i.e., $\eta = 0$) the azimuthal velocity must vanish, i.e.,

$$\frac{\partial \psi}{\partial r} = 0 = \frac{\partial \psi_I(1)}{\partial r} + \frac{C_4}{l}, \tag{4.13.34}$$

so that

$$\psi = \psi_I(r) - \frac{E_H^{1/2}}{E_V^{1/4}} \psi_I(1) e^{-\eta}. \tag{4.13.35}$$

Since $\psi_I(r) e^{-\eta}$ is effectively zero for all $r$ different from one, (4.13.35) may be approximated by

$$\psi = \psi_I(r)\left\{1 - \frac{E_H^{1/2}}{E_V^{1/4}} e^{-\eta}\right\}$$

$$= \psi_I(r)\left\{1 - \frac{E_H^{1/2}}{E_V^{1/4}} \exp\left(-[1 - r]\frac{E_V^{1/4}}{E_H^{1/2}}\right)\right\}. \tag{4.13.36}$$

The azimuthal velocity is similarly

$$u_\theta = 2\tau(r)\{1 - e^{-\eta}\} = 2\tau(r)\left[1 - \exp\left((1 - r)\frac{E_V^{1/4}}{E_H^{1/2}}\right)\right] \tag{4.13.37}$$

and is shown in Figure 4.13.4. In contrast to the Ekman layer, the boundary-layer velocity is monotonically increasing from the boundary to the interior.

This vertically standing boundary layer at $r = 1$ intersects the upper and lower Ekman layers in the crosshatched regions of Figure 4.13.5. In this intersection region of the boundary layers, say at $z = 0$, the velocity fields will be functions of $\zeta = z/E_V^{1/2}$ and $\eta = (1 - r)E_V^{1/4}/E_H^{1/2}$. The frictional terms in the horizontal equations will be of the form

$$\frac{E_V}{2} \frac{\partial^2 u_\theta}{\partial z^2} + \frac{E_H}{2} \frac{\partial^2 u_\theta}{\partial r^2} + \cdots = \frac{1}{2} \frac{\partial^2 u_\theta}{\partial \zeta^2} + \frac{E_V^{1/2}}{2} \frac{\partial^2 u_\theta}{\partial \eta^2}, \tag{4.13.38}$$

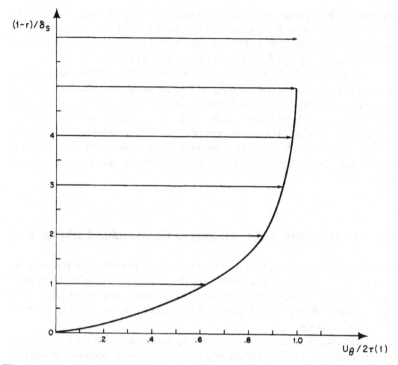

**Figure 4.13.4** Profile of $U_\theta(r)$ near the side boundary. The characteristic width of the side-wall layer is $\delta_s = (2A_H/f)^{1/2}(D/\delta_E)^{1/2}$.

so that in the crosshatched region the important frictional terms are those which produce the Ekman-layer balance. *The side-wall layer is so wide that for the frictional region below (and above) it the Ekman-layer analysis applies directly to it as if it were an ordinary geostrophic interior flow.* In particular the Ekman-layer boundary conditions (4.5.39) and (4.10.20) apply to the solution (4.13.35) at $z = 0$ and $z = 1$. At $z = 1$ the vertical velocity is again given by (4.13.21), since the velocity in the side-wall layer has no vertical shear. This means that the Ekman flux impinging on the side boundary does

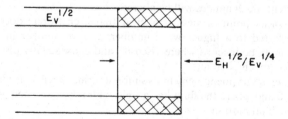

**Figure 4.13.5** The intersection region of the side-wall layers and Ekman layers is crosshatched. Within these regions the Ekman-layer dynamics still apply.

not descend in the layer of width $\delta_S$. Thus the side-wall layer described by (4.13.37) is capable only of satisfying the no-slip condition. The additional requirement that the small radial mass flux be turned to descend to the lower Ekman layer is not satisfied by this geostrophic layer.

In order to satisfy the mass-flux balance an additional boundary layer, *thinner than the layer of thickness* $\delta_S$, is also required. In order for this layer to have a different dynamical character it must contain either non-geostrophic or nonhydrostatic effects at the *lowest* order, for otherwise the $\delta_S$-layer will reappear. The discussion of these thin *upwelling* layers required to close the small $O(E_V^{1/2})$ mass balance is deferred to Section 5.12 and Section 8.3, where such nongeostrophic effects are described in greater detail.

## 4.14  The Dissipation of Enstrophy in Geostrophic Turbulence

When the mixing coefficients are sufficiently small, the effects of viscosity are confined to boundary layer zones where the spatial variation of the velocity is rapid enough to make the frictional forces locally important. Generally, when $E_V$ and $1/R_e$ are small, these considerations are sufficient to understand the role of viscosity.

However, there are occasions when the spatial scale *within* the fluid, distant from boundaries, shrinks sufficiently so that viscosity becomes important within the fluid to alter the scale of the vorticity, for example, from that which would be set by either the scale of the forcing or the geometry of the container. In such cases, our choice of scaling length $L$ becomes artificial. The fluid begins to manifest its own scale, which scale may not be evident *a priori*.

A particularly interesting example of this transfer to viscous scales occurs in the case of geostrophically controlled turbulent flow. In Section 3.28, we saw that the inviscid quasigeostrophic dynamics acted nonlinearly to effect scale alterations on the energy and enstrophy. Energy cascaded to larger scales while enstrophy was transferred to smaller scales. These are both manifestations of inviscid conservation principles and in the case of enstrophy also of the property that identifies lines of constant vorticity with material lines. The presence of viscosity will obviously have some effect, but the central issue is to what extent the dynamics will be altered.

From another point of view, it is clear that if enstrophy does continue to be transferred to a higher wave number, a wave number high enough must eventually be reached where viscosity and consequently dissipation of vorticity become important.

To investigate the phenomena in its simplest form, we will consider the case where $\beta = 0$ and where the fluid has a rigid lid (or $F = 0$). Then (4.6.14) will apply to the fluid motion.

For the moment, let us take $\zeta_T = 0$ so that the turbulent flow is freely evolving from some initially forced state. There is no difficulty in considering

$\zeta_T \neq 0$ and the reader is encouraged to paraphrase the arguments given below with that restriction removed.

If (4.6.13) is multiplied by $\psi = p_0$ and integrated over the infinite domain on whose boundaries tends to zero, we obtain the energy equation.

$$\frac{\partial}{\partial t} \iint \left(\frac{\nabla \psi}{2}\right)^2 dx\, dy = -r \iint (\nabla \psi)^2\, dx\, dy - \frac{1}{R_e} \iint \zeta_0^2\, dx\, dy, \quad (4.14.1)$$

where

$$(\nabla \psi)^2 = u_0^2 + v_0^2,$$
$$\nabla^2 \psi = \zeta_0. \quad (4.14.2)$$

If (4.6.13) is now multiplied by $\zeta_0$ and integrated over the plane, an equation for the enstrophy is obtained, i.e.,

$$\frac{\partial}{\partial t} \iint \frac{\zeta_0^2}{2} dx\, dy = -r \iint \zeta_0^2\, dx\, dy - \frac{1}{R_e} \iint (\nabla \zeta_0)^2\, dx\, dy. \quad (4.14.3)$$

Now (4.14.3) implies that the total enstrophy must decay, although rather slowly when $r$ is small and $R_e$ is large. In any case, (4.14.3) implies that the area integral of $\zeta_0^2$ never exceeds its initial value and (4.14.1) implies the same for the energy. Therefore, each of the integrals on the right-hand side of (4.14.1) is bounded and thus as $r \to 0$ and $R_e \to \infty$, the right-hand side of (4.14.1) vanishes. Thus in the limit of nearly frictionless flow, the energy is essentially conserved. In a three-dimensional, nongeostrophic flow, this would not be the case, for then the vorticity could locally become large due to the tilting and twisting of vortex lines, which mechanism is absent in this geostrophic flow.

The situation is less clear for the enstrophy, for its dissipation depends on the *gradients* of the vorticity field. We saw in Section 3.28 that the turbulent stretching of material lines would lead to a continuous increase in $\nabla \zeta_0$. The weaker the viscosity is, i.e., the larger is $R_e$, the more completely vorticity is frozen to material lines and the more efficient will be the inertial process of increasing $\nabla \zeta_0$. Hence, the size of the product $R_e^{-1}(\nabla \zeta_0)^2$ is problematic as $R_e \to \infty$, i.e., it may remain nonzero in the limit.

An equation for $\nabla \zeta_0$ can be obtained by differentiating (4.6.13) with respect to $x$ and $y$. After several obvious steps we obtain

$$\frac{1}{2} \frac{\partial}{\partial t} \iint \left[ \left(\frac{\partial \zeta_0}{\partial x}\right)^2 + \left(\frac{\partial \zeta_0}{\partial y}\right)^2 \right] dx\, dy$$

$$= -\iint \left[ \frac{\partial v_0}{\partial y} \left(\frac{\partial \zeta_0}{\partial y}\right)^2 + \frac{\partial u_0}{\partial x} \left(\frac{\partial \zeta_0}{\partial x}\right)^2 + \left(\frac{\partial v_0}{\partial x} + \frac{\partial u_0}{\partial y}\right) \frac{\partial \zeta_0}{\partial x} \frac{\partial \zeta_0}{\partial y} \right] dx\, dy \quad (4.14.4)$$

$$- r \iint (\nabla \zeta_0)^2\, dx\, dy - \frac{1}{R_e} \iint \left[ \left(\nabla \frac{\partial \zeta}{\partial x}\right)^2 + \left(\nabla \frac{\partial \zeta}{\partial y}\right)^2 \right] dx\, dy.$$

The first term on the right-hand side of (4.14.4) represents an inertial process by which the vorticity gradient can be sharpened by the turbulent velocity field. To see how this works, examine Figure 4.14.1. At an arbitrary point in

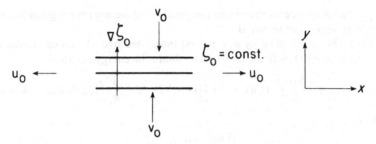

**Figure 4.14.1**   The magnification of the vorticity gradient by the turbulent velocity field.

the fluid, the $y$-axis has been aligned with the gradient of $\zeta_0$. Then locally the integrand of the first term on the right-hand side of (4.14.4) becomes simply

$$-\frac{\partial v_0}{\partial y}\left(\frac{\partial \zeta_0}{\partial y}\right)^2$$

which is independent of the *sign* of $\partial \zeta_0/\partial y$. Now if, as previously argued, the effect of the turbulence is to lead statistically to a stretching out of material lines, $\partial u_0/\partial x$ will be $>0$ as shown schematically in the figure. Since the geostrophic velocities are now divergent, this yields $\partial v_0/\partial y < 0$. The material lines crowd closer together and the integral we are considering on the right-hand side becomes positive. This by (4.14.4), tends to produce a growth of the area integrated squared vorticity gradient. The process will continue untill $\zeta_0$ reaches small enough scales for the viscous term in (4.14.4) to become important. This inertial magnification of $(\nabla \zeta_0)^2$ means that in the limit $R_e \to \infty$, the drain of enstrophy represented by the second term on the right-hand side of (4.14.3) may continue. Hence, there is no reason to believe that *enstrophy* will continue to be preserved if sufficient time has elapsed for the spectrum of the enstrophy spectrum to shift to high enough wave numbers.

Let us now repeat the steps leading to (3.28.4). With $K_1$ defined as in (3.28.1) it follows that

$$\frac{\partial}{\partial t}K_1^2 \int_0^\infty \varepsilon(K)\,dK = -\frac{\partial}{\partial t}\int_0^\infty (K - K_1)^2 \varepsilon(K)\,dK + \frac{\partial}{\partial t}\int_0^\infty K^2 \varepsilon(K)\,dK$$

$$= -\frac{\partial}{\partial t}\int_0^\infty (K - K_1)^2 \varepsilon(K)\,dK - r\iint \zeta_0^2\,dx\,dy \quad (4.14.5)$$

$$-\frac{1}{R_e}\iint (\nabla \zeta_0)^2\,dx\,dy,$$

where we have used the fact that $\int_0^\infty K^2 \varepsilon(K)\,dK$ is the total enstrophy and where (4.14.3) has been used to evaluate its rate of change.

Now in the limit $r \to 0$, $R_e \to \infty$, we have seen that: (1) the total energy will be conserved, and (2) that $\zeta_0^2$ will remain finite although $\iint (\nabla \zeta_0)^2\,dx\,dy$ may not. Hence in that limit

$$\frac{\partial K_1^2}{dt} = -\frac{\frac{\partial}{dt}\iint(K - K_1)^2\varepsilon(K)\,dK}{\int_0^\infty \varepsilon(K)\,dK} - \frac{1}{R_e}\frac{\iint(\nabla\zeta_0)^2\,dx\,dy}{\int_0^\infty \varepsilon(K)\,dK}. \qquad (4.14.6)$$

The last term in (4.14.5) representing the enstrophy dissipation is never positive. The first term on the right-hand side of (4.14.6) will be negative under the conditions of spectral broadening as discussed in Section 3.28. Thus any dissipation of enstrophy will *enhance* the consequences of spectral broadening and augment the transfer of energy to larger scales. Since the energy is even less vulnerable to dissipation on large scales, the unexpected result is that viscous dissipation of enstrophy locks up even more tightly the energy at lower wave numbers.

As energy moves to larger scales, eventually, should no other process intervene, enstrophy will be continuously drained from the fluid. Batchelor (1969) argued that if the viscosity was small enough, i.e., if the Reynolds number based on the scales of the energy containing eddies were large enough, then the viscous sink of enstrophy would be at wave numbers very much greater than the scale of the energy input. Suppose the enstrophy moves through the spectrum by a series of transfers up the spectrum, each step of which involves only the interaction of eddies of nearly that scale. That is, suppose the dynamics of the spectral transfer is *local* in the spectrum. Then a great number of transfers would be required to transfer the enstrophy any great distance up the spectrum and it is therefore plausible that as a consequence of this multiple-step process, information about the state at lower wave numbers might eventually be lost. In particular, it might be expected that the eddies becomes statistically isotropic at high wave number. It is unclear that the requirements of spectral localness of the transfer and isotropy will really be met, but if they are, a simple argument suffices to describe the nature of the resulting enstrophy cascading region of an *equilibrium* spectrum. That is, if we imagine the spectrum to be either steady or nearly so, then at least in the enstrophy cascading region, a constant flux of enstrophy moves through the enstrophy spectrum. The flux is generated at lower wave numbers and then finally dissipated at higher wave numbers.

Let $\chi$ be the dimensional rate at which enstrophy is drained by friction. That is, let $\chi$ be the dimensional equivalent to the final term in (4.14.5). Then $\chi$ is also the dimensional flux of enstrophy through the enstrophy spectrum in the equilibrium spectrum. The flux must be constant for any variation of the flux with $K$ would lead to a piling up of enstrophy at some $K$. Since the flux is constant, it must equal the rate at which enstrophy spills into the viscous sink at high $K$. We can imagine that $\chi$ is an external parameter, at least as far as the enstrophy cascading region is concerned and only lateral mixing becomes important due to the shrinking lateral scale as $R_e \to \infty$ and $r \to 0$.

Let $\Omega_*$ be the dimensional equilibrium spectrum for the enstrophy. Our choice of $U$ and $L$ for scaling parameters implies that then

$$\Omega_*(K_*) = \frac{U^2}{L}\Omega(K), \qquad (4.14.7)$$

where $\Omega$ is the nondimensional enstrophy spectrum and where $K_*L = K$. We are confronted next with the task of specifying $U$ and $L$ in terms of the relevant external parameters. We have argued that in the equilibrium enstrophy cascade domain these can be only $\chi$ and $A_H$. Now the dimensions of $\chi$ are equal to the those of the rate of decrease of enstrophy, hence

$$[\chi] = (\text{time})^{-3},$$

while

$$[A_H] = (\text{length})^2(\text{time})^{-1},$$

where the bracket notation $[\ ]$ stands for the specification of the dimensional equivalent of the function within the bracket. Hence, if we write

$$U = \chi^\alpha A_H^\beta, \qquad (4.14.8)$$

dimensional arguments alone imply that $\alpha = \frac{1}{6}$ and $\beta = \frac{1}{2}$, so

$$U = A_H^{1/2}\chi^{1/6}, \qquad (4.14.9)$$

while similarly the scaling length $L$ can only be chosen as

$$L = A_H^{1/2}\chi^{-1/6}. \qquad (4.14.10)$$

Thus

$$\Omega_*(K_*) = A_H^{1/2}\chi^{1/2}\Omega(K), \qquad (4.14.11)$$

where

$$K = K_* A_H^{1/2}\chi^{-1/6}. \qquad (4.14.12)$$

Note that as either the mixing coefficient $A_H$, is reduced or the flux rate is increased, the characteristic scale $L$, given by (4.14.10) decreases. This scaling length is the length on which horizontal friction becomes as important as inertial transfer in the spectral evolution of the enstrophy.

Now suppose that $A_H^{1/2}\chi^{-1/6}$ is very large compared to either the scale of the energy input or the scale of the energy containing eddies. In that case, there will be a large intermediate cascade zone satisfying (4.14.7) in which the direct effect of viscosity is unimportant. In this inertial range, in which $K_*$ is larger than the wave numbers of the energy containing scales but still small compared to $A_H^{1/2}\chi^{-1/6}$, $\Omega_*$ should be independent of $A_H^{1/2}$. This can only occur if in this inertial equilibrium zone,

$$\Omega(K) = \frac{\Omega_0}{K}, \qquad (4.14.13)$$

so that

$$\Omega_* = \Omega_0\chi^{2/3}K_*^{-1},$$

where $\Omega_0$ is a nondimensional constant.

The energy spectrum at any $K$ is related to the enstrophy spectrum by the relation

$$\varepsilon = \frac{\Omega}{K^2}, \tag{4.14.14}$$

so that in dimensional units, the energy in the enstrophy cascade region is

$$\varepsilon_* = \frac{\Omega_*}{K_*^2} = A_H^{3/2} \chi^{1/6} \frac{\Omega(K)}{K^2}, \tag{4.14.15}$$

so that the choice (4.14.13) yields

$$\varepsilon_*(K_*) = \Omega_0 \chi^{2/3} K_*^{-3}. \tag{4.14.16}$$

In the inertial cascade region the energy is expected to fall off like $K_*^{-3}$.

The arguments of self-similarity would also imply that if there were an energy flux in this part of the spectrum, it must be related to $\chi$. This would imply, by dimensional arguments alone, that the energy flux through the spectrum would be proportional to $K_*^{-2}$. But this would yield an energy flux varying with $K_*$ and be inconsistent with the notion of viscous equilibrium at these scales. Hence the only logical conclusion is that if there is indeed an equilibrium region determined by a constant enstrophy flux the spectral *energy* flux must vanish in that region of the spectrum.

This seems sensible for the flux of energy is in fact occurring at much lower wave numbers rather than in the enstrophy cascade region. In fact, the same arguments leading to (4.14.14) can be applied to a hypothetical domain at *low* wave numbers in which energy is imagined to be uniformly flowing through the spectrum from the wave number of energy input to some hypothetical sink at the largest scales. If $\mu$ is the proposed energy cascade rate [with dimensions (length)$^2$/(time)$^3$] then the energy spectrum in this domain would be, from dimensional arguments

$$\varepsilon_*(K_*) = C_0 \mu^{2/3} K_*^{-5/3}. \tag{4.14.17}$$

The enstrophy spectrum here would be

$$\Omega_*(K_*) = C_0 \mu^{2/3} K_*^{2/3} \tag{4.14.18}$$

and vanishes as $K_* \to 0$.

It seems unlikely that the requirements for the validity of (4.14.8) can easily be met. It is hard to imagine a spectrally localized sink at low wave numbers able to a continuously accept the energy cascaded to small $K_*$. At the same time the introduction of the $\beta$ effect will certainly eliminate the tendency for isotropy, and the arrest of the energy cascade by the transformation of turbulence into Rossby waves as described in Section 3.28 will severely limit the region of validity of (4.14.17).

In fact, although the presence of the $\beta$-effect is unlikely to be important at the smaller scales where the enstrophy cascade is expected, there are, nevertheless, important difficulties associated even with (4.14.14) and hence (4.14.16). Suppose we consider the contribution to the total enstrophy from any wave

number band within the enstrophy inertial range where (4.14.14) is supposed
to apply. This contribution is

$$\int_{K_{1*}}^{K_{2*}} \Omega_* \, dK_* = \Omega_0 \chi^{2/3} \ln\left(\frac{K_{2*}}{K_{1*}}\right). \qquad (4.14.19)$$

Consider the eddies with wave numbers $\sim K_{2*}$. The shear acting on eddies
of this scale by larger eddies is of the order of

$$\left\{\int_{K_{1*}}^{K_{2*}} \Omega_* \, dK_*\right\}^{1/2}$$

if $K_{1*}$ is the scale of the eddy of maximum size considered. Eddies at scales
much smaller than $K_{2*}^{-1}$ can be expected to have their shearing effects cancel
out. Now imagine the interval between $K_{1*}$ and $K_{2*}$ partitioned into octaves,
i.e., into segments whose end members $K_{j*}$ and $K_{j+1*}$ satisfy $K_{j+1*}/K_{j*} =$
constant. Then from (4.14.19), it follows that each such segment will make an
equal contribution to the shear acting on the eddy with scale $K_{2*}^{-1}$. This implies,
contrary to our hypothesis of local interactions in the spectrum, that the
enstrophy cascade links widely differing scales and this casts some doubt on the
validity of the assumptions leading to (4.14.14) and (4.14.15). Indeed, Kraichnan
(1971) has argued that (4.14.14), in fact, implies a nonuniform enstrophy flux
through the spectrum and suggested a correction to (4.14.15), namely,

$$\varepsilon_* = \text{constant} = K_*^{-3}\left(\ln\left(\frac{K_*}{K_{1*}}\right)\right)^{1/3},$$

where $K_{1*}$ is the wave number of the lower limit of the putative enstrophy
cascade domain. However, even in this case, the spectral dynamics may be
inferred to be nonlocal.

Nevertheless, the qualitative nature of the enstrophy cascading region has
received considerable verification from a number of numerical experiments
which satisfy the overall physical assumptions required by theory, i.e., of
homogeneous flow under essentially quasigeostrophic conditions at high
Reynolds number. Figure 4.14.2 shows the result of a calculation by Lilly
(1971) for which $r = 0$, $\beta = 0$ and $R_e^{-1} = 2.5 \times 10^{-4}$. One notes that after the
initiation of the experiment the region $1 \leq K \leq K_e = 8$ (where $K_e$ is the scale
in which energy enters the system due to external forcing) roughly matches
the prediction of (4.14.17) while at longer times, the energy begins to pile up
at wave number 1 invalidating the assumptions leading to (4.14.17). The
spectrum then rises more steeply to larger scale.

At the higher wave numbers, the enstrophy cascading domain with $\varepsilon \sim K^{-3}$
appears to exist at least for a few octaves. More recent calculations by
Basdevant et al. (1981) in which the effect of $\beta$ is included also clearly demonstrate the existence of an enstrophy cascading subrange in the calculated
spectrum in which there is nearly a uniform transport of enstrophy to higher
wave numbers and no transport of energy in that spectral region. However,
the calculated spectral slope is much steeper than the $K^{-3}$ law of (4.14.15)

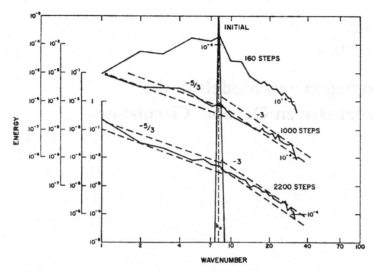

**Figure 4.14.2**   The energy spectrum $\varepsilon(k)$ calculated by Lilly (1969). The solid lines give the energy spectrum after the individual number of time steps in the calculation. The spectrum is forced at wave number = 8. Curves with slope $-\frac{5}{3}$ and $-3$ are shown for comparison.

approaching a $K^{-4}$ or even $K^{-6}$ behavior. This steeper slope has been attributed to the presence of spatial and temporal intermittence of the turbulent field which tends to reduce the efficiency of the enstrophy transfer to large wave number.

It is important to keep in mind that in any case, the *energy* of the turbulent flow is in all cases observed to remain locked in the largest scales. Thus even the most chaotic of flows will, because of the geostrophic constraint, keep its energy relatively immune from the small-scale horizontal mixing process. This implies that in most cases of oceanographic or meteorological interest, the attribution of single gross scales $U$ and $L$ to the most energetic portions of the spectrum remains sensible and useful and dissipation of energy will continue to be dominated by the behavior of the large-scale field of motion.

CHAPTER 5

# Homogeneous Models of the Wind-Driven Oceanic Circulation

## 5.1 Introduction

From ancient times descriptions of the movement of the sea have emphasized both change and continuity. Sudden, violent change and hypnotic
steadiness are the dramatic elements of many a sea story. With the advent of
the great age of navigation and exploration a more systematic description of
the general pattern of the oceanic circulation began to emerge. It is clear
now that a time-averaged circulation pattern exists in all oceans, although
its observations may be significantly distorted by the presence of energetic
fluctuations.

It would be a mistake to imagine that even now the time-averaged pattern
of the general circulation is known with great accuracy. Important observational and descriptive lacunae still exist. Nevertheless certain basic features
of the oceanic circulation are common to all of the world's oceans, and the
explanation of these features sets a problem in geophysical fluid dynamics of
great interest and beauty.

Figure 5.1.1 shows a schematic rendering of the circulation pattern of the
world's oceans as manifested by surface currents. Perhaps the most striking
feature of the pattern in each of the world's oceans is the remarkable western
intensification of the circulation. In the Atlantic Ocean of the northern
hemisphere, for example, a generally clockwise gyre possesses modest speeds
of the order of 1 to 10 cm/s except in the intense and narrow Gulf Stream
current pressed against the western boundary from Florida to Cape Hatteras, where it then rejoins the interior circulation. The intensity of this

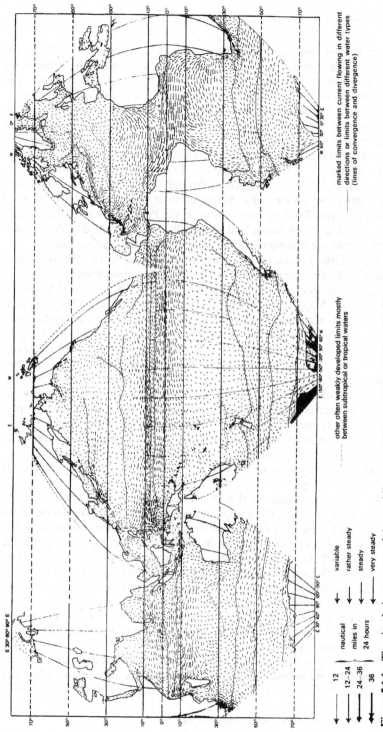

**Figure 5.1.1** The circulation pattern of the world's oceans. This map (after Defant (1961)) represents a long-term compilation of measurements and ship reports. Although it is schematic rather than precise, it reveals the westward intensification of the circulation in the major ocean gyres.

255

current is truly awesome. The current velocity is typically 100 cm/s, and its maximum may be double that. Its width is not unambiguously defined, but the region of strong flow is of the order of 50 to 100 km in width. The momentum per unit volume of an element of fluid in the Stream, $\rho\mathbf{u}$, is therefore of $O(100 \text{ gram cm}^{-2} \text{ s}^{-1})$, which is about an order of magnitude greater than the momentum of an element of air in the atmospheric jet stream, even though the speed of the latter is one hundred times greater. The immensity of the Gulf Stream current is also demonstrated by the magnitude of its total mass transport. Estimates still vary, but a typical estimate of the rate with which mass is transported through the Stream's cross section is on the order of 90 million cubic meters of water per second. Nor is the Gulf Stream unique. The Kuroshio Current in the Pacific Ocean, the Brazil Current in the South Atlantic, and the Agulhas Current off East Africa each demonstrate the dramatic western intensification of the circulation.

A particularly illuminating phenomenon occurs in the Indian Ocean, where in the wintertime a western current flows southward along the horn of Africa. In the summer period after the onset of the Southwest Monsoon (usually in May), the current reverses in direction and the strong, northward-flowing Somali Current appears with a structure and speed similar to both the Kuroshio and the Gulf Stream. The remarkable sensitivity to the shift in the winds of the Somali Current suggests that the ocean circulation can be attributed to the action of the wind. Indeed, with the important and singular exception of the western intensification, the pattern of the oceanic circulation reflects the pattern of the winds, i.e., generally westward flow in equatorial regions under the influence of the trade winds and predominantly eastward flow in mid-latitudes, the region of the westerly winds.* Although differential heating of the sea surface can produce motion by buoyant forces, it is possible to show (see Section 6.19) that over most of the ocean only the applied wind stress can produce a net (vertically integrated) horizontal transport.

In this chapter simple models of the oceanic general circulation are discussed which attribute the motion of the ocean entirely to the action of the wind, in particular to the stress exerted by the wind on the sea surface. These models are homogeneous in density and therefore completely ignore the dynamical effects of stratification, and most often the models ignore both the complex topography of the ocean floor and the complexity of the shape of the perimeter of each ocean basin. Nevertheless the models are remarkably successful in describing the general nature of the horizontal circulation, at least in plan view, while of course they make no pretense of adequately describing the vertical structure of the flow, which will be sensitive to the density field.

Each of the oceans manifests the westward intensification, although the basins differ in the details of shape, topography, and stratification (the latter to a more minor degree) and the pattern of the wind stress. This suggests

---

* Recall that westerly winds are winds *from* the west.

that the overall circulation moves according to powerful, yet simple constraints. The advantage of the homogeneous models is that they clearly expose these constraints in the most elementary way. The success of the models is a fair sign that they grasp much of the essential physics of the general oceanic circulation. The general success of the homogeneous models furthermore means that where they do fail, for example in adequately predicting the nature of oceanic fluctuations, the physical cause of the failure is transparently related to the absence of stratification.

## 5.2 The Homogeneous Model

In this section the mathematical model for the circulation of a homogeneous ocean will be developed in a heuristic fashion, making free use of the systematic scaling arguments and detailed calculations of Chapters 3 and 4. This is done to emphasize the basic and simple physical foundation of the model. The essential aspects of the model are depicted in Figure 5.2.1. The

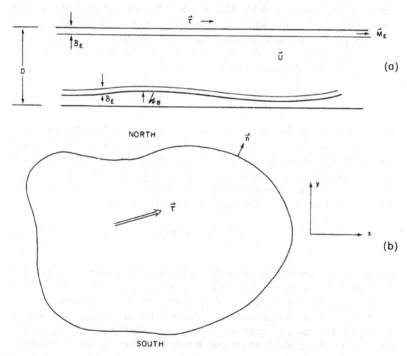

**Figure 5.2.1** (a) The essential ingredients of the homogeneous model. $\tau$ is the applied stress, and the surface layer has a thickness $\delta_E \ll D$. The mass flux in the surface layer is $\mathbf{M}_E$. There is a thin bottom layer over a bottom of variable depth $h_B$. (b) In plan view, the boundedness of the oceanic basin becomes an important element in the dynamics. The direction of north is dynamically revealed by the increase of the Coriolis parameter with latitude.

fluid layer consists in the main of three regions: a thin surface Ekman layer, the bulk interior of typical depth $D$, and a thin bottom frictional Ekman layer over a sloping bottom. As described in Section 4.10, the applied wind stress directly produces a horizontal volume flux in the upper Ekman layer:

$$\mathbf{M}_E = \frac{\tau}{\rho f} \times \mathbf{k}, \tag{5.2.1}$$

where $\tau$ is the wind stress, $\rho$ the water density, $f$ the Coriolis parameter, and $\mathbf{k}$ a unit vertical vector. In (5.2.1) and in all subsequent formulae of this section, all variables are *dimensional*. Recall that the Coriolis parameter is

$$f = 2\Omega \sin \theta, \tag{5.2.2}$$

where $\theta$ is the latitude.

The interior is sandwiched between the two thin friction layers. In previous chapters it was systematically demonstrated that for low Rossby numbers the basic momentum balance for the interior is given by the geostrophic approximation while the pressure field is determined by the dynamics of the vorticity field. Also for low Rossby numbers the vertical component of *absolute vorticity* was shown, in (4.6.7), to change as a consequence of: (1) vortex-tube stretching of the planetary vorticity filaments and (2) the generally weak effect of horizontal diffusion of vorticity. In dimensional units this statement is

$$\frac{d}{dt}(\zeta + f) = f\frac{\partial w}{\partial z} + A_H \nabla^2 \zeta. \tag{5.2.3}$$

There is one essential difference between (5.2.3) and (4.6.7). In the derivation of (4.6.7) $f$ was considered a constant. In (5.2.3) the variation of the Coriolis parameter with latitude is retained. The heuristic discussion of the $\beta$-plane in Section 3.17 suggests that for scales of motion $L$ for which $L < O(r_0)$, where $r_0$ is the earth's radius, the dynamically significant effect of the earth's sphericity is the introduction of the planetary vorticity gradient

$$\beta_0 = \frac{df}{dy} = \frac{2\Omega \cos \theta_0}{r_0}, \tag{5.2.4}$$

where $\theta_0$ is the central latitude of the region. Apart from this contribution to the vorticity dynamics, the sphericity of the earth can be neglected. This intuitive concept is developed in a rigorous method by asymptotic analysis in Chapter 6. The detailed analysis supports the intuitive picture already developed. The reader concerned at this point with the $\beta$-plane approximation may turn temporarily to Chapter 6 before returning to the development of the model as presented here.

The Ekman layer on the lower, sloping surface has already been discussed in Section 4.9. The frictional retardation of the interior flow produces a suction velocity into (or out of) the interior in addition to the vertical velocity produced by the topographic lifting of the flow. When (4.9.36) is written in dimensional units, this vertical velocity entering the interior region on its lower edge is

$$w(x, y, h_B) = \mathbf{u} \cdot \nabla h_B + \frac{\delta_E}{2} \zeta. \tag{5.2.5}$$

Since the interior is homogeneous and geostrophic, $u$, $v$, and therefore $\zeta$ must be independent of $z$, so that (5.2.3) may be easily integrated over the thickness of the geostrophic region to yield

$$[D - h_B]\left|\frac{d\zeta}{dt} + v\beta_0 - A_H \nabla^2\zeta\right| = f\{w(x, y, D) - w(x, y, h_B)\}, \tag{5.2.6}$$

where the relation

$$\frac{df}{dt} = v\beta_0 \tag{5.2.7}$$

has been used.

For steady flows, the vertical velocity sucked into the upper Ekman layer is given entirely by the divergence of the Ekman mass flux as described in Section 4.10, i.e.,

$$w(x, y, D) = \nabla \cdot \mathbf{M}_E$$

$$= \nabla \cdot \left(\frac{\tau \times k}{\rho f}\right) = \mathbf{k} \cdot \text{curl}\,\frac{\tau}{\rho f}. \tag{5.2.8}$$

If (5.2.5) and (5.2.8) are used in (5.2.6),

$$D\left[1 - \frac{h_B}{D}\right]\left|\frac{d\zeta}{dt} + v\beta_0 - A_H\nabla^2\zeta\right|$$

$$= f\left|\mathbf{k} \cdot \text{curl}\,\frac{\tau}{f\rho} - \frac{\delta_E}{2}\zeta - \mathbf{u} \cdot \nabla h_B\right|. \tag{5.2.9}$$

This is the mathematical expression of the simple physical model described earlier. Simple as it is, it is still a difficult differential equation to deal with as it stands. Further simplification, necessary for any progress to be made, is considerably eased by first nondimensionalizing the variables. In the usual way, introduce $L$, $U$, $L/U$, and $\tau_0$ as characteristic scales for horizontal scale of the *motion*, the horizontal velocity, time, and the stress— i.e., if dimensionless parameters are temporarily denoted by primes,

$$(x, y) = L(x', y'), \qquad \zeta = \frac{U}{L}\zeta'$$

$$(u, v) = U(u', v'), \tag{5.2.10}$$

$$\tau = \tau_0 \tau'.$$

Note that within the context of the $\beta$-plane approximation

$$f = f_0 + \beta_0 y$$

$$= f_0\left(1 + \frac{\beta_0 L}{f_0} y'\right), \tag{5.2.11}$$

where

$$\frac{\beta_0 L}{f_0} = (\cot \theta_0)\frac{L}{r_0}$$

$$= O\left(\frac{L}{r_0}\right) < O(1). \tag{5.2.12}$$

Then (5.2.9) becomes

$$\left[1 - \frac{h_B}{D}\right]\left|\frac{d\zeta}{dt} + \beta v - \frac{\nabla^2\zeta}{\text{Re}}\right|$$

$$= \frac{\tau_0 L}{\rho D U^2}\left[\text{curl } \tau + \tau^{(x)}\frac{\beta_0 L}{f}\right] - \frac{\delta_E}{2D}\frac{f_0 L}{U}\left(1 + \frac{\beta_0 L}{f_0}y\right)\zeta$$

$$- \mathbf{u} \cdot \nabla\left(\frac{h_B}{D}\right)\frac{f_0 L}{U}\left(1 + \frac{\beta_0 L}{f_0}y\right), \tag{5.2.13}$$

where

$$\text{Re} = \frac{UL}{A_H},$$

$$\beta = \frac{\beta_0 L^2}{U}, \tag{5.2.14}$$

$$\text{curl } \tau = \frac{\partial\tau^{(y)}}{\partial x} - \frac{\partial\tau^{(x)}}{\partial y},$$

and where *unprimed* variables are now *dimensionless*. Dimensional variables are denoted by an asterisk, e.g.,

$$u_* = Uu, \tag{5.2.15}$$

etc.

Since $\beta_0 L/f_0$ is necessarily small for the $\beta$-plane approximation to be consistent, (5.2.13) may be further simplified to

$$\frac{d\zeta}{dt} + \beta v = \frac{\tau_0 L}{\rho D U^2}\text{curl } \tau - r\zeta - \mathbf{u} \cdot \nabla\eta_B + \frac{\nabla^2\zeta}{\text{Re}}, \tag{5.2.16}$$

where

$$r = \frac{\delta_E}{2D\varepsilon},$$

$$\eta_B = \frac{h_B/D}{\varepsilon}, \tag{5.2.17}$$

$$\varepsilon = \frac{U}{f_0 L},$$

and the fact that $h_B/D \ll 1$ has been used.

In any even partly realistic model for the oceanic circulation Re is large and $r$ is $O(1)$ or less. $\eta_B$ may be large if actual topographic slopes are considered, due to the smallness of $\varepsilon$, but the homogeneous model which has velocities which extend uniformly to the bottom considerably overestimates the effect of topography on the vorticity balance. The topographic contribution to the vortex stretching depends in reality on the size of the *bottom* velocity, and in the ocean the time-averaged bottom velocity is usually considerably less than the average velocity over the whole depth. In the following analysis the ocean bottom will be considered to be flat, but if topographic effects are to be considered the size of $\eta_B$ should probably be reduced from the value obtained from the use of actual bottom slopes to something like the product of $\eta_B$ and the ratio of the bottom velocity to the average velocity. Similar considerations should also apply to $r$, which represents the effect of bottom friction. On the other hand $\beta$ is a very large number for all realistic mid-ocean flows. If $L$ is $10^3$ km and $U$ is even as large as 10 cm/s, with $\beta_0 = 10^{-13}$ cm$^{-1}$ s$^{-1}$, then $\beta = O(10^2)$. In the mid-ocean the dominant term in the vorticity balance is the increase of absolute vorticity by the northward motion of fluid in the planetary vorticity gradient. As vorticity is added to the fluid at a rate proportional to curl $\tau$, the fluid increases its absolute vorticity by languidly moving to a higher latitude, where it takes on a larger value of planetary vorticity. This balance suggests that the appropriate choice for the scaling velocity $U$ should be such as to balance the wind-stress curl and the $\beta$-term, i.e.,

$$\frac{\tau_0 L}{\rho D U^2} = \beta = \frac{\beta_0 L^2}{U} \tag{5.2.18}$$

or

$$\boxed{U = \frac{\tau_0}{\rho D \beta_0 L}}. \tag{5.2.19}$$

If $\tau_0$ is 1 dyne/cm$^2$ (a typical value of the wind stress) and $D$ is 5 km, this yields a velocity $U$ of $O(0.2$ cm/s$)$, which should be interpreted as the average of the mid-ocean velocity over the entire oceanic depth, or equivalently a mid-ocean transport per unit *width* of $O(10^5$ cm$^2$/s$)$.

If $D$ is interpreted as the depth of the main thermocline in the ocean (i.e., the depth of the relatively warm surface layer of the ocean in which the predominant large-scale currents are observed), then $D$ is O(1 km) and $U$ would be O(1 cm/s).

With this choice of $U$, (5.2.16) may be written in its final form:

$$\frac{1}{\beta}\frac{d\zeta}{dt} + v = \text{curl } \tau - \mathbf{u} \cdot \frac{\nabla \eta_B}{\beta} - \frac{r}{\beta}\zeta + \frac{\nabla^2 \zeta}{\beta \text{ Re}}. \qquad (5.2.20)$$

Since the velocity is geostrophic, $u$, $v$, and $\zeta$ may be written in terms of a geostrophic stream function $\psi$:

$$u = -\frac{\partial \psi}{\partial y}, \qquad v = \frac{\partial \psi}{\partial x}, \qquad \zeta = \nabla^2 \psi, \qquad (5.2.21)$$

in terms of which (5.2.20) is, for steady flow,

$$\frac{1}{\beta}\left\{\frac{\partial \psi}{\partial x}\frac{\partial}{\partial y}[\nabla^2 \psi + \eta_B] - \frac{\partial \psi}{\partial y}\frac{\partial}{\partial x}[\nabla^2 \psi + \eta_B]\right\} + \frac{\partial \psi}{\partial x}$$
$$= \text{curl } \tau - \frac{r}{\beta}\nabla^2 \psi + \frac{\nabla^4 \psi}{\beta \text{ Re}}. \qquad (5.2.22)$$

On the perimeter of the basin, $C$, the total *transport* normal to the boundary, i.e., the integrated horizontal velocity, must certainly vanish. The transport by the geostrophic velocity is, from (5.2.19), of the order

$$\rho U D = \left(\frac{\tau_0}{L}\right)\frac{1}{\beta_0}$$

(i.e., of the order of the wind-stress curl divided by $\beta_0$), while the *Ekman* transport in the upper Ekman layer is

$$\mathbf{M}_{*E} = O\left(\frac{\tau_0}{f_0}\right), \qquad (5.2.23)$$

so that

$$\frac{\mathbf{M}_{*E}}{\rho U D} = O\left(\frac{\beta_0 L}{f_0}\right) \ll 1. \qquad (5.2.24)$$

Thus to the lowest order in $L/r_0$ the dominant transport is due to the geostrophic flow. Note that the transport in the lower Ekman layer is only $O(U\delta_E)$. In order that the normal component of the transport may vanish on the basin perimeter, the geostrophic velocity normal to the boundary must vanish. Equivalently, the perimeter, in steady flow, must be a line of constant $\psi$. If the transport is calculated to higher order to include the effect of the Ekman transport, the geostrophic velocity at *higher* order need not vanish at the coast. This interesting question is discussed in Section 5.12. However at *lowest* order the above considerations show that on the boundary of the basin the condition of no normal transport requires $\psi$ to be constant, or

$$\psi = 0 \quad \text{on } C, \tag{5.2.25}$$

while the no-slip condition demands

$$\mathbf{n} \cdot \nabla\psi = 0 \quad \text{on } C, \tag{5.2.26}$$

where $\mathbf{n}$ is an outward normal vector on $C$.

Once $\tau$ has been specified, the problem for the circulation is completely posed. In fact, however, it is no trivial task to accurately determine the wind stress from observations of the wind. Apart from the imperfect knowledge of the surface wind field over the ocean, the relationship between the wind and the stress is itself a difficult question. Usually an empirical formula of the form

$$\tau_* = C_D \rho_a \mathbf{U}_a |\mathbf{U}_a|, \tag{5.2.27}$$

where $\rho_a$, $\mathbf{U}_a$, and $|\mathbf{U}_a|$ are the air density, air velocity, and air speed measured (or extrapolated to) some small distance above the sea surface, usually about 10 meters. The empirical coefficient $C_D$ is the *drag coefficient*. It is of the order of $2 \times 10^{-3}$, but like all empirical coefficients it may itself vary (usually increase) with wind speed. It is clear from (5.2.27) that the stress will be particularly sensitive to intervals of high wind speed. Therefore, even accepting (5.2.27) as correct, the difficulties of obtaining $\mathbf{U}_a$ accurately over short intervals of large $\mathbf{U}_a$ render the task of calculating $\tau_*$ difficult and delicate.

In this chapter the problem will be avoided altogether. Only the grossest features of the stress field will be needed for the purposes of the theoretical development. The general tendency of the stress to follow the pattern of the major wind systems will suffice to expose the basic dynamical issues. This view is consistent with the earlier observation that the overall pattern of the circulation (e.g., the western itensification) does not seem to depend on the *details* of the wind forcing.

## 5.3 The Sverdrup Relation

If the ocean bottom is flat, or $\eta_B \ll \beta$, and

$$\beta^{-1} = \frac{U}{\beta_0 L^2} \ll 1,$$

$$(\beta \, \mathrm{Re})^{-1} = \frac{A_H}{\beta_0 L^3} \ll 1, \tag{5.3.1}$$

$$\frac{r}{\beta} = \frac{\delta_E \, f/2D\beta_0}{L} \ll 1,$$

then to lowest order (5.2.20) reduces to

$$\boxed{v = \operatorname{curl} \tau}. \tag{5.3.2}$$

The conditions (5.3.1) are generally appropriate for the large-scale motion in the oceanic interior, while the condition $\eta_B \ll \beta$ reflects the fact, discussed in (5.2), that the effect of topography should be deemphasized to account for the relative weakness of the bottom velocity in the real ocean.

The condition (5.3.2) is the *Sverdrup relation*. It was shown in Section 3.12 that when the ambient potential vorticity gradient is much larger than the relative vorticity gradient, *free* geostrophic flow must move along the isolines of ambient potential vorticity. In the present case, $\beta \gg 1$ is just that condition, and the isolines of ambient potential vorticity, $f/D$, are the lines of constant $y$. Hence for free flow we have $\tau = 0$, and (5.3.2) shows that (consistent with the vorticity constraint) $v$ must be zero. Further, the Sverdrup relation shows that the fluid can only cross the isolines of $f$ to the extent that vorticity is fed into the fluid column by the curl of the wind stress. As positive (negative) vorticity is imparted by a positive (negative) wind-stress curl, the fluid must flow northward (southward). No information is obtained directly about the flow along the isolines of ambient potential vorticity. This is a reflection again of the degeneracy of geostrophic motion for as long as the flow is along the isolines of ambient potential vorticity any geostrophic flow is admissible. Only the small vorticity sources and sinks, neglected in the derivation of the Sverdrup relation can determine the geostrophic flow and by (5.3.1) these effects are utterly negligible in the oceanic interior.

If terms of $O(\beta_0 L/f_0)$ are *not* neglected in (5.2.9) but the inequalities (5.3.1) hold, (5.2.9) can still be approximated by

$$
\begin{aligned}
Dv_* \beta_0 &= f \mathbf{k} \cdot \operatorname{curl} \frac{\tau_*}{\rho f} \\
&= \mathbf{k} \cdot \operatorname{curl} \frac{\tau_*}{\rho} + \tau_*^{(y)} \frac{\beta_0}{\rho f}.
\end{aligned}
\tag{5.3.3}
$$

Since $-\tau_*^{(y)}/\rho f$ is the northward Ekman transport, (5.3.3) is simply

$$\beta_0 [Dv_* + M_{*E}^{(y)}] \equiv \beta_0 M_{*s}^{(y)} = \mathbf{k} \cdot \operatorname{curl} \frac{\tau_*}{\rho}. \tag{5.3.4}$$

The *total northward transport* $M_s^{(y)}$ is therefore given by the curl of the wind stress, and (5.3.2) is the approximation to this more general transport result in the $\beta$-plane limit in which the geostrophic transport $Dv_*$ exceeds the Ekman transport by $O(f/\beta_0 L)$ according to (5.2.24).

Consider an oceanic basin bounded by two meridional coasts at $x = X_E(y)$ and $x = X_W(y)$ as shown in Figure 5.3.1. On each boundary the normal flow (i.e., the normal transport) must vanish. This implies that the

**Figure 5.3.1**    A portion of an ocean bounded on the east by a coast at $X_E(y)$ and on the west by a coast at $X_W(y)$.

O(1) inviscid interior flow must satisfy

$$u = v \frac{\partial X_E}{\partial y}, \qquad x = X_E(y), \tag{5.3.5a}$$

$$u = v \frac{\partial X_W}{\partial y}, \qquad x = X_W(y), \tag{5.3.5b}$$

where $u$ and $v$ are the inviscid geostrophic velocities adjacent to the boundary but just outside whatever friction layers are required to satisfy the no-slip condition on the coasts. Now $v$, at least as given by the Sverdrup relation, is completely given in terms of curl $\tau$. Although the zonal flow is not determined, its derivative in the $x$-direction will be determined by considerations of mass balance, i.e., for the geostrophic flow

$$\frac{\partial u}{\partial x} = -\frac{\partial v}{\partial y}, \tag{5.3.6a}$$

so that with (5.3.2)

$$\frac{\partial u}{\partial x} = -\frac{\partial}{\partial y} \text{ curl } \tau. \tag{5.3.6b}$$

This can be integrated to yield

$$u(x, y) = -\int_{x_0}^{x} \frac{\partial}{\partial y} \text{ curl } \tau(x', y) \, dx' + U(y), \tag{5.3.7}$$

where the lower limit of integration is arbitrary and $U(y)$ is the arbitrary zonal flow admitted by the Sverdrup relation.

Suppose we try to use the Sverdrup solution to satisfy the inviscid conditions (5.3.5a). On $x = X_E(y)$, (5.3.5) implies that

$$-\int_{X_0}^{X_E(y)} \frac{\partial}{\partial y} \text{curl } \tau(x', y) \, dx' + U(y) = \text{curl } \tau(X_E, y) \frac{\partial X_E}{\partial y},$$

or

$$U(y) = \int_{X_0}^{X_E(y)} \frac{\partial}{\partial y} \text{curl } \tau(x', y) \, dx' + \text{curl } \tau(X_E, y) \frac{\partial X_E}{\partial y}$$

$$= \frac{\partial}{\partial y} \int_{X_0}^{X_E(y)} \text{curl } \tau(x', y) \, dx', \qquad (5.3.8)$$

so that (5.3.7) becomes

$$u(x, y) = -\int_{X_0}^{x} \frac{\partial}{\partial y} \text{curl } \tau(x', y) \, dx' + \frac{\partial}{\partial y} \int_{X_0}^{X_E(y)} \text{curl } \tau(x', y) \, dx'$$

or

$$u(x, y) = \int_{x}^{X_E(y)} \frac{\partial}{\partial y} \text{curl } \tau(x', y) \, dx' + \text{curl } \tau(X_E, y) \frac{dX_E}{dy}. \qquad (5.3.9)$$

On the western boundary (5.3.5) applies:

$$u(X_W, y) = \int_{X_W(y)}^{X_E(y)} \frac{\partial}{\partial y} \text{curl } \tau(x', y) \, dx' + \text{curl } \tau(X_E, y) \frac{\partial X_E}{\partial y}$$

$$= \text{curl } \tau(X_W, y) \frac{dX_W}{dy}. \qquad (5.3.10)$$

In order for (5.3.10) to be satisfied,

$$\frac{\partial}{\partial y} \int_{X_W(y)}^{X_E(y)} \text{curl } \tau(x', y) \, dx' = 0 = \frac{\partial}{\partial y} \int_{X_W}^{X_E} v(x', y) \, dx'. \qquad (5.3.11)$$

The Sverdrup solution (5.3.2) and (5.3.9) will satisfy the boundary conditions on both eastern and western boundaries *only* if the total southward flux of the Sverdrup flow is zero. Since this is given unambiguously in terms of the wind-stress curl, only very special stress curls will satisfy (5.3.11). Of course the stress field is completely free to be specified, and the condition (5.3.11) is usually not satisfied. For example, consider the idealized stress field shown in Figure 5.3.2. The stress is zonal and a function of latitude only. This can be thought of as a simple model of the stress field due to the westerly winds in mid-latitudes and the easterly trade winds in the south. The total Sverdrup transport to the north is, in this case,

$$\int_{X_W}^{X_E} v(x', y) \, dx' = (X_E - X_W) \text{curl } \tau. \qquad (5.3.12)$$

It is clear from Figure 5.3.2 that the transport vanishes only at the latitudes $y_N$ and $y_S$ where the curl vanishes, so that (5.3.11) is obviously violated. It is

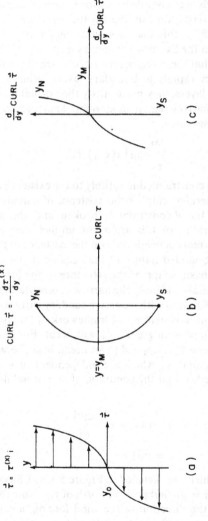

**Figure 5.3.2** (a) An idealized stress distribution with latitude which models the action of the westerlies in middle latitudes and the easterly trade winds in lower latitudes. (b) The wind-stress curl of (a). The curl has its maximum in magnitude at $y_M$. Note that the Sverdrup flow is everywhere southward between $y_N$ and $y_S$. (c) The derivative with respect to latitude, $y$, of curl $\tau$, to which the *zonal* Sverdrup transport is proportional. Note that $y_M$ is not in general coincident with $y_0$.

also clear that the Sverdrup solution does not in general satisfy mass conservation for the basin as a whole, since everywhere the Sverdrup transport is southward. Between $y_N$ and $y_M$ the southward transport increases as $y$ goes from $y_N$ to $y_M$, the latitude of maximum curl, while a convergence of mass flux occurs between $y_M$ and $y_S$. Although the Sverdrup relation (5.3.2) is correct, it is clearly not complete. It alone cannot satisfy the boundary conditions on both coasts, nor can it satisfy the mass balance for the basin *as a whole*. It follows from this that *somewhere* in the basin a region whose dynamics differs from the Sverdrup dynamics must occur. The discussion in Chapter 4 suggests that these regions may occur near boundaries where the dynamical fields vary rapidly in boundary layers. Whatever the nature of these thin boundary layers, they must satisfy the normal flow condition and must transport in a narrow region an amount of mass equal and opposite to the total Sverdrup mass flux,

$$\int_{X_W}^{X_E} \text{curl } \tau(x' \ y) \ dx'.$$

The strong Sverdrup constraint, due entirely to the existence of the planetary vorticity gradient, therefore implies the existence of narrow intense boundary currents where the $\beta$-constraint is broken and the mass flux compensated. The generality of this argument implies that the existence of intense boundary currents depends only on the existence of $\beta$ (i.e., the earth's sphericity) and the bounded nature of the oceanic basins, and not on the details of either the basin shape or the structure of the forcing.

Although it is natural to identify the narrow compensating currents called for by the Sverdrup solution with western boundary currents, it is important to note that there is no way, within the framework of the Sverdrup relation, to tell where the compensating currents will occur. For example, consider the stress field of Figure 5.3.2 applied to an ocean basin between two meridional boundaries $X_W$ and $X_E$ which are independent of $y$. If the Sverdrup solution is determined so that the condition of no normal flow on $x = X_E$ is satisfied, then

$$u = (X_E - x)\frac{\partial \text{ curl } \tau}{\partial y},$$
$$v = \text{curl } \tau,$$

(5.3.13)

the streamlines of which are sketched in Figure 5.3.3(a). Since $X_E - x > 0$, $u$ is negative south of $y = y_M$ and positive north of $y_M$. Note that although the flow is generally in the direction of the wind forcing, $u$ can be oppositely directed to $\tau$, since $y_M$ need not coincide with $y_0$. That is, $u$ depends on $\partial^2 \tau^{(x)}/\partial y^2$ and not simply $\tau^{(x)}$. Further, $v$ is always negative, and it is clear from Figure 5.3.3(a) that if (5.3.13) is correct, a narrow return current is required near the western boundary. However, the decision to use the Sverdrup solution to satisfy the flux condition on $x = X_E$ is completely arbitrary. It is equally valid, *a priori*, to choose $U(y)$ in (5.3.7) so that the normal flow

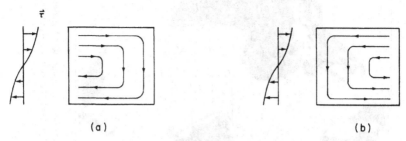

(a)                                                   (b)

**Figure 5.3.3**   (a) The streamlines of the Sverdrup transport for the stress distribution shown at left *if* the interior geostrophic transport is made to satisfy the zero-normal-flow condition at the eastern boundary. Note the implied requirement of a *western* boundary current to complete the flow. (b) The Sverdrup streamlines for the *same* stress distribution as (a), where now the Sverdrup transport has been made to satisfy the no-normal-flow condition at the western boundary. The validity of this choice implicitly requires the existence of an eastern boundary current.

condition is satisfied on $x = X_W$, in which case instead of (5.3.13), for the same $\tau$ as in Figure 5.3.2 we have

$$u = (X_W - x) \frac{\partial}{\partial y} \text{ curl } \tau,$$

$$v = \text{curl } \tau.$$

(5.3.14)

The northward component $v$ is of course unchanged, but since $X_W - x \leq 0$, the direction of $u$ as determined from (5.3.14) is precisely reversed from that given by (5.3.13). The resulting streamlines are shown in Figure 5.3.3(b). Although the flow appears uncomfortably at odds with the natural intuitive feeling that the ocean as a whole should circulate in the same sense as the wind-stress curl (i.e., clockwise in this case), it is vital to realize that (5.3.14) is just as good a solution as (5.3.13) and cannot by the Sverdrup relation alone be considered inferior to the flow in Figure 5.3.3(a). Indeed, as was pointed out above, the fact that the latitudes where $\tau^{(x)}$ and $\partial^2 \tau^{(x)} / \partial y^2$ vanish do not in general coincide means that in *both* solutions (5.3.13) and (5.3.14) the flow is generally flowing against the wind somewhere. Sverdrup (1947) was led to derive the relation which now bears his name precisely to explain the existence of the observed surface countercurrent visible in Figure 5.1.1 in the Pacific at a latitude between 5° and 10° N, where it flows against the trade winds. In order to obtain good agreement between the theory and observations Sverdrup arbitrarily chose the transport to satisfy the boundary condition at the eastern oceanic boundary.

Welander (1959) carried these ideas to their logical end point and calculated the oceanic transport for the entire world oceans using the Sverdrup theory and estimates of the stress field then available. He specified that the Sverdrup flow satisfy the flux condition on oceanic *eastern* boundaries and indicated where western boundary currents would be necessary to satisfy mass conservation. The qualitative similarity between Welander's result

**Figure 5.3.4** The Sverdrup transport and the implied western boundary currents in the oceans corresponding to the annual mean wind-stress field if the Sverdrup transport is required to satisfy the normal-flow condition on oceanic eastern boundaries. Figures represent the volume transports in Sverdrups (millions of m³ s⁻¹). (Reprinted courtesy Welander, 1959.)

shown in Figure 5.3.4 and the circulation shown in Figure 5.1.1 strongly supports the validity of the Sverdrup solution and the western boundary as the presumptive location of boundary currents.

Recent comparisons (Leetmaa, Niiler, and Stommel 1977) of the calculated total southward Sverdrup transport in the Atlantic with observations of that transport strongly support the quantitative validity of the Sverdrup theory for the mid-ocean. The calculations at 16° N, 24° N, and 32° N yield values of 12, 25, and 27 sverdrups (one sverdrup = $10^6$ m³/s), while the respective transports they inferred from direct observation were 14, 24, and 24 sverdrups respectively. This is astonishingly close agreement.

Nevertheless, at this point these results are only presumptive, for there is no *a priori* justification for choosing between the eastern and western boundaries to close the Sverdrup flow. Indeed, *a priori* there is no reason to believe that the Sverdrup solution need satisfy the no-normal-flow condition on *either* boundary. In order to remove the ambiguity inherent in the Sverdrup solution the boundary layers required on $X_E$ and $X_W$ must be examined in greater detail and their implications for the interior flow derived. It is important to keep in mind the essential fact that the total mass flux that must be returned in the boundary-current region is

$$M_B = \int_{X_W}^{X_E} \text{curl } \tau(x', y) \, dx' \tag{5.3.15}$$

and is *independent* of the detailed structure of the boundary current but depends only on its existence.

## 5.4 Meridional Boundary Layers: the Munk Layer

The degeneracy in the Sverdrup solution requires that a detailed analysis of the nature of possible boundary layers adjacent to $X_E$ and $X_W$ be examined. The existence of a boundary layer implies the local breakdown of the dynamical balances of the Sverdrup relation, i.e., that terms in the vorticity equation which are negligible in the open ocean become important near the boundaries. The inequalities (5.3.1) which measure these possible additional contributors to the vorticity balance may be interpreted in terms of ratios of length scales.

Define the lengths

$$\delta_I = \left(\frac{U}{\beta_0}\right)^{1/2},$$

$$\delta_M = \left(\frac{A_H}{\beta_0}\right)^{1/3},$$

and

$$\delta_S = \frac{\delta_E f}{2D\beta_0} \tag{5.4.1}$$

where $U$ is given by (5.2.19). Then the conditions (5.3.1) are equivalent to

$$L \gg \delta_I, \delta_M, \delta_S, \tag{5.4.2}$$

where $L$ is the scale of the motion. It is natural to anticipate, then, that for those regions of motion sufficiently narrow that the scale of the motion is of the same order as one of these three scales, the relevant term in the vorticity equation will be of the same order as the planetary-vorticity-gradient term. Depending on the relative size of $\delta_I$, $\delta_M$, and $\delta_S$, either inertia, horizontal friction, or bottom friction will enter the vorticity balance. Note that each of these scales goes inversely with $\beta_0$, so that just as the presence of $\beta_0$ produced the ambiguity associated with the Sverdrup constraint, it simultaneously introduces the possibility of the existence of narrow regions in which the constraint may be broken and the ambiguity removed.

To give concrete expression to these ideas, consider the case where each meridional boundary $X_E$ and $X_W$ is independent of $y$ (the more general case can be similarly described). As in the preceding section, the ocean bottom will be considered flat ($\eta_B = 0$). In the interior of the ocean, away from the meridional boundaries, the *interior stream function* $\psi_I$ satisfies (5.3.2):

$$\frac{\partial \psi_I}{\partial x} = \text{curl } \tau, \tag{5.4.3}$$

so that

$$\psi_I(x, y) = \int_{x_0}^{x} \text{curl } \tau \, dx' + \Psi_0(y), \tag{5.4.4}$$

where $\Psi_0(y)$ is arbitrary and to be determined.

Consider the possibility of a boundary layer on the *eastern* meridional boundary. Introduce the stretched boundary-layer coordinate

$$\lambda = \frac{X_E - x}{l}, \tag{5.4.5}$$

where $l$ is an unknown, nondimensional boundary-layer scale, i.e., the boundary-layer width is

$$l_* = Ll. \tag{5.4.6}$$

The definition (5.4.5) is chosen so that the range of $\lambda$ is $(0, \infty)$. In the boundary layer, the stream function for the flow is

$$\psi = \psi_B(\lambda, y), \tag{5.4.7}$$

and it is important to note, since the boundary current can carry no more mass than the interior, that the *magnitude* of the stream function in the boundary layer is of the same order as the magnitude of the interior stream function. If (5.4.6) and (5.4.7) are substituted into (5.2.22), we obtain

$$-\left(\frac{\delta_I}{l_*}\right)^2 \left[ \frac{\partial \psi_B}{\partial \lambda} \frac{\partial}{\partial y}\left(\frac{\partial^2 \psi_B}{\partial \lambda^2} + \frac{l_*^2}{L^2}\frac{\partial^2 \psi_B}{\partial y^2}\right) - \frac{\partial \psi_B}{\partial y}\frac{\partial}{\partial \lambda}\left(\frac{\partial^2 \psi_B}{\partial \lambda^2} + \frac{l_*^2}{L^2}\frac{\partial^2 \psi_B}{\partial y^2}\right) \right] - \frac{\partial \psi_B}{\partial \lambda}$$

$$= \frac{l_*}{L}\text{curl } \tau - \left(\frac{\delta_S}{l_*}\right)\left(\frac{\partial^2 \psi_B}{\partial \lambda^2} + \frac{l_*^2}{L^2}\frac{\partial^2 \psi_B}{\partial y^2}\right) + \left(\frac{\delta_M}{l_*}\right)^3 \left[\frac{\partial^2}{\partial \lambda^2} + \frac{l_*^2}{L^2}\frac{\partial^2}{\partial y^2}\right]^2 \psi_B. \tag{5.4.8}$$

The size of each term in the vorticity equation is therefore measured in terms of the boundary-layer width $l_*$ and $\delta_I$, $\delta_M$, and $\delta_S$, as expected. For $\psi = O(1)$ (i.e., for a boundary layer that can satisfy the condition of no normal flow and close the mass of the Sverdrup flow), $l_*$ must be chosen to be the *largest* of $\delta_I$, $\delta_M$, and $\delta_S$. If that choice is made, then that particular term will balance the $\beta$-term while the others will be small and negligible. Now each of $\delta_I$, $\delta_M$, and $\delta_S$ depends on independent parameters, and *a priori* any ordering relation between them is, in principle, possible. Since $\delta_S$ and $\delta_M$ depend on the turbulent mixing coefficients, it is well nigh impossible to determine their sizes with any precision. Therefore each of the various major possibilities will be considered separately and in turn to examine to what degree the major results are sensitive to this ordering. First note however that *regardless* of whether $l_*$ is equal to $\delta_M$, $\delta_I$, or $\delta_S$, the direct contribution of the wind-stress curl to the dynamics of the boundary layer is negligible, i.e., $O(l_*/L)$. The stress curl produces $\psi_I$, and it is the discrepancy of $\psi_I$ and the boundary values it must assume, given by (5.2.25) and (5.2.26), that produces the boundary layer.

The first possibility that will be examined is that

$$\delta_M = \left(\frac{A_H}{\beta_0}\right)^{1/3} \gg \delta_I, \delta_S. \tag{5.4.9}$$

Then $l_*$ must be chosen as

$$l_* = \delta_M. \tag{5.4.10}$$

The size of the inertial and bottom friction terms in (5.4.8) are then $(\delta_I/\delta_M)^2$ and $\delta_S/\delta_M$, respectively, and to lowest order can be neglected.

It is convenient to represent $\psi_B$ as the sum of the interior stream function $\psi_I$ plus a boundary-layer correction function $\phi_B(\lambda, y)$ required to allow $\psi_B$ to meet the boundary conditions, i.e.,

$$\psi_B(\lambda, y) = \psi_I(x, y) + \phi_B(\lambda, y). \tag{5.4.11}$$

In order for $\psi_B$ to merge smoothly into $\psi_I$ for large $\lambda$, $\phi_B$ must go to zero as $\lambda \to \infty$. If (5.4.11) is inserted into (5.4.8), and the fact that

$$\frac{\partial \psi_I}{\partial \lambda} = -\frac{\delta_M}{L}\frac{\partial \psi_I}{\partial x} = -\frac{\delta_M}{L}\text{ curl } \tau \tag{5.4.12}$$

is used, the resulting lowest-order terms in (5.4.8) yield

$$\frac{\partial^4 \phi_B}{\partial \lambda^4} + \frac{\partial \phi_B}{\partial \lambda} = 0. \tag{5.4.13}$$

Note that only the derivatives in the longitudinal direction are retained in the friction term as a consequence of the boundary-layer approximation. The general solution of (5.4.13) is

$$\phi_B = C_1 + C_2 e^{-\lambda} + C_3 e^{+\lambda/2} \cos \frac{\sqrt{3}}{2}\lambda + C_4 e^{+\lambda/2} \sin \frac{\sqrt{3}}{2}\lambda, \tag{5.4.14}$$

where $C_1, C_2, C_3$, and $C_4$ are arbitrary functions of $y$. For $\phi_B$ to vanish for large $\lambda$, we must set $C_1, C_3$, and $C_4$ equal to zero. Thus in the boundary layer on the eastern side of the basin,

$$\psi_B(\lambda, y) = \psi_I(x, y) + C_2 e^{-\lambda}. \qquad (5.4.15)$$

The boundary conditions on $x = X_E$, i.e., $\lambda = 0$ are the condition of no normal flow,

$$u = -\frac{\partial \psi_B}{\partial y} = 0 = \frac{\partial \psi_I}{\partial y}(X_E, y) + \frac{\partial C_2(y)}{\partial y}, \qquad (5.4.16)$$

and the no-slip condition

$$v = \frac{\partial \psi_B}{\partial x} = 0 = \frac{\partial \psi_I}{\partial x} - \frac{L}{\delta_M} \frac{\partial \phi_B}{\partial \lambda} \quad \text{on } \lambda = 0$$

$$= \frac{\partial \psi_I}{\partial x}(X_E, y) + \frac{L}{\delta_M} C_2(y). \qquad (5.4.17)$$

From (5.4.17) it follows that

$$C_2 = -\frac{\delta_M}{L} \frac{\partial \psi_I}{\partial x}(X_E, y) = -\frac{\delta_M}{L} \text{curl } \tau(X_E, y), \qquad (5.4.18)$$

so that to $O(\delta_M/L)$, (5.4.16) demands that

$$\boxed{\frac{\partial \psi_I}{\partial y}(X_E, y) = 0}. \qquad (5.4.19)$$

The Sverdrup solution must therefore itself satisfy the condition of no normal flow on the oceanic *eastern* boundary. This completely determines the interior Sverdrup flow so that (5.4.4) may be written

$$\boxed{\psi_I(x, y) = \int_{X_E}^{x} \text{curl } \tau \, dx'}, \qquad (5.4.20)$$

which corresponds to the choice of Figure 5.3.3(a) rather than 5.3.3(b) in the example described above and is identical to the condition applied by Welander in his calculation. Note that it is the inability of the eastern boundary to support a boundary layer capable of absorbing an $O(1)$ on-shore flow that specifies the interior solution. It remains to be seen whether the western boundary can support such a layer. It must if an inconsistency is to be avoided, for now the interior field is fixed in the neighborhood of $X_W$, i.e., near the western boundary the interior flow in the $x$-direction is given as

$$-\frac{\partial \psi_I}{\partial y} = -\int_{X_E}^{X_W} \frac{\partial}{\partial y} \text{curl } \tau \, dx'. \qquad (5.4.21)$$

The sole effect of the eastern boundary layer is to satisfy the no-slip condition. As illustrated in Figure 5.4.1 this affects the transport by only a tiny

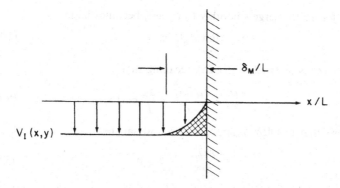

**Figure 5.4.1** The boundary layer at the eastern boundary acts only to satisfy the no-slip condition. The effect is to diminish the southward transport by the amount indicated by the crosshatched area and is $O(\delta_M/L)$, and hence negligible in proportion to the $O(1)$ Sverdrup transport.

amount, $O(\delta_M/L)$, a fact which is also apparent by combining (5.4.15) and (5.4.18) to yield

$$\psi_B = \psi_I(x, y) - \frac{\delta_M}{L} \frac{\partial \psi_I}{\partial x}(X_E, y)e^{-\lambda}, \qquad (5.4.22)$$

so that the correction in this layer to the interior mass flow is $O(\delta_M/L)$.

The situation is considerably different in the boundary layer on the western side of the ocean. A stretched boundary layer variable $\xi$ is defined as

$$\xi = \frac{(x - X_W)}{l} = \frac{(x - X_W)L}{l_*} \qquad (5.4.23)$$

so that the range of $\xi$ is $(0, \infty)$. The boundary-layer stream function is now

$$\psi_B = \psi_B(\xi, y). \qquad (5.4.24)$$

If (5.4.23) and (5.4.24) are substituted into (5.2.22) we obtain, instead of (5.4.8),

$$\left(\frac{\delta_I}{l_*}\right)^2 \left[ \frac{\partial \psi_B}{\partial \xi} \frac{\partial}{\partial y} \left( \frac{\partial^2 \psi_B}{\partial \xi^2} + \frac{l_*^2}{L^2} \frac{\partial^2 \psi_B}{\partial y^2} \right) - \frac{\partial \psi_B}{\partial y} \frac{\partial}{\partial \xi} \left( \frac{\partial^2 \psi_B}{\partial \xi^2} + \frac{l_*^2}{L^2} \frac{\partial^2 \psi_B}{\partial y^2} \right) \right] + \frac{\partial \psi_B}{\partial \xi}$$

$$= \frac{l_*}{L} \operatorname{curl} \tau - \frac{\delta_S}{l_*} \left( \frac{\partial^2 \psi_B}{\partial \xi^2} + \frac{l_*^2}{L^2} \frac{\partial^2 \psi_B}{\partial y^2} \right) + \left(\frac{\delta_M}{l_*}\right)^3 \left[ \frac{\partial^2}{\partial \xi^2} + \frac{l_*^2}{L^2} \frac{\partial^2}{\partial y^2} \right]^2 \psi_B. \qquad (5.4.25)$$

In the present case $l_* = \delta_M$, so that in the western boundary layer the inertial advection of relative vorticity and the effect of bottom friction are negligible, since (5.4.9) applies. It is again convenient to represent $\psi_B(\xi, y)$ as

$$\psi_B = \psi_I(x, y) + \tilde{\phi}_B(\xi, y). \qquad (5.4.26)$$

Again, for $\psi_B$ to merge smoothly to $\psi_I$ as $\xi$ becomes large,

$$\lim_{\xi \to \infty} \tilde{\phi}_B(\zeta, y) = 0. \tag{5.4.27}$$

If (5.4.26) is inserted in (5.4.25), recalling that

$$\frac{\partial \psi_I}{\partial \xi} = \frac{\delta_M}{L} \frac{\partial \psi_I}{\partial x} = \frac{\delta_M}{L} \text{ curl } \tau, \tag{5.4.28}$$

it follows that to $O[(\delta_I/\delta_M)^2, \delta_S/\delta_M] \, \tilde{\phi}$ now satisfies

$$\frac{\partial^4 \tilde{\phi}_B}{\partial \xi^4} - \frac{\partial \tilde{\phi}_B}{\partial \xi} = 0. \tag{5.4.29}$$

It is apparent that (5.4.29) differs from its eastern-boundary equivalent, (5.4.13), by a crucial change in sign before the first derivative, i.e., the term which represents the effect of the planetary vorticity gradient. The transformations (5.4.13) and (5.4.29) introduce the sign change when applied to an *odd*-ordered derivative only, and in particular to the planetary-vorticity gradient term. The presence of the $\beta$-effect destroys the invariance the complete vorticity equation otherwise would have under the transformation $x \to -x$, $y \to -y$. This loss of invariance simply reflects the fact that physical space is no longer dynamically isotropic in the presence of $\beta$. The northward increase of the ambient planetary vorticity picks out, *a priori*, the northward direction as dynamically significant. It has been already seen in the discussion of Rossby waves, their propagation and reflection properties, that the presence of the planetary vorticity gradient leads to an inherent anisotropy in the wave dynamics. It is the identical, dynamically profound anisotropy in the physics of the vorticity in the presence of $\beta$, rather than an artifice of the mathematics, that leads to the altered form of the boundary-layer equation at the western boundary.

The solution of (5.4.29) is

$$\tilde{\phi}_B = \tilde{C}_1(y) + \tilde{C}_2(y)e^\xi + \tilde{C}_3 e^{-\xi/2} \cos \frac{\sqrt{3}}{2} \xi + \tilde{C}_4 e^{-\xi/2} \sin \frac{\sqrt{3}}{2} \xi. \tag{5.4.30}$$

$\tilde{C}_1(y)$ and $\tilde{C}_2(y)$ must be zero so that (5.4.27) can be satisfied, which implies that in the western boundary layer

$$\psi_B(\xi, y) = \psi_I(x, y) + \tilde{C}_3 e^{-\xi/2} \cos \frac{\sqrt{3}}{2} \xi + \tilde{C}_4 e^{-\xi/2} \sin \frac{\sqrt{3}}{2} \xi, \tag{5.4.31}$$

which should be compared with (5.4.15). The conditions of no normal flow and no slip on $x = X_W$ are, respectively

$$0 = \frac{\partial \psi_I}{\partial y}(X_W, y) + \frac{\partial}{\partial y} \tilde{C}_3, \tag{5.4.32}$$

$$0 = \frac{\partial \psi_I}{\partial x} + \frac{L}{\delta_M} \left[ -\frac{\tilde{C}_3}{2} + \frac{\sqrt{3}}{2} \tilde{C}_4 \right]. \tag{5.4.33}$$

From (5.4.33) it follows that to $O(\delta_M/L)$,

$$\tilde{C}_4 = \frac{\tilde{C}_3}{\sqrt{3}}, \qquad (5.4.34)$$

while (5.4.32) implies that

$$\tilde{C}_3 = -\psi_I(X_W, y) + K, \qquad (5.4.35)$$

where $K$ is a constant. Hence

$$\psi_B = \psi_I(x, y)\left\{1 - e^{-\xi/2}\left(\cos\frac{\sqrt{3}}{2}\xi + \frac{1}{\sqrt{3}}\sin\frac{\sqrt{3}}{2}\xi\right)\right\}$$

$$+ Ke^{-\xi/2}\left[\cos\frac{\sqrt{3}}{2}\xi + \frac{1}{\sqrt{3}}\sin\frac{\sqrt{3}}{2}\xi\right]. \qquad (5.4.36)$$

The value of the stream function on $x = X_E$ is zero. In order for the *total* mass flux in the meridional direction to balance,

$$0 = \int_{X_W}^{X_E} v\, dx = \int_{X_W}^{X_E}\frac{\partial\psi}{\partial x}\, dx = \psi(X_E) - \psi(X_W), \qquad (5.4.37)$$

so that $\psi(X_W)$ must also vanish. The condition of no *net* mass flux is the condition that the boundary current returns northward the same amount of mass flux as the interior. In order that $\psi(X_W)$ may vanish, $K$ must be chosen to be zero, so that in the western boundary region

$$\psi_B = \psi_I(x, y)\left[1 - e^{-\xi/2}\left(\cos\frac{\sqrt{3}}{2}\xi + \frac{1}{\sqrt{3}}\sin\frac{\sqrt{3}}{2}\xi\right)\right]. \qquad (5.4.38)$$

The northward velocity in the western boundary current is, to $O(\delta_M/L)$,

$$v_B = \psi_I(X_W, y)\left(\frac{L}{\delta_M}\right)\frac{2e^{-\xi/2}}{\sqrt{3}}\sin\frac{\sqrt{3}}{2}\xi. \qquad (5.4.39)$$

Figure 5.4.2 shows $\psi_B$ and $v_B$ as a function of $\xi$, and Figure 5.4.3 shows a sketch of $v$ across the entire basin. A uniformly valid approximation for the entire flow consists of a composite of the interior solution plus each of the boundary-layer correction solutions on the eastern and western boundaries, i.e.,

$$\psi = \psi_I(x, y)\left[1 - e^{-\xi/2}\left(\cos\frac{\sqrt{3}}{2}\xi + \frac{1}{\sqrt{3}}\sin\frac{\sqrt{3}}{2}\xi\right)\right] - \frac{\delta_M}{L}\frac{\partial\psi_I}{\partial x}(X_E, y)e^{-\lambda},$$

$$(5.4.40)$$

where $\lambda$ and $\xi$ are given by (5.4.5) and (5.4.23) respectively, while $\psi_I$ is given by (5.4.20).

There are several important facts to note about the western boundary current. In contrast to the eastern boundary layer, the western boundary

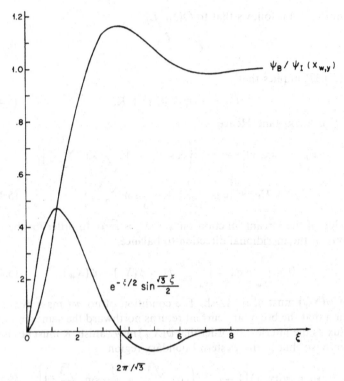

**Figure 5.4.2**   The structure of the northward velocity in the western boundary layer. Note the recirculation region for $\xi > 2\pi/\sqrt{3}$. The curve labeled $\psi_B/\psi_I$ shows the distribution of transport measured from the western boundary. There is a rapid increase of $\psi_B$ with $\xi$ until the recirculation region, and then $\psi_B$ attains its asymptotic value $\psi_I(x_W, y)$ for $\xi \approx 6$.

current makes an O(1) contribution to the mass flux as previously noted. Since in fact the mass flux in the boundary current region precisely balances the entire interior mass flux, the *northward* velocity in the boundary current must be very much larger than the interior velocity in order that the same mass flux can be realized in the narrow return current. In fact, as (5.4.39) shows, this velocity must be $O(L/\delta_M)$ times the interior flow. Thus in this model (and, as we shall see, in general) the presence of the constraint of the planetary vorticity gradient produces both a weak Sverdrup interior *and* an intense western boundary current to return the Sverdrup mass flux, while the dynamical balance of the Sverdrup interior holds, to lowest order, right up to the eastern boundary.

This particular model of the western intensification was first discussed by Munk (1950). Munk's boundary layer exists as a balance between the accession of relative vorticity produced in a column as it is shoved northward in

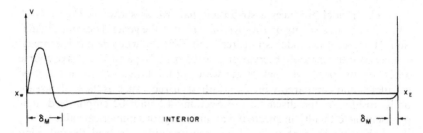

**Figure 5.4.3**   A schematic of the distribution of $v$ across the entire basin, indicating the intense return flow in the western boundary current, the weak interior Sverdrup flow, and the eastern boundary-layer region.

the western boundary current and the lateral diffusion of this additional relative vorticity into the western wall. A fluid element entering the western boundary layer has a negligible relative vorticity and possesses mostly planetary vorticity. If it moves northward a distance $\Delta y_*$ in the boundary layer, its planetary vorticity is increased an amount $\beta_0 \, \Delta y_*$ and in the absence of friction would suffer a change in relative vorticity $\zeta_*$ by an amount $O(-\beta_0 \, \Delta y_*)$. Frictional forces must however drain this vorticity from the element and diffuse it out the boundary, so that the fluid element can smoothly rejoin the Sverdrup interior, where fluid elements must have negligible relative vorticity. Consequently, the fluid element must stay in contact with the wall for at least a characteristic friction diffusion time for vorticity, i.e.,

$$t_{*D} = \frac{\delta_M^2}{A_H} \tag{5.4.41}$$

where $\delta_M$ is the layer width. On the other hand, this residence time must also equal

$$t_{*D} = \frac{\Delta y_*}{v_{B*}} \tag{5.4.42}$$

where $v_{B*}$ is the northward velocity of the current. Since $\zeta_*$ is $O(v_{B*}/\delta_M)$, it follows that in order for friction to drain the excess relative vorticity produced by $\beta_0$, we must have

$$\frac{\delta_M^2}{A_H} = O\left(\frac{\Delta y}{v_{B*}}\right) = O\left(\frac{\Delta y}{\delta_M \zeta_*}\right) = O\left(\frac{1}{\beta_0 \, \delta_M}\right), \tag{5.4.43}$$

or

$$\delta_M = \left(\frac{A_H}{\beta_0}\right)^{1/3}, \tag{5.4.44}$$

in agreement with (5.4.1).

Munk's model produces a streamline pattern, as shown in Figure 5.4.4, which possesses a strong qualitative similarity to the general oceanic circulation. However, the model is not free from difficulty when *detailed* comparison between theory and observation is attempted. In particular it is of course necessary to specify $A_H$ before the width of the boundary current can be calculated and compared with the width of currents such as the Gulf Stream or Kuroshio. As the discussion of Sections 4.1 and 4.2 emphasized, it is precisely the difficulty in specifying an unambiguous numerical value for $A_H$ that is the weak element in turbulent-viscosity models. Instead, the following tack is taken. For the essential validity of Munk's model, the friction terms must dominate the nonlinear term, i.e., $\delta_M \gg \delta_I$, or equivalently

$$A_H \gg \frac{U^{3/2}}{\beta_0^{1/2}} = U \delta_I, \tag{5.4.45}$$

or, in turn, the Reynolds number $\mathrm{Re}_\delta$ based on $U$ and $\delta_I$ must satisfy

$$\mathrm{Re}_\delta \equiv U \frac{\delta_I}{A_H} \ll 1. \tag{5.4.46}$$

If $U$ is $O(1 \text{ cm/s})$ and $\beta_0$ is $10^{-13} \text{ cm}^{-1} \text{ s}^{-1}$, $A_H$ must be far greater than $3 \times 10^6 \text{ cm}^2/\text{s}$ in order that $\delta_M \gg \delta_I$. This is probably too large a value for an $A_H$ attributable to momentum mixing by small-scale turbulence, although it *is* very difficult to truly estimate $A_H$ by direct observation. However, the width of the observed northward boundary currents is a readily accessible measurement. As Figure 5.4.2 shows, the natural definition of the current width is given by the point where $\sin[(\sqrt{3}/2)\xi]$ has its first nontrivial zero, i.e., at $\zeta = 2\pi/\sqrt{3}$. This yields a dimensional current width

$$\delta_{*0} = \frac{2\pi \, \delta_M}{\sqrt{3}}$$

$$= \mathrm{Re}_\delta^{-1/3} \left(\frac{U}{3\beta_0}\right)^{1/2} 2\pi. \tag{5.4.47}$$

With the same values as before, this yields a current width of $\mathrm{Re}_\delta^{-1/3} \times 120$ km. Since $\mathrm{Re}_\delta$ must be much less than one for Munk's model of the current to be accurate, we must conclude that this friction model cannot realistically model the observed currents whose widths are *narrower* than 100 km.

However, although the frictional model fails to describe the details of the boundary current accurately, the model possesses full internal consistency. It very clearly demonstrates the relationship between the oceanic western intensification and the planetary vorticity gradient. In the following sections other models will be examined, and in each case the cause of the western intensification is explicable, as in Munk's model, by the effect of $\beta$.

Finally, it is interesting to note that the effect of lateral friction in Munk's model produces a *counterflow* on the seaward side of the boundary current. The presence of the counterflow *increases* the magnitude of the transport of

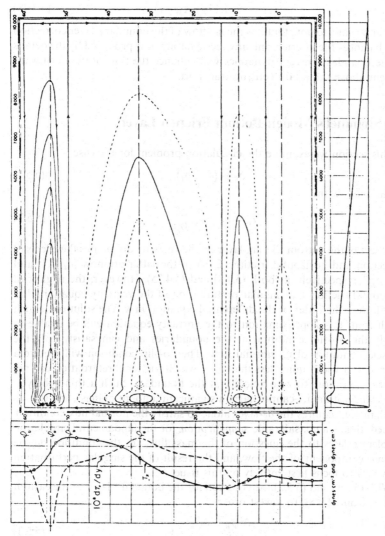

**Figure 5.4.4**  The streamlines of the circulation pattern, as given by Munk (1950), for a rectangular ocean driven by the stress distribution for the Pacific Ocean, shown in the panel at left.

the northward jet (i.e., the flow between the boundary and $\delta_{*0}$) *above* the value required to balance the interior flow—i.e., by (5.4.39),

$$\int_0^{.2\pi/\sqrt{3}} v_B \, dx = \psi_I(X_W, y)\{1 + e^{-\pi/\sqrt{3}}\}$$

$$\approx 1.17\psi_I(X_W, y). \tag{5.4.48}$$

It is important to note that it is the *net* flow of the boundary-layer correction that balances the interior, and any recirculating flow produced by the details of the boundary-current dynamics will enhance the flow of the northward current above the total Sverdrup transport.

## 5.5 Stommel's Model: Bottom Friction Layer

In this section we examine the circulation problem for the case

$$\delta_S \gg (\delta_M, \delta_I). \tag{5.5.1}$$

In this case, with

$$l_* = \delta_S = \frac{\delta_E}{2} \frac{f}{D\beta_0}, \tag{5.5.2}$$

the effect of the bottom Ekman layer on the vorticity dominates the effects of advection and horizontal friction. Each of the latter are now $(\delta_I/\delta_S)^2$ and $(\delta_M/\delta_S)^3$ respectively in both (5.4.8) and (5.4.25). However, the neglect of horizontal friction is a singular perturbation of the vorticity equation, that is, it lowers the order of the differential equation in $x$. The solutions which result from this approximation to the vorticity equation will be unable to satisfy the no-slip condition at the boundaries and can satisfy only the inviscid condition of no normal flow. The no-slip condition will require a layer with a width much less than $\delta_S$ in which frictional retardation can act to bring the flow to zero velocity at the boundary. Such a thin sublayer, embedded within $\delta_S$, cannot carry a significant amount of mass, and this, as we shall see, requires the $\delta_S$-layer alone to satisfy the normal flux condition. Indeed, this is a considerable *simplification*, for the fundamental circulation problem relates to the question of the mass flux, i.e., the condition on the normal component of the flow, and the result of the singular perturbation allows us to focus first entirely on this issue.

With (5.5.1) and (5.5.2) the equation for the boundary layer at the eastern wall is, from (5.4.8), to lowest order,

$$\frac{\partial^2 \psi_B}{\partial \lambda^2} - \frac{\partial \psi_B}{\partial \lambda} = 0. \tag{5.5.3}$$

If as before $\psi_B$ is written in terms of the interior $\psi$ plus a boundary-layer correction function, i.e.,

$$\psi_B = \psi_I(x, y) + \phi_B(\lambda, y), \tag{5.5.4}$$

then $\phi_B$ satisfies

$$\frac{\partial^2 \phi_B}{\partial \lambda^2} - \frac{\partial \phi_B}{\partial \lambda} = 0. \tag{5.5.5}$$

The general solution of (5.5.5) is

$$\phi_B = C_1(y) + C_2(y)e^\lambda. \tag{5.5.6}$$

Since $\phi_B$ must vanish as $\lambda \to \infty$, both $C_1$ and $C_2$ must be zero. Thus in order for $\psi_B$ to satisfy the condition of no normal flow on $x = X_E$, we must have

$$\frac{\partial \psi_I}{\partial y}(X_E, y) = 0, \tag{5.5.7}$$

so that again, *the interior Sverdrup flow must satisfy the normal-flow condition on the eastern boundary*. Hence, (5.4.20) applies regardless of whether $\delta_M > \delta_S$ or vice versa. In either case the intensification must occur on the western boundary.

To satisfy the no-slip condition on $x = X_E$, a region sufficiently narrow to reintroduce the effect of lateral friction must exist. In the Munk model horizontal friction balances the $\beta$-term directly, and all conditions are satisfied by one layer. In the present case bottom friction is as large as the $\beta$-term for $l_* = \delta_S$. If $l_*$ is even *less* than $\delta_S$, then (5.4.8) shows that the bottom friction will then dominate the $\beta$-term. If $l_*$ is sufficiently small, then both the bottom and lateral friction will dominate and balance when

$$\left(\frac{\delta_M}{l_*}\right)^3 = \frac{\delta_S}{l_*}, \tag{5.5.8}$$

or

$$l_* = \delta_S \left(\frac{\delta_M}{\delta_S}\right)^{3/2} \ll \delta_S. \tag{5.5.9}$$

If the definitions of $\delta_M$ and $\delta_S$ are used,

$$\frac{l_*}{L} = \frac{E_H^{1/2}}{E_V^{1/4}}, \tag{5.5.10}$$

so that this sublayer is identical to the boundary-layer region discussed in Section 4.13. Note that $l_*$ is independent of $\beta_0$, since the physics of this layer does *not* depend on the existence of the planetary-vorticity gradient. If $\delta_I/l_*$ is small, i.e., if

$$\delta_I < \delta_S \left(\frac{\delta_M}{\delta_S}\right)^{3/2}, \tag{5.5.11}$$

then the equation for this sublayer is, from (5.4.8),

$$\frac{\partial^4 \psi_B}{\partial \lambda^4} - \frac{\partial^2 \psi_B}{\partial \lambda^2} = 0, \tag{5.5.12}$$

where now

$$\lambda = (X_E - x) \frac{L}{\delta_S} \left( \frac{\delta_S}{\delta_M} \right)^{3/2}. \tag{5.5.13}$$

Note that (5.5.12) is identical in form to (4.13.26). The solution of (5.5.12) which merges with $\psi_I$ for large $\lambda$ and which satisfies the no-slip condition on $x = X_E$ is

$$\psi_B = \psi_I(x, y) - \frac{\partial \psi_I}{\partial x} (X_E, y) \frac{l_*}{L} e^{-(X_E - x)L/l_*}, \tag{5.5.14}$$

where $l_*$ is given by (5.5.9). The correction to the Sverdrup mass flux produced by this layer is utterly negligible. The layer has in fact little dynamical significance with regard to the circulation problem. If (5.5.11) is not satisfied, but $\delta_S > \delta_I$, the vorticity balance in this no-slip sublayer is nonlinear and the analysis of the sublayer is exceedingly difficult. It is important and indeed fortunate that this issue does not affect the question of the overall circulation.

On the *western boundary*, the choice $l_* = \delta_S$ yields, at lowest order,

$$\frac{\partial^2 \psi_B}{\partial \xi^2} + \frac{\partial \psi_B}{\partial \xi} = 0. \tag{5.5.15}$$

Again writing

$$\psi_B = \psi_I(x, y) + \phi_B(\xi, y), \tag{5.5.16}$$

it quickly follows that

$$\phi_B(\xi, y) = C(y) e^{-\xi}, \tag{5.5.17}$$

so that the solution which satisfies the normal-flow condition is

$$\boxed{\psi_B = \psi_I(x, y)[1 - e^{-\xi}]} \tag{5.5.18}$$

and

$$\boxed{v_B = \frac{L}{\delta_S} \psi_I(x, y) e^{-\xi}}. \tag{5.5.19}$$

Again, the mass flux in the Sverdrup flow is returned in a *western* boundary current. In this case the width of the intense western current is $O(\delta_S)$ and there is no seaward countercurrent. As in the case of the Munk layer, $\delta_S$ depends inversely on $\beta_0$, i.e., the existence of the layer depends crucially on the same planetary-vorticity gradient that produces the Sverdrup constraint on the interior. The northward velocity in the return current is order $L/\delta_S$ larger than the interior flow and, as (5.5.18) shows, is unable to satisfy the no-slip condition on $x = X_W$. If (5.5.11) holds, a boundary layer of width $\delta_S(\delta_M/\delta_S)^{3/2}$ exists within the outer layer of thickness $\delta_S$, as shown in Figure

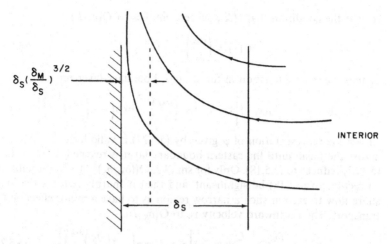

**Figure 5.5.1**   A schematic view of the western boundary current when friction dominates. The transport occurs mostly in the wide $\delta_s$-layer, while a sublayer of width $\delta_s(\delta_M/\delta_s)^{3/2}$ is required to satisfy the no-slip condition. Most streamlines do *not* enter the sublayer.

5.5.1. The outer layer turns the streamlines so that the normal-flow condition is satisfied, while the sole function of the inner layer is to frictionally retard the flow nearest to the boundary to satisfy the no-slip condition. The sublayer width implies that in the sublayer the stream function is a function of a new variable

$$\eta = (x - X_W)\frac{L}{\delta_s}\left(\frac{\delta_s}{\delta_M}\right)^{3/2} = \xi\left(\frac{\delta_s}{\delta_M}\right)^{3/2}, \tag{5.5.20}$$

while the stream function may be written as (5.5.18) plus a correction—i.e., in the sublayer,

$$\psi = \tilde{\psi}_B = \psi_B(\xi, y) + \tilde{\phi}_B(\eta, y). \tag{5.5.21}$$

If (5.5.2), (5.5.20), and (5.5.21) are substituted into (5.4.25), it follows that to lowest order

$$\frac{\partial^4 \tilde{\phi}_B}{\partial \eta^4} - \frac{\partial^2 \tilde{\phi}_B}{\partial \eta^2} = 0 \tag{5.5.22}$$

as long as (5.5.11) holds. The solution for $\tilde{\phi}_B$ which vanishes for large $\eta$ is

$$\tilde{\phi}_B = C(y)e^{-\eta}, \tag{5.5.23}$$

so that

$$\tilde{\psi}_B = \psi_B(\xi, y) + Ce^{-\eta}. \tag{5.5.24}$$

The northward velocity $v$ is given by

$$v = \frac{\partial \psi}{\partial x} = \frac{L}{\delta_s}\frac{\partial \psi_B}{\partial \xi} + \frac{L}{\delta_s}\left(\frac{\delta_s}{\delta_M}\right)^{3/2}\frac{\partial \tilde{\phi}_B}{\partial \eta}, \tag{5.5.25}$$

so that the condition that $v(X_W)$ is zero implies to $O(\delta_S/L)$

$$C(y) = \left(\frac{\delta_M}{\delta_S}\right)^{3/2} \psi_I(X_W, y) \qquad (5.5.26)$$

or that the stream function in the western boundary-layer region is

$$\psi = \psi_B = \psi_I(x, y)\left[1 - e^{-\xi} + \left(\frac{\delta_M}{\delta_S}\right)^{3/2} \exp\left(-\xi\left(\frac{\delta_S}{\delta_M}\right)^{3/2}\right)\right] \quad (5.5.27)$$

In fact the representation of $\psi$ given by (5.5.27) is valid from $x = X_W$ clear across the basin until the eastern boundary layer is reached. For $\xi = O(1)$ (5.5.27) reduces to (5.5.18). Only for small $\xi$, $O((\delta_M/\delta_S)^{3/2})$, will the sublayer correction to (5.5.18) be significant, and then it sharply reduces the long-shore flow to zero in such a narrow region as to have a minor effect on the transport. The northward velocity is, to $O(\delta_S/L)$,

$$v_B = \frac{\partial \psi_B}{\partial \xi} = \frac{L}{\delta_S} \psi_I(x, y)\left[e^{-\xi} - \exp\left(-\xi\left(\frac{\delta_S}{\delta_M}\right)^{3/2}\right)\right] \qquad (5.5.28)$$

and is illustrated in Figure 5.5.2, where (5.5.19) is shown for comparison.

The bottom-friction model is *mathematically* identical to the model introduced by Stommel (1948) in his epochal paper on the oceanic westward intensification. Stommel used a highly artificial drag law for fluid columns by which each column was retarded by a frictional force linearly proportional to the fluid velocity. Such a model introduces a decay term in the vorticity equation which is mathematically indistinguishable from the effect of the bottom Ekman layer in the model discussed in this section. With his model Stommel was the first to suggest that the westward intensification of the oceanic circulation is a consequence of the planetary vorticity gradient. We shall call the $\delta_S$-layer the Stommel layer.

The mathematical relationship between the Stommel layer, the Munk layer, and the frictional sublayer can be illustrated schematically as follows. The vorticity equation when nonlinear inertial effects can be ignored is

Munk layer      Sverdrup interior

$$\left(\frac{\delta_M}{L}\right)^3 \nabla^4\psi \;-\; \frac{\delta_S}{L}\nabla^2\psi \;-\; \frac{\partial \psi}{\partial x} \;=\; -\,\text{curl}\ \tau \qquad (5.5.29)$$

(lateral friction)  (bottom friction)  (planetary-vorticity gradient)  (wind-stress curl)

frictional sublayer  Stommel layer  Sverdrup interior

When $\delta_M > \delta_S$ the balances outlined above the equation hold and bottom friction is negligible everywhere. When $\delta_M < \delta_S$ the Munk layer splits into *two* layers, Stommel's layer and the sublayer. The sublayer, *together with Stommel's layer*, maintains the order of the differential equation in the boundary layer to be the same as Munk's layer. In all cases the vorticity balance in the interior is the Sverdrup balance. It is only the detailed balances in the

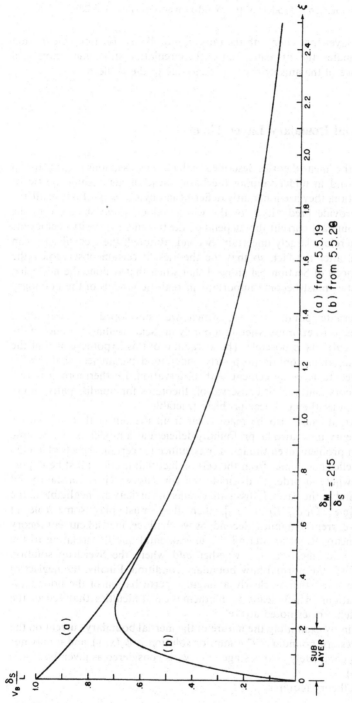

**Figure 5.5.2** The northward velocity in the bottom-friction-dominated western boundary current. Curve (a) shows the distribution of $v_B$ when the no-slip condition is relaxed, while (b) shows the distribution of $v_B$ when the lateral friction sublayer is included.

$$\frac{\delta_M}{\delta_S} = .215$$

(a) from 5.5.19
(b) from 5.5.20

boundary layer that alter with the ratio $\delta_S/\delta_M$. *Whichever* isotropic friction term dominates, the intensification of the circulation still comes about as a consequence of the anisotropy in (5.5.29) due to the $\beta$-effect.


## 5.6  Inertial Boundary-Layer Theory

Although the linear theories described in the last two sections provide useful models to aid in understanding the basic cause of the oceanic westward intensification, they are inherently deficient in obvious ways. First of all, the theories provide predictions for the width, velocity, and structure of the western boundary current that depend on the turbulent viscosity coefficients and are therefore largely uncertain. Second, although these coefficients can be chosen, after the fact, so that the theories fit certain observables, the considerations of Section 5.4 showed that when that is done the neglected nonlinearities must become important in realistic models of the boundary currents.

An alternate point of view, which was developed to meet these deficiencies, is to examine whether a purely inviscid, nonlinear model of the boundary currents is possible. The attraction of this hypothesis is that the resulting solutions contain no poorly understood parameters that can be manipulated to force agreement with observation. Furthermore, a purely inertial theory can exploit conservation theorems for nondissipative flows that can render the nonlinear problem tractable.

However, it must also be made clear from the outset that any *purely* inertial theory must also be profoundly deficient as a model for the oceanic circulation problem, even though it may suffice for certain aspects of it. One obvious deficiency comes from the certain fact that friction must be important somewhere in order to dissipate the vorticity which is constantly fed into the fluid in the basin. If dissipative sinks of vorticity are negligible in the interior, [e.g., $O((\delta_M/L)^3, \delta_S/L)$], then they must play some role in boundary-current dynamics. Second, as we shall see, inertial current theory alone is incapable of answering the fundamental question relating to the westward intensification, i.e., whether and where the Sverdrup solution should satisfy the normal-flow boundary condition. Finally, the neglect of dissipation altogether is clearly a singular perturbation of the boundary-layer equations, which leads to mathematical difficulties that reflect the physical deficiencies noted above.

We begin by examining the nature of the inertial boundary current on the oceanic western boundary. The interior solution is $\psi_I(x, y)$, and it may not be known completely at this stage—i.e., $\psi_I$ is considered as given by (5.4.3) and (5.4.4).

Inertial theory requires

$$\delta_I \gg \delta_M, \delta_S,\qquad\qquad(5.6.1)$$

so that

$$l_* = \delta_I. \tag{5.6.2}$$

Then to $O((\delta_M/\delta_I)^3, \delta_S/\delta_I, \delta_I/L)$, (5.4.25) yields at lowest order

$$\frac{\partial \psi_B}{\partial \xi} \frac{\partial}{\partial y} \frac{\partial^2 \psi_B}{\partial \xi^2} - \frac{\partial \psi_B}{\partial y} \frac{\partial^3 \psi_B}{\partial \xi^3} + \frac{\partial \psi_B}{\partial \xi} = 0, \tag{5.6.3}$$

or

$$J\left(\psi_B, \frac{\partial^2 \psi_B}{\partial \xi^2} + y\right) = 0, \tag{5.6.4}$$

where $J$ is the Jacobian with respect to $\xi$ and $y$. This is merely the statement of conservation of potential vorticity, which in the present case is the sum of the relative vorticity and the planetary vorticity. Since in the boundary layer

$$v_B \gg u_B,$$

$$\frac{\partial}{\partial x} \gg \frac{\partial}{\partial y}, \tag{5.6.5}$$

the vorticity is

$$\zeta = \frac{\partial v}{\partial x} - \frac{\partial u}{\partial y} \approx \frac{\partial v}{\partial x} = \frac{\partial^2 \psi}{\partial x^2}. \tag{5.6.6}$$

The conservation equation (5.6.4) may be integrated to yield

$$\frac{\partial^2 \psi_B}{\partial \xi^2} + y = G(\psi_B), \tag{5.6.7}$$

where $G$ is an arbitrary function of $\psi_B$. Since the motion in the boundary layer is nondissipative, the sum of the relative and planetary vorticities must be constant on a streamline. This immediately presents a fundamental difficulty. Consider Figure 5.6.1, where a hypothetical streamline is shown entering the inertial layer at $y = y_1$ and rejoins the interior flow at $y = y_2$. In the interior the relative vorticity of the fluid is negligible compared to the planetary vorticity. Thus at $y_1$, the total vorticity of the fluid column is simply $y_1$ (or $\beta_0 y_1$. in dimensional units). When the column is speeding northward in the boundary current it conserves total vorticity, so that at $y_2$ it possesses the negative relative vorticity

$$\zeta(y_2) = y_1 - y_2 \tag{5.6.8}$$

but still has total vorticity equal to $y_1$. In order to join the interior smoothly, however, it should have essentially no relative vorticity and the total vorticity of the ambient fluid, $y_2$. It is clearly impossible then for (5.6.7) to apply over the whole path, and we shall prove first that only over the $y$-interval for which $u_I(X_W, y)$ is *westward* will the inertial theory be valid. That is, we can expect the theory to hold only in the southern formative

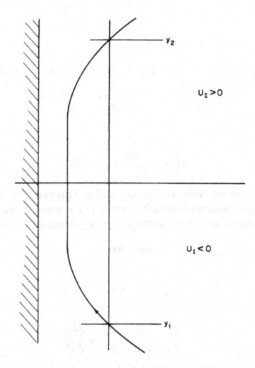

**Figure 5.6.1** A streamline entering and leaving an inertial western boundary current. At $y_1$ the relative vorticity of an element entering the region is nearly zero, and its (nondimensional) absolute vorticity is $y_1$. At $y_2$ its relative vorticity is $-(y_2 - y_1)$, and its absolute vorticity is still $y_1$. It cannot merge smoothly with the interior at $y_2$, where the relative vorticity is again nearly zero and the absolute vorticity is $y_2$.

region of the current, while dissipation must be crucial in the northern reentry region to dissipate the excess relative vorticity of the flow.

As in earlier sections, write $\psi_B$ as the sum of the interior stream function plus a boundary-layer correction,

$$\psi_B(\xi, y) = \psi_I(x, y) + \phi_B(\xi, y). \tag{5.6.9}$$

In particular, it is crucial to recall that $\phi_B$ must satisfy

$$\lim_{\xi \to \infty} \phi_B(\xi, y) = 0 \tag{5.6.10}$$

in order that $\phi_B$ may merge smoothly with the interior. Consider (5.6.3) in the region of large $\xi$. $\phi_B$ must be small in this transition region from the boundary layer into the interior, i.e., for large $\xi$

$$\phi_B \ll \psi_I. \tag{5.6.11}$$

If (5.6.9) is substituted into (5.6.3) and only *linear terms* in $\phi_B$ are retained, as a consequence of (5.6.11) we obtain

$$-\frac{\partial \psi_I}{\partial y}(X_W, y)\frac{\partial^3 \phi_B}{\partial \xi^3} + \frac{\partial \phi_B}{\partial \xi} = 0, \tag{5.6.12}$$

where terms of the form

$$\frac{\partial \phi_B}{\partial \xi}\frac{\partial}{\partial y}\frac{\partial^2 \psi_I}{\partial \xi^2} = \left(\frac{\delta_I}{L}\right)^2 \frac{\partial \phi_B}{\partial \xi}\frac{\partial}{\partial y}\frac{\partial^2 \psi_I}{\partial x^2} \tag{5.6.13}$$

have been neglected, since they are $O(\delta_I/L)^2$. Also we have used

$$-\frac{\partial \psi_I}{\partial y}(x, y) = -\frac{\partial \psi_I}{\partial y}\left(X_W + \frac{\delta_I}{L}\xi, Y\right) \approx -\frac{\partial \psi_I}{\partial y}(X_W, y) + O\left(\frac{\delta_I}{L}\right)$$

$$= u_I(X_W, y). \tag{5.6.14}$$

Integration of (5.6.12) once with respect to $\xi$ yields, with the aid of (5.6.10),

$$\boxed{u_I(X_W, y)\frac{\partial^2 \phi_B}{\partial \xi^2} + \phi_B = 0}. \tag{5.6.15}$$

The general solution of (5.6.15) will be *oscillatory* in $\xi$ for all $y$ for which $u_I(X_W, y) > 0$. The oscillations are undamped, and consequently (5.6.10) cannot be satisfied. A necessary condition for the existence of an inertial boundary current at $y$ is that

$$u_I(X_W, y) < 0. \tag{5.6.16}$$

North of the point where $u_I$ vanishes, a *purely* inertial current is no longer possible (Greenspan 1962). Hence (5.6.7) will be an adequate model of the western boundary current only in its southern sector. Where $u_I(X_W, y) < 0$ the decaying solution of (5.6.15) is of the form

$$\phi_B = C(y)\exp[-\xi[-u_I(X_W, y)]^{-1/2}] \tag{5.6.17}$$

Since (5.6.15) applies only far from the coast, we cannot determine $C$ through the application of the boundary condition there. However, (5.6.17) reveals an important fact. The decay rate for $\phi_B$ increases inversely with $u_I^{1/2}$. That is, at those $y$ where $u_I$ is on shore and small, the width of the inertial current is small. In fact, by combining the definition of $\delta_I$ with the result of (5.6.17), it is clear that the *local* width of the inertial current is

$$\delta_{I \text{ local}} = \sqrt{-\frac{U_*(y)}{\beta_0}}, \tag{5.6.18}$$

where $U_*(y)$ is the dimensional interior zonal flow. As the point $y_0$ is approached where $U_*$ vanishes, $\delta_{I \text{ local}}$ will shrink and finally be swallowed up within the thickness of a friction layer, such as Munk's which for $U_* = O(1)$

occupies only a small fraction of the inertial layer. Again, from this viewpoint friction is an essential ingredient of the dynamics in the boundary layer whenever *either* $u_I > 0$ or $\delta_{I\,\text{local}}$ becomes small with respect to $\delta_M$ or $\delta_S$.

For those $y$ for which $u_I(X_W, y)$ is $O(1)$ and westward, (5.6.7) may be used to find $\phi_B$ by the method described in Section 3.14. Thus for large $\xi$ where $\partial^2 \psi_B / \partial \xi^2$ is zero,

$$G(\psi_{B\infty}) = y, \tag{5.6.19a}$$

while at infinity

$$\psi_{B\infty} = \psi_I(X_W, y). \tag{5.6.19b}$$

If (5.6.19a,b) is inverted, then

$$y = \psi_I^{-1}(\psi_{B\infty}), \tag{5.6.20}$$

where $\psi_I^{-1}$ is the functional inverse to $\psi_I$. Thus

$$G(\psi_{B\infty}) = \psi_I^{-1}(\psi_{B\infty}), \tag{5.6.21a}$$

but since the dependence of $G$ on $\psi$ depends only on $\psi$ and not otherwise on location in the $x$, $y$ plane, the functional relation (5.6.21a) must hold on all streamlines which originate in the interior. Thus

$$G(\psi_B) = \psi_I^{-1}(\psi_B), \tag{5.6.21b}$$

so that (5.6.7) becomes

$$\frac{\partial^2 \psi_B}{\partial \xi^2} + y = \psi_I^{-1}(\psi_B). \tag{5.6.22}$$

In general the process of inverting (5.6.19) renders $\psi_I^{-1}(\psi_B)$ a highly nonlinear function of $\psi_B$, and (5.6.22) is rarely tractable analytically. In the simple case when

$$\psi_I = y = 1 \cdot y, \tag{5.6.23}$$

the inverse $\psi_I^{-1}$ is simple multiplication by 1, so that $\psi_B$ satisfies

$$\frac{\partial^2 \psi_B}{\partial \xi^2} - \psi_B = -y, \tag{5.6.24}$$

the solution for which is

$$\psi_B = y[1 - e^{-\xi}], \tag{5.6.25}$$

or exactly the solution found in Section 3.14. The streamlines of the flow given by (5.6.25) are therefore those of Figure 3.14.2. Note that in this case the asymptotic, linearized equation (5.6.15) is identical with the full equation

(5.6.22), which gives confidence to the qualitative conclusions inferred from the former.

The nature of the inertial boundary layer is also of little direct help in the determination of the overall circulation pattern, e.g., the choice between the flows of Figure 5.3.3(a,b). The form of (5.6.15) is invariant under the transformation $\xi \rightarrow -\xi$, so that it follows immediately that on $x = X_E$ the analogous equation there is

$$u_I(X_E, y)\frac{\partial^2\phi_B}{\partial\lambda^2} + \phi_B = 0 \qquad (5.6.26)$$

for the correction function in a possible eastern layer. Thus if $u_I(X_E, y)$ is also westward, an inertial boundary layer is possible on $x = X_E$. Figure 5.6.2 shows the two choices again, with the regions where inertial layers are possible indicated by crosshatching. Note that in Figure 5.6.2(a) the inertial

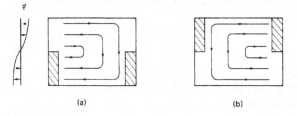

(a)                              (b)

**Figure 5.6.2**    The crosshatched regions in (a), (b) show the possible sites of inertial boundary layers corresponding to the choices of Figure 5.3.3 (a), (b) respectively. Note that in (b) the inertial layer is possible on the northwest wall, where it is not required, or on the northeast wall, where the flow exits from the presumed boundary current.

layer is possible in the formation region of the western boundary current, but in Figure 5.6.2(b) an inertial layer is not possible on the southeast wall of the basin, which would be the formation region of an eastern current. It seems natural to suppose that inertial currents should exist in formation regions, so that Figure 5.6.2(a) seems more intuitively acceptable than 5.6.2(b). However, it is important to realize that there is no proof of this hypothesis on the basis of inertial theory alone. Without dissipation it is not possible in the steady problem to rigorously exclude Figure 5.6.2(b). This question is considered in more detail in the next section.

If we take the point of view that the interior circulation is fixed by the usual choice (5.4.20), then it is clear that inertial theory gives a parameter-free model for the formation regions of western currents. The unambiguous prediction of the width is given by (5.6.18). For values of $U_*(y)$ between 1 and 10 cm/s, (5.6.18) yields current widths of the order of 30 to 100 km, which are quite realistic values.

## 5.7 Inertial Currents in the Presence of Friction

Although the inertial theory is adequate for some purposes, its limitation to regions where $u_I < 0$ prevents its use as the sole mechanism to close the oceanic mass flux, i.e., to complete the streamline as attempted in Figure 5.6.1. In this section we shall examine the behavior of the inertial theory in the presence of small lateral friction in much the same manner as Moore (1963). Again, for $\delta_I \gg \delta_M$, $\delta_S$, we choose $l_*$ to be $\delta_I$ in (5.4.25) and we represent $\psi_B$ as

$$\psi_B(\xi, y) = \psi_I(x, y) + \phi_B(\xi, y). \tag{5.7.1}$$

Consider the case where

$$\left(\frac{\delta_M}{\delta_I}\right)^3 \gg \frac{\delta_S}{\delta_I}, \tag{5.7.2}$$

so that the dominant frictional effect is due to lateral friction. Then from (5.4.25) the lowest-order equation which includes friction is

$$\frac{\partial \phi_B}{\partial \xi} \frac{\partial}{\partial y} \frac{\partial^2 \phi_B}{\partial \xi^2} - \left(\frac{\partial \psi_I}{\partial y} + \frac{\partial \phi_B}{\partial y}\right)\left(\frac{\partial^3 \phi_B}{\partial \xi^3}\right) + \frac{\partial \phi_B}{\partial \xi} = \left(\frac{\delta_M}{\delta_I}\right)^3 \frac{\partial^4 \phi_B}{\partial \xi^4}, \tag{5.7.3}$$

where again the fact that $\partial \psi_I / \partial \xi$ is $O(\delta_I/L)$ has been used throughout to neglect certain small terms in (5.4.25). Now consider (5.7.3) in the transition region of large $\xi$ where

$$\phi_B \ll \psi_I.$$

In this region (5.7.3) can be linearized in the same manner as in Section 5.6 to yield

$$u_I(X_W, y)\frac{\partial^3 \phi_B}{\partial \xi^3} + \frac{\partial \phi_B}{\partial \xi} = \left(\frac{\delta_M}{\delta_I}\right)^3 \frac{\partial^4 \phi_B}{\partial \xi^4}, \tag{5.7.4}$$

or since $\phi_B$ must vanish as $\xi \to \infty$,

$$u_I(X_W, y)\frac{\partial^2 \phi_B}{\partial \xi^2} + \phi_B = \left(\frac{\delta_M}{\delta_I}\right)^3 \frac{\partial^3 \phi_B}{\partial \xi^3}. \tag{5.7.5}$$

Note that when (5.7.5) is compared with (5.6.15), the term on the right-hand side, due to friction, introduces an asymmetry between $\pm \xi$. Solutions for $\phi_B$ can be found in the form

$$\phi_B = Ce^{\alpha\xi}, \tag{5.7.6}$$

where $\alpha$ satisfies the cubic

$$\left(\frac{\delta_M}{\delta_I}\right)^3 \alpha^3 - u_I\alpha^2 - 1 = 0. \tag{5.7.7}$$

For small values of $\delta_M/\delta_I$, solutions for $\alpha$ can be found in the form

$$\alpha = \alpha_0 + \left(\frac{\delta_M}{\delta_I}\right)^3 \alpha_1 + \cdots . \tag{5.7.8}$$

Other solutions give length scales small compared to $\delta_I$ and are irrelevant. If (5.7.8) is substituted into (5.7.7) and like powers of the small parameter $(\delta_M/\delta_I)^3$ are equated, we find that

$$\alpha_0^2 = -\frac{1}{u_I(y)}, \tag{5.7.9a}$$

$$\alpha_1 = \frac{\alpha_0^2}{2u_I(y)} = -\frac{1}{2u_I^2(y)}; \tag{5.7.9b}$$

hence

$$\alpha = \left|(-u_I)^{-1/2} - \frac{1}{2}\left(\frac{\delta_M}{\delta_I}\right)^3 u_I^{-2}\right|. \tag{5.7.10}$$

When $u_I < 0$, $\alpha_0$ is real and $\alpha_1$ represents an inconsequential correction to the decay scale of the inertial layer. Thus friction, as supposed earlier, can fundamentally be ignored in the formation region of the western boundary current. Consider however the case where $u_I > 0$, i.e., the "northern" reentry region. $\alpha_0$ is purely imaginary, and the solution (5.7.6) is

$$\phi_B = C \exp i[u_I^{-1/2}\xi]\exp\left(-\left|\frac{1}{2}\left(\frac{\delta_M}{\delta_I}\right)^3 u_I^{-2}\xi\right|\right) \tag{5.7.11}$$

and represents a *standing* Rossby wave with dimensional wave number

$$k_* = \frac{1}{\delta_I}u^{-1/2} = \left(\frac{\beta_0}{u_{I_*}}\right)^{1/2} \tag{5.7.12}$$

and wavelength

$$\lambda_* = 2\pi\left(\frac{u_{I_*}}{\beta_0}\right)^{1/2}, \tag{5.7.13}$$

in agreement with the results of Section 3.18. This standing Rossby wave, whose existence invalidated the purely inertial theory when $u_I > 0$, is nonetheless slowly damped by friction on the scale $u_I^2\delta_I^3/\delta_M^3$, or in *dimensional units* on the scale

$$l_A = \delta_I\left(u_I^2\frac{\delta_I^3}{\delta_M^3}\right) = \frac{u_{I_*}^2}{\beta_0 A_H}. \tag{5.7.14}$$

The number of oscillations before decay is therefore $O((\delta_I/\delta_M)^3)$. If $u_{I_*}$ is

O(1 cm/s) and $A_H$ is $O(10^6 \text{ cm}^2 \text{ s}^{-1})$, then $l_A$ is O(100 km). If $U_I$ is larger or $A_H$ smaller, the scale will be correspondingly larger. It will generally be small, though, compared with the extent of the basin, so that (5.7.11) implies that a complex inertial–viscous *boundary layer* is possible in the northern part of the basin on the *western* wall. On the other hand, the corresponding equation to (5.7.4) on the eastern wall yields solutions which exponentially increase seaward, so that the inertial viscous layer, whose decay scale depends again inversely on $\beta_0$, can exist only on the western boundary layer. Hence the presence of a small amount of *friction* in conjunction with the $\beta$-effect once more chooses Figure 5.6.2(a) as the proper general circulation pattern. The circulation pattern for $\delta_I > \delta_M \gg \delta_s$ is shown schematically in Figure 5.7.1.

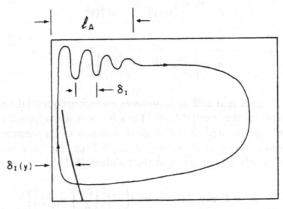

**Figure 5.7.1** A schematic of the inferred circulation pattern showing a Sverdrup interior, an inertial boundary layer in the western formation region, and a region of damped Rossby waves in the northwest, inertial-frictional region.

## 5.8 Rossby Waves and the Westward Intensification of the Oceanic Circulation

Each of the dynamically different models of the oceanic circulation shares the common feature of westward intensification in spite of considerable differences in their treatment of the western boundary layer. A simple physical explanation can be found by considering the fundamental character of Rossby waves. These waves act as the messengers of energy in quasi-geostrophic motion and strongly manifest the spatial anisotropy produced by the planetary-vorticity gradient.

The frequency of a Rossby wave in dimensional units* is (3.15.11):

$$\sigma = -\frac{\beta_0 k}{k^2 + l^2 + \mathcal{R}^{-2}}, \qquad (5.8.1)$$

---

\* In this section all variables are dimensional, so the asterisk convention for dimensional variables is temporarily suspended.

where $k$ and $l$ are the eastward and northward components of the wave vector, while $\mathscr{R}$ is the deformation radius,

$$\mathscr{R} = \frac{(gD)^{1/2}}{f}, \tag{5.8.2}$$

which is $O(2{,}000 \text{ km})$ for an ocean of depth 4 km in mid-latitudes where $f = 10^{-4}\ \text{s}^{-1}$. The planetary-vorticity gradient is $\beta_0$. The group velocity in the eastward direction is

$$C_{gx} = \beta_0 \frac{k^2 - (l^2 + \mathscr{R}^{-2})}{(k^2 + l^2 + \mathscr{R}^{-2})^2}, \tag{5.8.3}$$

in accordance with (3.19.13a). Energy with small $x$-scales ($k^2 > l^2 + \mathscr{R}^{-2}$) will be transmitted eastward, while energy with large $x$-scales ($k^2 < l^2 + \mathscr{R}^{-2}$) will move to the west with the appropriate group speed.

Now suppose that at some time, energy of varying scales is put into the ocean by the wind stress. The small-scale components will move to the eastern boundary of the ocean, where they will be reflected as components with large east–west scale in accord with the notions of Section 3.23. On the other hand, the large-scale components will move towards the western boundary, where they will be reflected as small-scale motions. The western boundary thus acts as a *source* of small-scale energy. This is the underlying physical basis for the preference for westward intensification. It depends only on the anisotropy in the energy transmission properties of planetary-vorticity waves.

This argument can be carried further to produce estimates of size of the region of western intensification. For example consider Munk's model, where lateral friction dominates both nonlinear advection and bottom dissipation of vorticity. The characteristic time for the decay of small scale ($k^2 \gg l^2 + \mathscr{R}^{-2}$) energy produced at the western boundary will be the viscous dissipation time, i.e.,

$$t_D = O(A_H k^2)^{-1}, \tag{5.8.4}$$

in which time the energy has moved eastward a distance

$$l = C_{gx} t_D = t_D \frac{\beta_0}{k^2}, \tag{5.8.5}$$

since for short-scale waves

$$C_{gx} \sim \frac{\beta_0}{k^2}. \tag{5.8.6}$$

The wave energy will be effectively trapped at the western boundary for scales sufficiently short that the distance $l$ is of the same order as the wavelength, i.e., for scales that are critically damped:

$$k \sim l^{-1}. \tag{5.8.7}$$

Combining (5.8.4), (5.8.5), and (5.8.7) leads to the estimate

$$l \sim \frac{\beta_0 \, l^4}{A_H}, \tag{5.8.8}$$

or

$$l = \left(\frac{A_H}{\beta_0}\right)^{1/3}, \tag{5.8.9}$$

in agreement with (5.4.1). In Stommel's model dissipation by bottom friction dominates, so that in that case, by (4.7.1),*

$$t_D \sim 2 \frac{D}{\delta_E} f^{-1} \tag{5.8.10}$$

which with (5.8.5) and (5.8.7) yields the trapping scale

$$l = \frac{\delta_E \, f/2}{D \beta_0}, \tag{5.8.11}$$

in agreement with (5.5.2).

In the inertial theory the advection of vorticity by the mean flow becomes significant, and as in Section 3.18, the mean flow alters the frequency and augments the group velocity. The short-wave components generated at the western boundary will now have a group velocity

$$C_{gx} = u_I + \frac{\beta_0}{k^2}, \tag{5.8.12}$$

where $u_I$ is the advection by the Sverdrup flow. Scales of motion for which $C_{gx} < 0$ will be trapped at the western boundary. This clearly requires $u_I < 0$ and $k$ large enough so that the intrinsic group velocity to the east becomes small enough to be compensated by the westward Sverdrup flow. Scales shorter than

$$l = \left(\frac{-u_I}{\beta_0}\right)^{1/2} \tag{5.8.13}$$

will be trapped. This estimate also reveals the basic difficulty of inertial theories where $u_I > 0$. The short-wave components generated at the western boundary, rather than being trapped there, are actually assisted by the Sverdrup flow and are radiated towards the oceanic interior as Rossby waves. As they radiate, they will slowly decay by dissipative processes. If lateral friction is the dominant dissipative mechanism, so that (5.8.4) applies, the scale over which Rossby waves of wavelength $k^{-1}$ will decay is, by (5.8.5),

$$l = \frac{\beta_0}{A_H} k^{-4}, \tag{5.8.14}$$

---

* The factor of 2 between (5.8.10) and (4.3.26) enters because the spin-up time is now twice as long, since only the lower Ekman layer acts to dissipate vorticity.

or, since $k \sim (\beta_0/u_I)^{1/2}$, from (3.18.11) with $u_I > 0$,

$$l = \frac{u_I^2}{\beta_0 A_H}, \tag{5.8.15}$$

in agreement with (5.7.14).

The various theories of the western boundary current can be classified, therefore, according to the method (viscous, inertial, or inertial–viscous) used to prevent motions of the boundary-layer scale from leaking into the oceanic interior from the western boundary where they are most naturally generated.

## 5.9 Dissipation Integrals for Steady Circulations

When the western boundary current is fundamentally inertial, analytical solutions for the complete circulation are not possible. Although the qualitative results already derived give a good picture of the nature of the circulation, numerical integration of the vorticity equation (5.2.20) is required to produce the explicit streamline structure. Before examining the numerical calculations that have been done, certain integral relations will be derived in this section that help in the understanding of the results of these necessarily intricate calculations.

In the steady-state circulation pattern calculated for a closed-basin ocean, all streamlines must be closed. Consider an *arbitrary* closed streamline which encloses an area $A_\psi$ and which consists of the curve $C_\psi$, as shown in Figure 5.9.1. The vorticity equation (5.2.20) may be written, for steady flow,

$$\mathbf{u} \cdot \nabla \left\{ \left( \frac{\delta_I^2}{L^2} \right) \zeta + y + \frac{\eta_B}{\beta} \right\} = \operatorname{curl} \tau - \frac{\delta_S}{L} \zeta + \left( \frac{\delta_M}{L} \right)^3 \nabla^2 \zeta \tag{5.9.1}$$

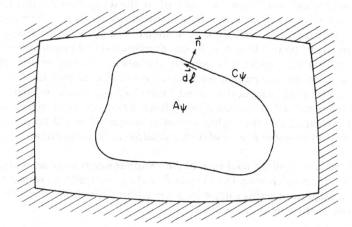

**Figure 5.9.1**  A contour $C_\psi$ coincident with a streamline of the steady circulation pattern encloses an area $A_\psi$.

if the definitions (5.4.1) are used. Integrate (5.9.1) over the area $A_\psi$. Since the velocity field is geostrophic,

$$\iint_{A_\psi} \mathbf{u} \cdot \nabla \left| \frac{\delta_I^2}{L^2} \zeta + y + \frac{\eta_B}{\beta} \right| dx\,dy = \iint \nabla \cdot \left[ \mathbf{u} \left( \frac{\delta_I^2}{L^2} \zeta + y + \frac{\eta_B}{\beta} \right) \right] dx\,dy, \quad (5.9.2)$$

so that with the divergence theorem,

$$\iint_{A_\psi} \mathbf{u} \cdot \nabla \left| \frac{\delta_I^2}{L^2} \zeta + y + \frac{\eta_B}{\beta} \right| dx\,dy = \oint_{C_\psi} (\mathbf{u} \cdot \mathbf{n}) \left| \frac{\delta_I^2}{L^2} \zeta + y + \frac{\eta_B}{\beta} \right| dl = 0,$$

$$(5.9.3)$$

since by definition $\mathbf{u}$ must be parallel to the streamline, while $\mathbf{n}$ is the unit vector normal to the streamline. Hence the integral of (5.9.1) over $A_\psi$ yields

$$\oint_{C_\psi} \boldsymbol{\tau} \cdot d\mathbf{l} = \frac{\delta_s}{L} \oint_{C_\psi} \mathbf{u} \cdot d\mathbf{l} - \left( \frac{\delta_M}{L} \right)^3 \oint_{C_\psi} (\nabla \zeta \cdot \mathbf{n})\,dl, \quad (5.9.4)$$

where $d\mathbf{l}$ is the infinitesimal vector line element tangent to $C_\psi$ whose magnitude is $dl$. In deriving (5.9.4) the following vector integral identities of Stokes's theorem and the divergence theorem have been used:

$$\iint_{A_\psi} \text{curl } \boldsymbol{\tau}\, dx\,dy = \oint_{C_\psi} \boldsymbol{\tau} \cdot d\mathbf{l} \quad (5.9.5a)$$

$$\iint_{A_\psi} \zeta\, dx\,dy = \oint_{C_\psi} \mathbf{u} \cdot d\mathbf{l} \quad (5.9.5b)$$

$$\iint_{A_\psi} \nabla^2 \zeta\, dx\,dy = \oint_{C_\psi} \nabla \zeta \cdot \mathbf{n}\, dl. \quad (5.9.5c)$$

The interpretation of (5.9.4) is simply that the total circulation produced by the wind stress on $C_\psi$ must be balanced by the dissipation of vorticity by bottom friction on $C_\psi$ and the diffusion of vorticity across $C_\psi$. The net input of vorticity is proportional to the circulation of $\tau$ on $C_\psi$, while the dissipation of vorticity by bottom friction is directly proportional to the circulation of the velocity field. It is important to note that the balance between the stress input and the frictional dissipation, as expressed by (5.9.4), must hold for *all* values of $\delta_I/\delta_S$ and $\delta_I/\delta_M$. That is, the inertial forces (5.9.3) are unable to balance the vorticity input and do not enter the balance. Friction must therefore act to yield a balance, and the neglect of friction altogether would be obviously inconsistent *no matter how small the frictional forces are compared with inertial effects.*

In the interior of the fluid each of the nondimensional variables is $O(1)$. It follows immediately from (5.9.4), since $\delta_S$ and $\delta_M$ are small with respect to $L$, that a balance of terms cannot be achieved on streamlines which remain entirely in the interior, since then

$$\oint \boldsymbol{\tau} \cdot d\mathbf{l} = O(1), \quad (5.9.6)$$

while the terms on the right-hand side are $O(\delta_S/L)$ and $O((\delta_M/L)^3)$ respectively. Therefore *every* streamline must go through a boundary-layer region where dissipation is important, as shown in Figure 5.9.2. Consider as a first example Munk's model. There is no effective bottom friction. In the boundary layer, from (5.4.39),

$$v_B = O\left(\frac{L}{\delta_M}\right),$$

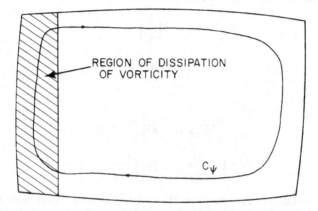

REGION OF DISSIPATION
OF VORTICITY

$C_\psi$

**Figure 5.9.2**    Each streamline must enter a region of dissipation of vorticity, and no streamline can close only in the interior, *if* the velocity in the interior is given by the Sverdrup relation.

and the vorticity gradient in the boundary layer is

$$\nabla\zeta \sim \frac{\partial\zeta}{\partial x} = O\left(v_B\frac{L^2}{\delta_M^2}\right) = O\left(\frac{L}{\delta_M}\right)^3 \tag{5.9.7}$$

so that

$$\left(\frac{\delta_M}{L}\right)^3 \oint \nabla\zeta \cdot n \, dl = O(1), \tag{5.9.8}$$

since each streamline remains in the layer over an $O(1)$ distance in $y$. Thus the vorticity diffusion by lateral friction in the boundary layer balances the wind-stress source of vorticity.

In Stommel's model, bottom friction dominates, and from (5.5.19)

$$v_B = O\left(\frac{L}{\delta_S}\right).$$

Thus since all streamlines thread through the western layer, the contribution to the first integral on the right-hand side of (5.9.4) from that part of the streamline in the Stommel layer is

$$\frac{\delta_S}{L} \int \mathbf{u} \cdot d\mathbf{l} = \frac{\delta_S}{L} \int v_B \, dy = O(1). \tag{5.9.9}$$

In each of the frictional cases the western layer which closes the mass flux is also capable of dissipating the vorticity input by the wind.

Consider however the inertial limit where $\delta_I \gg \delta_M$ or $\delta_S$. The contribution to the vorticity balance from dissipation acting on the flow as it travels in the inertial current can be easily estimated, since

$$v_B = O\left(\frac{L}{\delta_I}\right)$$

$$\zeta_B = O\left(\left(\frac{L}{\delta_I}\right)^2\right). \tag{5.9.10}$$

Thus

$$\frac{\delta_S}{L} \int v_B \, dy = O\left(\frac{\delta_S}{\delta_I}\right) \ll 1,$$

$$\left(\frac{\delta_M}{L}\right)^3 \int \frac{\partial \zeta_B}{\partial x} \, dy = O\left(\left(\frac{\delta_M}{\delta_I}\right)^3\right) \ll 1. \tag{5.9.11}$$

There is insufficient dissipation in the inertial boundary layer alone to balance the input of the wind stress. The resolution of this apparent paradox is subtle but important. For all $A_H \neq 0$, no matter how small, a frictional sublayer must exist within the inertial boundary layer to satisfy the no-slip condition. It is analogous to the sublayer of Section 5.5. However, since now $\delta_I > (\delta_M, \delta_S)$, the sublayer must be characterized by a balance between horizontal friction and nonlinear advection, i.e.,

$$\left(\frac{\delta_I}{L}\right)^2 \mathbf{u} \cdot \nabla \zeta = O\left(\left(\frac{\delta_M}{L}\right)^3\right) \nabla^2 \zeta. \tag{5.9.12}$$

The nondimensional width of the sublayer, $l$, can be estimated as follows. In the sublayer,

$$\left(\frac{\delta_M}{L}\right)^3 \nabla^2 \zeta = O\left(\left(\frac{\delta_M}{L}\right)^3\right) \frac{\partial^2 \zeta}{\partial x^2} = O\left(\left(\frac{\delta_M}{L}\right)^3 \frac{\zeta_B}{l^2}\right), \tag{5.9.13}$$

while

$$\left(\frac{\delta_I}{L}\right)^2 \mathbf{u} \cdot \nabla \zeta = \left(\frac{\delta_I}{L}\right)^2 \left(u \frac{\partial \zeta}{\partial x} + v \frac{\partial \zeta}{\partial y}\right) = O\left(\left(\frac{\delta_I}{L}\right)^2\right) v_B \zeta_B. \tag{5.9.14}$$

Hence (5.9.12) implies that

$$l = \left(\frac{\delta_M^3}{\delta_I^2 L v_B}\right)^{1/2} = \frac{1}{L} \left(\frac{\delta_M^3}{\delta_I}\right)^{1/2}, \tag{5.9.15}$$

or in dimensional units

$$l_* = lL = \sqrt{\frac{A_H \delta_I}{U}} = \frac{\delta_I}{\text{Re}_\delta^{1/2}} \ll \delta_I, \tag{5.9.16}$$

where

$$\text{Re}_\delta = \frac{U \delta_I}{A_H}$$

is the boundary-layer Reynolds number.

Consider now the streamline $C_0$ which coincides with the perimeter of the basin. For this outer streamline the vorticity balance (5.9.4) represents the vorticity balance for the basin as a whole. Since $C_0$ is immediately adjacent to the boundary, it must pass through the frictional sublayer and the contribution of the lateral diffusion of vorticity, for this streamline is, using (5.9.10) and (5.9.15),

$$\left(\frac{\delta_M}{L}\right)^3 \int \nabla \zeta \cdot n \, dl = O\left(\left(\frac{\delta_M}{L}\right)^3 \frac{v_B}{l^2}\right)$$

$$= O\left(\frac{\delta_M^3}{L^3} \frac{L}{\delta_I} \frac{L^2}{\delta_M^3} \delta_I\right) = O(1), \tag{5.9.17}$$

and the vorticity balance for the basin is restored. Thus no matter how small $A_H$ is, only the consideration of horizontal diffusion will allow a steady solution with a Sverdrup interior (so that the variables in (5.9.11) are truly $O(1)$ in the interior). The neglect of lateral diffusion is a singular perturbation which will completely alter the solution when nonlinear inertial terms become dominant.

However, not all streamlines will pass through the sublayer (else the sublayer itself would have emerged as the appropriate western boundary current). Consider, as shown in Figure 5.9.3, a streamline outside the sublayer which passes north through the inertial layer and rejoins the interior via the wavy northern inertial–viscous region. We have already seen that the vorticity dissipation in the formation region of the western boundary current is insufficient to balance the vorticity input by the wind stress.

Consider now the diffusion that occurs in the viscous–inertial region in the northwest part of the basin. In this region the vorticity is $O(v_B L/\delta_I)$ in nondimensional units, so that

$$\left(\frac{\delta_M}{L}\right)^3 \int \nabla \zeta \cdot \mathbf{n} \, dl = O\left(v_B \frac{L^2}{\delta_I^2} \cdot \left(\frac{\delta_M}{L}\right)^3\right) \mathscr{L}, \tag{5.9.18}$$

where $\mathscr{L}$ is the path length along the undulating streamline in the viscous–inertial region. The path length for each undulation is $O(1)$ but the number of undulations, as can be seen from (5.7.13) (after use is made of (5.7.14)) is

$$N = \frac{\delta_I^4/\delta_M^3}{\delta_I} = \frac{\delta_I^3}{\delta_M^3}. \tag{5.9.19}$$

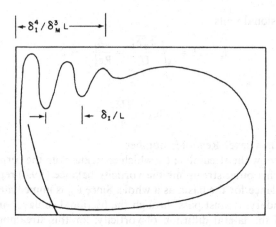

**Figure 5.9.3** The steady streamline pattern inferred from considerations of dissipation.

Since

$$v_B = O\left(\frac{L}{\delta_I}\right),\tag{5.9.20}$$

it follows that the size of the integral in (5.9.18) is

$$\frac{L}{\delta_I}\frac{L^2}{\delta_I^2}\frac{\delta_M^3}{L^3}N = O(1).\tag{5.9.21}$$

Thus another perspective on the undulations of the streamlines in the viscous inertial region is that the undulations serve as a baffle to allow sufficient time for vorticity to diffuse across the undulations into the frictional sublayer and out of the basin, so that (5.9.4) can be satisfied.

If $A_H$ is arbitrarily set to zero and only bottom friction is retained, the resulting singular perturbation does not lead to a vorticity balance with an $O(1)$ interior, as we have seen, even in the presence of bottom friction. What is the resolution in that case? Clearly, the velocity in the interior must be greater than $O(1)$ to achieve the vorticity-dissipation balance even though the *forcing* by the wind-stress curl is fixed at $O(1)$. This occurs in simple mechanical systems such as the mass–spring oscillator at *resonance* when the forcing frequency coincides with the natural frequency of the system. The amplitude of the response is then limited only by the friction in the system, and if the friction is small, the final amplitude can be considerably greater than the amplitude forced at nonresonant frequencies. This naturally leads to the question of whether a zero-frequency, inviscid free mode exists in the circulation problem which can resonate in the forced problem when $A_H$ is neglected. The question is treated in the following section.

## 5.10 Free Inertial Modes

Consider the free, steady inertial motions possible in the rectangular basin shown in Figure 5.10.1. For $\tau = \delta_M = \delta_S = 0$, (5.9.1) becomes for a flat-bottom ocean

$$\mathbf{u} \cdot \nabla \left\{ \left( \frac{\delta_I}{L} \right)^2 \zeta + y \right\} = 0, \tag{5.10.1}$$

or equivalently,

$$\left( \frac{\delta_I}{L} \right)^2 \nabla^2 \psi + y = G(\psi), \tag{5.10.2}$$

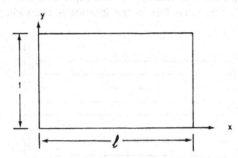

**Figure 5.10.1**   A closed rectangular basin used to calculate a zero-frequency inertial mode.

where $\psi$ is the geostrophic streamfunction. Since the basin is closed, all streamlines are also closed, so that $G(\psi)$ cannot be determined by the method of Section 5.6, since the flow is not known *a priori* on any streamline. Nothing prevents us, though, from *choosing*

$$G(\psi) = \frac{\psi}{A^2} \tag{5.10.3}$$

and observing the consequences. The constant $A^2$ is arbitrary. Then (5.10.2) becomes

$$A^2 \left( \frac{\delta_I}{L} \right)^2 \nabla^2 \psi - \psi = -y A^2 \tag{5.10.4}$$

with boundary conditions

$$\psi = \text{constant} \quad \text{on} \quad \begin{cases} x = 0, \\ x = l, \\ y = 0, \\ y = 1. \end{cases} \tag{5.10.5}$$

Since $(\delta_1/L)^2$ is a small parameter, an approximate solution to (5.10.4) which satisfies (5.10.5) can be found by boundary-layer methods. In the interior the solution is

$$\psi = \psi_I = yA^2 \qquad (5.10.6)$$

and represents a uniform westward flow. On $x = 0$ and $x = l$, boundary layers are required to bring the normal velocity to zero. Thus excluding the boundaries at $y = 0, 1$ from consideration momentarily, the solution to (5.10.4) which has zero normal velocity at $x = 0$ and $x = l$ is

$$\frac{\psi}{A^2} = y - (y - y_r)[e^{-xL/\delta_1 A} + e^{-(l-x)L/\delta_1 A}], \qquad (5.10.7)$$

where $y_r$ is an arbitrary constant. The flow represented by (5.10.7) is shown in Figure 5.10.2. The mass flux in the interior is everywhere westward. It

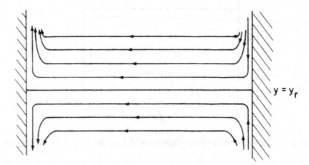

**Figure 5.10.2**   The flow pattern in the inertial mode, excluding the northern and southern boundaries.

impinges on the western boundary, where it splits at $y = y_r$, a portion then flowing northward, the remainder southward. Boundary layers are required on $y = 0$ and $y = 1$ to return the mass flux eastward, i.e., homogeneous solutions to (5.10.4) of the form

$$\psi_N = C_N e^{-(1-y)L/\delta_1 A},$$
$$\psi_S = C_S e^{-yL/\delta_1 A} \qquad (5.10.8)$$

must be added to (5.10.7) to account for the return mass flux. The total eastward mass flux in the northern boundary layer is $-C_N$, and this must be equal to the northward mass flux in the western layer when it reaches the northern boundary, i.e.,

$$-\frac{C_N}{A^2} = (1 - y_r). \qquad (5.10.9)$$

Similarly

$$-\frac{C_S}{A^2} = y_r, \qquad (5.10.10)$$

so that the complete solution for small $(\delta_I/L)^2$ is

$$\frac{\psi}{A^2} = y - (y - y_r)[e^{-xL/\delta_I A} + e^{-(I-x)L/\delta_I A}]$$

$$- (1 - y_r)e^{-(1-y)L/\delta_I A} + y_r e^{-yL/\delta_I A}, \qquad (5.10.11)$$

the streamlines for which are shown schematically in Figure 5.10.3. This free mode was first discussed by Fofonoff (1954). It is a fundamentally non-linear

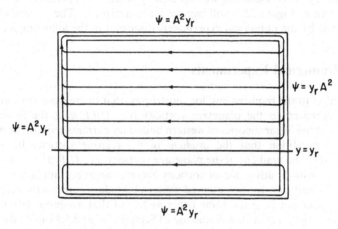

**Figure 5.10.3**  The Fofonoff free inertial mode. Note that the value of $y_r$ is arbitrary and may be shifted to any latitude in the basin.

mode since strictly linear, geostrophic steady flow is impossible on the geostrophic contours of constant $y$ when interrupted by walls at $x = 0$ or 1. It is clear from Figure 5.10.3 that inertial boundary currents are required to complete the mass flux. It is important to note that the amplitude of the mode is arbitrary.

Now consider the presence of a small amount of bottom friction (i.e., $\delta_S \ll \delta_I$) while neglecting lateral friction altogether. The dissipation in the Fofonoff mode is then, from (5.9.4),

$$\frac{\delta_S}{L} \oint \mathbf{u} \cdot d\mathbf{l} = O\left(\frac{\delta_S}{L} v_B\right), \qquad (5.10.12)$$

since the dissipation will be accomplished primarily in the high-speed boundary currents. In these currents the speed, from (5.10.11), is (say in the western layer)

$$\frac{\partial \psi}{\partial x} = \frac{L}{\delta_I} A(y - y_r)e^{-xL/\delta_I A} = O\left(\frac{AL}{\delta_I}\right), \qquad (5.10.13)$$

so that

$$\frac{\delta_S}{L} \oint \mathbf{u} \cdot d\mathbf{l} = O\left(A\left(\frac{\delta_S}{\delta_I}\right)\right). \qquad (5.10.14)$$

This will give rise to an O(1) contribution to the dissipation of vorticity if

$$A = O\left(\frac{\delta_I}{\delta_S}\right) \gg 1. \tag{5.10.15}$$

It can be expected then that in the presence of O(1) wind stress a resonance with the Fofonoff free mode can occur producing an interior flow $O(\delta_I/\delta_S)$—i.e., greater by this factor than the O(1) flow predicted by Sverdrup theory. This resonance will occur only when $A_H$ is precisely zero, else the pattern of Figure 5.9.3 will emerge for nonzero $A_H$. These remarks are illustrated by the numerical experiments discussed in the following section.

## 5.11 Numerical Experiments

In addition to the requirement, for geostrophy, that the relative vorticity be small compared to the planetary vorticity (i.e., $U/fL \ll 1$), the Sverdrup theory and the linear theories of western boundary currents require the more stringent condition that the *gradient* of the relative vorticity be small compared to the gradient of the planetary vorticity, i.e., $U/\beta_0 L^2 \ll 1$. Otherwise the nonlinear advection of vorticity becomes an important factor in the balance of vorticity. In the observed western boundary currents the relative-vorticity gradient is of the same order as $\beta_0$, so that nonlinear effects are important. However, as the discussion of Sections 5.7 and 5.8 shows, friction must be included in the total circulation problem to achieve a steady solution. This leads to a mathematical and physical problem of great difficulty, and although the analytical arguments given above indicate the qualitative nature of the circulation to be expected, explicit solutions require the use of computer-generated numerical solutions. It is important to bear in mind that like analytical treatments, numerical calculations have their own limitations determined by considerations of scale resolution, economy, and the nature of the generated information. The last consideration is particularly important. By their very nature numerical solutions present on each occasion a single realization of a dynamical situation rather than general relationships between dynamical variables. This is completely analogous to the information received in a laboratory experiment. In each case the results require that additional information be brought to bear for useful physical interpretation of the results. When numerical calculations take as their starting point the relevant equations of motion rather than simply evaluating an analytically derived formula, the calculation is usually termed a *numerical experiment*. It is beyond the scope of this book to describe the techniques of numerical experimentation and the accompanying numerical analysis except to note that in most cases the relevant partial differential equations are solved by representing derivatives by finite-difference approximations of varying degrees of accuracy. These finite-difference approximations require the repetitive evaluation of the dynamic variables on a spatial grid whose spacing between the points must be fine enough to resolve the fields properly. The grid cannot be too fine, however, for then the computation burden becomes excessive, both technologically and economically. There have been several important numerical experiments which have generated solutions of the vorticity equation (5.2.22), and the results of these experiments follow.

Bryan (1963) integrated the vorticity equation for a rectangular, flat-bottomed ocean in the absence of bottom friction, i.e., he set $r = 0$ and $\eta_B = 0$ in (5.2.22). The interior velocity was scaled as in (5.2.19), and the calculations were done for the most part with a fixed value of

$$\beta^{-1} = \frac{U}{\beta_0 L^2} = 1.28 \times 10^{-3}, \tag{5.11.1}$$

while the Reynolds number

$$\mathrm{Re} = \frac{UL}{A_H} \tag{5.11.2}$$

was varied from 5 to 60. Beyond $\mathrm{Re} = 60$, only unsteady solutions were obtained; they are not discussed here. Since

$$\frac{\delta_I}{\delta_M} = \left(\frac{\mathrm{Re}}{\beta^{1/2}}\right)^{1/3}, \tag{5.11.3}$$

the experiments ranged over values of $\delta_I/\delta_M$ from about 0.56 at $\mathrm{Re} = 5$, to 1.29 for $\mathrm{Re} = 60$. In the latter limit the boundary layer is certainly strongly nonlinear, but friction is nearly as important. Indeed, the limit $\delta_M/\delta_I \ll 1$ may be largely irrelevant for the steady problem, for the experiments indicate that in that limit the steady solutions are undoubtedly unstable and unrealizable in practice.

Figure 5.11.1 shows Bryan's calculated streamline patterns for four different Reynolds numbers, i.e., four values of $\delta_I/\delta_M$. The applied wind stress had the form

$$\mathrm{curl}\ \tau = -\sin \pi y. \tag{5.11.4}$$

Figure 5.11.1(a) shows the computed pattern for $\mathrm{Re} = 5$, which is nearly indistinguishable from the pattern predicted by Munk's linear model, i.e., by (5.4.40). As Re increases the effects of nonlinearity become increasingly evident, until with $\mathrm{Re} = 60$ the pattern achieves the general nature of the flow discussed in Section 5.7. The northern part of the basin is the site of an undulating stream field of decaying Rossby waves in which the relative vorticity is dissipated by horizontal diffusion as described in Section 5.9. The rough estimate (5.9.19) of the number of undulations in the stream pattern predicts $N \approx 2$ for $\mathrm{Re} = 60$ and is in good agreement with the pattern of 5.11.1(d). Figure 5.11.2 shows the stream-function field as a function of $x$ in the northern part of the basin. The figures show an increase in $\psi$ above the linear solution as Re increases, reflecting the increased transport required to compensate for the countercurrents produced by nonlinearity. Munk's

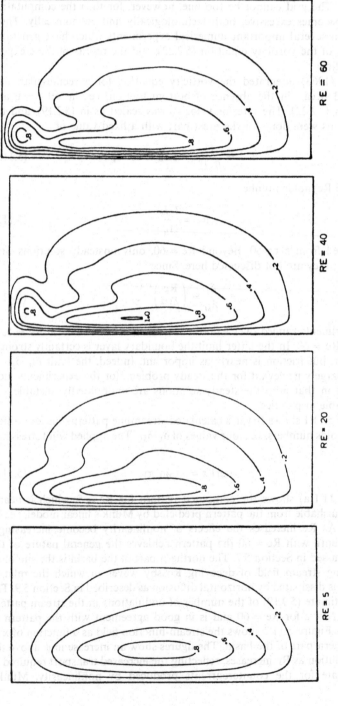

**Figure 5.11.1** Streamlines of the steady flow pattern. (a) Re = 5, (b) Re = 20, (c) Re = 40, (d) Re = 50; each calculation done for $\beta^{-1} = 1.28 \times 10^{-3}$ (Bryan 1963).

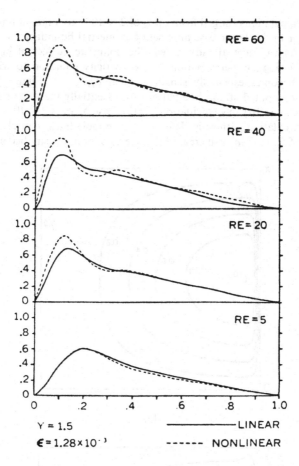

Figure 5.11.2 The profile with longitude of the stream function for the four Reynolds numbers of (5.11.1) shown as dashed curves. The solid curve is Munk's linear solution (Bryan 1963).

linear model also possessed a slight frictionally driven countercurrent and hence an enhancement of the transport of the boundary current. In the present calculation the enhancement is produced by the standing Rossby wave driven by the inertial–viscous balance in the northern part of the basin.

A very interesting numerical experiment was carried out by Veronis (1966). He set $A_H = 0$ and so relaxed the condition of no slip on the boundary of a rectangular basin and considered only bottom friction as a dissipation mechanism. In this case, in which the ocean again had a flat bottom, the relevant measure of nonlinearity in the western boundary current region is

$$\frac{\delta_I}{\delta_S} = \frac{\beta^{1/2}}{r}. \qquad (5.11.5)$$

Veronis's calculations ranged from $\delta_I/\delta_S$ of $2 \times 10^{-2}$ to a maximum of 8. It is clear that in the latter case an essentially inertial boundary current was achieved. As we have already noted, the complete neglect of horizontal friction is a singular perturbation of the vorticity equation to which the dynamics becomes especially sensitive when $\delta_I/\delta_S \gg 1$. Figure 5.11.3(a) shows the computed streamline pattern for essentially the same pattern of wind stress as in Bryan's calculation. In this case $\delta_I/\delta_S$ is $2 \times 10^{-2}$ and the resulting circulation pattern is identical to Stommel's linear model discussed in Section 5.5. As $\delta_I/\delta_S$ is increased the effects of nonlinearity become more

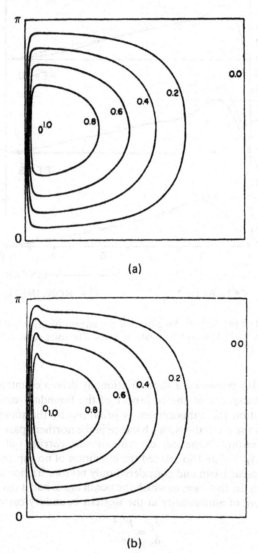

(a)

(b)

See legend on page 314.

apparent. A north–south asymmetry in the flow develops, and in Figure 5.11.3(c), for which $\delta_I/\delta_S$ is 1, the familiar undulation of the stream field in the northern portion of the basin appears. As $\delta_I/\delta_S$ increases beyond unity the dissipation considerations of Section 5.9 become increasingly significant. Figure 5.11.3(e), for example, shows the result of the calculation for $\delta_I/\delta_S = 4$. The undulations have *disappeared*. Instead, a strong northern boundary current has formed which shoots across the entire northern wall, so that fluid reemerges into the interior from the *east*. In the extreme limit, shown in Figure 5.11.3(f), for which $\delta_I/\delta_S$ is 8, the very western

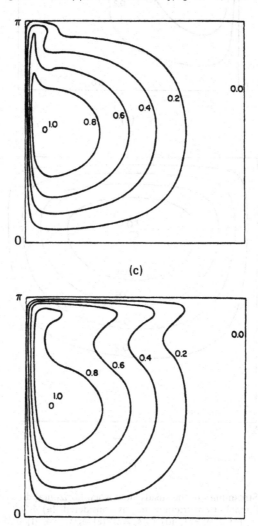

(c)

(d)

See legend on page 314.

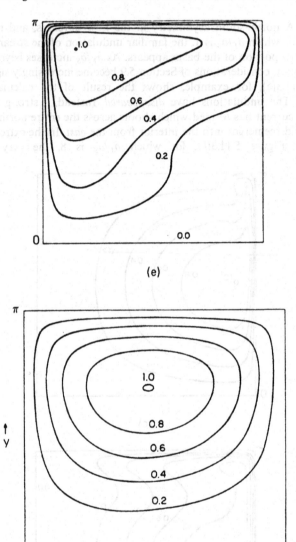

**Figure 5.11.3**   Streamlines of the steady flow when horizontal diffusion of momentum is ignored and bottom friction only is considered. (a) $\delta_I/\delta_s = 2 \times 10^{-2}$, (b) $\delta_I/\delta_s = 6 \times 10^{-2}$, (c) $\delta_I/\delta_s = 1$, (d) $\delta_I/\delta_s = 2$, (e) $\delta_I/\delta_s = 4$, (f) $\delta_I/\delta = 8$ (Veronis 1966). Contours of $\psi$ are shown spaced in increments of 0.2 from $\psi/\psi_{max} = 0$ to $\psi/\psi_{max} = 1.0$. (a) $\psi_{max} = 1.841$, (b) $\psi_{max} = 1.836$, (c) $\psi_{max} = 1.829$, (d) $\psi_{max} = 1.797$, (e) $\psi_{max} = 3.164$, (f) $\psi_{max} = 12.81$.

intensification itself of the circulation is entirely absent, and the pattern is strikingly similar to the Fofonoff free mode of Section 5.10. *The absence of lateral friction has led to resonance with the Fofonoff mode.* In the linear solution of Figure 5.11.3(a) the maximum value of $\psi$ is 1.84, while in the resonant case the maximum value of $\psi$ has jumped by about an order of magnitude to 12.81. This is consistent with the estimate (5.10.15), which shows that in the absence of lateral friction the interior stream function, rather than being given by the Sverdrup theory, is $O(\delta_I/\delta_S)$ when $\delta_I \gg \delta_S$. This would lead to a prediction of $(1.84) \times 8 \sim 14.7$ for the amplitude of the Fofonoff mode, in fair agreement with the calculations. These very important qualitative differences between Bryan's and Veronis's calculations strikingly illustrate the results of the singular perturbation involved in ignoring lateral friction.

## 5.12 Ekman Upwelling Circulations

The theory of the wind-driven circulation developed in this chapter has made explicit use of the $\beta$-plane approximation. In particular the vorticity equation was simplified throughout by the explicit use of the condition (5.2.12), i.e.,

$$\frac{\beta_0 L}{f_0} < O(1). \tag{5.12.1}$$

Among other implications of (5.12.1), this condition allows us to ignore to $O(\beta_0 L/f_0)$ the Ekman-layer transport in comparison with the interior geostrophic transport as shown by (5.2.24), so that the total transport is approximated by its geostrophic component. Yet $\beta_0 L/f_0$, while small—e.g., $O(10^{-1})$ for most oceanic gyres—is not completely negligible. Indeed, the only justification for neglecting terms of this order while retaining the generally smaller terms of $O(\delta_I^2/L^2, \delta_M^3/L^3, \delta_S/L)$ in the vorticity equation is that these latter terms, while utterly negligible in the interior, become $O(1)$ in the boundary-layer regions where the scale of motion $L$ shrinks to either $\delta_I, \delta_M$, or $\delta_S$.

In this section the $O(\beta L/f_0)$ circulation produced by the Ekman-layer transport is reexamined. More than a desire for a quantitative improvement of the circulation theory is involved. More fundamental issues are at stake. The transport in the Ekman layer is determined by the local value of the wind stress and is perpendicular to it. What happens to this Ekman-layer flux, for example, when it impinges on the ocean boundary? The geostrophic flow is driven by the divergence of the Ekman flux. What happens when there is a nondivergent Ekman flux? In such cases is the flow limited to the Ekman layer? Are western boundary currents required? Must the geostrophic velocity in the interior still vanish on an oceanic eastern boundary? These fundamental questions are beyond the capacity of the simple model (5.2.22) to answer, and yet a generalization of that model, unless carefully restricted, will be intractable. To answer the fundamental questions asked above we shall consider a very simple flat-bottom ocean model in which the nonlinear terms are ignored *a priori*. It will be apparent from their simplicity that the answers we derive are more general than the model.

The dimensional equations of motion for the homogeneous layer are given by (4.3.1a,b,c). If the Coriolis parameter is linearized about the central latitude $\theta_0$, then

$$f = f_0\left(1 + \frac{\beta_0}{f_0} y_*\right) = f_0 \tilde{f} \tag{5.12.2}$$

where

$$f_0 = 2\Omega \sin \theta_0,$$

$$\beta_0 = \frac{2\Omega \cos \theta_0}{r_0},$$

$$\tilde{f} = 1 + \frac{\beta_0 y_*}{f_0},$$

where $r_0$ is the earth's radius and $y_*$ is the *dimensional* northward coordinate. If $U$ as given by (5.2.19) is used to scale horizontal velocities while $(D/L)U$, $L$, $D$, $\rho L f_0 U$, and $\tau_0$ are used to scale the vertical velocity, horizontal length, vertical length, horizontally variable pressure, and wind stress, the linearized equations of motion in nondimensional form follow directly from (4.3.1a,b,c) as

$$-\tilde{f}v = -\frac{\partial p}{\partial x} + \frac{E_V}{2}\frac{\partial^2 u}{\partial z^2} + \frac{E_H}{2}\left(\frac{\partial^2 u}{\partial x^2} + \frac{\partial^2 u}{\partial y^2}\right), \tag{5.12.3a}$$

$$\tilde{f}u = -\frac{\partial p}{\partial y} + \frac{E_V}{2}\frac{\partial^2 v}{\partial z^2} + \frac{E_H}{2}\left(\frac{\partial^2 v}{\partial x^2} + \frac{\partial^2 v}{\partial y^2}\right), \tag{5.12.3b}$$

$$0 = -\frac{\partial p}{\partial z} + \frac{\delta^2}{2}\left[\frac{E_V}{2}\frac{\partial^2 w}{\partial z^2} + \frac{E_H}{2}\left(\frac{\partial^2 w}{\partial x^2} + \frac{\partial^2 w}{\partial y^2}\right)\right], \tag{5.12.3c}$$

while the continuity equation remains

$$\frac{\partial u}{\partial x} + \frac{\partial v}{\partial y} + \frac{\partial w}{\partial z} = 0. \tag{5.12.4}$$

On the upper surface at $z = 1$, the presence of the wind stress requires the existence of an Ekman layer of thickness $\delta_E$, where

$$\frac{\delta_E}{D} = E_V^{1/2} \tag{5.12.5}$$

as discussed in Section 4.10. The action of the stress produces an Ekman flux which in *nondimensional* units is

$$\mathbf{M}_E = \left(\frac{\beta_0 L}{f_0}\right)\frac{\boldsymbol{\tau} \times \mathbf{k}}{\tilde{f}} \tag{5.12.6}$$

where **k** is a unit vector in the $z$-direction. Since

$$\nabla \cdot \mathbf{M}_E = \frac{\beta_0 L}{f_0} \text{ curl } \frac{\tau}{\tilde{f}} \qquad (5.12.7)$$

the vertical velocity pumped into the upper Ekman layer from the interior is

$$w(x, y, 1) = \frac{\beta_0 L}{f_0} \text{ curl } \frac{\tau}{\tilde{f}}. \qquad (5.12.8)$$

Now in the interior the terms in (5.12.3a,b,c) proportional to $E_H$ and $E_V$ are completely negligible, so that for the interior

$$\tilde{f}v_I = + \frac{\partial p_I}{\partial x}, \qquad (5.12.9a)$$

$$\tilde{f}u_I = - \frac{\partial p_I}{\partial y}, \qquad (5.12.9b)$$

$$0 = - \frac{\partial p_I}{\partial z}, \qquad (5.12.9c)$$

where the subscript $I$ labels the interior representation of the dynamic variables. Eliminating $p_I$ between (5.12.9a,b) yields

$$\left( \frac{\beta_0 L}{f_0} \right) v_I = \tilde{f} \frac{\partial w_I}{\partial z}. \qquad (5.12.10)$$

Since $u_I$ and $v_I$ are independent by $z$ by (5.12.9c), the integral of (5.12.10) yields

$$\frac{\beta_0 L}{f_0} v_I = \tilde{f}(w_I(x, y, 1) - w_I(x, y, 0)). \qquad (5.12.11)$$

Since, by (4.5.39),

$$w_I(x, y, 0) = \frac{E_V^{1/2}}{2} \left| \frac{\partial v_I}{\partial x} - \frac{\partial u_I}{\partial y} \right|, \qquad (5.12.12)$$

to $O(E_V^{1/2})$ (5.12.11) becomes

$$\boxed{v_I = \tilde{f} \text{ curl } \frac{\tau}{\tilde{f}}}, \qquad (5.12.13)$$

or

$$v_I = \text{curl } \tau + \frac{\tau^{(x)}}{\tilde{f}} \frac{\beta_0 L}{f_0}, \qquad (5.12.14)$$

where $\tau^{(x)}$ is the $x$-component of $\tau$. Note that (5.12.14) will reduce to (5.3.2) when terms of $O(\beta_0 L/f_0)$ are neglected. Since the total northward transport is the sum of the geostrophic plus Ekman transport,

$$M_s^{(y)} = v_I + M_E^{(y)} = \text{curl } \tau \qquad (5.12.15)$$

if (5.12.6) is used. $M_s^{(y)}$ is the total northward Sverdrup transport and is still given by the curl of the wind stress when terms of $O(\beta_0 L/f_0)$ are retained. From (5.12.9) and (5.12.13) it follows that

$$p_I = -\int_x^{X_E} \bar{f}^2 \operatorname{curl} \frac{\tau}{\bar{f}} \, dx' + P(y), \qquad (5.12.16)$$

where $P(y)$ is *arbitrary* and $X_E$ is the position of the eastern boundary. The eastward geostrophic velocity is therefore

$$u_I = \frac{1}{\bar{f}} \frac{\partial}{\partial y} \int_x^{X_E} \bar{f}^2 \operatorname{curl} \frac{\tau}{\bar{f}} \, dx' - \frac{1}{\bar{f}} \frac{\partial P(y)}{\partial y}. \qquad (5.12.17)$$

As before, the eastward *geostrophic* velocity contains an arbitrary function of latitude, i.e., a purely zonal flow along the geostrophic contours. Note however, that the Ekman transport in the $x$-direction,

$$M_E^{(x)} = \frac{\beta_0 L}{f_0} \frac{\tau^{(y)}}{\bar{f}} (x, y),$$

is completely determined in terms of the local value of the $y$-component of the wind stress. In order to determine $P(y)$, the fate of the Ekman flux impinging on the eastern boundary must be studied. Consider Figure 5.12.1. The Ekman flux $M_E^{(x)}$ in the upper layer, of depth $E_V^{1/2}$, impinges on the boundary, where according to Ekman theory it will be nonzero if the wind stress along the coast, $\tau^{(y)}(X_E, y)$, is nonzero. Nevertheless the actual transport in the upper Ekman layer must vanish on $x = X_E$. Clearly in some small region near the coast, forces which are elsewhere negligible in the Ekman layer must here arise to invalidate (5.12.6). The only terms in (5.12.3a,b) that can increase in magnitude locally so as to alter the balance in

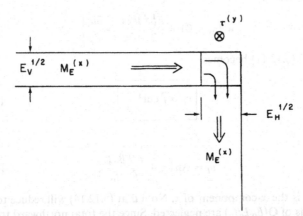

**Figure 5.12.1** A wind stress directed into the paper, as shown, drives an Ekman mass flux $\mathbf{M}_E$ towards the boundary, where it must descend. The depth (non-dimensional) of the Ekman layer is $E_V^{1/2}$, while the turning region has a width $E_H^{1/2}$.

the upper layer are the horizontal friction terms, and these will be of the same order in the Ekman layer as the vertical friction terms only if each $x$ derivative is of order $E_H^{-1/2}$. This leads us to examine the corner region of depth $E_V^{1/2}$ and width $E_H^{1/2}$. To do this we introduce the boundary-layer coordinates

$$\zeta = \frac{1-z}{E_V^{1/2}}$$

$$\xi = \frac{X_E - x}{E_H^{1/2}},$$

(5.12.18)

and write each of the dynamical fields as a sum of their interior values plus the correction in the corner layer—i.e., in the corner,

$$u = u_I(x, y) + u_c(\xi, y, \zeta)E_V^{-1/2},$$

$$v = v_I(x, y) + v_c(\xi, y, \zeta)E_V^{-1/2},$$

$$w = w_I(x, y) + w_c(\xi, y, \zeta)E_H^{-1/2},$$

$$p = p_I(x, y) + p_c(\xi, y, \zeta)\delta^2\left(\frac{E_V}{E_H}\right)^{1/2}.$$

(5.12.19)

The magnitudes of the correction functions are determined by the requirement that the horizontal *transport* in the corner be of the same order as the Ekman transport, i.e., $u \times E_V^{1/2} = O(1)$, from which it follows that $u$ and $v$ are $O(E_V^{-1/2})$ in the corner. The scaling for $w_c$ follows from mass conservation, i.e., $w \times E_H^{1/2}$ must be $O(1)$ if the transport impinging on the coast is to be turned downward in the corner, while the scaling for the correction to $p_I$ is determined by (5.12.3c) now that the scale of $w$ is determined. If (5.12.19) is substituted into (5.12.3) and only $O(1)$ terms are retained, we obtain equations for $u_c$, $v_c$, and $w_c$ after explicit use is made of the relations (5.12.9a,b,c) for the interior variables, i.e.,

$$-\tilde{f}v_c = \frac{1}{2}\left(\frac{\partial^2 u_c}{\partial \xi^2} + \frac{\partial^2 u_c}{\partial \zeta^2}\right),$$

(5.12.20a)

$$\tilde{f}u_c = \frac{1}{2}\left(\frac{\partial^2 v_c}{\partial \xi^2} + \frac{\partial^2 v_c}{\partial \zeta^2}\right),$$

(5.12.20b)

$$\frac{\partial w_c}{\partial \zeta} + \frac{\partial u_c}{\partial \xi} = 0.$$

(5.12.20c)

On the upper surface the fluid stress must match the applied stress as expressed by (4.10.2), which in dimensionless units becomes

$$\frac{\partial u_c}{\partial \zeta} = -2\frac{\beta_0 L}{f_0}\tau^{(x)},$$

(5.12.21a)

$$\frac{\partial v_c}{\partial \zeta} = -2\frac{\beta_0 L}{f_0}\tau^{(y)},$$

(5.12.21b)

while far beneath the corner stress region $\partial u_c/\partial \zeta$ and $\partial v_c/\partial \zeta$ must vanish. Integrating (5.12.20a,b,c) from $\zeta = 0$ to $\zeta = \infty$ yields

$$-\tilde{f} V_c = \frac{1}{2} \frac{\partial^2 U_c}{\partial \xi^2} - \frac{\beta_0 L}{f_0} \tau^{(x)}, \tag{5.12.22a}$$

$$\tilde{f} U_c = \frac{1}{2} \frac{\partial^2 V_c}{\partial \xi^2} - \frac{\beta_0 L}{f_0} \tau^{(y)}, \tag{5.12.22b}$$

where

$$\begin{bmatrix} U_c \\ V_c \end{bmatrix} = \int_0^\infty d\zeta \begin{bmatrix} u_c \\ v_c \end{bmatrix}. \tag{5.12.23}$$

$U_c$ and $V_c$ are the vertically integrated velocities in the corner region. Since in this region $dz = -d\zeta\, E_V^{1/2}$ while $u$ and $v$ are given at lowest order by $u_c E_V^{-1/2}$ and $v_c E_V^{-1/2}$, it follows that $U_c$ and $V_c$ are the horizontal *transports* in the upper corner region. The set (5.12.22a,b) are *ordinary* differential equations for the transports. Since $U_c$ and $V_c$ must vanish on $x = X_E$ (i.e., $\xi = 0$), by virtue of the normal flow and no-slip conditions, the solutions of (5.12.22a,b) are immediately determined as

$$U_c = -\frac{\beta_0 L}{f_0} \frac{\tau^{(y)}}{\tilde{f}} + \frac{\beta_0 L}{f_0} e^{-\tilde{f}^{1/2}\xi}$$
$$\times \left[ \frac{\tau^{(y)}}{\tilde{f}} \cos \tilde{f}^{1/2}\xi + \frac{-\tau^{(x)}}{\tilde{f}} \sin \tilde{f}^{1/2}\xi \right],$$

$$V_c = +\frac{\beta_0 L}{f_0} \frac{\tau^{(x)}}{\tilde{f}} + \frac{\beta_0 L}{f_0} e^{-\tilde{f}^{1/2}\xi}$$
$$\times \left[ -\frac{\tau^{(y)}}{\tilde{f}} \sin \tilde{f}^{1/2}\xi + \frac{-\tau^{(x)}}{\tilde{f}} \cos \tilde{f}^{1/2}\xi \right]. \tag{5.12.24}$$

Note that for large $\xi$ the transports $U_c$ and $V_c$ reduce, as required, to the Ekman transports in agreement with (5.12.6). The integral of (5.12.20c) yields the vertical velocity forced out of the corner and pumped into the fluid below, i.e.,

$$-w_c(\xi, y, \infty) = -\frac{\partial U_c}{\partial \xi}(\xi, y)$$

$$= \frac{\beta L}{f_0} \tilde{f}^{1/2} e^{-\tilde{f}^{1/2}\xi} \left[ \frac{\tau^{(y)}}{\tilde{f}} \{\cos \tilde{f}^{1/2}\,\xi + \sin \tilde{f}^{1/2}\xi\} \right.$$

$$\left. + \frac{\tau^{(x)}}{\tilde{f}} \{\cos \tilde{f}^{1/2}\xi - \sin \tilde{f}^{1/2}\xi\} \right]. \tag{5.12.25a}$$

The total mass pumped *downward* is

$$-\int_0^\infty w_c(\xi, y, \infty)\, d\xi = +\frac{\beta L}{f_0} \frac{\tau^{(y)}}{\tilde{f}}(x, y), \tag{5.12.25b}$$

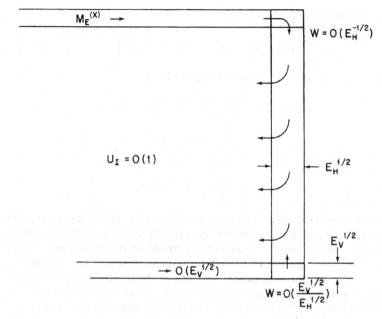

**Figure 5.12.2** A schematic of the vertical circulation at the eastern boundary. The flux in the upper Ekman layer descends in the side-wall layer and exits into the interior as a geostrophic zonal flow.

which is simply the eastward Ekman flux impinging on the coast. The corner flow turns the Ekman flux downwards into a coastal layer beneath the corner. In this coastal layer, shown in Figure 5.12.2, the dynamical variables will be rapidly varying functions of $x$, but will vary with $z$ on the O(1) depth scale appropriate for matching to the interior. Hence the dominant terms in (5.12.3a,b,c) will be

$$-\tilde{f}v_B = -\frac{\partial p_B}{\partial x} + \frac{E_H}{2}\frac{\partial^2 u_B}{\partial x^2}, \qquad (5.12.26a)$$

$$\tilde{f}u_B = -\frac{\partial p_B}{\partial y} + \frac{E_H}{2}\frac{\partial^2 v_B}{\partial x^2}, \qquad (5.12.26b)$$

$$0 = -\frac{\partial p_B}{\partial z} + \delta^2\frac{E_H}{2}\frac{\partial^2 w_B}{\partial x^2}, \qquad (5.12.26c)$$

where the subscript $B$ refers to the representation of the variables in the coastal layer, whose width must now be determined. Now the flow in the interior, into which the $B$ subscript variables must smoothly merge, is independent of $z$. Let us provisionally assume that $u_B$ and $v_B$ are also independent of $z$ and check after the fact the validity of this presumption. If $u_B$ and $v_B$ *are* independent of $z$, elimination of $p_B$ from (5.12.26a,b) yields

$$\frac{\beta_0 L}{f_0}v_B = \tilde{f}\frac{\partial w_B}{\partial z} + \frac{E_H}{2}\frac{\partial^2}{\partial x^2}\left(\frac{\partial v_B}{\partial x} - \frac{\partial u_B}{\partial y}\right), \qquad (5.12.27)$$

the vertical integral of which is

$$\frac{\beta_0 L}{f_0} v_B = \tilde{f}[w_B(x, y, 1) - w_B(x, y, 0)] + \frac{E_H}{2} \frac{\partial^2}{\partial x^2}\left(\frac{\partial v_B}{\partial x} - \frac{\partial u_B}{\partial y}\right). \quad (5.12.28)$$

At $z = 1$, $w_B$ is given by (5.12.19) and (5.12.25), i.e.,

$$w_B(x, y, 1) = E_H^{-1/2} \frac{\beta_0 L}{f_0} \tilde{f}^{1/2} \exp(-\tilde{f}^{1/2}\xi)$$

$$\times \left[ -\frac{\tau^{(y)}}{\tilde{f}} \{\cos \tilde{f}^{1/2}\xi + \sin \tilde{f}^{1/2}\xi\} \right.$$

$$\left. + \frac{\tau^{(x)}}{\tilde{f}} \{\sin \tilde{f}^{1/2}\xi - \cos \tilde{f}^{1/2}\xi\} \right]. \quad (5.12.29)$$

On the other hand, since the *interior* velocity is $O(1)$, the Ekman transport in the lower layer impinging on the coast is only $O(E_V^{1/2})$, so that the vertical velocity pumped into the coastal layer from the *lower* corner is $O(E_V^{1/2}/E_H^{1/2})$, as shown in Figure 5.12.2, so that it is inconsequential compared to $w_B(x, y, 1)$ and can be neglected. Further, in the coastal layer $u_B$ and $v_B$ are $O(1)$ so as to match to the interior velocities, so that $\partial u_B/\partial y$ can be neglected in comparison with $\partial v_B/\partial x$ in (5.12.28). That equation then becomes an ordinary differential equation for $v_B$, forced by the inhomogeneous term $w_B(x, y, 1)$.

Since the forcing has an $x$-scale of $O(E_H^{1/2})$, the solution for $v_B$ will also vary on that scale. Thus the ratio

$$\frac{v_B \beta_0 L/f_0}{(E_H/2)(\partial^3 v_B/\partial x^3)} = O(E_H^{1/2}) \ll 1, \quad (5.12.30)$$

so that (5.12.28) reduces to

$$\frac{\partial^3 v_B}{\partial \xi^3} = -E_H^{-3/2} \frac{\partial^3 v_B}{\partial x^3} = +\tilde{f}^{3/2} \frac{2\beta_0 L}{f_0}$$

$$\times \left[ -\frac{\tau^{(y)}}{\tilde{f}} \{\cos \tilde{f}^{1/2}\xi + \sin \tilde{f}^{1/2}\xi\} \right.$$

$$\left. + \frac{\tau^{(x)}}{\tilde{f}} \{\sin \tilde{f}^{1/2}\xi - \cos \tilde{f}^{1/2}\xi\} \right] e^{-\tilde{f}^{1/2}\xi}, \quad (5.12.31)$$

the first integral of which is

$$\frac{\partial^2 v_B}{\partial \xi^2} = 2\frac{\beta_0 L}{f_0}[+\tau^{(y)}\{\cos \tilde{f}^{1/2}\xi\} - \tau^{(x)}\{\sin \tilde{f}^{1/2}\xi\}]e^{-\tilde{f}^{1/2}\xi}, \quad (5.12.32)$$

since $\partial^2 v_B/\partial \xi^2$ must vanish for large $\xi$, where $v$ is a function of $x$ and not $\xi$. But from (5.12.26b),

$$u_B = -\frac{\partial p_B}{\partial y} + \frac{1}{2}\frac{\partial^2 v_B}{\partial \xi^2}. \quad (5.12.33)$$

Since $u_B$ must vanish on $x = X_E$ (i.e., $\xi = 0$),

$$\left(\frac{\partial p_B}{\partial y}\right)_{\xi=0} = \frac{\beta_0 L}{f_0} \tau^{(y)}(X_E, y).$$  (5.12.34)

However, $u_B$, $v_B$, and $p_B$ are still $O(1)$, so that (5.12.26a) implies that to $O(E_H^{1/2})$

$$\frac{\partial p_B}{\partial \xi} = 0.$$  (5.12.35)

Therefore the pressure in the boundary layer is, by the matching principle, the pressure of the geostrophic flow immediately outside the coastal boundary layer. Now on the oceanic *eastern* boundary the results of earlier sections of this chapter have demonstrated that no *geostrophic* boundary layer can exist which corrects the *interior* on-shore zonal flow by an $O(1)$ amount. Hence the zonal flow immediately outside the coastal layer is, to $O(1)$, the *interior* geostrophic flow. In turn that implies that with (5.12.35),

$$\left(\frac{\partial p_B}{\partial y}\right)_{\xi=0} = \left(\frac{\partial p_B}{\partial y}\right)_{\xi=\infty} = \frac{\partial p_I}{\partial y}(X_E, y) = \frac{\beta_0 L}{f_0} \tau^{(y)}(X_E, y).$$  (5.12.36)

This condition then completely determines the arbitrary zonal flow in (5.12.17), since from (5.12.16)

$$\frac{1}{\tilde{f}} \frac{\partial p_I}{\partial y}(X_E, y) = \frac{\partial P}{\partial y} \frac{1}{\tilde{f}} = \frac{\beta_0 L}{f_0} \frac{\tau^{(y)}}{\tilde{f}}(X_E, y),$$  (5.12.37)

or

$$\boxed{u_I = \frac{1}{\tilde{f}} \frac{\partial}{\partial y} \int_x^{X_E} \tilde{f}^2 \, \text{curl} \, \frac{\tau}{\tilde{f}} \, dx' - \frac{\beta_0 L}{f_0} \frac{\tau^{(y)}}{\tilde{f}}(X_E, y).}$$  (5.12.38)

Thus on $x = X_E$ the geostrophic flow is zero *only* to $O(1)$, for at $O(\beta_0 L/f_0)$ a geostrophic flow must exist given by

$$u_I(X_E, y) = -\frac{\beta_0 L}{f_0} \frac{\tau^{(y)}}{\tilde{f}}(X_E, y) = -M_E^{(y)}(X_E, y).$$  (5.12.39)

This geostrophic mass flux is equal and *opposite* to the Ekman-layer mass flux in the surface layer at the coast, so that the *total* transport—i.e., the sum of the geostrophic transport and the Ekman transport—vanishes on $x = X_E$. After a little manipulation (5.12.38) may be rewritten as

$$u_I = \frac{\partial}{\partial y} \int_x^{X_E} \text{curl} \, \tau \, dx' - \frac{\beta_0 L}{f_0} \frac{\tau^{(y)}(x, y)}{\tilde{f}}.$$  (5.12.40)

On the oceanic western boundary, the analysis of the upper corner region is identical with the analysis for the eastern side. Again the Ekman mass flux impinging on the coast plunges downward in a layer of width $E_H^{1/2}$ and enters

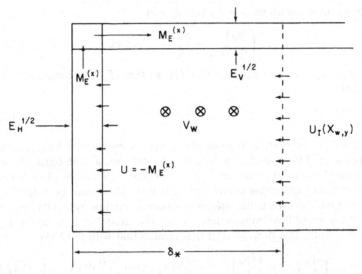

**Figure 5.12.3** A schematic of the circulation near a western boundary. A mass flux $M_E^{(x)}$ is driven off shore by the wind and is replenished by vertical motion in the upwelling layer of width $E_H^{1/2}$. The interior flow is $U_I(X_W, y)$. The discrepancy between $U_I(x_W, y)$ and $M_E^{(x)}$ represents the transport that must be deflected laterally and flow northward in the western boundary current of width $\delta_*$.

the geostrophic region, as shown in Figure 5.12.3. Consider the case, as in Figure 5.12.3, where the Ekman flux at the western boundary is eastward. The upwelling of this flux in the corner requires a geostrophic flux of fluid to enter the upwelling layer at $x = X_W$, so that the geostrophic velocity at $x = X_W$ is

$$u(X_W, y) = -M_E^{(x)}(X_W, y) = -\frac{\beta_0 L}{f_0} \frac{\tau^{(y)}(X_W, y)}{\tilde{f}}. \qquad (5.12.41)$$

However, (5.12.40) demonstrates that at $x = X_W$

$$u_I(X_W, y) = \frac{\partial}{\partial y} \int_{X_W}^{X_E} \operatorname{curl} \tau \, dx' - \frac{\beta_0 L}{f_0} \frac{\tau^{(y)}}{\tilde{f}}(X_W, y), \qquad (5.12.42)$$

so that a discrepancy exists in the magnitude of the geostrophic flow at $X_W$ between its *interior* value and the value it must assume upon entering the upwelling layer, by an amount

$$u_I(X_W, y) - u(X_W, y) = \frac{\partial}{\partial y} \int_{X_W}^{X_E} \operatorname{curl} \tau \, dx'. \qquad (5.12.43)$$

This mass flux must therefore be absorbed by the western Gulf Stream–like $\beta$-layer and turned northward. Whether the layer is the Munk, Stommel, or inertial layer is basically irrelevant. The $\beta$-layer is horizontally nondivergent, so that its eastward and northward velocities satisfy

$$\frac{\partial v_W}{\partial y} = -\frac{\partial u_W}{\partial x}, \tag{5.12.44}$$

where the $W$ subscript refers to the flow in the western boundary layer. Integration across the layer yields

$$\frac{\partial}{\partial y} \int_{X_W}^{X_W + \delta_*} v_W \, dx' = u_W(X_W, y) - u_W(X_W + \delta_*, y)$$

$$= -\frac{\partial}{\partial y} \int_{X_W}^{X_E} \text{curl } \tau \, dx' \tag{5.12.45}$$

or

$$\int_{X_W}^{X_W + \delta_*} v_W \, dx' = -\int_{X_W}^{X_E} \text{curl } \tau \, dx', \tag{5.12.46}$$

where the constant of integration is set equal to zero, so that the *total* southward transport (i.e., the sum of the Sverdrup and boundary layer transports) balances. Note that the transport of the western boundary current is identical to the earlier results; it depends only on the curl of the wind stress and balances the total Sverdrup transport of the interior as given by (5.12.15).

Before proceeding we must briefly return to the derivation and check whether $u_B$ and $v_B$ are indeed consistently independent of $z$. Eliminating $p_B$ between (5.12.26a) and (5.12.26c) yields

$$\bar{f} \frac{\partial v_B}{\partial z} + \frac{E_H}{2} \frac{\partial^2}{\partial x^2} \frac{\partial u_B}{\partial z} = \delta^2 \frac{E_H}{2} \frac{\partial^3 w_B}{\partial x^3}. \tag{5.12.47}$$

Since $w_B$ is $O(E_H^{-1/2})$, while $\partial/\partial x$ is also $O(E_H^{-1/2})$ in the upwelling coastal layer, $v_B$ and $u_B$, which are $O(1)$, may be consistently considered independent of $z$ if

$$\delta^2 \ll E_H,$$

or in dimensional units, if

$$\frac{D}{\delta_u} \ll 1, \tag{5.12.48}$$

where $\delta_u$ is the width of the coastal upwelling layer:

$$\delta_u = LE_H^{1/2} = \left(\frac{2A_H}{f_0}\right)^{1/2}. \tag{5.12.49}$$

The condition (5.12.48) is simply the familiar condition that the validity of the hydrostatic approximation is assured if the ratio of the vertical to horizontal scales of motion is small.

To illustrate these results, consider the following examples. First suppose that the wind stress is given by

$$\tau = \tau^{(y)}(y)\mathbf{j}, \qquad \tau^{(y)}(y) > 0, \qquad (5.12.50)$$

i.e., is strictly northward and independent of longitude. Then both curl $\tau$ and curl $\tau/f$ are identically zero. There is no net transport, i.e., the sum of the geostrophic and Ekman transports in the interior is identically zero and there is no vertical motion in the interior: the Ekman transport is non-divergent. An Ekman flux $(\beta L/f_0)\tau^{(y)}/f$ is directed eastward in the upper Ekman layer. It impinges on the eastern boundary where the flow downwells (i.e., sinks) and returns westward in the geostrophic interior, where $u_I$, by (5.12.40), is simply $-(\beta L/f_0)\tau^{(y)}/f$. This geostrophic flow is completely absorbed into the upwelling layer on the western boundary, where fluid elements rise and enter the Ekman layer at $x = X_W$. There are no $\beta$-layers, since curl $\tau$ is zero and the entire flux circuit is confined to the $x, z$ plane. There is no *westward intensification*, for that would require curl $\tau \neq 0$.

As a second example, consider the wind stress

$$\tau = A(x)\mathbf{j}\mathbf{i}, \qquad A > 0. \qquad (5.12.51)$$

In this case the wind stress is zonal and $v_I$, from (5.12.13), is *zero*. The Sverdrup transport, by (5.12.15), is given by the northward Ekman transport,

$$M_s^{(y)} = M_E^{(y)} = \text{curl }\tau = -\frac{\tau^{(x)}(x)}{f}\frac{\beta_0 L}{f_0} = -A(x)\frac{\beta_0 L}{f_0}. \qquad (5.12.52)$$

Thus in the region outside the western boundary current the meridional transport is confined to the upper Ekman layer. Yet by (5.12.46) this mass flux will be balanced by a western geostrophic boundary current extending to the ocean bottom.

Since $\beta_0 L/f_0$ is small, these circulations are generally weaker than the circulation patterns driven by the wind stress curl and discussed in Sections 5.3–5.10. However, consideration of these circulations and the upwelling layers is required in order to understand how the mass circuit of the fluid, initiated by the Ekman flux, is closed. It is especially important to keep in mind that the geostrophic velocity normal to the boundaries need not vanish if upwelling or downwelling occurs, but that the geostrophic velocity normal to the boundary must balance the on-shore Ekman transports so that the *total* on-shore transport vanishes at the boundary.

## 5.13 The Effect of Bottom Topography

When the ocean bottom is not flat, (5.9.1) shows that the planetary-vorticity gradient must be supplemented by the ambient potential vorticity associated with bottom variations, $\eta_B$. Geostrophic contours, i.e., lines of constant ambient potential vorticity, which previously were the lines of constant $y$, are now distorted to coincide with the lines

$$B(x, y) = \beta y + \eta_B = \text{constant}, \qquad (5.13.1)$$

which are the lines of constant $f/H$, where $H$ is the variable ambient depth of the fluid.

The Sverdrup interior equation now becomes, from (5.2.22),

$$\frac{\partial \psi}{\partial x}\left(\beta + \frac{\partial \eta_B}{\partial y}\right) - \frac{\partial \psi}{\partial y}\frac{\partial \eta_B}{\partial x} = \beta \text{ curl } \tau, \qquad (5.13.2)$$

or

$$\nabla \psi \cdot [\nabla B \times \mathbf{k}] = \beta \text{ curl } \tau. \qquad (5.13.3)$$

Thus it is the flow across the isolines of $B$, the ambient potential vorticity, that is determined by the wind-stress curl.

Introduce the new, natural coordinates

$$\theta = \theta(x, y),$$
$$s = s(x, y), \qquad (5.13.4)$$

or equivalently

$$x = x(\theta, s),$$
$$y = y(\theta, s), \qquad (5.13.5)$$

where

$$\left(\frac{\partial x}{\partial \theta}\right)_s = 1 + \beta^{-1}\frac{\partial \eta_B}{\partial y} = \beta^{-1}\frac{\partial B}{\partial y},$$
$$\left(\frac{\partial y}{\partial \theta}\right)_s = -\beta^{-1}\frac{\partial \eta_B}{\partial x} = -\beta^{-1}\frac{\partial B}{\partial x}. \qquad (5.13.6)$$

Thus

$$\left(\frac{dy}{dx}\right)_s = \left(\frac{\partial y}{\partial \theta}\right)_s \bigg/ \left(\frac{\partial x}{\partial \theta}\right)_s = -\frac{\partial B}{\partial x}\bigg/\frac{\partial B}{\partial y} = \left(\frac{\partial y}{\partial x}\right)_B, \qquad (5.13.7)$$

so that lines of constant $s$ in the $x$, $y$ plane coincide, as shown in Figure 5.13.1, with the isolines of $B$, while (5.13.6) determines the relationship between $x$, $y$, and $\theta$. In terms of $\theta$ and $s$, the Sverdrup relation is now

$$\left(\frac{\partial \psi}{\partial \theta}\right)_s = \text{curl } \tau \equiv T(\theta, s), \qquad (5.13.8)$$

since curl $\tau$ can be written in terms of $\theta$ and $s$ as well as $x$ and $y$. Integrating (5.13.8) yields

$$\psi(\theta, s) = \int^{\theta} T(\theta', s)\, d\theta' + \psi_0(s), \qquad (5.13.9)$$

where $\psi_0(s)$ represents an undetermined geostrophic flow along the isolines

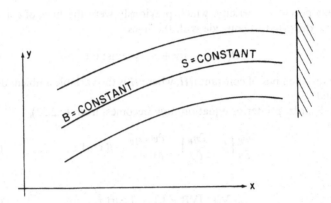

**Figure 5.13.1**   Isolines of constant $f/H$ are given by $B =$ constant. The natural coordinate $s$ is chosen so as to be constant on the isolines of $B$, which are assumed to intersect the oceanic eastern boundary.

of $B$, i.e., the isolines of $f/H$. If $\eta_B$ is zero, then $x$ corresponds to $\theta$ and $s$ to $y$, and the flat-bottom ocean model is recovered. To determine $\psi_0(s)$, the boundary layer on $x = X_E$, where the geostrophic contours intersect the eastern boundary, must be examined. Let us restrict our attention to the simplest such model, i.e., Stommel's, where the bottom friction plays the dominant role in the boundary-layer dynamics. Retaining only the bottom-friction terms in (5.2.22) leads to

$$\frac{\partial \psi}{\partial x}\left(1 + \frac{1}{\beta}\frac{\partial \eta_B}{\partial y}\right) - \frac{1}{\beta}\frac{\partial \psi}{\partial y}\frac{\partial \eta_B}{\partial x} = \text{curl } \tau - \frac{\delta_S}{L}\nabla^2\psi. \qquad (5.13.10)$$

On the oceanic eastern boundary let

$$\psi = \psi_B = \psi_I(x, y) + \tilde{\psi}_B(\xi, y), \qquad (5.13.11)$$

where

$$\xi = \frac{L(X_E - x)}{\delta_S}.$$

Then to $O(\delta_S/L)$,

$$\frac{\partial^2 \tilde{\psi}_B}{\partial \xi^2} - \frac{\partial \tilde{\psi}_B}{\partial \xi}\frac{\partial B}{\partial y}(X_E, y) = 0, \qquad (5.13.12)$$

so that a boundary layer on the eastern side of the ocean is possible only if $\partial B/\partial y < 0$, i.e., only if the topographic slope is sufficiently strong to reverse the sign of the ambient potential-vorticity gradient from its value in the flat-bottom case. Otherwise $\psi_I$ must satisfy the no-normal-flow condition on the eastern boundary. A similar analysis on the western boundary shows that if $\partial B/\partial y > 0$, a western boundary current is, as before, possible and westward intensification occurs. Thus if $\partial B/\partial y(X_E, y) > 0$,

$$\psi_I = \int_{\theta_E(X_E, s)}^{\theta} T(\theta', s)\, d\theta', \tag{5.13.13}$$

where $\theta_E(X_E, s)$ is the value of $\theta$ on each isoline of $B$ which intersects the boundary at $x = X_E$.

In (5.13.9)

$$\int T(\theta', \zeta)\, d\theta' \rightarrow \int^{\theta} T(\theta', \zeta)\, d\theta'.$$

Of course, the possibility exists that the topography may distort the isolines of $f/H$ sufficiently so that rather than intersecting the eastern boundary, they intersect the northern or southern boundary of the basin. Then we must investigate the possibility of boundary currents on these boundaries.

On an oceanic northern boundary a Stommel boundary layer may exist only if $\partial \eta_B/\partial x > 0$. To see this let

$$\psi = \psi_B = \psi_I(x, y) + \tilde{\psi}_B(x, \eta) \tag{5.13.14}$$

near the northern boundary at $y = 1$, where

$$\eta = \frac{L(1 - y)}{\delta_S}. \tag{5.13.15}$$

Then to $O(\delta_S/L)$ (5.13.10) becomes

$$\frac{\partial^2 \tilde{\psi}_B}{\partial \eta^2} + \frac{\partial \tilde{\psi}_B}{\partial \eta} \frac{\partial B}{\partial x}(1, x) = 0, \tag{5.13.16}$$

so that a northern Stommel-like layer is possible only if $\partial B/\partial x > 0$. Similarly, a Stommel-like layer is possible on the southern only if $\partial B/\partial x < 0$.

As an example consider the simple case where $\eta_B = \alpha x$ so that the depth of the gyre becomes shallow to the east. The Sverdrup interior equation (5.13.2) becomes

$$\frac{\partial \psi}{\partial x} - \frac{\alpha}{\beta} \frac{\partial \psi}{\partial y} = \text{curl } \tau. \tag{5.13.17}$$

The lines of constant $B$ satisfy

$$y = \frac{\alpha}{\beta}(X_E - x) + \text{constant}, \tag{5.13.18}$$

where $X_E$ is the coordinate of the eastern boundary. The contours of constant $B$ are shown in Figure 5.13.2. There is a critical contour given by

$$y = \frac{\alpha}{\beta}(X_E - x),$$

which interacts the eastern wall at $y = 0$. The region south of this critical line

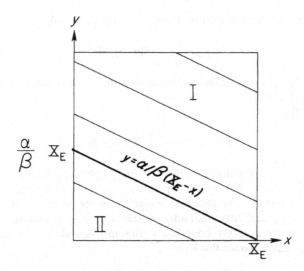

**Figure 5.13.2**   The contours of constant $B$ for the case $\eta_B = \alpha x$. The heavy line is the $B$ isopleth which intersects $x = x_E$ at $y = 0$. This isopleth intersects the western boundary at $y = \alpha x_E/\beta$. The flow in regions I and II are determined by boundary conditions at the eastern and southern boundaries, respectively.

is spanned by isolines of $B$ which strike the southern rather than the eastern boundary.

From (5.13.9) it is clear that the constant of integration required to complete the Sverdrup solution is determined on that boundary, struck by an isoline of $B$, which cannot support a boundary layer. Thus, in the southern region shown in Figure 5.13.2, this boundary will be the line $y = 0$ and will determine the solution in the entire region labeled II, since no boundary layer is possible on $y = 0$ for $\alpha > 0$. In region I the isolines of $B$ strike the eastern boundary on which again, no boundary layer is possible. This shift of the intersection of the $B$ contours from one boundary to another can have an important effect on the solution.

Suppose that

$$\text{curl } \tau = -\sin \pi y. \tag{5.13.19}$$

Then in both region I and II the solution may be written:

$$\psi = \frac{\beta}{\alpha\pi} \cos \pi y + \Phi\left(y + \frac{\alpha x}{\beta}\right), \tag{5.13.20}$$

where $\Phi$ is an arbitrary function of its argument. In region I, no boundary layer is allowed on $x = X_E$ so that $\psi$ must vanish there. Thus

$$\Phi\left(y + \frac{\alpha X_E}{\beta}\right) = -\frac{\beta}{\alpha\pi} \cos \pi y, \tag{5.13.21}$$

which implies that

$$\Phi\left(y + \frac{\alpha x}{\beta}\right) = -\frac{\beta}{\alpha\pi}\cos\pi\left[y + \frac{\alpha}{\beta}(x - X_E)\right], \qquad (5.13.22)$$

so that in region I

$$\psi = -\frac{\beta}{\alpha\pi}\left\{\cos\pi y - \cos\pi\left[y + \frac{\alpha}{\beta}(x - X_E)\right]\right\}. \qquad (5.13.23)$$

In the limit $\alpha/\beta \to 0$, region I occupies the entire basin and in that limit

$$\psi \to -(x - X_E)\sin\pi y$$

in agreement with (5.4.20).

In region II on the other hand, the condition that $\psi$ vanish on $y = 0$ implies that

$$\Phi\left(\frac{\alpha x}{\beta}\right) = -\frac{\beta}{\alpha\pi}, \qquad (5.13.24)$$

i.e., that the function $\Phi$ must be a constant since, in distinction ot the case for region I, the lines of constant $B$ here intersect a boundary upon which curl $\tau$ is constant. Thus in region II

$$\psi = \frac{\beta}{\alpha\pi}(1 - \cos\pi y). \qquad (5.13.25)$$

The reader may verify that on the boundary between regions I and II, the stream function $\psi$ and each of the velocities are continuous.

The striking feature of the solution is that in region II the transport is strictly zonal. The meridional transport is identically zero in all of zone II and this will occur in all such regions, isolated from the eastern boundary if curl $\tau/(\partial\eta_B/\partial x)$ is independent of $x$.

Figure 5.13.3(a) shows the interior circulation pattern and, schematically, the required boundary layer for the case $\alpha/\beta = 0.25$ and $X_E = 2$. Note that if curl $\tau/(\partial\eta_B/\partial x)$ were independent of $x$ a solution for $\psi$ which is dependent only on $y$, i.e., a strictly zonal flow, would be possible everywhere were it not for the meridional boundary at $x = X_E$. The presence of the meridional boundary introduces longitudinal pressure gradients and a meridional flow in region I alone. The circulation pattern requires boundary layers on both western and northern boundaries (where they are possible). Figure 5.13.3(b) shows the Sverdrup solution for the same wind stress curl in the absence of topography. In the latter case, no boundary layer is required on $y = 1$ and the solution is symmetric about $y = 0.5$. Topography destroys that symmetry and squeezes the center of the interior circulation northward while producing an eastward

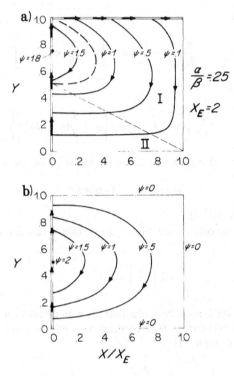

**Figure 5.13.3**   (a) The circulation pattern for curl $\tau = \sin \pi y$ and $\eta_B = \alpha x$ for $\alpha = \beta/4$. Note that in region II the flow is strictly zonal. The dotted curve is the streamline ($\psi = 1.27$) which passed through the point $(0, 1)$. Note the existence of boundary layers on the northern as well as western boundary. The maximum $\psi$ occurs at $x = 0$, $y = 0.75$, and is $\psi = 1.8$. (b) The circulation pattern for $\eta_B = 0$. Note, the north–south symmetry and the enhanced value of the transport; the maximum value of $\psi$ now equals 2.

jet along the northern boundary of the gyre. The dotted curve in Figure 5.13.3(a) is the boundary between regions in which streamlines thread through the northern jet and streamlines which close entirely through the western boundary layer. Since the maximum $\psi$ in this case is 1.8 and the value of $\psi$ on the boundary is 1.27, it follows that most of the flow ($\sim 70$ percent) flows through the northern boundary layer.

The overall value of the interior transport is also reduced by the topography in this case. While the maximum Sverdrup transport corresponds to $\psi = 2$ in the case of $\alpha = 0$, it is reduced to 1.8 by topography. Thus both the structure and the magnitude of the interior transport can be affected by topography and it follows that the structure and transport of the boundary currents will similarly be affected.

Generally speaking, although the solution structure may differ as a consequence of topography, the fundamental dynamical considerations remain the same with the role of the planetary vorticity gradient simply generalized to include the potential vorticity gradient of the bottom slope. One noteworthy exception to this qualitative remark can occur in the exceptional situation when the contours of constant potential vorticity close on themselves entirely *within* the interior. When this occurs the Sverdrup solution fails within the domain girdled by this closed contour for then an integral of (5.13.8) around such a closed curve $C$ would lead to an inconsistency if

$$\oint_C T(\theta, S) \, d\theta \neq 0. \tag{5.13.26}$$

In the likely event that (5.13.26) occurs, the dynamics within the closed contour alters. Consider the case shown in Figure 5.13.4. Although most of

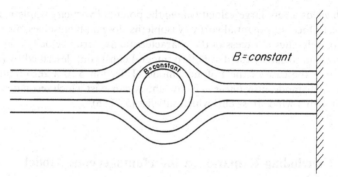

**Figure 5.13.4**   A schematic of a domain in which the $B$ contours close on themselves. Contours external to this domain tread back and intersect the eastern boundary.

the $B$ contours in the figure strike the eastern boundary, a domain of closed contours exists in the interior whose isolines of potential vorticity close on themselves and, for simplicity, are in this example, taken to be circular. Outside this closed region the solution is determined as previously described by integrating along the lines of constant $B$ from their intersection with the eastern wall. In the enclosed domain, where $B$ is a function only of $r = (x^2 + y^2)^{1/2}$, equation (5.13.10) written in polar coordinates becomes

$$\frac{\delta_S}{L} \left( \frac{1}{r} \frac{\partial}{\partial r} r \frac{\partial \psi}{\partial r} + \frac{1}{r^2} \frac{\partial^2 \psi}{\partial \theta^2} \right) + \frac{\partial B}{\partial r} \frac{\partial \psi}{\partial \theta} = \text{curl } \tau = T(r, \theta). \tag{5.13.27}$$

Now partition $T(r, \theta)$ into a part $T_0(r)$ which is independent of $\theta$ and the remainder, $T'(r, \theta)$ where

$$\int_0^{2\pi} T'(r, \theta) \, d\theta' = 0. \tag{5.13.28}$$

Similarly, $\psi$ may be partitioned as

$$\psi = \psi_0(r) + \psi'(r, \theta). \tag{5.13.29}$$

Then $\psi'$, to $O(\delta_S/L)$, is simply given by

$$\psi'(r, \theta) = \frac{\int_0^\theta T'(r, \theta') \, d\theta'}{\partial B/\partial r}, \tag{5.13.30}$$

and is qualitatively the same as the Sverdrup solution on nonclosed $B$ contours. However, the total solution is dominated by its $\theta$-independent portion, $\psi_0(r)$, since from (5.13.27) the only possible solution is given by

$$\frac{\partial \psi}{\partial r} = \frac{1}{r} \frac{L}{\delta_S} \int_0^r r' T_0(r') \, dr', \tag{5.13.31}$$

which yields a very large velocity *along* the potential vorticity isopleth, i.e., of $O(L/\delta_S)$. Once the potential vorticity isopleths close on themselves, only large interior velocities in excess of the characteristic Sverdrup velocity will allow the solution to satisfy the dissipation integral constraint described by (5.9.4). Just as in the case of the Fofonoff mode of Section 5.10 a zero-frequency geostrophic mode, now *linear* in its dynamics, can exist which resonates when forced with a wind stress curl which satisfies (5.13.26).

## 5.14 Concluding Remarks on the Homogeneous Model

The homogeneous model of the ocean circulation is, in many ways, archetypical of geophysical fluid dynamics. The model is physically extremely crude and, *a priori*, makes no attempt to include the description of the vertical structure of the circulation it predicts, which, being geostrophic, depends completely on horizontal density gradients. Further, the model deliberately excludes external surface heating as a source of large-scale motion. Certainly in the nonlinear regime near the western boundary the motions produced by wind and heating cannot be decoupled and linearly superposed. Finally the absence of stratification also implies the absence of potential energy available for the growth of fluctuation eddy motions of the type described in Chapter 7. These fluctuations could well affect the structure of the mean flow by producing rectified transports of heat and momentum, and it is not at all clear that they can be parametrized in terms of simple turbulent mixing coefficients.

Nevertheless it is striking how successful the homogeneous models are in offering a compelling simple explanation for the overall pattern of the oceanic circulation and in particular its western intensification. The crucial elements of the theory, i.e., the wind-stress curl, the planetary-vorticity gradient, and the presence of meridional boundaries, operate in such a clear and powerful way that the qualitative predictions of the model evidently transcend the deficiencies associated with many of the more detailed features of the various assumptions that are required to pose the complete problem. What is particularly characteristic of geophysical fluid dynamics is the need for careful and complete analysis of physically simple and idealized models in order to produce a trustworthy *qualitative* picture of the phenomenon.

CHAPTER 6

# Quasigeostrophic Motion of a Stratified Fluid on a Sphere

## 6.1 Introduction

In Chapter 1 it was noted that variations in density which arise from the differential heating of the atmosphere and oceans are responsible for the circulations of both systems. Even the wind-driven oceanic circulation implicitly requires atmospheric buoyancy forces to produce the wind. In addition, the presence of a stable stratification in which heavier fluid underlies lighter, characteristic of both the atmosphere and the oceans, inhibits vertical motion and strongly affects the nature of the dynamics. It is also clear from the discussion in Section 2.9 that the vertical structure of the winds and currents is directly related to the presence and strength of horizontal density gradients by the thermal wind relation.

In Chapters 3–5 it was possible to discuss several interesting phenomena of geophysical interest with homogeneous models which completely ignore stratification and, in consequence of the Taylor–Proudman theorem, must therefore have horizontal currents independent of height. In this chapter the Taylor–Proudman constraint will be removed by the introduction of stratification and baroclinicity. We will now observe not only the relationship between homogeneous and stratified models of the same phenomena, such as Rossby waves, but will also examine the possibility of new modes of motion whose very existence depends on stratification.

Now the motion of a stratified and (especially in the case of the atmosphere) compressible, rotating fluid is no easy matter to describe and calculate. Progress in understanding that motion requires that we considerably simplify the equations of motion so that they describe only the essential aspects of large scale motions. The method of simplification, by the systematic use of scaling arguments, is similar to the scaling analysis of Chapters 3 and 4, but the presence of stratification introduces some new considerations that require special attention. At the same time, the absence of the Taylor–Proudman theorem eliminates the direct dynamical analogy between bottom-slope variations and the variation of the Coriolis parameter with latitude. The analogy was based to a large extent on the principle of potential-vorticity conservation (3.17.1) for a homogeneous fluid. Thus, it is essential now to go beyond the heuristic arguments of Section 3.17 and derive the $\beta$-plane model, for both homogeneous and stratified fluids, from the same systematic scaling considerations used to derive the general quasigeostrophic equations.

It is important to keep in mind that the scaling analysis which follows is a consequence of an explicit *choice* to describe a particular class of motions, i.e., the large-scale, essentially geostrophic motion. The scaling arguments and results do not follow automatically from the equations of motion. Rather, we must bring to the equations, and systematically use, a preconception of the qualitative nature of the motions in terms of their scales and amplitude. All we can expect the equations to help us decide is whether our preconception is consistent (rather than apt), and if so, which simplified equations consistently describe the dynamics of the assumed motion.

## 6.2 The Equations of Motion in Spherical Coordinates: Scaling

The situation to be described is schematically depicted in Figure 6.2.1(a,b). We consider motions on a sphere of radius $r_0$. That is, we will ignore *ab initio* the slight departures of the figure of the earth from sphericity. The characteristic vertical scale of the motion, $D$, is in all cases of interest so small compared to $r_0$ that the effective gravitational acceleration $g$ can be considered constant over the depth of the fluid. The horizontal scale of the motion $L$ is large in the sense described in Section 2.1 (i.e., $L$ is large enough so that the Rossby number is small), but first we will focus our attention on the situation where $L$ is considerably smaller than $r_0$. The reason for this is threefold. First, it is characteristic of many important geophysical phenomena of interest, including cyclone waves in the atmosphere and mesoscale eddies in the oceans, and describes well the structure of the general circulation of both systems. Secondly we anticipate *a priori* that the geostrophic approximation, about which most of the subsequent development will pivot, must fail in the equatorial regions, where the Coriolis force on the horizontal currents is extremely feeble. Hence by its nature quasigeostrophic theory must be less than global. Finally, the restriction to small $L/r_0$ allows considerable simplification of the equations of motion and is a particular requirement for the validity of the $\beta$-plane approximation. The case where $L/r_0$ is O(1) is

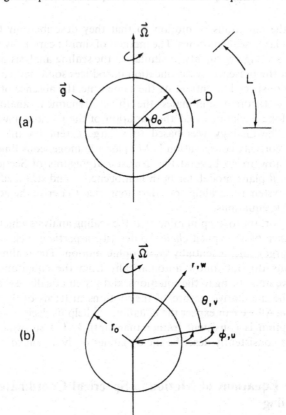

**Figure 6.2.1** (a) The motion to be studied has a horizontal scale $L$ and a vertical scale $D$, and is centered at latitude $\theta_0$. (b) The spherical polar coordinate system. Longitude is $\phi$, latitude is $\theta$, and $r$ is distance from the earth's center. $u$, $v$, and $w$ are the velocities in the eastward, northward, and vertical directions.

described in Sections 6.21 and 6.24. The coordinate system to be used in the spherical system is shown in Figure 6.2.1(b). The position of any point in the fluid is fixed by $r$, $\theta$, and $\phi$, which are the distance from the earth's center, the latitude, and the longitude respectively. The velocities in the eastward, northward, and vertical directions are $u$, $v$, and $w$, as shown. The equation for conservation of mass (1.4.2) in these spherical coordinates is*

$$\frac{d\rho}{dt} + \rho\left\{\frac{\partial w}{\partial r} + \frac{2w}{r} + \frac{1}{r\cos\theta}\frac{\partial(v\cos\theta)}{\partial\theta} + \frac{1}{r\cos\theta}\frac{\partial u}{\partial\phi}\right\} = 0, \quad (6.2.1)$$

---

* The equations of motion in several useful curvilinear coordinate frames can be found in many fluid-dynamics texts, e.g., Batchelor's *Fluid Dynamics*.

where

$$\frac{d}{dt} \equiv \frac{\partial}{\partial t} + \frac{u}{r \cos \theta} \frac{\partial}{\partial \phi} + \frac{v}{r} \frac{\partial}{\partial \theta} + w \frac{\partial}{\partial r}. \tag{6.2.2}$$

The momentum equations are

$$\frac{du}{dt} + \frac{uw}{r} - \frac{uv}{r} \tan \theta - 2\Omega \sin \theta v + 2\Omega \cos \theta w$$

$$= -\frac{1}{\rho r \cos \theta} \frac{\partial p}{\partial \phi} + \frac{\mathscr{F}_\phi}{\rho}, \tag{6.2.3a}$$

$$\frac{dv}{dt} + \frac{wv}{r} + \frac{u^2}{r} \tan \theta + 2\Omega \sin \theta u$$

$$= -\frac{1}{\rho r} \frac{\partial p}{\partial \theta} + \frac{\mathscr{F}_\theta}{\rho}, \tag{6.2.3b}$$

$$\frac{dw}{dt} - \frac{u^2 + v^2}{r} - 2\Omega \cos \theta u$$

$$= -\frac{1}{\rho} \frac{\partial p}{\partial r} - g + \frac{\mathscr{F}_r}{\rho}, \tag{6.2.3c}$$

where $\mathscr{F}_\phi$, $\mathscr{F}_\theta$, $\mathscr{F}_r$ are the three components of the frictional forces acting on the fluid. The equations of motion must be completed with the addition of a thermodynamic equation, either (1.4.18) for the atmosphere or an equation of the form (1.4.22) for the oceans. That is, for the atmosphere we have

$$\frac{d\theta}{dt} = \frac{\theta}{C_p T} \left( \frac{k}{\rho} \nabla^2 T + Q \right), \tag{6.2.4}$$

where $\theta$ is the potential temperature, i.e.,

$$\theta = T \left( \frac{p_0}{p} \right)^{R/C_p}, \tag{6.2.5}$$

while $p$, $\rho$, and the temperature $T$ are related by

$$p = \rho R T, \tag{6.2.6}$$

where $R$ is the gas constant for air. On the other hand for the oceans we will use the simplified thermodynamic equation

$$\frac{d\rho}{dt} = \kappa \nabla^2 \rho - \frac{\alpha \rho_0}{C_p} Q, \tag{6.2.7}$$

where $\kappa$ is the thermal diffusivity and $\alpha$ the coefficient of thermal expansion. The state equation is (1.4.20). In both (6.2.4) and (6.2.7), $Q$ is the rate of internal heating and $C_p$ is the specific heat at constant pressure for air or water as the case may be.

Consider now the description of a motion, in either the ocean or the atmosphere, whose horizontal spatial scale of variation is given by the length scale $L$ and whose horizontal velocities are characterized by the velocity scale $U$. Indeed, the fundamental assumption of this chapter is that a single well-defined scale for the velocity and its derivatives exists such that the magnitude of the terms in the equations of motion can be systematically estimated in terms of these scales. We further suppose that the motion occurs in a mid-latitude region, distant from the equator, around some central latitude $\theta_0$. It then becomes convenient to introduce new longitude and latitude coordinates as follows. Define $x$ and $y$ by

$$
\begin{aligned}
x &= \phi r_0 \cos \theta_0, \\
y &= (\theta - \theta_0) r_0.
\end{aligned}
\tag{6.2.8}
$$

The variables $x$ and $y$ have dimensions of length. They are however exact measures of eastward and northward distance only at the earth's surface $(r = r_0)$ and at the central latitude $\theta_0$. Although $x$ and $y$ are in principle simply new longitude and latitude coordinates in terms of which the equations of motion may be rewritten without approximation, they are obviously introduced in the expectation that for small $L/r_0$ and $D/r_0$ they will be the Cartesian coordinates of the $\beta$-plane approximation. Note that

$$
\begin{aligned}
\frac{\partial}{\partial \phi} &= r_0 \cos \theta_0 \frac{\partial}{\partial x}, \\
\frac{\partial}{\partial \theta} &= r_0 \frac{\partial}{\partial y}.
\end{aligned}
\tag{6.2.9}
$$

It is similarly convenient to introduce

$$
z = r - r_0,
\tag{6.2.10}
$$

so that

$$
\frac{\partial}{\partial r} = \frac{\partial}{\partial z}.
\tag{6.2.11}
$$

The existence of characteristic scales is exploited by the introduction of nondimensional variables denoted by primes, i.e.,

$$
x = L x',
\tag{6.2.12a}
$$

$$
y = L y',
\tag{6.2.12b}
$$

and

$$
z = D z',
\tag{6.2.12c}
$$

while the time is scaled by the advective time $L/U$, i.e.,

$$
t = \frac{L}{U} t'
\tag{6.2.12d}
$$

in accordance with the remarks of Section 3.12. For the horizontal velocity components,

$$u = Uu',$$
$$v = Uv'.$$
(6.2.13)

Geometrical considerations imply that if the vertical scale of motion is $D$ and its horizontal scale is $L$, the corresponding slope of a fluid element's trajectory will not exceed $D/L$, so that the appropriate scaling for $w$ is

$$w = \frac{D}{L} Uw'.$$
(6.2.14)

The actual scale of $w$ may be less than $DU/L$ if other dynamical constraints act to reduce the vertical motion, and the scaling (6.2.14) is best thought of as an upper bound on the size of $w$.

The scaling of the pressure and density is more subtle. If the relative velocities are small (i.e., for small Rossby number), the pressure will be only slightly disturbed from the value it would have in the absence of motion, $p_s(z)$, defined by the relation

$$\frac{\partial p_s(z)}{\partial z} = -\rho_s(z)g,$$
(6.2.15)

which is what (6.2.3a,b,c) reduce to if $u$, $v$, and $w$ are zero. We can think of $p_s(z)$ and $\rho_s(z)$ as defining a "standard" atmosphere or ocean, i.e., a basic state upon which fluctuations due to the motion occur. The basic state is assumed known, although in fact its determination from first principles requires the consideration of mechanisms such as radiative transfer in the atmosphere, etc. Alternatively, $p_s$ and $\rho_s$ may be defined as the global average of $p$ and $\rho$ at each level $z$. We then write

$$p = p_s(z) + \tilde{p}(x, y, z, t),$$
$$\rho = \rho_s(z) + \tilde{\rho}(x, y, z, t),$$
(6.2.16)

where $\tilde{p}$ and $\tilde{\rho}$ are spatially and temporally varying departures from the standard values $p_s$ and $\rho_s$. How shall $\tilde{p}$ and $\tilde{\rho}$ be scaled? We anticipate that for the motions of interest to us the horizontal pressure gradient will be of the same order as the Coriolis acceleration. The order of the Coriolis acceleration at the latitude $\theta_0$ is given by

$$\rho \, 2\Omega u \sin \theta_0 = O(2\Omega \sin \theta_0 \, U\rho_s),$$

while the magnitude of the pressure gradient is $\tilde{p}/L$. Thus we anticipate that $\tilde{p} = O(\rho_s \, Uf_0 \, L)$, where

$$f_0 = 2\Omega \sin \theta_0$$
(6.2.17)

is the Coriolis parameter at the central latitude $\theta_0$. These considerations imply that the pressure should be written

$$p = p_s(z) + \rho_s Uf_0 Lp'$$
(6.2.18)

with the expectation that $p'$ is an $O(1)$ quantity with an $O(1)$ variation over distances of $O(L)$. Note that $\rho_s$ in (6.2.18) is a function of $z$ alone.

Similarly we may anticipate that the buoyancy force per unit mass due to $\tilde{p}$ will be of the same order as the vertical pressure gradient, since an observed feature of large-scale motions is the excellence of the hydrostatic approximation. The vertical pressure gradient associated with $\tilde{p}$ is

$$\frac{\partial \tilde{p}}{\partial z} = O\left(\frac{\tilde{p}}{D}\right) = O\left(\frac{\rho_s U f_0 L}{D}\right), \tag{6.2.19}$$

and hence if $\tilde{\rho} g$ is to be of this same order,

$$\tilde{\rho} = O\left(\rho_s U \frac{f_0 L}{gD}\right), \tag{6.2.20}$$

so that the density should be written as

$$\rho = \rho_s(z)[1 + \varepsilon F \rho'], \tag{6.2.21}$$

where

$$\varepsilon = \frac{U}{f_0 L} \tag{6.2.22}$$

is the Rossby number and

$$F = \frac{f_0^2 L^2}{gD}. \tag{6.2.23}$$

If (6.2.12), (6.2.13), (6.2.14), (6.2.15), (6.2.18), and (6.2.21) are substituted into (6.2.3a,b,c), we obtain, after division by the constants multiplying the pressure terms,

$$\varepsilon\left\{\frac{du}{dt} + \frac{L}{r_*}[\delta uw - uv \tan \theta]\right\} - v\frac{\sin \theta}{\sin \theta_0} + \delta w\frac{\cos \theta}{\sin \theta_0}$$
$$= \frac{-\cos \theta_0}{\cos \theta}\frac{r_0}{r_*}\frac{\partial p}{\partial x}\frac{1}{1 + \varepsilon F \rho} + \frac{\mathscr{F}_{*\phi}}{\rho_* U f_0}, \tag{6.2.24a}$$

$$\varepsilon\left\{\frac{dv}{dt} + \frac{L}{r_*}[\delta vw + u^2 \tan \theta]\right\} + u\frac{\sin \theta}{\sin \theta_0}$$
$$= -\frac{r_0}{r_*}\frac{\partial p}{\partial y}\frac{1}{1 + \varepsilon F \rho} + \frac{\mathscr{F}_{*\theta}}{\rho_* U f_0}, \tag{6.2.24b}$$

$$(1 + \varepsilon F \rho)\left[\varepsilon \delta^2 \frac{dw}{dt} - \frac{\varepsilon \delta L}{r_*}(u^2 + v^2) - \frac{\delta u \cos \theta}{\sin \theta_0}\right]$$
$$= -\frac{\partial}{\rho_s \partial z}(p\rho_s) - \rho + \frac{\mathscr{F}_{*z}}{\rho_s U f_0}\delta, \tag{6.2.24c}$$

where

$$\frac{d}{dt} = \frac{\partial}{\partial t} + u \frac{\cos \theta_0}{\cos \theta} \frac{r_0}{r_*} \frac{\partial}{\partial x} + v \frac{r_0}{r_*} \frac{\partial}{\partial y} + w \frac{\partial}{\partial z} \qquad (6.2.25a)$$

and

$$\delta = \frac{D}{L}. \qquad (6.2.25b)$$

Unprimed variables are nondimensional, while dimensional variables are henceforth denoted by asterisks, so that (6.2.12a,b) become, for example,

$$x_* = Lx,$$

$$y_* = Ly,$$

and

$$p_* = p_s(z) + \rho_s U f_0 L p,$$

etc. Note that, in particular

$$\frac{r_*}{r_0} = 1 + \delta \left(\frac{L}{r_0}\right) z. \qquad (6.2.26)$$

The continuity equation (6.2.1) becomes

$$\varepsilon F \frac{d\rho}{dt} + (1 + \varepsilon F \rho)$$

$$\times \left[ \frac{w}{\rho_s} \frac{\partial \rho_s}{\partial z} + \frac{\partial w}{\partial z} + 2 \frac{D}{r_*} w + \frac{\partial v}{\partial y} \frac{r_0}{r_*} - \frac{L}{r_*} v \tan \theta + \frac{r_0}{r_*} \frac{\cos \theta_0}{\cos \theta} \frac{\partial u}{\partial x} \right] = 0. \qquad (6.2.27)$$

To this point no approximations have been made. The equations have simply been scaled so that the relative order of each term is clearly measured by the nondimensional parameter multiplying it. This is the basic presumption of the scaling method. It is now possible to systematically exploit the smallness of the parameters $\varepsilon$, $\delta$, $L/r_0$, and $F$. Some preliminary remarks must be made at this point. The parameters $\varepsilon$, $\delta$, $L/r_0$, and $F$ are all independent, and their relative orders will vary somewhat from phenomenon to phenomenon. The nature of the approximations to be derived will depend on these relative orders, in particular on the ratio $L/\varepsilon r_0$. The treatment which follows will not exhaustively cover all possible relative magnitudes of the parameters. Only those cases of particular significance will be examined. The methods of approach should become sufficiently clear that the reader may apply the same method to other parameter relationships.

The trigonometric functions will be expanded about the latitude $\theta_0$, e.g., with (6.2.8) and (6.2.12):

$$\sin \theta = \sin \theta_0 + \frac{L}{r_0} y \cos \theta_0 - \left(\frac{L}{r_0}\right)^2 \frac{y^2}{2} \sin \theta_0 + \cdots, \qquad (6.2.28a)$$

$$\cos \theta = \cos \theta_0 - \frac{L}{r_0} y \sin \theta_0 - \left(\frac{L}{r_0}\right)^2 \frac{y^2}{2} \cos \theta_0 + \cdots, \qquad (6.2.28b)$$

$$\tan \theta = \tan \theta_0 + \frac{L}{r_0} y \cos^{-2} \theta_0 + \left(\frac{L}{r_0}\right)^2 y^2 \frac{\tan \theta_0}{\cos^2 \theta_0}. \qquad (6.2.28c)$$

The magnitude of the friction forces will be estimated as follows. The *order* of the horizontal friction forces, e.g., $\mathscr{F}_{*\theta}/\rho_*$ will be estimated in terms of characteristic turbulent mixing coefficients, as in Chapter 4. That is,

$$\frac{\mathscr{F}_{*\theta}}{\rho_*} = O\left(A_H \frac{U}{L^2}, A_V \frac{U}{D^2}\right), \qquad (6.2.29)$$

so that

$$\frac{\mathscr{F}_{*\theta}}{\rho_* U f_0} = O\left|\frac{E_H}{2}, \frac{E_V}{2}\right|, \qquad (6.2.30)$$

where

$$E_H = 2 \frac{A_H}{f_0 L^2},$$
$$\qquad\qquad\qquad\qquad (6.2.31)$$
$$E_V = 2 \frac{A_V}{f_0 D^2},$$

are the Ekman numbers associated with the horizontal and vertical mixing of momentum by turbulence at scales smaller than $O(L)$ and $O(D)$ respectively. Except in boundary layers, these frictional terms will *usually* be neglected; in the boundary layers the equations must be rescaled to take explicit note of the change of length scales.

Finally we define

$$\beta_0 = \frac{2\Omega}{r_0} \cos \theta_0 = \left(\frac{1}{r_0} \frac{df}{d\theta}\right)_{\theta=\theta_0} \qquad (6.2.32)$$

as the northward gradient of the Coriolis parameter at the latitude $\theta_0$. Note that

$$\frac{\beta_0 L}{f_0} = \frac{L}{r_0} \cot \theta_0 = O\left(\frac{L}{r_0}\right) \qquad (6.2.33)$$

so that the ratio

$$\frac{\beta_0 L/f_0}{\varepsilon} = \frac{\beta_0 L^2}{U} = O\left|\frac{L}{r_0 \varepsilon}\right|. \qquad (6.2.34)$$

Thus while $\varepsilon$ measures the ratio of the relative vorticity and the planetary vorticity normal to the sphere at $\theta_0$, the magnitude of the relative-vorticity *gradient* and the planetary vorticity *gradient* is measured by the parameter

$$\beta^{-1} = \frac{U}{\beta_0 L^2} = O\left(\varepsilon \frac{r_0}{L}\right). \tag{6.2.35}$$

Since the relative vorticity varies on the scale $L$ while the planetary vorticity varies on the scale $r_0$, the smallness of $\varepsilon$ does not imply the smallness of $U/\beta_0 L^2$. Hence while $\varepsilon$ may be small, $\beta^{-1}$ may be large, order one, or small, and each of these possibilities gives rise to quite different quasigeostrophic dynamical systems.

The consideration of the thermodynamic equations in scaled form will be deferred to later sections as needed.

# 6.3 Geostrophic Approximation: $\varepsilon = O\left(\dfrac{L}{r_0}\right) \ll 1$

The first case we will examine is the case where

$$\varepsilon = O\left(\frac{L}{r_0}\right) \ll 1, \tag{6.3.1}$$

or equivalently, with (6.2.35), where

$$\frac{U}{\beta_0 L^2} \equiv \frac{1}{\beta} = O(1). \tag{6.3.2}$$

This parameter setting is of particular interest in both meteorology and oceanography. Large-scale waves in the atmospheric westerly winds are characterized by $U = O(10 \text{ m/s})$, $L = O(10^3 \text{ km})$ while in mid-latitudes $f_0 = O(10^{-4} \text{ s}^{-1})$ and $\beta_0 = O(10^{-13} \text{ cm}^{-1} \text{ s}^{-1})$. Hence

$$\varepsilon = O(10^{-1}), \qquad \beta = O(1)$$

if these scales are used. For the mesoscale or synoptic eddies that have been observed in the western Atlantic, $U$ is $O(5 \text{ cm/s})$, $L = O(100 \text{ km})$, so that for these scales

$$\varepsilon = O(5 \times 10^{-3}), \qquad \beta = O(0.5).$$

In the former case, where $D$ is typically 10 km (the depth of the troposphere), the parameter $F$ is $O(10^{-1})$—i.e., $O(\varepsilon)$—while in the oceanic case just mentioned, where $D$ is $O(4 \text{ km})$, $F$ is much smaller (since $L$ is only 100 km): $F = O(0.2 \times 10^{-2})$, which is also $O(\varepsilon)$. Therefore in analyzing the approximate form of (6.2.24) and (6.2.27) we shall be particularly interested in the case

$$\frac{L}{r_0} = O(F) = O(\varepsilon). \tag{6.3.3}$$

For the case of large-scale atmospheric waves,

$$\delta = \frac{D}{L} = O(10^{-2}) = O(\varepsilon^2),$$

while for the oceanic synoptic scale

$$\delta = O(2 \times 10^{-2}) = O(\varepsilon).$$

This distinction however leads to no significant difference in the following analysis.

The frictional terms are more difficult to estimate, due to the difficult and vexing uncertainty in the value to attribute to $A_H$ and $A_V$. On the basis of the studies of Chapters 4 and 5 it seems plausible that in geophysically relevant situations the *direct* effect of friction is felt only in narrow boundary layers adjacent to the fluid boundaries. The approximate equations to be derived will be valid only outside such layers, and we therefore assume that

$$(E_V, E_H) < O(\varepsilon). \tag{6.3.4}$$

Finally we note that, from (6.2.26),

$$\frac{r_*}{r_0} - 1 = O\left(\frac{\delta L}{r_0}\right) \leq O(\varepsilon^2).$$

Each dependent variable, for example $u$, which is a solution of the equations of motion, will be a function not only of $x$, $y$, $z$, and $t$, but also of the nondimensional parameters $\varepsilon$, $L/r_0$, $F$, $\delta$, etc. In order to make progress we must exploit the smallness of each of the parameters. The results will depend on the relative orders of each of the parameters and in expanding the equations and the solutions we must, of necessity, choose particular ordering relations between the parameters. We may choose, for example, to let $\varepsilon r_0 /L$ remain of order one in the limit $\varepsilon \to 0$. To some extent this is suggested by the observational evidence described above. However, such information hardly suffices to prove that if $\varepsilon$ were indeed to become ever smaller that $L/r_0$ would shrink correspondingly; in particular, linearly. We must, in addition, bring to the scaling analysis an intuitive appreciation that certain ordering relationships are highlighted by the way they distinguish important and natural physical balances that are physically relevant to the phenomena of interest. The limit $\varepsilon \to 0$, with $\varepsilon r_0 /L$ chosen to be of order one, has special significance because it examines geostrophic dynamics when the planetary vorticity gradient contributes equally with the relative vorticity gradient to the overall vorticity balance. This recognition of the particular meaning of this ordering relation then gives confidence that the observed numerical relations between the parameters are not merely fortuitous.

Once the ordering relationships are chosen in this way, each variable may be considered a function of $\varepsilon$ with the other parameters directly related to $\varepsilon$ by the preceding considerations. Since $\varepsilon$ is small, we write, for example:

$$u(x, y, z, t, \varepsilon) = u_0(x, y, z, t) + \varepsilon u_1(x, y, z, t) + \cdots, \tag{6.3.5}$$

where the $u_K$ are independent of $\varepsilon$ and $O(1)$. The $u_K$ are also now implicitly functions of the parameters $L/r_0\,\varepsilon$, $F/\varepsilon$, $\delta/\varepsilon$, etc., which are here assumed to be order one. If the expansion (6.3.5) is applied to each dependent variable and that expansion is substituted into the equations of motion, terms of like order in $\varepsilon$ which result must balance, since $\varepsilon$, while small, is also arbitrary, and the equations must hold for all orders in $\varepsilon$. If the trigonometric terms are expanded as in (6.2.28), the $O(1)$ terms in the $\varepsilon$-expansion of (6.2.24) and (6.2.27) are

$$v_0 = \frac{\partial p_0}{\partial x}, \tag{6.3.6a}$$

$$u_0 = -\frac{\partial p_0}{\partial y}, \tag{6.3.6b}$$

$$\rho_0 = \frac{-1}{\rho_s}\frac{\partial}{\partial z}(\rho_s p_0), \tag{6.3.6c}$$

and

$$\frac{1}{\rho_s}\frac{\partial}{\partial z}\{w_0\rho_s\} + \frac{\partial u_0}{\partial x} + \frac{\partial v_0}{\partial y} = 0. \tag{6.3.7}$$

Equations (6.3.6a,b) are the geostrophic approximation. Note that the equations are indistinguishable from the geostrophic relations appropriate to horizontal flow over a *flat* region for which $x$ and $y$ are the horizontal cartesian coordinates. The $\beta$-plane approximation first emerges at this point. The relations between $u_0$, $v_0$, and $p_0$ are not only in cartesian coordinate form, but more significantly, the Coriolis parameter relating the $O(1)$ velocity and the $O(1)$ horizontal pressure gradient is given by the *constant value*, $f_0$, *it assumes at* $\theta = \theta_0$, since (6.3.6a) is simply

$$v_{*0} = \frac{1}{\rho_s f_0}\frac{\partial p_{*0}}{\partial x_*} \tag{6.3.8}$$

if the dimensional units are restored. This crucial simplification further implies that the horizontal, $O(1)$ geostrophic velocities are horizontally non-divergent, since from (6.3.6a,b),

$$\frac{\partial v_0}{\partial y} + \frac{\partial u_0}{\partial x} = 0, \tag{6.3.9}$$

so that (6.3.7) implies

$$\frac{\partial}{\partial z}(\rho_s(z)w_0) = 0. \tag{6.3.10}$$

Hence $\rho_s w_0$ must be independent of $z$. If $w_0$ vanishes for any $z$ it will be zero for all $z$. If the region under consideration is bounded above or below by a horizontal solid surface, then the $O(1)$ vertical velocity on that surface must vanish. The existence of the viscous boundary layer on such a surface will give rise to a small vertical velocity, by (4.5.39), of $O(\delta_E/D)$. Hence the $O(1)$ vertical velocity, $w_0$, must vanish on such surfaces. Now indeed we must later examine whether the presence of stratification seriously affects the Ekman-layer structure. We shall see in Section 6.6 that the answer is essentially negative, but the agreement here really only depends on the simple fact that by mass conservation the vertical velocity pumped out of the layer on a solid surface must be, when compared to the horizontal velocity, of the order of the aspect ratio of the boundary layer. Hence as long as the boundary-layer thickness is less than $D$, its contribution to $w$ will be negligible to the order of the layer thickness divided by $D$. Similarly, if the solid surface is not flat but is characterized by variations in height of $O(h_B)$, a vertical velocity will be produced of order

$$w_* = O\left(U\frac{h_B}{L}\right),$$

or in dimensionless units,

$$w = O\left(\frac{h_B}{D}\right).$$

As in Section 3.12, we limit our attention to those situations where

$$\frac{h_B}{D} = O(\varepsilon), \tag{6.3.11}$$

since the case of larger $h_B/D$ can be recovered, as in (3.12), as a limiting case of the theory to be derived.

With (6.3.11), we can then specify that $w_0$ must vanish for at least one value of $z$, so that from (6.3.10)

$$w_0 = 0, \tag{6.3.12a}$$

and hence

$$w = \varepsilon w_1 + \varepsilon^2 w_2 + \cdots, \tag{6.3.12b}$$

which, it is important to note, is a direct consequence of the effect of rotation, i.e., of the geostrophic approximation. The geostrophic approximation (6.3.6a,b) for the $O(1)$ flow leads to a (by now) familiar difficulty of geostrophic degeneracy, i.e., the inability of $O(1)$ geostrophic approximation to determine $p_0$ and thence $u_0$ and $v_0$. As in Chapter 3, it is necessary to consider the higher-order dynamics, i.e., the $O(\varepsilon)$ terms in the equations of motion. With careful attention to the expansion of the trigonometric terms and the use of (6.3.12), we obtain from (6.2.24a,b)

$$\frac{\partial u_0}{\partial t} + u_0\frac{\partial u_0}{\partial x} + v_0\frac{\partial u_0}{\partial y} - v_1 - v_0\left(\frac{L}{r_0\varepsilon}\right)y\cot\theta_0$$

$$= -\frac{\partial p_1}{\partial x} - \frac{Ly}{\varepsilon r_0}\tan\theta_0\frac{\partial p_0}{\partial x}, \tag{6.3.13a}$$

$$\frac{\partial v_0}{\partial t} + u_0\frac{\partial v_0}{\partial x} + v_0\frac{\partial v_0}{\partial y} + u_1 + u_0\left(\frac{L}{r_0\varepsilon}\right)y\cot\theta_0 = -\frac{\partial p_1}{\partial y}. \tag{6.3.13b}$$

The terms proportional to $L/r_0 \varepsilon$ on the left-hand side of (6.3.13a,b) are due to the variation of the Coriolis parameter with latitude. The presence of these terms is to be expected on the basis of the heuristic $\beta$-plane argument. The term proportional to $L/\varepsilon r_0$ on the *right*-hand side of (6.3.13a) arises, however, from the variation of the metric term $\cos \theta$ relating longitude variations and eastward length changes. If $\tan \theta_0$ were small, this term would be negligible. In that case (6.3.13) would reduce to the momentum equation, at $O(\varepsilon)$, for a *flat* earth with a Coriolis parameter linearly varying in the northward direction. This would validate the $\beta$-plane idealization immediately, on the level of the momentum balance. However the restriction to small $\tan \theta_0$ is quite a strong one and indeed would tend to restrict the $\beta$-plane approximation to latitude regions uncomfortably near the equator where the geostrophic approximation is weakest. We must conclude that the model of a flat earth with sphericity accounted for only by a varying $f$ is not valid for the consideration of the $O(\varepsilon)$ momentum balance. However, we must recall that the $\beta$-plane approximation used earlier only required that the *vorticity* equation satisfy the requirement of the $\beta$-plane idealization. Before forming the vorticity equation, though, it is useful to first examine the $O(\varepsilon)$ balance in the continuity equation (6.2.27), namely

$$\frac{\partial u_1}{\partial x} + \frac{\partial v_1}{\partial y} - \frac{L}{\varepsilon r_0} v_0 \tan \theta_0 + \frac{L}{\varepsilon r_0} y \tan \theta_0 \frac{\partial u_0}{\partial x} + \frac{1}{\rho_s} \frac{\partial}{\partial z} (\rho_s w_1) = 0. \quad (6.3.14)$$

The vorticity equation for

$$\zeta_0 = \frac{\partial v_0}{\partial x} - \frac{\partial u_0}{\partial y} \qquad (6.3.15)$$

can be obtained by cross differentiation of (6.3.13a,b) with respect to $x$ and $y$ to yield

$$\frac{\partial \zeta_0}{\partial t} + u_0 \frac{\partial \zeta_0}{\partial x} + v_0 \frac{\partial \zeta_0}{\partial y} + \beta v_0$$
$$= \frac{L}{\varepsilon r_0} \tan \theta_0 \frac{\partial p_0}{\partial x} + \frac{L}{\varepsilon r_0} y \tan \theta_0 \frac{\partial^2 p_0}{\partial x\, \partial y} - \left( \frac{\partial u_1}{\partial x} + \frac{\partial v_1}{\partial y} \right) \qquad (6.3.16)$$

with the aid of (6.2.33) and (6.3.12). The continuity equation then yields, with the aid of (6.3.6a,b),

$$\boxed{\frac{d_0}{dt} \{\zeta_0 + \beta y\} = \frac{1}{\rho_s} \frac{\partial}{\partial z} (\rho_s w_1)} \qquad (6.3.17)$$

where

$$\frac{d_0}{dt} \equiv \frac{\partial}{\partial t} + u_0 \frac{\partial}{\partial x} + v_0 \frac{\partial}{\partial y}. \qquad (6.3.18)$$

This is *precisely* the form the $O(1)$ vorticity equation assumes for a flat-earth model with a Coriolis parameter which varies linearly with $y$. Note that $u_0$

and $v_0$ are determined directly in terms of $p_0$ by the "flat" geostrophic approximation (6.3.6a,b), so that

$$\zeta_0 = \frac{\partial^2 p_0}{\partial x^2} + \frac{\partial^2 p_0}{\partial y^2}. \tag{6.3.19}$$

The theory is not complete, however, until $w_1$ is related to $p_0$. In general this requires the consideration of the thermodynamic equation, i.e., either (6.2.4) or (6.2.7). This consideration is taken up and the derivation completed in the following sections. At this point we can, however, directly obtain the governing vorticity equation for a *homogeneous* fluid layer of depth $D$. If the fluid is homogeneous, $\rho_s$ is constant and $\rho$ is identically zero, and this implies from (6.3.6c) that $p_0$ is independent of $z$, so that $u_0$ and $v_0$ must be also $z$-independent. In that case (6.3.17) may be integrated over the depth of the fluid (i.e., from $z = 0$ to $z = 1$) to yield

$$\frac{d_0}{dt}(\zeta_0 + \beta y) = w_1(z = 1) - w_1(z = 0) \tag{6.3.20}$$

to $O(h_B/D)$. If the lower layer has a slope $h_B$, then (4.11.4) applies; if the upper surface is a free surface, (4.11.10) applies at $z = 1$, so that (6.3.20) becomes

$$\frac{d_0}{dt}\{\zeta_0 + \beta y + \eta_B - F\eta_0\} = \left[\frac{\tau_0}{\rho f_0 U D \varepsilon}\right]\mathbf{k}\cdot\operatorname{curl}\tau - \frac{r}{2}\zeta_0, \tag{6.3.21}$$

where

$$\eta_0 = p_0,$$

$$\eta_B = \frac{h_B}{D\varepsilon},$$

and $r = E_V^{1/2}/\varepsilon$. We have retained the free-surface term of $O(F)$, since for a homogeneous fluid the condition $F \ll 1$ is not crucial to simplify the continuity equation as it is when $\rho \neq 0$. The important point is that (6.3.21) is precisely the $\beta$-plane model for the vorticity equation for a homogeneous fluid, as comparison with (5.2.16) shows.* That is, in terms of the geostrophic dynamics, the geostrophic approximation (6.3.6a,b) and the potential-vorticity equation (6.3.21), which is the governing equation for the determination of the $O(1)$ variables, are precisely the equations for motion on a plane with a linearly varying Coriolis parameter. This is furthermore true even when $\tan\theta_0$ is $O(1)$, i.e., when the $O(\varepsilon)$ momentum equation (6.3.13a,b) is not modeled accurately by the $\beta$-plane model. So at least for a homogeneous model we have shown that the sole effect of the earth's spheri-

---

* In the present case the term in (5.2.16) proportional to $\mathrm{Re}^{-1}$ has been neglected, but this is not an essential difference. The same viscous term could be included in (6.3.21) if the parameter scaling $E_H = O(\varepsilon)$ were chosen.

city on the O(1) fields is due entirely to the $\beta$-term, i.e., the variation of $f$ with latitude. In the next sections we extend this result to the case of stratified fluids.

## 6.4 The Concept of Static Stability

The completion of the derivation of the quasigeostrophic equations of motion requires the representation of the small vertical velocity, $\varepsilon w_1$, in terms of the O(1) geostrophic fields. This is a subtle issue, and before proceeding directly to that question we consider in this section the concept of *static stability*, with which the vertical velocity is intimately connected.

Consider first the atmosphere in its rest state, i.e., the lowest-order equilibrium state in which the pressure surfaces and density surfaces are surfaces of constant $z$.* Then, as shown in Figure 6.4.1, consider the displacement of

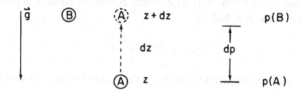

**Figure 6.4.1**   A fluid element $A$ at $z$ is slowly raised to $z + dz$ at the same level as element $B$. The element preserves potential temperature during the displacement. Its temperature and density are altered, since $p(B) \neq p(A)$. That change of density, when added to the difference $\rho(A) - \rho(B)$, yields the buoyancy force on $A$ as given by (6.4.7).

fluid element $A$ upwards to the level occupied by fluid element $B$. Let the displacement be slow enough so that the pressure of $A$ continuously adjusts to its surroundings (this only requires that $A$ move slowly compared with the local speed of sound), but rapid enough so that thermal dissipation or heat addition is negligible during the displacement, i.e., that during the displacement $A$ undergoes an adiabatic (constant-entropy) transition to the new level. According to (1.4.18) this implies that the potential temperature (1.4.19),

$$\theta = T\left(\frac{p_0}{p}\right)^{R/C_p}, \tag{6.4.1}$$

is preserved during the displacement. With the perfect-gas law (1.4.16), (6.4.1) may be rewritten as

$$\rho = \frac{p_0}{R\theta}\left|\frac{p}{p_0}\right|^{+1/\gamma}, \tag{6.4.2}$$

* In this section variables are in *dimensional* units unless otherwise specified.

where

$$\gamma = \frac{C_p}{C_v}$$

is the ratio of the specific heats of air at constant pressure and density respectively. For air $\gamma = 1.4$. Then at $z + dz$ the density of element $A$ will have changed by an amount

$$\Delta\rho_A = \frac{1}{\gamma}\frac{p_0}{R\theta}\left(\frac{p}{p_0}\right)^{+1/\gamma}\frac{\partial p}{\partial z}\frac{dz}{p} \qquad (6.4.3)$$

due to the decrease of pressure experienced by the element in its displacement, so that its new density at $z + dz$ is

$$\rho_A + \Delta\rho_A = \rho_A(z) + \frac{1}{\gamma}\frac{\rho}{p}\frac{\partial p}{\partial z}dz. \qquad (6.4.4)$$

On the other hand the density of element $B$ may be written in terms of the undisturbed density $A$ had at $z$, i.e.,

$$\rho_B = \rho_A(z) + \frac{\partial\rho}{\partial z}dz. \qquad (6.4.5)$$

Thus the excess density element $A$ has when judged by its new surroundings at $z + dz$ is

$$(\rho_A + \Delta\rho_A) - \rho_B = \left(\frac{1}{\gamma}\frac{\rho}{p}\frac{\partial p}{\partial z} - \frac{\partial\rho}{\partial z}\right)dz. \qquad (6.4.6)$$

This gives rise to a *restoring force per unit mass* due to the excess density of an amount

$$\frac{g}{\rho}\{\rho_A + \Delta\rho_A - \rho_B\} = g\left|\frac{1}{\gamma p}\frac{\partial p}{\partial z} - \frac{1}{\rho}\frac{\partial\rho}{\partial z}\right|dz$$
$$= \frac{g}{\theta}\frac{\partial\theta}{\partial z}dz. \qquad (6.4.7)$$

Thus if $\partial\theta/\partial z > 0$, the buoyancy force is restoring and the static state is stable with respect to small, constant-entropy, vertical displacements. The quantity $(1/\theta)\,\partial\theta/\partial z$ is called the *static stability*. Since the restoring force in (6.4.7) is proportional to the displacement, an oscillation of the fluid element about the equilibrium level is implied with frequency

$$N = \left(\frac{g}{\theta}\frac{\partial\theta}{\partial z}\right)^{1/2}. \qquad (6.4.8)$$

$N$ is called the Brunt–Väisälä frequency. It is in general a function of height. In terms of the temperature $T$, from (6.4.1),

$$\frac{1}{\theta}\frac{\partial\theta}{\partial z} = \frac{1}{T}\frac{\partial T}{\partial z} - \frac{R}{C_p}\frac{1}{p}\frac{\partial p}{\partial z}, \qquad (6.4.9)$$

and since for the *static* atmospheric state

$$\frac{\partial p}{\partial z} = -\rho g, \tag{6.4.10}$$

this becomes

$$\frac{1}{\theta}\frac{\partial \theta}{\partial z} = \frac{1}{T}\left[\frac{\partial T}{\partial z} + \frac{g}{C_p}\right]. \tag{6.4.11}$$

Hence even if $\partial T/\partial z < 0$, the atmosphere will be statically stable as long as the *lapse rate*, $-\partial T/\partial z$, does not exceed $g/C_p$. Figure 6.4.2 shows typical

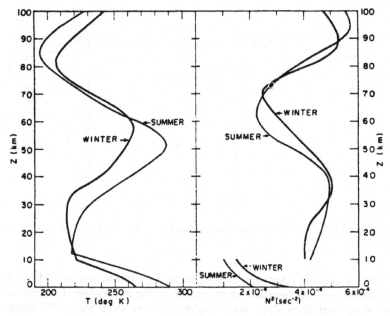

**Figure 6.4.2**  (a) Vertical distribution of temperature in the atmosphere in winter and summer. (b) The corresponding distribution of $N^2(z)$. (Reprinted from Charney and Drazin (1961).)

profiles of $T(z)$ and $N^2(z)$ for the summer and winter atmospheres. In the troposphere, where $\partial T/\partial z < 0$ (and where "weather" phenomena occur), $N$ is $O(10^{-2}\ \mathrm{s}^{-1})$ and sharply increases as the stratosphere, at about 10 km, is entered, due to the sudden decrease of the lapse rate at the lower boundary of the stratosphere. (This upper boundary of the troposphere is called the tropopause.) Since $N \gg f$, the oscillations indicated by (6.4.7) and (6.4.8) form a high-frequency oscillation around the more ponderous large-scale geostrophic motion, and the geostrophic theory, limited to the more energetic motions with characteristic frequencies $\ll f$, will not explicitly deal with this oscillation. Yet as we shall see, the presence of $N \neq 0$ strongly inhibits the large-scale vertical motion and couples in an essential way the small, large-scale, vertical velocity to the O(1) geostrophic fields.

For the oceans, where the compressibility effect on fluid elements undergoing small displacements is negligible, the Brunt–Väisälä frequency is given simply in terms of the equilibrium density distribution, i.e., for the oceans

$$N = \left| -\frac{g}{\rho}\frac{\partial\rho}{\partial z}\right|^{1/2}. \qquad (6.4.12)$$

Figure 6.4.3 shows a schematic profile of the density with depth in the ocean.

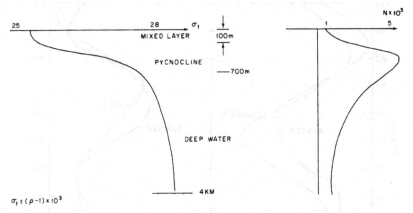

$\sigma_t \equiv (\rho - 1) \times 10^3$

**Figure 6.4.3** A typical distribution of density in the ocean and the corresponding $N(z)$.

Beneath a shallow, essentially homogeneous wind-mixed layer of O(100 m), the density sharply increases in the region of the pycnocline at a depth of O(700 m) and then stays fairly uniform in the deep water. $N$, correspondingly, is small near the surface and at great depth, with a sharp maximum O($10^{-3}$ s$^{-1}$) at the depth of the pycnocline.

Note that for the atmosphere

$$\frac{D}{\theta}\frac{\partial\theta}{\partial z} = \frac{N^2 D}{g} = O(10^{-1}) \qquad (6.4.13)$$

if $D$ is 10 km, while for the oceans

$$-\frac{D}{\rho}\frac{\partial\rho}{\partial z} = \frac{N^2 D}{g} = O(10^{-3}). \qquad (6.4.14)$$

In the atmospheric case the variation of $\theta$ over $D$, while important, is small (though $(D/\rho)\,\partial\rho/\partial z$ is O(1)), while for the oceans it is the total change of the density itself which is small over the oceanic depth. Nevertheless, the stratification, while small, is crucial in providing gravitational stability to the fluid.

## 6.5 Quasigeostrophic Potential-Vorticity Equation for Atmospheric Synoptic Scales

We return now to the completion of the scaling for the case $\beta = O(1)$, $\varepsilon \ll 1$. In order to complete the dynamical description, $w_1$ in (6.3.17) must be related to $O(1)$ geostrophic dynamic variables. We first consider the case of motions in the atmosphere and consider the proper scaling for the potential temperature $\theta_*$.* Since by (6.4.2)

$$\ln \theta_* = \frac{1}{\gamma} \ln p_* - \ln \rho_*, \tag{6.5.1}$$

(6.2.18) and (6.2.21) yield

$$\ln \theta_* = \left(\frac{1}{\gamma} \ln p_s - \ln \rho_s\right) + \frac{1}{\gamma} \ln\left[1 + \varepsilon \frac{f_0^2 L^2 p}{p_s/\rho_s}\right] - \ln[1 + \varepsilon F \rho] \tag{6.5.2}$$

$$= \frac{1}{\gamma} \ln p_s(z) - \ln \rho_s(z) + \frac{1}{\gamma} \varepsilon \frac{f_0^2 L^2}{p_s/\rho_s} p - \varepsilon F \rho + O(\varepsilon^2 F).$$

Therefore, guided by (6.5.2), we express $\theta_*$ (as we did $\rho_*$) as the sum of the rest-state value $\theta_s(z)$ and a small motion-dependent deviation of $O(\varepsilon F)$, i.e.,

$$\theta_* = \theta_s(z)(1 + \varepsilon F \theta(x, y, z, t)), \tag{6.5.3}$$

where

$$\ln \theta_s(z) \equiv \frac{1}{\gamma} \ln p_s(z) - \ln \rho_s(z) + \text{const.}$$

Thus if $\theta$ is expanded in an $\varepsilon$-series

$$\theta = \theta_0 + \varepsilon \theta_1 + \cdots, \tag{6.5.4}$$

it follows from (6.5.2) that

$$\theta_0 = -\rho_0 + \frac{1}{\gamma}\left(\frac{\rho_s g D}{p_s}\right) p_0. \tag{6.5.5}$$

However, from (6.3.6c) and the fact that (using (6.2.15) with the dimensionless variable)

$$g D \rho_s = -\frac{\partial p_s}{\partial z},$$

---

* In this section we restore the asterisk notation for dimensional variables.

we find

$$\theta_0 = \frac{\partial p_0}{\partial z} - p_0 \frac{1}{\theta_s} \frac{\partial \theta_s}{\partial z}. \tag{6.5.6}$$

Since from (6.4.13)

$$\frac{1}{\theta_s} \frac{\partial \theta_s}{\partial z} = \frac{D}{\theta_s} \frac{\partial \theta_s}{\partial z_*} = O(10^{-1}) = O(\varepsilon), \tag{6.5.7}$$

it follows that

$$\theta_0 = \frac{\partial p_0}{\partial z}. \tag{6.5.8}$$

Now, (6.3.6c) is obviously the hydrostatic approximation, which is valid whenever $\delta \ll 1$,* and it is clear that it is more concisely expressed, in (6.5.8) in terms of the potential temperature deviation $\theta_0$.

If (6.5.3) is substituted into (6.2.4), the resulting *non-dimensional* equation is

$$\frac{d\theta}{dt} + \frac{w}{\varepsilon F \theta_s} \frac{\partial \theta_s}{\partial z}(1 + \varepsilon F \theta) = \frac{\theta_*}{\theta_s} \left(\frac{\mathscr{H}_*}{C_p T_*}\right) \frac{gD}{U^2 f_0}, \tag{6.5.9}$$

where $\mathscr{H}_*$ is the total heating rate of each fluid element, i.e., by (6.2.4),

$$\mathscr{H}_* = \frac{k}{\rho_*} \nabla^2 T_* + Q_*. \tag{6.5.10}$$

For the troposphere $C_p T_* = O(gD) = O(200 \text{ cm}^2/\text{s}^2)$; hence (6.5.9) is consistent with heating rates $\mathscr{H}_* \le O(U^2 f_0) = 10^2 \text{ cm}^2/\text{s}^3$ if $U = O(10 \text{ m/s})$. Define

$$\mathscr{H} = \mathscr{H}_* \frac{gD}{C_p T_* f_0 U^2}; \tag{6.5.11}$$

then to lowest order in $\varepsilon$, after explicit use is made of (6.3.12), (6.5.9) becomes

$$\frac{d_0 \theta_0}{dt} + w_1 S = \mathscr{H}, \tag{6.5.12}$$

where the stratification parameter

$$S(z) = \frac{F^{-1}}{\theta_s} \frac{\partial \theta_s}{\partial z} = \frac{N_s^2 D^2}{f_0^2 L^2} \tag{6.5.13}$$

---

* Note, from (6.2.24c) that since $w$ is $O(\varepsilon)$, the hydrostatic approximation in a *flat* geometry requires only that $\varepsilon^2 \delta^2 \ll 1$, so that (6.3.6c) remains valid for $\delta = O(1)$ for small $\varepsilon$ in flat geometries. This has practical implications for laboratory analogues of atmospheric motions, i.e., they need not be confined to small-aspect-ratio systems. However, in order to be able to ignore the effect of the vertical component of the Coriolis acceleration on the sphere, (6.3.6c) requires $\delta \ll 1$.

is $O(1)$, and $N_s$ is the Brunt–Väisälä frequency of the rest-state atmosphere, i.e.,

$$N_s^2 = \frac{g}{\theta_s}\frac{d\theta_s}{dz_*} = \frac{1}{D}\frac{g}{\theta_s}\frac{d\theta_s}{dz}. \qquad (6.5.14)$$

The operator $d_0/dt$ is given by (6.3.18). For atmospheric motions of synoptic scale, $S(z)$ is $O(1)$, since $F$ is $O(\varepsilon)$, as is $(1/\theta_s)\,\partial\theta_s/\partial z$. The heating $\mathscr{H}$ either may be considered a given forcing function for (6.5.12) or may depend on $\theta_0$ and the $O(1)$ motion field. In most of the problems to be discussed the accession of heat over the advective time $L/U$ will be considered small and the heating will be generally ignored.

In general (6.5.12) may be solved for $w_1$:

$$w_1 = \left\{ \mathscr{H} - \frac{d_0}{dt}\theta_0 \right\}\frac{1}{S} \qquad (6.5.15)$$

and substituted into (6.3.17) to evaluate the vortex-tube stretching due to $w_1$, i.e.,

$$
\begin{aligned}
\frac{1}{\rho_s}\frac{\partial}{\partial z}\rho_s w_1 &= \frac{1}{\rho_s}\frac{\partial}{\partial z}\left\{ \rho_s\left[ \frac{\mathscr{H}}{S} - \frac{1}{S}\frac{d_0\theta_0}{dt} \right] \right\} \\
&= \frac{1}{\rho_s}\frac{\partial}{\partial z}\frac{\rho_s\mathscr{H}}{S} - \frac{1}{\rho_s}\frac{\partial}{\partial z}\left( \frac{\rho_s}{S}\frac{d_0\theta_0}{dt} \right) \\
&= \frac{1}{\rho_s}\frac{\partial}{\partial z}\left[ \frac{\rho_s\mathscr{H}}{S} \right] - \frac{1}{\rho_s}\frac{d_0}{dt}\left( \frac{\partial}{\partial z}\left[ \frac{\rho_s}{S}\theta_0 \right] \right) \\
&\quad + S^{-1}\left( \frac{\partial u_0}{\partial z}\frac{\partial\theta_0}{\partial x} + \frac{\partial v_0}{\partial z}\frac{\partial\theta_0}{\partial y} \right).
\end{aligned}
\qquad (6.5.16)
$$

The geostrophic approximation (6.3.6a,b), when combined with the hydrostatic approximation (6.5.8), yields the thermal wind equation

$$
\begin{aligned}
\frac{\partial u_0}{\partial z} &= -\frac{\partial\theta_0}{\partial y}, \\
\frac{\partial v_0}{\partial z} &= \frac{\partial\theta_0}{\partial x},
\end{aligned}
\qquad (6.5.17)
$$

and the application of these relations to the last term on the right-hand side of (6.5.16) shows that the term vanishes identically. Thus the vorticity equation (6.3.17) may be written

$$\frac{d_0}{dt}\left[ \zeta_0 + \beta y + \frac{1}{\rho_s}\frac{\partial}{\partial z}\left( \frac{\rho_s}{S}\theta_0 \right) \right] = \frac{1}{\rho_s}\frac{\partial}{\partial z}\left[ \frac{\rho_s\mathscr{H}}{S} \right] \qquad (6.5.18)$$

which in the absence of heating becomes the conservation law

$$\frac{d_0}{dt}\left[ \zeta_0 + \beta y + \frac{1}{\rho_s}\frac{\partial}{\partial z}\left( \frac{\rho_s}{S}\theta_0 \right) \right] = 0. \qquad (6.5.19)$$

The geostrophic and hydrostatic approximations allow us to express each dependent variable in (6.5.19) in terms of

$$\psi = p_0, \tag{6.5.20}$$

so that (6.5.19) becomes the governing equation of motion, i.e.,

$$\left[ \frac{\partial}{\partial t} + \frac{\partial \psi}{\partial x}\frac{\partial}{\partial y} - \frac{\partial \psi}{\partial y}\frac{\partial}{\partial x} \right] \left[ \frac{\partial^2 \psi}{\partial x^2} + \frac{\partial^2 \psi}{\partial y^2} + \frac{1}{\rho_s}\frac{\partial}{\partial z}\left( \frac{\rho_s}{S}\frac{\partial \psi}{\partial z} \right) + \beta y \right] = 0. \tag{6.5.21}$$

A comparison of (6.5.21) with (3.12.30) shows that (6.5.21) is the generalization to a stratified fluid on a sphere of the quasigeostrophic potential-vorticity equation (3.12.30) for a homogeneous layer of fluid. The generalization is further sharpened if we note that $S^{1/2}$ is the ratio of the geometric scale $L$ and the intrinsic length

$$L_D = \frac{N_s D}{f_0} = \frac{(g'D)^{1/2}}{f_0}, \tag{6.5.22}$$

where

$$g' = \frac{gD}{\theta_s}\frac{\partial \theta_s}{\partial z_*}$$

is the "reduced gravity," i.e., the effective gravitational acceleration for buoyant fluid elements. The length $L_D$ is called the *internal Rossby radius of deformation*, or when the context is clear, simply the deformation radius. Note that the ratio of $L_D$ to the external deformation radius, $R$, is

$$\frac{L_D}{R} = \left( \frac{1}{\theta_s}\frac{\partial \theta_s}{\partial z} \right)^{1/2} \ll 1. \tag{6.5.23}$$

The governing equation for synoptic-scale atmospheric motion, (6.5.21) (or the generalization (6.5.18)), is completely written in terms of the $O(1)$ pressure field $p_0$. Once $p_0$, or $\psi$, has been determined, $u_0$, $v_0$, $\rho_0$, $\theta_0$, and $w_1$ follow directly from (6.3.6a,b,c), (6.5.8), and (6.5.15).

Before discussing the appropriate boundary conditions for (6.5.21), the relationship with Ertel's theorem (2.5.8), (2.5.9) will be established. In the case $\mathcal{H} = 0$, the potential temperature is a conserved quantity, so that $\lambda$ in (2.5.8) may be identified with $\theta_*$. Hence the *dimensional* potential vorticity is

$$\Pi_* = \frac{\boldsymbol{\omega}_{*a} \cdot \nabla \theta_*}{\rho_*}, \tag{6.5.24}$$

where

$$\boldsymbol{\omega}_{*a} = (f_0 + \beta_0 y_* + \zeta_*)\mathbf{k} + \mathbf{i}\xi_* + \mathbf{j}(\eta_* + f_0 \cot \theta_0). \tag{6.5.25}$$

The unit vectors $\mathbf{i}$, $\mathbf{j}$, and $\mathbf{k}$ are eastward, northward, and upward respectively, and $\zeta_*$, $\xi_*$, and $\eta_*$ are the vertical and horizontal components of vorticity. The first term in (6.5.25) is the vertical component of vorticity, while the second and third terms are the horizontal components (including the northward component of the planetary vorticity). If $\Pi_*$ is expressed in nondimensional units, with the aid of (6.5.3) we obtain

$$
\Pi_* = \left[\frac{f_0/D}{\rho_s(1 + \varepsilon F\rho)}\right] \left[\{1 + \varepsilon(\zeta + \beta y)\}\left\{\frac{\partial\theta_s}{\partial z} + \varepsilon F\frac{\partial}{\partial z}(\theta\theta_s)\right\}\right.
$$
$$
\left. + \varepsilon F\left\{\varepsilon\zeta\frac{\partial\theta}{\partial x} + \varepsilon\eta\frac{\partial\theta}{\partial y} + \varepsilon\delta\cot\theta_0\frac{\partial\theta}{\partial y}\right\}\right]
\tag{6.5.26}
$$

while the operator $d/dt_*$ in dimensionless units is

$$
\frac{d}{dt_*} = \frac{U}{L}\left[\frac{\partial}{\partial t} + u\frac{\partial}{\partial x} + v\frac{\partial}{\partial y} + w\frac{\partial}{\partial z}\right].
\tag{6.5.27}
$$

If only terms of $O(\varepsilon^2, \varepsilon F, \varepsilon \, \partial\theta_s/\partial z)$ are retained, the conservation statement

$$
\frac{d}{dt_*}\Pi_* = 0
\tag{6.5.28}
$$

becomes

$$
\frac{d_0}{dt}\{\zeta_0 + \beta y\} + \frac{F}{(\rho_s/\theta_s)\,\partial\theta_s/\partial z}\frac{d_0}{dt}\frac{\partial\theta_0}{\partial z} + \frac{\rho_s w_1}{\partial\theta_s/\partial z}\frac{\partial}{\partial z}\left(\frac{\partial\theta_s/\partial z}{\rho_s}\right) = 0, \quad (6.5.29)
$$

or with the adiabatic form of (6.5.15),

$$
\frac{d_0}{dt}\{\zeta_0 + \beta y\} + \frac{1}{S}\frac{d_0}{dt}\frac{\partial\theta_0}{\partial z} - \frac{\rho_s}{S^2}\left(\frac{d_0}{dt}\theta_0\right)\frac{\partial}{\partial z}\frac{S}{\rho_s} = 0,
\tag{6.5.30}
$$

or finally, after combining the last two terms,

$$
\frac{d_0}{dt}\left\{\zeta_0 + \beta y + \frac{1}{\rho_s}\frac{\partial}{\partial z}\frac{\rho_s}{S}\theta_0\right\} = 0,
\tag{6.5.31}
$$

which is identical to (6.5.19).

Therefore, the governing equation of motion for synoptic-scale motions in the atmosphere is the quasigeostrophic potential-vorticity equation, i.e., it is the statement of conservation of potential vorticity in which the geostrophic and hydrostatic approximations are used to evaluate horizontal velocities and potential temperature in terms of the pressure. Further, the $\beta$-plane approximation may be used, in which the *only* $O(1)$ effect of sphericity is to introduce a linear variation in the planetary-vorticity contribution to the potential vorticity. It is extremely important to note that although the horizontal components of vorticity are large compared to the vertical component of the relative vorticity (by a factor $L/D$), it is the vertical component of relative vorticity, by virtue of the basic stratification, that enters the potential-vorticity equation as the controlling influence on the dynamics. It is also important to note that geostrophy does not require $\theta_s^{-1}(d\theta_s/dz) = O(\varepsilon)$, i.e., $S = O(1)$. If $S$ is very much less than one (6.5.15) merely becomes an inappropriate equation for $w_1$. Indeed, if $S \to 0$, the homogeneous form of the potential-vorticity equation is to be used instead. On the other hand, if $S$ is very large (i.e., if the stratification is very large), then $w_1 \to 0$, the mechanism of vortex stretching is essentially eliminated, and (6.5.21) reduces to the statement of the conservation of absolute vorticity, $\zeta_0 + \beta y$.

Finally, for the sake of further reference (6.5.21) may be expressed in terms of *dimensional* units as

$$\left[\frac{\partial}{\partial t_*} + u_* \frac{\partial}{\partial x_*} + v_* \frac{\partial}{\partial y_*}\right]\left[\zeta_* + \beta_0 y_* + \frac{1}{\rho_s}\frac{\partial}{\partial z_*}\left|\frac{f_0^2}{N_s^2}\rho_s \frac{\partial}{\partial z}\frac{\delta p_*}{\rho_s}\right|\right] = 0, \quad (6.5.32)$$

where

$$\delta p_* = p_*(x, y, z, t) - p_s(z) \quad (6.5.33)$$

and

$$u_* = -\frac{1}{\rho_s f_0}\frac{\partial \delta p_*}{\partial y_*},$$

$$v_* = \frac{1}{\rho_s f_0}\frac{\partial}{\partial x_*}\delta p_*, \quad (6.5.34)$$

$$\zeta_* = \frac{1}{\rho_s f_0}\left(\frac{\partial^2}{\partial x_*^2} + \frac{\partial^2}{\partial y_*^2}\right)\delta p_*.$$

Note again that the geostrophic relations (6.5.34) involve *only* $f_0$, i.e., only the constant part of the Coriolis parameter.

## 6.6 The Ekman Layer in a Stratified Fluid

For a homogeneous fluid the vertical velocity pumped out of the Ekman layer formed an essential component in the specification of the boundary condition on the interior vertical velocity at solid surfaces. In this section we examine what effect, if any, the presence of stratification will have on the Ekman layer and in particular on the Ekman vertical velocity.

We will use the $\beta$-plane approximation for the $O(1)$ momentum equations as discussed in Section 6.2. In the vicinity of the lower boundary we further assume, as in Chapter 4, that the dominant contributions to the frictional forces can be represented as

$$\frac{\mathscr{F}_{*\phi}}{\rho_* U f_0} = \frac{A_V}{f_0 D^2}\frac{\partial^2 u}{\partial z^2} \equiv \frac{E_V}{2}\frac{\partial^2 u}{\partial z^2}$$

$$(6.6.1)$$

$$\frac{\mathscr{F}_{*\theta}}{\rho_* U f_0} = \frac{A_V}{f_0 D^2}\frac{\partial^2 v}{\partial z^2} \equiv \frac{E_V}{2}\frac{\partial^2 v}{\partial z^2}.$$

Let us suppose that an ordinary Ekman layer exists on the surface $z = 0$, i.e., the lower boundary of the atmosphere. Introducing the boundary-layer coordinate

$$\zeta = \frac{z}{E_V^{1/2}}$$

and writing all variables in the boundary layer as functions of $x$, $y$, and $\zeta$ yields, from (6.2.24a,b), the following $O(1)$ equations, to $O(\varepsilon, \delta, L/r_0)$:

$$-\tilde{v}_0 = -\frac{\partial \tilde{p}_0}{\partial x} + \frac{1}{2}\frac{\partial^2 \tilde{u}_0}{\partial \zeta^2}, \tag{6.6.2a}$$

$$\tilde{u}_0 = -\frac{\partial \tilde{p}_0}{\partial y} + \frac{1}{2}\frac{\partial^2 \tilde{v}_0}{\partial \zeta^2}, \tag{6.6.2b}$$

where the tilde $\tilde{\ }$ reminds us that these variables are the representation of the dynamic fields in the presumed Ekman layer near $z = 0$. The equations (6.6.2a,b) are identical to (4.5.25a,b), i.e., the Ekman-layer equations for the homogeneous fluid. However, the crucial question is whether $\tilde{p}_0$ is independent of $\zeta$ as in the homogeneous case or whether the presence of density variations produces an $O(1)$ $\partial \tilde{p}_0/\partial \zeta$. The continuity equation becomes

$$\frac{\partial \tilde{w}}{\partial \zeta} = -E_V^{1/2}\left(\frac{\partial \tilde{u}_0}{\partial x} + \frac{\partial \tilde{v}_0}{\partial y}\right) \tag{6.6.3}$$

which implies that $\tilde{w}$ is $O(E_V^{1/2})$ in the Ekman layer. This allows an estimate of $\partial \tilde{p}_0/\partial \zeta$ from the hydrostatic approximation, which is certainly valid in the Ekman layer, whose aspect ratio is even smaller than the interior's. Hence

$$\frac{\partial \tilde{p}_0}{\partial \zeta} = E_V^{1/2}\tilde{\theta}_0. \tag{6.6.4}$$

On the other hand, from (6.5.12), (6.3.12),

$$\tilde{\theta}_0 = O(w_1 S) = O\left(\frac{w}{\varepsilon}S\right) = O\left(E_V^{1/2}\frac{S}{\varepsilon}\right), \tag{6.6.5}$$

so that

$$\frac{\partial \tilde{p}_0}{\partial \zeta} = O\left(E_V\frac{S}{\varepsilon}\right). \tag{6.6.6}$$

Thus as long as

$$E_V < O\left(\frac{\varepsilon}{S}\right), \tag{6.6.7}$$

$\tilde{p}_0$ will be constant over the Ekman-layer depth and the homogeneous model of the Ekman layer remains valid for the stratified fluid. In particular, from (4.5.39)

$$w(x, y, 0) = \varepsilon w_1(x, y, 0) = \frac{E_V^{1/2}}{2}\zeta_0(x, y, 0),$$

or, at the lower boundary,

$$w_1(x, y, 0) = \frac{E_V^{1/2}}{2\varepsilon} \zeta_0(x, y, 0).$$ 

(6.6.8)

In the presence of a topographic slope, from (4.9.36)

$$\boxed{w_1(x, y, 0) = \frac{E_V^{1/2}}{2\varepsilon} \zeta_0(x, y, 0) + \mathbf{u}_0 \cdot \nabla\left(\frac{h_B}{\varepsilon D}\right),}$$

(6.6.9)

where $h_B(x, y)$ is the elevation of the lower boundary above the reference level $z = 0$. This is the lower boundary condition for (6.5.21) or (6.5.18), and written in terms of $\psi$ it becomes, with (6.5.12),

$$\boxed{S^{-1}\left(\frac{d_0}{dt}\frac{\partial \psi}{\partial z}\right)_{z=0} = -\frac{\partial \psi}{\partial x}\frac{\partial \eta_B}{\partial y} + \frac{\partial \psi}{\partial y}\frac{\partial \eta_B}{\partial x} - \frac{E_V^{1/2}}{2\varepsilon}\left(\frac{\partial^2 \psi}{\partial x^2} + \frac{\partial^2 \psi}{\partial y^2}\right) + \frac{\mathcal{H}(x, y, 0)}{S}.}$$

(6.6.10)

## 6.7 Boundary Conditions for the Potential-Vorticity Equation: the Atmosphere

On the earth's surface, (6.6.10) provides the boundary condition for the quasigeostrophic potential vorticity equation. The upper boundary condition is a more difficult problem. The atmosphere is not bounded above, although the most energetic "weather"-related motions of synoptic scale are to a great extent confined to the troposphere. We will defer the mathematical formulation of the upper boundary condition to Section 6.12, but in physical terms it can be stated as follows. For the motions to be discussed in this book, either the amplitude of the energy of the motion must decay to zero for large $z$, or if it remains finite but nonzero, the flux of motion energy must be directed *upwards*.

## 6.8 Quasigeostrophic Potential-Vorticity Equation for Oceanic Synoptic Scales

For oceanic motions for which $\beta = O(1)$ and $\varepsilon \ll 1$, (6.3.17), as in the atmospheric counterpart, is the $O(1)$ vorticity equation. However, the thermodynamic equation for oceanic motions is (6.2.7) rather than (6.2.4), and is written entirely in terms of the density. Since by (6.2.21)

$$\rho_* = \rho_s(z)\{1 + \varepsilon F\rho\},$$

(6.8.1)

(6.2.7) becomes, in nondimensional form,

$$\varepsilon F \frac{d\rho}{dt} + \frac{w}{\rho_s}\frac{\partial \rho_s}{\partial z}(1 + \varepsilon F\rho) = -\frac{\mathcal{H}_* L}{U},$$

(6.8.2)

where

$$\mathcal{H}_* = -\frac{\kappa \nabla^2 \rho_*}{\rho_s} + \frac{\alpha \rho_0}{\rho_s C_p} Q.$$

In the oceans the density field changes by only a few parts per thousand over the entire depth of the ocean, and therefore, as shown by (6.4.14),

$$\frac{D}{\rho_s} \frac{\partial \rho_s}{\partial z_*} = \frac{1}{\rho_s} \frac{\partial \rho_s}{\partial z} = O(10^{-3}) = O(F) = O(\varepsilon).$$

Hence to *lowest* order in the $\varepsilon$-expansion (6.8.2) becomes

$$-\frac{d_0 \rho_0}{dt} + w_1 S = \mathcal{H}, \tag{6.8.3}$$

where

$$\mathcal{H} = \frac{\mathcal{H}_*}{(U^2/gD)f_0} \tag{6.8.4}$$

and

$$S \doteq \frac{N_s^2 D^2}{f_0^2 L^2} = O(1).$$

In this case the Brunt–Väisälä frequency $N$ is given by (6.4.12), i.e.,

$$N_s = \left| -\frac{g}{\rho_s} \frac{\partial \rho_s}{\partial z_*} \right|^{1/2}. \tag{6.8.5}$$

The similarity of (6.8.3) to (6.5.12) is striking. Since $(1/\rho_s) \partial \rho_s / \partial z$ is small, the hydrostatic approximation (6.3.6c) becomes

$$\rho_0 = -\frac{\partial p_0}{\partial z}, \tag{6.8.6}$$

which is the oceanic counterpart of (6.5.8). The thermal wind equations become

$$\frac{\partial u_0}{\partial z} = \frac{\partial \rho_0}{\partial y},$$

$$\frac{\partial v_0}{\partial z} = -\frac{\partial \rho_0}{\partial x}, \tag{6.8.7}$$

which are precisely the same as (6.5.17) with $\theta_0$ replaced by $-\rho_0$. If (6.8.3) and (6.8.7) are used to evaluate the vortex-tube stretching term in (6.3.17), after again using the smallness of $(1/\rho_s) \partial \rho_s / \partial z$, we obtain

$$\frac{d_0}{dt} \left| \zeta_0 + \beta y - \frac{\partial}{\partial z} \left( \frac{\rho_0}{S} \right) \right| = \frac{\partial}{\partial z} \left[ \frac{\mathcal{H}}{S} \right]. \tag{6.8.8}$$

In the absence of heating, (6.8.8) becomes the conservation law

$$\frac{d_0}{dt}\left\{\zeta_0 + \beta y - \frac{\partial}{\partial z}\left(\frac{\rho_0}{S}\right)\right\} = 0, \tag{6.8.9}$$

which, upon use of the geostrophic and hydrostatic approximations, can be written entirely in terms of the geostrophic stream function

$$\psi = p_0, \tag{6.8.10}$$

i.e.,

$$\left[\frac{\partial}{\partial t} + \frac{\partial \psi}{\partial x}\frac{\partial}{\partial y} - \frac{\partial \psi}{\partial y}\frac{\partial}{\partial x}\right]\left[\frac{\partial^2 \psi}{\partial x^2} + \frac{\partial^2 \psi}{\partial y^2} + \frac{\partial}{\partial z}\left(\frac{1}{S}\frac{\partial \psi}{\partial z}\right) + \beta y\right] = 0. \tag{6.8.11}$$

A comparison of (6.8.11) with (6.5.21) shows that the governing equation of motion for oceanic synoptic-scale motions is precisely the *same* as for atmospheric motions, with the single further simplification that the *basic* density field may be considered constant except insofar as it contributes to the static-stability parameter $S$. This striking similarity between the governing equations of motion for synoptic-scale motions in the atmosphere and oceans is the crucial fact that provides a single, unified conceptual framework in geophysical fluid dynamics for the description and understanding of both atmospheric and oceanic synoptic-scale motions. In both fluids the dynamical process is determined by potential-vorticity conservation, in which horizontal velocities and the thermodynamic variable are consistently evaluated geostrophically and hydrostatically and in which the Coriolis parameter is considered constant except insofar as it produces a linearly varying planetary-vorticity contribution to the total potential vorticity. In both systems the effect of stratification is measured by $S$, which can be written

$$S = \left(\frac{L_D}{L}\right)^2,$$

where

$$L_D = \frac{N_s D}{f_0}$$

and $L$ is the spatial scale of the motion. For the atmospheric synoptic scale $L_D = O(10^3 \text{ km})$, while for the oceans $L_D = O(100 \text{ km})$. The condition $\beta = O(1)$ in both fluids requires $L$ to be $O(1000 \text{ km})$ in the atmosphere and $O(100 \text{ km})$ for the ocean. Thus, for synoptic-scale motions in both the atmosphere and the ocean, $L$ is $O(L_D)$. This implies that the variable planetary vorticity, the relative vorticity, and the potential vorticity contribution due to the thermodynamic field are each of the *same* order. Indeed, the requirement that the relative vorticity dynamics be linked to the thermodynamics is precisely the condition $S = O(1)$. It is important to keep in mind, though, that $\beta = O(1)$ and $S = O(1)$ are both observational facts and are not required *a priori*. In principle, synoptic-scale motions in either the atmosphere or the ocean could have $\beta = O(1)$ and $S$ small or large, or $S = O(1)$ and $\beta$ different from $O(1)$. The basic mechanisms which favor motions on these scales have yet to be discussed. What we have shown to this point is that whatever the mechanism is to produce synoptic-scale motions in the oceans and atmosphere, their dynamics are described in the same theoretical framework.

Finally it should be noted that although the potential-vorticity equation is a fully three-dimensional partial differential equation (i.e., $\psi$ is a function of all three space dimensions), Equations (6.5.21) and (6.8.11) still have a strongly two-dimensional character. Only the advection of the geostrophic potential vorticity by the horizontal velocity explicitly enters the equations, due to the smallness of $w$.

## 6.9 Boundary Conditions for the Potential-Vorticity Equation: the Oceans

On the lower boundary of the ocean (6.6.9) and (6.6.10) apply as in the atmospheric case. The boundary condition at the upper boundary is of course different, because the oceanic height is finite and the extremely sharp density jump, between the ocean and the air above, acts as a very effective barrier to the upward transfer of mechanical energy. As in Section 6.6, it may easily be shown that the upper Ekman layer is unaffected by stratification. Consider the upper surface at $z_* = h_* = O(D)$. Then in our dimensionless units, the upper boundary condition is given by (4.10.18), i.e.,

$$w(x, y, h_*/D) = \frac{d}{dt}\left(\frac{h_*}{D}\right) + \frac{\tau_0}{\rho_s f_0 UD} \mathbf{k} \cdot \operatorname{curl} \boldsymbol{\tau}, \qquad (6.9.1)$$

where the applied stress at the upper boundary is

$$\boldsymbol{\tau}_* = \tau_0 \boldsymbol{\tau}. \qquad (6.9.2)$$

Now at the upper boundary the pressure of the ocean must match the external pressure imposed by the atmosphere, $p_{*a}$, i.e.,

$$p_* = p_s(z) + \rho_s U f_0 Lp = p_{*a}(x, y, t), \quad z = \frac{h_*}{D}. \qquad (6.9.3)$$

Write

$$h_* = D(1 + \mu\eta), \qquad (6.9.4)$$

where $\eta(x, y, t)$ is the nondimensional deviation of the upper surface and $\mu$ is an (as yet) undetermined scaling constant. However, we expect that $\mu$ will be small. The upper boundary is therefore at $z = 1 + \mu\eta(x, y, t)$, so that (6.9.3) may be rewritten at $z = 1$, by the use of a Taylor series expansion, i.e., keeping only *linear* first order terms in the motion fields,

$$p_*\left(x, y, \frac{h_*}{D}\right) = p_s(1) + \frac{\partial p_s}{\partial z}\mu\eta + \cdots + \rho_s Uf_0 Lp(x, y, 1) + \cdots = p_{*a}.$$

$$(6.9.5)$$

But since

$$\frac{1}{D}\frac{\partial p_s}{\partial z} = -\rho_s g, \qquad (6.9.6)$$

it follows that at $z = 1$,

$$\mu\eta = \frac{Uf_0 L}{gD}p(x, y, 1) - \frac{p_{*a} - p_s(1)}{\rho_s gD}. \qquad (6.9.7)$$

Thus the appropriate *choice* for $\mu$ is

$$\mu = \frac{Uf_0 L}{gD} = \varepsilon F. \qquad (6.9.8)$$

Thus the free surface is written as

$$h_* = D(1 + \varepsilon F\eta), \qquad (6.9.9)$$

where

$$\eta = p(x, y, 1, t) + \frac{p_s(1) - p_{*a}}{\rho_s f_0 UL}. \qquad (6.9.10)$$

Therefore, the upper boundary condition, to lowest order in $\varepsilon$, becomes

$$w(x, y, 1, t) = \varepsilon w_1(x, y, 1, t) = \varepsilon F\frac{d_0}{dt}p_0(x, y, 1, t)$$

$$(6.9.11)$$

$$+ \frac{\tau_0}{\rho_s f_0 UD}\mathbf{k}\cdot\text{curl }\tau + \frac{1}{gD\rho_s}\frac{d_0}{dt}[p_s(1) - p_{*a}],$$

or

$$w_1(x, y, 1, t) = F\frac{d_0\psi}{dt} + \frac{\tau_0}{\varepsilon\rho_s f_0 UD}\mathbf{k}\cdot\text{curl }\tau$$

$$(6.9.12)$$

$$- \frac{1}{gD\varepsilon}\frac{d_0}{dt}\frac{p_{*a}}{\rho_s}.$$

Now, on the basis of our earlier atmospheric analysis we can estimate that

$$p_{*a} = O(\rho_{*a}U_a f_0 L_a),$$

where the subscript $a$ refers to atmospheric scales. Furthermore the horizontal scale for the variable $p_{*a}$ is on a scale $L_a = O(10^3 \text{ km})$ rather than $L = O(100 \text{ km})$, so that

$$\frac{1}{gD\varepsilon}\frac{d}{dt}\frac{o p_{*a}}{\rho_s} = O\left(\frac{\rho_{*a}U_a}{\rho_s U}\right)\frac{f_0^2 L^2}{gD} = O(10^{-3}). \qquad (6.9.13)$$

The ratio of the momentum per unit volume of the atmosphere to that of the ocean is $O(1)$, the smaller density of the air compensating its larger velocity. Hence the last term on the right of (6.9.12) is $O(F)$, like the first term, and can be neglected. The term involving the wind-stress curl is $O(\tau_0/\rho_s f_0 UD)L/L_a$ after account is taken of the fact that $\tau$, like $p_{*a}$, varies on the larger scale $L_a$ rather than $L$. If $\tau_0 = 1$ dyne/cm$^2$, $D = 4$ km, $U = 5$ cm/s, and $L/L_a = 10^{-1}$, then the parameter

$$\frac{\tau_0}{\varepsilon\rho_s f_0 UD} \frac{L}{L_a} = 10^{-1},$$

which while small is not utterly negligible. In fact, during the passage of more intense atmospheric storms it can be $O(1)$. Thus while the wind-stress curl on the synoptic scale is generally a minor contribution to $w_1$ and hence to the potential-vorticity balance, it is reasonable to retain it for the sake of generality. Thus on $z = 1$, the boundary condition for (6.8.11) becomes

$$w_1 = -S^{-1}\left(\frac{d_0}{dt}\frac{\partial\psi}{\partial z}\right)_{z=1} = \left[\frac{\tau_0 L^2}{\rho_s U^2 DL_a}\right]\mathbf{k}\cdot\text{curl }\tau . \qquad (6.9.14)$$

It is important to note that the motion of the free surface, $F d_0\psi/dt$, plays no role in the dynamics of the synoptic scale. That is, as far as the quasigeostrophic dynamics are concerned, the free surface can be thought of as rigid. The reason for this is that $F$ is small while $S$ is $O(1)$, which implies that

$$L_D = O(L) \ll R = \frac{(gD)^{1/2}}{f_0}. \qquad (6.9.15)$$

The length scale of the motion is of the order of the *internal* deformation radius and is far smaller than $R$, which is the deformation radius of the free surface. As was noted in Chapter 3, whenever $L \ll R$ the free surface acts as a rigid lid. The isopycnal surfaces within the ocean deform far more easily than the free surface, since less potential energy is associated with a vertical displacement of the former, and therefore it is the deformation of the isopycnals rather than the free surface that produces vorticity changes on the synoptic scale by vortex-tube stretching if $L \ll R$.

On the vertical boundaries that bound the ocean laterally, the $O(1)$ geostrophic motion must vanish, i.e.,

$$\mathbf{n} \times \nabla\psi = 0, \qquad (6.9.16)$$

where $\mathbf{n}$ is a horizontal unit vector normal to the basin perimeter.

In a manner completely similar to the derivation of Equation (3.25.7), it also follows that the integral condition

$$\frac{\partial}{\partial t}\oint_c (\nabla\psi \cdot \mathbf{n})\, dl = 0 \qquad (6.9.17)$$

must be satisfied on each closed, solid boundary $C$ bounding the fluid, in order that the $O(\varepsilon)$ horizontal normal velocities may also vanish on the perimeter. Equations (6.9.16) and (6.9.17) must be satisfied *for all z on the boundary.*

## 6.10  Geostrophic Energy Equation and Available Potential Energy

The quasigeostrophic potential-vorticity equation (6.5.21) or (6.8.11) with the appropriate boundary and initial conditions forms the complete dynamical basis for the calculation of the fields of motion. Other mathematical statements concerning energy are derivable *directly* from the potential-vorticity equation. However, a satisfying interpretation of the results requires consideration of the momentum and thermodynamic equations. In this section the energy equation for synoptic-scale atmospheric motions is derived; the oceanic counterpart follows trivially by a mere redefinition of certain terms and the simplification inherent in the smallness of the variation of $\rho_s$.

If (6.5.21) is multiplied by $-\rho_s \psi$, it follows that

$$\frac{\partial}{\partial t} \frac{\rho_s}{2} \left\{ \left(\frac{\partial \psi}{\partial x}\right)^2 + \left(\frac{\partial \psi}{\partial y}\right)^2 + \frac{1}{S}\left(\frac{\partial \psi}{\partial z}\right)^2 \right\} + \nabla \cdot \mathbf{\bar S} = 0, \qquad (6.10.1)$$

where $\mathbf{\bar S}$ is the vector

$$\begin{aligned}
\mathbf{\bar S} = &\; \mathbf{i}\left\{ -\rho_s \psi \frac{\partial^2 \psi}{\partial x\, \partial t} - \frac{\rho_s \beta \psi^2}{2} - \rho_s u_0 \psi \Pi_0 \right\} \\
&+ \mathbf{j}\left\{ -\rho_s \psi \frac{\partial^2 \psi}{\partial y\, \partial t} - \rho_s v_0 \psi \Pi_0 \right\} \\
&+ \mathbf{k}\left\{ -\frac{\rho_s \psi}{S} \frac{\partial^2 \psi}{\partial z\, \partial t} \right\}.
\end{aligned} \qquad (6.10.2)$$

Here $u_0$, $v_0$, and $\Pi_0$ are given by

$$u_0 = -\frac{\partial \psi}{\partial y}, \qquad (6.10.3a)$$

$$v_0 = \frac{\partial \psi}{\partial x}, \qquad (6.10.3b)$$

$$\Pi_0 = \frac{\partial^2 \psi}{\partial x^2} + \frac{\partial^2 \psi}{\partial y^2} + \frac{1}{\rho_s}\frac{\partial}{\partial z}\frac{\rho_s}{S}\frac{\partial \psi}{\partial z}. \qquad (6.10.3c)$$

Consider the motion in a region $R$ which is bounded on $z = 0$ by a rigid boundary where (6.6.10) applies. Since we have assumed $\mathcal{H} = 0$ to derive (6.10.1) (the generalization is trivial), (6.6.10) when multiplied by $\rho_s \psi$

becomes

$$\frac{\rho_s \psi}{S} \frac{\partial^2 \psi}{\partial z \, \partial t} = - \frac{\partial}{\partial x} \left\{ \rho_s u_0 \left[ S^{-1} \frac{\partial \psi}{\partial z} + \eta_B \right] \right\} - \frac{\partial}{\partial y} \left\{ \rho_s v_0 \left[ S^{-1} \frac{\partial \psi}{\partial z} + \eta_B \right] \right\}$$

$$- \frac{E_V^{1/2}}{2\varepsilon} \rho_s \psi \left[ \frac{\partial^2 \psi}{\partial x^2} + \frac{\partial^2 \psi}{\partial y^2} \right]. \tag{6.10.4}$$

The perimeter of $R$ may be composed of either closed or open vertical boundaries. If the boundary is considered closed, then (6.9.16) and (6.9.17) must apply. If the boundary is open, the fields are generally undetermined on the boundary. In the case of the oceans (6.9.16) and (6.9.17) are certainly appropriate. For the atmosphere we often consider a latitude band, i.e., an open $x$-interval on which the fields are periodic with a spatial period equal to the distance $2\pi(\cos \theta_0) r_0 / L$, which is the nondimensional distance around the earth at the central latitude $\theta_0$. The region must however be finite in $y$ for the $\beta$-plane approximation to be sensible, and the specification of the fields at the boundaries of the latitude band, $y_1 \leq y \leq y_2$, is somewhat artificial. The $\beta$-plane idea *implicitly* assumes that the dynamical processes occurring in a region around $\theta_0$ are essentially comprehensible in terms of the dynamics of the fluid occupying that band of latitude in isolation from the fluid at more distant regions of the globe. This implicit notion is given mathematical expression by the boundary conditions

$$v_1 = v_0 = 0 \quad \text{at } y = y_1 \text{ and } y = y_2, \tag{6.10.5}$$

which acts to isolate energetically the fluid in the latitude band from its surroundings and is equivalent to the application of (6.9.16) and (6.9.17).

If (6.10.1) is integrated over the volume $R$ and (6.9.16), (6.9.17), and (6.10.4) are applied on the boundaries of $R$, we obtain

$$\frac{\partial}{\partial t} \iiint_R \frac{\rho_s}{2} \left[ \left( \frac{\partial \psi}{\partial x} \right)^2 + \left( \frac{\partial \psi}{\partial y} \right)^2 + \frac{1}{S} \left( \frac{\partial \psi}{\partial z} \right)^2 \right] dx \, dy \, dz$$

$$= + \iint_A dx \, dy \frac{\rho_s \psi}{S} \frac{\partial^2 \psi}{\partial z \, \partial t} \bigg|_{z = z_T} \tag{6.10.6}$$

$$- \frac{E_V^{1/2}}{2\varepsilon} \iint_A dx \, dy \frac{\rho_s}{2} \left| \left( \frac{\partial \psi}{\partial x} \right)^2 + \left( \frac{\partial \psi}{\partial y} \right)^2 \right| \bigg|_{z = 0},$$

where $z_T$ is the upper boundary of the region and $A$ is the area on the $x, y$ plane occupied by $R$.

The integrand of the integral on the left-hand side of (6.10.6) is the positive definite scalar quantity

$$E = \frac{\rho_s}{2} \left[ \left( \frac{\partial \psi}{\partial x} \right)^2 + \left( \frac{\partial \psi}{\partial y} \right)^2 + \frac{1}{S} \left( \frac{\partial \psi}{\partial z} \right)^2 \right]$$

$$= \frac{\rho_s}{2} \left[ u_0^2 + v_0^2 + \frac{\theta_0^2}{S} \right]. \tag{6.10.7}$$

The term $\rho_s(u_0^2 + v_0^2)/2$ is clearly the kinetic energy per unit volume of the fluid. The vertical velocity makes a contribution to the kinetic energy of $O(\delta^2 \varepsilon^2)$ and is therefore utterly negligible. The remaining term $\rho_s \theta_0^2/2S$ is called the *available potential energy* and has the following interpretation. Consider the atmosphere (or ocean) in a state of rest with absolutely level density and pressure surfaces. The potential energy of this equilibrium state, in which no horizontal variations of $p$, $\rho$, and $\theta$ exist, can be defined as the ground state of potential energy, for clearly no potential energy in this state can spontaneously be liberated to yield kinetic energy. Now consider a disturbed state of the atmosphere or ocean characterized by departures from the equilibrium state. This state can be achieved by considering the process by which fluid elements are displaced a distance $\delta z_*$ vertically, where $\delta z_*$ is a function of $x_*$, $y_*$, and $z_*$. Then the restoring force per unit volume felt by the fluid element is, by (6.4.7),

$$F_{z_*} = \rho_s \frac{g}{\theta_s} \frac{\partial \theta_s}{\partial z_*} \delta z_*, \qquad (6.10.8)$$

which is linear in $\delta z_*$, and hence the increase in potential energy of the element is

$$\mathscr{A}_* = \rho_s \frac{g}{\theta_s} \frac{\partial \theta_s}{\partial z_*} \frac{\delta z_*^2}{2}. \qquad (6.10.9)$$

However, the vertical displacement $\delta z_*$ can be directly related to the departures of $\theta_*$ from $\theta_s$ by

$$\delta \theta_* = + \frac{\partial \theta_s}{\partial z_*} \delta z_* + O(\varepsilon^2). \qquad (6.10.10)$$

Thus

$$\mathscr{A}_* = \frac{\rho_s}{2} g \frac{\theta_s}{\partial \theta_s/\partial z_*} \left(\frac{\delta \theta_*}{\theta_s}\right)^2, \qquad (6.10.11)$$

so that to lowest order in $\varepsilon$, after use is made of the relation

$$\frac{\delta \theta_*}{\theta_s} = \varepsilon F \theta_0, \qquad (6.10.12)$$

we have

$$\mathscr{A}_* = \frac{\rho_s U^2 \theta_0^2}{2S} \equiv U^2 \mathscr{A}. \qquad (6.10.13)$$

Thus the *available* potential energy is directly proportional to the departure of the $\theta$-surfaces from their equilibrium, horizontal configuration, and is a quadratic, positive definite function of these departures. It is this portion alone of the *total* potential energy which is available for transformation into kinetic energy. The *total* energy in the geostrophic approximation is therefore the sum of the kinetic energy of the horizontal motion plus the available potential energy. In dimensional units

$$E_* = \frac{\rho_s}{2}\left(u_*^2 + v_*^2 + \frac{g}{N_s^2}\left(\frac{\delta\theta_*}{\theta_s}\right)^2\right),$$
(6.10.14)

or in dimensionless units,

$$E \equiv \frac{E_*}{U^2} = \frac{\rho_s}{2}\left(u_0^2 + v_0^2 + \frac{\theta_0^2}{S}\right).$$
(6.10.15)

The energy equation (6.10.6) therefore states that the rate of change of the sum of the kinetic energy and available potential energy in a closed region $R$ will be altered only by the two terms on the right-hand side. The second term, $O(E_V^{1/2}/\varepsilon)$, represents the dissipation of energy in the surface Ekman layer and is negative definite. The first term is

$$\iint_A dx\, dy \left(\frac{\rho_s\psi}{S}\frac{\partial^2\psi}{\partial z\,\partial t}\right)_{z=z_T} = -\iint_A dx\, dy (\rho_s p_0 w_1)_{z=z_T}.$$
(6.10.16)

If on the upper surface $p_0 w_1$ has a positive average when integrated over $A$ then the fluid in $R$ is doing work, by exerting pressure forces on the surface $z_T$, in which case the energy in $R$ will correspondingly decrease. When dimensional units are used to represent (6.10.16), the energy equation becomes

$$\frac{\partial}{\partial t_*}\iiint_R E_*\, dx_*\, dy_*\, dz_* = -\iint_A dx_*\, dy_*(\delta p_* w_*)_{z=z_T}$$
$$- \frac{\delta_E f_0}{2}\iint_A dx_*\, dy_*\left(\frac{u_*^2 + v_*^2}{2}\right)_{z=0},$$

(6.10.17)

where

$$\delta p_* = p_* - p_s = \rho_s f_0 U L p_0,$$
$$\delta_E = D E_V^{1/2}.$$
(6.10.18)

Thus in the absence of viscous dissipation or work done by (or on) the fluid at the upper surface, the sum of the kinetic energy plus the available potential energy is conserved. The pressure work on $z_T$ will vanish if $w_1$ is zero there or, in the case $z_T \to \infty$, if the product $\rho_s p_0 w_1$ vanishes as $z \to \infty$.

A differential statement of the energy balance is given by (6.10.1) in terms of the energy flux vector $\check{S}$. This differential statement can be rewritten in a more physically meaningful way if use is made of the following identity:

$$\rho_s u_0 \psi \frac{\partial \Pi_0}{\partial x} + \rho_s v_0 \psi \frac{\partial \Pi_0}{\partial y} = \frac{\partial}{\partial x}(\rho_s u_0 \psi \Pi_0) + \frac{\partial}{\partial y}(\rho_s v_0 \psi \Pi_0)$$

$$= \frac{\partial}{\partial x}\left\{ u_0 \psi \rho_s \frac{\partial^2 \psi}{\partial x^2} + v_0 \psi \rho_s \frac{\partial^2 \psi}{\partial x \, \partial y} \right\}$$

$$+ \frac{\partial}{\partial y}\left\{ u_0 \psi \rho_s \frac{\partial^2 \psi}{\partial x \, \partial y} + v_0 \psi \rho_s \frac{\partial^2 \psi}{\partial y^2} \right\} \quad (6.10.19)$$

$$+ \frac{\partial}{\partial z}\left\{ \frac{u_0 \psi \rho_s}{S} \frac{\partial^2 \psi}{\partial z \, \partial x} + \frac{v_0 \psi \rho_s}{S} \frac{\partial^2 \psi}{\partial z \, \partial y} \right\}$$

$$- \frac{\partial}{\partial x}(u_0 E) - \frac{\partial}{\partial y}(v_0 E).$$

Thus $\nabla \cdot \bar{\mathbf{S}} \equiv \nabla \cdot \mathbf{S}$ where

$$\mathbf{S} = \mathbf{i}\left[ u_0 E - \rho_s \psi \frac{d_0 v_0}{dt} - \rho_s \frac{\beta \psi^2}{2} \right]$$

$$+ \mathbf{j}\left[ v_0 E + \rho_s \psi \frac{d_0 u_0}{dt} \right] \quad (6.10.20)$$

$$+ \mathbf{k}\left[ -\frac{\rho_s \psi}{S} \frac{d_0 \theta_0}{dt} \right].$$

If the $O(\varepsilon)$ momentum equations (6.3.13a,b) are used, $\mathbf{S}$ may be further rewritten as

$$\mathbf{S} = \mathbf{i}\left[ u_0 E + \rho_s \psi \left( u_1 + u_0 \frac{L}{r_0 \varepsilon} y \cot \theta_0 + \frac{\partial p_1}{\partial y} \right) - \rho_s \beta \frac{\psi^2}{2} \right]$$

$$+ \mathbf{j}\left[ v_0 E + \rho_s \psi \left( v_1 + v_0 \frac{L}{r_0 \varepsilon} y \cot \theta_0 - \frac{\partial p_1}{\partial x} \right) - \frac{\psi L y}{\varepsilon r_0} \tan \theta_0 \frac{\partial p_0}{\partial x} \right]$$

$$+ \mathbf{k}[\rho_s \psi w_1], \quad (6.10.21)$$

where the thermodynamic equation has been used to evaluate the vertical component of $\mathbf{S}$. Now the vector

$$\mathbf{i}\left\{ \rho_s \frac{\partial}{\partial y} p_1 \psi + \rho_s \psi \frac{u_0 L}{r_0 \varepsilon} y \cot \theta_0 - \rho_s \frac{\beta \psi^2}{2} - \rho_s p_0 \frac{u_0}{\varepsilon} \right\}$$

$$+ \mathbf{j}\left\{ -\rho_s \frac{\partial}{\partial x} p_1 \psi + \rho_s \psi \frac{v_0 L}{r_0 \varepsilon} y \cot \theta_0 - \rho_s p_0 \frac{v_0}{\varepsilon} \right\}$$

is *trivially* nondivergent, so that

$$\nabla \cdot \mathbf{S} \equiv \nabla \cdot \mathbf{J}, \quad (6.10.22)$$

where

$$
\mathbf{J} = \mathbf{i}\left[ u_0 E + \rho_s\left[ \psi u_1 - p_1 \frac{\partial \psi}{\partial y} \right] + \rho_s p_0 \frac{u_0}{\varepsilon} \right]
$$

$$
+ \mathbf{j}\left[ v_0 E + \rho_s\left[ \psi v_1 + p_1 \frac{\partial \psi_1}{\partial x} - \frac{Ly}{\varepsilon r_0} \tan \theta_0 \ \psi v_0 \right] + \rho_s p_0 v_0 \right]
$$

$$
+ \mathbf{k}[\rho_s \psi w_1],
$$

or finally

$$
\mathbf{J} = \mathbf{i}\left[ u_0 E + \rho_s\left| \frac{(p_0 + \varepsilon p_1)(u_0 + \varepsilon u_1)}{\varepsilon} + O(\varepsilon) \right| \right]
$$

$$
+ \mathbf{j}\left[ v_0 E + \rho_s(p_0 + \varepsilon p_1)(v_0 + \varepsilon v_1)\frac{\cos \theta}{\varepsilon} + O(\varepsilon) \right] \qquad (6.10.23)
$$

$$
+ \mathbf{k}\left[ \rho_s p_0 \frac{w}{\varepsilon} \right],
$$

where (6.2.28b) has been used. Hence to $O(\varepsilon)$ the energy flux vector $\mathbf{J}$ is the sum of the flux of energy carried horizontally by the flow itself plus the rate of pressure work done by the fluid element on its surroundings, and

$$
\boxed{\frac{\partial E}{\partial t} + \nabla \cdot \mathbf{J} = 0}. \qquad (6.10.24)
$$

There are several remarks to be made about the form of $\mathbf{J}$. The advantage of writing the energy flux vector in the form (6.10.23) is that its interpretation in terms of the material flux of energy and the pressure work is most intuitive. However, the dominant pressure-work term $p_0 u_0 \mathbf{i} + p_0 v_0 \mathbf{j}$ is trivially non-divergent and does no net work on any fluid element. Because of this fact the $O(\varepsilon)$ corrections to the pressure and velocity field are crucial in correctly determining the net pressure work done on each fluid element. However, the solutions to the potential-vorticity equation yield only the $O(1)$ fields, and $p_1$, $u_1$, and $v_1$ are not directly accessible (although in principle they may be calculated). Therefore S, as given by (6.10.20), is clearly preferable for the actual calculation of the energy-flux divergence. It is important to note that S and J are *not* equal. They differ by a nondivergent vector whose addition, while leaving the physically meaningful statement (6.10.24) unaltered, renders the definition of energy flux *necessarily* ambiguous, since the energy flux vector can always be altered by a nondivergent vector without affecting any observable variable or its rate of change. The vector S is also a preferable definition of energy flux for two further reasons. They are basically aesthetic. First, the vector S *naturally* arises when reference is made only to the *governing* equation of motion, i.e., to the potential-vorticity equation—which, as we have noted, is sufficient unto itself for the determination of the fields of motion. Second S is clearly the generalization of the energy flux vector of

Section 3.21 to nonlinear quasigeostrophic motion of a stratified fluid, and it was seen there that **S** had the pretty property that, in the limit of small-amplitude motions, it reduced to the product of the energy and the group velocity. To preserve that property, **S**, as defined by (6.10.20), must be used as the energy flux vector, and it is apparent from (6.10.20) that one effect of nonlinearity therefore will be to augment the group velocity with the material velocity $iu_0 + jv_0$.

Finally we note that the ratio of the available potential energy

$$\mathscr{A} = \frac{\rho_s}{2} \frac{\theta_0^2}{S} \tag{6.10.25}$$

to the kinetic energy

$$K = \frac{\rho_s}{2} (u_0^2 + v_0^2) \tag{6.10.26}$$

is

$$\frac{\mathscr{A}}{K} = \frac{\theta_0^2}{(u_0^2 + v_0^2)S} = O\left(\frac{L}{L_D}\right)^2. \tag{6.10.27}$$

Thus another interpretation of the Rossby deformation radius $L_D$ is that when the scale of motion, $L$, is of the order of the deformation radius, the available potential energy is of the same order as the kinetic energy. If $L \ll L_D$, then $K \gg \mathscr{A}$, while if $L \gg L_D$, then $\mathscr{A} \gg K$.

In the next few sections several applications of the quasigeostrophic theory for the case $\beta = O(1)$ will be discussed before proceeding to the alternative case $\beta \gg 1$. The limit $\beta \ll 1$ (i.e., where the spatial extent of the motion is small compared to $(U/\beta_0)^{1/2}$), is obtained by merely setting $\beta = 0$ in (6.5.21) and (6.8.11). The resulting equations are the so-called *f-plane* equations, in which the earth's sphericity plays no dynamic role except the specification of the appropriate constant value of $f_0$, the Coriolis parameter.

## 6.11 Rossby Waves in a Stratified Fluid

In Chapter 3 it was shown that a homogeneous fluid, in the presence of the planetary-vorticity gradient, supported an intriguing wave motion, the Rossby wave. In this section we extend the theory of the Rossby wave to a stratified fluid.

Consider first the question of the existence of wave solutions in an unbounded medium. This is relevant only to the situation where horizontal boundaries are distant from the region of the wave disturbance. This is a restrictive assumption, but forms a useful first step in our discussion. In addition, to simplify the analysis, suppose that $S$ (i.e., $N^2$) is constant (i.e., over the scale of the wave) and that

$$\frac{-1}{\rho_s} \frac{\partial \rho_s}{\partial z} = H^{-1} \tag{6.11.1}$$

is constant (and of course greater than zero). The nondimensional number $H$ is the *density scale height* divided by $D$. For the ocean $H$ is very large. For the atmosphere, $H$ is $O(1)$ and is related to the temperature of the rest state by the perfect-gas law (6.2.6) and the hydrostatic equation (6.2.15). Thus in our nondimensional units

$$H^{-1} = \frac{1}{T_s} \frac{dT_s}{dz} + \frac{gD}{RT_s} \tag{6.11.2}$$

and $H$ will be exactly constant for an isothermal atmosphere, or for one where

$$T_s = T_0 + \mathcal{T}_0 \, e^{z/H} \tag{6.11.3}$$

for the arbitrary constants $\mathcal{T}_0$ and $T_0$; $H^{-1}$ is then $gD/RT_0$. Although the temperature in the troposphere is actually a linearly decreasing function of $z$, the assumption of constant $H$ is adequate for our purposes at this point. The potential-vorticity equation (6.5.21) then becomes

$$\left| \frac{\partial}{\partial t} + \frac{\partial \psi}{\partial x} \frac{\partial}{\partial y} - \frac{\partial \psi}{\partial y} \frac{\partial}{\partial x} \right| \left| S^{-1} \frac{\partial^2 \psi}{\partial z^2} - (SH)^{-1} \frac{\partial \psi}{\partial z} + \frac{\partial^2 \psi}{\partial x^2} + \frac{\partial^2 \psi}{\partial y^2} \right| + \beta \frac{\partial \psi}{\partial x} = 0. \tag{6.11.4}$$

A plane-wave solution of the form

$$\psi = A e^{z/2H} \cos(kx + ly + mz - \sigma t) \tag{6.11.5}$$

is an exact solution of (6.11.4) if

$$\sigma = \frac{-\beta k}{k^2 + l^2 + S^{-1}(m^2 + 1/4H^2)}, \tag{6.11.6}$$

since

$$S^{-1} \frac{\partial^2 \psi}{\partial z^2} - (SH)^{-1} \frac{\partial \psi}{\partial z} + \frac{\partial^2 \psi}{\partial x^2} + \frac{\partial^2 \psi}{\partial y^2} = -\left( k^2 + l^2 + \frac{m^2}{S} + \frac{1}{4H^2 S} \right) \psi, \tag{6.11.7}$$

so that the Jacobian of the potential vorticity and $\psi$ vanishes identically. If the dispersion relation (6.11.6) is compared with the dispersion relation (3.15.11) for Rossby waves in a homogeneous fluid, it is immediately apparent that the frequency relations are identical if the factor $F$ in (3.15.11) is identified with the modified vertical wave-number factor $S^{-1}(m^2 + 1/4H^2)$ in (6.11.6). Indeed, since we have implicitly assumed that the vertical scale of the wave is much less than $D$ (and therefore $H$), the factor $(4H^2)^{-1}$ can be ignored in (6.11.6), so that

$$\sigma = -\frac{\beta k}{(k^2 + l^2 + m^2 S^{-1})}. \tag{6.11.8}$$

The relationship between (3.15.11) and (6.11.8) is made more apparent if we recall that

$$F = \left(\frac{L}{R}\right)^2,$$

where $R = (gD)^{1/2}/f_0$ is the deformation radius of the free surface of the homogeneous fluid, while

$$\frac{m^2}{S} = m^2\left(\frac{L}{L_D}\right)^2 = \frac{L^2}{(L_D/m)^2}, \qquad (6.11.9)$$

where $L_D$ is the deformation radius of the isopycnal surfaces, so that

$$\frac{L_D}{m} = \frac{N_s(D/m)}{f_0}. \qquad (6.11.10)$$

Since $m$ is the vertical wave number, $D/m$ is the vertical distance between two nodes of the wave, and so the factor $m^2S^{-1}$ is equivalent to an $F$ based on the reduced gravity $(g/\theta_s)\,\partial\theta_s/\partial z$ [or $g(-(1/\rho_s)\,\partial\rho_s/\partial z)$ for the ocean] and the vertical distance $d = D/m$ instead of $D$. The vertical velocity in the Rossby wave is given by (6.5.15), viz.

$$w_1 = -\frac{m\sigma A}{S}e^{z/2H}\cos(kx + ly + mz - \sigma t), \qquad (6.11.11)$$

where again a factor $O((mH)^{-1})$ has been ignored. The vortex-tube stretching in the wave is simply

$$\frac{\partial w_1}{\partial z} = +\frac{m^2}{S}\frac{d\psi}{dt}, \qquad (6.11.12)$$

so that the vorticity equation (6.3.17) reduces to

$$\frac{d}{dt}\left|\zeta_0 - \frac{m^2}{S}\psi + \beta y\right| = 0, \qquad (6.11.13)$$

which should be compared with (3.12.24). Note that when $m = 0$ the wave phase is independent of $z$, and aside from the factor $e^{z/2H}$, the wave is *barotropic*. This follows from the fact that for $m = 0$, the vertical velocity vanishes from (6.11.11), and hence the equilibrium density surfaces are *undisturbed* by the wave motion. The trajectories of the fluid elements are horizontal, and the fluid does not experience any buoyancy forces.

The group-velocity components in the $x$-, $y$-, and $z$-directions are respectively

$$C_{gx} = \frac{\beta(k^2 - (l^2 + m^2/S))}{(k^2 + l^2 + m^2/S)^2},$$

$$C_{gy} = \frac{2\beta kl}{(k^2 + l^2 + m^2/S)^2}, \qquad (6.11.14)$$

$$C_{gz} = \frac{2\beta km/S}{(k^2 + l^2 + m^2/S)^2},$$

which should be compared with (3.19.13). Note that

$$\frac{\sigma/m}{C_{gz}} = -\frac{(k^2 + l^2 + m^2/S)^2}{2m^2/S} < 0, \qquad (6.11.15)$$

so that the vertical group velocity and phase velocity are *oppositely* directed. If the lines of constant phase propagates upwards (downwards), the energy in the wave propagates downwards (upwards).

The kinetic energy of the wave averaged over a wave period is

$$\langle K \rangle = \frac{\sigma}{2\pi} \int_0^{2\pi/\sigma} \frac{1}{2} \left\{ \left( \frac{\partial \psi}{\partial x} \right)^2 + \left( \frac{\partial \psi}{\partial y} \right)^2 \right\} \rho_s \, dt = \rho_0 (k^2 + l^2) \frac{A^2}{4}, \qquad (6.11.16)$$

where $\rho_0 = \rho_s e^{z/H}$ is the surface density, i.e., $\rho_s(0)$. The similarly averaged available potential energy is

$$\langle A \rangle = \frac{\sigma}{2\pi} \int_0^{2\pi/\sigma} \frac{\rho_s}{2S} \left( \frac{\partial \psi}{\partial z} \right)^2 \, dt = \rho_0 \frac{m^2}{S} \frac{A^2}{4}, \qquad (6.11.17)$$

so that the total wave energy

$$\langle E \rangle = \rho_0 \left( k^2 + l^2 + \frac{m^2}{S} \right) \frac{A^2}{4}, \qquad (6.11.18)$$

which should be compared with (3.21.8a).

The energy flux, **S**, as defined by (6.10.20) can be calculated using (6.11.5) and (6.11.14). Again, neglecting terms of $O(mH)^{-1}$:

$$\begin{aligned} \mathbf{S} = \hat{\imath} 2 \langle E \rangle \{ \cos^2 \phi C_{gx} + \sin^2 \phi u_0 \} \\ + \hat{\jmath} 2 \langle E \rangle \{ \cos^2 \phi C_{gy} + \sin^2 \phi v_0 \} + \hat{k} 2 \langle E \rangle C_{gz} \cos^2 \phi, \end{aligned} \qquad (6.11.19)$$

where $\phi$ is the wave phase,

$$\phi = kx + ly + mz - \sigma t, \qquad (6.11.20)$$

and $u_0$ and $v_0$ are the horizontal velocity components of the fluid, i.e.,

$$u_0 = -\frac{\partial \psi}{\partial y} = Al \sin \phi,$$

$$v_0 = \frac{\partial \psi}{\partial x} = Ak \sin \phi. \qquad (6.11.21)$$

If **S** is averaged over a wave period, the nonlinear terms, involving the advection of wave energy by the *fluid* velocity vanish identically so that

$$\langle \mathbf{S} \rangle = \mathbf{C}_g \langle E \rangle, \qquad (6.11.22)$$

where

$$\mathbf{C}_g = \hat{\imath} C_{gx} + \hat{\jmath} C_{gy} + \hat{k} C_{gz}. \qquad (6.11.23)$$

It is clear from a comparison of these results with the results for the Rossby wave in a homogeneous fluid of Chapter 3, that the concepts appropriate to the Rossby wave in the simple homogeneous case can be directly taken over to the more physically complex case of a stratified fluid.

## 6.12  Rossby-Wave Normal Modes: the Vertical Structure Equation

Although the plane-wave solution of the preceding section is a useful example to demonstrate the vorticity dynamics of a stratified fluid on the $\beta$-plane, the neglect of horizontal boundaries is a serious restriction. In this section that restriction will be removed. In addition, only linear, small-amplitude motions will be considered, for as we saw in Chapter 3, although a single plane wave is a nonlinear solution, a wave packet is not, and the latter is a more realistic description of the wave field. The linearized form of (6.5.21) is

$$\frac{\partial}{\partial t}\left|\frac{\partial^2\psi}{\partial x^2} + \frac{\partial^2\psi}{\partial y^2} + \frac{1}{\rho_s}\frac{\partial}{\partial z}\left(\frac{\rho_s}{S}\frac{\partial\psi}{\partial z}\right)\right| + \beta\frac{\partial\psi}{\partial x} = 0. \qquad (6.12.1)$$

The oceanographic case is realized by simply letting $\rho_s$ be a constant. Consider now the possible Rossby waves in the region, unbounded laterally but limited in the vertical to the interval

$$0 \le z \le z_T. \qquad (6.12.2)$$

For the oceanic case we may take $z_T = 1$, while for the atmosphere, which is unbounded above, $z_T \to \infty$. The boundary conditions at $z = 0$ are

$$\frac{\partial^2\psi}{\partial t\,\partial z} = 0, \qquad z = 0, \qquad (6.12.3)$$

where we have neglected dissipation and bottom slope. At the upper surface (6.9.14) will apply in the oceanic case, where in the absence of applied stress, the linearized form of (6.9.14) becomes

$$\frac{\partial^2\psi}{\partial t\,\partial z} = 0, \qquad z = 1. \qquad (6.12.4)$$

In the case of the atmosphere. where the upper surface is at infinity, we must require that the energy be finite as $z \to \infty$, or equivalently that

$$\rho_s\psi^2 \text{ remains finite} \quad \text{as } z \to \infty. \qquad (6.12.5)$$

It is not clear at this stage whether (6.12.5) is sufficient to determine the motion. It is certainly necessary, and we defer further discussion of this point until later.

First we examine the possibility that separable solutions of the form

$$\psi = \text{Re } e^{i(kx+ly-\sigma t)}\Phi(z) \qquad (6.12.6)$$

are possible, where $\Phi$ is the vertical structure function which must satisfy

$$\frac{1}{\rho_s}\frac{d}{dz}\frac{\rho_s}{S}\frac{d\Phi}{dz} = -\lambda\Phi. \tag{6.12.7a}$$

Where $\lambda$, from (6.12.1), is given by

$$\lambda = -\left|\frac{\beta k}{\sigma} + k^2 + l^2\right|. \tag{6.12.7b}$$

The boundary conditions for $\Phi$ are determined from (6.12.3) and (6.12.4) or (6.12.5):

$$\frac{d\Phi}{dz} = 0, \qquad z = 0, \tag{6.12.8a}$$

and *either*, for the ocean,

$$\frac{d\Phi}{dz} = 0, \qquad z = 1, \tag{6.12.8b}$$

*or*, for the atmosphere

$$\rho_s |\Phi|^2 \text{ finite } \quad \text{as } z \to \infty. \tag{6.12.8c}$$

Consider the oceanic case first. Then (6.12.7a,b) and (6.12.8a,b) constitute an eigenvalue problem for the eigenvalue $\lambda$. Since $\rho_s$ and $S$ are always positive definite in the interval $(0, 1)$, we are assured that there are an infinite number of solutions $\Phi_n(z)$, $n = 1, 2, 3 \ldots$, each one associated with a real, discrete eigenvalue $\lambda_n$, $n = 1, 2, 3, \ldots$. Furthermore, since it is easily seen, after integration by parts, that

$$\lambda = \frac{\int_0^1 \frac{\rho_s}{S}\left|\frac{d\Phi}{dz}\right|^2 dz}{\int_0^1 \rho_s |\Phi|^2 dz}, \tag{6.12.9}$$

it follows that all the nonzero $\lambda$'s are positive. It is also clear that $\lambda = 0$ is an eigenvalue for arbitrary $\rho_s(z)$ and $S(z)$, since (6.12.7), (6.12.8a), and (6.12.8b) are satisfied by

$$\lambda = 0, \qquad \Phi = 1. \tag{6.12.10a}$$

This mode is the *barotropic mode*. Its $\psi$-field and hence its horizontal velocities are independent of depth, and its vertical velocity and density perturbation are identically zero. It follows that for this $\lambda = 0$ mode

$$\sigma = \sigma_0 = -\frac{\beta k}{k^2 + l^2}, \tag{6.12.10b}$$

which is identical to the Rossby-wave frequency of a homogeneous fluid with a rigid top (i.e., $F = 0$). It is important to emphasize that this barotropic mode is possible regardless of the detailed structure of $\rho_s$ and $S$. Thus all the results previously derived pertaining to barotropic Rossby waves apply to the barotropic *mode* in a stratified fluid.

For $\lambda \neq 0$ the integral of (6.12.7a) from $z = 0$ to $z = 1$ yields, after application of the boundary conditions at the end points,

$$\int_0^1 \rho_s \Phi(z) \, dz = 0, \qquad \lambda \neq 0, \tag{6.12.11}$$

which with (6.12.6) implies that all the solutions corresponding to nonzero $\lambda$ have zero vertically integrated horizontal mass flux, i.e., for these modes

$$\int_0^1 \rho_s u_0 \, dz = \int_0^1 \rho_s v_0 \, dz = 0. \tag{6.12.12}$$

These modes are called the baroclinic modes, since they deform the density surfaces, have a nonzero vertical velocity, and depend for their existence on the presence of a the basic stratification. As an example consider the simple case where $\rho_s$ and $S$ are both constant. Then

$$\frac{d^2\Phi}{dz^2} = -\lambda S\Phi, \tag{6.12.13}$$

whose solution satisfying (6.12.8a) is

$$\Phi = \cos(\lambda S)^{1/2} z. \tag{6.12.14}$$

The condition (6.12.8b) yields the eigenvalue relation

$$\sin(\lambda S)^{1/2} = 0, \tag{6.12.15}$$

or

$$\lambda = \lambda_n = \frac{n^2 \pi^2}{S}, \qquad n = 0, 1, 2, \dots. \tag{6.12.16}$$

The $n = 0$ mode is the barotropic mode previously discussed. For $n > 0$ the solutions are the set

$$\Phi_n(z) = \cos n\pi z, \qquad n = 1, 2, \dots, \tag{6.12.17}$$

each corresponding to the eigenvalue $\lambda_n$ as given by (6.12.16). It is a simple matter to verify (6.12.11) and in addition verify the following general feature of the eigenvalue problem. If the $\lambda_n$'s are arranged to form an increasing sequence,

$$\lambda_0 = 0 < \lambda_1 < \lambda_2 < \lambda_3 < \cdots < \lambda_{n-1} < \lambda_n < \lambda_{n+1} \cdots, \tag{6.12.18}$$

then for any $n > 0$, $\Phi_n$ has one more zero in the interval $(0, 1)$ than does $\Phi_{n-1}$. The higher modes are more "wiggly" in $z$. If $S(z)$ is a more complicated function of $z$, the numerical values of the $\lambda_n$ will be altered as well as the structure of the *baroclinic* modes $\Phi_n$. Their general character will remain unaltered. Figure 6.12.1 shows the vertical modal structure for the first four modes, i.e., the barotropic and first three baroclinic modes as calculated by

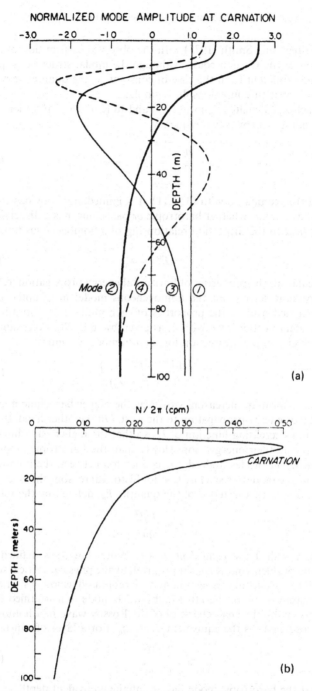

**Figure 6.12.1** (a) The barotropic and first three baroclinic modes, as calculated by Kundu, Allen, and Smith (1975) for (b) the distribution of $N$ observed at ocean station Carnation, off the Oregon coast.

Kundu, Allen, and Smith (1975) with the observed distribution of $N^2(z)$ in the region of the Oregon coast. Although the modal structure is quantitatively quite different from the case of constant $N$, the qualitative nature of the vertical structure functions $\Phi_n$ is similar.

For each $\lambda_n$ calculated from (6.12.7) and (6.12.9), (6.12.10), there exists a corresponding Rossby-wave-frequency,

$$\boxed{\sigma_n = -\frac{\beta k}{k^2 + l^2 + \lambda_n}}, \qquad n = 0, 1, 2, \dots . \tag{6.12.19}$$

If (6.12.19) is compared with (3.15.11), it is immediately obvious that each Rossby-wave mode, whether barotropic *or* baroclinic, has a dispersion relation *identical* to the dispersion relationship of a homogeneous fluid with a value of

$$F = F^{(n)} = \lambda_n. \tag{6.12.20}$$

In particular all the properties of horizontal energy propagation, reflection, and dispersion derived for the homogeneous model in Chapter 3 can be directly carried over to the properties of each mode in the stratified fluid with the identification of $F$ with $\lambda_n$ as given by (6.12.20). For example, the group velocity in the $x$-direction for the $n$th mode is simply

$$C_{gx} = \frac{\beta(k^2 - [l^2 + \lambda_n])}{(k^2 + l^2 + \lambda_n)^2}. \tag{6.12.21}$$

Since the $\lambda_n$ form an increasing sequence, the higher baroclinic modes will tend to favor energy propagation to the west. On the other hand, the group velocity is a decreasing function of $\lambda_n$, so that the higher baroclinic modes will propagate their energy more slowly than the barotropic mode or the lower baroclinic modes. Indeed, the connection between the homogeneous and baroclinic models is so close that for historical reasons the eigenvalue $\lambda_n$ is sometimes written in terms of the quantity $h_n$, defined by the relation

$$\lambda_n = \frac{f_0^2 L^2}{g h_n}, \tag{6.12.22}$$

where $h_n$ is called the *equivalent depth*.* Note that $h_n$ is defined by the eigenvalue problem and is *not* simply given by the physical vertical scales. In this form the results of this section may be rephrased as follows. The propagation characteristic of the $n$th Rossby-wave mode in a stratified fluid are given entirely by the characteristics of the Rossby wave in a homogeneous layer whose depth is the equivalent depth $h_n$. For a layer of constant $N$,

$$h_n = \frac{N_s^2 D^2}{g n^2 \pi^2} \tag{6.12.23}$$

Note that the barotropic mode has an infinite equivalent depth.

_____
* As defined here, $h_n$ is *dimensional*.

The problem for the Rossby normal modes in the atmosphere is rendered considerably more complex by the lack of lid at the top of the atmosphere. Of course, as mentioned previously, the barotropic mode, $\lambda = 0$, is still possible for arbitrary $\rho_s(z)$ and $S(z)$. To examine the possibility of nontrivial baroclinic modes it is useful to reduce (6.12.7a) to standard form by the transformation

$$\Phi = \left(\frac{S}{\rho_s}\right)^{1/2} Z(z), \tag{6.12.24}$$

where $Z$ satisfies

$$\frac{d^2}{dz^2} Z + \left[\lambda S - \left(\frac{S}{\rho_s}\right)^{1/2} \frac{d^2}{dz^2} \left(\frac{\rho_s}{S}\right)^{1/2}\right] Z = 0 \tag{6.12.25a}$$

and

$$\frac{1}{Z} \frac{dZ}{dz} + \left(\frac{\rho_s}{S}\right)^{1/2} \frac{d}{dz} \left(\frac{S}{\rho_s}\right)^{1/2} = 0, \qquad z = 0. \tag{6.12.25b}$$

If $S$ is finite and nonzero at infinity, (6.12.8c) merely requires $Z$ to be finite as $z \to \infty$. The essential nature of the problem may be illustrated by the example of the isothermal atmosphere. If $T_s(z)$ is a constant, then from (6.4.8), (6.4.11), and (6.5.13) $N_s$ and hence $S$ are constant, while

$$\rho_s(z) = \rho_s(0)e^{-z/H}, \tag{6.12.26}$$

where, by (6.11.2),

$$H = \frac{RT_s}{gD}, \tag{6.12.27}$$

so that (6.12.25a) becomes

$$\frac{d^2Z}{dz^2} + q^2Z = 0, \tag{6.12.28}$$

where

$$q^2 = \lambda S - \frac{1}{4H^2}, \tag{6.12.29}$$

while (6.12.25b) is

$$\frac{dZ}{dz} + \frac{1}{2H} Z = 0, \qquad z = 0. \tag{6.12.30}$$

If $q^2 < 0$, i.e., if $\lambda < (4H^2S)^{-1}$, the only solution for $Z$ which remains finite at infinity is

$$Z = Z_0 e^{-|q|z}, \tag{6.12.31}$$

which will satisfy (6.12.30) only if

$$|q|^2 = \frac{1}{4H^2} - \lambda S = \frac{1}{4H^2}. \tag{6.12.32}$$

This implies that $\lambda = 0$, which is the barotropic mode previously described. On the other hand, if $q^2 > 0$, both homogeneous solutions to (6.12.28) are finite at infinity, i.e.,

$$Z = Ae^{iqz} + Be^{-iqz}. \tag{6.12.33}$$

The first solution, proportional to $A$, corresponds by (6.12.6) to a wave

$$\psi = \text{Re}\left(\frac{S}{\rho_s}\right)^{1/2} Ae^{i(kx+ly+qz-\sigma t)}, \tag{6.12.34}$$

where by (6.12.7b)

$$\sigma = \frac{-\beta k}{k^2 + l^2 + S^{-1}(q^2 + 1/4H^2)}. \tag{6.12.35}$$

If $k > 0$, then $\sigma < 0$ and the wave given by (6.12.34) has a phase speed in the $z$-direction which is negative. Similarly the wave with amplitude $B$ has a positive phase speed in the $z$-direction. Since the group and phase speeds in the $z$-direction are *oppositely* directed (6.11.15), the "$A$" wave has an upward energy flux, while the "$B$" wave has a downward energy flux. If there are no sources of wave energy at infinity, we must impose the radiation condition, viz., that all waves have an outward-directed energy flux. This radiation condition implies that $B$ must be zero, and it then follows as before that (6.12.33), with $B = 0$, can satisfy (6.12.30) only if $\lambda = 0$. Thus for the unbounded *isothermal* atmosphere only the barotropic free mode is possible. Of course the atmosphere is not isothermal, and the problem (6.12.25a,b) is in general quite complex. Chapman and Lindzen (1970) have made detailed calculations and have shown that for the realistic standard atmosphere, no baroclinic modes are possible.* This does not mean baroclinic modes are impossible in principle. Had the earth a different thermal structure, it would be possible in principle for nonzero, positive $\lambda$'s to yield nontrivial solutions to (6.12.25).

In the case of the atmosphere, however, the relevance of the pure Rossby wave in an atmosphere at rest is problematic. The potential-vorticity gradient of the mean winds is of $O(\beta_0)$, i.e., $\beta = O(1)$, so that the vertical structure and the dispersion relation for free atmospheric waves can be expected to be strongly influenced by the presence of the mean zonal winds in which the wave is embedded. This problem is discussed at length in Chapter 7.

Over much of the ocean the relative vorticity gradient of the *mean* flow is small (i.e., $\beta \gg 1$). Indeed, this formed the basis of the Sverdrup theory of the mean flow discussed in Chapter 5. In this case we might expect the results of the present section to apply to Rossby-wave propagation. However, the contribution to the ambient potential vorticity gradient due to the horizontal density gradients may not be negligible. From (6.8.9) we note that, for

---

* Their lowest eigenvalue, $\lambda_0$, is $O(F)$, which in the theory presented here is a small parameter, indistinguishable from zero.

example, the ambient potential vorticity gradient in the $y$-direction, ignoring the relative-vorticity gradient of the mean flow, is

$$\beta - \frac{\partial}{\partial z}\frac{1}{S}\frac{\partial \rho_0}{\partial y} = \beta - \frac{\partial}{\partial z}\frac{1}{S}\frac{\partial u_0}{\partial z}. \tag{6.12.36}$$

The ratio of the second to the first term in (6.12.36) is

$$\frac{\dfrac{\partial}{\partial z}\dfrac{1}{S}\dfrac{\partial u_0}{\partial z}}{\beta} = O\!\left(\frac{\partial^2 u_*/\partial z_*^2}{\beta_0 N_s^2/f_0^2}\right) = O\!\left(\frac{\Delta u_*}{\beta_0 N_s^2 d_*^2/f_0^2}\right), \tag{6.12.37}$$

where $\Delta u_*$ is the characteristic magnitude of the velocity variation over the depth and $d_*$ is the vertical scale of variation of $u_*$. In mid-latitudes, where $\beta_0 \sim 10^{-13}\,\mathrm{cm}^{-1}\,\mathrm{s}^{-1}$, $N_s^2 \sim 5 \times 10^{-5}\,\mathrm{s}^{-2}$, $f_0^2 \sim 10^{-8}\,\mathrm{s}^{-2}$, and $d_* \sim 10^5\,\mathrm{cm}$, the ratio given by (6.12.37) will be less than one unless $\Delta u_*$ is greater than or equal to 5 cm/s. Thus in regions of the ocean where the mean currents are much less than this value, the Rossby-wave theory developed here is directly applicable. For regions where $\Delta u_*$ is larger, the effect of the mean currents must be included in the discussion of the wave dynamics, as in the case of the atmosphere.

Since for the first baroclinic mode $\lambda_1 = O(S^{-1})$, the maximum phase speed for the first baroclinic mode in the $x$-direction is

$$C_{1x} = \max\text{-}\lim_{\substack{k\to 0 \\ l\to 0}} \frac{\sigma_1}{k} = -\frac{\beta}{\lambda_1} = O(\beta S) = \frac{\beta_0 N_s^2 D^2/f_0^2}{U}, \tag{6.12.38}$$

so that the condition that the potential-vorticity gradients of the mean currents can be ignored in the theory of baroclinic Rossby wave propagation may equally well be written

$$\Delta u_* < C_{1x*} \tag{6.12.39}$$

i.e., that the thermal wind of the mean flow is small compared to the phase speed of the first baroclinic modes. Since the higher baroclinic modes have slower phase speeds, the effect of the mean flow will become increasingly more critical for the higher modes.

## 6.13 Forced Stationary Waves in the Atmosphere

The vertical structure of free motions in a resting fluid over a flat surface depends only on the solutions of (6.12.7) and therefore only on $\rho_s(z)$ and $S(z)$. However, forced motions possess a vertical structure more intimately connected with the structure of the forcing and the complete potential-vorticity dynamics of the response. An example of great importance is the response of the atmosphere to forcing at or near its lower boundary by heating and topography. Only the simplest example will be treated explicitly.

In the absence of ocean–continent contrasts in temperature and atmospheric heating, and in the absence of topographic variations in the longitudinal direction, the response of the atmosphere to the meridional heating gradient from equator to pole could lead, at least in principle, to a zonally symmetric flow (i.e., a flow independent of longitude). Whether such a flow would be actually realized is another question to be taken up in Chapter 7. This question aside, it is clear that the differences of the thermal properties of continents and oceans and the existence of mountain ranges might seriously perturb the zonally symmetric state. To examine this question further, the following model is examined.

Consider the zonal flow, independent of longitude,

$$u = u_0(y, z) \qquad (6.13.1)$$

corresponding to the stream function

$$\psi = \psi_0(y, z) = -\int^y u_0(y', z)\, dy'. \qquad (6.13.2)$$

Next consider the nature in which this stream field is perturbed by the existence of a *small* topographic variation

$$\eta_B = \eta_B(x, y) \qquad (6.13.3)$$

and an atmospheric heating

$$\mathscr{H} = \mathscr{H}(x, y, z). \qquad (6.13.4)$$

Let the new stream function be given by

$$\psi = \psi_0(y, z) + \phi(x, y, z), \qquad (6.13.5)$$

where $\phi(x, y, z)$ is the small disturbance produced by the heating and topography. If (6.13.5) is substituted into (6.5.18) and only linear terms in $\phi$ are retained, we find that for steady flow, $\phi$ satisfies

$$u_0 \frac{\partial}{\partial x} \left| \frac{1}{\rho_s} \frac{\partial}{\partial z} \frac{\rho_s}{S} \frac{\partial \phi}{\partial z} + \frac{\partial^2 \phi}{\partial x^2} + \frac{\partial^2 \phi}{\partial y^2} \right|$$

$$+ \left| \beta - \frac{\partial^2 u_0}{\partial y^2} - \frac{1}{\rho_s} \frac{\partial}{\partial z} \frac{\rho_s}{S} \frac{\partial u_0}{\partial z} \right| \frac{\partial \phi}{\partial x} = \frac{1}{\rho_s} \frac{\partial}{\partial z} \left| \frac{\rho_s \mathscr{H}}{S} \right|, \qquad (6.13.6)$$

while on $z = 0$, the linearization of (6.6.10) yields

$$S^{-1} \left| u_0 \frac{\partial^2 \phi}{\partial z\, \partial x} - \frac{\partial \phi}{\partial x} \frac{\partial u_0}{\partial z} \right| = -u_0 \frac{\partial \eta_B}{\partial x} + S^{-1} \mathscr{H}(x, y, 0), \qquad (6.13.7)$$

where friction has been neglected. Terms of $O(\phi^2, \phi\eta_B)$ have been also neglected as small. It is clear from the form of (6.13.6) and (6.13.7) that $\eta_B$ and $\mathscr{H}$ will produce a forced, stationary response for $\phi$. A simple but illuminating example is the case where $u_0$ is a *constant* and $\eta_B$ and $\mathscr{H}$ have the form

$$\eta_B = \eta_0 \cos kx \sin \pi y, \qquad (6.13.8a)$$

$$\mathscr{H} = \mathscr{H}_0 e^{-zz} \cos(kx + \theta_H) \sin \pi y \qquad (6.13.8b)$$

in the latitude band $|y| \leq 1$. The phase of the heating with respect to the topography is the constant, $\theta_H$. The heating is a maximum at the earth's surface and decreases with altitude. The topography and heating are periodic in longitude, and this may be considered as a rough model of the recurrent forcing of the flow of the mid-latitude atmospheric westerly winds as they travel around the earth.

The solution for $\phi$ may be sought in the form*

$$\phi = \operatorname{Re} \, \Phi(z) e^{ikx} \sin \pi y, \tag{6.13.9}$$

where $\Phi(z)$ satisfies

$$\frac{1}{\rho_s} \frac{d}{dz} \frac{\rho_s}{S} \frac{d\Phi}{dz} + \left| \frac{\beta}{u_0} - K^2 \right| \Phi = \frac{1}{\rho_s} \frac{\partial}{\partial z} \frac{\rho_s \mathscr{H}_0 e^{-\alpha z} e^{i\theta_H}}{iku_0 S} \tag{6.13.10}$$

and

$$S^{-1} \frac{d\Phi}{dz} = -\eta_0 + S^{-1} \frac{\mathscr{H}_0 e^{i\theta_H}}{iku_0}, \qquad z = 0 \tag{6.13.11}$$

where

$$K^2 = k^2 + \pi^2. \tag{6.13.12}$$

For large $z$, (6.12.5) applies, so that $z \to \infty$

$$\rho_s |\Phi|^2 \text{ remains finite.} \tag{6.13.13}$$

The condition (6.13.13) may need to be supplemented by the radiation condition, i.e., that the *wave* energy flux at $z \to \infty$ is directed outward, i.e., that for the wave field

$$\overline{\rho_s p_0 w_1}^{x,y} = -\overline{\rho_s \frac{\phi}{S} u_0 \frac{\partial \phi}{\partial x \, \partial z}}^{x,y} \geq 0, \tag{6.13.14}$$

since $-(u_0/S)(\partial^2 \phi / \partial x \, \partial z)$ is the vertical velocity of wave field and where the overbar $\overline{\phantom{x}}^{x,y}$ over any expression denotes the averaging operation

$$\overline{(\phantom{x})}^{x,y} \equiv \frac{k}{4\pi} \int_0^{2\pi,k} dx \int_{-1}^1 dy \, (\phantom{x}). \tag{6.13.15}$$

In terms of $\Phi$, the radiation condition becomes

$$\overline{\rho_s p_0 w_1}^{x,y} = \frac{ik\rho_s u_0}{8S} \left| \Phi \frac{d\Phi^*}{dz} - \Phi^* \frac{d\Phi}{dz} \right| \geq 0 \tag{6.13.16}$$

where $\Phi^*$ is the complex conjugate of $\Phi$.

To further simplify the analysis, assume that $S$ is constant and the non-dimensional density scale height

$$H = -\left( \frac{1}{\rho_s} \frac{\partial \rho_s}{\partial z} \right)^{-1} \tag{6.13.17}$$

---

* It is left to the reader to verify that, in this case, $\psi_0 + \phi$ is a solution to the complete, non-linear potential vorticity equation.

is also constant, so that (6.13.10) may be written

$$\frac{d^2\Phi}{dz^2} - \frac{1}{H}\frac{d\Phi}{dz} + \left\{\frac{\beta}{u_0} - K^2\right\}S\Phi = -\frac{(1 + (\alpha H)^{-1})\alpha \mathcal{H}_0 e^{-\alpha z + i\theta_H}}{iku_0}. \quad (6.13.18)$$

A *particular* solution to (6.13.18) may be easily found and is given by

$$\Phi = \Phi_p(z) = -\frac{\alpha \mathcal{H}_0(1 + (\alpha H)^{-1})}{iku_0} \frac{e^{-\alpha z}e^{i\theta_H}}{\{(\beta/u_0 - K^2)S + \alpha^2(1 + (\alpha H)^{-1})\}}.$$

$$(6.13.19)$$

This solution alone, however, will not satisfy (6.13.11), so that $\Phi_p(z)$ must be supplemented by the homogeneous solutions of (6.13.18). The homogeneous solutions are of the form

$$\Phi_H(z) = e^{z/2H}\{Ae^{imz} + Be^{-imz}\}, \quad (6.13.20a)$$

where

$$m = \left|\left(\frac{\beta}{u_0} - K^2\right)S - \frac{1}{4H^2}\right|^{1/2}. \quad (6.13.20b)$$

There are two important and distinct cases. If

$$\frac{\beta}{u_0} > K^2 + (4H^2S)^{-1}, \quad (6.13.21)$$

$m^2$ will be positive and the solution (6.13.20) will be oscillatory in $z$. If the inequality in (6.13.21) is reversed, then $m^2$ is negative and the solutions are an exponentially growing and decaying pair, without oscillation, of the form

$$\Phi_{II} = e^{z/2H}\{Ce^{-qz} + De^{qz}\}, \quad (6.13.22)$$

where

$$q = S^{1/2}\left|K^2 + (4H^2S)^{-1} - \frac{\beta}{u_0}\right|^{1/2}. \quad (6.13.23)$$

Note that the existence of oscillatory solutions requires

$$0 > -u_0 k > -\frac{\beta k}{K^2 + (4H^2S)^{-1}}, \quad (6.13.24)$$

which has the following interesting interpretation in terms of the Rossby waves of Section 6.11. There it was shown that for $k > 0$ the Rossby-wave frequency was negative and, as a function of vertical wave number, achieves an algebraic minimum* when $m^2 = 0$, for which the frequency in a medium otherwise at rest is

$$\sigma_0 = -\frac{\beta k}{K^2 + (4H^2S)^{-1}}. \quad (6.13.25)$$

---

* But a maximum in absolute numerical value.

In the present case the waves are steady in a frame fixed to the earth's surface. If we move to a frame traveling with the uniform speed $u_0$, so that the mean flow is zero, we observe the wave with the frequency

$$\sigma = -u_0 k. \tag{6.13.26}$$

The condition (6.13.24) is thus the condition that $u_0$ is small enough so that the forcing frequency, $-u_0 k$, *lies within the range of vertically propagating Rossby waves*. Otherwise, $m^2 < 0$, no oscillations in $z$ are possible, and the motions are trapped near the surface.

Since

$$\rho_s(z) = \rho_s(0)e^{-z/H}, \tag{6.13.27}$$

the condition (6.13.13) implies that

$$\lim_{z \to \infty} e^{-z/2H}\Phi(z) \text{ must be finite.} \tag{6.13.28}$$

If $m^2 < 0$, the second solution in (6.13.22) must therefore be neglected, so that $D = 0$. On the other hand, if $m^2 > 0$, both solutions in (6.13.20) will satisfy (6.13.28), i.e., both solutions will have finite energy flux at infinity. The energy flux of the first solution in the $z$-direction, by (6.13.14), is

$$\overline{\rho_s p_0 w_1}^{x,y} = \frac{|A|^2 km\rho_s(0)}{4S}, \tag{6.13.29}$$

while for the second solution, proportional to $B$,

$$\overline{\rho_s p_0 w_1}^{x,y} = \frac{-|B|^2 km\rho_s(0)}{4S}. \tag{6.13.30}$$

The first solution therefore represents a wave with upward energy flux, while the second solution represents a wave with a flux of energy directed towards the surface. To satisfy the radiation condition the wave-disturbance energy flux must be directed, at infinity, *away* from the energy source, and this requires that only the first solution be chosen. Consequently we choose $B = 0$.

A comparison with (6.11.14) shows that this condition is precisely equivalent to the choice of a wave whose vertical group velocity, proportional to the product of the $x$ and $z$ wave numbers, is *positive*. Thus for $m^2$ either greater or less than zero, the homogeneous solution may be written

$$\Phi_H = Ae^{z/2H}e^{imz} \tag{6.13.31}$$

with the convention that the square root $m$ in (6.13.20b) is interpreted as

$$im = i\{(K_s^2 - K^2)S - 1/4H^2\}^{1/2}, \qquad m^2 > 0, \tag{6.13.32a}$$

$$im = -\left|\frac{1}{4H^2} + (K^2 - K_s^2)S\right|^{1/2}, \qquad m^2 < 0, \tag{6.13.32b}$$

where the positive square root is chosen in each instance and where

$$K_s^2 = \frac{\beta}{u_0} \qquad (6.13.33)$$

is the stationary Rossby wave number for a *barotropic* Rossby wave (3.18.10). With this convention, the total solution is

$$\Phi = \Phi_p + \Phi_H, \qquad (6.13.34)$$

where $\Phi_p$ and $\Phi_H$ are given by (6.13.19) and (6.13.31). The constant $A$ is determined by substituting (6.13.34) into the lower boundary condition (6.13.11), which yields

$$A = - \frac{\eta_0(1/2H - im)}{K_s^2 - K^2} + \frac{\mathcal{H}_0 e^{i\theta_H}}{iku_0} \frac{1/2H - im}{(K_s^2 - K^2)S + \alpha^2(1 + \mu)},$$

where

$$\mu \equiv (\alpha H)^{-1}. \qquad (6.13.35)$$

The solution may then be written for $\phi$ itself as

$$\phi = - \frac{\mathcal{H}_0}{ku_0} \left\{ \alpha(1 + \mu)e^{-\alpha z} - \left| \frac{1}{2H} + q \right| e^{(1/2H - q)z} \right\}$$
$$\times \frac{\sin(kx + \theta_H)\sin \pi y}{[K_s^2 - K^2]S + \alpha^2(1 + \mu)} - \frac{\eta_0(1/2H + q)}{K_s^2 - K^2} e^{z(1/2H - q)} \cos kx \sin \pi y$$

$$(6.13.36)$$

if $m^2 < 0$, while for $m^2 > 0$

$$\phi = \frac{\sin \pi y \mathcal{H}_0}{ku_0} \left| -\sin(kx + \theta_H)e^{-\alpha z}\alpha(1 + \mu) \right.$$

$$+ \frac{1}{2H} \sin(kx + mz + \theta_H)e^{z/2H}$$

$$\left. - m \cos(kx + mz + \theta_H)e^{z/2H} \right|$$

$$\times ((K_s^2 - K^2)S + \alpha^2(1 + \mu))^{-1}$$

$$- \frac{\sin \pi y \eta_0 e^{z/2H}}{K_s^2 - K^2} \left| \frac{1}{2H} \cos(kx + mz) + m \sin(kx + mz) \right|,$$

$$(6.13.37)$$

where $q$ and $m$ are given by (6.13.23) and (6.13.20b) respectively. If $m^2 < 0$, the total response of the atmosphere is trapped near the surface, while if $m^2 > 0$ the perturbation due to *both* heating and topography will radiate to great heights. Such radiating waves, initiated in the lower troposphere, are commonly considered (e.g., Holton 1975) to play a significant role in the dynamics of the stratosphere. According to (6.13.21) radiating waves are favored in the presence of weak, westerly winds in the stratosphere, which occur mainly in the spring and autumn. It is also clear from (6.13.36) that for $K^2 \to K_s^2$ (for which $m^2 = -1/4H^2$ and $q = 1/2H$), resonance occurs for the topographically induced wave. At this particular wavelength, the forcing resonates with the *barotropic* Rossby wave of Section 6.12. The amplitude of the *topographically* forced wave becomes infinite as $K^2 \to K_s^2$ while the response due to the heating remains finite at this wavelength. Naturally, dissipation or nonlinear effects will limit the amplitude of the topographically forced wave as $K^2 \to K_s^2$.

The denominator of the response term due to the heating vanishes when $K^2 = K_s^2 + (\alpha^2/S)(1 + \mu)$. At this value of $K^2$, $m^2$ is negative and

$$\frac{1}{2H} - q = -\alpha,$$

so that the forcing has the same vertical structure as the homogeneous solution. No true resonance occurs at this wave number in the strict sense since no free mode other than the barotropic mode exists which satisfies (6.13.13) or (6.13.14).

As $K^2 \to K_s^2 + \alpha^2/S(1 + \mu)$, both the numerator and the denominator vanish. However, since the forcing has the same $z$-structure as the homogeneous solution, the response now contains an algebraic term in $z$, i.e.,

$$\phi \to \frac{\mathscr{H}_0 e^{-\alpha z}}{kU_0(2 + \mu)} \sin(kx + \theta_H)\sin \pi y[z(1 + \mu) - \alpha^{-1}],$$

as

$$K^2 \to K_s^2 + \frac{\alpha^2}{S}(1 + \mu).$$

Thus, in contrast with the topographically forced solution, the response amplitude remains bounded. There is a quasi-resonance which occurs in the sense that the ratio of the amplitude of the response to the amplitude of the forcing linearly increases with height. However, at large $z$ where this effect is most pronounced, both the forcing and the response are exponentially small.

If the atmospheric heating is assumed to occur in a thin layer near the earth's surface, where effects of condensation and evaporation occur, then $\alpha$ in (6.13.36) and (6.13.37) will be large compared with unity and, as (6.13.36) and (6.13.37) indicate the atmospheric response, especially at large $z$, will tend to be small compared to the topographic response. For this reason the remaining discussion will be limited to the topographic wave, which is

$$\phi = \phi_T = \frac{-\eta_0\{q + 1/2H\}e^{z\{1/2H - q\}}}{(K_s^2 - K^2)} \cos kx \sin \pi y, \qquad m^2 < 0$$

$$= \frac{-\eta_0 e^{z/2H}}{K_s^2 - K^2} \left\{ \frac{1}{2H} \cos(kx + mz) + m \sin(kx + mz)\right\} \sin \pi y, \qquad m^2 > 0.$$

$$(6.13.38)$$

Associated with this stream-function perturbation are the dynamical wave fields, which in the case $m^2 > 0$ are

$$\tilde{\theta}_0 = \frac{\partial \phi_T}{\partial z} = -\eta_0 S e^{z/2H} \cos(kx + mz)\sin \pi y, \qquad (6.13.39a)$$

$$\tilde{v}_0 = \frac{\partial \phi_T}{\partial x} = \frac{\eta_0 e^{z/2H}}{K_s^2 - K^2} \qquad\qquad (6.13.39b)$$

$$\times \left|\frac{k}{2H} \sin(kx + mz) - mk \cos(kx + mz)\right| \sin \pi y,$$

$$\tilde{u}_0 = -\frac{\partial \phi_T}{\partial y} = +\frac{\eta_0 e^{z/2H}}{(K_s^2 - K^2)} \qquad (6.13.39c)$$

$$\times \pi \left|\frac{1}{2H} \cos(kx + mz) + m \sin(kx + mz)\right| \cos \pi y,$$

$$\tilde{w}_1 = -\frac{u_0}{S}\frac{\partial^2 \phi_T}{\partial x\,\partial z} = -u_0 \eta_0 k e^{z/2H} \sin(kx + mz)\sin \pi y, \quad (6.13.39d)$$

where the tildes denote the perturbation dynamical fields.

Note that the perturbation in the potential temperature is 180° out of phase with the surface variation, for when $\eta_B(x, y)$ is positive, fluid elements with lower values of potential temperature are lifted. The vertical velocity produced by topographic slope, $\tilde{w}_1$, yields a vortex-tube stretching, which in turn yields a simple expression for $\tilde{v}_0$, viz.

$$\tilde{v}_0 = \frac{1}{\beta - u_0 K^2}\frac{1}{\rho_s}\frac{\partial}{\partial z}\rho_s \tilde{w}_1 \qquad (6.13.40)$$

which may be verified from (6.13.39b,d). The northward velocity is correlated with the potential-temperature fluctuation as follows. As the flow goes over a rise in $\eta_B$, the total potential temperature decreases as remarked earlier. The resulting vortex-tube compression produces a southward flow due to the influence of the planetary-vorticity gradient. Therefore on the *average*, negative $\tilde{\theta}_0$ is associated with southward flow, and equivalently, positive $\tilde{\theta}_0$ is correlated with northward flow, so that

$$\overline{\rho_s \tilde{\theta}_0 \tilde{v}_0}^{x, y} = \frac{\rho_s \eta_0^2 S}{K_s^2 - K^2}\frac{km}{4}. \qquad (6.13.41)$$

Thus, if $K_s^2 > K^2$, as it must be for radiation, the potential temperature flux associated with the radiating topographically induced stationary wave is northward. The average eastward flux $\overline{\tilde{\theta}_0 \tilde{u}_0}^{x, y}$ is identically zero. The vertical wave-energy flux

$$\overline{\rho_s p_0 w_1}^{x, y} = u_0 \frac{\rho_s \eta_0^2}{K_s^2 - K^2}\frac{km}{4} \qquad (6.13.42)$$

is proportional to the northward heat flux. The existence of the upward radiation of energy implies that the topography in the presence of the mean flow generates waves which carry energy upwards. This implies a momentum and energy exchange with the topography. In particular, the wave drag force, exerted by the topography on the fluid, is

$$\overline{p_0\left(\frac{\partial \eta}{\partial x}\right)_{z=0}}^{x,\,y} = \frac{\eta_0^2}{K_s^2 - K^2} \frac{km}{4},$$

(6.13.43)

while the rate at which the wave drag force does work on the fluid, $u_0 \times$ drag, is seen to be precisely equal to the energy flux radiated to great heights.

When $m^2 < 0$ no vertically propagating wave is possible, and it is a simple matter, left for the reader to verify, that the vertical propagation of energy is identically zero, as are both the northward heat flux $\overline{\tilde{v}_0 \tilde{\theta}_0}^{x,\,y}$ and the wave drag exerted by the topography on the fluid. Thus the radiating wave is significant, not only for its transfer of energy to the stratosphere, but also for its contribution to the wave-induced heat flux and momentum balance of the lower atmosphere. The relationship

$$\overline{\tilde{p}_0 \tilde{w}_1}^{x,\,y} = \frac{u_0}{S} \overline{\tilde{\theta}_0 \tilde{v}_0}^{x,\,y}$$

(6.13.44)

is true for an arbitrary $u_0(z)$ and can be proved in steady flow by a direct consideration of the linearized potential-temperature equation (6.5.12) in the absence of heating, i.e., from (6.5.12),

$$u_0 \frac{\partial}{\partial x} \tilde{\theta}_0 - \frac{\partial u_0}{\partial z} \frac{\partial \phi}{\partial x} = -\tilde{w}_1 S.$$

(6.13.45)

Multiplication by $p_0 = \phi$ and integration over $x$ and $y$ immediately yields (6.13.44).

To emphasize the importance of the upper boundary condition, consider the case where the region is limited in the vertical rather than being unbounded, i.e., let $0 \le z \le 1$. If the upper boundary is flat, then (6.13.13) and (6.13.14) are replaced by the condition.

$$\frac{\partial \Phi}{\partial z} = 0, \qquad z = 1.$$

(6.13.46)

For simplicity, consider the case where $\rho_s$ is constant (as might be appropriate for an oceanic model to which the bounded interval is natural).

It is useful to represent $\mathcal{H}$ in terms of a Fourier cosine series, i.e.,

$$= \sum_{n=1}^{\infty} \mathcal{H}_n \sin n\pi z \cos(kx + \theta_H)\sin \pi y.$$

(6.13.47)

Then the solution to (6.13.10) subject to (6.13.11) and (6.13.46) can be written

$$\Phi = \sum_{n=1}^{\infty} \frac{n\pi \mathcal{H}_n}{iku_0} \frac{\cos n\pi z e^{i\theta_H}}{[S(K_s^2 - K^2) - n^2\pi^2]} - \frac{S^{1/2}\eta_0 \cos m(z-1)}{(K_s^2 - K^2)^{1/2} \sin m},$$

(6.13.48)

where
$$m^2 = S(K_s^2 - K^2). \qquad (6.13.49)$$

Note that the sum in (6.13.48) runs from $n = 1$, i.e., heating always produces a response which has no barotropic component.

The response to *both* heating and topographic forcing may now suffer resonance in distinction to the case of the unbounded domain. The resonance corresponding to the excitation of the barotropic Rossby wave is identical to the unbounded case and involves only the topographic forcing. Since the vertical velocity in the barotropic mode is identically zero, this resonance must be unaffected by the presence of the upper rigid lid which only acts to suppress the vertical velocity associated with baroclinic motions. However, in addition, the denominators of both the heating and topographic terms vanish to give resonance whenever

$$K_s^2 \equiv \frac{\beta}{u_0} = K^2 + \frac{n^2\pi^2}{S}, \qquad n = 1, 2, 3, \dots. \qquad (6.13.50)$$

Comparison with (6.12.16) and (6.12.19) shows that this resonance occurs when the forcing resonates with the free Rossby *internal* modes. The resonance will occur for *dimensional* values of the flow speed given by

$$u_{0_*} = \frac{\beta_0}{K_*^2 + n^2\pi^2/L_D^2}, \qquad n = 1, 2, 3, \dots \qquad (6.13.51)$$

where $L_D$ is the Rossby deformation radius.

It follows from (6.13.48) that

$$\phi = \sum_{n=1}^{\infty} \frac{n\pi}{ku_0} \mathcal{H}_n \frac{\cos n\pi z \, \sin(kx + \theta_H)}{[S(K_s^2 - K^2) - n^2\pi^2]} \sin \pi y$$

$$- \frac{S^{1/2}\eta_0}{(K_s^2 - K^2)^{1/2}} \frac{\cos m(z-1)}{\sin m} \cos kx \, \sin \pi y. \qquad (6.13.52)$$

It is important to note, as the reader may verify, that the topographic drag due to the *topographically induced flow* is now identically zero. This can be easily understood if we note that the topographic term in (6.13.52) can be written as a sum of two plane waves, one radiating energy upwards, while the other with equal amplitude radiates downwards. A glance at (6.13.43) shows that the wave drag of each component is proportional to its assoicated vertical wave number. Since these are equal and opposite for the two waves which comprise the standing wave, the total sums to zero.

For functions which are limited to large horizontal and vertical scales so that

$$K_s^2 \gg K^2 + \frac{n^2\pi^2}{S}$$

the first term in (6.13.52) can be approximated by

$$\phi = \frac{1}{k\beta S} \frac{\partial}{\partial z} \sin(kx + \theta_H) \sin \pi y, \qquad (6.13.53)$$

which is equivalent to a potential vorticity balance

$$\beta v_0 = \frac{1}{S} \frac{\partial \mathcal{H}}{\partial z} \qquad (6.13.54)$$

in which the advection of planetary vorticity just balances the vortex tube stretching induced by internal heating.

## 6.14 Wave-Zonal Flow Interactions

It was apparent in the previous section that waves produced by heating and/or topography will generally possess a *nonzero* zonally averaged northward flux of heat and a vertical energy flux. If these fluxes produced by the waves are convergent, there is the *a priori* possibility that the convergence of the fluxes may alter both the zonally averaged temperature field and the zonally averaged zonal velocity. However, the relationship between the wave fluxes and the alteration of the mean temperature and zonal velocity fields is not entirely straightforward because of the presence in the momentum equation, for example, of the Coriolis term due to the ageostrophic meridional velocity.

To make this point clear, consider each variable (e.g., $u_0$) partitioned on the infinite $x$-interval between its zonal average,

$$\overline{u_0}^x = \lim_{X \to \infty} \frac{1}{2X} \int_{-X}^{X} u_0 \, dx \qquad (6.14.1)$$

and a longitudinal fluctuation around the mean, $u_0'$, where

$$u_0' = u_0 - \overline{u_0}^x, \qquad (6.14.2)$$

and thus

$$\overline{u_0'}^x \equiv 0. \qquad (6.14.3)$$

We may consider the primed fields as due to waves or eddies superimposed on the zonal mean flow.

Since $u_0$ and $v_0$ are geostrophic, it follows that

$$\overline{v_0}^x = 0 \qquad (6.14.4)$$

assuming $p_0$ remains finite as $|x| \to \infty$.

Each variable may be written

$$u_0 = \overline{u_0}^x + u_0',$$
$$\theta_0 = \overline{\theta_0}^x + \theta_0',$$
$$v_0 = v_0', \qquad (6.14.5)$$
$$v_1 = \overline{v_1}^x + v_1',$$
$$w_1 = \overline{w_1}^x + w_1'.$$

The essence of the wave-zonal flow interaction problem is the fact that quadratic products of terms, each of which have zero x-average, as might a wave, nevertheless may jointly yield a nonzero average, e.g.,

$$\overline{u_0 v_0}^x = \overline{u_0' v_0'}^x \neq 0. \tag{6.14.6}$$

If the decomposition indicated by (6.14.5) is used in (6.3.13a) and the averaging operation of (6.14.1) is applied to the result, an equation for the rate of change of the mean zonal velocity results:

$$\frac{\partial \overline{u_0}^x}{\partial t} = \overline{v_1}^x - \frac{\partial}{\partial y}(\overline{u_0' v_0'}^x), \tag{6.14.7}$$

while the same averaging operator applied to (6.5.12) yields

$$\frac{\partial \overline{\theta_0}^x}{\partial t} = -\overline{w_1}^x S - \frac{\partial}{\partial y}(\overline{\theta_0' v_0'}^x) + \mathscr{H}. \tag{6.14.8}$$

For the purposes of the present discussion, it is possible to slightly generalize the scaling discussion of Section 6.3 and allow for the presence of a force, due to dissipation, say, in the horizontal momentum equations. In that case (6.14.7) becomes,

$$\frac{\partial \overline{u_0}^x}{dt} = \overline{v_1}^x - \frac{\partial}{\partial y}(\overline{u_0' v_0'}^x) + \overline{\mathscr{F}_x}^x, \tag{6.14.7'}$$

where in the notation of Section 6.2

$$\mathscr{F}_x = \frac{\mathscr{F}_* \phi}{\rho_* \varepsilon U f_0}. \tag{6.14.9}$$

This permits a dynamical symmetry between the form of (6.14.7') and (6.14.8). The x-average of the equation for mass conservation, (6.3.14) yields

$$\frac{\partial}{\partial z}(\rho_s \overline{w_1}^x) + \frac{\partial}{\partial y}(\rho_s \overline{v_1}^x) = 0. \tag{6.14.10}$$

Note that while the $O(\varepsilon)$ momentum and mass equations generally contain metric terms that are absent in a $\beta$-plane mode, the equations for the *zonally-averaged* flow to $O(\varepsilon)$ are precisely the equations that would obtain in a $\beta$-plane model using Cartesian coordinates.

The zonal momentum equation (6.14.7') shows that the mean zonal velocity may be altered by two physically distinct processes, in addition to the direct action of the force $\mathscr{F}_x$. The first is due to the small ageostrophic, mean northward velocity, $\overline{v_1}^x$. In the presence of the earth's rotation the Coriolis force acting on northward moving fluid elements will lead to an average eastward acceleration of the zonal flow. The second term on the right-hand side of (6.14.7) is the *Reynolds stress gradient*. If $\overline{u_0' v_0'}^x$ is different from zero, the departures from the mean, which we may interpret as due to the eddy or wave field, will be transporting positive (or westerly) zonal momentum northward. If this northward flux of eastward momentum is divergent, i.e., if $\partial/\partial y (\overline{u_0' v_0'}^x) > 0$, then more zonal momentum, on average, leaves the local latitude interval than enters. The mean zonal momentum then tends to decrease in accordance with (6.14.7'). Similarly, (6.14.8) describes the change of the mean potential temperature. A northward heat flux, $\overline{v_0'\theta_0'}^x$ due to the wave field which varies meridionally, will tend to lead to local cooling if more heat flux leaves the region than enters, i.e., if $\partial/\partial y \; \overline{v_0'\theta_0'}^x > 0$. A slow, ageostrophic upward motion, $\overline{w_1}^x$, in the presence of the overall static stability $S$ will tend to lower the local $\overline{\theta_0}^x$ as fluid of lower potential temperature is raised replacing warmer fluid.

The chief difficulty with the prediction and interpretation of the effect of the eddy or wave fluxes on the mean fields is that the zonally averaged meridional circulation given by $\overline{v_1}^x$ and $\overline{w_1}^x$ is not itself independent of the eddy fluxes. In general $\overline{v_1}^x$ and $\overline{w_1}^x$ are themselves functionals of the eddy fields and the overall effect of the eddies cannot be isolated in the flux terms on the right-hand sides of (6.14.7') and (6.14.8). We might, in fact, ask whether the presence of the eddy fluxes might not be identically balanced by the effects of the mean meridional circulation forced by the eddy fluxes and so yield no alteration of $\overline{u_0}^x$ and $\overline{\theta_0}^x$. Although this may seem unlikely, it is worth pursuing the question further. Momentarily setting $\mathcal{H}^*$ and $\mathcal{F}_x^*$ to zero, so as to isolate the role of the eddy field, the condition that $\partial \overline{u_0}^x/\partial t$ and $\partial/\partial t \, \overline{\theta_0}^x$ both vanish requires

$$\overline{w_1}^x = -\frac{\partial}{\partial y} \frac{\overline{(\theta_0' v_0')}^x}{S}, \tag{6.14.11a}$$

$$\overline{v_1}^x = \frac{\partial}{\partial y} \overline{(u_0' v_0')}^x, \tag{6.14.11b}$$

while mass conservation, (6.14.10), in turn implies that

$$\frac{\partial}{\partial y}\left\{ -\frac{\partial}{\partial y}\rho_s\overline{(u_0' v_0')}^x + \frac{\partial}{\partial z}\rho_s\frac{\overline{(\theta_0' v_0')}^x}{S}\right\} = 0. \tag{6.14.12}$$

If $v_0'$ vanishes at some latitude it follows that for all $y$ the condition that $\partial \overline{u_0}^x/\partial t$ and $\partial \overline{\theta_0}^x/\partial t$ both vanish is that

$$-\frac{\partial}{\partial y}\rho_s\overline{(u_0' v_0')}^x + \frac{\partial}{\partial z}\rho_s\frac{\overline{\theta_0' v_0'}^x}{S} = 0. \tag{6.14.13}$$

This condition has a simple interpretation. First note that the zonal average of the thermal wind relation, (6.5.17), implies that

$$\frac{\partial \overline{u_0}^x}{\partial z} = -\frac{\partial \overline{\theta_0}^x}{\partial y}. \tag{6.14.14}$$

The potential vorticity equation for the *mean* flow can be obtained by using (6.14.10) to eliminate $\overline{v_1}^x$ and $\overline{w_1}^x$ between (6.14.7') and (6.14.8) to yield

$$\frac{\partial}{\partial t}\overline{\Pi_0}^x = -\frac{\partial}{\partial y}\overline{v_0'\Pi_0'}^x + \frac{1}{\rho_s}\frac{\partial}{\partial z}\left(\frac{\rho_s\overline{\mathcal{H}}^x}{S}\right) - \frac{\partial \overline{\mathcal{F}_x}^x}{\partial y}, \tag{6.14.15}$$

where

$$\overline{\Pi_0}^x = -\frac{\overline{\partial u_0}^x}{\partial y} + \frac{1}{\rho_s} \frac{\partial}{\partial z} \left( \frac{\rho_s \overline{\theta_0}^x}{S} \right), \tag{6.14.16a}$$

and

$$\Pi_0' = \frac{\partial v_0'}{\partial x} - \frac{\partial u_0'}{\partial y} + \frac{1}{\rho_s} \frac{\partial}{\partial z} \left( \frac{\rho_s \theta_0'}{S} \right). \tag{6.14.16b}$$

Since $u_0'$, $v_0'$, and $\theta_0'$ are in geostrophic and hydrostatic balance, it follows that

$$\overline{v_0' \Pi_0'}^x = -\frac{\partial}{\partial y} \overline{u_0' v_0'}^x + \frac{1}{\rho_s} \frac{\partial}{\partial z} \left( \frac{\rho_s \overline{v_0' \theta_0'}^x}{S} \right). \tag{6.14.17}$$

Thus the condition (6.14.13) that the mean fields remain stationary, is equivalent to the condition that the meridional eddy flux of potential vorticity vanish and this may, in fact, occur as we shall see.

This result is not surprising if we keep in mind that within quasigeostrophic theory all the dynamical fields can be obtained via the potential vorticity equation. Hence, were the eddy flux of potential vorticity to vanish, the eddies would be entirely unable to affect $\overline{\Pi_0}^x$. Since

$$\overline{\Pi_0}^x = \frac{\partial^2 \overline{\psi}^x}{\partial y^2} + \frac{1}{\rho_s} \frac{\partial}{\partial z} \frac{\rho_s}{S} \frac{\partial \overline{\psi}^x}{\partial z}, \tag{6.14.18}$$

once $\overline{\Pi_0}^x$ is known, $\overline{\psi}^x$ is determined by solving (6.14.18) and hence $\overline{u_0}$ and $\overline{\theta_0}$ can be determined. Obviously, if $\overline{\Pi_0}^x$ were independent of $t$, so would be $\overline{\psi}^x$ (unless there is time-dependent boundary forcing). By fixing attention on the evolution of the potential vorticity, which depends directly on only the eddy flux term, the difficulties associated with the role of the ageostrophic velocities in determining the individual evolutions of $\overline{u_0}^x$ and $\overline{\theta_0}^x$ could be avoided.

However, there are certain conceptual and diagnostic advantages in discussing how each of the elements of $\overline{\Pi_0}^x$, i.e., $\overline{u_0}^x$ and $\overline{\theta_0}^x$, evolve although, as noted above, their evolutions are hardly independent.

We have already seen that when the flux of potential vorticity vanishes, the effect of both the momentum and heat fluxes on $\overline{u_0}^x$ and $\overline{\theta_0}^x$ are exactly canceled by the effects of the mean meridional circulation. Or, stated equivalently, the eddy fluxes force a meridional circulation whose effect on the mean field is vitiated entirely by the eddy fluxes. In that case, the zonally averaged vertical velocity is given directly in terms of the divergence of the meridional heat flux (6.14.11a). In the more general case, there will be an imbalance in (6.14.8) between the heat flux divergence and the mean vertical velocity. Following Andrews and McIntyre (1976) and Edmond, Hoskins, and McIntyre (1980), we can express this imbalance in terms of a "residual" meridional circulation, i.e., as the difference between the actual vertical and meridional velocity and the portion whose effect on the mean field is canceled by eddy effects. Thus, we define†

$$\overline{w_{1*}}^x = \overline{w_1}^x + \frac{\partial}{\partial y} \frac{\overline{\theta_0' v_0'}^x}{S}. \tag{6.14.19a}$$

When (6.14.11a) is satisfied $\overline{w_{1*}}^x$ is identically zero. In order that the residual circulation also satisfy the continuity condition (6.14.10) we define

$$\overline{v_{1*}}^x = \overline{v_1}^x - \frac{1}{\rho_s} \frac{\partial}{\partial z} \frac{\rho_s \overline{\theta_0' v_0'}^x}{S}. \tag{6.14.19b}$$

If the mean momentum and heat equations are rewritten in terms of the residual vertical and meridional velocities (6.14.7′) and (6.14.8) become

$$\frac{\partial \overline{u_0}^x}{\partial t} = \overline{v_{1*}}^x + \frac{1}{\rho_s} \nabla \cdot (\rho_s \mathbf{F}) + \overline{\mathscr{F}_*}^x, \tag{6.14.20a}$$

$$\frac{\partial \overline{\theta_0}^x}{\partial t} = -\overline{w_{1*}}^x S + \overline{\mathscr{H}}^x, \tag{6.14.20b}$$

$$\frac{1}{\rho_s} \frac{\partial \rho_s \overline{w_{1*}}^x}{\partial z} + \frac{\partial}{\partial y} \overline{v_{1*}}^x = 0, \tag{6.14.20c}$$

where the vector $\mathbf{F} = \hat{j} F_y + \hat{k} F_z$, which lies in the meridional plane, is defined in terms of its components

$$F_y = -\overline{u_0' v_0'}^x, \qquad (6.14.21)$$
$$F_z = \frac{\overline{\theta_0' v_0'}^x}{S},$$

while

$$\frac{1}{\rho_s} \nabla \cdot \rho_s \mathbf{F} = -\frac{\partial}{\partial y} (\overline{u_0' v_0'}^x) + \frac{1}{\rho_s} \frac{\partial}{\partial z} \frac{\overline{\rho_s \theta_0' v_0'}^x}{S}. \tag{6.14.22}$$

The vector $\mathbf{F}$ is called the Eliassen–Palm flux and its density-weighted divergence is the *only* term in the transformed equations which explicitly depends on the eddy field. The set (6.14.20a, b, c) is called the Transformed Eulerian Mean equations (TEM) and this transformed set has several attractive conceptual properties although, of course, it is entirely equivalent to the original set (6.11.7′), (6.14.8), and (6.14.10).

The first feature of the TEM equations, as noted above, is that the sole *explicit* eddy term occurs in the momentum equation in the form of the divergence of the Eliassen–Palm flux vector, which is, as (6.14.17) shows the same as the eddy potential vorticity flux. Since the dynamics of the quasi-geostrophic system is determined entirely in terms of the potential vorticity, it is clearly conceptually an improvement to be able to rewrite the eddy forcing

---

† The asterisk subscript in this section refers to the "residual" dependent variables and not to restoration of dimensions for the labeled variable, i.e., $\overline{w_{1*}}^x$ is *nondimensional*.

in terms of the potential vorticity flux which is dynamically more fundamental than either the momentum or heat fluxes separately. Note that in this formulation, the heat equation is entirely free of eddy terms and the difference between the heating and the rate of change of $\overline{\theta_0}^x$ must be balanced entirely by $\overline{w_*}^x$ acting on the basic state's vertical temperature gradient. The residual circulation then comes closer to acting like the actually zonally averaged Lagrangian meridional circulation of fluid elements.

Nevertheless, in order to understand how $\overline{u_0}^x$ and $\overline{\theta_0}^x$ change with time it is still necessary to know the residual circulation. In this sense, the TEM equations share the feature of the original Eulerian set that the effect of the eddies cannot be ascertained until the ageostrophic or meridional circulation is calculated. However, the second attractive feature of the TEM equations is that the problem for the residual circulation can itself be written in terms of $(1/\rho_s)\nabla \cdot (\rho_s\mathbf{F})$.

Since the residual circulation is nondivergent by (6.14.20c), we may define a residual stream function for the meridional velocities, e.g.,

$$\rho_s\overline{w_{1*}}^x = -\frac{\partial \chi}{\partial y},$$

$$\rho_s\overline{v_{1*}}^x = \frac{\partial \chi}{\partial z}.$$

(6.14.23)

Using the thermal wind relation to eliminate the time derivatives between (6.14.20a) and (6.14.20b) yields

$$\frac{\partial}{\partial z}\frac{1}{\rho_s}\frac{\partial \chi}{\partial z} + \frac{S}{\rho_s}\frac{\partial^2 \chi}{\partial y^2} = -\frac{\partial}{\partial z}\left(\frac{1}{\rho_s}\nabla \cdot \rho_s\mathbf{F}\right) - \frac{\partial \overline{\mathscr{H}}^x}{\partial y} - \frac{\partial}{\partial z}\frac{\overline{\mathscr{F}_*}^x}{\rho_s}.$$ (6.14.24)

Thus the residual circulation produced by the eddies is due only to the flux of potential vorticity (the Eliassen–Palm flux divergence). It is important to discuss the appropriate boundary conditions for (6.14.24) but note first that if $\nabla \cdot \rho_s\mathbf{F}$ should be independent of $y$, a particular solution of (6.14.24) would have as the residual circulation due to the eddy field

$$\overline{v_{1*}}^x \equiv \frac{1}{\rho_s}\frac{\partial \chi}{\partial z} = -\frac{1}{\rho_s}\nabla \cdot (\rho_s\mathbf{F})$$

in which case, by (6.14.20a), there would be no alteration of the zonal velocity. Again, this apparently fortuitous cancellation can be easily explained in terms of potential vorticity dynamics. If $\overline{v_0'\Pi_0'}^x$ is independent of $y$, (6.14.15) shows that the eddies will leave $\overline{\Pi_0}^x$ and hence $\overline{u_0}^x$ independent of time (again assuming consistency with the boundary conditions).

Suppose on the latitude lines $y = y_1, y_2$, that $v_0$ and $v_1$ vanish (where $y_1$ and $y_2$ may be at $\pm\infty$). Then from their definitions, it follows that

$$\frac{\partial \chi}{\partial z} = 0, \qquad y = y_1, y_2.$$ (6.14.25)

On $z = 0$, supposing it to be a rigid lower boundary, we have, from (6.6.9) and (6.14.19a)

$$\overline{w_{1*}}^x = -\frac{1}{\rho_s}\frac{\partial \chi}{\partial y} = \frac{E_V^{1/2}}{2\varepsilon}\overline{\zeta_0}^x + \frac{\partial}{\partial y}\overline{(v_0'\eta_B)}^x - \frac{\partial}{\partial y}\overline{\frac{v_0'\theta_0'}{S}}^x, \qquad (6.14.26)$$

where

$$\overline{\zeta_0}^x = -\frac{\partial}{\partial y}\overline{u_0}^x,$$

and where it has been assumed that $\overline{\eta_B}^x = 0$.

It is important to note that the problem for $\chi$ involves two further eddy terms in addition to the Eliassen–Palm flux. They are the divergence of the eddy heat flux at the surface $z = 0$ and the term

$$\overline{v_0'\eta_B}^x = -p_0'\frac{\overline{\partial \eta_B}^x}{\partial x}, \qquad (6.14.27)$$

which is the force in the $x$-direction exerted on the fluid by the sloping lower boundary.

The upper boundary condition will depend on the nature of the boundary. If it too is rigid, the boundary condition at $z = z_T$ will be similar to (6.14.26).

So far no restriction has been placed on the amplitude of the eddy field. However, in most cases, further theoretical progress has been limited to the case where the amplitude of the eddy field is small.

If (6.5.18) is then linearized about the zonal mean state $\overline{u_0}^x$ under the presumption that

$$(u_0', v_0') \ll \overline{u_0}^x, \qquad (6.14.28)$$

we obtain the linearized potential vorticity equation

$$\frac{\partial \Pi_0'}{\partial t} + \overline{u_0}^x\frac{\partial \Pi_0'}{\partial x} + v_0'\frac{\partial \overline{\Pi_0}^x}{\partial y} = \frac{1}{\rho_s}\frac{\partial}{\partial z}\left(\frac{\rho_s\mathcal{H}'}{S}\right) + \frac{\partial}{\partial x}\frac{\mathcal{F}_y'}{\rho_s} - \frac{\partial}{\partial y}\frac{\mathcal{F}_x'}{\rho_s}, \qquad (6.14.29)$$

where $\Pi_0'$, given by (6.14.16b), may be written in term of a stream function $\psi'$ as

$$\Pi_0' = \frac{\partial^2\psi'}{\partial x^2} + \frac{\partial^2\psi'}{\partial y^2} + \frac{1}{\rho_s}\frac{\partial}{\partial z}\frac{\rho_s}{S}\frac{\partial\psi'}{\partial z}, \qquad (6.14.30)$$

and where

$$\frac{\partial \overline{\Pi_0}^x}{\partial y} = \beta - \frac{\partial^2\overline{u_0}^x}{\partial y^2} - \frac{1}{\rho_s}\frac{\partial}{\partial z}\frac{\rho_s}{S}\frac{\partial\overline{u_0}^x}{\partial z} \qquad (6.14.31)$$

is the meridional gradient of the potential vorticity of the state associated with the zonally averaged flow. $\mathcal{H}'$, $\mathcal{F}_y'$, and $\mathcal{F}_x'$ are the departures from the zonal mean of the heating and frictional forcing in the $y$- and $x$-directions, respectively.

The linearization of the heat equation yields

$$\frac{\partial \theta_0}{\partial t} + \overline{u_0}^x \frac{\partial \theta_0'}{\partial x} + v_0' \frac{\partial \overline{\theta_0}}{\partial y} + w_1' S = \mathcal{H}'. \qquad (6.14.32)$$

If (6.14.29) is multiplied by $\Pi_0'$ and averaged in $x$, we obtain an explicit expression for $\overline{v_0' \Pi_0'}^x$, i.e.,

$$\overline{v_0' \Pi_0'}^x \doteq -\frac{\partial}{\partial t}\left(\frac{\overline{\Pi_0'^2}^x}{\partial \overline{\Pi_0}^x / \partial y}\right) + \mathcal{D}, \qquad (6.14.33)$$

where

$$\mathcal{D} = \overline{\frac{\Pi_0'}{\rho_s}\left\{\frac{\partial}{\partial z}\frac{\rho_s \mathcal{H}'}{S} + \frac{\partial}{\partial x}\frac{\mathcal{F}_y'}{\rho_s} - \frac{\partial}{\partial y}\frac{\mathcal{F}_x'}{\rho_s}\right\}}^x \Big/ \frac{\partial \overline{\Pi_0}^x}{\partial y}. \qquad (6.14.34)$$

If $\partial \overline{\Pi_0}/\partial y^x$ should vanish at some point in the $(y, z)$-plane, the apparent singularity in (6.14.33) can be avoided by multiplying (6.14.30) by $\psi'$ instead, to obtain an equivalent expression for $\overline{v_0' \Pi_0'}^x$. Should both $\overline{u_0}^x$ and $\partial \overline{\Pi_0}/\partial y^x$ vanish for the same $y$ and $z$, it can be shown that the flow $\overline{u_0}^x$ is inherently unstable (see Chapter 7).

Similarly, multiplication of (6.14.32) by $\psi'$ and $\theta_0'$ allows (6.14.26) to be rewritten as either

$$-\frac{1}{\rho}\frac{\partial \chi}{\partial y} = \frac{\partial}{\partial y}\left[\overline{\frac{\psi'}{S}\frac{\partial \theta_0'}{\partial t}}^x + \frac{Ev^{1/2}}{2\varepsilon}\overline{\psi' \zeta_0'}^x - \overline{\frac{\psi' \mathcal{H}'}{S}}^x\right]\Big/ \overline{u_0}^x + \frac{Ev^{1/2}}{2\varepsilon}\overline{\zeta_0}, \qquad (6.14.35a)$$

or

$$-\frac{1}{\rho_s}\frac{\partial \chi}{\partial y} = \frac{\partial}{\partial y}\left[\overline{(v_0' \eta_B)}^x - \left\{\frac{\partial}{\partial t}\frac{\overline{\theta_0'^2}^x}{2S} + \overline{u_0}^x \theta_0'\frac{\partial \eta_B}{\partial x} + \frac{Ev^{1/2}}{2\varepsilon}\overline{\theta_0' \zeta_0'} - \overline{\frac{\theta' \mathcal{H}'}{S}}\right\}\Big/ \frac{\partial \overline{\theta_0}^x}{\partial y}\right]$$
$$+ \frac{Ev^{1/2}}{2\varepsilon}\overline{\zeta_0}. \qquad (6.14.35b)$$

Each of these representations is useful in separate circumstances. The two topographic terms in (6.14.35b) may be re-expressed using (6.14.32), i.e.,

$$\overline{v_0' \eta_B}^x - \overline{u_0}^x \theta_0'\frac{\partial \eta_B}{\partial x} \Big/ \frac{\partial \overline{\theta_0}^x}{\partial y} = \overline{\eta_B\left\{\frac{\partial \theta_0'}{\partial t} - S\frac{Ev^{1/2}}{2\varepsilon}\zeta_0' - \mathcal{H}'\right\}}^x \Big/ \frac{\partial \overline{\theta_0}^x}{\partial y}. \qquad (6.14.35c)$$

These explicit expressions for the potential vorticity flux and the boundary heat flux allow the proof of a far-reaching theorem due originally to Charney and Drazin (1961). Suppose that the wave field is conservative, i.e., that all dissipative terms acting on the eddy field are negligible, i.e., that $\mathcal{H}'$, $\mathcal{F}_x'$, and $\mathcal{F}_y'$ vanish and that terms in $E_V^{1/2}$ may be neglected at $z = 0$ (and $z_T$). Furthermore, suppose the eddy field has an intensity that is constant in time so that

$$\frac{\partial}{\partial t}\overline{\Pi_0'^2}^x \quad \text{and} \quad \frac{\partial}{\partial t}\overline{\theta_0'^2}^x = 0. \qquad (6.14.36)$$

Note that this still allows the motion itself to be unsteady, that is, a uniformly propagating wave of fixed amplitude would also satisfy (6.14.36).

It then follows from (6.14.33) that: (1) $\nabla \cdot \rho_s \mathbf{F}$ vanishes, (2) that there is no eddy forcing in the differential equation for $\chi$, and (3) that the boundary condition at $z = z_T$ contains no eddy term except a term proportional to $\overline{\eta_b \, \partial \theta_0'/\partial t}^x$. This latter term identically vanishes if either (a) there is no topography or (b) the eddy field is precisely steady [as the fields calculated in Section 6.13]. If neither (a) nor (b) is satisfied, there may be an eddy forcing of $\chi$, but (6.14.36) implies that it will oscillate with the period of the wave field with a zero time average. It seems sensible then to momentarily disregard such forcings as irrelevant for the long-term dynamics of the mean flow. In which case the problem for $\chi$ is completely homogeneous and its solution is simply $\chi = 0$.

With $\chi$ and $\nabla \cdot \rho_s \mathbf{F}$ identically zero, it follows directly from (6.14.20a,b) that the eddies do not alter the mean zonal flow at all. *Thus in the absence of heating and friction in a stationary wave field, the fluxes of heat and momentum produce mean meridional circulations which precisely cancel the tendency of the fluxes to alter the mean state. The mean state is unaffected by the wave fluxes.*

If the wave field is conservative but (6.14.36) is not satisfied, there will be an alteration of the mean field. However, if $\overline{\Pi_0'^2}^x$ and $\overline{\theta_0'^2}^x$ eventually return to their initial values then so will the mean field. Thus the passage of a wave packet through a mean field will give rise to a temporary alteration of the zonal flow. However, if the wave field is conservative, the mean field will be restored to its original state after the wave packet has departed, aside from such changes of $\overline{u_0}^x$ that are forced by $\overline{\mathscr{H}}^x$ and $\overline{\mathscr{F}_x}^x$ which, of course, are not associated with the eddy field.

Thus changes in the mean zonal state induced by the eddy field imply either that the wave field is nonconservative or nonstationary in time. Otherwise, the Eliassen–Palm flux will remain nondivergent.

There is a particularly intriguing correction between the Eliassen–Palm flux vector and the Rossby waves discussed in Section 6.11.1. First, we define the wave activity, $A_w$,

$$A_w = \frac{\rho_s \overline{\Pi_0'^2}^x}{\partial \Pi_0 / \partial y}. \tag{6.14.37}$$

Then (6.14.33) may be written

$$\frac{\partial}{\partial t} A_w + \nabla \cdot (\rho_s \mathbf{F}) = \rho_s \mathscr{D}. \tag{6.14.38}$$

Thus the wave activity satisfies a simple budget equation in which the flux of wave activity is given by the Eliassen–Palm flux, $\rho_0 \mathbf{F}$, while the non-conservative source term $\rho_s \mathscr{D}$ acts to increase the wave activity.

Suppose the wave field under consideration is a wave packet carrying a wave with a wavelength short enough so that *locally* the fields $\overline{u_0}^x$ and $\overline{\partial \Pi_0}^x/\partial y$ appear *spatially uniform*. Then to a good approximation $\psi'$ can be written

$$\psi' = a(y, z, t)\cos(kx + ly + mz - \sigma t), \tag{6.14.39}$$

where $a(y, z, t)$ is a slowly varying function of time, and latitude and height reflecting the structure of the mean field in which it is embedded.

Then it is a simple matter to show, using the ideas of Section 6.11, that

$$\mathbf{F} = (\hat{\mathbf{j}} C_{gy} + \hat{\mathbf{k}} C_{gz}) A_w, \tag{6.14.40a}$$

where

$$C_{gy} = \frac{\partial \overline{\Pi}_0^x}{\partial y} \frac{2kl}{(k^2 + l^2 + m^2/S)^2}, \tag{6.14.40b}$$

$$C_{gz} = \frac{\partial \overline{\Pi}_0^x}{\partial y} \frac{2km/S}{(k^2 + l^2 + m^2/S)^2}. \tag{6.14.40c}$$

Hence the Eliassen–Palm flux in such a case can be thought of as a propagation, by Rossby wave packets, of wave activity from one region to another. Comparison of (6.14.40a) to (6.14.21) shows that an upward flux of wave activity ($F_z > 0$) corresponds to a *northward* eddy heat flux while a poleward flux of wave activity ($F_y > 0$) corresponds to a *southward* flux of westerly momentum. Of course, the relations between the signs of $F_y$ and $F_z$ and the signs of the heat and momentum fluxes are independent of the assumptions required for the validity of (6.14.40a,b,c). Hence the orientation of $\mathbf{F}$ in the $(y, z)$-plane more generally indicates the relative magnitude of the eddy heat flux and momentum flux.

Edmond, Hoskins, and McIntyre (1980) have presented a vivid demonstration of the diagnostic value of these ideas in interpreting atmospheric observations of eddy-induced accelerations of the mean zonal winds. The reader is referred to their paper for a more complete discussion than there is space to present here.

It would, however, be an oversimplification of the eddy-mean flow interaction dynamics to concentrate entirely on the spatial structure of the Eliassen–Palm flux as an indicator of the structure of the forcing of the mean flow. It must be remembered that the mean flow is just as strongly affected by the residual meridional circulation. Even though this circulation is driven by $\nabla \cdot (\rho_s \mathbf{F})$ the elliptic character of (6.14.24) suggests that the resulting circulation can penetrate well beyond the region of the source terms on the right-hand side.

As an example, consider the case where $\overline{\mathscr{H}}^x$ and $\overline{\mathscr{F}}^x$ are identically zero. For simplicity, let $\rho_s$ and $S$ be taken as constant and let us suppose that $\nabla \cdot (\rho_s \mathbf{F})$ and all other eddy forcing is limited to a region so far away from any boundary that the domain may be considered infinite in $y$ and $z$. In this case it is convenient to choose the horizontal scaling length $L$ to be the deformation radius, $L_D$ so that $S = 1$. Suppose now that the potential vorticity flux has the form

$$\overline{u_0' \Pi_0'}^x = \nabla \cdot \mathbf{F} = G_0 e^{-\alpha r^2}, \qquad r = (y^2 + z^2)^{1/2}, \tag{6.14.41}$$

i.e., that the divergence of the Eliassen–Palm flux is a maximum at the origin of the meridional domain $-\infty \le y \le +\infty$, $-\infty \le z \le \infty$. The contours of constant potential vorticity flux are circles in the $(y, z)$-plane.† The characteristic scale for the flux divergence is $\alpha^{-1/2}$ and it decays rapidly with increasing $r$. In order to produce the steady distribution specified in (6.14.41), a particular combination of heating and friction implied by (6.14.34) are required. The potential vorticity flux in (6.14.41) will contribute to a mean flow acceleration which by (6.14.20a) will have the same spatial structure as (6.14.41). However, to compute the rate of change of $\overline{u_0}^x$, the meridional residual circulation $\overline{v_{1*}}^x$ must be found.

It follows from (6.14.24) that if $\tilde{\chi}_\rho$ is a solution of

$$\frac{\partial^2 \tilde{\chi}}{\partial y^2} + \frac{\partial^2 \tilde{\chi}}{\partial z^2} = -\nabla \cdot \mathbf{F} \tag{6.14.42}$$

that

$$\chi = \frac{\partial \tilde{\chi}}{\partial z} \tag{6.14.43}$$

will be a solution of (6.14.24). The solution of (6.14.42) which is regular at $r = 0$ when (6.14.41) is the prescribed Eliassen–Palm flux is simply

$$\tilde{\chi} = -\frac{G_0}{2\alpha^2} \int_0^r \left( \frac{1 - e^{-\alpha^2 \zeta^2}}{\zeta^2} \right) d\zeta, \tag{6.14.44}$$

so that

$$\chi = \frac{\partial \tilde{\chi}}{\partial z} = \frac{z}{r} \frac{\partial \tilde{\chi}}{\partial r},$$

or

$$\chi = -\frac{G_0}{2\alpha^2} \frac{z}{r^2} (1 - e^{-\alpha^2 r^2}). \tag{6.14.45}$$

Note that for large $r$, $\chi$, reflecting the elliptic nature of (6.14.24) decays only slowly like $r^{-1}$ rather than like $e^{-\alpha^2 r^2}$ as does the Eliassen–Palm flux. The resulting circulation is depicted in Figure 6.14.1. Note the extensive vertical and meridional scale of the circulation. The meridional and vertical residual velocities are given by (6.14.23) and (6.14.44), i.e.,

$$\overline{v_{1*}}^x = -\frac{G_0}{2\alpha^2} \left[ \frac{(1 - e^{-\alpha^2 r^2})}{r^2} \left( 1 - \frac{2z^2}{r^2} \right) + \frac{2^2 \alpha z^2}{r^2} e^{-\alpha^2 r^2} \right], \tag{6.14.46a}$$

$$\overline{w_{1*}}^x = -\frac{G_0}{2\alpha^2} \frac{yz}{r^2} \left[ -\frac{(1 - e^{-\alpha^2 r^2})}{r^2} + 2\alpha^2 e^{-\alpha^2 r^2} \right]. \tag{6.14.46b}$$

Figure 6.14.2 shows $\overline{v_{1*}}^x$ as a function of $z$ through the center line of the meridionally cell at $y = 0$. In the region $|\alpha z| < 1.122$, the meridional $\overline{v_{1*}}^x$ is

---

† Since $D$ has been used as a vertical scale and $L_D$ as a horizontal scale, the isolines of $\nabla \cdot \mathbf{F}$ are, in fact, ellipses in the $(y_*, z_*)$ plane. The ellipse "flattens" for larger static stability, i.e., for larger $L_D$.

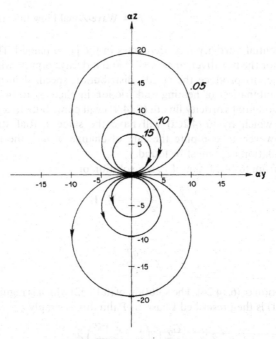

**Figure 6.14.1** Streamline pattern for the residual meridional circulation driven by the Eliassen and Palm flux divergence $\nabla \cdot \mathbf{F} = G_0 \exp[-\alpha^2(y^2 + z^2)]$. The contours are labeled in units of $-G_0/2\alpha$ for $\chi$. Note that the streamlines extend far beyond $\alpha(y^2 + z^2)^{1/2} = 1$.

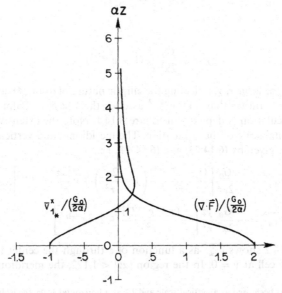

**Figure 6.14.2** The residual meridional velocity $\overline{v_{1_*}}^x$ is shown on the line $y = 0$ for $z > 0$. For $z < 0$, $\overline{v_{1_*}}^x(z) = \overline{v_{1_*}}^x(-z)$. Also shown is the potential vorticity flux $(\nabla \cdot \mathbf{F})$. Note that for $\alpha z > 1.5$ $\overline{v_{1_*}}^x > \nabla \cdot \mathbf{F}$.

negative and *opposes* the tendency of $\nabla \cdot \mathbf{F}$ to accelerate the zonal flow. For $|\alpha z| > 1.122$, $\overline{v_{1*}}^x$ is positive and the mean flow is accelerated by the residual mean meridional circulation. Figure 6.14.2 also shows $(\nabla \cdot \mathbf{F})$ along $y = 0$ and it is apparent that for $\alpha r > 1.5$ the acceleration of the zonal flow is dominated by the effect of the eddy-driven meridional circulation and that this accelera-tion term decays far more slowly than the potential vorticity flux which is ultimately responsible for the circulation. In much the same way that electric charges produce electric fields which penetrate space well beyond the charge distribution, the presence of the meridional circulation introduces an "action-at-a-distance" character to the relationship between the acceleration of the zonal velocity and $\nabla \cdot \mathbf{F}$. Indeed, on $y = 0$ (6.14.20a), (6.14.42), and (6.14.47a) imply that

$$\frac{\partial \overline{u_0}^x}{\partial t} = \frac{G_0}{2\alpha^2 z^2}(1 - e^{-\alpha^2 z^2}). \tag{6.14.47}$$

For $\alpha z \to 0$, $\partial \overline{u_0}^x/\partial t \to G_0/2$ (independent of the scale $\alpha^{-1}$) while for large $z$ $\partial \overline{u_0}^x/\partial t \to G_0/(2\alpha^2 z^2)$, i.e., the acceleration falls off only algebraically with height.

If the eddy field consists of a uniformly propagating field which satisfies

$$\frac{\partial \Pi'}{\partial t} = -c\frac{\partial \Pi'}{\partial x},$$

then (6.14.29) becomes

$$(\overline{u_0}^x - c)\frac{\partial}{\partial x}\left\{\frac{\partial^2 \psi'}{\partial x^2} + \frac{\partial^2 \psi'}{\partial y^2} + \frac{1}{\rho_s}\frac{\partial}{\partial z}\frac{\rho_s}{s}\frac{\partial \psi'}{\partial z}\right\} + \frac{\partial \overline{\Pi_0}^x}{\partial y}\frac{\partial \psi_0'}{\partial x} = \mathscr{D}', \tag{6.14.48}$$

where $\mathscr{D}'$ signifies the right-hand side of (6.14.29). Points in the $(y, z)$-plane where $\overline{u_0}^x - c$ vanishes have special significance since (6.14.48) is *singular* at those points where the wave speed coincides with the local zonal velocity. Due to that singularity rapid spatial variations in $\psi'$ can be anticipated in the vicinity of these points and frictional and nonlinear effects which might be elsewhere negligible can play a significant note in the vicinity of the region where $\overline{u_0}^x - c$ becomes small. If friction and thermal dissipation are significant near the critical points, i.e., in the so-called *critical* layer, then nonzero flux convergence of potential vorticity and associated alterations of the zonal flow can be expected.

In the case where the eddy field can be described locally as a Rossby wave packet, then the same approximations that lead to (6.14.40b,c) yield the dispersion relation

$$c = \overline{u_0}^x - \frac{\partial \overline{\Pi_0}^x/\partial y}{k^2 + l^2 + m^2/S}. \tag{6.14.49}$$

As the wave packet approaches the critical layer where $c \to \overline{u_0}^x$, the wave numbers must become very large unless, fortuitously, $\partial \overline{\Pi_0}/\partial y^x$ vanishes there.

In general, that will not happen and

$$k^2 + l^2 + \frac{m^2}{S} \to \infty \quad \text{as} \quad c \to \overline{u}_0^{\,x}.$$

However, since the coefficients of (6.14.48) are independent of $x$, the wave number in the $x$-direction, $k$, remains fixed and thus $l$ and $m$ must increase. It follows from (6.14.40b,c) that as the packet approaches the critical layer, its group velocity in the meridional plane must then go to zero. The packet moves very slowly and, subject to dissipation, it is natural to anticipate the complete dissipation of wave energy and wave activity. If the critical layer does act as such a graveyard for wave activity, i.e., if the layer is absorbing, the associated spatial gradients of potential vorticity flux will produce a locally strong divergence of Eliassen–Palm flux.

If, however, nonlinearities enter strongly into the dynamics of the critical layer, the situation is somewhat obscure. Analysis beyond the scope of this book (Benney and Bergeron, 1969) indicates the possibility that if dissipation is negligible compared to nonlinear effects in the critical layer, reflection of wave activity rather than absorption will take place. If the reflection is complete so that there is no net flux of potential vorticity, the mean flow would remain unaltered.

## 6.15 Topographic Waves in a Stratified Ocean

In a homogeneous fluid layer the potential-vorticity gradient associated with a sloping bottom is dynamically equivalent to the planetary-vorticity gradient. The dynamic similarity of the two effects depends crucially on the Taylor–Proudman theorem, i.e., that in a *homogeneous* fluid at small Rossby number the horizontal velocity must be independent of depth. This in turn implies that vortex stretching, initiated at the lower boundary by motion over the slope, will be felt throughout the layer, and the consequent change in relative vorticity will be indistinguishable from the changes produced by motion in the field of the planetary-vorticity gradient. Obviously the presence of stratification which inhibits the vertical velocity and also allows the horizontal velocity to be depth dependent can be expected to significantly alter the dynamical relationship between bottom slope and the $\beta$-effect. At the same time the introduction of a sloping bottom will alter the eigenvalue problem of Section 6.12 and render the simple decomposition of the wave motion into a barotropic mode plus zero-mass-transport baroclinic internal modes no longer strictly possible. The introduction of a sloping bottom will, in general, tend to force a vertical motion, and hence baroclinic structure, into *all* modes. To examine the nature of the motion in the presence of $\beta$ and a bottom slope, we will study the following problem, after Rhines (1970). Consider the motion of a stratified fluid layer with constant $S$ (i.e., constant Brunt–Väisälä frequency $N_s$) over a bottom sloping in the $y$-direction such that

$$\eta_B = \alpha y. \tag{6.15.1}$$

The slope parameter $\alpha$ is related to the actual bottom slope by (3.12.18), i.e., in dimensional units

$$\alpha = \left(\frac{f_0 L^2}{UD}\right)\frac{\partial h_B}{\partial y_*}. \tag{6.15.2}$$

In addition, the density scale height will be assumed to be large compared to the depth of the fluid, so that $\rho_s$ may be considered constant. The model is therefore directly relevant to the oceanic case, but it is also useful qualitatively for the atmosphere.

Small oscillations in a fluid otherwise at rest must satisfy the linearized potential-vorticity equation

$$\frac{\partial}{\partial t}\left|S^{-1}\frac{\partial^2\psi}{\partial z^2} + \frac{\partial^2\psi}{\partial x^2} + \frac{\partial^2\psi}{\partial y^2}\right| + \beta\frac{\partial\psi}{\partial x} = 0, \tag{6.15.3}$$

subject to

$$\frac{\partial}{\partial t}\frac{\partial\psi}{\partial z} = 0, \qquad z = 1, \tag{6.15.4}$$

and

$$S^{-1}\frac{\partial}{\partial t}\frac{\partial\psi}{\partial z} = -\alpha\frac{\partial\psi}{\partial x}, \qquad z = 0, \tag{6.15.5}$$

which follow from the application of (6.6.10) and (6.9.14) in their linearized forms. Note from (6.15.5) that if $\alpha \neq 0$, no nontrivial solution with zero vertical variation can exist unless $\partial\psi/\partial x$ is zero on $z = 0$. If it vanishes on $z = 0$ and $\psi$ is independent of $z$, (6.15.3) shows that no nontrivial wave motion is possible. Only the steady geostrophic streaming flow along lines of constant $y$ is then allowed.

Plane-wave solutions of the form

$$\psi = \text{Re } \Phi(z)e^{i(kx + ly - \sigma t)} \tag{6.15.6}$$

are possible if nontrivial solutions can be found for $\Phi(z)$ which satisfies

$$\frac{d^2\Phi}{dz^2} + m^2\Phi = 0, \tag{6.15.7}$$

$$\frac{d\Phi}{dz} = 0, \qquad z = 1, \tag{6.15.8a}$$

$$\frac{d\Phi}{dz} = \frac{\alpha k}{\sigma}S\Phi, \qquad z = 0, \tag{6.15.8b}$$

where

$$m^2 = -S\left|\frac{\beta k}{\sigma} + K^2\right|, \tag{6.15.9a}$$

$$K^2 = k^2 + l^2. \tag{6.15.9b}$$

The problem (6.15.7), (6.15.8a,b) is an eigenvalue problem for $m$, or equivalently for $\sigma$. If $\beta$ is zero, i.e., if the horizontal wavelength is sufficiently small for the earth's sphericity to be neglected, then $m$ is independent of $\sigma$, but the satisfaction of (6.15.8a,b) will determine $\sigma(k, l)$. Let us consider this case first.

If $\beta = 0$,

$$m^2 = -\mu^2 = -SK^2 < 0. \tag{6.15.10}$$

The solution for $\Phi$ which satisfies (6.15.8a) is

$$\Phi = \frac{A \cosh S^{1/2} K(z - 1)}{\cosh S^{1/2} K}. \tag{6.15.11}$$

In this case the potential-vorticity equation with $\beta = 0$ reduces to

$$\frac{\partial}{\partial t} \left( S^{-1} \frac{\partial^2 \psi}{\partial z^2} + \frac{\partial^2 \psi}{\partial x^2} + \frac{\partial^2 \psi}{\partial y^2} \right) = 0 \tag{6.15.12}$$

and only determines the relationship between vertical and horizontal scales. The solution (6.15.11) is a maximum at $z = 0$ and decays away from the lower boundary. It is bottom trapped, and its vertical $e$-folding scale is

$$d = \frac{1}{S^{1/2} K} = \frac{\lambda}{2\pi S^{1/2}}, \tag{6.15.13}$$

where $\lambda$ is the horizontal wavelength,

$$\lambda = \frac{2\pi}{(k^2 + l^2)^{1/2}}. \tag{6.15.14}$$

In dimensional units

$$d_* = Dd = \frac{\lambda_* \, f_0}{2\pi \, N_s}. \tag{6.15.15}$$

Stratification enhances the trapping, while rotation increases $d_*$. Naturally, as $N_s \to 0$ and the fluid becomes homogeneous, $d_* \to \infty$, for this is the content of the Taylor–Proudman theorem. If (6.15.11) is substituted into (6.15.8b), the eigenvalue relation for $\sigma$ is obtained:

$$\sigma = -\frac{\alpha S^{1/2} k}{K \tanh K S^{1/2}}. \tag{6.15.16}$$

For each wave number there is a single mode of oscillation, whose frequency is given by (6.15.16). In the limit $S \to 0$,

$$\sigma \sim -\alpha \frac{k}{K^2}, \tag{6.15.17}$$

or in dimensional units

$$\sigma_* = -\frac{f_0}{D} \frac{\partial h_B}{\partial y_*} \frac{k_*}{k_*^2 + l_*^2},$$

which yields the Rossby wave in a homogeneous fluid with a rigid upper lid ($F = 0$). In this limit $\Phi(z)$ becomes independent of $z$. The homogeneous limit is in fact recovered whenever $K S^{1/2} \to 0$, i.e., whenever the horizontal wavelength considerably exceeds the Rossby internal deformation radius. Thus even if $S = O(1)$, sufficiently *long* waves will appear barotropic, and topography will produce an oscillation dynamically similar to Rossby $\beta$-waves. On the other hand, when $KS^{1/2}$ exceeds unity there is a significant change, as shown in Figure 6.15.1. For large $KS^{1/2}$ the wave is increasingly affected by stratification and increasingly bottom trapped, and in the limit of large $SK^{1/2}$

$$\sigma \sim -\frac{\alpha S^{1/2} k}{K}$$

$$\Phi \sim A e^{-KS^{1/2}z}, \qquad KS^{1/2} \to \infty. \tag{6.15.18}$$

Note that in this limit the frequency is independent of wavelength and depends only on the *orientation* of the wave vector. The dimensional frequency is

$$\sigma_* = \frac{U}{L}\sigma = -\frac{\partial h_B}{\partial y_*} N_S\left(\frac{k_*}{K_*}\right) \tag{6.15.19}$$

and is independent of the rotation. The exponential structure of the solution does depend on $f_0$. Indeed, in this limit the wave is reminiscent of the Kelvin wave of Section 3.9, whose frequency was also independent of $f_0$ but whose spatial structure similarly depended on $f_0$. In each case the relationship between vertical and horizontal scale is determined by the condition of zero perturbation potential vorticity. Note that for the topographic wave $\sigma/k < 0$ if $\partial h_B/\partial y > 0$.

The presence of $\beta$ leads to a somewhat more complex situation. The parameter $m^2$ in (6.15.9a) now depends on the eigenvalue $\sigma$. The solutions of (6.15.7) then fall into two categories. If $m^2 > 0$ the solution to (6.15.7), subject to (6.15.8a), is

$$\Phi(z) = A \cos m(z - 1), \tag{6.15.20}$$

while the application of (6.15.8b) yields the eigenvalue relation

$$m \tan m = \frac{\alpha k S}{\sigma}. \tag{6.15.21}$$

Since by (6.15.9a)

$$\sigma = -\frac{\beta k}{K^2 + m^2/S}, \tag{6.15.22}$$

(6.15.21) can be written

$$\boxed{\tan m = -\frac{\alpha}{\beta}\left\{m + \frac{SK^2}{m}\right\}.} \tag{6.15.23}$$

On the other hand, if $m^2 = -\mu^2 < 0$, then

$$\Phi(z) = A \cosh \mu(z - 1), \tag{6.15.24}$$

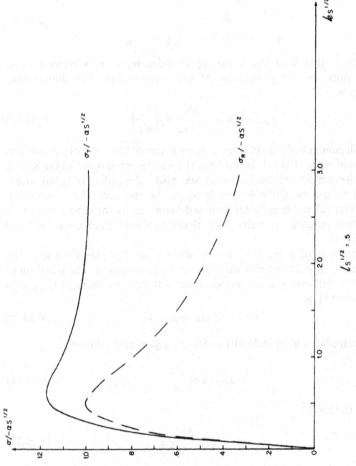

**Figure 6.15.1** The dispersion relation for quasigeostrophic, topographic Rossby waves in a stratified fluid. The parameter $KS^{1/2}$ is proportional to the deformation radius divided by the wavelength $\lambda_*$, namely, $KS^{1/2} = 2\pi L_D/\lambda_*$.

and (6.15.8b) yields

$$\mu \tanh \mu = -\frac{\alpha k}{\sigma} S, \qquad (6.15.25a)$$

and

$$\tanh \mu = -\frac{\alpha}{\beta}\left\{\mu - \frac{SK^2}{\mu}\right\}, \qquad (6.15.25b)$$

which may be obtained directly from (6.15.23) by the substitution $m = i\mu$. The eigenvalue equations (6.15.23) and (6.15.25) depend only on the magnitude of the wave vector (i.e., on the wavelength) and *not* on the orientation of the wave vector. The frequency, which is given by (6.15.22) once $m$ (or $\mu$) has been determined, *does* depend on the direction of the wave vector.

Consider the relation (6.15.25b) first. Without loss of generality $\mu$ can be considered positive, and there is a single $\mu > 0$ for which (6.15.25) may be satisfied, as can be verified by plotting the left- and right-hand sides of the equation separately as shown in Figure 6.15.2. Since

$$\mu^2 = S\left(K^2 + \frac{k\beta}{\sigma}\right), \qquad (6.15.26)$$

small $S$ will imply $\mu \ll 1$, so that (6.15.25) may be rewritten

$$\mu \tanh \mu \simeq \mu^2 = S\left(K^2 + \frac{\beta k}{\sigma}\right) = -\frac{\alpha}{\beta}\{\mu^2 - K^2 S\}$$

$$= -\frac{\alpha k S}{\sigma}, \qquad (6.15.27)$$

or

$$\sigma = -\frac{(\alpha + \beta)k}{K^2} \qquad (6.15.28)$$

corresponding to

$$\mu = K\left(\frac{\alpha S}{\alpha + \beta}\right)^{1/2}. \qquad (6.15.29)$$

Thus in the limit $S \to 0$ the dispersion relation becomes precisely the Rossby-wave frequency relation with the planetary-vorticity gradient and the potential-vorticity gradient of the bottom slope combining to form the total potential-vorticity gradient in the manner described in Section 3.17. For the homogeneous fluid the two effects are indistinguishable. As the strength of the stratification increases, or for increasing $\alpha/\beta$, the single root of (6.15.25) approaches $\mu = KS^{1/2}$, and from (6.15.25a)

$$\sigma \to -\frac{\alpha k S^{1/2}}{K \tanh KS^{1/2}}, \qquad (6.15.30)$$

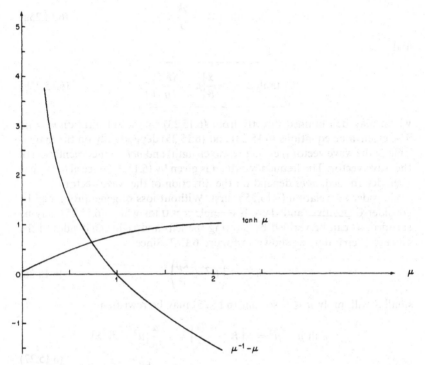

**Figure 6.15.2**   The eigenvalue relation (6.15.25b) for bottom-trapped modes, showing the single solution possible.

which is the dispersion relation for the $\beta = 0$ case. The mode structure given by $\Phi(z)$ becomes increasingly bottom trapped. Thus for small $S$ (nearly homogeneous fluid), topography and $\beta$ combine linearly for a barotropic oscillation. For large $S$ (or large $\alpha$ and/or $K$) the mode given by (6.15.25b) becomes so limited in vertical extent that the influence of $\beta$ becomes negligible compared to that of the bottom slope. The latter influence is, after all, measured by

$$\frac{f_0}{D}\frac{dh_B}{dy_*},$$

where $D$ is the vertical scale of the *motion*. As the mode becomes bottom trapped its vertical scale decreases, so that the potential-vorticity gradient of the bottom becomes dominant.

The eigenvalue relation (6.15.23) has an infinite number of solutions due to the periodicity of $\tan m$, as shown in Figure 6.15.3. Note, however, that no

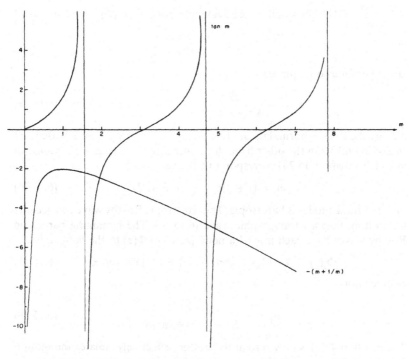

**Figure 6.15.3**  The eigenvalue relation (6.15.23) demonstrates a multiplicity of solutions, each referring to a different vertical mode structure.

intersection with the first branch of tan $m$ is possible. That "missing" mode is of course the bottom-trapped mode discussed above. As $S \to 0$, either $m \to 0$ (in which case the solution becomes the barotropic mode) or, for $m = O(1)$, $\sigma$ becomes $O(S)$ by (6.15.22). Unless $K^2$ also is $O(S^{-1})$, the wave becomes nondispersive in this limit. Thus, baroclinic modes exist for small but nonzero $S$ of the form

$$\Phi(z) = \cos m_0(z - 1), \tag{6.15.31}$$

where $m_0$ is the solution of

$$\tan m_0 = -\frac{\alpha}{\beta} m_0 \tag{6.15.32}$$

and

$$\sigma = -\frac{\beta S k}{m^2}. \tag{6.15.33}$$

For $S = O(1)$ but small $\alpha$ (i.e., negligible bottom slope), the solutions of (6.15.23) are

$$m_j = j\pi, \qquad j = 0, 1, 2, \tag{6.15.34}$$

corresponding to frequencies

$$\sigma_j = -\frac{\beta k}{K^2 + j^2\pi^2/S}, \qquad j = 0, 1, 2, \tag{6.15.35}$$

and are the Rossby normal modes (including the barotropic mode) discussed in Section 6.12. On the other hand, for *either* large $\alpha$ or large $S$ the values of $m$ which satisfy (6.15.23) correspond to the set

$$m_j = (j + \tfrac{1}{2})\pi, \qquad j = 0, 1, 2. \tag{6.15.36}$$

None of these modes is barotropic. The barotropic Rossby wave has become the bottom-trapped topographic wave (6.15.30). The remaining baroclinic Rossby waves have each moved a node (zero) of $\Phi(z)$ to the *bottom*, i.e.,

$$\Phi_j(z) = A \cos[(j + \tfrac{1}{2})\pi(z - 1)] = (-1)^j A \sin j\pi z. \tag{6.15.37}$$

In this limit

$$\sigma_j = -\frac{\beta k}{K^2 + (j + \tfrac{1}{2})^2\pi^2/S}. \tag{6.15.38}$$

The fact that $\Phi_j(z)$ is now zero at the bottom effectively isolates the motion from the otherwise overwhelming effect of the bottom slope, since now the horizontal velocity in these modes at $z = 0$ identically *vanishes*. This is evident in the absence of an explicit dependence of $\sigma_j$ on $\alpha$ in (6.15.38). The effect of large $\alpha$ *or* large $S$ is felt implicitly in the consequent change in the structure of $\Phi$ and the change in the vertical wave number from $j\pi$ to $(j + \tfrac{1}{2})\pi$.

The presence of stratification has substantially altered the structure of the motion due to bottom topography, yet even in a stratified fluid, the topography yields free oscillations recognizably similar in their dynamics to the Rossby topographic waves of Chapter 3.

## 6.16   Layer Models

Although the quasigeostrophic approximation considerably simplifies the analysis of the dynamics of large-scale motions, the potential-vorticity equation, (6.5.21) or (6.8.11), remains a nonlinear partial differential equation in the four independent variables $x$, $y$, $z$, and $t$. Multilayer models, in which the fluid consists of a finite number of homogeneous layers of uniform but distinct densities, provide a useful intermediate system between the single-layer barotropic model of Chapter 3 and the continuously stratified, baroclinic model of this chapter. Baroclinic effects can often be modeled with striking simplicity in the layer models. Of course, the simplicity is purchased at the cost of a reduction in the model's ability to resolve the vertical structure of the motion, so that care and experience are required in the interpretation of the finite layered system.

Consider the motion of the fluid system shown in Figure 6.16.1. The fluid is composed of $N$ layers, each of which consists of fluid of uniform, constant density $\rho_n$, $n = 1, 2, \ldots, N$, where

$$\rho_N > \rho_{N-1} > \cdots \rho_{n+1} > \rho_n > \rho_{n-1} > \cdots > \rho_1.$$

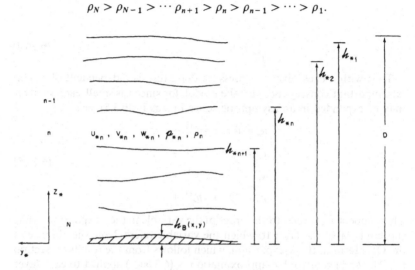

**Figure 6.16.1**   The $N$-layer model.

Let $D$ be the overall vertical scale for the fluid height, and let $h_{*n}$, $n = 1$, $2, \ldots, N$, represent the instantaneous elevation of the upper surface of the $n$th layer above the reference level at $z_* = 0$. As in Section 6.12, we suppose that $U$, $L$, $D$, and $UD/L$ are characteristic scales for the horizontal velocity, the horizontal length and vertical length of the motion, and the vertical velocity respectively. Further, we also assume that as in (6.12)

$$\varepsilon = \frac{U}{f_0 L} = O\left(\frac{L}{r_0}\right) \ll 1,$$

$$\delta = \frac{D}{L} \ll 1, \tag{6.16.1}$$

where $r_0$ is the earth's radius. This suggests, in analogy with (6.2.18), that the pressure in each layer be written

$$P_{*n} = \sum_{K=1}^{n-1} \rho_K g D_K + \rho_n g(H_n - z_*) + \rho_n f_0 U L p_n(x, y, z, t), \tag{6.16.2}$$

where the constant $D_K$ is the dimensional thickness of the $K$th layer in the absence of motion, and $H_n$ is the constant dimensional elevation of the upper surface of the $n$th layer, also in the absence of motion. Thus $\rho_n f_0 UL p_n$ represents the departure of the pressure field from the hydrostatic value it would have in the absence of motion. Within each layer the hydrostatic approximation, valid as a result of (6.16.1), implies

$$\frac{\partial p_{*n}}{\partial z_*} = -\rho_n g, \tag{6.16.3}$$

or

$$\frac{\partial p_n}{\partial z} = 0. \tag{6.16.4}$$

Thus within each layer the pressure departure is independent of $z$. This has important consequences for the model, for since $\varepsilon$ is small, each variable may be expanded in an asymptotic series in $\varepsilon$ as in (6.3.5), i.e.,

$$
\begin{aligned}
u_n &= u_n^{(0)} + \varepsilon u_n^{(1)} + \cdots, \\
v_n &= v_n^{(0)} + \varepsilon v_n^{(1)} + \cdots, \\
w_n &= \phantom{v_n^{(0)}} + \varepsilon w_n^{(1)} + \cdots, \\
p_n &= p_n^{(0)} + \varepsilon p_n^{(1)} + \cdots,
\end{aligned}
\tag{6.16.5}
$$

where superscripts refer to the order of the variable in the $\varepsilon$-expansion, while subscripts label the layer to which the variable refers. Note the absence of the $O(1)$ term in the $w$-expansion, which follows from the results of Section 6.3. The $O(1)$ geostrophic approximation is (6.3.6a,b) applied to each layer in turn, i.e.,

$$
\begin{aligned}
v_n^{(0)} &= \frac{\partial p_n^{(0)}}{\partial x}, \\
u_n^{(0)} &= -\frac{\partial p_n^{(0)}}{\partial y},
\end{aligned}
\tag{6.16.6}
$$

which, with (6.16.4), implies that $u_n^{(0)}$ and $v_n^{(0)}$ are independent of $z$ within each layer. However, the horizontal velocities may vary from one layer to another. Within each layer the Taylor–Proudman theorem applies, but the presence of sharp density jumps at the interfaces will permit accompanying jumps in the horizontal velocities from layer to layer. To determine this variation, consider first the $n$th interface, whose height may be written

$$h_{*n} = Dh_n(x, y, t) = H_n + R_n \eta_n(x, y, t), \tag{6.16.7}$$

where $R_n$ is a yet unspecified scaling constant, while $\eta_n$ is the $O(1)$ nondimensional deviation of the $n$th surface from its undisturbed value. Both $R_n$ and $\eta_n$ are determined by the condition that at each interface the pressure must be continuous. For example at $z_* = h_n$,

$$p_{*n-1} = p_{*n}, \tag{6.16.8}$$

or, using (6.16.2) and the fact that $D_{K-1} = H_{K-1} - H_K$,

$$\sum_{K=1}^{n-2} \rho_K g D_K + \rho_{n-1} g\{H_{n-1} - H_n - R_n\eta_n\} + \rho_{n-1} f_0 ULp_{n-1}(x, y, t)$$

$$= \sum_{K=1}^{n-1} \rho_K g D_K + \rho_n g\{H_n - H_n - R_n\eta_n\} + \rho_n f_0 ULp_n(x, y, t),$$

(6.16.9)

from which it follows that

$$R_n g(\rho_n - \rho_{n-1})\eta_n = \rho_n f_0 ULp_n - \rho_{n-1} f_0 ULp_{n-1}. \qquad (6.16.10)$$

In order to ensure that $\eta_n$ is an $O(1)$ dimensionless variable, $R_n$ is *chosen* as

$$R_n = \frac{\rho_0 f_0 UL}{g\{\rho_n - \rho_{n-1}\}} = \varepsilon FD \frac{\rho_0}{\rho_n - \rho_{n-1}}, \qquad (6.16.11)$$

where $\rho_0$ is a characteristic (constant) value for the density of the fluid, and as before

$$F \equiv \frac{f_0^2 L^2}{gD}. \qquad (6.16.12)$$

Thus (6.16.7) becomes

$$h_n = \frac{H_n}{D} + \frac{\varepsilon F\eta_n(x, y, t)}{(\rho_n - \rho_{n-1})/\rho_0}. \qquad (6.16.13)$$

In analogy with (6.5.13) we assume that

$$\frac{F\rho_0}{\rho_n - \rho_{n-1}} = O(1), \qquad (6.16.14)$$

while

$$\frac{\Delta\rho}{\rho} = \frac{\rho_n - \rho_{n-1}}{\rho_0} \ll 1 \qquad (6.16.15)$$

in analogy with (6.4.13) and (6.4.14), although the restriction (6.16.15) is not necessary for the validity of the layer model and is an assumption easily relaxed. It is convenient for our purposes however to require (6.16.15). Then to $O(\Delta\rho/\rho)$, (6.16.10) becomes

$$\boxed{\eta_n = p_n - p_{n-1}}. \qquad (6.16.16)$$

Variations of the interface are therefore directly related to the differences in the pressure deviations from layer to layer. In particular, (6.16.6) with (6.16.16) implies that

$$u_n^{(0)} - u_{n-1}^{(0)} = -\frac{\partial \eta_n^{(0)}}{\partial y},$$

$$v_n^{(0)} - v_{n-1}^{(0)} = \frac{\partial \eta_n^{(0)}}{\partial x}.$$

(6.16.17)

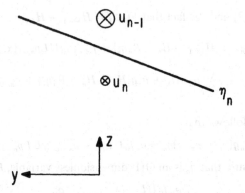

**Figure 6.16.2**  The interface slope is directly proportional, at each horizontal position, to the velocity difference between two adjacent layers. Here, $U_{n-1}^{(0)} > U_n^{(0)}$.

Thus the *difference* in the velocity from one layer to the next depends directly on the slope of the interface between the two layers as shown in Figure 6.16.2. Horizontal density gradients are confined to the interface surfaces, where the density changes discontinuously, leading, by an obvious limit of the thermal wind balance, to discontinuous changes in $u$ and $v$. The sloping interfaces play the role of the sloping density or potential-temperature surfaces in the continuous model. Note from (6.16.13) that a given velocity difference between two layers requires an interface deformation that depends *inversely* upon the density difference between the layers. It is left to the reader to demonstrate that

$$\eta_1 = p_1, \qquad R_1 = \varepsilon F D, \qquad (6.16.18)$$

so that the upper free surface deforms proportionally less, by a factor of $O(\Delta\rho/\rho)$, than the other interfaces, due to the $O(1)$ density jump between the first layer and the essentially zero-density region above it. At each interface, the vertical velocity must satisfy

$$w_{*n} = \frac{D}{L} U w_n = \frac{D}{L} U \varepsilon w_n^{(1)} + \cdots$$

$$= \frac{d}{dt_*} h_{*n} = \frac{U}{L} R_n \frac{d}{dt} \eta_n^{(0)} + \cdots, \qquad (6.16.19)$$

or

$$w_n^{(1)} = \begin{cases} \dfrac{F}{(\rho_n - \rho_{n-1})/\rho_0} \dfrac{d\eta_n^{(0)}}{dt}, & \text{at } z = h_n,\ n > 1, \qquad (6.16.20a) \\[3mm] F \dfrac{d\eta_1^{(0)}}{dt} & \text{at } z = h_1. \qquad (6.16.20b) \end{cases}$$

The $O(1)$ vorticity balance is given by (6.3.17) for each layer. Since the density is constant *within* each layer, (6.3.17) may be written, in the present context, as

$$\frac{d_0}{dt}\{\zeta_n^{(0)} + \beta y\} = \frac{\partial}{\partial z} w_n^{(1)}, \tag{6.16.21}$$

where

$$\zeta_n^{(0)} = \frac{\partial v_n^{(0)}}{\partial x} - \frac{\partial u_n^{(0)}}{\partial y} = \left(\frac{\partial^2}{\partial x^2} + \frac{\partial^2}{\partial y^2}\right) p_n^{(0)} \tag{6.16.22}$$

and

$$\frac{d_0}{dt} = \frac{\partial}{\partial t} + u_n^{(0)}\frac{\partial}{\partial x} + v_n^{(0)}\frac{\partial}{\partial y}. \tag{6.16.23}$$

Since $u_n^{(0)}$, $v_n^{(0)}$, and consequently $\zeta_n^{(0)}$ are independent of $z$ within each layer, (6.16.21) may be integrated over the depth of the $n$th layer to yield

$$\left|\frac{D_n}{D} + \frac{O(\varepsilon F)}{\Delta\rho/\rho_0}\right|\frac{d_0}{dt}\{\zeta_n^{(0)} + \beta y\} = w_n^{(1)}(x, y, h_n) - w_n^{(1)}(x, y, h_{n+1}). \tag{6.16.24}$$

For all $n$ except $n = 1$ and $n = N$, the right-hand side of (6.16.24) may be evaluated with the aid of (6.16.20a) to yield

$$\frac{d_0}{dt}[\zeta_n^{(0)} + \beta y] = \rho_0 \frac{DF}{D_n}\frac{d_0}{dt}\left[\frac{\eta_n^{(0)}}{\rho_n - \rho_{n-1}} - \frac{\eta_{n+1}^{(0)}}{\rho_{n+1} - \rho_n}\right], \tag{6.16.25a}$$

or with (6.16.16), for $n \neq 1, N$,

$$\frac{d_0}{dt_0}\left[\zeta_n^{(0)} + \beta y - \frac{DF}{D_n}\left\{\frac{p_n^{(0)} - p_{n-1}^{(0)}}{(\rho_n - \rho_{n-1})/\rho_0} - \frac{p_{n+1}^{(0)} - p_n^{(0)}}{(\rho_{n+1} - \rho_n)/\rho_0}\right\}\right] = 0. \tag{6.16.25b}$$

If each of the density jumps is equal, then (6.16.25b) reduces, for $n \neq 1, N$, to

$$\boxed{\frac{d_0}{dt_0}\left[\zeta_n^{(0)} + \beta y - \frac{DF}{D_n \Delta\rho/\rho_0}\{2p_n^{(0)} - p_{n+1}^{(0)} - p_{n-1}^{(0)}\}\right] = 0}, \tag{6.16.26}$$

where

$$\frac{\Delta\rho}{\rho_0} = \frac{\rho_n - \rho_{n-1}}{\rho_0} = \frac{\rho_{n+1} - \rho_n}{\rho_0}.$$

The lowest layer must be treated specially. At its lower boundary, i.e., at $z_* = h_{*B}$, the vertical velocity is given by (6.6.9):

$$w_N^{(1)}(x, y, h_{*B}) = \frac{E_V^{1/2}}{2\varepsilon}\zeta_N^{(0)}(x, y, t) + \mathbf{u}_N^{(0)} \cdot \nabla\eta_B \frac{D_N}{D}, \tag{6.16.27}$$

where

$$\eta_B = \frac{h_{*B}}{\varepsilon D_N}. \tag{6.16.28}$$

The vorticity equation, when integrated over the lowest layer, yields

$$\frac{d_0}{dt_0}\left[\zeta_N^{(0)} + \beta y - \frac{DF}{D_N}\left\{\frac{p_N^{(0)} - p_{N-1}^{(0)}}{(\rho_N - \rho_{N-1})/\rho_0}\right\} + \eta_B(x, y)\right] = -\frac{E_V^{1/2}}{2\varepsilon}\frac{D}{D_N}\zeta_N^{(0)},$$

(6.16.29)

where

$$E_V = \frac{2A_V}{f_0 D^2}.$$

The equation for the uppermost layer follows similarly. If the upper surface is *free*, then (6.16.20b) is appropriate. To order $\Delta\rho/\rho$, then, the motion of the upper surface becomes negligible compared to the motion of the interface $h_2$, and the upper surface (to $O(\Delta\rho/\rho)$) appears rigid, insofar as the vertical velocity is concerned. Thus it easily follows for $n = 1$ that

$$\frac{d_0}{dt}\left[\zeta_1^{(0)} + \beta y - \frac{DF}{D_1}\left\{\frac{p_1^{(0)} - p_2^{(0)}}{(\rho_2 - \rho_1)/\rho_0}\right\}\right] = 0.$$

(6.16.30)

It is sometimes useful to consider models where the upper surface is truly rigid, so that (4.5.50) applies. Then (6.16.30) can easily be shown to become

$$\frac{d_0}{dt}\left[\zeta_1^{(0)} + \beta y - \frac{DF}{D_1}\left\{\frac{p_1^{(0)} - p_2^{(0)}}{(\rho_2 - \rho_1)/\rho_0}\right\}\right] = \frac{E_V^{1/2}}{2\varepsilon}\frac{D}{D_1}\{\zeta_T - \zeta_1^{(0)}\},\quad (6.16.31)$$

where $\zeta_T$ is the applied vorticity of the rigid upper boundary.

The simplest model which retains baroclinic features is the two-layer model, i.e., $N = 2$. If we define

$$\psi_1 = p_1^{(0)},$$
$$\psi_2 = p_2^{(0)}$$

(6.16.32)

and

$$F_1 = \frac{f_0^2 L^2}{g(\Delta\rho/\rho)D_1},$$

$$F_2 = \frac{f_0^2 L^2}{g(\Delta\rho/\rho)D_2},$$

(6.16.33)

$$\frac{\Delta\rho}{\rho} = \frac{\rho_2 - \rho_1}{\rho_0},$$

then (6.16.29) and (6.16.30) become

$$\left[\frac{\partial}{\partial t} + \frac{\partial \psi_1}{\partial x}\frac{\partial}{\partial y} - \frac{\partial \psi_1}{\partial y}\frac{\partial}{\partial x}\right]$$

$$\times \left[\frac{\partial^2 \psi_1}{\partial x^2} + \frac{\partial^2 \psi_1}{\partial y^2} - F_1(\psi_1 - \psi_2) + \beta y\right] = 0, \quad (6.16.34a)$$

$$\left[\frac{\partial}{\partial t} + \frac{\partial \psi_2}{\partial x}\frac{\partial}{\partial y} - \frac{\partial \psi_2}{\partial y}\frac{\partial}{\partial x}\right]\left[\frac{\partial^2 \psi_2}{\partial x^2} + \frac{\partial^2 \psi_2}{\partial y^2} - F_2(\psi_2 - \psi_1) + \beta y + \eta_B\right]$$

$$= -\frac{r_2}{2}\left|\frac{\partial^2 \psi_2}{\partial x^2} + \frac{\partial^2 \psi_2}{\partial y^2}\right|, \quad (6.16.34b)$$

where

$$r_2 = \frac{(2A_V f_0)^{1/2}}{U}\frac{L}{D_2}. \quad (6.16.35)$$

The system (6.16.34a,b) is an obvious generalization of the single-layer, quasigeostrophic potential-vorticity equation (4.11.12). The baroclinic motion of the two layers is coupled in (6.16.34a,b) by the motion of the interface,

$$h_2 = \frac{H_2}{D} + \varepsilon\frac{D_2}{D}F_2(\psi_2 - \psi_1), \quad (6.16.36)$$

which produces vortex-tube compression in one layer and stretching in the other. This baroclinic system is a considerable simplification over the continuous model. There are now only three independent variables $(x, y, t)$ for the two coupled equations.

An energy equation may be easily derived by multiplying (6.16.34a,b) by $-\psi_1 D_1/D$ and $-\psi_2 D_2/D$ respectively to obtain, after addition,

$$\frac{\partial}{\partial t}\left|\frac{1}{2}\sum_{K=1}^{2}\left|\left(\frac{\partial \psi_K}{\partial x}\right)^2 + \left(\frac{\partial \psi_K}{\partial y}\right)^2\right|\frac{D_K}{D} + \frac{1}{2}\left(\frac{F_1 D_1 + F_2 D_2}{D}\right)(\psi_1 - \psi_2)^2\right|$$

$$+ \nabla \cdot \mathbf{S} = -\frac{r_2}{2}\left|\left(\frac{\partial \psi_2}{\partial x}\right)^2 + \left(\frac{\partial \psi_2}{\partial y}\right)^2\right|\frac{D_2}{D}, \quad (6.16.37)$$

where

$$\mathbf{S} = \mathbf{i}\left[\sum_{K=1}^{2}\left\{-\psi_K\frac{\partial^2 \psi_K}{\partial x \,\partial t} - u_K^{(0)}\psi_K\Pi_K^{(0)}\right\}\frac{D_K}{D} - \beta\frac{\psi_K^2}{2}\frac{D_K}{D} - \frac{r_2}{2}\psi_2\frac{\partial \psi_2}{\partial x}\frac{D_2}{D}\right]$$

$$+ \mathbf{j}\left[\sum_{K=1}^{2}\left\{-\psi_K\frac{\partial^2 \psi_K}{\partial y \,\partial t} - v_K^{(0)}\psi_K\Pi_K^{(0)}\right\}\frac{D_K}{D} - \frac{r_2}{2}\psi_2\frac{\partial \psi_2}{\partial y}\frac{D_2}{D}\right], \quad (6.16.38)$$

where

$$\Pi_K^{(0)} = \frac{\partial^2 \psi_K}{\partial x^2} + \frac{\partial^2 \psi_K}{\partial y^2} + (-1)^K F_K(\psi_1 - \psi_2). \quad (6.16.39)$$

This conservation statement implies that in the absence of dissipation ($r_2 = 0$), the sum of the kinetic energy and the available potential energy, which for the two-layer model is

$$\mathscr{A} = \frac{1}{2} \frac{F_1 D_1 + F_2 D_2}{D} (\psi_1 - \psi_2)^2, \tag{6.16.40}$$

is conserved if S vanishes on the lateral boundary of the region containing the fluid. In a manner analogous to the argument of Section 6.10, it can easily be shown that indeed the integral of the normal component of S around a closed, rigid boundary vanishes.

## 6.17 Rossby Waves in the Two-Layer Model

As an example of the application of the two-layer model to the description of dynamical phenomena, consider the dynamics of linear Rossby waves. If dissipation and bottom slope are ignored, the two-layer equations for small-amplitude motions about a state of rest are, from (6.16.34a,b),

$$\frac{\partial}{\partial t} \left[ \frac{\partial^2 \psi_1}{\partial x^2} + \frac{\partial^2 \psi_1}{\partial y^2} - F_1(\psi_1 - \psi_2) \right] + \beta \frac{\partial \psi_1}{\partial x} = 0, \tag{6.17.1a}$$

$$\frac{\partial}{\partial t} \left[ \frac{\partial^2 \psi_2}{\partial x^2} + \frac{\partial^2 \psi_2}{\partial y^2} - F_2(\psi_2 - \psi_1) \right] + \beta \frac{\partial \psi_2}{\partial x} = 0. \tag{6.17.1b}$$

On the infinite $x$, $y$ plane, solutions in the form of plane waves may be sought in the form

$$\psi_1 = \text{Re } A_1 e^{i(kx + ly - \sigma t)},$$
$$\psi_2 = \text{Re } A_2 e^{i(kx + ly - \sigma t)}, \tag{6.17.2}$$

where $A_1$ and $A_2$ are complex constants whose moduli and phases yield the magnitude and relative phase of the motion in each layer. If (6.17.2) is substituted into (6.17.1a,b), we obtain two, homogeneous, coupled algebraic equations for $A_1$ and $A_2$:

$$A_1\{\sigma(K^2 + F_1) + \beta k\} + \{-\sigma F_1\}A_2 = 0, \tag{6.17.3a}$$

$$A_1\{-\sigma F_2\} + \{\sigma(K^2 + F_2) + \beta k\}A_2 = 0, \tag{6.17.3b}$$

where

$$K^2 = k^2 + l^2.$$

Nontrivial solutions for $A_1$ and $A_2$ are possible only if the determinant of their coefficients vanishes, i.e., only if $\sigma$ is a root of the quadratic

$$\sigma^2 K^2(K^2 + F_1 + F_2) + \sigma \beta k(2K^2 + F_1 + F_2) + \beta^2 k^2 = 0, \tag{6.17.4}$$

whose two roots are

$$\sigma_1 = -\frac{\beta k}{K^2},\tag{6.17.5a}$$

$$\sigma_2 = -\frac{\beta k}{K^2 + F_1 + F_2}.\tag{6.17.5b}$$

Corresponding to each root, either (6.17.3a) or (6.17.3b) may be used to obtain the vertical-structure ratio $A_1/A_2$. Thus

$$A_1 = A_2, \qquad \sigma = \sigma_1,\tag{6.17.6}$$

while

$$A_1 F_2 = -A_2 F_1, \qquad \sigma = \sigma_2,$$

which may also be written

$$A_1 D_1 = -A_2 D_2, \qquad \sigma = \sigma_2.\tag{6.17.7}$$

The first root corresponds to the barotropic mode, as the comparison of (6.17.5a) with the identical result (6.12.10) for the continuous model shows. Since $A_1 = A_2$ in this mode, the motion in both layers is identical, the velocity is independent of depth, and the displacement of the interface, by (6.16.36), is identically zero. This reinforces the conclusion drawn earlier that the barotropic mode, when it exists, is independent of the detailed nature of the stratification. On the other hand, the dependence (6.17.5b) of $\sigma_2$ on the parameters $F_1$ and $F_2$ demonstrates that the second mode depends on the density difference between the two layers. More precisely, $\sigma_2$ depends on the ratio of the length scale $L$ (which may be thought of as the wavelength) to the two deformation radii, i.e.,

$$F_n = \frac{L^2}{R_n^2}, \qquad n = 1, 2\tag{6.17.8a}$$

where

$$R_n = \frac{1}{f_0}\left(g\frac{\Delta\rho}{\rho}D_n\right)^{1/2}, \qquad n = 1, 2.\tag{6.17.8b}$$

The total, instantaneous horizontal transport in the second mode is identically zero, as shown by (6.17.7). The interface deformation is given by

$$h_2 = \frac{H_2}{D} - \varepsilon F_2\,\mathrm{Re}\,A_1 e^{i(kx+ly-\sigma t)}, \qquad \sigma = \sigma_2,\tag{6.17.9}$$

Therefore, the nature of the second mode corresponds to the baroclinic modes of Section 6.12, which by (6.12.11) also satisfy the zero-transport condition. In addition, the structure of the two modes, as shown in Figure 6.17.1, when compared with the modal structure of the first two modes of the continuous model, shows that the two-layer model can be thought of as representing the barotropic mode and the *first* baroclinic mode (the mode

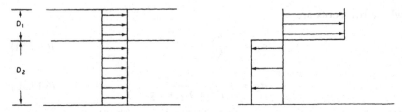

**Figure 6.17.1** The structure of the barotropic mode and baroclinic mode in the two-layer model.

with a single internal node) of the continuous model. The description of the structure and dynamics of higher modes is beyond the capacity of the two-layer model. In general an $N$-layer model can only model the first $N$ modes of the continuous model.

## 6.18 The Relationship of the Layer Models to the "Level" Models

The relative simplicity of the layer models, insofar as analysis is concerned, is an extremely attractive feature. Moreover, from the discussion of Section 6.16 the model is, at least within the quasigeostrophic context, an accurate description of the dynamics of an idealized but physically realizable fluid system. To the degree that the geostrophic approximation is valid, the results of the layer models can therefore be applied with confidence to the dynamics of a real physical system. This allows the formulation of simple but mathematically and physically well-posed problems of geophysical interest. However it is also clear that while physically meaningful in their own right, the layer models have been introduced as analogues for the continuous model. Indeed, in the previous section it was apparent from the results of the special calculation for Rossby waves that motions in the layer models bear a close equivalence to a subclass of the possible motions of the continuous model. In this section the relationship is explored between the layer models, which are accurate models of very simple, idealized systems, and the so-called "level" models, which are finite-difference approximations to the potential-vorticity equation for a continuously stratified fluid.

Consider the continuously stratified fluid shown in Figure 6.18.1. At levels of *fixed* height $Z_n$, the vorticity equation (6.3.17) may be written

$$\frac{d_0}{dt}\{\zeta_0(x, y, Z_n, t) + \beta y\} = \frac{1}{\rho_s(Z_n)}\left(\frac{\partial}{\partial z}\rho_s w_1\right)_{z=Z_n}. \qquad (6.18.1)$$

The derivative in (6.181) may be approximated by the finite-difference form

$$\left(\frac{\partial}{\partial z}\rho_s w_1\right)_{z=Z_n} = \frac{\rho_s(h_n)w_1(h_n) - \rho_s(h_{n+1})w(h_{n+1})}{d_n} + O(d_n), \qquad (6.18.2)$$

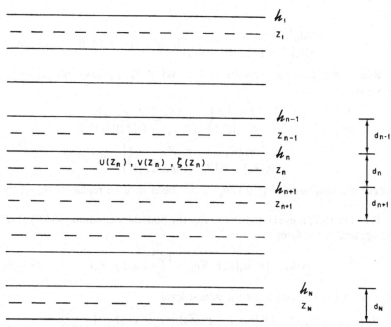

**Figure 6.18.1**   The levels $Z_n$ at which the potential-vorticity equation is applied in a continuously stratified fluid, and the intermediate levels $h_n$ where the thermodynamic variables and the vertical velocity are evaluated.

where $h_n$ is the level intermediate (though not necessarily equidistant) between $Z_n$ and $Z_{n-1}$, and $d_n \equiv h_n - h_{n+1}$.

In the absence of heating (6.5.15) yields

$$w_1(h_n) = -\frac{1}{S(h_n)}\left[\frac{d_0}{dt}\theta_0(h_n)\right]$$   (6.18.3)

where

$$S(h_n) = \frac{F^{-1}}{\theta_s(h_n)}\frac{\partial\theta_s(h_n)}{\partial z},$$   (6.18.4)

$$\theta_0(h_n) = \frac{\partial p_0}{\partial z}(h_n),$$   (6.18.5)

from (6.5.8) and (6.5.13) respectively. Using, again, finite-difference approximations

$$\theta_0(h_n) \approx \frac{p_0(Z_{n-1}) - p_0(Z_n)}{Z_{n-1} - Z_n},$$   (6.18.6a)

and

$$S(h_n) \approx \frac{F^{-1}}{\theta_s(h_n)}\frac{\theta_s(Z_{n-1}) - \theta_s(Z_n)}{Z_{n-1} - Z_n},$$   (6.18.6b)

or

$$\frac{\theta_0(h_n)}{S(h_n)} \sim \frac{F\theta_s(h_n)}{\theta_s(Z_{n-1}) - \theta_s(Z_n)} [p_0(Z_{n-1}) - p_0(Z_n)]. \tag{6.18.7}$$

Reapplication of these results at the level at $h_{n+1}$ allows (6.18.1) to be rewritten

$$\frac{d_0}{dt} \left[ \zeta_0(Z_n) + \beta y - \frac{F\theta_s(h_n)}{d_n \rho_s(Z_n)} \left| \frac{p_0(Z_n) - p_0(Z_{n-1})}{[\theta_s(Z_{n-1}) - \theta_s(Z_n)]/\rho_s(h_n)} \right. \right.$$
$$\left. \left. - \frac{p_0(Z_{n+1}) - p_0(Z_n)}{[\theta_s(Z_n) - \theta_s(Z_{n+1})]/\rho_s(h_{n+1})} \right| \right] = 0, \tag{6.18.8}$$

where the smallness of $(1/\theta_s)\, \partial\theta_s/\partial z$ has been used to replace $\theta_s(h_{n+1})$ by $\theta_s(h_n)$.

When (6.18.1) is evaluated at $z = Z_N$, the boundary condition (6.6.9) must be applied in the form

$$w(h_{N+1}) = \mathbf{u}_0(Z_n) \cdot \nabla\eta_B + \frac{E_V^{1/2}}{2\varepsilon} \zeta_0(x, y, Z_n), \tag{6.18.9}$$

so that (6.18.1) becomes, for the lowest level,

$$\frac{d_0}{dt} \left[ \zeta_0(Z_N) + \beta y - \frac{F\theta_s(h_n)}{d_N \rho_s(Z_n)} \left| \frac{p_0(Z_n) - p_0(Z_{N-1})}{[\theta_s(Z_{n-1}) - \theta_s(Z_N)]/\rho_s(h_n)} \right| + \eta_B \right]$$
$$= -\frac{E_V^{1/2}}{2\varepsilon d_N} \zeta_0(Z_N). \tag{6.18.10}$$

At $z = Z_1$, the vorticity equation may be written, using $\rho_s(h_1) = 0$, as

$$\frac{d_0}{dt} \left[ \zeta_0(Z_1) + \beta y - \frac{F\theta_s(h_1)}{d_1 \rho_s(Z_1)} \left| \frac{p_0(Z_1) - p_0(Z_2)}{[\theta_s(Z_1) - \theta_s(Z_2)]/\rho_s(h_2)} \right| \right] = 0. \tag{6.18.11}$$

If (6.18.8), (6.18.10), (6.18.11) are compared with (6.16.25a), (6.16.25b), and (6.16.30) respectively, it is clear that the equations for the $N$-layer model are equivalent to an $N$-level finite-difference approximation to the continuously stratified fluid model in which the dynamical fields, $u$, $v$, and $p$, are evaluated at $N$ *fixed* points, $Z_n$, in the vertical. The equivalence between the two sets of equations requires the identification

| (level model) | (layer model) |

$$\frac{\theta_s(h_n)\rho_s(h_n)/\rho_s(Z_n)}{d_n[\theta_s(Z_{n-1}) - \theta_s(Z_n)]} \leftrightarrow \frac{1}{\rho_n - \rho_{n-1}} \frac{\rho_0}{D_n/D}, \tag{6.18.12}$$

$$\frac{\theta_s(h_n)\rho_s(h_{n+1})/\rho_s(Z_n)}{d_n[\theta_s(Z_n) - \theta_s(Z_{n+1})]} \leftrightarrow \frac{1}{\rho_{n+1} - \rho_n} \frac{\rho_0}{D_n/D}.$$

In the oceanic case where (6.8.11) applies, the equivalence required is even simpler, for then $\rho_s(h_n)/\rho_s(Z_n)$ is unity while in (6.18.12) $\theta_s(Z_n) \to \rho_s^{-1}(Z_n)$.

The simplest *level* model is again the two-level model. Then, with (6.5.20), we obtain

$$\left[\frac{\partial}{\partial t} + \frac{\partial \psi}{\partial x}(Z_1)\frac{\partial}{\partial y} - \frac{\partial \psi}{\partial y}(Z_1)\frac{\partial}{\partial x}\right]\left[\frac{\partial^2 \psi(Z_1)}{\partial x^2} + \frac{\partial^2 \psi}{\partial y^2}(Z_1)\right.$$

$$\left. - \tilde{F}_1(\psi(Z_1) - \psi(Z_2)) + \beta y\right] = 0, \tag{6.18.13a}$$

$$\left[\frac{\partial}{\partial t} + \frac{\partial \psi(Z_2)}{\partial y}\frac{\partial}{\partial y} - \frac{\partial \psi(Z_2)}{\partial y}\frac{\partial}{\partial x}\right]\left[\frac{\partial^2 \psi(Z_2)}{\partial x^2} + \frac{\partial^2 \psi(Z_2)}{\partial y^2}\right.$$

$$\left. - \tilde{F}_2(\psi(Z_2) - \psi(Z_1)) + \beta y + \eta_B\right] \tag{6.18.13b}$$

$$= -\frac{\tilde{r}_2}{2}\left[\frac{\partial^2 \psi(Z_2)}{\partial x^2} + \frac{\partial^2 \psi(Z_2)}{\partial y^2}\right],$$

where

$$\tilde{F}_1 = \frac{f_0^2 L^2}{gDd_1}\frac{\rho_s(h_2)}{\rho_s(Z_1)}\frac{\theta_s(h_2)}{\theta_s(Z_1) - \theta_s(Z_2)},$$

$$\tilde{F}_2 = \frac{f_0^2 L^2}{gDd_2}\frac{\rho_s(h_2)}{\rho_s(Z_2)}\frac{\theta_s(h_2)}{\theta_s(Z_1) - \theta_s(Z_2)}, \tag{6.18.14}$$

$$\tilde{r}_2 = \frac{(2A_V f_0)^{1/2}}{U}\frac{L}{Dd_2}.$$

The equivalence between (6.18.13a,b) and (6.16.34a,b) is now *exact* if $\tilde{F}_n$ is identified with $F_n$. If (6.8.11) applies, then the equations (6.18.13a,b) are unchanged, while

$$F_n = \frac{f_0^2 L^2}{gDd_n}\frac{\rho_0}{(\rho(Z_2) - \rho(Z_1))}. \tag{6.18.15}$$

Thus the two-layer model, in which the fluid layers are immiscible, incompressible fluids of constant density, bears a remarkable one-to-one relationship with the simplest finite-difference approximation to the continuous model for both the atmosphere (where potential temperature is conserved and defines the deformation radius) and the ocean (where the density is conserved). While the finite-level models are gross mathematical approximations to an accurate portrayal of the true physical systems, the layer models invert this relationship and serve as accurate mathematical representations of a crude physical representation of actually more complex geophysical systems. It is consequently useful and reassuring when applying the level models, especially with low $N$, to realize that the governing equations for the crudest finite-difference approximation do represent the dynamics of a real, albeit simplified, physical system.

In the atmospheric case, the expressions for the $\tilde{F}_n$ in (6.18.4) may be considerably simplified if the hydrostatic equation

$$\frac{\partial p_s}{\partial z} = -\rho_s g D$$

is used in finite-difference form as

$$
\begin{aligned}
[p_s(h_2) - p_s(0)] &= -\rho_s(Z_2)gDd_2, \\
[p_s(h_1) - p_s(h_2)] &= -\rho_s(Z_1)gDd_1, \\
[p_s(Z_1) - p_s(Z_2)] &= -\rho_s(h_2)gD(Z_1 - Z_2),
\end{aligned}
\tag{6.18.16}
$$

where $p_s(0)$ is the surface pressure in the basic state. If the two levels are chosen so that $h_2$ separates two regions of equal mass, the hydrostatic equation implies that

$$p_s(h_2) = \frac{p_s(0)}{2},$$

$$p_s(Z_1) - p_s(Z_2) = \frac{-p_s(0)}{2}, \tag{6.18.17}$$

$$p_s(h_1) = 0.$$

Then

$$\tilde{F}_1 = \frac{f_0^2 L^2}{(\Delta\theta_s/\theta_s)g(Z_1 - Z_2)} = \tilde{F}_2, \tag{6.18.18}$$

where

$$\frac{\Delta\theta_s}{\theta_s} = \frac{\theta_s(Z_1) - \theta_s(Z_2)}{\theta_s(h_1)}, \tag{6.18.19}$$

which then corresponds precisely to (6.16.33) in the case $D_1 = D_2$.

In subsequent chapters, especially in the discussion of stability theory in Chapter 7, the layer (or level) models provide an especially useful simplified setting for the discussion of complex physical phenomena.

## 6.19 Geostrophic Approximation $\varepsilon \ll L/r_0 < 1$; the Sverdrup Relation

In the preceding sections geostrophic motion, for which $\varepsilon \ll 1$, was examined for the case $\beta = \beta_0 L^2/U = O(1)$, i.e., for situations where the scale of the motion is of the order of the stationary Rossby wavelength $(U/\beta_0)^{1/2}$. If $\beta$ is small, the resulting geostrophic dynamics can be obtained directly from (6.5.21) or (6.8.11) by taking $\beta = 0$. The resulting equations correspond to those appropriate for a *flat* earth rotating with angular velocity $f_0/2$.

When the relative-vorticity *gradient* is small compared to the planetary-vorticity *gradient*, $\beta$ is large and the nature of the resulting dynamics requires careful consideration of three further parameters, namely

$$\beta S = \frac{\beta_0 L_D^2}{U}, \tag{6.19.1a}$$

$$\beta \varepsilon = \frac{\beta_0 L}{f_0} = O\left(\frac{L}{r_0}\right), \tag{6.19.1b}$$

$$\frac{S}{\varepsilon} = (\beta S)\frac{f_0}{\beta_0 L} = O\left(\beta S \frac{r_0}{L}\right). \tag{6.19.1c}$$

The case $\beta \gg 1$ is of particular relevance for large-scale oceanic motions, since $(U/\beta_0)^{1/2}$ is only $O(100 \text{ km})$.

The *nature* of the dynamics for large $\beta$ depends on the order of these additional parameters, roughly as follows. If $\beta$ is large and if $\beta S$ (which is independent of $L$) is order one (i.e., if the deformation radius $L_D$ is of the same order as the stationary Rossby wavelength), then the vorticity advection in the planetary-vorticity gradient is balanced only by vortex-tube stretching, and from (6.3.17) this implies that in nondimensional units

$$\beta v_0 = O\left(\frac{\partial w_1}{\partial z}\right) = O\left(\varepsilon^{-1}\frac{\partial w}{\partial z}\right), \qquad \beta \gg 1. \tag{6.19.2}$$

Thus for $\beta \gg 1$, $\beta S = O(1)$; and where $L/r_0 \ll 1$ (so that the $\beta$-plane approximation used here is valid),

$$w = O(\beta \varepsilon) = O\left(\frac{L}{r_0}\right) = O\left(\beta_0 \frac{L}{f_0}\right). \tag{6.19.3}$$

This implies, for $\beta S = O(1)$ and $\beta \varepsilon \ll 1$, that the scaling for $w$ should be

$$w_* = \frac{UD}{L} w = \frac{UD}{L}\left\{\beta_0 \frac{L}{f_0} w_1 + \cdots\right\} \tag{6.19.4}$$

instead of the scaling given by (6.3.12). If $u$, $v$, $p$, and $\rho$ are scaled as in (6.2.13), (6.2.18), and (6.2.21), then the vorticity equation becomes, to lowest order in $L/r_0$, simply

$$\frac{\partial w_1}{\partial z} = v_0, \tag{6.19.5}$$

where $\rho_s(z)$ has been assumed essentially constant over the vertical scale of the motion. Northward motion is therefore possible only to the extent that it is balanced by vortex-tube stretching.

The density equation, (6.2.7), now becomes

$$\frac{\partial \rho_0}{\partial t} + u_0 \frac{\partial \rho_0}{\partial x} + v_0 \frac{\partial \rho_0}{\partial y} + \frac{\beta_0 L w_1}{\varepsilon f_0 F}\left\{\frac{1}{\rho_s}\frac{\partial \rho_s}{\partial z} + \varepsilon F \frac{\partial \rho_0}{\partial z}\right\} = -\mathcal{H}, \tag{6.19.6}$$

where $\mathscr{H}$ is defined by (6.8.3) and (6.8.4). Note that

$$\frac{\beta_0 L}{f_0 \varepsilon F} \frac{1}{\rho_s} \frac{\partial \rho_s}{\partial z} = -\beta S = O(1),$$
(6.19.7a)

while

$$\frac{1}{\rho_s} \frac{\partial \rho_s / \partial z}{\varepsilon F} = \frac{S}{\varepsilon} = O\left(\beta S \frac{r_0}{L}\right) \gg 1.$$
(6.19.7b)

Hence the density equation to lowest order becomes, in the absence of heating,

$$\frac{d_0}{dt} \rho_0 - \beta S w_1 = 0.$$
(6.19.8)

Using

$$v_0 = \frac{\partial \psi}{\partial x},$$

$$\rho_0 = -\frac{\partial \psi}{\partial z},$$
(6.19.9)

which follow directly from geostrophy and hydrostatic balance, (6.19.5) and (6.19.8) may be combined to yield

$$\frac{d_0}{dt} \left[ \frac{\partial}{\partial z} \frac{1}{S\beta} \frac{\partial \psi}{\partial z} + y \right] = 0.$$
(6.19.10)

This is precisely the limiting form of (6.8.11) for large $\beta$, when $\beta S = O(1)$. Thus as long as $L/r_0$ is small, the *synoptic*-scale equations of motion (6.5.21) and (6.8.11) remain valid for $\beta \gg 1$.

The vertical integral of (6.19.5) over the depth of the geostrophic region yields

$$M_S^{(y)} = \int_0^1 v_0 \, dz = w_1(x, y, 1) - w_1(x, y, 0).$$
(6.19.11)

At the upper surface of the ocean where the wind stress acts, (5.2.8) applies, i.e.,

$$w_*(x, y, 1) = U \frac{D \beta_0 L}{L f_0} w_1 = \mathbf{k} \cdot \text{curl} \frac{\tau_*}{\rho_0 f}$$
(6.19.12)

$$= \frac{\tau_0}{\rho_0 L f_0} \text{curl } \tau + O(L/r_0),$$

or

$$w_1(x, y, 1) = \left| \frac{\tau_0}{\rho_0 U D \beta_0 L} \right| \text{curl } \tau,$$
(6.19.13)

where curl $\tau$ is defined as in (5.2.14). On the lower boundary (4.9.36) applies, so that $w_1$ at the lower boundary depends directly on the magnitude of the bottom horizontal velocity and vorticity. If the horizontal velocity and hence the vorticity are negligible at the ocean bottom:

$$M_S^{(y)} = \int_0^1 v_0 \, dz = w_1(x, y, 1)$$

$$= \frac{\tau_0}{\rho_0 U D \beta_0 L} \, \text{curl } \tau,$$

(6.19.14)

or if, as given by (5.2.19), the scaling velocity $U$ is chosen as $\tau_0 / \rho_0 D \beta_0 L$, then

$$\boxed{M_S^{(y)} = \int_0^1 v_0 \, dz = \text{curl } \tau}\,,$$

(6.19.15)

which is identical to the Sverdrup relation derived in Chapter 5 for the *homogeneous* model of the wind-driven oceanic circulation. The present derivation, which is similar to Sverdrup's original argument (1947), in fact relies crucially on the *baroclinic* nature of the fluid to allow the neglect of fluid interaction with the lower boundary. The fact is, of course, that the result (6.19.15) is clearly *independent* of the detailed nature of the basic stratification. The fact that the Sverdrup relation for the wind-driven transport in a baroclinic ocean satisfies the same relationship to the curl of the stress as the homogeneous model of the oceanic circulation is one of the primary reasons why the homogeneous models have been thought to yield useful information about the vertically integrated transport fields. It is beyond the scope of this book to recapitulate the study of the question of western boundary currents in a stratified fluid. In brief, the synoptic-scale dynamics of Section 6.8—supplemented, as in Chapter 5, by at least rudimentary models of turbulent mixing processes—must be applied for models of the narrow boundary currents. Linear theories, such as Munk's (Section 5.4) can be shown to apply precisely and without change to the description of the integrated transport. The inertial models become considerably more complex, but retain strong qualitative similarities with the inertial models of Chapter 5.

The vertical structure of the motion for $\beta \gg 1$, $\beta S = O(1)$ is described by (6.19.10). As remarked above, the equation is merely a limiting form of the synoptic-scale potential-vorticity equation. In particular, as long as $L/r_0 \ll 1$ and $\beta S = O(1)$, (6.19.7b) shows that the vertical advection of density remains dominated by the fluid motion in the average density field $\rho_s(z)$. Thus motions for which $L_D \sim (U/\beta_0)^{1/2}$ have the important property that the linearization of the density field about its mean value remains valid to the same extent as the $\beta$-plane approximation itself retains its validity. For motions of extremely large scale, such that $L/r_0 \sim 1$, *both* the $\beta$-plane approximation and the useful partitioning of the density field lose their relevance. This limiting case is of great importance in oceanography and is discussed in the following section.

## 6.20 Geostrophic Approximation $\varepsilon \ll 1$, $L/r_0 = O(1)$

When the horizontal length scale of the motion becomes very large, such that $L/r_0 = O(1)$, several fundamental changes occur in the formulation of the quasigeostrophic dynamics. Most obviously, a reduction to a flat geometry is no longer a natural approximation. Equally important, given the observed values of key parameters such as the Rossby deformation radius $L_D$ in the ocean, where $L_D \sim 50$ km $\ll r_0$, the horizontal variation of the basic density field can no longer be ignored. For then, by (6.19.7b), the vertical density gradient in the basic state is no larger than the vertical density gradient associated with the density changes due to the large-scale motion field. In order to adequately describe the dynamics on these scales it is necessary to return to the fundamental equations of motion in Section 6.2 and reconsider the appropriate scaling relations. The argument will be explicitly carried through for the "oceanic" case for which (6.2.7) applies and where the density scale height can be considered large compared to the vertical scale of motion.

Let the dimensional variables, denoted by asterisks, be written in terms of unsubscripted dimensionless variables as

$$r_* = r_0\left(1 + \frac{D}{r_0}z\right),$$

$$u_* = Uu,$$

$$v_* = Uv,$$

$$w_* = U\frac{D}{r_0}w = Ww, \tag{6.20.1}$$

$$t_* = \frac{r_0}{U}t,$$

$$p_* = -\rho_0 gDz + \rho_0 2\Omega Ur_0 p,$$

$$\rho_* = \rho_0 + \frac{\rho_0 2\Omega Ur_0}{gD}\rho.$$

Here $r_0$ is the earth's radius, $D$ the vertical scale of the motion, $U$ the scale of the horizontal velocity, and $W = UD/r_0$ the scale of the vertical velocity. Since the horizontal scale of the motion is assumed $O(r_0)$, $r_0$ has been used to scale horizontal lengths. The advective time $r_0/U$ has been used to non-dimensionalize the time. The density has been partitioned between a constant value $\rho_0$ and a remainder, $[\rho_0 2\Omega U r_0 /gD]\rho$, which describes the *complete* variation of the density field in space (and time). The scaling for the density field has been chosen, as in (6.2), in anticipation of the fact that for small Rossby numbers the horizontal pressure gradient will be of order of the Coriolis acceleration. This sets the pressure scale as in (6.20.1), while the expectation that the buoyancy forces will be of the same order as the *vertical* pressure gradient determines the density scaling in (6.20.1). If the relations (6.20.1) are used to rewrite the equations of motion (6.2.1), (6.2.3a,b,c), and (6.2.7), we obtain, after certain obvious manipulations,

$$\varepsilon F \frac{d\rho}{dt} + [1 + \varepsilon F \rho]\left[\frac{\partial w}{\partial z} + \frac{2D}{r_*} w + \left(\frac{r_0}{r_*}\right)\frac{1}{\cos \theta}\frac{\partial(v \cos \theta)}{\partial \theta} + \frac{1}{\cos \theta}\frac{\partial u}{\partial \phi}\right] = 0,$$

(6.20.2a)

$$\varepsilon\left[\frac{du}{dt} + uw\frac{D}{r_*} - uv\frac{r_0}{r_*}\tan \theta\right] - \sin \theta\, v + \cos \theta\left(\frac{D}{r_0}\right)w$$

$$= -\frac{1}{1 + \varepsilon F \rho}\frac{r_0}{r_*}\frac{\partial p}{\cos \theta\, \partial \phi} + \frac{\mathscr{F}_{*\phi}}{\rho_* U 2\Omega},$$

(6.20.2b)

$$\varepsilon\left[\frac{dv}{dt} + vw\frac{D}{r_*} + u^2\frac{r_0}{r_*}\tan \theta\right] + \sin \theta\, u = -\frac{1}{1 + \varepsilon F \rho}\frac{r_0}{r_*}\frac{\partial p}{\partial \theta} + \frac{\mathscr{F}_{*\theta}}{\rho_* U 2\Omega},$$

(6.20.2c)

$$(1 + \varepsilon F \rho)\left[\varepsilon\frac{D^2}{r_0^2}\frac{dw}{dt} - \frac{\varepsilon D}{r_*}(u^2 + v^2) - \frac{D}{r_0}\cos \theta\, u\right] = -\frac{\partial p}{\partial z} - \rho + \frac{\mathscr{F}_{*z}}{\rho_* U 2\Omega},$$

(6.20.2d)

$$\frac{d\rho}{dt} = \frac{K_V r_0}{UD^2}\frac{\partial^2 \rho}{\partial z^2} + \frac{K_H}{Ur_0}\nabla_H^2 \rho,$$

(6.20.2e)

where in the present context

$$\varepsilon = \frac{U}{2\Omega r_0},$$

$$F = \frac{4\Omega^2 r_0^2}{gD},$$

(6.20.3)

$$\frac{r_*}{r_0} = 1 + \frac{D}{r_0}z.$$

The frictional terms in (6.20.2b,c,d) have been left unspecified in detail, though as in Section 6.2 we estimate their magnitude as

$$\frac{\mathscr{F}_{*\theta}}{\rho_* 2\Omega U} = O\left|\frac{A_H}{2\Omega r_0^2}, \frac{A_V}{2\Omega D^2}\right| \ll 1$$

(6.20.4)

etc. Outside of Ekman layers on the horizontal boundary surfaces we can ignore these friction terms. In the equation for the density, (6.20.2e), different turbulent diffusivities in the vertical and horizontal directions are assumed in analogy with the formulation of the turbulent mixing of momentum described in Section 4.2. The operators $d/dt$ and $\nabla_H^2$ are defined as

$$\frac{d}{dt} = \frac{\partial}{\partial t} + \frac{r_0}{r_*}\left(\frac{u}{\cos\theta}\frac{\partial}{\partial\phi} + v\frac{\partial}{\partial\theta}\right) + w\frac{\partial}{\partial z},$$

$$\nabla_H^2 = \frac{r_0^2}{r_*^2}\left[\frac{1}{\cos\theta}\frac{\partial}{\partial\theta}\cos\theta\frac{\partial}{\partial\theta} + \frac{1}{\cos^2\theta}\frac{\partial}{\partial\phi^2}\right].$$

(6.20.5)

For $U = O(1\ \text{cm/s})$, with $r_0 = O(6\times10^8)$ cm, $D = O(1\ \text{km})$, and $2\Omega \approx 1.4 \times 10^{-4}\ \text{s}^{-1}$, the parameters $\varepsilon$, $F$, and $D/r_0$ are respectively

$$\varepsilon = O(10^{-5}),$$

$$F = O(70),$$

(6.20.6)

$$\frac{D}{r_0} = O(1.6\times10^{-4}).$$

Hence to $O(1)$, the leading terms in the equations of motion are

$$\frac{1}{\cos\theta}\frac{\partial}{\partial\theta}(v\cos\theta) + \frac{1}{\cos\theta}\frac{\partial u}{\partial\phi} + \frac{\partial w}{\partial z} = 0,$$

(6.20.7a)

$$\sin\theta\, v = \frac{1}{\cos\theta}\frac{\partial p}{\partial\phi},$$

(6.20.7b)

$$\sin\theta\, u = -\frac{\partial p}{\partial\theta},$$

(6.20.7c)

$$\rho = -\frac{\partial p}{\partial z},$$

(6.20.7d)

$$\frac{\partial\rho}{\partial t} + \frac{u}{\cos\theta}\frac{\partial\rho}{\partial\phi} + v\frac{\partial\rho}{\partial\theta} + w\frac{\partial\rho}{\partial z} = \lambda\frac{\partial^2\rho}{\partial z^2}.$$

(6.20.7e)

If (6.20.7) is compared with the equivalent approximation for synoptic-scale motions, (6.3.6) and (6.3.7), it is evident that in both cases the velocity field is incompressible,* geostrophic, and hydrostatic. However, in the present case the full variation in the metric terms must be retained in the expression for the divergence and, more significantly, the Coriolis parameter is *variable at lowest order*. The scale of the motion now being considered is so vast that over that scale the local normal component of the earth's rotation varies by $O(1)$. In the equation for the density field the vertical density gradient is now determined by the motion itself rather than being preset in terms of a *horizontally* uniform standard density. As in the case of the Coriolis parameter, the horizontal scale of motion is large enough so that the lateral variations in the static stability, $\rho^{-1}\,\partial\rho/\partial z$, become significant.

---

* That is, if $\rho_s$ in (6.3.7) has a scale height in excess of $D$.

The parameter $\lambda$ is given by

$$\lambda = \frac{K_V r_0}{UD^2}. \tag{6.20.8}$$

The term proportional to $\lambda$ is a rough model for the vertical diffusion of density by smaller-scale motions. This term, while most likely small by any realistic estimate, is retained to allow thermal (density) boundary-layer behavior to be uniformly represented by the set (6.20.7). If $K_V$ is $O(1 \text{ cm}^2/\text{s})$, $\lambda$ is $O(6 \times 10^{-2})$ if $U = 1$ cm/s and $D = 1$ km.

The truly crucial difference in the geostrophic dynamics on this scale and on synoptic scales is that the lowest-order geostrophic approximation is no longer dynamically degenerate. That is, if the pressure is eliminated between (6.20.7b,c) and the continuity equation (6.20.7a) is used, we easily obtain

$$\boxed{\cos\theta\, v = \sin\theta\, \frac{\partial w}{\partial z},} \tag{6.20.9}$$

which corresponds in dimensional form to the rudimentary vorticity equation

$$\beta_* v_* = f \frac{\partial w_*}{\partial z_*}, \tag{6.20.10}$$

where

$$f = 2\Omega \sin\theta$$

$$\beta_* = 2\Omega \frac{\cos\theta}{r_0}.$$

The vorticity equation (6.20.9) is the generalization of (6.19.5) to $L \sim r_0$, where $O(1)$ variations of $f_*$ and $\beta_*$ are explicitly retained. Due to the $O(1)$ variation of the Coriolis parameter, the lowest-order geostrophic velocities possess an $O(1)$ horizontal divergence, and the consequent production of vorticity by vortex-tube stretching in the planetary vorticity field is balanced by north–south motion in the field of the planetary vorticity gradient. The thermal wind relations take the form

$$\sin\theta \frac{\partial u}{\partial z} = \frac{\partial \rho}{\partial \theta} \tag{6.20.11a}$$

$$\sin\theta \frac{\partial v}{\partial z} = -\frac{1}{\cos\theta} \frac{\partial \rho}{\partial \phi}. \tag{6.20.11b}$$

If the density equation is differentiated once with respect to $z$, then with the aid of (6.20.11) we obtain

$$\frac{d}{dt} \frac{\partial \rho}{\partial z} + \frac{\partial w}{\partial z} \frac{\partial \rho}{\partial z} = \lambda \frac{\partial^3 \rho}{\partial z^3}. \tag{6.20.12}$$

The term in $\partial w/\partial z$ may be eliminated with the aid of the vorticity equation (6.20.9) to yield

$$\frac{d}{dt}\left|\sin\theta\,\frac{\partial\rho}{\partial z}\right| = \lambda\frac{\partial^2}{\partial z^2}\left|\sin\theta\,\frac{\partial\rho}{\partial z}\right| \tag{6.20.13}$$

This is the potential vorticity equation for geostrophic motion for which $L \sim r_0$ and $\varepsilon \ll 1$. On these scales the vorticity dynamics of the fluid involves only the planetary vorticity. The vertical component of the density gradient is the largest component, since $D/r_0$ is small; hence in dimensional units the potential vorticity is

$$\frac{f}{\rho_0}\frac{\partial\rho_*}{\partial z_*} = \frac{2\Omega}{D}\,\varepsilon F\left(\frac{\partial\rho}{\partial z}\sin\theta\right), \tag{6.20.14}$$

which, aside from a multiplicative constant, is the quantity appearing in (6.20.13). Note that in the absence of diffusion ($\lambda = 0$) the potential vorticity is conserved. If $\lambda \neq 0$, the potential vorticity itself satisfies the same diffusion equation as the density field.

If diffusion can be neglected, the potential vorticity and the density are both conserved. Further, the Bernoulli function

$$B = p + \rho z \tag{6.20.15}$$

satisfies

$$\begin{aligned}\frac{dB}{dt} &= \frac{dp}{dt} + z\frac{d\rho}{dt} + \rho\frac{dz}{dt}, \\[4pt] &= \frac{dp}{dt} + z\lambda\frac{\partial^2\rho}{\partial z^2} + \rho w.\end{aligned} \tag{6.20.16}$$

However,

$$\begin{aligned}\frac{dp}{dt} &= \frac{\partial p}{\partial t} + \frac{u}{\cos\theta}\frac{\partial p}{\partial\phi} + v\frac{\partial p}{\partial\theta} + w\frac{\partial p}{\partial z} \\[4pt] &= \frac{\partial p}{\partial t} - w\rho\end{aligned} \tag{6.20.17}$$

if (6.20.7b.c.d) are used. Hence

$$\frac{dB}{dt} = \frac{\partial p}{\partial t} + z\lambda\frac{\partial^2\rho}{\partial z^2} \tag{6.20.18}$$

If the flow is *steady* and the diffusion of density can be ignored, then $B$ is conserved following a fluid element. Thus in the steady state the three functions

$$\Pi = \sin\theta\,\frac{\partial\rho}{\partial z}, \tag{6.20.19a}$$

$$B = p + \rho z, \tag{6.20.19b}$$

$$\rho = \rho \tag{6.20.19c}$$

are *each* conserved. When the flow is steady, the surfaces of each conserved property are fixed in space, and a fluid element must move so as to remain on the surface it started on. Consider a fluid element on the surfaces $B = B_1$, $\rho = \rho_1$. As shown in Figure 6.20.1, it must flow along the *intersection* of these

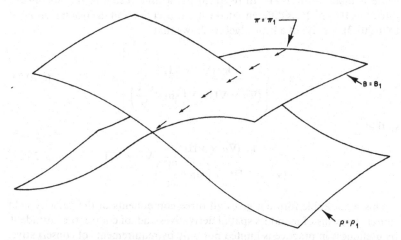

**Figure 6.20.1**  The intersection of the surfaces of constant $B = p + \rho z$ and constant $\rho$ yields a line along which $\Pi$ is constant and which is coincident with a streamline for nondissipative flow when $U/\beta L^2 \ll 1$.

two surfaces so as to remain on both. Since $\Pi$ is constant along this trajectory, in the steady state $\Pi$ must depend only on the intersection of the surfaces of constant $\rho$ and $B$. That is, given a particular value of $\rho$ and $B$, $\Pi$ must be constant for that combination of $\rho$ and $B$, or equivalently, $\Pi$ must be a function of $\rho$ and $B$ only:

$$\Pi = \Pi(\rho, B) \tag{6.20.20}$$

for steady, nondiffusive flow.

Since both $\rho$ and $\Pi$ are conserved if $\lambda = 0$, it follows that for steady, conservative flow **u** must be perpendicular to both $\nabla\rho$ and $\nabla\Pi$, i.e.,

$$\mathbf{u} = \alpha\nabla\rho \times \nabla\Pi, \tag{6.20.21}$$

where $\alpha$ is an arbitrary scalar function of $\phi$, $\theta$, and $z$.

As noted by Needler (1985) $\alpha$ may in fact be determined as follows. By definition

$$\frac{\partial\mathbf{u}}{\partial z} = \frac{\partial\mathbf{u}_H}{\partial z} + \hat{\mathbf{k}}\frac{\partial w}{\partial z}, \tag{6.20.22}$$

where $\mathbf{u}_H$ is the horizontal velocity, $\mathbf{u} - \hat{\mathbf{k}}w$. From (6.20.9) and (6.20.11a,b) it follows that

$$\frac{\partial \mathbf{u}}{\partial z} = -\frac{\hat{\mathbf{k}} \times \nabla\rho}{\sin\theta} + \cot\theta\hat{\mathbf{k}}v$$

$$= -\frac{\hat{\mathbf{k}} \times \nabla\rho}{\sin\theta} + \alpha\cot\theta\hat{\mathbf{k}}(\hat{\mathbf{j}}\cdot[\nabla\rho \times \nabla\Pi]),$$

(6.20.23)

if use is made of (6.20.21). In (6.20.23) $\hat{\mathbf{j}}$ is a unit vector in the northward direction. If (6.20.21) is differentiated with respect to $z$ and the result is equated to (6.20.23), the use of a little algebra shows that

$$\alpha = \left\{ \frac{\hat{\mathbf{k}}\cdot(\nabla\rho \times \nabla\Pi)}{(\nabla\rho \times \nabla\Pi)\cdot\nabla\left(\sin\theta\dfrac{\partial\Pi}{\partial z}\right)} \right\},$$

(6.20.24)

so that

$$\mathbf{u} = \frac{\hat{\mathbf{k}}\cdot(\nabla\rho \times \nabla\Pi)}{(\nabla\rho \times \nabla\Pi)\cdot\nabla\left(\sin\theta\dfrac{\partial\Pi}{\partial z}\right)}\nabla\rho \times \nabla\Pi.$$

(6.20.25)

This remarkable formula gives all three components of the velocity field *entirely* in terms of $\rho$ and its spatial derivatives and, of course, the latitude $\theta$. Its usefulness in practice is limited not only by requirements of conservative, steady flow but also by the dependence of the formula on rather high derivatives (third) of the density field. Also, the formula clearly fails when surfaces of constant $\rho$ coincide with constant potential vorticity surfaces, and it is difficult to apply in the case of near coincidence. Nevertheless, the result is conceptually important because it illustrates that in principle the determination of the density field for an ideal fluid motion on global scales completely determines the flow field.

Another useful diagnostic relation follows from the direct application of (6.20.21) to the determination of $w$ in terms of $v$. It is left to the reader to demonstrate that (6.20.21) along with (6.20.19a) implies that

$$w = -\frac{v\hat{\mathbf{k}}\cdot(\nabla\rho \times \nabla\Pi)}{\tan\theta\left(\dfrac{\partial\rho}{\partial z}\right)^2 \dfrac{\partial}{\partial z}\left(\dfrac{\partial z}{\partial\phi}\right)_\rho}$$

(6.20.26)

where $(\partial z/\partial\phi)_\rho$ is the increase of elevation with longitude of a constant density surface. This formula allows the vertical velocity to be determined in terms of the meridional velocity and gradients of the density and potential vorticity.

Finally, we note one more diagnostic relation. The steady, conservative form of (6.20.12), along with the thermal wind relation implies that

$$\hat{\mathbf{k}}\cdot\left(\mathbf{u}_H \times \frac{\partial\mathbf{u}_H}{\partial z}\right) = \frac{w}{\sin\theta}\frac{\partial\rho}{\partial z}.$$

(6.20.27)

If we write

$$\mathbf{u}_H = U_H \cos v \hat{\mathbf{i}} + U_H \sin v \hat{\mathbf{j}}, \qquad (6.20.28)$$

where $v$ is the angle $\mathbf{u}_H$ makes with latitude and $U_H$ is its magnitude, it follows that

$$w = \frac{U_H^2 \sin \theta (\partial v / \partial z)}{\partial \rho / \partial z}. \qquad (6.20.29)$$

Since $\partial \rho / \partial z < 0$ it follows that in regions where $w > 0 \, (<0)$ the velocity vector must rotate clockwise (counterclockwise) with increasing $z$ (decreasing depth).

## 6.21 The Thermocline Problem

A fundamental feature of the temperature, density, and salinity structure of the world's ocean is the existence of a fairly narrow zone of rapid variation of these properties with depth. A typical density profile was shown in Section 6.4. Figure 6.21.1(a) shows a schematic, three-dimensional temperature distribution in the North Atlantic, while (b) demonstrates the ubiquity of this phenomenon. This region of rapid vertical variation, the *thermocline*, occurs at different depths depending on latitude and longitude; it is deepest in mid-latitudes, and considerably shallower at low and high latitudes. The variable depth of the thermocline implies the existence of strong horizontal density gradients associated with its slope, with concomitant "thermal wind" currents. The currents in turn influence the density field according to (6.20.7e); hence the structure of the density field is intimately related to the structure of all the dynamic fields. The fundamental question these observations pose is quite simply stated. Why is there a relatively sharp region of density change instead of a smooth and gradual variation from the ocean's surface value to its value at great depths? In the atmosphere the basic temperature and density structure is determined by a complex interaction between the dynamics and the process of radiative transfer through an atmosphere containing nonhomogeneous distributions of thermally absorbing and emitting constituents. It seems likely, on the other hand, that the temperature distribution of the ocean, heated and cooled on its upper surface, is determined primarily by the advection of heat by the oceanic circulation, affected to an uncertain degree by the turbulent diffusion of heat (and hence density) by smaller-scale motions. Furthermore, the global scale observed for the variation of the thermocline structure makes plausible the notion that the dynamical processes of significance will be of large scale, i.e., described by the dynamical model derived in the preceding section. From a somewhat more general viewpoint the question can be reformulated as a need to understand how surface forcing by heating and wind stress penetrates to

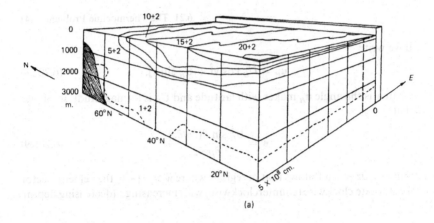

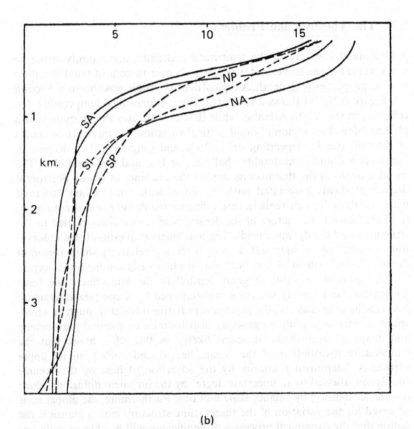

**Figure 6.21.1** (a) A schematic view of the temperature distribution in the North Atlantic. (b) Characteristic distributions of temperature with depth in the world's oceans. S = south, N = north, A = Atlantic, P = Pacific, I = Indian (from Robinson and Stommel 1959).

great depths in the oceanic gyres. The results of Chapter 5 suggest that Ekman pumping from the upper, turbulent boundary layer is important but the interplay of this mechanism with the constraints associated with stratification produces a dynamical problem of considerable complexity.

The theories of the thermocline to be described explicitly assume that the mid-ocean thermocline depends on the large-scale dynamical balances discussed in Section 6.20. As we shall see, however, even such theories at subtle points require a statement about the role of mixing of potential vorticity within the oceanic gyres. The degree of mixing (in its simplest form represented by $\lambda$ in (6.20.7e)) is uncertain. It is unclear, in fact, whether such a representation for the turbulent diffusion of density is apt at all. In addition, one weakness shared by all current theories is the incompleteness of the models with regard to western boundary currents. The oceanic circulation is a closed circulation but most theories aim at a stratified extension of the description of the Sverdrup interior flow described in Chapter 5 and it remains to be seen whether such partial treatments can be successfully closed *via* western boundary currents.

Nevertheless, the problem to be posed now must be understood before more complex and sophisticated models can be investigated. Simply put, we will ask what is the response to large-scale heating and wind stress of an ocean governed by the dynamical relations of the preceding section? In particular, we will continue to ignore lateral diffusion of density since the observed density distribution is relatively sharp in $z$, i.e., we are implicitly making a boundary layer approximation in the density equation, although subsequently we will relax this restriction when needed. Additionally, we will restrict our attention entirely to steady situations, i.e., we assume that the time-averaged density structure can be obtained without explicitly dealing with temporal variations in the forcing fields.

For steady circulations, the vorticity (6.20.9) and density (6.20.7e) equations become

$$\cos \theta v = \sin \theta \frac{\partial w}{\partial z}, \tag{6.21.1a}$$

$$\frac{u}{\cos \theta} \frac{\partial \rho}{\partial \phi} + v \frac{\partial \rho}{\partial \theta} + \frac{\partial \rho}{\partial z} = \lambda \frac{\partial^2 \rho}{\partial z^2}, \tag{6.21.1b}$$

while $u$ and $v$ are related to $p$ by (6.20.7b,c) and their vertical derivatives to $\rho$ by (6.20.11a,b).

At the upper boundary of the geostrophic region described by the above equations, the *vertical* velocity must be specified. The geostrophic dynamics are inviscid and therefore only the normal velocity can be specified on the upper boundary. Now, as described in Chapter 4, this implies that the dynamics described here are valid *beneath* the upper Ekman layer. Hence the appropriate value for the $w$ which appears in (6.21.1) to attain as the upper surface is approached is the velocity pumped into or out of the upper Ekman layer. The vertical velocity pumped into the Ekman layer at the upper surface is given in dimensional units by (4.10.21), which in the steady state is

$$w_*(\text{surface}) = \hat{k} \cdot \text{curl}\left(\frac{\tau_*}{\rho_0 f}\right), \tag{6.21.2}$$

where $f = 2\Omega \sin\theta$. Hence $w$ at the upper boundary is determined by the wind stress. Define the Ekman velocity as

$$w_{e*} \equiv \hat{k} \cdot \text{curl}\left(\frac{\tau_*}{\rho_0 f}\right) = W_e w_e(\theta, \phi), \tag{6.21.3}$$

where $W_e$ is the characteristic magnitude of $w_{e*}$. Then, using (6.20.1) at the upper surface (i.e., at $z = 1$)

$$w_*(\phi, \theta, 1) = Ww(\phi, \theta, 1) = W_e w_e(\theta, \phi). \tag{6.21.4}$$

It is convenient to choose

$$W = W_e,$$

so that (6.21.4) becomes

$$w(\phi, \theta, 1) = w_e, \qquad z = 1. \tag{6.21.5}$$

The choice $W = w_e$ reflects an *a priori* belief that the circulation is largely driven by the wind stress although we will examine this presumption more carefully below.

On the upper surface, some condition on the density must be satisfied. This may either be related to the flux of heat through the surface, the surface density, or some combination of both. For simplicity, we shall assume that the surface density is given and that this is impressed through the thin Ekman layer to the geostrophic region below.

In dimensional form, this becomes the condition

$$\rho_*(\phi, \theta, 1) = \rho_0 + (\Delta\rho)\rho_s(\phi, \theta)$$
$$= \rho_0 + \rho_0 \frac{2\Omega U r_0}{gD}\rho, \tag{6.21.6}$$

where $\Delta\rho$ is the magnitude of the surface density variation and the function $\rho_s(\phi, \theta)$ is its geographical structure. The second equality follows from (6.20.1) and it therefore follows that on $z = 1$

$$\rho = \left(\frac{D}{\delta_a}\right)^2 \rho_s(\phi, \theta), \tag{6.21.7}$$

where

$$\delta_a = \left(\frac{2\Omega W_e r_0^2}{g \, \Delta\rho/\rho_0}\right)^{1/2}, \tag{6.21.8}$$

and where the relations

$$U = \frac{r_0}{D}W = \frac{r_0}{D}W_e$$

have been used.

The length scale, $\delta_a$, is called the *advective scale* and we shall shortly see its relevance to thermocline theory.

The boundary conditions (6.21.5) and (6.21.7) must be satisfied by solutions of (6.21.1a,b). In principle, boundary conditions must also be specified along the lateral boundaries to complete the solution. These conditions are not the physical conditions externally imposed along the rim of the basin because those boundaries are shielded from the interior dynamics by boundary current regions in which advection of heat and relative vorticity are important. To get the proper boundary conditions for the interior problem formulated here, the boundary layer regions must first be analyzed for the boundary-layer fields in terms of unspecified interior fields and then the boundary conditions for the interior fields must be determined by matching the interior to the boundary layers. The general nature of this process is described in Chapter 5 were it is shown how the boundary-layer analysis on the eastern oceanic boundary, for example, determines the boundary condition for the Sverdrup interior flow. Although similar in conception, the process in the present case for nonlinear, stratified dynamics is far too difficult to actually be carried out. Thus we shall usually be content with specifying plausible conditions on the eastern boundary, e.g., $u = 0$ if the boundary is a meridian. This leaves uncertain the appropriate boundary conditions on, particularly, the western boundary. Theoretical progress has only been made by searching for solutions without regard for their ability to be matched to western boundaries and the reader must be aware of that weakness shared by all the solutions discussed below.

Before considering specific solutions, it is useful to examine the structure of the problem posed by (6.21.1a,b), (6.21.50), and (6.21.7) as a function of $\lambda$. Now $\lambda$, as given by (6.20.8) can be rewritten as a ratio of length scales, i.e.,

$$\lambda = \frac{\delta_D}{D}, \tag{6.21.9}$$

where $\delta_D$ is the *diffusive scale* defined by

$$\delta_D = \frac{K_V}{W_e}. \tag{6.21.10}$$

Now for the realistic oceanic values, $W_e = 10^{-4}$ cm/s, $\Delta\rho/\rho_0 = 10^{-3}$ and $2\Omega = 1.4 \times 10^{-4}$, $\delta_a$ is about 700 m. This is about the right order of magnitude for the thermocline scale and it would be natural therefore to choose $D$, the vertical scale of the motion to be $\delta_a$, in which case $\lambda$ would $\delta_D/\delta_a$. The estimates of Section 6.20 would then imply that $\lambda \ll 1$. This is the limit and the scaling for $D$ to which we shall shortly return. However, it is first illuminating to consider the alternative, less realistic case first. That is, imagine a case where the Ekman pumping is sufficiently weak so that $\delta_a \ll \delta_D$. Then the choice $D = \delta_a$ would make $\lambda \gg 1$ and nothing in (6.21.16) could balance the vertical diffusion term. On the other hand, if we chose $D = \delta_D$, the coefficient $(D/\delta_a)^2$ in (6.21.7) would be greater than O(1). This suggests rescaling $\rho$ in this limit by writing

$$\rho = \left(\frac{D}{\delta_a}\right)^2 \tilde{\rho}(\phi, \theta, z), \qquad (6.21.11a)$$

which from the thermal wind balance (6.20.11a,b) requires a similar rescaling of $u$ and $v$, i.e.,

$$(u, v) = \left(\frac{D}{\delta_a}\right)^2 \tilde{\rho}(\tilde{u}, \tilde{v}), \qquad (6.21.11b)$$

where the tilde variables are O(1). However, the rescaling of $v$ requires, by (6.21.1a) a rescaling of $w$, i.e.,

$$w = \left(\frac{D}{\delta_a}\right)^2 \tilde{w}, \qquad (6.21.11c)$$

so that the dynamical equations now become

$$\left(\frac{D}{\delta_a}\right)^2 \left[\frac{\tilde{\delta}}{\cos\theta}\frac{\partial\tilde{\rho}}{\partial\theta} + v\frac{\partial\tilde{\rho}}{\partial\theta} + w\frac{\partial\tilde{\rho}}{\partial z}\right] = \frac{\delta_D}{D}\frac{\partial^2\tilde{\rho}}{\partial z^2}, \qquad (6.21.12)$$

while the boundary conditions on $z = 1$ become

$$\tilde{w} = \left(\frac{\delta_a}{D}\right)^2 w_e(\phi, \theta), \qquad (6.21.13a)$$

$$\tilde{\rho} = \rho_s(\phi, \theta). \qquad (6.21.13b)$$

For large $\lambda$, or for $\delta_D/D \gg 1$, a balance between diffusion and advection is required, else the diffusion term, unbalanced, would allow the surface density field to permeate completely are oceanic bottom. If we reject this possibility this requires that $(D/\delta_a)^2 \sim \delta_D/D$, or that in this limit the vertical scale predicted for the thermocline is

$$D = (\delta_a^2 \delta_D)^{1/3} = \left(\frac{2\Omega r_0^2 K_v}{g \, \Delta\rho/\rho_0}\right)^{1/3}. \qquad (6.21.14)$$

Note that $D$ is independent of $w_E$ and that

$$\frac{D}{\delta_a} = \left(\frac{\delta_D}{\delta_a}\right)^{1/3} \gg 1$$

if $\lambda \gg 1$. Since $w = O(D/\delta_a)^2$ this implies that the vertical velocity in this limit is much larger than the Ekman pumping. Indeed, to lowest order, *in this diffusive limit* (6.21.13a) reduces to $\tilde{w} = 0$ on $z = 1$, i.e., the problem becomes *independent of the Ekman pumping*. The downward diffusion of the applied surface density is so strong in this limit that the resulting interior density gradients force vertical motions which are large compared to $w_e$ and so Ekman pumping becomes irrelevant. The fluid is driven primarily by buoyancy gradients diffusing into the fluid and the vertical velocity is *internally* generated in this limit. Thus for $\delta_D/\delta_a \gg 1$, the dynamics becomes a balance between advection and diffusion and the motion is primarily thermally driven with the wind forcing playing negligible role in determining the structure and dynamics of the thermocline.

Now let us turn our attention back to the opposite limit $\delta_D/\delta_a \ll 1$ which we have already noted is probably the more realistic limit in the natural ocean.

In this limit, $\delta_D/\delta_a \ll 1$, and one balance occurs for $D = \delta_a$. Since then $\lambda = \delta_d/\delta_a \ll 1$, the obvious approximation to (6.21.6) is

$$\frac{u}{\cos\theta}\frac{\partial\rho}{\partial\phi} + v\frac{\partial\rho}{\partial\theta} + w\frac{\partial\rho}{\partial z} = 0. \tag{6.21.15}$$

The density balance becomes strictly advective, hence the name advective scale for $\delta_a$. The vorticity equation (6.21.1a) and the boundary conditions remain unchanged although (6.21.7) becomes simply the condition

$$\rho(\phi, \theta, 1) = \rho_s(\phi, \theta). \tag{6.21.16}$$

The approximation of (6.21.15) to (6.21.16) is clearly a singular perturbation since the highest $z$-derivative is dropped and the ability of (6.21.15) to satisfy conditions on $w$ and $\rho$ at the surface must be examined. If flow enters the geostrophic region from the mixed layer, it is clear on physical grounds that $\rho$ must be specified at this inflow region to determine solutions of (6.21.15). Or, phrased differently, we can expect that if the flow is inward ($w_e < 0$) that solutions of (6.21.15) will satisfy (6.21.16). Thus in the oceanic subtropical gyres where $w_e < 0$ a purely advective balance can exist. In northern regions of the subpolar gyres where $w_e > 0$, fluid leaves the geostrophic domain and enters the Ekman layer carrying fluid whose density has been determined within the gyre and the density of this exiting fluid cannot be expected to match the arbitrarily specified surface density. How can this mismatch be corrected?

We can see from (6.21.16) that in the limit $\delta_D/\delta_a \ll 1$, a second dynamical balance can be obtained in which $D = \delta_D$. Then (6.21.7) implies that

$$\rho = \left(\frac{\delta_D}{\delta_a}\right)^2 \hat{\rho} \tag{6.21.17}$$

and hence from the thermal wind balance $u$ and $v$ become $O(\delta_D/\delta_a)^2$. Thus the density equation reduces to

$$w\frac{\partial\hat{\rho}}{\partial z} = \frac{\partial^2\hat{\rho}}{\partial z^2}, \tag{6.21.18}$$

while, to lowest order the vorticity equation becomes

$$\frac{\partial w}{\partial z} = 0. \tag{6.21.19}$$

The boundary condition on $w$ at $z = 1$, with (6.21.19) implies that in the region with length scale $\delta_D$

$$w = w_e$$

and is thus known. For $\delta_D \ll \delta_a$, a motion on the $\delta_D$ scale is so shallow that no thermal wind of any magnitude can develop. The absence of a meridional velocity then implies a $z$-independent $w$ from the vorticity balance. It is clear that (6.21.18) will only admit solutions of a boundary layer type when $w_e > 0$. Only then will the density signal decay away from the upper surface. Hence, this solution is not possible in the subtropical gyres ($w_e < 0$) and, as we have already noted, it is not needed there since when $w_e < 0$, the purely advective solution can satisfy the surface density condition. On the other hand, in regions where $w_e > 0$, the advective solution cannot satisfy the surface condition on $\rho$ and the secondary region governed by (6.21.18) must be used. When that happens, the thin diffusive layer with $D = \delta_D$ becomes sandwiched between the deeper advective domain and the upper surface. In each case the overall depth of the thermocline region is the advective scale $\delta_a$ in the limit $\delta_D/\delta_a \ll 1$.

In the case $\lambda \ll 1$, $w_e > 0$, the solution to (6.21.18) would represent a shallow solution for the thermocline complete in itself if we were willing to accept an undiminished Ekman velocity with depth and were not concerned about matching the thermocline solution $w_e > 0$, with scale $\delta_D$ to the advective thermocline that would exist south of the line $w_e = 0$ whose scale is $\delta_a$. Thus solutions of (6.21.1a,b,) without regard to appropriate conditions at great depth or on lateral boundaries can be nonunique if insufficiently constrained.

Several special solutions to (6.21.1a,b) have been found for cases in which the constraints associated with the boundary conditions are somewhat relaxed. In particular, these solution are of *similarity form*, i.e., the vertical form of the solution is imagined, *a priori*, to be essentially the same everywhere except insofar as it must be stretched by a geography dependent scale factor. There are numerous boundary layer problems in fluid mechanics where this, in fact, naturally occurs. Usually, the absence of any intrinsic scale, geographic or otherwise is required for the validity of such solutions.

One of the most interesting of the class of similarity solutions that have been found is the one discussed by Needler (1967). He proposed a trial solution of the form

$$p = m(\phi, \theta)\exp[k(\theta, \phi)z] \tag{6.21.20}$$

from which $\rho$, $u$, and $v$ can be determined from (6.20.7b,c,d). It is convenient in discussing this solution to change the origin of $z$ of coordinates to the sea surface and measure $z$ from the sea surface. Since $\delta_a$ is much less than the oceanic depth, the interval in $z$ effectively runs from zero to large negative $z$, i.e., $-\infty \leq z \leq 0$. The pressure field decays exponentially downward at a rate $k$ which depends on $\theta$ and $\phi$. Obviously $k$ must be positive. In this trial solution, the vertical profiles of density

$$\rho = -\frac{\partial p}{\partial z} = -kme^{kz} \tag{6.21.21}$$

are everywhere *similar*, i.e., everywhere the same except for a laterally variable stretching factor. Since (6.21.1a) may be written

$$\frac{\partial w}{\partial z} = (\sin \theta)^{-2} \frac{\partial p}{\partial \phi} = (\sin \theta)^{-2} e^{kz} \left( \frac{\partial m}{\partial \phi} + zm \frac{\partial k}{\partial \phi} \right) \qquad (6.21.22)$$

an integration with respect to $z$ yields

$$w = (\sin \theta)^{-2} \left[ k^{-1} \frac{\partial m}{\partial \phi} (e^{kz-1}) + k^{-1} m \frac{\partial k}{\partial \phi} (ze^{kz} - k^{-1}[e^{kz} - 1]) \right] + w_e,$$

$$(6.21.23)$$

where the constants of integration have been chosen to satisfy (6.21.5). The density equation may be rewritten in terms of $p$ and $w$ alone using (6.20.7b,c,d) as

$$(\sin \theta \cos \theta)^{-1} \left( \frac{\partial p}{\partial \phi} \frac{\partial^2 p}{\partial \theta \, \partial z} - \frac{\partial p}{\partial \theta} \frac{\partial^2 p}{\partial \phi \, \partial z} \right) + w \frac{\partial^2 p}{\partial z^2} = \lambda \frac{\partial^3 p}{\partial z^3}. \qquad (6.21.24)$$

If (6.21.20) and (6.21.23) are substituted into (6.21.24), after a little algebra we obtain

$$e^{kz} \left[ \frac{\left( m^2 \frac{\partial k}{\partial \phi} - mk \frac{\partial m}{\partial \phi} \right)}{\sin^2 \theta} + mk^2(w_e - \lambda k) \right]$$

$$(6.21.25)$$

$$+ e^{2kz} \left[ \frac{m \left( \frac{\partial m}{\partial \phi} \frac{\partial k}{\partial \theta} - \frac{\partial m}{\partial \theta} \frac{\partial k}{\partial \phi} \right)}{\sin \theta \cos \theta} + \frac{mk}{\sin^2 \theta} \left( \frac{\partial m}{\partial \phi} + m \frac{\partial k}{\partial \phi} [z - k^{-1}] \right) \right]$$

In order tnat (6.20.21) may truly be a solution each coefficient in (6.21.25) of $e^{kz}$, $ze^{2kz}$, and $e^{2kz}$ must separately vanish. This condition applied to the term in $ze^{2kz}$ implies that

$$m \frac{\partial k}{\partial \phi} = 0. \qquad (6.21.26)$$

If $m \neq 0$, i.e., for a nontrivial solution, $k$ must be independent of longitude. Thus $k = k(\theta)$. The vanishing of the coefficient of $e^{2kz}$ then implies that

$$m \frac{\partial m}{\partial \phi} \left[ \frac{\partial k}{\partial \theta} + \frac{\cos \theta}{\sin \theta} k \right] = 0 \qquad (6.21.27)$$

from which it follows that either $m$ is independent of $\phi$ or

$$k = \frac{C}{\sin \theta}, \qquad (6.21.28)$$

where $C$ is a free constant. A glance at (6.21.23) shows that the choice of $m$ independent of $\phi$ renders $w$ independent of $z$ [if $p$ is independent of $\phi$, $v = 0$ and $\partial w/\partial z$ must vanish by (6.21.1a)]. This if we reject this solution which has

$w$ constant, it follows that the vertical scale for the thermocline is $k^{-1} = \sin \theta / C$, which becomes increasingly shallow as the equator is approached in qualitative agreement with observations.

The remaining condition is the vanishing of the coefficient of $e^{kz}$ which yields

$$\lambda C^3 - C^2 w_e \sin \theta + k \frac{\partial m}{\partial \phi} \sin \theta = 0. \tag{6.21.29}$$

On $z = 0$, $\rho$ must equal $\theta_s$, hence $mk = -\rho_s(\phi, \theta)$. Thus (6.21.29) becomes

$$\lambda C^3 - C^2 w_e \sin \theta - \frac{\partial \rho_s}{\partial \phi} \sin \theta = 0. \tag{6.21.30}$$

In principle, (6.21.30) is the equation which determines $C$. However, since $C$ is a constant, this requires a special and rather artificial relationship between $\rho_s$ and $w_e$. This is one of the chief deficiencies of the similarity form, i.e., the inability to match arbitrary boundary data. Nevertheless, it is interesting to examine the nature of the solutions for $C$ arbitrary $\lambda$. In doing this, we will assume that $\rho_s$ and $w_e$ are always chosen to make all solutions for $C$ independent of $\theta$ and $\phi$. Figure 6.21.2(a) shows a plot of $C$ versus $\partial \rho_s / \partial \phi$ for the case $w_e < 0$. For moderate values of $\partial \rho_s / \partial \phi$, the solution for $C$ is the advective solution

$$C = \left( -\frac{\partial \rho_s / \partial \phi}{w_e} \right)^{1/2}, \tag{6.21.31}$$

which requires $\partial \rho_s / \partial \phi > 0$ if $w_e < 0$. For large $\partial \rho_s / \partial \phi$ (or small $w_e$) the solution alters and the advective-diffusive solution

$$C = \left( \frac{\partial \rho_s / \partial \phi}{\lambda} \right)^{1/3} \tag{6.21.32}$$

corresponding to (6.21.14) is obtained. In this case the thermocline scale is independent of $w_e$.

An interesting situation arises for the case $w_e > 0$. In that case the graph of $C$ versus $\partial \rho_s / \partial \theta$ is shown in Figure 6.21.2(b). For large $\partial \rho_s / \partial \phi$ (or large $\lambda$), the solution is once again (6.21.32) and is independent of $w_e$. Thus in this limit the solutions for the subpolar and subtropical gyres are essentially the same. However, when $\partial \rho_s / \partial \phi < 0$ there are *two* solutions in the range

$$-\frac{4}{27} \frac{w_e^3}{\lambda^3} \sin^2 \theta < \frac{\partial \rho_s}{\partial \phi} < 0. \tag{6.21.33}$$

This possibility was noted by Blandford (1965). The lower branch with smaller $C$ (hence with a deeper thermocline) is given by the advective balance (6.21.31). The second and additional solution is the diffusive solution. On this branch

$$C \sim w_e \sin \frac{\theta}{\lambda} \tag{6.21.34}$$

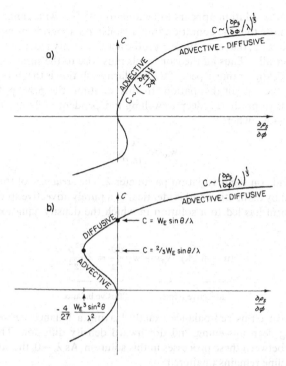

**Figure 6.21.2** (a) A schematic graph of the solution for $C$ as given by (6.21.30) for the case $w_e < 0$. Only $C > 0$ is relevant. (b) The solution for $C$ for $w_e > 0$. Note the multiple solutions for $\partial \rho_s / \partial \phi < 0$.

corresponding to a shallow thermocline of depth $\delta_D$. Hence the similarity solution retrieves each of the cases anticipated by scaling analysis but rather than presenting them as portions of a total, composite solution, the similarity form treats each balance as an independent possibility for the thermocline.

The dynamical fields implied by (6.21.20) are of the form

$$p = -\frac{\rho_s}{k} e^{kz},$$

$$\rho = \rho_s e^{kz},$$

$$u = -\left[\frac{\partial}{\partial \theta}\left(\frac{\rho_s}{k}\right) + z \frac{\partial k}{\partial \theta} \frac{\rho_s}{k}\right] \frac{e^{kz}}{\sin \theta}, \qquad (6.21.35)$$

$$v = -\left[\frac{1}{k} \frac{\partial \rho_s}{\partial \phi}\right] \frac{e^{kz}}{\sin \theta \cos \theta},$$

$$w = -\frac{1}{C^2} \frac{\partial \rho_s}{\partial \phi} e^{kz} + \frac{\lambda C}{\sin \theta}.$$

Although the solution appears to be able to satisfy a wide range of conditions on $\rho_s$, note that the condition that $u$ vanish on a meridian requires that $\rho_s$ vanish there, i.e., that at some $\phi = \phi_E$, the actual density must be the abyssal density $\rho_0$ for all $z$. Thus all the isopycnals must rise to the surface at $\phi = \phi_E$.

Another disappointing aspect of the similarity solution is that it tells us very little about the role of dissipation. In this solution, the principal effect of dissipation is to produce a deep upwelling independent of depth. That is, as $z \to -\infty$, $w \to w_\infty$ where

$$w_\infty = \frac{\lambda C}{\sin \theta}, \tag{6.21.36}$$

which depends on the dissipation parameter $\lambda$. The *structure* of the solution is unaffected by dissipation and is identical to a purely advective solution. The similarity form has led to a solution in which the density equation may be written

$$\underbrace{(\mathbf{u} - w_\infty \hat{\mathbf{k}}) \cdot \nabla \rho + w_\infty \frac{\partial \rho}{\partial z}}_{\text{advective terms}} = \underbrace{\lambda \frac{\partial^2 \rho}{\partial z^2}}_{\text{diffusive balance}}. \tag{6.21.37}$$

The advective terms self-balance exactly leaving a balance between terms representing deep upwelling and downward density diffusion. There is no interaction between these processes in this solution. As $\lambda \to 0$, the structure of the thermocline remains unaltered.

This is connected to another special feature of the similarity solution. The potential vorticity

$$\Pi = \sin \theta \frac{\partial \rho}{\partial z} = C\rho. \tag{6.21.38}$$

Hence the potential vorticity is constant on density surfaces. We have already noted that for a nondissipative flow any $\Pi = \Pi(\rho, B)$ will satisfy the dissipation-free dynamics. Hence the present case is a rather degenerate example of that class in which $\Pi$ is independent of the Bernoulli function and depends only on $\rho$. This is what allows the exponential solution to "work," i.e., the nonlinearities associated with advection cancel since $\Pi = \Pi(\rho)$ must always yields a solution to the dissipationless thermocline dynamics.

Figure 6.21.3 shows a meridional cross section of temperature as calculated by Needler (1967) from the similarity solution with the arbitrary *choice*

$$\frac{\delta_a}{C} = 1500 \text{ m},$$

and a surface temperature proportional to $\cos(\theta + 10°)$. The temperature anomaly is convected to $\rho$ by

$$T_* = -\mathcal{T}_0 \rho(\theta, \phi, z),$$

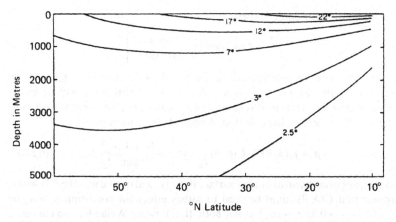

**Figure 6.21.3** North–south section of the temperature field with the surface temperature proportional to $\cos(\theta + 10°)$ and $C/\delta_a = (1500\ \text{m})^{-1}$ (from Needler 1967).

where $\mathcal{T}_0$ is the surface temperature amplitude. Recall though that the solution requires a special meridionally varying surface density.

Since the structure of the Needler solution is largely independent of $\lambda$ and has the advective structure as long as $\delta_D < \delta_a$ it might be natural to ask what solutions could be found ignoring $\lambda$ altogether. If dissipation is unimportant a first integral of the equations is given by (6.20.20), i.e.,

$$\sin\theta\frac{\partial\rho}{\partial z} = \Pi(\rho, p + \rho z). \tag{6.21.39}$$

A particularly simple example, suggested by Welander (1971a) occurs when the functional form of (6.21.39) is *chosen* to be

$$\Pi = a_0\rho + b_0(p + \rho_z) = \sin\frac{\partial\rho}{\partial z}. \tag{6.21.40}$$

Whereas Needler's solution has a linear function of $\rho$, this choice (6.21.40) generalizes that by adding a linear dependence on $B$. There is no reason *a priori* why this form should be particularly relevant to a problem with given surface conditions and we therefore cannot expect the solution to satisfy arbitrary conditions on $\rho_s$ and $w_e$.

If (6.21.40) is differentiated with respect to $z$, the hydrostatic balance yields

$$\frac{\partial^2\rho}{\partial z^2} = a\frac{\partial\rho}{\partial z} + bz\frac{\partial\rho}{\partial z} = -2\frac{z + z_0}{D^2\sin\theta}\frac{\partial\rho}{\partial z}, \tag{6.21.41}$$

where

$$z_0 \equiv \frac{a}{b}, \qquad D^2 \equiv -\frac{2}{b}.$$

The solution for $\partial\rho/\partial z$ is then

$$\frac{\partial\rho}{\partial z} = C(\phi, \theta)\exp\left[-\frac{(z + z_0)^2}{D^2 \sin\theta}\right]. \tag{6.21.42}$$

Clearly $D^2$ must be positive and hence $b < 0$. Note that $\partial\rho/\partial z$ takes on its maximum value at $z = -z_0$. If $a < 0$, $z_0 > 0$ and this point will lie in the physical domain. Thus in this case the maximum of $\partial\rho/\partial z$ is not forced to lie at $z = 0$ as in the Needler solution. A second integration yields

$$\rho = \rho_s(\phi, \theta) + C(\phi, \theta)\int_0^z \exp\left[-\frac{(\zeta + z_0)^2}{D^2 \sin^2\theta}\right]d\zeta, \tag{6.21.43}$$

so that, apparently, an arbitrary surface density field can be matched. It would appear that $C(\phi, \theta)$ could be used to satisfy either the condition $w = w_e$ on $z = 0$, or $\rho \to 0$ as $z \to \infty$, but *not both*. If, following Welander, we choose $C$ so that $\rho$ goes to zero (i.e., $\rho_* \to \rho_0$) as $z \to -\infty$, then

$$C = \frac{\rho_s}{\displaystyle\int_0^{-\infty} \exp\left[-\frac{(\zeta + z_0)^2}{D^2 \sin^2\theta}\right]d\zeta}. \tag{6.21.44}$$

The pressure $p$ and hence $w$ can now be calculated at the surface. The hydrostatic balance yields

$$p = p_0(\phi, \theta) - z\rho_s - C\int_0^z \int_0^\zeta \exp\left[-\frac{(\zeta' + z_0)^2}{D^2 \sin^2\theta}\right]d\zeta' \, d\zeta. \tag{6.21.45}$$

However, the function $p_0$ is not arbitrary since $p$ and $\rho$ must satisfy (6.21.40). This implies that

$$bp_0 + a\rho_s = \sin\theta\hat{C} = \sin\theta C \exp\left[-\frac{z_0^2}{D^2 \sin^2\theta}\right], \tag{6.21.46}$$

and this determines $p_0$. On $z = 0$

$$w = w_e = \frac{-\dfrac{u}{\cos\theta}\dfrac{\partial\rho_s}{\partial\phi} - v\dfrac{\partial\rho_s}{\partial\phi}}{\partial\rho/\partial z}, \tag{6.21.47}$$

which may be rewritten with the use of geostrophy and (6.21.46) in the form

$$w_e = \frac{\left\{\dfrac{\partial\rho_s}{\partial\phi}\dfrac{\partial}{\partial\phi}\hat{C}\sin\theta - \dfrac{\partial\rho_s}{\partial\theta}\dfrac{\partial}{\partial\phi}\hat{C}\sin\theta\right\}}{b\hat{C}\sin\theta\cos\theta}. \tag{6.21.48}$$

If $\rho_s$ were a function only of latitude, $\rho_s$ and $\hat{C}$ would both be independent of $\phi$ and $w_e$ would be forced to be zero. Hence Welander's solution shares with Needler's the need for a special form of the surface density field for nontrivial solutions. It is not clear that $C$ should be chosen as described above. An alternative viewpoint might be that the prescription (6.21.40) holds only over

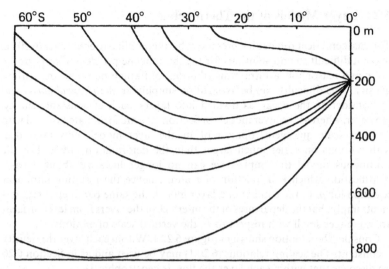

**Figure 6.21.4**   Thermocline structure for an ideal-fluid thermocline in which $\Pi$ is a linear function of $B$ and $\rho$ (from Welander 1971a).

a given density or depth range below which the fluid might be at rest or possess an entirely different $\Pi(\rho, B)$ relation. In that case (6.21.48) becomes a differential equation for $C$ allowing a match to arbitrary $w_e$. This, in turn would lead to density discontinuities between moving and stagnant fluid. There seems no simple way to assure both continuity of solutions and the satisfaction of arbitrary boundary conditions as long as the prescription for $\Pi(\rho, B)$ is determined *a priori*. In principle, we should expect this relationship to be determined by the solution rather than being free to be chosen *a priori*.

Figure 6.21.4 shows the thermocline structure given by (6.21.40) as presented by Welander. He chose $\rho_s$ to match observed surface density in the South Pacific and then calculated $C$ from (6.21.43). The figure shows the meridional cross section for the case where $z_0 \delta_a = 200$ m. Again, we note the tendency for isopycnals to rise toward the equator in response to potential vorticity conservation.

Both Needler's and Welander's solutions share the character that they are sophisticated "guesses." That is, the form of the solution is first proposed and the equations place certain constraints on free constants or free functions in the solution. It is no surprise then that the solutions are unable to satisfy general boundary conditions. A more fundamental problem with these solutions is their implicit insistence that the form of the solution hold throughout the domain of the flow, for there is no reason why that should be true. Nevertheless, the solutions presented above represent almost all of the solutions of the continuously stratified models of the thermocline. The thermocline models presented in the next section are able to break away from the similarity form and satisfy surface data directly, but only by treating the thermocline problem in a layered model.

## 6.22  Layer Models of the Thermocline

The mathematical difficulties presented by the nonlinear equations (6.20.7c) make it difficult to find solutions for the thermocline problem in response to given and reasonably arbitrary distributions of Ekman pumping and surface density. This difficulty can be avoided by simplifying the *physical model* and by considering the ocean as divided into layers, each of constant density. However, whereas the layer models of Section 6.16 are formulated to deal with synoptic scale quasigeostrophic motion, the layer models now must deal with motions on scales much larger than the deformation radius. For the continuous model this implies we can no longer linearize about a basic stratification which is a function of $z$ alone, hence the resulting nonlinear equation for $\rho$. In the context of a layer model, the same parametric requirements imply that the departures of the interface of the layers from level surfaces are no longer small with respect to $D$, the vertical scale of motion.

Consider the situation shown in Figure 6.22.1. Within each layer the density is constant. The scaling relations (6.20.1) may be used and as in Section 6.20 we discover that within each layer the flow is geostrophic, i.e.,

$$\sin \theta v_n = \frac{1}{\cos \theta} \frac{\partial p_n}{\partial \phi}, \tag{6.22.1a}$$

$$\sin \theta u_n = -\frac{\partial p_n}{\partial \theta}, \tag{6.22.1b}$$

where subscripts label the layer. The continuity equation is

$$\frac{1}{\cos \theta} \frac{\partial}{\partial \theta}(v_n \cos \theta) + \frac{1}{\cos \theta} \frac{\partial u_n}{\partial \phi} + \frac{\partial w_n}{\partial z} = 0, \tag{6.22.2}$$

while the vorticity equation in this limit is again

$$\cos \theta v_n = \sin \theta \frac{\partial w_n}{\partial z}, \tag{6.22.3}$$

i.e., the relative vorticity variations are negligible with respect to the planetary vorticity gradient. Since the density is constant within each layer it follows that $u_n$ and $v_n$ are independent of $z$ within each layer but, of course, may change discontinuously from layer to layer. It is convenient in the present case to rewrite the pressure in each layer as

$$p_{n*} = -\rho_n g D z + \rho_0 2\Omega U r_0 p_n, \tag{6.22.4}$$

where $\rho_0$ is the characteristic value for the overall density field. Usually $|\rho_n - \rho_0| \ll \rho_0$. The hydrostatic balance

$$\frac{\partial p_{n*}}{\partial z_*} = -\rho_n g$$

then implies that $p_n$ in (6.22.4) is independent of $z$ within each layer. On the

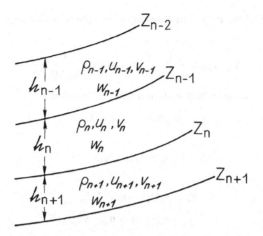

**Figure 6.22.1**   The layer model. The thickness of the $n$th layer is $h_n$ and its density is $\rho_n$. The vertical coordinate of the base of the $n$th layer is $z_n(\phi, \theta)$. The density is constant within each layer.

interface between layers $n$ and $n + 1$, the pressure must be continuous, hence on $z = z_n(\phi, \theta)$ it follows from (6.22.4) that

$$\rho_0 2\Omega U r_0(p_{n+1} - p_n) = (\rho_{n+1} - \rho_n)gDz_n. \qquad (6.22.5)$$

Now, as in Section 6.21, $D$ is the vertical scale of the motion. If we imagine that it is $W = r_0/DU$ whose magnitude is known (e.g., from the Ekman pumping) an ordering equality in (6.22.5) is attained with (see (6.21.8))

$$D = \left(\frac{2\Omega W r_0^2}{g\dfrac{\Delta\rho}{\rho_0}}\right)^{1/2} \equiv \delta_a, \qquad (6.22.6)$$

where $\Delta\rho$ is a characteristic value of $(\rho_{n+1} - \rho_n)$.

If this advective scale is chosen for $D$, (6.22.5) becomes

$$p_{n+1} - p_n = \frac{(\rho_{n+1} - \rho_n)}{\Delta\rho} z_n, \qquad (6.22.7)$$

while the height of any interface (from an as yet unspecified origin) is

$$z_{n*} = \delta_a z_n.$$

Note that $z_n$ may depart significantly from a level surface. In fact, we will shortly consider interfaces which, modeling isopycnal surfaces in the thermocline, start from the sea surface and reach great depths some distance from their outcrop position.

The horizontal velocities are independent of depth and this allows (6.22.2) to be integrated in $z$ from $z_n$ to $z_{n-1}$ to yield

$$(z_{n-1} - z_n)\nabla_H \cdot \mathbf{u}_{Hn} + w_n(z_{n-1}) - w_n(z_n) = 0, \qquad (6.22.8)$$

where $\mathbf{u}_H$ is the velocity vector tangent to the sphere with components $(u_n, v_n)$ and while

$$\nabla_H \cdot \mathbf{u}_{nH} \equiv \frac{1}{\cos\theta} \frac{\partial}{\partial\theta}(v_n \cos\theta) + \frac{1}{\cos\theta} \frac{\partial u_n}{\partial\phi}.$$

The integrated continuity equation can be rewritten

$$\nabla_H \cdot (\mathbf{u}_{nH} h_n) + V_{n-1} - V_n = 0, \qquad (6.22.9)$$

where

$$h_n \equiv z_{n-1} - z_n, \qquad (6.22.10)$$

and

$$V_n \equiv w_n(z_n) - \mathbf{u}_{nH} \cdot \nabla z_n,$$
$$V_{n-1} = w_n(z_{n-1}) - \mathbf{u}_{nH} \cdot \nabla z_{n-1}. \qquad (6.22.11)$$

In steady flow $V_n$ is the fluid velocity normal to the $n$th interface. If each fluid element preserves its density it is unable to cross the interface and $V_n$ must be zero. If $V_n$ is different from zero in steady flow it implies that fluid of density $\rho_{n+1}$ is being converted to fluid of density $\rho_n$ (or vice-versa). Such a conversion represents a *nonconservative process acting on the density field* and in the layer model is the analogue of the diffusion term in (6.20.7e). For unsteady flow $V_n$ need not be zero, even for density conserving fluids since then the normal velocity at the $n$th interface is

$$U_n = w_n(z_n) - \mathbf{u}_{nH} \cdot \nabla z_n - \frac{\partial z_n}{\partial t}$$
$$= V_n - \frac{\partial z_n}{\partial t}. \qquad (6.22.12)$$

Hence the rate at which fluid crosses the $n$th interface for a general *unsteady* flow is given by $U_n$. $U_n$ is the velocity with which fluid of density $\rho_{n+1}$ in the $(n + 1)$st layer is *entrained* into the $n$th layer and is incorporated into that layer as additional fluid with density $\rho_n$. Normally the entrainment process is produced by smaller scale turbulence although since we are here considering motion on the very largest planetary scale, the entrainment velocity could be a manifestation of eddy activity on scales of the order of the deformation radius. It is physically obvious that the entrainment velocity, by mass conservation must be continuous, i.e.,

$$U_n(z_{n-1}) = U_{n-1}(z_{n-1}). \qquad (6.22.13)$$

The vorticity equation can be integrated over layer $n$ and combined with (6.22.8) and (6.22.12) to yield

$$\cos\theta v_n h_n = \sin\theta [w_n(z_{n-1}) - w_n(z_n)]$$
$$= \sin\theta \left[ \frac{\partial h_n}{\partial t} + \mathbf{u}_{nH} \cdot \nabla h_n + U_{n-1} - U_n \right], \qquad (6.22.14)$$

or

$$\left(\frac{\partial}{\partial t} + \mathbf{u}_{nH} \cdot \nabla\right)\left(\frac{\sin \theta}{h_n}\right) = \frac{\sin \theta}{h_n^2}(U_{n-1} - U_n). \tag{6.22.15}$$

If fluid parcels conserve their density, i.e., if no fluid crosses the interfaces defining the $n$th layer, then $U_{n-1}$ and $U_n$ are both zero and the quantity

$$\Pi_n = \frac{\sin \theta}{h_n} \tag{6.22.16}$$

is conserved following the motion. This form of $\Pi_n$ is the layer equivalent of (6.20.19a) and is the obvious generalization of (3.4.7) for the case of planetary motions for which the relative vorticity is negligible. Note, however, that in contrast to the synoptic scale, quasigeostrophic motions $h_n$ cannot be linearized about a given level surface as in Section 6.16.

In dimensional units the potential vorticity is just $f/h_{n*}$ where $f$ is $2\Omega \sin \theta$ and $h_{n*}$ is $\delta_a h_n$, the dimensional layer thickness.

Since the entrainment velocities are continuous at each interface, it follows that

$$U_n = w_n(z_n) - \mathbf{u}_{nH} \cdot \nabla z_n - \frac{\partial z_n}{\partial t}$$

$$= w_{n+1}(z_n) - \mathbf{u}_{n+1H} \cdot \nabla z_n - \frac{\partial z_n}{\partial t}, \tag{6.22.17}$$

or

$$w_{n+1}(z_n) - w_n(z_n) = (\mathbf{u}_{n+1H} - \mathbf{u}_{nH}) \cdot \nabla z_n, \tag{6.22.18}$$

but from (6.22.1a,b)

$$\mathbf{u}_{nH} = \hat{\mathbf{k}} \times \nabla p_n$$

from which it follows, with the aid of (6.22.7), that

$$w_{n+1}(z_n) - w_n(z_n) = \frac{\rho_{n+1} - \rho_n}{\Delta \rho}(\hat{\mathbf{k}} \times \nabla z_n) \cdot \nabla z_n \equiv 0. \tag{6.22.19}$$

Hence, *because of geostrophy* the vertical velocity *is also continuous at each interface* even though the horizontal velocities are discontinuous.

This allows (6.22.14) to be summed over all the moving layers to yield

$$\cos \theta \sum_n v_n h_n = \sin \theta[w_T - w_B], \tag{6.22.20}$$

where $w_T$ is the fluid pumped out of the upper geostrophic layer while $w_B$ is the fluid pumped into the lowest moving geostrophic layer. If the upper layer is capped by an Ekman layer exposed to a wind stress, then $w_T$ would be the (nondimensional) Ekman velocity $w_e$ whose scale has been used to define $W$. If the fluid motion in the layer in contact with the oceanic bottom is at rest, or if the bottom is flat and frictional effects ignored, then $w_B$ vanishes, leading to

$$\cos \theta \sum_n v_n h_n = \sin \theta w_e, \tag{6.22.21}$$

which is the Sverdrup relation for the layer model. Of course, as long as there is no interaction with the bottom (6.20.9) can be integrated directly to yield (6.22.21), i.e., the Sverdrup relation is independent of the particular structure whether continuous or layered, by which the ocean is represented in the vertical.

As an application of these ideas, we now examine the problem of the thermocline. To keep matters as simple as possible, we will consider, at first, only the most rudimentary model, e.g., one with two moving layers. Figure 6.22.2 shows the model schematically. Attention is focused on the region of the subtropical gyre where $w_e$ is negative and we will assume that $w_e(\phi, \theta)$ vanishes on a line of constant latitude, $\theta_0$. The two moving layers have density $\rho_1$ and $\rho_2$. The depth of the upper layer is $h_1$ and that of the lower layer is $h_2$.

The interface between layers 1 and 2 outcrops within the subtropical gyre so that layer 2 is directly exposed to Ekman pumping north of the outcrop line. In just the same way that the surface density may be prescribed in the continuous model, the equivalent surface condition in the layer model is the specification of the line along which the density interface outcrops. This line then determines the surface density distribution. We will choose to study the simplest case where the outcrop line is a latitude circle. This is equivalent to specifying a surface density which is a function only of latitude and we noted in Section 6.21 that neither solution described there could accept such a condition.

The deeper layers in this model are assumed to be at rest since they are nowhere forced by Ekman pumping. This is a consistent choice but it will be

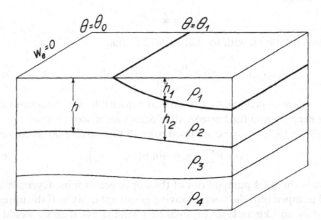

**Figure 6.22.2** A schematic view of the two-moving-layer model of the thermocline in the subtropical gyre. The Ekman pumping vanishes on $\theta = \theta_0$ and is downward (negative) for $\theta < \theta_0$.

necessary to examine later to what extent the solution so obtained is unique. We will discuss this point further in Section 6.23.

If layer 3 is at rest $\nabla p_3$ must be zero, from which it follows, using (6.22.7) that

$$\nabla p_2 = -\frac{\rho_3 - \rho_2}{\Delta \rho} \nabla z_2.$$

$$(6.22.22)$$

Now with an error of $O(\Delta\rho/\rho_0)\nabla z_2$ is just $-\nabla h$ where $h$ is the combined thickness of the two layers, i.e.,

$$h = h_1 + h_2.$$

$$(6.22.23)$$

This ignores the contribution the free surface makes to the thickness of the exposed layer but as shown in Section 6.16 this produces only an $O(\Delta\rho/\rho_0)$ error. Thus

$$\nabla p_2 = \gamma_2 \nabla h,$$

$$(6.22.24)$$

where

$$\gamma_n \equiv \frac{\rho_{n+1} - \rho_1}{\Delta \rho}.$$

$$(6.22.25)$$

In the region $\theta_1 < \theta < \theta_0$ only layer 2 is in motion (layer 1 does not exist in this region) and the Sverdrup relation, (6.22.21), reduces to

$$\cos \theta v_2 h_2 = \sin \theta w_e.$$

$$(6.22.26)$$

However, from (6.22.1a), (6.22.24) and the fact that $h_2 = h$ in this region, (6.22.26) becomes

$$\cos \theta \frac{\partial}{\partial \phi} h^2 = 2 \frac{\sin^2 \theta}{\gamma_2 \cos \theta} w_e,$$

$$(6.22.27)$$

which can be integrated to yield

$$h^2 = h_2^2 = D_0^2(\phi, \theta) + H_2^2,$$

$$(6.22.28)$$

where

$$D_0^2 = -\frac{2 \sin^2 \theta}{\gamma_2} \int_\phi^{\phi_E} w_e(\theta, \phi') \, d\phi'.$$

$$(6.22.29)$$

Note that $D_0^2$ vanishes on the eastern boundary.

Since $w_e < 0$ in the subtropical gyre, $D_0^2$ is positive and an increasing function of $(\phi_E - \phi)$ where $\phi_E$ is the longitude of the eastern boundary. $H_2$ is the thickness of layer 2 on the eastern boundary. If we assume that the geostrophic velocity vanishes on the eastern boundary, then (6.22.1b) and (6.22.24) imply that $H_2$, which in principle could be a function of $\theta$, must, in fact, be a constant.

$H_2$ is thus the depth of layer 2 on the eastern boundary where $D_0^2$ vanishes and on the northern boundary of the gyre where $w_e$ and hence $D_0^2$ is zero. The value of $H_2$ is arbitrary and cannot be predicted from the ideal fluid thermocline model. Thus we will obtain a family of solutions which depend parametrically on $H_2$. In actuality, $H_2$ is probably set by physical processes like small-scale convection along the northern boundary of the gyre; processes not described in the present theory.

Once $h$ is known, both $u_2$ and $v_2$ can be determined geostrophically. In particular,

$$v_2 = \frac{\tan \theta w_e}{(D_0^2 + H_2^2)^{1/2}}, \tag{6.22.30}$$

so that $v_2$ is everywhere negative, i.e., the flow in layer 2 is southward. The potential vorticity is

$$\Pi_2 = \frac{\sin \theta}{h_2} = \frac{\sin \theta}{(D_0^2 + H_2^2)^{1/2}}, \tag{6.22.31}$$

and, at a fixed latitude, decreases monotonically westward since $D_0^2$ and hence $h_2$ must increase westward.

Since the fluid in layer 2 flows southward, it will eventually reach the outcrop line, $\theta = \theta_1$. At that point, as the fluid in layer 2 flows further southward, it becomes shielded from the Ekman pumping by the presence of layer 1 and the solution (6.22.28) is no longer valid. However, for all stream-lines on which fluid in layer 2 flows under layer 1, the potential vorticity is conserved in the absence of density entrainment. That is, for $U_1$ and $U_2$ equal to zero, $\Pi_2$ is conserved. Furthermore, the value of $\Pi_2$ on each streamline in the region south of $\theta_1$ is set by the value of potential vorticity in the fluid column possessed at the outcrop line and that is known from (6.22.31).

South of the outcrop line, in layer 2

$$\mathbf{u}_{2H} \cdot \nabla \Pi_2 = 0, \tag{6.22.32}$$

or, using geostrophy and (6.22.24)

$$\hat{\mathbf{k}} \cdot (\nabla h \times \nabla \Pi_2) = 0. \tag{6.22.33}$$

Thus lines of constant $\Pi_2$ and $h$ must coincide or equivalently

$$\Pi_2 \equiv \frac{\sin \theta}{h_2} = G(h), \tag{6.22.34}$$

where $G$ is an arbitrary function. The relation (6.22.34) holds on every stream-line in layer 2. To determine $G(h)$, we first examine those streamlines which originate from the outcrop line.

On the outcrop line, $\theta = \theta_1$ and $h_1$, by definition, is zero. Thus on $\theta = \theta_1$, $h_2 = h$ and

$$G(h) = \frac{\sin \theta_1}{h}, \qquad \theta = \theta_1. \tag{6.22.35}$$

Although the potential vorticity is generally a complicated function of longitude on $\theta = \theta_1$ (6.22.31), it is a very simple function of $h$. As the stream-lines enter there region where $\Pi_2$ is conserved, the relation (6.22.35) between

$\Pi_2$ and $h$ is preserved on these streamlines exactly as it was set at the outcrop line. Hence, for all the subducted fluid in the region, $\theta \leq \theta_1$,

$$\Pi_2(h) = \frac{\sin \theta}{h_2} = G(h) = \frac{\sin \theta_1}{h}. \qquad (6.22.36)$$

Thus

$$h_2 = \frac{\sin \theta}{\sin \theta_1} h = \frac{f}{f_1} h \qquad (6.22.37a)$$

$$h_1 = \left(1 - \frac{\sin \theta}{\sin \theta_1}\right) h = \left(1 - \frac{f}{f_1}\right) h \qquad (6.22.37b)$$

where $f/f_1 = \sin \theta / \sin \theta_1$, is the ratio of the Coriolis parameter at $\theta$ to its value at the outcrop line. Note that the ratio of the layer depths is determined entirely by the requirement of potential vorticity conservation.

The remaining unknown is $h$ which can be determined by applying (6.22.21). In layer 1, (6.22.7) implies that

$$\begin{aligned} p_1 &= p_2 + \gamma_1 h_1 \\ &= \gamma_2 h + \gamma_1 h_1, \end{aligned} \qquad (6.22.38)$$

so that the Sverdrup relation, (6.22.21) becomes

$$\cos \theta (v_1 h_1 + v_2 h_2) = \sin \theta w_e,$$

or

$$\gamma_2 \frac{\partial}{\partial \phi}\left(h^2 + \frac{\gamma_1}{\gamma_2} h_1^2\right) = 2 \sin^2 \theta w_e / \gamma_2. \qquad (6.22.39)$$

The integral of (6.22.39) yields

$$h^2 + \frac{\gamma_1}{\gamma_2} h_1^2 = D_0^2 + H^2 + \frac{\gamma_1}{\gamma_2} H_1^2, \qquad (6.22.40)$$

where $H$ and $H_1$ are the constant values of $h$ and $h_1$ on $\phi = \phi_E$ where $D_0$ vanishes. Both $h$ and $h_1$ must be independent of $\theta$ to make $u_1$ and $u_2$ zero on $\phi = \phi_E$. However, $h_1$ is zero for $\theta > \theta_1$ and hence must *remain* zero on the eastern boundary or else a zonal velocity will be induced there. Thus $H_1$ is zero and $H = H_2$ which is the same constant which occurs in (6.22.28). It is important to note that this argument is entirely independent of potential vorticity conservation and the arguments leading to (6.22.37a,b) and hence (6.22.40) holds everywhere in the region $\theta < \theta_1$.

However, on the streamlines in layer 2 which *can* be traced back to the outcrop line conservation of $\Pi_2$ yields (6.22.37a,b) the latter of which allows us to use (6.22.40) to solve for $h$, i.e., in $\theta \leq \theta_1$

$$h = \frac{D_0^2 + H_2^2}{\left\{1 + \frac{\gamma_1}{\gamma_2}\left(1 - \frac{f}{f_1}\right)^2\right\}}^{1/2}. \qquad (6.22.41)$$

With $h$ determined, $h_1$ and $h_2$ are known and so are the geostrophic velocities. Before discussing the qualitative nature of the solution, we must first ask about the domain in which it is valid. First of all (6.22.37a,b) and hence (6.22.41) cannot apply up to the eastern wall if the geostrophic zonal velocity must vanish there. For if $u_1$ and $u_2$ both vanish on $\phi = \phi_E$, then both $h_1$ and $h_2$ (and hence $h = h_1 + h_2$) must be independent of $\theta$ on $\phi = \phi_E$. However, a $\theta$ derivative of (6.22.37a,b) shows that

$$\frac{\partial h_2}{\partial \theta} = \frac{h}{f_1}\frac{\partial f}{\partial \theta} + \frac{f}{f_1}\frac{\partial h}{\partial \theta},$$

$$\frac{\partial h_2}{\partial \theta} = -\frac{h}{f_1}\frac{\partial f}{\partial \theta} + \left(1 - \frac{f}{f_1}\right)\frac{\partial h}{\partial \theta}.$$

Thus if $\partial h/\partial \theta$ vanishes at $\phi = \phi_E$, the only way $h_1$ and $h_2$ can each be independent of $\theta$ is if $h$ vanishes at $\phi = \phi_E$. This, in turn, implies that all moving layers must have zero thickness on the eastern boundary. This violates the condition $h = H_2 \neq 0$ on $\theta = \theta_1$. If we wish to have $H_2 \neq 0$ it then must follow that the solution (6.22.41) *cannot hold right up to the eastern boundary.* The physical reason is quite simple. A fluid column in layer 2 simply cannot flow parallel to the eastern wall with a constant value of $h_2$, so as to avoid a nonzero zonal geostrophic velocity, and at the same time conserve potential vorticity, $\sin \theta/h_2$.

Consider a trajectory in layer 2 south of the outcrop. The trajectory is a curve of constant $h$, thus the trajectory or streamline is given parametrically by the condition

$$\frac{D_0^2(\phi, \theta) + H_2^2}{\left[1 + \dfrac{\gamma_1}{\gamma_2}\left(1 - \dfrac{f}{f_1}\right)^2\right]} = \text{constant}.$$

We can evaluate the constant by tracing the trajectory back to the outcrop latitude at its point of origin, e.g., $(\phi_0, \theta_1)$. Thus the parametric equation for the trajectory originating at the point $\phi_0, \theta_1$ is

$$D_0^2(\phi, \theta) = H_2^2 \frac{\gamma_1}{\gamma_2}\left(1 - \frac{f}{f_1}\right)^2 + D_0^2(\phi_0, \theta_1)\left[1 + \frac{\gamma_1}{\gamma_2}\left(1 - \frac{f}{f_1}\right)^2\right]. \quad (6.22.42)$$

Now consider the trajectory emanating from the point on the outcrop latitude at the eastern wall, i.e., let $\phi_0 = \phi_E$. $D_0^2(\phi_E, \theta_1)$ vanishes there so that such a limiting trajectory, $\Phi_2(\theta)$ is determined by the condition

$$D_0^2(\Phi_2, \theta) = H_2^2 \frac{\gamma_1}{\gamma_2}\left(1 - \frac{f}{f_1}\right)^2 \quad (6.22.43)$$

and is shown in Figure 6.22.3(a). Since $D_0^2$ is an increasing function of $\phi_E - \phi$, and since the right-hand side of (6.22.43) increases as $\theta$ decreases south of the outcrop line, the trajectory $\Phi_2(\theta)$ must curve westward as the fluid moves southward. Thus as long as $H_2$ differs from zero, there must be fluid in layer 2 that cannot be reached by geostrophic flow from the outcrop line. Thus layer

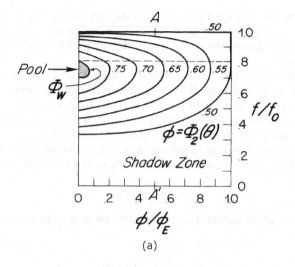

(a)

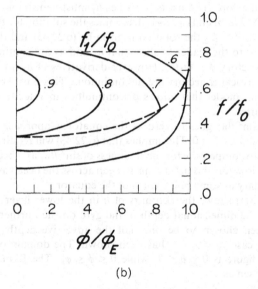

(b)

**Figure 6.22.3** (a) Contours of constant $h$ for the case $\gamma_1 = \gamma_2 = 1$, $\theta_0 = 45°$, $H_2 = 0.5$, $\theta_1 = 35°$, and $w_e = -\phi_E^{-1} \sin \pi f/f_0$. Since the flow is geostrophic these also show the trajectories of the flow. The outcrop line along which $h_1$ vanishes is shown by the horizontal dotted line. The boundary of the shadow zone is labeled $\Phi_2(\theta)$. The pool of constant potential vorticity is shaded and is bounded by the curve $\Phi_w(\theta)$. (b) The contours of $h + \gamma_1/\gamma_2 \, h_1$ which are the trajectories of the geostrophic flow in layer 1 are shown. The dotted curve is $\Phi_2(\theta)$. Note the abrupt change in the trajectories of the flow across $\Phi_2$.

2 can be ventilated by fluid which has felt the free surface only in the region west of the critical trajectory $\Phi_2(\theta)$. East of this critical line there is fluid in a zone, called the *shadow zone*, which cannot be refreshed by fluid flowing from the outcrop line.

The eastern boundary of this region in layer 2 is $\phi_E$ across which no geostrophic flow is allowed. Its western boundary is $\Phi_2(\theta)$ which, being a streamline also allows no fluid to cross it. Since $\Phi_2$ meets $\phi_E$ at $\theta = \theta_1$, the fluid in the shadow zone is completely unforced by Ekman pumping either directly or indirectly. Hence eastward of $\phi = \Phi_2$ only the fluid in the upper layer can be in motion and in this region it must carry the entire Sverdrup transport. Thus for $\theta < \theta_1$ and $\phi_E < \phi < \Phi_2(\theta)$, $u_2$ and $v_2$ vanish and (6.22.21) becomes

$$\cos\theta v_1 h_1 = \sin\theta w_e, \tag{6.22.44}$$

or, since $h$ must be constant in this region, we may obtain directly from (6.22.40)

$$h_1 = \left(\frac{\gamma_2}{\gamma_1}\right)^{1/2} D_0(\phi, \theta), \tag{6.22.45}$$

since as argued before $H_1$ must be zero, i.e., $h_1$ must vanish on $\phi = \phi_E$. Thus for $\phi < \Phi_2$ (6.22.37a,b) and (6.22.41) determines the solution for the isopycnal surface while for $\phi > \Phi_2$ the solution is given by (6.22.45) and the condition $h = H_2$. It is left to the reader to show that $h_1$ and $h_2$ are continuous across the critical trajectory $\phi = \Phi_2(\theta)$. However derivatives of $h_1$ and $h_2$ perpendicular to the critical trajectory are discontinuous. This must be anticipated in an ideal fluid model that allows discontinuities in velocities *parallel* to region boundaries.

As long as $\sin^2\theta w_e$ goes to zero as the equator is approached, $D_0^2$ must vanish, for fixed $\phi$, as $\theta \to 0$. This implies that (6.22.43) will require $\phi_E - \Phi_2$ to become large, to compensate for the smallness of $\sin^2\theta w_e$ as $\theta$ becomes small. Hence the shadow zone boundary must sweep across the ocean and strike the western boundary at some point north of the equator.

Figure 6.22.3(a) shows the contours of $h$ in the lower layer for the case $H_2 = 0.5$ (i.e., the dimensional depth of the gyre on the northern boundary $\theta = \theta_0$ has been chosen to be one half the advective depth $\delta_a$), and for simplicity, the case $\gamma_1 = \gamma_2 = 1$ has been shown. The domain of the region shown in the figure is $0 \le \theta \le \theta_0$ while $0 \le \phi \le \phi_E$. The Ekman pumping velocity $w_e$ chosen as

$$w_e = -\frac{\sin\pi f/f_0}{\phi_E},$$

so that

$$D_0^2 = (1 - \phi)\sin^2\theta \sin\pi f/f_0.$$

where $f/f_0 = \sin\theta/\sin\theta_0$. For the example shown $\theta_0$ has been chosen to be $45°$.

As shown in the figure the fluid in layer 2 flows in a clockwise gyre. Over a relatively small portion of its journey the fluid is directly forced by the Ekman

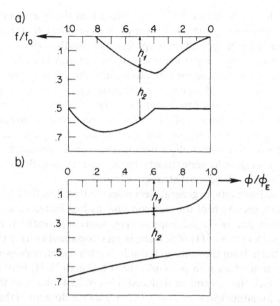

**Figure 6.22.4** (a) The configuration of the layer thicknesses for the circulation shown in Figure 6.22.3 along the line $\phi = 0.5$. Note that $h$ is flat in the shadow zone. (b) As in (a) along the line $\sin \theta / \sin \theta_0 = 0.5$, i.e., $\theta \sim 20°$. Note the shadow zone where $h$ is flat and $h_1$ varies rapidly with longitude.

pumping while after it crossed the outcrop line it is shielded from the Ekman pumping and preserves its potential vorticity. Hence in the region south of $\theta_1$ but north of the shadow zone boundary the flow lines are also contours of constant potential vorticity and it is important to note how altered the contours have become when compared to lines of constant latitude.

Figure 6.22.3(b) shows the trajectories of the flow in the upper layer, i.e., contours of $h + \gamma_1/\gamma_2 h_1$. The trajectories cross the line $\Phi_2(\theta)$ since in the shadow zone, layer 1 must carry the entire Sverdrup transport. The contours in the figure are, of course, only the projection of the paths of the flow on the horizontal plane. Note that there is a discontinuity in direction of the upper layer flow across $\Phi_2(\theta)$. Although the layer thicknesses are continuous across $\Phi_2$, their normal gradients are not. Most of the trajectories in layer 1 have their point of origin at the outcrop line rather than at a point in the western boundary. This implies that most of the fluid in layer 1 (i.e., the fluid between the 0.5 and the 0.8 contours of $h + \gamma_1/\gamma_2 h_1$ comes directly from the upper Ekman layer. Fluid in the region west of the 0.8 contour must contain some fluid coming from the western boundary current region which, rather than being ventilated from the Ekman layer, continuously recirculates through the gyre and western boundary current.

Figure 6.22.4(a), (b) shows cross sections which demonstrate the shaped imposed on the isopycnal surfaces by the circulation. The first, (6.22.4a) shows a north–south section at $\phi = 0.5$. Note that the overall depth of the thermocline, $h$, deepens south of $\theta = \theta_0$ and then rises back to its initial depth, $H_2$, as the shadow zone boundary is approached. South of that point $h$ is flat. The depth of the upper layer vanishes at the outcrop latitude, reaches its deepest point at the shadow zone boundary and then rises to the surfaces as the equator is approached. Naturally, the geostrophic dynamics used in the theory is not valid near the equator and the region of validity of the solution must exclude a narrow zone of at least a few degrees about the equator. Figure 6.22.4(b) shows the layer configuration as a function of longitude at $f/f_0 = 0.5$. Again, $h$ becomes flat east of $\Phi_2$ where layer 1 carries the entire transport. Note that in this region $h_1$ varies most strongly and it is here that the potential vorticity in layer 2 will be most strongly variable.

The potential vorticity, in general, decreases westward, as the layers thicken, and also decreases south of the outcrop line, as $|w_e|$ increases in size. Thus near the western part of the subtropical gyre, pools of potential vorticity may be obtained with values of $\Pi_2$ which are, in fact, too small to have their origin in fluid subducted from the outcrop latitude. Such a region is shown in Figure 6.22.3(a) as a shaded area. In principle, the solution (6.22.41) may be applied there but, in fact, the streamline deduced from (6.22.41) have as their origin along $\theta = \theta_1$ longitudes *west* of $\phi = 0$, i.e., *outside* the domain of the problem. Hence it is not legitimate to apply the ventilated solution to this domain since it contains fluid which is not subject to (6.12.35). To find the boundary of this region, we need to consider the critical curve $\Phi_w(\theta)$ which is the trajectory of the flow which emanates from the western line. Since it is a line of constant $h$, it may be determined from (6.22.42) by setting $\phi_0 = 0$, i.e., $\Phi_w(\theta)$ as determined from the relation

$$D_0^2(\Phi_w, \theta) = D_0^2(0, \theta)\left[1 + \frac{\gamma_1}{\gamma_2}\left(1 - \frac{f}{f_1}\right)^2\right] + H_2^2\frac{\gamma_1}{\gamma_2}\left(1 - \frac{f}{f_1}\right)^2. \quad (6.22.46)$$

In the vicinity of $\theta_1$ the terms $(1 - f/f_1)^2$ are negligible. If $D_0^2$ increases southward, at fixed longitude, due to an increase in the Ekman downwelling, $\phi_E - \Phi_w$ must decrease to compensate and $\Phi_w(\theta)$ must bend eastward. As $(1 - f/f_1)^2$ increases and $|w_e|\sin^2\theta$ decreases, inevitably $\Phi_w(\theta)$ will turn westward carving out an isolated zone near the western boundary. This zone will exist as long as the outcrop line lies north of the maximum of $|w_e|\sin^2\theta$.

Within this pool the solution is not fixed by conditions at the subduction latitude. Following the suggestion by Rhines and Young (1982), Luyten, Pedlosky, and Stommel (1983) assumed that within the pool potential vorticity would be mixed to uniformity by even the smallest degree of turbulent eddy stirring. The rationale for this choice is developed in Section 6.23. For now we only wish to apply the idea to determine the solution in the western pool. The constant region is chosen to be equal to the potential vorticity on its boundary contour and this is known since it must equal

$$\frac{\sin\theta_1}{h_2(0, \theta_1)} = \frac{\sin\theta_1}{(D_0^2(0, \theta_1) + H_2^2)^{1/2}}, \quad (6.22.47)$$

so that within the pool

$$\frac{\sin \theta}{h_2} = \frac{\sin \theta_1}{h_2(0, \theta_1)}. \tag{6.22.48}$$

With $h_2$ determined by (6.22.48), $h_1$ can be determined through application of the Sverdrup condition (6.22.21). It can be shown that the layer depths so determined match continuously the layer depths in the ventilated region across the pool boundary.

The physical model of the thermocline structure described by the layer model is relatively speaking, quite crude. Greater vertical resolution can, of course, be achieved by adding additional layers. Nevertheless, the most striking characteristic of the solution apparent even here is its decidedly *non-self-similar character*. The simplicity of the model allows us to explicitly consider the role of the meridional boundaries and more arbitrary boundary conditions and we see that the character of the solution changes fundamentally across internally generated critical lines. Even in this simple model, restricted to the subtropical gyre there appear three separate zones; the shadow zone, the region of ventilation, and the pool of constant potential vorticity. In each of these zones the structure of the solution is different.

## 6.23 Flow in Unventilated Layers: Potential Vorticity Homogenization

The thermocline model described in the previous section assumed that fluid in layers which were not at some point exposed to Ekman pumping would remain at rest. This is clearly a consistent assumption when there is no frictional coupling or mass exchange between the layers. However, we must ask whether the theory *requires* such layers to be at rest and whether alternative solutions are possible. Stated another way, we must examine with some care the nondissipative limit of the thermocline equations to see if the limit is regular or singular, i.e., whether motions produced as a consequence of frictional coupling truly vanish as the dissipation itself goes to zero.

Consider once again the model pictured in Figure 6.22.2. If the fluid in all layers beneath layer 2 are at rest, then the base of each of these layers must be flat. For layers 4, 5, 6, ..., etc., the potential vorticity in each layer will then be simply $\sin \theta / H_n$ where $H_n$ is the constant thickness of the layer on the eastern boundary. Hence in each of these layers, the isolines of potential vorticity strike the eastern boundary. Since potential vorticity is conserved to within the order of the small dissipation, the flow in these layers, were it to exist, must essentially lie *along* latitude circles. However, the presence of the eastern boundary on which the interior flow must satisfy the condition of no normal flow effectively blocks such a zonal flow in turn forcing the fluid in the layer to remain everywhere at rest. This will be true of each layer which is covered by a resting layer.

Now consider layer 3, which we have assumed is also at rest. It is covered by layer 2 which is in motion so that the base of layer 2 and hence the thickness of layer 3 is variable. The potential vorticity layer 3, when layer 3 is motionless, is

$$\Pi_3 = \frac{\sin \theta}{h_3}. \tag{6.23.1}$$

Consider the region north of the outcrop latitude where layer 2 is the only layer supposed in motion. Then

$$h_3 = H_2 + H_3 - h_2(\phi, \theta), \tag{6.23.2}$$

where $H_2$ and $H_3$ are the constant thicknesses of layers 2 and 3 on the eastern and northern boundaries. The layer thickness, $h_2$, is given by (6.22.28) so that

$$\Pi_3 = \frac{\sin \theta}{h_3} = \frac{\sin \theta}{[H_2 + H_3 - (D_0^2 + H_2^2)^{1/2}]}. \tag{6.23.3}$$

Now near the eastern boundary $D_0^2 \to 0$ and $\Pi_3$ approaches $\sin \theta / H_3$. Consider the contour of constant potential vorticity which emanates from the point $\theta_*$ on the eastern wall. Solving (6.23.3) for $D_0^2$ along such a contour, on which $\Pi_3 = \sin \theta_* / H_3$, yields

$$D_0^2(\phi, \theta) = 2H_2 H_3 \left(1 - \frac{f}{f_*}\right) + H_3^2 \left(1 - \frac{f}{f_*}\right)^2, \tag{6.23.4}$$

where, as before, $f/f_* = \sin \theta / \sin \theta_*$. The relation for $D_0^2$ is an implicit equation for the potential vorticity isoline in layer 3 which intersects the eastern wall at $\theta = \theta_*$. Note that as the forcing goes to zero, $D_0^2$ vanishes and the solution of (6.23.4) is $\theta = \theta_*$, i.e., the isolines of potential vorticity become latitude circles. In fact, the presence of the circulation in layer 2 can considerably distort the isopleths of $\Pi_3$ from latitude circles. Figure 6.23.1 shows the isopleths of $\Pi_3$ for the case where $D_0^2$ is given by

$$D_0^2 = H_2^2(1 - \phi/\phi_E) \sin \pi \frac{f}{f_0} \tag{6.23.5}$$

and where, for simplicity, layer 1 has been removed (by letting $\theta_1 \to 0$) so that layer 2 is the only moving layer. This allows (6.23.3) to apply everywhere. In the eastern and southern parts of the basin where $D_0^2$ is small, the potential vorticity isolines, while significantly distorted, still strike the eastern boundary. Hence flow along these contours must be zero. The figure shows though that not all the contours strike the eastern boundary. To see how this happens let us examine (6.23.4) more carefully. As $f_* \to f_0$, i.e., for isolines emanating from latitudes very near the northern wall, the right-hand side, of course, vanishes for all $\phi$ as $f \to f_*$. However, since $D_0^2$ (i.e., $w_e$) vanishes for all $\phi$ as $f \to f_0$, the left-hand side also exactly vanishes as $f \to f_* \to f_0$, placing no constraint on $\phi$. Suppose $w_e$ is a function only of $\phi$, then (6.23.4) may be rewritten

$$\phi_E - \phi = -\frac{\gamma_2}{\sin^2 \theta w_e(\theta)} \left[ 2H_2 H_3 \left(1 - \frac{f}{f_*}\right) + H_3^2 \left(1 - \frac{f}{f_*}\right)^2 \right]. \tag{6.23.6}$$

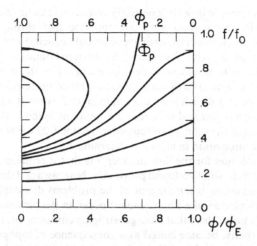

**Figure 6.23.1** Isolines of potential vorticity in the unventilated layer beneath the moving surface layer. $\Phi_p(\theta)$ is the critical curve separating isolines of potential vorticity which thread back to the eastern boundary from those that fold back to strike the western boundary. $\phi_p$ is the intersection of $\Phi_p$ with the zero wind stress curl line.

Now as both $\theta_*$ and $\theta$ approach $\theta_0$, the ratio only becomes defined by L'Hopital's rule. Hence in that limit

$$\phi_E - \phi_* = \frac{2\gamma_2 H_2 H_3}{(\sin\theta_0)^3 \left(\dfrac{\partial w_e}{\partial \sin\theta}\right)_{\theta=\theta_0}},$$ (6.23.7a)

or

$$\phi_E - \phi_* = \frac{2\gamma_2 H_2 H_3}{(\sin\theta_0)^3 \left(\dfrac{\partial w_e}{\partial \theta}\right)_{\theta=\theta_0}} \cos\theta_0,$$ (6.23.7b)

which is shown in Figure 6.23.1. The longitude $\phi_{p*}$ marks the interaction of the critical potential vorticity contour $\Phi_p(\theta)$ which strikes the *northern* rather than the eastern boundary. Along $\Phi_p$, $\Pi_3 = \sin\theta_0/H_3$. Since this is also the potential vorticity along the latitude circle $\theta = \theta_0$, the curve $\Phi_p$ can be thought of as continuing westward along $\theta_0$ to the western boundary. Within the pool region defined by $\Phi_p(\theta)$, there exist additional contours of constant $\Pi_3$. For these contours, which originate from the western wall, (6.23.3) can be solved for $D_0^2$ for a given $\Pi_3 = \sin\theta_0/(H_2 + H_3 - h_{2w})$. $h_{2w}$ is the value of $h_2$ on the western wall at $\theta = \theta_*$, i.e.,

$$h_{2w} = (D_0^2(0, \theta_*) + H_2^2)^{1/2}.$$

Two of these contours are shown in the figure within the sector carved out by $\Phi_p(\theta)$.

For all contours within the outermost contour, $\Phi_p(\theta)$, there is no prohibition against geostrophic flow. It is perfectly conceivable to imagine a dissipative free flow circulating about such contours and then through a western boundary current (if it does not dissipate potential vorticity) and rejoining its contour to complete a circuit. Whether such a western boundary current exists to allow the circuit to be closed is another matter. For the moment, let us concentrate on the issue of how such a free flow could be determined. This is an important issue for if such a circulation in layer 3 exists, it will carry part of the total Sverdrup flow and thus alter the solutions found in the preceding section. In this sense the solutions discussed in Section 6.22 are not completely unique until the circulation in layer 3 is determined.

Clearly one solution for the flow in layer 3 is that it remains zero. Hence the consistency of the solution found previously. Now we ask if there are other solutions. The situation is reminiscent of the problems discussed in Section 5.13, e.g., the circulation of a homogeneous ocean in the presence of bottom topography. We noted there that if the geostrophic contours, i.e., the isopleths of potential vorticity, became closed as a consequence of topography, large circulations could be produced along the contours of potential vorticity by $O(1)$ forcing. In the case at hand the "topography" is due to the distortion of the base of layer 2. Now the forcing is hypothesized to be weak and due to interlayer coupling due to friction, and we are asking whether a weak forcing will produce an $O(1)$ response. In Section 5.13 the along-isopleth flow was found by integrating the potential vorticity equation around the closed contour of potential vorticity and just such an approach will be useful in the present case.

Following Rhines and Young (1982a,b), we first consider the idealized case where the outermost contour $\Phi_p(\theta)$ closes on itself entirely within the oceanic interior. This requires $w_e$ to be positive as well as negative, but this artifice allows us first to consider the determination of the flow within $\Phi_p(\theta)$ without consideration of boundary layer dynamics. Thus consider the contour $C_3$ as shown in Figure 6.23.2 formed by the girdling isopleth of $\Pi_3$ determined by $\Phi_p(\theta)$.

We next consider the potential vorticity equation for steady flow and we append to it the result of adding a small dissipative term associated with the flux of potential vorticity by small scale eddy processes, i.e., we rewrite (6.22.15) as

$$\mathbf{u}_{n_H} \cdot \nabla \left( \frac{\sin \theta}{h_n} \right) = \frac{\sin \theta}{h_n^2} (U_{n-1} - U_n) + \frac{\mathscr{D}_n}{h_n}, \qquad (6.23.8)$$

where $\mathscr{D}_n$ represents a small-scale dissipative process acting on the vorticity. If (6.23.8) is multiplied by $h_n$ and use is made of (6.22.9), we obtain for layer 3

$$\nabla \cdot \{\mathbf{u}_{3H} h_3 \Pi_3\} = \mathscr{D}_3. \qquad (6.23.9)$$

Although $\mathscr{D}_3$ represents terms which are locally small and locally negligible, it becomes significant when (6.23.9) is integrated over the area, $A_3$, bounded by $C_3$ for then the advective terms precisely cancel. The reason for this cancellation is that for small dissipation $C_3$ is a streamline of the flow as well

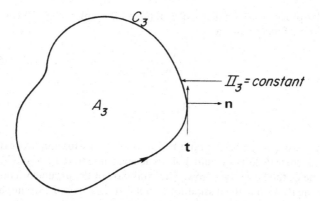

**Figure 6.23.2**  The domain $A_3$ in layer 3 bounded by an isoline of $\Pi_3$ in which potential vorticity is homogenized.

as an isoline of $\Pi_3$. Thus, as in Section 5.9, the locally advective terms integrate to zero leaving

$$\iint_{A_3} dx \, dy \, \mathcal{D}_3 = 0. \tag{6.23.10}$$

In fact, (6.23.10) must apply to *each* closed contour within $C_3$ as well as the boundary contour determined by $\Phi_p(\theta)$.

Consider now the important case where $\mathcal{D}_3$ can be represented in terms of a potential vorticity flux vector $\mathcal{F}_3$ so that

$$\mathcal{D}_3 = \nabla_H \cdot \mathcal{F}_3. \tag{6.23.11}$$

The justification for (6.23.11) essentially relies on the notion that within $A_3$ there are no true sources or sinks for potential vorticity and that the eddy mixing terms whose stirring effects are represented by $\mathcal{D}_3$ can only cooperatively transport potential vorticity from one place to another. If, in fact, this is so, (6.21.1a) becomes

$$\oint_{C_3} \mathcal{F}_3 \cdot \mathbf{n} \, dl = 0, \tag{6.23.12}$$

where $\mathbf{n}$ is the outward normal to $C_3$ and $dl$ is the line element along $C_3$.

The simplest conceivable representation for the flux vector $\mathcal{F}_3$ in terms of the mean field is the flux law

$$\mathcal{F}_3 = -\kappa \nabla_H \Pi_3, \tag{6.23.13}$$

where $\kappa$ is a mixing coefficient for potential vorticity. The circumstances where (6.23.1) might be expected to hold are discussed in detail by Rhines and Young (1982a) and are beyond the scope of the present treatment. However, the representation (6.23.13) fundamentally requires the characteristic displacements in the eddy stirring field to be short compared to the gyre scale of motion so that the mixing of potential vorticity mimics the mixing accomplished by

molecular processes at smaller scales. If (6.23.13) is apt, then (6.23.12) reduces
to the simple flux condition

$$\oint_{C_3} \frac{\partial \Pi_3}{\partial n} \, dl = 0, \tag{6.23.14}$$

where

$$\frac{\partial \Pi_3}{\partial n} \equiv \mathbf{n} \cdot \nabla \Pi_3.$$

Assuming our hypotheses about the form of the dissipation process are, in
fact valid, then (6.23.14) would hold whenever integrated around a closed
streamline of the large-scale flow. The vital part of the argument consists in
recognizing that when the dissipation is small, (6.22.15) or (6.23.9) implies that
*locally* potential vorticity is conserved, or equivalently that locally on each
contour of $p_3$

$$\Pi_3 = \Pi_3(p_3), \tag{6.23.15}$$

i.e., that the isolines of $\Pi_3$ are $p_3$ coincide and so $\Pi_3$ is a function of only $p_3$.
For a ventilated layer the functional form of $\Pi_3(p_3)$ is determined by the
process of subduction as described in Section 6.22. For an unventilated layer,
the function $\Pi_3(p_3)$ is irrelevant unless the fluid is in motion, i.e., within $C_3$.
If (6.23.15) is used in (6.23.14) and we recall that $p_3$ and hence $\Pi_3$ is constant
on $C_3$, we obtain

$$\frac{\partial \Pi_3}{\partial p_3} \oint_{C_3} \kappa \mathbf{u}_{3H} \cdot \mathbf{t} \, dl = 0, \tag{6.23.16}$$

where $\mathbf{t}$ is the unit tangent vector, $\hat{\mathbf{k}} \times \mathbf{n}$. Unless the integral in (6.23.16)
vanishes (and this is unlikely since, for constant $\kappa$, the integral is proportional
to the fluid circulation around $C_3$) it must follow that

$$\frac{\partial \Pi_3}{\partial p_3} = 0 \tag{6.23.17}$$

on $C_3$ and by repetition of the argument, on each closed contour within $C_3$.
Thus, if the dissipation potential vorticity is indeed representable in terms of
a downgradient, flux of mean potential vorticity, the effect of weak mixing is
to render the potential vorticity uniform, independent of streamline, *i.e., to
homogenize the potential vorticity within such closed contours.* Then $\Pi_3$ must
take on the value of potential vorticity which occurs on $C_3$ as specified by the
flow field external to the circuit.

It is very important to note that the process of homogenization depends
crucially on the *smallness* of the dissipation. If the dissipation were large and
locally dominated the advection, the effect of nonuniform distributions of
potential vorticity on the gyre boundaries would diffuse throughout the region
leading to a nonuniform distribution of $\Pi_3$. However, when the dissipation
is weak, advection can first encircle a domain with a streamline along which
$\Pi_3$ is very nearly constant and then the effects of dissipation, however weak,

can gradually homogenize the domain encircled by the flow. Dominant advection is required to form a region bounded by uniform potential vorticity in which the dissipation can inexorably act.

For the problem originally under discussion, the situation is even more complex since the circuit cannot be closed without at least a portion of it entering the western boundary current. For a barotropic flow, as discussed in Chapter 5, the western boundary current is generally a zone of significant vorticity dissipation. It is less clear that this must be true layer by layer in a baroclinic model. Conceivably, if the western boundary current and its extension is sufficiently inertial, the arguments presented above can be extended to contours threading through such a current. It must be admitted that the justification for this hypothesis remains largely heuristic at this point although there is evidence in numerical experiments (see Rhines and Young (1982b), for a discussion of this point) to support it.

If these generalized homogenization arguments are accepted as a reasonable premise, the flow in layer 3 may be easily calculated. Thus, outside $\Phi_p$, i.e., to the east and south of the critical contour, layer 3 is at rest. Within $\Phi_p$, layer 3 is in motion and the fluid in layer 3 *within* $\Phi_p$ has uniform potential vorticity. The value of $\Pi_3$ must be the magnitude of $\Pi_3$ on $\Phi_p$. Since $\Phi_p$ intersects the northern boundary of the gyre where $h_3 = H_3$ it follows that

$$\Pi_3 = \frac{\sin \theta}{h_3} = \frac{\sin \theta_0}{H_3}, \tag{6.23.18}$$

or that everywhere within $\Phi_p$,

$$h_3 = \frac{f}{f_0} H_3. \tag{6.23.19}$$

Thus, within $\Phi_p$, the thickness of layer 3 is a function only of latitude.

With $h_3$ known, $h_2$ can be determined from the Sverdrup relation (6.22.21), i.e.,

$$\cos \theta [v_2 h_2 + v_3 h_3] = \sin \theta w_e. \tag{6.23.20}$$

However,

$$v_3 = \frac{\gamma_3}{\sin \theta \cos \theta} \frac{\partial h}{\partial \phi}, \tag{6.23.21a}$$

$$v_2 = \frac{\gamma_3}{\sin \theta \cos \theta} \frac{\partial}{\partial \phi} \left( h + \frac{\gamma_2}{\gamma_3} h_2 \right), \tag{6.23.21b}$$

where $h = h_2 + h_3$. Thus (6.23.20) becomes

$$\frac{\partial}{\partial \phi} \left[ h^2 + h_2^2 \frac{\gamma_2}{\gamma_3} \right] = \frac{\gamma_2}{\gamma_3} \frac{\partial D_0^2}{\partial \phi}, \tag{6.23.22}$$

or

$$h^2 + h_2^2 \frac{\gamma_2}{\gamma_3} = \frac{\gamma_2}{\gamma_3} D_0^2 + H^2 + H_2^2 \frac{\gamma_2}{\gamma_3}, \tag{6.23.23}$$

where $H$ and $H_2$ are the constant values of $h$ and $h_2$ on $\phi = \phi_E$. Since

$$h = h_2 + h_3 = h_2 + \frac{f}{f_0} H_3, \tag{6.23.24}$$

(6.24.23) may easily be solved for $h_2$. Note that on $C_3$, $h = H$ and continuity of $h_2$ and $h_3$ with the values of $h_2$ and $h_3$ external to $C_3$ requires that on $C_3$

$$h_2 = H_2 + H_3 \left( 1 - \frac{f}{f_0} \right),$$

which when used in (6.24.33) recovers (6.23.4). Hence continuity of layer depths and thickness is assured across $C_3$.

Figure 6.23.3 shows the pattern of the flow for the wind forcing given by (6.23.5) and for $\gamma_2/\gamma_3 = 1.25$. In (a) the upper layer thickness, when *only* the upper layer is in motion, is shown and the thickness contours are coincident with fluid trajectories. In this case there is no flow in layer 3. In (b) the upper layer thickness is shown for the case where $H_3/H_2 = 0.5$ and flow occurs in layer 3 within $\Phi_p$ and the moving fluid in layer 3 has uniform potential vorticity. Qualitatively the thickness pattern in the upper layer is the same as (a) and only a relatively small diminution in the thickness gradient can be perceived. In (c) the thickness of the lower layer is presented when layer 3 is in motion. When layer 3 is at rest, the thickness pattern of layer 3 is, of course, a mirror of that in layer 2. However, when layer 3 is in motion with constant potential vorticity, the thickness of layer 3 within the homogenized pool of potential vorticity is independent of longitude and the extent of the zone of homogenization can be seen as the boundary of region of longitude independent of $h_3$. Figure 6.23.3(d) shows the upper layer flow paths when layer 3 is moving while Figure 6.23.3(e) shows the pattern of flow in the lower layer within the pool defined by $\Phi_p$.

When $H_3$ is increased, the relative distortion of the isolines of $\Pi_3$ becomes smaller and $\phi_{p*}$ moves westward (6.23.7) and the pool of homogenized potential vorticity shrinks. Figure 6.23.3(f) shows the domain of flow in layer 3 when $H_3$ is doubled so that $H_3 = H_2$. A similar effect, e.g., a shrinking of the pool occurs when the Ekman pumping is reduced.

Now that layer 3 has been set into motion, it becomes necessary to reexamine the justification for the necessity for layer 4 to be at rest. Thus within $\Phi_p(\theta)$ the thickness of layer 4 is no longer uniform and the arguments applied to layer 3 can now be applied to layer 4. Pedlosky and Young (1983) show that a nested set of curves $\Phi_p^{(n)}$ will exist such that $\Phi_p^{(n)} < \Phi_p^{(n-1)}$ where the superscript refers to the layer label. Thus as each layer is set into motion as a

**Figure 6.23.3**  The thickness of the upper layer when (a) all unventilated layers are ▶ motionless and (b) when the uppermost unventilated layer is in motion. (c) The thickness of the uppermost unventilated layer. Its domain of motion is easily seen as the region where its thickness is independent of longitude. (d) The upper-layer [compare with (a)] and (e) the lower-layer streamline pattern. (f) The lower-layer streamline when $H_3$ is doubled over its value used in (e).

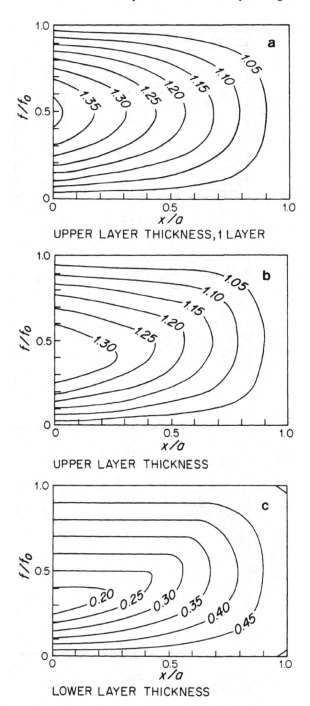

UPPER LAYER THICKNESS, 1 LAYER

UPPER LAYER THICKNESS

LOWER LAYER THICKNESS

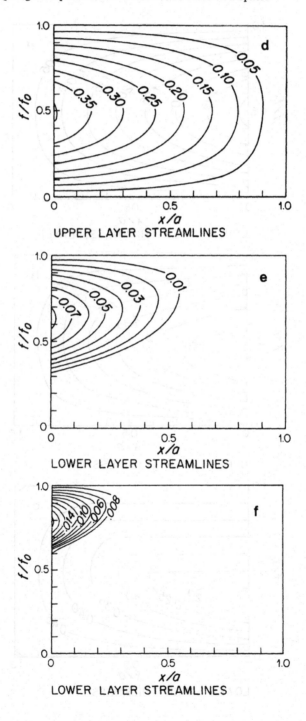

UPPER LAYER STREAMLINES

LOWER LAYER STREAMLINES

LOWER LAYER STREAMLINES

consequence of potential vorticity homogenization, the layer beneath it contains a smaller domain of closed potential vorticity contours in which motion and homogenization can occur.

Of course, it is also possible to add additional ventilated layers which outcrop within the gyre as in Section 6.22. Pedlosky and Young (1983) have presented a synthesis of the thermocline models driven by ventilation processes and the unventilated models drived by weak eddy processes and potential vorticity homogenization. The general picture that emerges is of the total Sverdrup transport at each geographical location being partitioned between flow in the deeper unventilated layers, driven by potential vorticity homogenization, endlessly recirculating in the northwest portion of the gyre and flow in shallower layers directly driven by Ekman pumping in which the potential vorticity is reset in each circuit by exposure to the atmosphere. The interaction between these layers is nonlinear since the Sverdrup relation (6.22.21), when used geostrophically, is quadratic in the layer thicknesses. Thus there is a subtle interplay between the ventilated and unventilated domains of the flow in which the character of the flow in each domain is affected by the other.

Naturally, it would be useful to re-examine the problem for a continuous rather than a layered model. This remains to be done. However, the results of the layer model forcefully demonstrate the important alterations of the flow structure between the different domains of the solution. This will undoubtedly make solutions of (6.21.1(a,b)) difficult to obtain which effectively describe the nonself-similar character of the flow as revealed by the layer models.

## 6.24 Quasigeostrophic Approximation: an Alternative Derivation

The geostrophic equations of motion on the synoptic scale and for the global scale have been derived separately in Sections 6.5 and 6.20. A peculiar feature of the derivation of the synoptic scale quasigeostrophic equations is the requirement that the buoyancy frequency be a given function independent of latitude and longitude. At the same time, the global scale equations of Section 6.20 describe how the static stability field (e.g., the vertical density gradient) varies on the global scale. There is no contradiction between these theories, they simply relate to spatial variations (or their absence) on supposedly distinct spatial scales. The synoptic scale equations describe motions on the scale of the Rossby deformation radius while the geostrophic equations of Section 6.20 describe motions of truly planetary scale.

In many cases of interest, the appropriate situation is one in which the synoptic scale motions are embedded within the larger scale, i.e., both dynamical systems are operating simultaneously. For example, in the ocean the meso-scale eddy field exists within the general oceanic circulation and the ratio of length scales of the two motions may be $O(10^{-2})$.

Formally, the derivation of the synoptic scale equations requires that the buoyancy frequency, $N^2$, be independent of horizontal coordinate. That limits the applicability of the synoptic scale equations, at least in principle, to a synoptic "patch" within the global system. A different synoptic patch might require a different, horizontally uniform $N^2$. In no case, formally, could the synoptic scale equation [e.g., (6.5.18)] be applied over, say, an entire oceanic gyre unless we accept the unrealistic limitation of constant $N^2$ over the gyre. Similarly, derivation of Section 6.5 has used the notion of a "central" latitude $\theta_0$ about which all metric terms in the dynamic equations are expanded. As a result, the final equations contains parameters, e.g., $\beta$ and $S$ which explicitly depend on the value of the "central" latitudes. Yet, the exact value of that central latitude is rather arbitrary. As in the choice of an appropriate value of $N^2$ this arbitrariness is computationally unimportant within a given synoptic patch because variations of the parameters on the synoptic scale are negligible. What the derivation of Section 6.5 lacks is a way to connect continuously the synoptic patches over global scales. The issues are probably more pressing in the oceanographic context where there is a great spatial separation of scales but even in the atmospheric case the notion of a separation between the synoptic scale and the truly planetary scale is a conceptually useful one.

In this section we will re-derive the geostrophic equation paying particular attention to the issues raised above. As an important by-product of the derivation, the connection between the equations on both scales will become clear. In fact, both will be derived *simultaneously*.

In order to keep the derivation as simple and as uncluttered as possible, our attention will be restricted to the case where dissipation may be neglected and where the density scale height is large (i.e., where formally the density itself is a conservative quantity). The necessary elaborations required to include dissipation and compressibility effects are straightforward and are left as an exercise for the reader.

The key element of the derivation is the intuitive statement that all dynamical fields may be variable on two separate scales. First on the planetary scale, i.e., on length scales comparable to the earth's radius $r_0$, and then also on the deformation radius $L_D$ which characterizes the synoptic scale. The independent horizontal coordinates which measure the global variations are the longitude $\phi$ and the latitude $\theta$. For the synoptic scale we introduce two *stretched* coordinates

$$\xi = \frac{r_0}{L}\phi, \qquad (6.24.1a)$$

$$\eta = \frac{r_0}{L}\theta, \qquad (6.24.1b)$$

to measure variations on the synoptic scale. Note that $\xi$ and $\eta$ are angles and not linear instances. $L$ is the characteristic length scale of the synoptic scale motions and is of the order of $L_D$. Our presumption is that

$$\gamma = \frac{L}{r_0} \ll 1. \qquad (6.24.2)$$

Hence an O(1) change in $\xi$ or $\eta$ is produced by a small O($\gamma$), change in $\phi$ or $\theta$, i.e., synoptic variations correspond to small global displacements.

All variables, e.g., the zonal velocity $u_*$ are considered functions of both sets of variables, i.e.,

$$u_* = Uu(\xi, \eta, \phi, \theta, z, \tau, T), \tag{6.24.3}$$

where $U$ is the characteristic horizontal velocity of the fluid. The dimensionless variables $z$, $\tau$, and $T$ are defined as

$$z = \frac{z_*}{D}, \tag{6.24.4a}$$

$$\tau = \frac{Ut_*}{L}, \tag{6.24.4b}$$

$$T = \frac{Ut_*}{r_0}. \tag{6.24.4c}$$

The two times $\tau$ and $T$ measure the evolution of each field on time scales appropriate to the synoptic and global scales, respectively.

The remaining variables are scaled in a manner similar to Section 6.2, namely

$$p_* = -\rho_0 gz + \rho_0 2\Omega ULp, \tag{6.24.5a}$$

$$\rho_* = \rho_0 + \rho_0 \frac{2\Omega UL}{gD}\rho. \tag{6.24.5b}$$

In (6.24.5a,b) $\rho_0$ is a constant. No separation into a basic $z$-dependent density field and a dynamic anomaly is assumed. Similarly, note that the pressure and density fields use $2\Omega$ rather than some prescribed value of $2\Omega \sin \theta$. No central latitude has been chosen. The essence of the method used in the derivation is the transformation of spatial derivatives now that all the fields are considered to be function of the quite distinct independent variables $(\phi, \theta, T)$ and $(\xi, \eta, \tau)$. For example, a derivative with respect to $\theta$ in the original equations, e.g., (6.2.3), must be rewritten

$$\frac{\partial}{\partial \theta} \rightarrow \frac{\partial}{\partial \theta} + \frac{r_0}{L}\frac{\partial}{\partial \eta}$$

by the chain rule and the relations (6.24.1a,b).

If (6.2.1) and (6.2.3) are rewritten, using these variables and the scaling relations of (6.24.4) and (6.24.5) are used, we obtain

$$\varepsilon\left[\frac{\partial u}{\partial \tau} + \frac{u}{\cos \theta}\frac{\partial u}{\partial \xi} + v\frac{\partial u}{\partial \eta} + w\frac{\partial u}{\partial z}\right] + \varepsilon\gamma\left[\frac{\partial u}{\partial T} + \frac{u}{\cos \theta}\frac{\partial u}{\partial \phi} + v\frac{\partial u}{\partial \theta} - uv\tan \theta\right]$$

$$- \sin \theta v = -\frac{1}{\cos \theta}\left[\frac{\partial p}{\partial \xi} + \gamma\frac{\partial p}{\partial \phi}\right] \tag{6.24.6a}$$

$$\varepsilon\left[\frac{\partial v}{\partial \tau} + \frac{u}{\cos\theta}\frac{\partial v}{\partial \xi} + v\frac{\partial v}{\partial \eta} + w\frac{\partial v}{\partial z}\right] + \varepsilon\gamma\left[\frac{\partial v}{\partial T} + \frac{u}{\cos\theta}\frac{\partial v}{\partial \phi} + v\frac{\partial v}{\partial \theta} + u^2\tan\theta\right]$$

$$+ \sin\theta u = -\left[\frac{\partial p}{\partial \eta} + \gamma\frac{\partial p}{\partial \phi}\right] \qquad (6.24.6b)$$

$$\rho = -\frac{\partial p}{\partial z} \qquad (6.24.6c)$$

$$\frac{\partial w}{\partial z} + \left(\frac{\partial v}{\partial \eta} + \frac{1}{\cos\theta}\frac{\partial u}{\partial \xi}\right) + \gamma\left(\frac{1}{\cos\theta}\frac{\partial(v\cos\theta)}{\partial\theta} + \frac{1}{\cos\theta}\frac{\partial u}{\partial \phi}\right) = 0, \quad (6.24.6d)$$

where

$$\varepsilon = \frac{U}{2\Omega L}.$$

The conservation of $\rho$ may be written

$$\varepsilon\left[\frac{\partial\rho}{\partial\tau} + \frac{u}{\cos\theta}\frac{\partial\rho}{\partial\xi} + v\frac{\partial\rho}{\partial\eta} + w\frac{\partial\rho}{\partial z}\right] + \varepsilon\gamma\left[\frac{\partial\rho}{\partial T} + \frac{u}{\cos\theta}\frac{\partial\rho}{\partial\phi} + v\frac{\partial\rho}{\partial\theta}\right] = 0.$$

$$(6.24.6e)$$

We have already made certain approximations, the justifications for which have been described in earlier sections of this chapter. For example, the hydrostatic approximation (6.24.6c) and the neglect of Coriolis accelerations proportional to $\cos\theta$ in (6.24.6a) and (6.24.6b) follow from the assumption $D/L \ll 1$. Similarly, the neglect of $w/r_0$ with respect to $\partial w/\partial z$ in (6.24.6d) and the replacement of the earth's radius $r$ by a constant value $r_0$ are justified by the fact that $D/r_0 \ll 1$.

It follows from (6.24.5b) that

$$\frac{1}{\rho_0}\frac{\partial\rho_*}{\partial z_*} = \varepsilon\frac{4\Omega^2 L^2}{gD^2}\frac{\partial\rho}{\partial z}, \qquad (6.24.7)$$

hence

$$\varepsilon\frac{\partial\rho}{\partial z} = -\frac{L_D^2}{L^2}, \qquad (6.24.8)$$

where $L_D$ is the internal Rossby deformation radius, i.e., in the present context

$$L_D = \left(-\frac{g}{\rho_0}\frac{\partial\rho_*}{\partial z_*}\frac{D^2}{4\Omega^2}\right)^{1/2}. \qquad (6.24.9)$$

However, since $L$ is $O(L_D)$, this implies, from (6.24.8) that $\rho$ must be $O(\varepsilon^{-1})$. The horizontal velocities on the other hand must be $O(1)$. This implies that density and pressure must be independent of $\xi$ and $\eta$ to lowest order or else the thermal wind equations and geostrophy would give rise to unacceptably large velocities. That is, if $\rho$ and $p$ are order $\varepsilon^{-1}$ and vary with $\xi$ and $\eta$ they would give rise to velocities of $O(\varepsilon^{-1})$ which would make the actual Rossby number $O(1)$. To satisfy the requirement that the flow be geostrophic it follows that to lowest order $\rho$ should depend only on $\theta$ and $\phi$. This suggests the following expansion for the dynamical fields:

$$\rho = \frac{1}{\varepsilon}(\rho^{(0)} + \varepsilon\rho^{(1)} + \varepsilon^2\rho^{(2)} + \cdots), \qquad \text{(a)}$$

$$p = \frac{1}{\varepsilon}(p^{(0)} + \varepsilon p^{(1)} + \varepsilon^2 p^{(2)} + \cdots), \qquad \text{(b)}$$

$$u = \frac{1}{\varepsilon}(0 + \varepsilon u^{(1)} + \varepsilon^2 u^{(2)} + \cdots), \qquad \text{(c)} \quad \text{(6.24.10)}$$

$$v = \frac{1}{\varepsilon}(0 + \varepsilon v^{(1)} + \varepsilon^2 v^{(2)} + \cdots), \qquad \text{(d)}$$

$$w = \frac{1}{\varepsilon}(0 + \varepsilon w^{(1)} + \varepsilon^2 w^{(2)} + \cdots), \qquad \text{(e)}$$

where $p^{(0)}$ and $\rho^{(0)}$ are independent of $\tau$, $\xi$, and $\eta$ while all other variables are functions of $\tau$, $\xi$, and $\eta$ as well as $T$, $\phi$, and $\theta$.

When (6.24.10a,b,c,d,e) are substituted into (6.24.6) and like orders in $\varepsilon$ are equated a series of equalities is obtained. For the moment we assume that $\gamma = O(\varepsilon)$.

At $O(1)$:

$$\sin\theta v^{(1)} = \frac{1}{\cos\theta}\frac{\partial p^{(1)}}{\partial\xi} + \frac{\gamma}{\varepsilon}\frac{1}{\cos\theta}\frac{\partial p^{(0)}}{\partial\phi}, \qquad \text{(6.24.11a)}$$

$$\sin\theta u^{(1)} = -\frac{\partial p^{(1)}}{\partial\eta} - \frac{\gamma}{\varepsilon}\frac{\partial p^{(0)}}{\partial\theta}, \qquad \text{(6.24.11b)}$$

$$\rho^{(0)} = -\frac{\partial p^{(0)}}{\partial z}, \qquad \text{(6.24.11c)}$$

$$\frac{\partial w^{(1)}}{\partial z} + \frac{\partial v^{(1)}}{\partial\eta} + \frac{1}{\cos\theta}\frac{\partial u^{(1)}}{\partial\xi} = 0, \qquad \text{(6.24.11d)}$$

while from (6.24.6e)

$$w^{(1)}\frac{\partial\rho^{(0)}}{\partial z} = 0. \qquad \text{(6.24.11e)}$$

Note that the geostrophic balance for $u^{(1)}$ and $v^{(1)}$ is formed from two distinct pressure fields. The more rapid variation of $p^{(1)}$ on the synoptic scale will be of the same order as the pressure gradient of the $p^{(0)}$ field on the larger scale. These two pressure gradients are of the same order as long as $\gamma/\varepsilon$ is $O(1)$. Now by definition

$$\frac{\gamma}{\varepsilon} = \frac{2\Omega\, L^2}{R\, U}, \qquad \text{(6.24.12)}$$

and comparison with (6.3.2) shows that this parameter is, in fact, $O(1)$ for synoptic scales as long as $\cos\theta$ is $O(1)$. Thus the pressure gradient of the large scale field combines with the synoptic scale pressure gradient to produce the lowest order velocity field. Equivalently, the lowest-order velocity field can be split into two parts, a synoptic field $u_s$ and $v_s$ which satisfies

$$\sin \theta(u_s, v_s) = \left( -\frac{\partial p^{(1)}}{\partial \eta}, \frac{1}{\cos \theta} \frac{\partial p^{(1)}}{\partial \xi} \right), \qquad (6.24.13a)$$

and a planetary scale field $(u_T, v_T)$ which satisfies

$$\sin \theta(u_T, v_T) = \frac{\gamma}{\varepsilon} \left( -\frac{\partial p^{(0)}}{\partial \theta}, \frac{1}{\cos \theta} \frac{\partial p^{(0)}}{\partial \phi} \right). \qquad (6.24.13b)$$

The density equation (6.24.11e) implies that

$$w^{(1)} = 0,$$

which is connected with (6.24.11d) since (6.24.11a,b) imply that

$$\frac{\partial v^{(1)}}{\partial \eta} + \frac{1}{\cos \theta} \frac{\partial u^{(1)}}{\partial \xi} \equiv 0,$$

as a consequence of geostrophy and the independence of $p^{(0)}$ of $\phi$ and $\theta$.

As $O(\varepsilon)$ the equations (6.24.6) yield for the momentum and continuity equation the relations:

$$\frac{\partial u^{(1)}}{\partial \tau} + \frac{u^{(1)}}{\cos \theta} \frac{\partial u^{(1)}}{\partial \xi} + v^{(1)} \frac{\partial u^{(1)}}{\partial \eta} - \sin \theta v^{(2)}$$

$$= \frac{1}{\cos \theta} \left\{ \frac{\partial p^{(2)}}{\partial \xi} + \frac{\gamma}{\varepsilon} \frac{\partial p^{(1)}}{\partial \phi} \right\}, \qquad (6.24.14a)$$

$$\frac{\partial v^{(1)}}{\partial \tau} + \frac{u^{(1)}}{\cos \theta} \frac{\partial v^{(1)}}{\partial \xi} + v^{(1)} \frac{\partial v^{(1)}}{\partial \eta} + \sin \theta u^{(2)} = -\left\{ \frac{\partial p^{(2)}}{\partial \eta} + \frac{\gamma}{\varepsilon} \frac{\partial p^{(1)}}{\partial \theta} \right\}, \qquad (6.24.14b)$$

$$\frac{\partial w^{(2)}}{\partial z} + \left\{ \frac{\partial v^{(2)}}{\partial \eta} + \frac{1}{\cos \theta} \frac{\partial u^{(2)}}{\partial \phi} \right\} + \frac{\gamma}{\varepsilon \cos \theta} \left\{ \frac{\partial}{\partial \theta} (v^{(1)} \cos \theta) + \frac{\partial u^{(1)}}{\partial \phi} \right\} = 0. \qquad (6.24.14c)$$

If (6.24.14a) is differentiated with respect to $\eta$ while (6.24.14b) is differentiated with respect to $\xi$ and multiplied by $(\cos \theta)^{-1}$, their difference yields

$$\left( \frac{\partial}{\partial t} + \frac{u^{(1)}}{\cos \theta} \frac{\partial}{\partial \xi} + v^{(1)} \frac{\partial}{\partial \eta} \right) \zeta^{(1)} + \sin \theta \left\{ \frac{1}{\cos \theta} \frac{\partial u^{(2)}}{\partial \xi} + \frac{\partial v^{(2)}}{\partial \eta} \right\}$$

$$= -\frac{\gamma}{\varepsilon} \left[ \frac{\partial^2 p^{(1)}}{\partial \xi \, \partial \theta} - \frac{\partial^2 p^{(1)}}{\partial \eta \, \partial \phi} \right], \qquad (6.24.15)$$

where

$$\zeta^{(1)} = \frac{1}{\cos \theta} \frac{\partial v^{(1)}}{\partial \xi} - \frac{\partial u^{(1)}}{\partial \eta}, \qquad (6.24.16)$$

i.e., $\zeta^{(1)}$ is the relative vorticity of the synoptic scale motion.

The continuity equation is used to eliminated the term in $u^{(2)}$ and $v^{(2)}$ in (6.24.15). We recognize from (6.24.11a,b) that

$$\frac{\sin\theta}{\cos\theta}\left[\frac{\partial v^{(1)}}{\partial\theta}\frac{\cos\theta}{}+\frac{\partial u^{(1)}}{\partial\phi}\right]+\cos\theta v^{(1)}=\frac{\gamma}{\varepsilon}\left[\frac{\partial^2 p^{(1)}}{\partial\theta\,\partial\xi}-\frac{\partial^2 p^{(1)}}{\partial\eta\,\partial\phi}\right]\Bigg/\cos\theta,$$

and thus it follows that

$$\frac{\partial\zeta^{(1)}}{\partial\tau}+\frac{u^{(1)}}{\cos\theta}\frac{\partial\zeta^{(1)}}{\partial\xi}+v^{(1)}\frac{\partial\zeta^{(1)}}{\partial\eta}+\frac{\gamma}{\varepsilon}\cos\theta v^{(1)}=\sin\theta\frac{\partial w^{(2)}}{\partial z}, \quad (6.24.17)$$

and the reader is recommended to compare (6.24.17) with (6.3.17). Note that

$$\frac{\gamma}{\varepsilon}\cos\theta=\frac{2\Omega\cos\theta L^2}{r_0}\frac{}{U}=\frac{\beta_* L^2}{U}\equiv\beta, \quad (6.24.18)$$

where $\beta_*$ is the dimensional northward gradient of planetary vorticity. In contrast to the discussion of Section 6.3 here $\beta$ is *a slowly varying function of latitude*. However, it is independent of the synoptic variables $\xi$ and $\eta$, i.e., it is *constant on the synoptic scale*.

The density equation at $O(\varepsilon)$ is

$$\frac{\partial\rho^{(1)}}{\partial\tau}+\frac{u^{(1)}}{\cos\theta}\frac{\partial\rho^{(1)}}{\partial\xi}+v^{(1)}\frac{\partial\rho^{(1)}}{\partial\eta}+w^{(2)}\frac{\partial\rho^{(0)}}{\partial z}$$

$$=-\frac{\gamma}{\varepsilon}\left[\frac{\partial\rho^{(0)}}{\partial T}+\frac{u^{(1)}}{\cos\theta}\frac{\partial\rho^{(0)}}{\partial\phi}+v^{(1)}\frac{\partial\rho^{(0)}}{\partial\theta}\right]. \quad (6.24.19)$$

We may now use (6.24.19) to solve for $w^{(2)}$ and use the result in (6.24.17). If repeated use is made of the fact that $\rho^{(0)}$ is independent of $\xi$ and $\eta$ and that, from (6.24.11a,b,c)

$$\frac{\partial u^{(1)}}{\partial z}\left[\frac{\partial\rho^{(1)}}{\partial\xi}+\frac{\gamma}{\varepsilon}\frac{\partial\rho^{(0)}}{\partial\phi}\right]\Bigg/\cos\theta+\frac{\partial v^{(1)}}{\partial z}\left[\frac{\partial\rho^{(1)}}{\partial\eta}+\frac{\gamma}{\varepsilon}\frac{\partial\rho^{(0)}}{\partial\theta}\right]=0, \quad (6.24.20)$$

it can easily be shown that (6.24.17) can be written,

$$\left[\frac{\partial}{\partial\tau}+\frac{u^{(1)}}{\cos\theta}\frac{\partial}{\partial\xi}+v^{(1)}\frac{\partial}{\partial\eta}\right]\Pi+\frac{\gamma}{\varepsilon}\cos\theta v^{(1)}$$

$$=-\frac{\gamma}{\varepsilon}\left[\frac{\partial}{\partial z}\left(\frac{\partial\rho^{(0)}}{\partial T}\Bigg/\frac{\partial\rho^{(0)}}{\partial z}\right)+\frac{u^{(1)}}{\cos\theta}\frac{\partial}{\partial z}\left(\frac{\partial\rho^{(0)}}{\partial\phi}\Bigg/\frac{\partial\rho^{(0)}}{\partial z}\right)\right. \quad (6.24.21)$$

$$\left.+v^{(1)}\frac{\partial}{\partial z}\left(\frac{\partial\rho^{(0)}}{\partial\theta}\Bigg/\frac{\partial\rho^{(0)}}{\partial z}\right)\right]\sin\theta,$$

where

$$\Pi=\zeta^{(1)}+\sin\theta\frac{\partial}{\partial z}\left(\frac{\rho^{(1)}}{\dfrac{\partial\rho^{(0)}}{\partial z}}\right) \quad (6.24.22)$$

is the quasigeostrophic potential vorticity. If we write

$$\psi = \frac{p^{(1)}}{\sin\theta},$$
(6.24.23)

we find that

$$\Pi = \left[\frac{1}{\cos^2\theta}\frac{\partial^2\psi}{\partial\xi^2} + \frac{\partial^2\psi}{\partial\eta^2} + \frac{\partial}{\partial z}\frac{1}{S}\frac{\partial\psi}{\partial z}\right],$$

where

$$S = -\frac{\sin^2\theta}{\dfrac{\partial\rho^{(0)}}{\partial z}} = \frac{f^2 L^2}{N^2 D^2}.$$
(6.24.24)

In deriving (6.24.24), we have used the fact that with an O($\varepsilon$) error it follows from (6.24.7) and (6.24.10a) that

$$\frac{1}{\rho_0}\frac{\partial\rho_*}{\partial z_*} = \frac{4\Omega^2 L^2}{gD^2}\frac{\partial\rho^{(0)}}{\partial z} + O(\varepsilon),$$
(6.24.25)

and thus

$$N^2 = -\frac{\partial\rho^{(0)}}{\partial z}\frac{4\Omega^2 L^2}{D^2}.$$

The parameter $S$, the Burger number, is the familiar measure of stratification on the synoptic scale introduced in Section 6.5. However, now $S$ is considered to be a slowly varying function of horizontal position as well as being a function of $z$. Although it is a constant with respect to the synoptic scale variables $\xi$ and $\eta$, the dependence of $f$ and $\partial\rho^{(0)}/\partial z$ on $\phi$ and $\theta$ will allow $S$ to smoothly change from one synoptic patch to another.

To complete the derivation it is necessary now to use the partitioning of the velocity field introduced in (6.24.13a,b).

Then (6.24.21) may be rewritten as

$$\left[\frac{\partial}{\partial\tau} + \frac{u_s}{\cos\theta}\frac{\partial}{\partial\xi} + v_s\frac{\partial}{\partial\eta}\right](\Pi + \beta\eta) + \frac{u_T}{\cos\theta}\frac{\partial}{\partial\xi}\Pi + v_T\frac{\partial}{\partial\eta}\Pi$$

$$+ \frac{\gamma}{\varepsilon}\left[\frac{u_s}{\cos\theta}\frac{\partial}{\partial z}\left(\frac{\partial\rho^{(0)}}{\partial\phi}\bigg/\frac{\partial\rho^{(0)}}{\partial z}\right) + v_s\frac{\partial}{\partial z}\left(\frac{\partial\rho^{(0)}}{\partial\theta}\bigg/\frac{\partial\rho^{(0)}}{\partial z}\right)\right]$$

$$= -\frac{\gamma}{\varepsilon}\left[v_T\cos\theta + \frac{\partial}{\partial z}\left(\frac{\partial\rho^{(0)}}{\partial T}\bigg/\frac{\partial\rho^{(0)}}{\partial z}\right) + \frac{u_T}{\cos\theta}\frac{\partial}{\partial z}\left(\frac{\partial\rho^{(0)}}{\partial\phi}\bigg/\frac{\partial\rho^{(0)}}{\partial z}\right)\right.$$

$$\left. + v_T\frac{\partial}{\partial z}\left(\frac{\partial\rho^{(0)}}{\partial\theta}\bigg/\frac{\partial\rho^{(0)}}{\partial z}\right)\right]\sin\theta.$$
(6.24.26)

The right-hand side of (6.24.26) depends on combinations of variables all of which are independent of $\tau$, $\xi$, and $\eta$, the synoptic scale independent variables. Now (6.24.26) can be interpreted as a budget equation for the synoptic scale potential vorticity, $\Pi + \beta \eta$. That is, the first term on the left-hand side of (6.24.26) is the rate of change of the synoptic scale potential vorticity following a fluid element moving on the synoptic scale. The right-hand side therefore represents a source term which is constant on the synoptic scale and hence would exorably add to the synoptic scale potential vorticity leading to a catastrophic increase in the synoptic scale fields. This can be avoided if the right-hand side is zero and we are forced to insist on that if our original expansion (6.24.10) is to remain valid.

To see this more formally, we note that (6.24.26) can be rewritten as

$$\frac{\partial \Pi}{\partial \tau} + \frac{\partial A}{\partial \xi} + \frac{\partial B}{\partial \eta} = -\frac{\gamma}{\varepsilon} C, \qquad (6.24.27)$$

where

$$A = \frac{u_s}{\cos \theta} \Pi + \frac{\beta p^{(1)}}{\cos \theta \sin \theta} + \frac{u_T \Pi}{\cos \theta} + \frac{\gamma}{\varepsilon} \frac{1}{\cos \theta \sin \theta} \frac{\partial}{\partial z} \left( \frac{\partial \rho^{(0)}}{\partial \theta} \bigg/ \frac{\partial \rho^{(0)}}{\partial z} \right),$$

$$B = v_s \Pi + u_T \Pi - \frac{\gamma}{\varepsilon} \frac{1}{\cos \theta \sin \theta} \frac{\partial}{\partial z} \left( \frac{\partial \rho^{(0)}}{\partial \theta} \bigg/ \frac{\partial \rho^{(0)}}{\partial z} \right),$$

while $\gamma / \varepsilon \ C$ is the right-hand side of (6.24.26).

Now imagine that (6.24.27) is integrated over an area that is large compared to the synoptic domain but small compared to the global scale. In dimensionful units this area is $L^2 A_0$ where $A_0$ is a large number but where

$$L^2 A_0 \ll r_0^2.$$

Then, since the second and third terms in (6.24.27) combine to form the divergence of the vector $(A, B)$, the size of the integral of these terms is of the order of the length of the perimeter of $A_0$, i.e., $O(A_0^{1/2})$ while the size of the integral of the remaining terms is $O(A_0)$.

Thus to $O(A_0^{-1/2})$

$$\frac{\partial}{\partial t} \iint_{A_0} \Pi \ d\xi \ d\eta = \frac{\gamma}{\varepsilon} A_0 C(\theta, \phi, T, z). \qquad (6.24.28)$$

If $C \neq 0$ the area average of $\Pi$ (and hence of $\rho^{(1)}$) will increase like $\gamma \tau / \varepsilon$, so that in a time of $O(\gamma^{-1}) \rho^{(1)}$ would be as large as $\rho^{(0)}$ in contradiction to the expansion (6.24.10). Thus a condition for the validity of the expansion leading to the development of the synoptic scale quasigeostrophic potential vorticity equation is the side condition $C = 0$ or equivalently,

$$\frac{\partial}{\partial z} \left( \frac{\partial \rho^{(0)}}{\partial T} \bigg/ \frac{\partial \rho^{(0)}}{\partial z} \right) + \frac{u_T}{\cos \theta} \frac{\partial}{\partial z} \left( \frac{\partial \rho^{(0)}}{\partial \phi} \bigg/ \frac{\partial \rho^{(0)}}{\partial z} \right)$$

$$+ v_T \frac{\partial}{\partial z} \left( \frac{\partial \rho^{(0)}}{\partial \theta} \bigg/ \frac{\partial \rho^{(0)}}{\partial z} \right) + v_T \frac{\cos \theta}{\sin \theta} = 0, \qquad (6.24.29)$$

which becomes with the hydrostatic relation between $\rho^{(0)}$ and $p^{(0)}$ and the geostrophic relation between $(u_T, v_T)$ and $p^{(0)}$, a partial differential equation in $p^{(0)}$ alone. The reader may wish to verify the equivalence of this equation with (6.21.1) and (6.21.2) for the case $\lambda = 0$. The solution of this equation yields $\rho^{(0)}(\phi, \theta, z, T)$ which is a required parameter in the synoptic scale equations, i.e., the solution of (6.24.29) determines the synoptic scale static stability.

The vertical velocity may also be partitioned between synoptic and global scales with the use of (6.24.19). If the synoptic scales vertical velocity is defined so that

$$w = w_s + \frac{\gamma}{\varepsilon} w_T, \tag{6.24.30a}$$

such that

$$\lim_{A_0 \to \infty}, \quad \frac{1}{A_0} \iint w_s \, d\xi \, d\eta \to 0, \tag{6.24.30b}$$

then

$$w_T = \left[ -\frac{\partial \rho^{(0)}}{\partial T} + \frac{u_T}{\cos \theta} \frac{\partial \rho^{(0)}}{\partial \phi} + v_T \frac{\partial \rho^{(0)}}{\partial \theta} \right] \bigg/ \frac{\partial \rho^{(0)}}{\partial z}. \tag{6.24.31}$$

Then (6.24.31) together with (6.24.29) yields

$$\left[ \frac{\partial}{\partial \tau} + \frac{u_T}{\cos \theta} \frac{\partial}{\partial \phi} + v_T \frac{\partial}{\partial \theta} + w_T \frac{\partial}{\partial z} \right] \left( \frac{\partial \rho^{(0)}}{\partial z} \sin \theta \right) = 0, \tag{6.24.32}$$

which we recognize [see (6.20.19a)] as the statement of potential vorticity conservation on the global scale. Thus this global scale conservation statement is a by-product of the requirement of a consistent potential vorticity balance on the synoptic scale.

Now that the consistency condition, $C = 0$, has been met leading to the derivation of the global scale geostrophic equations, we return to the synoptic scale equation which may be written in the form

$$\left( \frac{\partial}{\partial \tau} + \frac{\partial \psi}{\partial x} \frac{\partial}{\partial y} - \frac{\partial \psi}{\partial y} \frac{\partial}{\partial x} \right) \left[ \frac{\partial^2 \psi}{\partial x^2} + \frac{\partial^2 \psi}{\partial y^2} + \frac{\partial}{\partial z} \frac{1}{S} \frac{\partial \psi}{\partial z} \right]$$

$$+ \left( u_T \frac{\partial}{\partial x} + v_T \frac{\partial}{\partial y} \right) \left[ \frac{\partial^2 \psi}{\partial x^2} + \frac{\partial \psi}{\partial y^2} + \frac{\partial}{\partial z} \frac{1}{S} \frac{\partial \psi}{\partial z} \right] \tag{6.24.33}$$

$$+ \frac{\partial \psi}{\partial x} \left[ \beta - \frac{\partial}{\partial z} \frac{1}{S} \frac{\partial u_T}{\partial z} \right] - \frac{\partial \psi}{\partial y} \left[ \frac{\partial}{\partial z} \frac{1}{S} \frac{\partial v_T}{\partial z} \right] = 0.$$

The new variables

$$x = \xi \cos \theta,$$
$$y = \eta, \tag{6.24.34}$$

have been introduced. On the synoptic scale $\cos \theta$ is a constant so that (6.24.34) represents a simple linear transformation. However, at this stage the introduction of the Cartesian $\beta$-plane approximation is complete without the necessity of expanding the metric terms about a "central" latitude.

In the limit $\gamma/\varepsilon \to 0$, $u_T$ and $v_T$ go to zero and (6.24.33) becomes identical to (6.5.21) (when $\rho_s$ is constant as we have assumed here) or to (6.8.11). When $\gamma/\varepsilon$ is nonzero, there are two additional terms present. The first additional term represents the advection of the synoptic scale potential vorticity by the large scale velocity field $u_T$ and $v_T$ while the last grouping of terms in (6.24.33) represents the advection of potential vorticity with gradients on the global scale by the synoptic scale velocity. In fact (6.24.33) can be derived directly from (6.8.11) recognizing that now the total stream function is the sum of the synoptic stream function, $\psi$, plus a portion representing a flow whose velocity is independent of $x$, $y$ and depends only on $z$, i.e.,

$$\psi \to \psi - u_T(z)y + v_T(z)x,$$

which when substituted into (6.8.11) yields (6.24.33). The $z$-variation of $u_T$ and $v_T$ gives rise to a global scale potential vorticity gradient.

The use of the formalism of multiple space scales has allowed us to re-derive the geostrophic equations of motion on both the synoptic and global scales without many of the artificial restrictions apparently placed on the synoptic scale. In particular, although the validity of the synoptic scale equations requires that the scale of variation of the synoptic fields be small compared to the scale of variation of $f$ and $N$ the present derivation illustrates that the geographic *extent* of the synoptic domain, in fact, need not be small.

# ıstability Theory

## 7.1 Introduction

Solar heating is the ultimate energy source for the motion of both the atmosphere and the oceans with the exception of the lunar forcing of the tides. The radiant energy emitted by the sun may vary somewhat over very long periods, but a sensible idealization for most meteorological and ocean-ographic purposes consists in considering the solar source strength as fixed. Temporal variations in the incident radiation (and its spatial distribution) are then fixed by the astronomical relation between the positions of the earth and sun, e.g., by the seasonal progress of the earth in its solar orbit. Quite clearly, though, the motions of both the atmosphere and the oceans exhibit fluctuations whose time scales are *not* directly related to the astronomical periodicities of the earth–sun system. The phenomenon of *weather* in the atmosphere is in fact nothing more than the existence of large-scale wavelike fluctuations in the circulation of the atmosphere whose occurrence cannot be predicted, as the tides can be, by a simple almanac of assured recurrence based on past experience. Observations of oceanic motions have also revealed fluctuations at periods which bear no evident relationship with the astronomical periods which characterize the externally imposed forces. Not only do the observed oceanic and meteorological fluctuations occur on time scales which do not match the periods of the external forcing, but in addition, any particular observation of the fluctuations in the circulation shows them to occur erratically if not randomly distributed in time.

490

It is possible, though, to imagine the atmosphere and the ocean in dynamical state which would be consistent with the external forcing ai boundary conditions in which all change *would* be predictable with t appropriate astronomical period, in which each season is identical to predecessor and such that an almanac of the past *would* serve as an accura predictor of the future. Such a physical system might be consistent wi every physical principle, but it is not the state realized in nature. Happil mankind instead experiences a rich variety of motions in the atmosphe and oceans which depart dramatically from a simple, repetitive recurrenc

The existence of fluctuations in the circulations of the atmosphere ar oceans can be attributed to the *instability* of the dynamical state withou fluctuations to very small wavelike disturbances. Such small disturbances ai inevitably present in any real system, but their effect on stable systems i ephemeral. If a state of flow, however, is *unstable* with respect to sma fluctuations, the fluctuations will grow in amplitude with time and spac scales determined by the dynamics of the interaction of the initial perturba tion and the structure of the original flow state. This at once leads to ā natural explanation, conceptually, for the inevitable presence of fluctuatior energy at nonastronomical periods. This hypothesis requires for its valida- tion, however, two quite formidable problem elements whose sequence forms a program of investigation and whose relationship is crucial. First, if the existence of fluctuations is to be demonstrated as due to the instability of the circulation which would occur in the *absence* of the fluctuations, it is first necessary to know what that fluctuation-free state would be. This first task is ordinarily very difficult. In some classical problems in hydrodynamic instab- ility, such as the instability of the conductive temperature gradient of a fluid layer heated uniformly from below, the calculation of the basic state is sufficiently simple that attention instinctively and immediately focuses on the second element of the program, i.e., the exposure of the initial state to small perturbations and their subsequent evolution. For the study of the stability of atmospheric and oceanic flows, the calculation of the physically and mathematically possible flows in the absence of fluctuations is itself so difficult that it is rarely possible to carry through even this first part of the program completely. It is then natural to ask whether there are alternatives to the detailed calculation of the fluctuation-free state of flow. Of course it is possible to observe the actual flow pattern and construct averages, in time say, and observationally produce a pattern which has filtered out the velo- city and temperature variations associated with the fluctuations. This *defines* an observed mean flow state. It is crucial to realize that this state can*not*, in general, be used as the flow whose stability or instability will determine whether the observed fluctuations can be attributed to an instability process. The structure of the observed mean flow will inevitably be affected by presence of the very fluctuations we seek to predict, since in general the fluctuations will give rise by nonlinear processes to fluxes of heat and momentum with nonzero time averages. The convergence of these fluxes must be balanced by dissipation or counterbalancing fluxes of the mean flow quantities if a time-averaged state is to exist. Consequently the structure of the observed mean flow already implicitly assumes the existence of fluctuations, and it is a generally misleading fiction to suppose that the stability of the averaged state accurately portrays the stability of the fluctuation-free states, since *in most cases the nature of the fluctuations alters the fluctuation-free state in the direction of stability.* That is, the time-averaged state, if considered as the initial state, is frequently found to be considerably more stable than the relevant initial state we should be examining.

The unknown nature of the precise fluctuation-free state required for the stability analysis may in fact be turned to advantage. Instead of precisely calculating the mean state, we may arbitrarily *prescribe* an initial state. For example, let us imagine a planet with no imposed longitudinal variations of flow produced by heating and topography. Any steady zonal flow, i.e., a flow independent of longitude, will then satisfy (6.5.21). We can imagine such a flow initially determined by the balance between friction and externally imposed heating, since these forces, while negligible for flows varying in $x$, become determining when the terms retained in (6.5.21) identically vanish, as they do for $x$-independent flows. Each imagined initial state will correspond to a particular distribution of heat sources and frictional forces which may be specified after the fact in order that the hypothesized flow may be a solution of the equations of motion. This allows the consideration of *classes* of initial states, each corresponding to a different constellation of forces, and the stability of each of these initial states may then be directly examined to see which feature of the initial states the instability can be attributed to. If the feature responsible for the instability is sufficiently general and robustly persistent in a variety of circumstances, and if the resulting instability can be identified as geophysically relevant, then the consideration of an imagined class of initial states, rather than a precisely calculated single example, serves to actually deepen our understanding of the nature of the instability process and the criteria for instability.

This in turn assumes that it is possible to judge whether the predicted mode of instability, i.e., the response of the initial state to a small perturbation, is indeed physically relevant to the ocean or the atmosphere. It is not evident *a priori* that the nature of the instability process will be clearly evident in the mature, finite-amplitude fluctuations that are realized in the ocean and atmosphere. A truly major contribution of the early pioneer workers in the field of atmospheric instability, such as Charney (1947) and Eady (1949), was their demonstration that the mode of instability of conceptually "reasonable" initial states possessed time and space scales and a physical structure remarkably close to the observed weather waves in the atmosphere. The notion that the observed fluctuations in the atmosphere could be explained in terms of the small-amplitude stability analysis of a highly idealized flow is not an obvious one, and its subsequent verification is a tribute to the profound physical insight of the early investigators.

The purpose of this chapter is to discuss the fundamentals of quasigeostrophic instability theory. Although the unperturbed state of the atmosphere should in fact be characterized by a longitudinally varying flow as a consequence of both continent–ocean variations of surface temperature and topographic forcing, most stability theories idealize the initial state as zonally uniform, i.e., longitudinally invariant. The study of instability of such idealized initial states reveals in the most straightforward way the mechanisms which give rise to the instability process, and is capable of predicting the general character of the observed fluctuations.

## 7.2 Formulation of the Instability Problem: the Continuously Stratified Model

The observed fluctuations to which the theory to be developed applies, in both the atmosphere and the ocean, are of synoptic scale, i.e., in the notation of Chapter 6,

$$\beta = \frac{\beta_0 L^2}{U} = O(1), \tag{7.2.1a}$$

$$S = \frac{N_s^2 D^2}{f_0^2 L^2} = O(1), \tag{7.2.1b}$$

where $N_s$ is given either by

$$N_s^2 = \frac{g}{\Theta_s} \frac{\partial \Theta_s}{\partial z_*}$$

for the case of the atmosphere, or

$$N_s^2 = -\frac{g}{\rho_s} \frac{\partial \rho_s}{\partial z_*}$$

for the case of the oceans. The parameters $\beta_0$ and $f_0$ are defined, as in Section 6.2, in terms of the central, mid-latitude, $\theta_0$, which in turn defines the geographical region under consideration. Clearly $\theta_0$ must be well away from the equator, where $f_0$ vanishes, for quasigeostrophic theory to apply.

Consider an initial state of flow, wherein the velocity is strictly zonal (i.e., along latitude circles) and given in terms of the geostrophic stream function

$$\psi = \Psi(y, z). \tag{7.2.2}$$

The initial, or *basic* state is therefore characterized by a nondimensional, zonal velocity

$$U_0 = -\frac{\partial \Psi}{\partial y}(y, z), \tag{7.2.3}$$

which in general is a function of latitude ($y$) and height ($z$). The *dimensional* velocity is related to $U_0$ by the constant scaling velocity, $U$, which typifies its magnitude. $U$ may be the average velocity over the meridional cross section, or equally well may characterize the range of the dimensional zonal velocity over $y$ and $z$. The only restriction on $U$ is that it be so chosen that the dimensional velocity is given by

$$U_{0_*} = UU_0(y, z)$$

and such that $U_0(y, z)$ is order unity.*

The formulation of the stability problem for this basic state will be given for the case of a zonal flow in the atmosphere. The oceanographic case can be considered as a special case of the atmospheric problem *insofar as the formulation is concerned* and is obtained by simply considering the standard density field $\rho_s(z)$ as a constant and replacing the atmospheric potential-temperature anomaly $\theta$ by the negative of the oceanic density anomaly, as discussed in Section 6.8.

Corresponding to the zonal flow (7.2.3), is a potential temperature

$$\theta_* = \theta_s(z)[1 + \varepsilon F \Theta(y, z)] \tag{7.2.4}$$

where to lowest order in Rossby number

$$\Theta_0(y, z) = \frac{\partial \Psi}{\partial z}. \tag{7.2.5}$$

From the thermal wind relation

$$\frac{\partial U_0}{\partial z} = -\frac{\partial \Theta_0}{\partial y}, \tag{7.2.6}$$

so that variations of the initial zonal wind with height are directly related to the existence of meridional potential temperature gradients, or by (6.10.13) to the existence of available potential energy in the basic flow.

The initial state, $\Psi(y, z)$, is a solution of (6.5.21). Now consider the evolution of the perturbed state, characterized by the stream function

$$\psi(x, y, z, t) = \Psi(y, z) + \phi(x, y, z, t) \tag{7.2.7}$$

where $\phi$ represents the perturbation of the initial state. The function $\phi$ represents the structure of the evolving perturbation field. If (7.2.7) is substituted into (6.5.21), a nonlinear problem for $\phi$ results, i.e.,

$$\left(\frac{\partial}{\partial t} + U_0 \frac{\partial}{\partial x}\right)q + \frac{\partial\phi}{\partial x}\frac{\partial\Pi_0}{\partial y} + \left|\frac{\partial\phi}{\partial x}\frac{\partial q}{\partial y} - \frac{\partial\phi}{\partial y}\frac{\partial q}{\partial x}\right| = 0, \tag{7.2.8}$$

where $q(x, y, z, t)$ is the perturbation potential vorticity defined by

$$q = \frac{\partial^2\phi}{\partial x^2} + \frac{\partial^2\phi}{\partial y^2} + \frac{1}{\rho_s}\frac{\partial}{\partial z}\left(\frac{\rho_s}{S}\frac{\partial\phi}{\partial z}\right), \tag{7.2.9}$$

---

* The notation of this chapter follows that of Chapter 6. Namely, dimensional *variables* are starred. The subscripts 0 and 1 refer to the first two orders of each variable in the Rossby-number expansion. Thus for example $u_* = U(u_0 + \varepsilon u_1 + \cdots) = u_{0_*} + \varepsilon u_{1_*} + \cdots$.

while $\partial \Pi_0 / \partial y$ is the meridional gradient of the potential vorticity of the basic state, viz.,

$$\Pi_0 = \beta y + \frac{\partial^2 \Psi}{\partial y^2} + \frac{1}{\rho_s} \frac{\partial}{\partial z}\left(\frac{\rho_s}{S} \frac{\partial \Psi}{\partial z}\right) \tag{7.2.10}$$

so that

$$\frac{\partial \Pi_0}{\partial y} = \beta - \frac{\partial^2 U_0}{\partial y^2} - \frac{1}{\rho_s} \frac{\partial}{\partial z}\left(\frac{\rho_s}{S} \frac{\partial U_0}{\partial z}\right). \tag{7.2.11}$$

If the mean flow were independent of $y$ and $z$, the ambient meridional potential vorticity gradient would reduce to $\beta$. The essential physical question is now to determine how the structure of $U_0$ determines the evolution of the perturbation field $\phi$. That is, given a particular initial state $U_0(y, z)$, will the perturbation field $\phi$ placed on the flow tend to grow or decay? If the former occurs, we may infer that the initial state is unstable with respect to the disturbance field $\phi$. To assert that $U_0$ is indeed stable, it is necessary to show the initial state is stable with respect to *all* possible initial disturbances, while instability may be demonstrated by the presence of a single perturbation $\phi$ to which the initial state is unstable.

To complete the problem for $\phi$, boundary conditions must be specified. Since the extent in latitude of the region must be less than the full sphere for the $\beta$-plane approximation to hold, boundary conditions in $y$ must be specified.

We assume that at $y = \pm 1$ rigid walls exist containing the region of the flow and the perturbations. Although clearly an artifice, this effectively isolates the region from its surroundings and assures that should an instability arise, its source must lie *within* the region under consideration. Under these conditions it follows that for $v_0$ to vanish on $y = \pm 1$,

$$\frac{\partial \phi}{\partial x} = 0, \qquad y = \pm 1. \tag{7.2.12}$$

If the zonal momentum equation (6.3.13a) is integrated in $x$ from $-\infty$ to $\infty$ (which is equivalent to integration around a complete latitude circle), it follows that

$$\frac{\partial \bar{u}_0}{\partial t} = \bar{v}_1 - \frac{\partial}{\partial y} \overline{v_0 u_0} \tag{7.2.13}$$

where an overbar represents the averaging process

$$\overline{(\ )} = \lim_{X \to \infty} \frac{1}{2X} \int_{-X}^{X} (\ ) \, dx. \tag{7.2.14}$$

Note that $\bar{v}_0$ must be zero, since $v_0$ is geostrophic. Since *both* $v_1$ and $v_0$ must vanish on $y = \pm 1$, it in turn follows that, in addition to (7.2.12), $\phi$ must satisfy

$$\frac{\partial^2}{\partial t \, \partial y} \bar{\phi} = 0, \qquad y = \pm 1, \tag{7.2.15}$$

which is a special case of (6.9.17). On $z = 0$, (6.6.10) applies. In order for the basic state to be $x$-independent, either $U_0(y, 0)$ must vanish, or more generally, $\partial \eta_B / \partial x$ must be zero. We choose the latter condition, in which case the boundary condition for $\phi$ becomes

$$
\left( \frac{\partial}{\partial t} + U_0 \frac{\partial}{\partial x} \right) \frac{\partial \phi}{\partial z} + \left( S \frac{\partial \eta_B}{\partial y} - \frac{\partial U_0}{\partial z} \right) \frac{\partial \phi}{\partial x} + \left| \frac{\partial \phi}{\partial x} \frac{\partial^2 \phi}{\partial y \, \partial z} - \frac{\partial \phi}{\partial y} \frac{\partial^2 \phi}{\partial x \, \partial z} \right|
$$
$$
= - \frac{E_V^{1/2}}{2\varepsilon} S \left| \frac{\partial^2 \phi}{\partial x^2} + \frac{\partial^2 \phi}{\partial y^2} \right| \quad \text{on } z = 0.
$$

(7.2.16)

At the upper surface a variety of boundary conditions are possible. If the upper surface at $z = z_T$ is a free surface, e.g., the surface of the ocean, then (6.9.14) applies, which for the perturbation implies that at $z = z_T$

$$
\left( \frac{\partial}{\partial t} + U_0 \frac{\partial}{\partial x} \right) \frac{\partial \phi}{\partial z} - \frac{\partial U_0}{\partial z} \frac{\partial \phi}{\partial x} + \left| \frac{\partial \phi}{\partial x} \frac{\partial^2 \phi}{\partial y \, \partial z} - \frac{\partial \phi}{\partial y} \frac{\partial^2 \phi}{\partial x \, \partial z} \right| = 0
$$
$$
\text{on } z = z_T.
$$

(7.2.17)

If, on the other hand, we wish to examine the stability problem of a flow bounded above by a *rigid* horizontal surface, then (4.6.11) will apply, so that for $\phi$,

$$
\left( \frac{\partial}{\partial t} + U_0 \frac{\partial}{\partial x} \right) \frac{\partial \phi}{\partial z} - \frac{\partial U_0}{\partial z} \frac{\partial \phi}{\partial x} + \left| \frac{\partial \phi}{\partial x} \frac{\partial^2 \phi}{\partial y \, \partial z} - \frac{\partial \phi}{\partial y} \frac{\partial^2 \phi}{\partial x \, \partial z} \right|
$$
$$
= \frac{E_V^{1/2}}{2\varepsilon} S \left| \frac{\partial^2 \phi}{\partial x^2} + \frac{\partial^2 \phi}{\partial y^2} \right| \quad \text{at } z = z_T.
$$

(7.2.18)

For atmospheric models, where the upper boundary is at $z_T = \infty$, the radiation condition

$$
\lim_{z \to \infty} \int_{-1}^{1} dy \, \rho_s \overline{w_1 \phi} \geq 0
$$

(7.2.19)

is required. In fact, as we shall see below, in all cases to be discussed the radiation condition for *growing* perturbations may be replaced by the stronger condition that

$$
\lim_{z \to \infty} \int_{-1}^{1} dy \, \rho_s \overline{w_1 \phi} = 0.
$$

(7.2.20)

The stability problem consists of first a specification of $U_0(y, z)$ and then the examination of the evolution of $\phi$, governed by (7.2.8), (7.2.12), (7.2.16), and either (7.2.17), (7.2.18), or (7.2.19), for arbitrary initial perturbation fields. The resulting nonlinear problem for $\phi$ is generally intractable, and progress usually requires that the perturbation be limited to small amplitudes, i.e., $\phi \ll 1$, so that the nonlinear terms proportional to $\phi^2$ can be neglected, at least during the earliest stages of the evolution of the perturbation. However, before proceeding to this linearization process, certain general remarks may be made about the nonlinear problem. The zonal average of the thermodynamic equation, (6.5.15), yields, in the absence of heating,

$$\frac{1}{S}\frac{\partial \bar{\theta}_0}{\partial t} = -\overline{w_1} - \frac{\partial}{\partial y}\frac{\overline{(v_0\theta_0)}}{S}. \tag{7.2.21}$$

If (7.2.13) is differentiated with respect to $y$, and if (7.2.21) is multiplied by $\rho_s$, then differentiated with respect to $z$, and then divided by $\rho_s$, the difference of the two resulting equations yields an equation for the evolution of the zonally averaged potential vorticity:

$$\frac{\partial \bar{q}}{\partial t} = +\frac{\partial^2}{\partial y^2}\overline{v_0 u_0} - \frac{1}{\rho_s}\frac{\partial^2}{\partial y\,\partial z}\frac{\rho_s}{S}\overline{v_0 \theta_0} \tag{7.2.22}$$

if use is made of the zonally averaged continuity equation. Since

$$\bar{q} = -\frac{\partial \bar{u}_0}{\partial y} + \frac{1}{\rho_s}\frac{\partial}{\partial z}\frac{\rho_s}{S}\bar{\theta}_0, \tag{7.2.23}$$

it follows from (7.2.22) that

$$\frac{\partial \bar{q}}{\partial t} = -\frac{\partial}{\partial y}\overline{(v_0 q)}. \tag{7.2.24}$$

Now $v_0$, by definition, is $O(\phi)$ and has a zero average in $x$, since $v_0$ is geostrophic. Thus the perturbation fluxes $\overline{u_0 v_0}$, $\overline{\theta_0 v_0}$, and $\overline{v_0 q}$ are no larger than $O(\phi^2)$. As the perturbation grows, the zonally averaged fields of $u_0$, $\theta_0$, and $q$ will alter with time by an amount $O(\phi^2)$. However, certain important *a priori* constraints may be placed on this evolution of the zonally averaged fields. If (7.2.13) is multiplied by $\rho_s$ and integrated over a meridional cross section, we obtain

$$\frac{\partial}{\partial t}\int_0^{z_T}\int_{-1}^{1}\rho_s\bar{u}_0\,dy\,dz = +\int_0^{z_T}\int_{-1}^{1}\rho_s\overline{v_1}\,dy\,dz, \tag{7.2.25}$$

since $v_0$ vanishes at $y = \pm 1$.

The term $\int_0^{z_T}\rho_s\bar{v}_1\,dz$ in (7.2.25) represents the total northward transport of mass by the $O(\varepsilon)$ nongeostrophic velocity in the inviscid interior of the flow. This must be balanced by the southward cross-isobar transport of mass in the lower Ekman layer on $z = 0$ and in the upper Ekman layer if the upper surface is rigid. Using the results of the Ekman-layer analysis of Section 4.5, it can then be shown that

$$\int_0^{z_T}\rho_s\bar{v}_1\,dz = -\frac{E_v^{1/2}}{2\varepsilon}\left[\rho_s(0)\bar{u}_0(y, 0) + \rho_s(z_T)\bar{u}_0(y, z_T)\right], \tag{7.2.26}$$

where the second term on the right-hand side is absent unless the surface at $z = z_T$ is indeed rigid and supports an Ekman layer. Thus in the *absence* of friction the total northward flow must vanish, and hence the total zonal momentum is preserved regardless of the nature of the perturbation. Except insofar as surface friction on $z = 0$ and $z = z_T$ retards the flow, the effect of the perturbation is to only redistribute the zonal momentum over the $y$, $z$ cross section while the *total* $x$-averaged zonal momentum is preserved, i.e.,

$$\frac{\partial}{\partial t}\int_0^{z_T}\int_{-1}^1 \rho_s \bar{u}_0 \, dy \, dz = O\left(\frac{E_v^{1/2}}{\varepsilon}\right), \qquad (7.2.27)$$

while similarly, the reader may easily verify that from (7.2.21)

$$\frac{\partial}{\partial t}\int_0^{z_T}\int_{-1}^1 \rho_s \frac{\bar{\theta}_0}{S} \, dy \, dz = O\left(\frac{E_v^{1/2}}{\varepsilon}\right). \qquad (7.2.28)$$

The total energy of the $x$-independent portion of the flow is defined as

$$\bar{E} = \int_{-1}^1 dy \int_0^{z_T} dz \, \rho_s \left\lvert\frac{(\bar{u}_0)^2 + (\bar{\theta}_0)^2/S}{2}\right\rvert, \qquad (7.2.29)$$

and it follows from (7.2.13) and (7.2.21) that

$$\frac{\partial \bar{E}}{\partial t} = -\int_0^{z_T}\int_{-1}^1 dy \, dz \, \rho_s[\bar{w}_1 \bar{\theta}_0 - \bar{v}_1 \bar{u}_0]$$
$$\qquad\qquad - \int_0^{z_T}\int_{-1}^1 dy \, dz \, \rho_s\left[\bar{u}_0 \frac{\partial}{\partial y}(\overline{v_0 u_0}) + \bar{\theta}_0 \frac{\partial}{\partial y}\frac{(\overline{v_0 \theta_0})}{S}\right]. \qquad (7.2.30)$$

Using the geostrophic relations to evaluate $\bar{\theta}_0$ and $\bar{u}_0$ in terms of $\bar{\psi}$, (7.2.30) yields, after integration by parts,

$$\frac{\partial \bar{E}}{\partial t} = -\int_0^{z_T}\int_{-1}^1 dy \, dz \, \rho_s\left[\bar{u}_0 \frac{\partial}{\partial y}(\overline{v_0 u_0}) + \bar{\theta}_0 \frac{\partial}{\partial y}\frac{(\overline{v_0 \theta_0})}{S}\right]$$
$$\qquad\qquad + \int_0^1 dy \, [\rho_s(0)(\overline{\bar{\psi} \bar{w}_1})_{z=0} - \rho_s(z_T)(\overline{\bar{\psi} \bar{w}_1})_{z_T}]. \qquad (7.2.31)$$

The last two terms in (7.2.31) represent the change in the energy of the zonally averaged component of the flow due to mean vertical energy fluxes across each of the horizontal boundaries. In the absence of viscosity both of these terms will vanish. However, the first two terms on the right-hand side of (7.2.31) represent the change of the energy of the zonal flow due to convergences of fluxes of momentum and heat in the perturbation fields, and in general these will not vanish. This is, in fact, the essence of the process of instability and its effect on the zonal flow. Although the zonal fields $\bar{u}_0$ and $\bar{\theta}_0$ can only be redistributed by the perturbations, the *energy* of the zonal fields, $\bar{E}$, will be altered by the perturbations, and the energy lost by the zonally averaged portion of the field becomes available for the further growth of wavelike perturbations. A further integration by parts in (7.2.31) yields, in the absence of friction,

$$\frac{\partial \bar{E}}{\partial t} = \int_0^{z_T}\int_{-1}^1 dy \, dz \, \rho_s\left[\overline{v_0 u_0}\frac{\partial \bar{u}_0}{\partial y} + \frac{\overline{v_0 \theta_0}}{S}\frac{\partial \bar{\theta}_0}{\partial y}\right]. \qquad (7.2.32)$$

If the perturbation heat flux, $\overline{v_0 \theta_0}$, is positive where $\partial\bar{\theta}_0/\partial y$ is negative, i.e., if the perturbation heat flux is directed down the gradient of the zonally averaged meridional temperature gradient, then this perturbation heat flux will decrease the energy of the $x$-independent fields. Similarly, if the perturbation momentum flux $v_0 u_0$ (the negative of the Reynolds stress) is directed down the gradient of the zonally averaged velocity $\bar{u}_0$, then the momentum flux also will decrease the energy of the $x$ − averaged flow. Thus the energy associated with the zonally averaged flow will be decreased by the perturbation field to the extent that the perturbations tend to smooth the meridional gradients of $\bar{u}_0$ and $\bar{\theta}_0$ while preserving their averages over the meridional plane.

## 7.3 The Linear Stability Problem: Conditions for Instability

The nonlinear stability problem, (7.2.8), is hardly tractable as it stands. While it would be certainly preferable to be able to examine the stability of the initial state to disturbances of arbitrary amplitude, the nonlinearity of (7.2.8) renders this impractical. Instead we assume that, at least initially, the amplitude of the disturbance is sufficiently small, i.e.,

$$\phi \ll 1, \tag{7.3.1}$$

so that the terms proportional to $\phi^2$ and $\phi q$ in (7.2.8), (7.2.16), (7.2.17), and (7.2.18) may be ignored. These terms are each quadratic in the amplitude of the disturbance. In ignoring the terms proportional to $\phi^2$ while retaining terms (for example) $O(E_v^{1/2}/\varepsilon)$, we are assuming that if the latter term is retained although small, the former is smaller still. Ignoring terms of order $\phi^2$ leads to the linear stability problem:

$$\left(\frac{\partial}{\partial t} + U_0 \frac{\partial}{\partial x}\right) q + \frac{\partial \phi}{\partial x}\frac{\partial \Pi_0}{\partial y} = 0, \tag{7.3.2}$$

where

$$q = \frac{\partial^2 \phi}{\partial x^2} + \frac{\partial^2 \phi}{\partial y^2} + \frac{1}{\rho_s}\frac{\partial}{\partial z}\frac{\rho_s}{S}\frac{\partial \phi}{\partial z}, \tag{7.3.3}$$

with boundary conditions

$$\frac{\partial \phi}{\partial x} = 0, \qquad y = \pm 1. \tag{7.3.4}$$

We then have

$$\left(\frac{\partial}{\partial t} + U_0 \frac{\partial}{\partial x}\right)\frac{\partial \phi}{\partial z} + \left(S\frac{\partial \eta_B}{\partial y} - \frac{\partial U_0}{\partial z}\right)\frac{\partial \phi}{\partial x}$$
$$= -\frac{E_v^{1/2}S}{2\varepsilon}\left(\frac{\partial^2 \phi}{\partial x^2} + \frac{\partial^2 \phi}{\partial y^2}\right) \quad \text{on } z = 0, \tag{7.3.5}$$

while at $z = z_T$,

$$\left(\frac{\partial}{\partial t} + U_0 \frac{\partial}{\partial x}\right)\frac{\partial \phi}{\partial z} - \frac{\partial U_0}{\partial z}\frac{\partial \phi}{\partial x} = \begin{cases} +\dfrac{E_V^{1/2}S}{2\varepsilon}\left(\dfrac{\partial^2 \phi}{\partial x^2} + \dfrac{\partial^2 \phi}{\partial y^2}\right) & \text{(rigid)}, \\[2mm] 0 & \text{(free)}, \end{cases} \tag{7.3.6}$$

depending on whether the horizontal upper surface is free or rigid. If $z_T \to \infty$, we choose (7.2.19) as the appropriate condition, i.e.,

$$\lim_{z \to \infty} \int_{-1}^{1} dy\, \rho_s \overline{w_1 \phi} \geq 0. \tag{7.3.7}$$

The linear problem formulated above inquires whether the flow field $U_0(y, z)$ is unstable when subjected to a *small* perturbation of essentially infinitesimal amplitude. If the flow is unstable to such disturbances and they grow, they will eventually attain an amplitude where nonlinear effects, ignored here, must be reconsidered. Or, if the flow is stable to infinitesimal disturbances, it is necessary to investigate whether disturbances sufficiently large will destabilize the flow. The second possibility will not present itself in the present context; the first will, and this will require, in Section 7.16, a discussion of the finite-amplitude problem. What is remarkable is that the linear problem does adequately describe to a large extent the structure (e.g., wavelength and vertical scale) of observed fluctuations in the atmosphere and oceans.

An energy equation for the disturbance field is obtained by multiplying (7.3.2) by $\rho_s \phi$ and integrating over the volume of the flow. After use is made of the periodicity of the disturbance field in $x$ and the boundary conditions on $y = \pm 1$ and on $z = 0, z_T$, we obtain after several integrations by parts

$$\frac{\partial}{\partial t}\int_0^{z_T}\int_{-1}^{1} dy\, dz \frac{\rho_s}{2}\left[\left(\frac{\partial \phi}{\partial x}\right)^2 + \left(\frac{\partial \phi}{\partial y}\right)^2 + S^{-1}\left(\frac{\partial \phi}{\partial z}\right)^2\right]$$

$$= \int_0^{z_T}\int_{-1}^{1} dy\, dz \left[\rho_s \frac{\partial \phi}{\partial x}\frac{\partial \phi}{\partial y}\frac{\partial U_0}{\partial y} + \rho_s S^{-1}\frac{\partial \phi}{\partial x}\frac{\partial \phi}{\partial z}\frac{\partial U_0}{\partial z}\right]$$

$$- \frac{E_V^{1/2}}{2\varepsilon}\int_{-1}^{1} dy\left[\rho_s\left|\left(\frac{\partial \phi}{\partial x}\right)^2 + \left(\frac{\partial \phi}{\partial y}\right)^2\right|\right]_{z=0}$$

$$+ \begin{cases} -\dfrac{E_V^{1/2}}{2\varepsilon}\displaystyle\int_{-1}^{1} dy\left[\rho_s\left|\left(\dfrac{\partial \phi}{\partial x}\right)^2 + \left(\dfrac{\partial \phi}{\partial y}\right)^2\right|\right]_{z=z_T} & \text{(rigid)}, \\[4mm] 0 & \text{(free)}, \\[4mm] -\displaystyle\int_{-1}^{1} dy\, (\rho_s \overline{\phi w_1})_{z \to \infty} & (z_T \to \infty), \end{cases} \tag{7.3.8}$$

where the overbar represents the $x$-average as defined by (7.2.14).

The volume integral on the left of (7.3.8) represents the time rate of change of the sum of the kinetic and available potential energies of the disturbance field. This growth or decay of the perturbation energy is given by the terms on the right-hand side. The first of these terms may be written

$$\int_0^{z_T} \int_{-1}^1 \rho_s \overline{\frac{\partial \phi}{\partial x} \frac{\partial \phi}{\partial y}} \frac{\partial U_0}{\partial y} \, dy \, dz = -\int_0^{z_T} \int_{-1}^1 \rho_s \overline{u_0 v_0} \frac{\partial U_0}{\partial y} \, dy \, dz \quad (7.3.9)$$

and is the integral over the meridional plane of the Reynolds stress $-\rho_s \overline{u_0 v_0}$ of the disturbance multiplied by the horizontal shear of the zonal flow, $U_0$. The second term on the right-hand side of (7.3.8) may be written

$$\int_0^{z_T} \int_{-1}^1 \rho_s \overline{\frac{\partial \phi}{\partial x} \frac{\partial \phi}{\partial z}} S^{-1} \frac{\partial U_0}{\partial z} \, dy \, dz = -\int_0^{z_T} \int_{-1}^1 \rho_s \overline{\theta_0 v_0} S^{-1} \frac{\partial \Theta_0}{\partial y} \, dy \, dz \quad (7.3.10)$$

after use is made of (7.2.6), and therefore represents the integral over the meridional plane of the product of the rectified northward heat flux of the fluctuations multiplied by $-\partial \Theta_0 / \partial y$, i.e., by the equatorial gradient of potential temperature. The remaining terms in (7.3.8) represent the sink of fluctuation energy due to frictional dissipation in the lower Ekman layer and to the dissipation in the upper Ekman layer (if $z_T$ is rigid) or to the vertical flux of energy out of the system (if $z_T \to \infty$). Thus the only possible *sources* of fluctuation energy are the terms described by (7.3.9) and (7.3.10) and depend crucially on the horizontal and vertical shear of $U_0$. In most meteorologically and oceanographically relevant situations the dissipation time scale due to the Ekman-layer friction is long compared to the advective time, i.e., $E_v^{1/2}/\varepsilon \ll 1$, and so the frictionless approximation for the initial evolution of the perturbations is valid. In the absence of friction and the vanishing of the energy flux at $z = z_T$,

$$\frac{\partial E(\phi)}{\partial t} = -\int_0^{z_T} \int_{-1}^1 dy \, dz \, \rho_s \left[ \overline{v_0 u_0} \frac{\partial U_0}{\partial y} + \frac{\overline{v_0 \theta_0}}{S} \frac{\partial \Theta_0}{\partial y} \right] \quad (7.3.11)$$

where $E(\phi)$ is the total perturbation energy, i.e.,

$$E(\phi) \equiv \int_0^{z_T} \int_{-1}^1 dy \, dz \, \frac{\rho_s}{2} \overline{\left[ \left( \frac{\partial \phi}{\partial x} \right)^2 + \left( \frac{\partial \phi}{\partial y} \right)^2 + \frac{1}{S} \left( \frac{\partial \phi}{\partial z} \right)^2 \right]}. \quad (7.3.12)$$

In (7.3.11) $u_0$, $v_0$, and $\theta_0$ are, of course, the perturbation fields defined by

$$u_0 - U_0 = -\frac{\partial \phi}{\partial y}, \quad (7.3.13a)$$

$$v_0 = \frac{\partial \phi}{\partial x}, \quad (7.3.13b)$$

$$\theta_0 - \Theta_0 = \frac{\partial \phi}{\partial z}. \quad (7.3.13c)$$

The equation for the zonally averaged kinetic energy in the linear theory may be derived directly from (7.2.32) when it is noted that for small amplitudes, to $O(\phi^2)$,

$$\overline{v_0 u_0}\frac{\partial \bar{u}_0}{\partial y} = \overline{v_0 u_0}\frac{\partial U_0}{\partial y} \tag{7.3.14a}$$

$$\overline{v_0 \theta_0}\frac{\partial \bar{\theta}_0}{\partial y} = \overline{v_0 \theta_0}\frac{\partial \Theta_0}{\partial y} \tag{7.3.14b}$$

so that (7.2.32) becomes, in the absence of friction,

$$\frac{\partial \bar{E}}{\partial t} = \int_0^{z_T} \int_{-1}^1 dy\, dz\, \rho_s\left[\overline{v_0 u_0}\frac{\partial U_0}{\partial y} + \frac{\overline{v_0 \theta_0}}{S}\frac{\partial \Theta_0}{\partial y}\right]. \tag{7.3.15}$$

If (7.3.11) is compared with (7.3.15), it is evident that the sum of the energy of the $x$-averaged flow* and the energy of fluctuations is preserved. The fluctuation momentum and heat fluxes acting on the mean gradients therefore represent *energy conversion mechanisms* between the mean and the fluctuating flow. Energy gained by the perturbation field must therefore be lost by the mean field through the instability process. The phenomenon of instability is the phenomenon of the preferential transfer of energy from the wave-free flow to the fluctuating flow.

The instability process which depends on the existence of the horizontal shear of the basic current is called *barotropic instability*, for it can occur in a homogeneous fluid in the absence of vertical shear. In order for the process of barotropic instability to transfer energy to the perturbations, it follows from (7.3.8) that on average over the meridional plane the product

$$\overline{\frac{\partial \phi}{\partial x}\frac{\partial \phi}{\partial y}}\frac{\partial U_0}{\partial y}$$

must be positive. Since

$$\overline{\frac{\partial \phi}{\partial x}\frac{\partial \phi}{\partial y}}\frac{\partial U_0}{\partial y} = \overline{\left(\frac{\partial \phi}{\partial x}\Big/\frac{\partial \phi}{\partial y}\right)\left(\frac{\partial \phi}{\partial y}\right)^2}\frac{\partial U_0}{\partial y}$$

$$= -\overline{\left(\frac{\partial y}{\partial x}\right)_\phi\left(\frac{\partial \phi}{\partial y}\right)^2}\frac{\partial U_0}{\partial y}, \tag{7.3.16}$$

lines of constant $\phi$ must be sloping (on average) northwest to southeast in regions where $\partial U_0/\partial y$ is positive if barotropic instability is to augment the energy of the disturbance, as shown in Figure 7.3.1(a). Were the lines of constant $\phi$ to slope in the opposite direction, the perturbations would be feeding energy into the mean flow. The rate of energy conversion depends on the size of the mean horizontal shear. If $\partial U_0/\partial y < 0$, then of course the directions of energy releasing streamlines of $\phi$ are reversed.

---

* We shall also refer to $U_0$ as the mean flow, *without* implying that $U_0$ is the $x$-average of the observed flow.

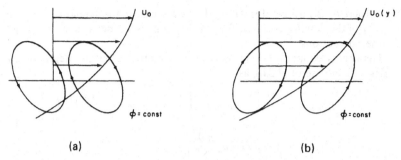

(a)                                         (b)

**Figure 7.3.1** (a) the slope of the perturbation streamlines for a disturbance whose Reynolds stress extracts energy from the horizontal shear of the basic current. Note the impression that the perturbation is "leaning" against the zonal flow so as to decelerate it. (b) The slope of the perturbation streamlines for a disturbance whose Reynolds stress transfers energy from the perturbations to the basic current.

The remaining instability process depends on the existence of the vertical shear of the basic current. Since the vertical shear implies a horizontal temperature gradient, this process is called *baroclinic instability*. The presence of horizontal temperature gradients implies the existence of *available potential energy* in the basic flow, and this is the energy source for baroclinic instability. The condition for the baroclinic conversion term to be positive is that, on average,

$$\frac{\partial \phi}{\partial x}\frac{\partial \phi}{\partial z}\frac{\partial U_0}{\partial z} = -\left(\frac{\partial z}{\partial x}\right)_\phi \left(\frac{\partial \phi}{\partial z}\right)^2 \frac{\partial U_0}{\partial z} > 0, \qquad (7.3.17)$$

so that in regions where $\partial U_0/\partial z > 0$ the lines of constant $\phi$ must be sloping upward and westward for the baroclinic conversion to be positive. The sensitivity of the energy conversion processes to the spatial structure of $\phi$ is a manifestation of the delicate process of instability in which the disturbance must be properly organized to release the available kinetic and potential energy of the mean flow. Note that in each case where the perturbations gain energy, the orientation of the lines of constant $\phi$ is precisely opposite to the directions a passive tracer (such as a dye streak) would be tilted by the shear flow.

The energy equation (7.3.11) may be rewritten in terms of the product of the flux convergences and the mean flow $U_0$ by an integration by parts, i.e.,

$$\frac{\partial E(\phi)}{\partial t} = \int_0^{z_T}\int_{-1}^1 dy\, dz\rho_s \left[ U_0 \frac{\partial}{\partial y}\overline{v_0 u_0} + \Theta_0 S^{-1}\frac{\partial}{\partial y}\overline{v_0 \theta_0} \right]. \qquad (7.3.18)$$

Note that if $(\partial/\partial y)(\overline{v_0 u_0}) > 0$, then (7.2.13) implies that the mean zonal flow decreases, i.e., the Reynolds stresses act as a retarding force on the mean flow. The product of this retarding force and a positive $U_0$ implies a rate of decrease of the energy of the mean flow and, as (7.3.18) shows, a consequent increase in $E(\phi)$. Similarly a divergence of the heat flux of the fluctuations, i.e., $(\partial/\partial y)(\overline{v_0 \theta_0}) > 0$, will tend to decrease the mean potential temperature

$\bar{\theta}_0$. Where $\bar{\theta}_0$ is positive this will lead to a decrease in the available potential energy of the zonally averaged flow, which implies an increase in the energy of the fluctuations.

The energy equation (7.3.11) may also be written, after an integration by parts and the use of (7.2.6), as

$$
\frac{\partial}{\partial t} E(\phi) = + \int_0^{z_T} \int_{-1}^1 dy \, dz \, U_0 \left[ \frac{\partial}{\partial y} \left( \overline{\rho_s u_0 v_0} \right) - \frac{\partial}{\partial z} \frac{\rho_s \overline{\theta_0 v_0}}{S} \right]
$$
$$
+ \int_{-1}^1 dy \, U_0 \rho_s \frac{\overline{v_0 \theta_0}}{S} \Big|_0^{z_T}
$$
(7.3.19)

where the notation

$$
(\ )\Big|_0^{z_T} = (\ )_{z=z_T} - (\ )_{z=0}
$$
(7.3.20)

has been used to write more compactly the boundary term which arises from the integration by parts of the second term. The identity

$$
\overline{\rho_s v_0 q} = -\frac{\partial}{\partial y} \overline{\rho_s v_0 u_0} + \frac{\partial}{\partial z} \frac{\rho_s}{S} \overline{v_0 \theta_0}
$$
(7.3.21)

allows the energy equation to be written in terms of the *potential vorticity flux* of the perturbation field, i.e.,

$$
\frac{\partial E(\phi)}{\partial t} = - \int_0^{z_T} \int_{-1}^1 dy \, dz \, U_0 \rho_s \overline{v_0 q} + \int_{-1}^1 dy \, U_0 \rho_s \frac{\overline{v_0 \theta_0}}{S} \Big|_0^{z_T}.
$$
(7.3.22)

It is convenient to introduce the function $\eta(x, y, z, t)$ defined by

$$
\frac{\partial \eta}{\partial t} + U_0 \frac{\partial \eta}{\partial x} = v_0 .
$$
(7.3.23)

$\eta$ is the displacement of fluid elements in the northward direction, and (7.3.23) is the linearized form of the more general relation between particle displacement and Eulerian velocity. The reason for introducing $\eta$ becomes clear when (7.3.2) is rewritten then as

$$
\left( \frac{\partial}{\partial t} + U_0 \frac{\partial}{\partial x} \right) q = - \left( \frac{\partial}{\partial t} + U_0 \frac{\partial}{\partial x} \right) \eta \frac{\partial \Pi_0}{\partial y} ,
$$
(7.3.24)

the particular solution of which is

$$
q = -\eta \frac{\partial \Pi_0}{\partial y}
$$
(7.3.25)

and corresponds to disturbances whose perturbation potential vorticity is due entirely to the advection of mean potential vorticity by the displacements. If (7.3.25) is used, it follows that

$$
\overline{v_0 q} = - \left( \frac{\partial}{\partial t} \frac{\overline{\eta^2}}{2} \right) \frac{\partial \Pi_0}{\partial y} ,
$$
(7.3.26)

which allows the energy equation (7.3.22) to be written

$$\frac{\partial}{\partial t}\left[E(\phi) - \int_0^{z_T}\int_{-1}^1 dy\,dz\,\frac{\rho_s\overline{\eta^2}}{2}\left(U_0\frac{\partial\Pi_0}{\partial y}\right)\right] = \int_{-1}^1 dy\,U_0\rho_s\frac{\overline{v_0\theta_0}}{S}\Big|_0^{z_T}. \qquad (7.3.27)$$

Similarly, on $z = 0$, it follows from (7.3.13c), (7.3.5), and (7.3.23) that, in the absence of friction

$$\overline{v_0\theta_0} = -\left|S\frac{\partial\eta_B}{\partial y} - \frac{\partial U_0}{\partial z}\right|\frac{\partial\overline{\eta^2}/2}{\partial t}, \qquad z = 0. \qquad (7.3.28)$$

If we assume for the moment that $z_T$ is finite so that (7.3.6) applies, then in the absence of friction

$$\overline{v_0\theta_0} = \frac{\partial U_0}{\partial z}\frac{\partial}{\partial t}\frac{\overline{\eta^2}}{2}, \qquad z = z_T \qquad (7.3.29)$$

for both free and rigid upper surfaces. It then follows that the energy equation for the fluctuations in the absence of dissipation may be written as the *conservation* statement

$$\frac{\partial}{\partial t}\left\{E(\phi) - \int_0^{z_T}\int_{-1}^1 dy\,dz\,\rho_s\frac{\overline{\eta^2}}{2}\left(U_0\frac{\partial\Pi_0}{\partial y}\right)\right.$$
$$\left. - \int_{-1}^1 dy\left(\frac{U_0}{S}\frac{\partial U_0}{\partial z}\frac{\rho_s\overline{\eta^2}}{2}\right)_{z=z_T} \qquad (7.3.30)$$
$$\left. + \int_{-1}^1 dy\left[U_0\left(S^{-1}\frac{\partial U_0}{\partial z} - \frac{\partial\eta_B}{\partial y}\right)\frac{\rho_s\overline{\eta^2}}{2}\right]_{z=0}\right\} = 0.$$

If $z_T$ is then allowed to become infinite, the contribution at $z = z_T$ will vanish if either the vertical shear of $U_0$ vanishes at infinity or if $\eta$ vanishes there.

If $U_0$ is unstable with respect to the perturbation field $\phi$, then obviously the energy $E(\phi)$ must increase with time. Similarly in a growing disturbance the mean square displacement in the $y$-direction of any line of fluid elements must also increase, i.e., the dispersion $\overline{\eta^2}$ must be increasing. This last requirement is a powerful one. It first of all, with (7.3.26) requires that the meridional potential vorticity flux be in the direction *opposite* to the basic potential vorticity gradient. That is, the fluctuations must transfer potential vorticity *down* the basic potential-vorticity gradient.

Furthermore, for flows for which the terms involving $\overline{\eta^2}$ in (7.3.30) are everywhere positive, an increase of $E(\phi)$ *and* $\overline{\eta^2}$ in time is inconsistent. Thus, if for all $y$ and $z$ in the meridional plane

$$U_0\frac{\partial\Pi_0}{\partial y} \le 0 \qquad (7.3.31a)$$

and

$$U_0\left(S^{-1}\frac{\partial U_0}{\partial z} - \frac{\partial\eta_B}{\partial y}\right) \ge 0, \qquad z = 0, \qquad (7.3.31b)$$

$$U_0 S^{-1} \frac{\partial U_0}{\partial z} \qquad \leq 0, \qquad z = z_T, \qquad (7.3.31c)$$

then the flow $U_0(y, z)$ is *stable* to infinitesimal disturbances. These conditions are therefore *sufficient* conditions for the stability of basic state. The violation of these conditions is a *necessary* condition for its instability. It is important to note that these conditions depend entirely on the character of $U_0(y, z)$.

Now by (7.2.6) and (6.5.3)

$$S^{-1} \frac{\partial U_0}{\partial z} = -\varepsilon^{-1} \frac{\partial \theta_*/\partial y}{\partial \theta_s/\partial z} = \frac{L}{\varepsilon D} \left( \frac{\partial z_*}{\partial y_*} \right)_{0_*}, \qquad (7.3.32)$$

where $(\partial z_*/\partial y_*)$ is, to $O(\varepsilon)$, the slope of the surfaces of potential temperature in the basic state. The boundary terms (7.3.31b,c) are therefore each proportional to the slope of the potential-temperature surfaces of the basic state relative to the *boundary slope*, for $\eta_B$ is related to the dimensional variation of the lower boundary by

$$\eta_B = \frac{h_B}{\varepsilon D}, \qquad (7.3.33)$$

so that

$$S^{-1} \frac{\partial U_0}{\partial z} - \frac{\partial \eta_B}{\partial y} = \frac{L}{\varepsilon D} \left[ \left( \frac{\partial z_*}{\partial y_*} \right)_{0_*} - \frac{\partial h_B}{\partial y_*} \right]. \qquad (7.3.34)$$

An additional, somewhat stronger condition for instability may be derived directly from the condition (7.2.27), i.e., that in the absence of dissipation $(E_V^{1/2}/\varepsilon \to 0)$ the effect of the fluctuations on the mean flow is merely to redistribute the $x$-averaged zonal momentum. This condition follows from (7.2.26) and the identity

$$0 = - \int_0^{z_T} \int_{-1}^1 \frac{\partial}{\partial y} \rho_s \overline{u_0 v_0} \, dy \, dz. \qquad (7.3.35)$$

With (7.3.21) this becomes

$$0 = \int_0^{z_T} \int_{-1}^1 \rho_s \overline{v_0 q} \, dy \, dz - \int_0^{z_T} \int_{-1}^1 \frac{\partial}{\partial z} \rho_s \frac{\overline{v_0 \theta_0}}{S} \, dz \, dy \qquad (7.3.36)$$

which with (7.3.26), (7.3.28), and (7.3.29) can be written

$$0 = \int_0^{z_T} \int_{-1}^1 \rho_s \frac{\partial \Pi_0}{\partial y} \left[ \frac{\partial}{\partial t} \overline{\eta^2} \right] dy \, dz + \int_{-1}^1 \left[ \frac{\rho_s}{S} \frac{\partial U_0}{\partial z} \frac{\partial}{\partial t} \overline{\eta^2} \right]_{z = z_T} dy$$

$$- \int_{-1}^1 \left[ \rho_s \left[ S^{-1} \frac{\partial U_0}{\partial z} - \frac{\partial \eta_B}{\partial y} \right] \frac{\partial \overline{\eta^2}}{\partial t} \right]_{z = 0} dy. \qquad (7.3.37)$$

The vanishing of this integral is a necessary condition for instability. As noted before, the dispersion of any fluid line of elements originally aligned along a latitude circle will increase in an unstable motion, i.e., $(\partial/\partial t)\overline{\eta^2} > 0$.

Therefore certain stringent conditions must be satisfied by the basic state in order that (7.3.37) may be satisfied. Consider, for example, the case where the surfaces of constant potential temperature are *parallel* to the lower boundary, i.e., $(\partial z_*/\partial y_*)_{\theta_*} = \partial h_B/\partial y_*$ at $z = 0$. In that case the final term in (7.3.37) will vanish. Further, let $\partial U_0/\partial z$ at the upper boundary $z_T$, whether finite or infinite, vanish also. Then the necessary condition for instability yields

$$\boxed{\int_0^{z_T} \int_{-1}^1 dy \, dz \, \rho_s \frac{\partial \Pi_0}{\partial y} \left( \frac{\partial}{\partial t} \overline{\eta^2} \right) = 0}.$$  (7.3.38)

Thus if $\partial \overline{\eta^2}/\partial t$ is to be positive, the potential-vorticity gradient of the basic current must be positive in some subregion of the meridional plane and negative in others. In the absence of friction the necessary condition for the instability of a zonal flow $U_0(y, z)$, each of whose horizontal boundaries is at constant potential temperature, is that the potential-vorticity gradient $\partial \Pi_0/\partial y$ must *vanish* on a line in the meridional cross section (Charney and Stern 1962). The archetypal case where this condition is not met occurs for a uniform current where $\partial \Pi_0/\partial y$ is equal to $\beta$. In that case only stable (nongrowing) Rossby waves are possible, as demonstrated in Section 6.12.

If $\partial \Pi_0/\partial y$ is of a single sign throughout the meridional cross section, instability (for inviscid perturbations) can only occur in the presence of meridional temperature gradients on either $z_T$ or $z = 0$. Consider the case where $\partial U_0/\partial z = 0$ on $z = z_T$ (an important example of which occurs when $z_T \to \infty$ and $(\partial U_0/\partial z)(\infty) = 0$). Then if $\partial \Pi_0/\partial y$ has the same sign throughout the meridional cross section of the current, instability can only arise if, for some $y$,

$$\left[ \left( \frac{\partial z_*}{\partial y_*} \right)_{\theta_*} - \frac{\partial h_B}{\partial y_*} \right]_{z=0} \frac{\partial \Pi_0}{\partial y} > 0.$$  (7.3.39)

If $\partial \Pi_0/\partial y$ is positive, i.e., if the basic potential vorticity gradient everywhere has the same sign as $\beta$, then the necessary condition for instability demands that at the lower boundary the northward slope of the surfaces of $\theta_*$ exceed the northward slope of the boundary. Note that sufficiently strong topographic slopes can *stabilize* a flow $U_0(y, z)$, regardless of the amount of available potential or kinetic energy in the basic state. If $z_T$ is finite and the contribution from $z = 0$ to (7.3.37) vanishes, then (7.3.39) is replaced by the condition

$$\left( \frac{\partial U_0}{\partial z} \right)_{z=z_T} \frac{\partial \Pi_0}{\partial y} < 0$$  (7.3.40)

for some $y$ in $(-1, 1)$ if $\partial \Pi_0/\partial y$ is of a single sign.

If (7.3.37) is indeed satisfied, a comparison with (7.3.30) shows that the latter may be written

$$
\int_0^{z_T} \int_{-1}^{1} dy\, dz\, \rho_s \left(\frac{\partial \overline{\eta^2}}{\partial t}\right)(U_0 - c_0)\frac{\partial \Pi_0}{\partial y}
$$

$$
+ \int_{-1}^{1} dy\left[(U_0 - c_0)S^{-1}\frac{\partial U_0}{\partial z}\, \rho_s \frac{\partial \overline{\eta^2}}{\partial t}\right]_{z=z_T}
$$

$$
- \int_{-1}^{1} dy\left[(U_0 - c_0)\left\{S^{-1}\frac{\partial U_0}{\partial z} - \frac{\partial \eta_B}{\partial y}\right\}\rho_s\left(\frac{\partial \overline{\eta^2}}{\partial t}\right)\right]_{z=0}
$$

$$
= 2\int_0^{z_T}\int_{-1}^{1} dy\, dz\, \frac{\partial E(\phi)}{\partial t}, \qquad (7.3.41)
$$

where $c_0$ is *any* constant. The combination of terms multiplied by $c_0$ vanishes identically by (7.3.37). Consider the case where $(\partial U_0/\partial z)(z_T)$ vanishes (which again may occur if $z_T \to \infty$), and where either the lower surface is at constant potential temperature *or* $U_0$ at the lower boundary is independent of $y$. If the former holds, the boundary term at $z = 0$ identically vanishes. If the latter holds (i.e., if $U_0$ is independent of $y$ at $z = 0$), the boundary term may be eliminated by choosing $c_0$ equal to $U_0(y, 0)$. Then (7.3.41) becomes

$$
\int_0^{z_T}\int_{-1}^{1} dy\, dz\, \rho_s\left(\frac{\partial \overline{\eta^2}}{\partial t}\right)[U_0(y, z) - U_0(y, 0)]\frac{\partial \Pi_0}{\partial y} > 0 \qquad (7.3.42)
$$

if $\partial E(\phi)/\partial t > 0$. Thus, in addition to the conditions required by (7.3.37), (7.3.42) demands that the product of the zonal velocity relative to its surface value times the meridional gradient of the potential vorticity must be somewhere positive for instability to result. Thus, while (7.3.38) would be satisfied if $\partial \Pi_0/\partial y$ were everywhere zero, (7.3.42) would clearly not be.

The sensitivity of the criteria for instability to the structure of the potential-vorticity distribution of the basic current is profoundly important. Although (7.3.11) shows that the existence of horizontal shear and horizontal temperature gradients in the basic state make energy available for the perturbation fields, the true availability of this energy depends on the ability of the perturbation fields to release that energy. This release, in turn, must be consistent with the underlying potential-vorticity dynamics of the perturbations. If the necessary conditions for instability (which derive directly from the potential-vorticity dynamics) are not met, the energy in the basic state will not be *dynamically* available for the growth of fluctuations.

## 7.4 Normal Modes

Although it is possible to discover sufficient conditions for stability by the general integral methods of Section 7.3, these lead only to *necessary* conditions for the instability of $U_0(y, z)$. If a flow field satisfies these necessary conditions, detailed calculations are still required to verify that the flow is in fact unstable. In addition, information concerning the structure of the unstable perturbations (which, we recall, are of geophysical interest themselves) is obtainable only by direct calculation.

A method of solution of the perturbation problem (7.3.2) which has been found fruitful is the method of normal modes. Since the *coefficients* of (7.3.2) as well as the coefficients of (7.3.5) and (7.3.6), depend only on $y$ and $z$, solutions may be sought in the form

$$\phi(x, y, z, t) = \text{Re } \Phi(y, z)e^{ik(x-ct)}, \tag{7.4.1}$$

where Re denotes the real part of the expression it prefaces. The zonal wave number $k$ must be real in order that $\phi$ may be finite for both large positive and negative $x$, i.e., in order that $\phi$ may be periodic in longitude. Without loss of generality we may consider $k$ positive. On the other hand, both the amplitude function $\Phi$ and the frequency $kc$ may be complex. In particular, the phase speed $c$ may be written in terms of its real and imaginary parts:

$$c = c_r + ic_i. \tag{7.4.2}$$

If solutions of the form (7.4.1) are found with $c_i > 0$, then $\phi$ will grow exponentially, since

$$\phi = \text{Re } \Phi e^{ik(x-c_r t)}e^{kc_i t}. \tag{7.4.3}$$

The growth rate of the disturbance is therefore $kc_i$. Naturally, if $kc_i > 0$, after a sufficient time has elapsed the perturbation will become large enough so that nonlinear effects, ignored in linear theory, will become important. Nevertheless, the possibility of explosive growth of the perturbations, at least initially, demonstrates clearly the instability of the basic state, and the relative growth rates of different perturbations yield a natural basis for deciding which perturbation is most favored by the instability process. It should be noted that the periodic form of the solution in $x$ imposes no loss of generality, since each wave number may be summed by Fourier integration to represent an arbitrary disturbance. The assumption of an exponential time factor, while consistent, is not the most general form the evolution of a disturbance may take. Algebraic growth or decay is also possible. However, if exponentially growing solutions are found, they grow faster than any algebraic power. Hence, solutions of the form (7.4.3), if they exist, are the most relevant. If the growth rate has a significant maximum at a particular wave number, then the solution also will naturally explain the observed wavelike nature of the fluctuations in both the atmosphere and the oceans.

The normal-mode problem for $\Phi$ is obtained by substituting (7.4.1) into (7.3.2) and the relevant boundary conditions to obtain

$$(U_0 - c)\left[\frac{1}{\rho_s}\frac{\partial}{\partial z}\frac{\rho_s}{S}\frac{\partial \Phi}{\partial z} + \frac{\partial^2 \Phi}{\partial y^2} - k^2\Phi\right] + \Phi\frac{\partial \Pi_0}{\partial y} = 0 \tag{7.4.4}$$

where, as before,

$$\frac{\partial \Pi_0}{\partial y} = \beta - \frac{\partial^2 U_0}{\partial y^2} - \frac{1}{\rho_s}\frac{\partial}{\partial z}\frac{\rho_s}{S}\frac{\partial U_0}{\partial z}. \tag{7.4.5}$$

The boundary conditions for $\Phi$ are, from (7.2.12),

$$\Phi = 0, \qquad y = \pm 1, \tag{7.4.6}$$

while (7.3.5), which holds at $z = 0$, becomes

$$(U_0 - c)\frac{\partial \Phi}{\partial z} + \left[S\frac{\partial \eta_B}{\partial y} - \frac{\partial U_0}{\partial z}\right]\Phi = i\frac{E_V^{1/2}}{2k\varepsilon}S\left[\frac{\partial^2 \Phi}{\partial y^2} - k^2\Phi\right]. \tag{7.4.7}$$

If $z_T$ is finite, (7.3.6) is relevant, which yields as the condition for $\Phi$

$$(U_0 - c)\frac{\partial \Phi}{\partial z} - \frac{\partial U_0}{\partial z}\Phi = \begin{cases} -\dfrac{iE_V^{1/2}}{2k\varepsilon}S\left[\dfrac{\partial^2 \Phi}{\partial y^2} - k^2\Phi\right] & \text{(rigid)} \\ 0 & \text{(free)} \end{cases} \tag{7.4.8}$$

depending on whether the upper surface is rigid or free. If $z_T$ is infinite, (7.3.7) applies. Now the vertical energy flux may be written

$$\rho_s\overline{\phi w_1} = -\frac{\rho_s}{S}\overline{[\text{Re }\Phi e^{ik(x-ct)}]}\,\text{Re}\left[\left\{(U_0 - c)\frac{\partial \Phi}{\partial z} - \frac{\partial U_0}{\partial z}\Phi\right\}ike^{ik(x-ct)}\right], \tag{7.4.9}$$

since for the perturbations

$$w_1 S = -\left[\left(\frac{\partial}{\partial t} + U_0\frac{\partial}{\partial x}\right)\frac{\partial \phi}{\partial z} - \frac{\partial U_0}{\partial z}\frac{\partial \phi}{\partial x}\right]. \tag{7.4.10}$$

Using the fact that

$$\text{Re}(\ ) = \frac{(\ ) + (\ )^*}{2} \tag{7.4.11}$$

where $(\ )^*$ represents the complex conjugate of $(\ )$, (7.4.9) becomes

$$\rho_s\overline{\phi w_1} = -\frac{S^{-1}}{4}\rho_s\left[(U_0 - c)\Phi^*\frac{\partial \Phi}{\partial z} - (U_0 - c^*)\Phi\frac{\partial \Phi^*}{\partial z}\right]ike^{2kc_it}. \tag{7.4.12}$$

If (7.4.4) is multiplied by $\rho_s\Phi^*$, a little manipulation of the result yields, with (7.4.12),

$$\begin{aligned}
\frac{\partial}{\partial z}\int_{-1}^{1} dy\,\rho_s\overline{\phi w_1} = -\int_{-1}^{1} dy\Bigg\{&\frac{kc_i}{2}\rho_s\left[S^{-1}\left|\frac{\partial \Phi}{\partial z}\right|^2 + \left|\frac{\partial \Phi}{\partial y}\right|^2 + k^2|\Phi|^2\right] \\
&+ \frac{\partial U_0}{\partial z}\frac{ik}{4}\frac{\rho_s}{S}\left[\frac{\partial \Phi}{\partial z}\Phi^* - \Phi\frac{\partial \Phi^*}{\partial z}\right] \\
&+ \frac{\partial U_0}{\partial y}\frac{ik}{4}\rho_s\left[\frac{\partial \Phi}{\partial y}\Phi^* - \Phi\frac{\partial \Phi^*}{\partial y}\right]\Bigg\}e^{2kc_it},
\end{aligned} \tag{7.4.13}$$

which is, in fact, the perturbation form of the energy equation written in terms of $\Phi$. The content of (7.4.13) is merely that the divergence of the energy flux must be balanced locally either by the decay of perturbation energy (the first term on the right of (7.4.13)) or by the local conversion of energy of the mean flow. This local conversion is proportional to the local values of $\partial U_0/\partial z$ and $\partial U_0/\partial y$. Consider the case where, for large $z$, $U_0$ is independent of $y$ and $z$. In such regions,

$$\frac{\partial}{\partial z}\rho_s\overline{\phi w_1} = -\frac{kc_i}{2}\int_{-1}^{1} dy\,\rho_s\left[S^{-1}\left|\frac{\partial\Phi}{\partial z}\right|^2 + \left|\frac{\partial\Phi}{\partial y}\right|^2 + k^2|\Phi|^2\right], \quad (7.4.14)$$

so that the energy flux must *decrease* with height for an unstable disturbance. However, from (7.4.12),

$$\rho_s\int_{-1}^{1}\overline{\phi w_1}\,dy$$

$$= -ikS^{-1}\frac{\rho_s}{4}\left[(U_0-c)\Phi^*\frac{\partial\Phi}{\partial z} - (U_0-c^*)\Phi\frac{\partial\Phi^*}{\partial z}\right]e^{2kc_it}$$

$$\le \tfrac{1}{4}S^{-1}\rho_s\left|2k\,|U_0-c|\,|\Phi|\left|\frac{\partial\Phi}{\partial z}\right|\right|e^{2kc_it}$$

$$= \tfrac{1}{4}S^{-1/2}\rho_s\left|2\,|U_0-c|\,|\Phi k|\left|S^{-1/2}\frac{\partial\Phi}{\partial z}\right|\right|e^{2kc_it}$$

$$\le \tfrac{1}{4}S^{-1/2}\rho_s|U_0-c|\left\{S^{-1}\left|\frac{\partial\Phi}{\partial z}\right|^2 + k^2|\Phi|^2\right\}e^{2kc_it}$$

$$\le \tfrac{1}{4}S^{-1/2}\rho_s\,|U_0-c|\left\{S^{-1}\left|\frac{\partial\Phi}{\partial z}\right|^2 + \left|\frac{\partial\Phi}{\partial y}\right|^2 + k^2|\Phi|^2\right\}e^{2kc_it}.$$

$$(7.4.15)$$

Thus, (7.4.14) implies that

$$-\frac{\partial}{\partial z}\int_{-1}^{1} dy\,\rho_s\overline{\phi w_1} \ge \frac{2kc_iS^{1/2}\rho_s}{|U_0-c|}\int_{-1}^{1} dy\,\overline{\phi w_1}. \quad (7.4.16)$$

Integrating from the level $z = z_1$ at the base of the region where $U_0$ is independent of $y$ and $z$ yields

$$\rho_s\int_{-1}^{1}\overline{\phi w_1}\,dy \le \left[\rho_s\int_{-1}^{1}\overline{\phi w_1}\,dy\right]_{z=z_1}\exp\left(-\int_{z_1}^{z}\frac{2kc_iS^{1/2}}{|U_0-c|}\,dz'\right) \quad (7.4.17)$$

As $z \to \infty$, (7.4.17) implies that

$$\lim_{z\to\infty}\rho_s\int_{-1}^{1}\overline{\phi w_1}\,dy = 0 \quad (7.4.18)$$

for unstable modes ($kc_i > 0$). In the absence of local energy production the growth of energy in the unstable mode must be balanced by a convergence of energy flux produced in regions where the instability has its sources. Since the energy flux is bounded by a multiple of the local energy density, this leads to an exponential decrease of the energy flux away from the source region. The technical consequence of this result is that the stronger condition (7.4.18) may be applied in place of the inequality (7.3.7) whenever the flow is bounded above by an infinite source-free region for the perturbations.

The stability problem can now be formally stated in the following way. Since both the perturbation equation (7.4.4) and the boundary conditions are linear and homogeneous, nontrivial solutions will exist for only certain values of $c$, all other parameters considered fixed. The complex phase speed $c$ is therefore the *eigenvalue* of the stability problem. If there exists a solution with an associated eigenvalue with a positive imaginary part, then the flow is unstable. For a particular $U_0(y, z)$ the solutions of the eigenvalue problem will yield the growth rate

$$kc_i = kc_i\left(k, \beta, S, \frac{E_v^{1/2}}{\varepsilon}\right). \qquad (7.4.19)$$

The wave number $k$ for which $kc_i$ is a maximum yields the zonal wavelength of the most unstable mode, and it is at least plausible that this mode will emerge first from the background of small disturbances, i.e., that it will be the most likely to be realized in finite amplitude. The program implied by the normal-mode analysis requires the calculation of the parametric domain of instability of $U_0(y, z)$ and the growth-rate dependence on $k$.

The necessary conditions for instability developed in Section 7.3 can be derived also directly from (7.4.4). If $c_i \neq 0$, then (7.4.4) can always be divided by $U_0 - c$, since then $U_0 - c$ is never zero. If the resulting equation is multiplied by $\rho_s \Phi^*$ and integrated over the meridional cross section, we obtain, in the absence of friction, after integration by parts,

$$\int_{-1}^{1} dy \int_{0}^{z_T} dz\, \rho_s \left\{ S^{-1} \left| \frac{\partial \Phi}{\partial z} \right|^2 + \left| \frac{\partial \Phi}{\partial y} \right|^2 + k^2 |\Phi|^2 \right\}$$

$$= \int_{-1}^{1} dy \int_{0}^{z_T} dz \frac{\rho_s |\Phi|^2}{U_0 - c} \frac{\partial \Pi_0}{\partial y} + \int_{-1}^{1} dy \left\{ \rho_s S^{-1} \frac{|\Phi|^2}{U_0 - c} \frac{\partial U_0}{\partial z} \right\}\bigg|_{z = z_T} \qquad (7.4.20)$$

$$- \int_{-1}^{1} dy \frac{\rho_s |\Phi|^2}{|U_0 - c|} \left[ S^{-1} \frac{\partial U_0}{\partial z} - \frac{\partial \eta_B}{\partial y} \right]\bigg|_{z = 0},$$

while if $z_T \to \infty$, there is no contribution from the integrated term at $z = z_T$ if (7.4.18) applies.

Since

$$\frac{1}{U_0 - c} = \frac{1}{|U_0 - c|^2}\{U_0 - c_r + ic_i\}, \qquad (7.4.21)$$

the imaginary part of (7.4.20) may be written

$$
c_i \left\{ \left| \int_{-1}^{1} dy \int_{0}^{z_T} dz \, \frac{\rho_s |\Phi|^2}{|U_0 - c|^2} \frac{\partial \Pi_0}{\partial y} + \int_{-1}^{1} dy \frac{|\rho_s S^{-1}|\Phi|^2}{||U_0 - c|^2} \frac{\partial U_0}{\partial z} \right|_{z=z_T} \right.
$$

$$
\left. - \int_{-1}^{1} dy \left| \rho_s \frac{|\Phi|^2}{|U_0 - c|^2} \left[ S^{-1} \frac{\partial U_0}{\partial z} - \frac{\partial \eta_B}{\partial y} \right] \right|_{z=0} \right\} = 0.
$$

(7.4.22)

If $c_i$ is not to equal zero, i.e., if the mode is to be unstable, then the integrals multiplied by $c_i$ must vanish. The resulting condition for instability is precisely the same as (7.3.37). If we write

$$
\eta = \mathrm{Re}\, \mathcal{N}(y, z) e^{ik(x - ct)},
$$

(7.4.23)

then by (7.3.23)

$$
\mathcal{N} = \frac{\Phi}{U_0 - c},
$$

(7.4.24)

or

$$
\overline{\eta^2} = \tfrac{1}{2} |\mathcal{N}|^2 e^{2kc_i t} = \frac{1}{2} \frac{|\Phi|^2}{|U_0 - c|^2} e^{2kc_i t}.
$$

(7.4.25)

Thus

$$
\frac{\partial}{\partial t} \overline{\eta^2} = \frac{kc_i |\Phi|^2}{|U_0 - c|^2} e^{2kc_i t},
$$

(7.4.26)

which allows (7.3.37) to be rewritten directly as (7.4.22). The real part of (7.4.20) yields, with (7.4.22) for $c_i \neq 0$, and $c_0$ any constant,

$$
\int_{-1}^{1} dy \int_{0}^{z_T} dz \, \rho_s \frac{|\Phi|^2}{|U_0 - c|^2} (U_0 - c_0) \frac{\partial \Pi_0}{\partial y}
$$

$$
+ \int_{-1}^{1} dy \left[ \frac{\rho_s}{S}(U_0 - c_0) \frac{|\Phi|^2}{|U_0 - c|^2} \frac{\partial U_0}{\partial z} \right]_{z=z_T}
$$

$$
- \int_{-1}^{1} dy \left[ \rho_s (U_0 - c_0) \left\{ S^{-1} \frac{\partial U_0}{\partial z} - \frac{\partial \eta_B}{\partial y} \right\} \frac{|\Phi|^2}{|U_0 - c|^2} \right]_{z=0}
$$

$$
= \int_{-1}^{1} dy \int_{0}^{z_T} dz \, \rho_s \left[ S^{-1} \left| \frac{\partial \Phi}{\partial z} \right|^2 + \left| \frac{\partial \Phi}{\partial y} \right|^2 + k^2 |\Phi|^2 \right] > 0,
$$

(7.4.27)

which, with (7.4.26), is identical with the condition (7.3.41). The implications of these necessary conditions for instability have been discussed in the previous section. It is important to note that the same conditions for instability arise whether the disturbance under consideration is assumed to be wavelike as in this section, or is allowed a general form as in Section 7.3.

## 7.5 Bounds on the Phase Speed and Growth Rate

The calculation of the phase speed $c$, and the growth rate $kc_i$ for a given $U_0(y, z)$ is a generally difficult task, and it is useful to have in hand certain *a priori* information about the range of allowable phase speeds and growth rates before the search for the complex eigenvalue $c$ is undertaken.

It is useful for this purpose to rewrite (7.4.4) in terms of $\mathcal{N}$, the amplitude of the northward displacement, which is related to $\Phi$ by (7.4.24). If this relation is substituted into (7.4.4), then with (7.4.5) we obtain

$$\frac{\partial}{\partial y}\left[(U_0 - c)^2 \frac{\partial \mathcal{N}}{\partial y}\right] + \frac{1}{\rho_s}\frac{\partial}{\partial z}\left[(U_0 - c)^2 \frac{\rho_s}{S}\frac{\partial \mathcal{N}}{\partial z}\right]$$

$$- k^2 \mathcal{N}(U_0 - c)^2 + \beta \mathcal{N}(U_0 - c) = 0 \quad (7.5.1)$$

while the boundary conditions become

$$\mathcal{N} = 0, \qquad y = \pm 1,$$

and in the absence of friction

$$(U_0 - c)^2 \frac{\partial \mathcal{N}}{\partial z} + (U_0 - c)S \frac{\partial \eta_B}{\partial y}\mathcal{N} = 0, \qquad z = 0, \qquad (7.5.2a)$$

$$\frac{\partial \mathcal{N}}{\partial z} = 0, \qquad z = z_T. \qquad (7.5.2b)$$

If (7.5.1) is multiplied by $\rho_s \mathcal{N}^*$ and integrated over the meridional plane, we obtain

$$\int_0^{z_T}\int_{-1}^1 (U_0 - c)^2 \rho_s \left[S^{-1}\left|\frac{\partial \mathcal{N}}{\partial z}\right|^2 + \left|\frac{\partial \mathcal{N}}{\partial y}\right|^2 + k^2 |\mathcal{N}|^2\right] dy\, dz$$

$$= \beta \int_0^z \int_{-1}^1 (U_0 - c)\rho_s |\mathcal{N}|^2 \, dy\, dz \qquad (7.5.3)$$

$$+ \int_{-1}^1 dy\, (U_0 - c)\frac{\partial \eta_B}{\partial y}\{\rho_s |\mathcal{N}|^2\}_{z=0},$$

whose imaginary part yields, if $c_i \neq 0$,

$$\boxed{\int_0^{z_T}\int_{-1}^1 U_0 P\, dy\, dz = c_r \int_0^{z_T}\int_{-1}^1 P\, dy\, dz + \frac{\beta}{2}\int_0^{z_T}\int_{-1}^1 J\, dy\, dz}$$

$$+ \int_{-1}^1 dy\, \frac{J(y, 0)}{2}\frac{\partial \eta_B}{\partial y}, \qquad (7.5.4)$$

where

$$P(y, z) \equiv \rho_s\left[S^{-1}\left|\frac{\partial \mathcal{N}}{\partial z}\right|^2 + \left|\frac{\partial \mathcal{N}}{\partial y}\right|^2 + k^2 |\mathcal{N}|^2\right],$$

$$J(y, z) = \rho_s |\mathcal{N}|^2. \qquad (7.5.5)$$

The real part of (7.5.3) in turn yields, with the aid of (7.5.4),

$$\int_0^{z_i} \int_{-1}^1 U_0^2 P \, dy \, dz = (c_r^2 + c_i^2) \int_0^{z_T} \int_{-1}^1 P \, dy \, dz$$
$$+ \beta \int_0^{z_T} \int_{-1}^1 U_0 J \, dy \, dz + \int_{-1}^1 dy \, J(y, 0) \frac{\partial \eta_B}{\partial y}. \tag{7.5.6}$$

In order to fully exploit the relations (7.5.4) and (7.5.6) certain preliminary results are required. Any function $\mathcal{N}$ which vanishes on $y = \pm 1$ can be expanded in the Fourier series

$$\mathcal{N} = \sum_{j=0}^{\infty} A_j \cos(j + \tfrac{1}{2})\pi y. \tag{7.5.7}$$

Thus

$$\frac{\partial \mathcal{N}}{\partial y} = - \sum_{j=0}^{\infty} A_j \pi (j + \tfrac{1}{2}) \sin(j + \tfrac{1}{2})\pi y, \tag{7.5.8}$$

and using the orthogonality of the $\sin(j + \tfrac{1}{2})\pi y$ over the interval $(-1, 1)$,

$$\int_{-1}^1 dy \left| \frac{\partial \mathcal{N}}{\partial y} \right|^2 = \sum_{j=0}^{\infty} |A_j|^2 \pi^2 (j + \tfrac{1}{2})^2 \geq \sum_{j=0}^{\infty} \pi^2 |A_j|^2 / 4. \tag{7.5.9}$$

But since

$$\int_{-1}^1 dy \, |\mathcal{N}|^2 = \sum_{j=0}^{\infty} |A_j|^2, \tag{7.5.10}$$

it immediately follows that

$$\frac{\pi^2}{4} \int_{-1}^1 dy \, |\mathcal{N}|^2 \leq \int_{-1}^1 dy \left| \frac{\partial \mathcal{N}}{\partial y} \right|^2. \tag{7.5.11}$$

Thus,

$$\int_0^{z_T} dz \int_{-1}^1 dy \, P \geq \left( k^2 + \frac{\pi^2}{4} \right) \int_0^{z_T} dz \int_{-1}^1 dy \, J. \tag{7.5.12}$$

Consider the case where $\partial \eta_B / \partial y$ is zero, i.e., where the bottom is flat; then from (7.5.4)

$$c_r = \frac{\int_0^{z_T} \int_{-1}^1 U_0 P \, dy \, dz}{\int_0^{z_T} \int_{-1}^1 P \, dy \, dz} - \frac{\beta \int_0^{z_T} \int_{-1}^1 J \, dy \, dz}{2 \int_0^{z_T} \int_{-1}^1 P \, dy \, dz}. \tag{7.5.13}$$

Let $U_{\text{MAX}}$ and $U_{\text{MIN}}$ be the maximum and minimum values attained by $U_0$ in the meridional plane. Using (7.5.13) and the inequality (7.5.11), it follows that

$$\boxed{U_{\text{MIN}} - \frac{\beta}{2(\pi^2/4 + k^2)} \leq c_r \leq U_{\text{MAX}}.} \tag{7.5.14}$$

Thus the real part of the phase speed of an *unstable* inviscid perturbation of a zonal flow on the $\beta$-plane must be within the range given by (7.5.14). Consider now the obvious inequality

$$0 \geq \int_0^{z_T} \int_{-1}^1 dy \, dz \, (U_0 - U_{MAX})(U_0 - U_{MIN})P$$

$$= \int_0^{z_T} \int_{-1}^1 dy \, dz \, \{U_0^2 P - (U_{MAX} + U_{MIN})U_0 P + U_{MAX} U_{MIN} P\},$$

(7.5.15)

which, with (7.5.4) and (7.5.6) yields, when $\partial \eta_B/\partial y = 0$,

$$0 \geq [c_r^2 - c_r(U_{MAX} + U_{MIN}) + c_i^2 + U_{MAX} U_{MIN}] \int_0^{z_T} \int_{-1}^1 dy \, dz \, P$$

$$+ \beta \int_0^{z_T} \int_{-1}^1 dy \, dz \, J \left| U_0 - \frac{U_{MAX} + U_{MIN}}{2} \right|$$

(7.5.16)

If $U_0$ is replaced by $U_{MIN}$ in the final integral in (7.5.16), it is easily seen that

$$\int_0^{z_T} \int_{-1}^1 dy \, dz \, J \left| U_0 - \frac{U_{MAX} + U_{MIN}}{2} \right|$$

$$\geq -\frac{U_{MAX} - U_{MIN}}{2} \int_0^{z_T} \int_{-1}^1 dy \, dz \, J$$

(7.5.17)

$$\geq -\frac{U_{MAX} - U_{MIN}}{2(\pi^2/4 + k^2)} \int_0^{z_T} \int_{-1}^1 dy \, dz \, P,$$

so that the inequality in (7.5.16) may be rewritten, after division by the positive definite integral of $P$ over the meridional plane, as

$$\left( \frac{U_{MAX} - U_{MIN}}{2} \right)^2 + \frac{\beta}{k^2 + \pi^2/4} \left( \frac{U_{MAX} - U_{MIN}}{2} \right)$$

$$\geq \left( c_r - \frac{U_{MAX} + U_{MIN}}{2} \right)^2 + c_i^2.$$

(7.5.18)

Thus for unstable waves the complex phase speed $c$ must lie within a semicircle in the $c$-plane whose radius is given by the square root of the left-hand side of (7.5.18) and whose center is on the real axis at the mean velocity, $(U_{MAX} + U_{MIN})/2$, as shown in Figure 7.5.1. Equation (7.5.14) reveals, however, that the shaded portion of the semicircle, for which $c_r > U_{MAX}$, is not an allowed region for the eigenvalue $c$. If $\beta$ were zero, (7.5.18) would reduce to the beautiful theorem of Howard's (1961) in which the radius of the semicircle is half the range of the zonal velocity. Note that

$$c_i^2 \leq \left( \frac{U_{MAX} - U_{MIN}}{2} \right)^2 + \frac{\beta}{k^2 + \pi^2/4} \frac{U_{MAX} - U_{MIN}}{2},$$

(7.5.19)

so that $c_i$ must vanish for uniform flows, where $U_{MAX} - U_{MIN}$ is zero.

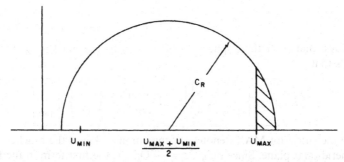

**Figure 7.5.1** The semicircle in the complex velocity plane in which the complex phase speed must lie. The shaded region is not a possible region for $C$. The radius, $C_R$, of the semicircle is given by

$$C_R^2 = \left(\frac{U_{MAX} - U_{MIN}}{2}\right)^2 + \frac{\beta}{k^2 + \pi^2/4}\left(\frac{U_{MAX} - U_{MIN}}{2}\right)$$

An alternative upper bound on the growth may be obtained by utilizing the transformation

$$\Phi(y, z) = (U_0 - c)^{1/2}\chi(y, z), \tag{7.5.20}$$

which, when substituted into (7.4.4), yields

$$\begin{aligned}
&\frac{1}{\rho_s}\frac{\partial}{\partial z}\left[(U_0 - c)S^{-1}\rho_s\frac{\partial \chi}{\partial z}\right] + \frac{\partial}{\partial y}\left[(U_0 - c)\frac{\partial \chi}{\partial y}\right] - k^2(U_0 - c)\chi \\
&\quad - \left[\left(\frac{\partial U_0}{\partial y}\right)^2 + S^{-1}\left(\frac{\partial U_0}{\partial z}\right)^2\right]\frac{\chi}{4(U_0 - c)} \\
&\quad + \left[\frac{1}{2}\frac{\partial^2 U_0}{\partial y^2} + \frac{1}{2\rho_s}\frac{\partial}{\partial z}\left(S^{-1}\rho_s\frac{\partial U_0}{\partial z}\right) + \frac{\partial \Pi_0}{\partial y}\right]\chi = 0
\end{aligned} \tag{7.5.21}$$

with boundary conditions

$$\chi = 0, \qquad y = \pm 1,$$

and in the absence of friction,

$$\begin{aligned}
(U_0 - c)\frac{\partial \chi}{\partial z} + \left[S\frac{\partial \eta_B}{\partial y} - \frac{\partial U_0}{\partial z}\right]\chi = 0, \qquad z = 0, \\
(U_0 - c)\frac{\partial \chi}{\partial z} - \frac{\partial U_0}{\partial z}\chi = 0, \qquad z = z_T.
\end{aligned} \tag{7.5.22}$$

If (7.5.21) is multiplied by $\rho_s\chi^*$ and the product is integrated over the meridional plane, the imaginary part of the result yields

$$\begin{aligned}
\int_{-1}^{1} dy \int_{0}^{z_T} dz\,\rho_s&\left[S^{-1}\left|\frac{\partial \chi}{\partial z}\right|^2 + \left|\frac{\partial \chi}{\partial y}\right|^2 + k^2|\chi|^2\right] \\
&= \int_{-1}^{1} dy \int_{0}^{z_T} dz\left[S^{-1}\left(\frac{\partial U_0}{\partial z}\right)^2 + \left(\frac{\partial U_0}{\partial y}\right)^2\right]\frac{\rho_s|\chi|^2}{4|U_0 - c|^2}.
\end{aligned} \tag{7.5.23}$$

Since

$$|U_0 - c|^2 \geq c_i^2,$$

it follows that with the application of (7.5.11) to the function $\chi$, (7.5.23) implies that

$$(2kc_i)^2 \leq \left[\frac{k^2}{k^2 + \pi^2/4}\right]\left[S^{-1}\left(\frac{\partial U_0}{\partial z}\right)^2 + \left(\frac{\partial U_0}{\partial y}\right)^2\right]_{MAX} \qquad (7.5.24)$$

where the subscript MAX denotes the maximum value of the bracket over the meridional plane. Since $\partial U_0/\partial z = -\partial\Theta_0/\partial y$, the first term in the final factor of the right-hand side of (7.5.24) represents the available potential energy, while the second term represents the available kinetic-energy density for the perturbation in the horizontal shear of the zonal flow. The sum of the two energy sources bounds the growth rate for the energy, $2kc_i$. It is important to note that the maximum growth rate diminishes with decreasing $k$. Waves with very long zonal wavelengths will have small northward velocities and will transport relatively little heat and momentum meridionally; as is clear from (7.3.11), this fact is essential for the growth of perturbation energy.

A final upper bound on the growth rate is derivable directly from (7.4.27). When $U_0$ at $z = 0$ and $z = z_T$ is such that the boundary terms in (7.4.27) can be made to vanish by a judicious choice of $c_0$, then with (7.5.11) it follows that

$$(kc_i)^2 \leq \left[(U_0 - c_0)\frac{\partial\Pi_0}{\partial y}\right]_{MAX}\frac{k^2}{k^2 + \pi^2/4} \qquad (7.5.25)$$

## 7.6 Baroclinic Instability: the Basic Mechanism

The energy equation for the perturbations (7.3.11) reveals two potential sources of instability in the basic current: the vertical and horizontal shear of $U_0$. The vertical shear implies horizontal temperature gradients, and therefore the presence of available potential energy which may be released and transferred to the perturbations by a process called baroclinic instability.

Although all currents in the atmosphere and oceans possess, to varying degrees, both horizontal and vertical shear, it is helpful to consider at first simplified situations in which one or the other is absent. This serves two purposes. From a mathematical viewpoint the assumption that $U_0$ is a function of $y$ or $z$, but not both, enables (7.4.4) to be immediately reduced from a partial to an ordinary differential equation. From a physical viewpoint the arbitrary suppression of first one and then the other fluctuation energy source enables us to focus on the fundamental physical character of each of the types of energy transfer. Once this has been done, it allows a more natural interpretation of the complex problem in which $U_0$ possesses vertical and horizontal shear simultaneously.

The pioneering studies of Charney (1947) and Eady (1949) showed that the disturbances observed in mid-latitude in the atmosphere could be explained as a manifestation of the baroclinic instability of the zonal winds, and we shall first study this mechanism in isolation.

The mechanism for baroclinic instability is easily described by an argument originally put forward by Eady (1949). Consider the situation depicted in Figure 7.6.1, where the surfaces of constant potential temperature tilt

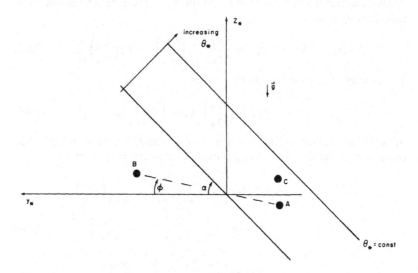

**Figure 7.6.1** The slope of the potential-temperature surface with respect to the horizontal opens a "wedge of instability" of angle $\tan^{-1}(\partial z_*/\partial y_*)_{\theta_*}$. Fluid trajectories within this wedge (e.g., the motion of element $A$ to the position of $B$) will release potential energy, and a fluid element on such a trajectory will be accelerated away from its initial position.

upward in the meridional plane at an angle $\alpha$ with respect to the horizontal. This is an equilibrium situation only because the Coriolis force balances the horizontal pressure gradient implied by the sloping $\theta$-surfaces. In a nonrotating flow the $\theta$-surfaces, if initially tilted, would immediately slump into a horizontal configuration.

Consider now the displacement of a fluid element from position $A$ in Figure 7.6.1 to a new position, say $B$.

Let the *dimensional* displacements in $y$ and $z$ of the fluid elements be $\eta_*$ and $\zeta_*$ respectively. Then by the same analysis as in Section 6.4, the density change experienced by element $A$ when moved to the position of element $B$ and adjusted to the ambient pressure is

$$\Delta\rho_{*A} = +\frac{1}{\gamma}\frac{p_{*0}}{R\theta_{*A}}\left(\frac{p_*}{p_{*0}}\right)^{1/\gamma}\left(\frac{\partial p_*}{\partial z_*}\frac{1}{p_*}\zeta_* + \frac{\partial p_*}{\partial y_*}\frac{1}{p_*}\eta_*\right), \qquad (7.6.1)$$

assuming that the potential temperature

$$\theta_* = \frac{p_{*0}}{R\rho_*}\left(\frac{p_*}{p_{*0}}\right)^{1/\gamma} \tag{7.6.2}$$

is preserved during the displacement. The constant $p_{*0}$ is a standard pressure, usually chosen to be typical of the pressure at the earth's surface. The expression (7.6.1) is a slight generalization of (6.4.3) to account for the density change due to the lateral variation of the pressure. The density now possessed by $A$ is

$$\rho_{*A} + \Delta\rho_{*A} = \rho_{*A} + \frac{1}{\gamma}\frac{\rho_{*A}}{p_{*A}}\left[\zeta_*\left(\frac{\partial p_*}{\partial z_*}\right)_A + \eta_*\left(\frac{\partial p_*}{\partial y_*}\right)_A\right]. \tag{7.6.3}$$

The density of element $B$ may be written

$$\rho_{*B} = \rho_{*A} + \left(\frac{\partial\rho_*}{\partial y_*}\right)_A \eta_* + \left(\frac{\partial\rho_*}{\partial z_*}\right)_A \zeta_* \tag{7.6.4}$$

by a simple Taylor expansion. The excess density element $A$ has, when compared to the density of the undisturbed element $B$, is given by

$$
\begin{aligned}
\rho_{*A} + \Delta\rho_{*A} - \rho_{*B} &= \rho_{*A}\left[\zeta_*\left|\frac{1}{\gamma p_{*A}}\left(\frac{\partial p_*}{\partial z_*}\right)_A - \frac{1}{\rho_{*A}}\left(\frac{\partial\rho_*}{\partial z_*}\right)_A\right.\right. \\
&\quad\left.\left. + \eta_*\left|\frac{1}{\gamma p_{*A}}\left(\frac{\partial p_*}{\partial y_*}\right)_A - \frac{1}{\rho_{*A}}\left(\frac{\partial\rho_*}{\partial y_*}\right)_A\right|\right] \\
&= \rho_{*A}\left[\frac{\zeta_*}{\theta_{*A}}\frac{\partial\theta_{*A}}{\partial z_*} + \frac{\eta_*}{\theta_{*A}}\frac{\partial\theta_{*A}}{\partial y_*}\right].
\end{aligned}
\tag{7.6.5}
$$

The *restoring* force per unit mass along the displacement direction, $AB$, due to the density excess is simply

$$E_* = \left(\frac{\rho_{*A} + \Delta\rho_{*A} - \rho_{*B}}{\rho_{*A}}\right)g\sin\phi, \tag{7.6.6}$$

where $\phi$ is the angle of the displacement defined by

$$\tan\phi = \frac{\zeta_*}{\eta_*}. \tag{7.6.7}$$

With (7.6.5) the restoring force can therefore be written

$$\boxed{E_* = \frac{g}{\theta_*}\frac{\partial\theta_*}{\partial z_*}\sin\phi\left[\zeta_* - \eta_*\left(\frac{\partial z_*}{\partial y_*}\right)_{\theta_*}\right],} \tag{7.6.8}$$

where the identity

$$\left(\frac{\partial z_*}{\partial y_*}\right)_{\theta_*} = -\frac{\partial\theta_*/\partial y_*}{\partial\theta_*/\partial z_*}$$

has been used. The subscript $A$ has been dropped in going from (7.6.6) to (7.6.8), since the expression for the restoring force clearly applies to the displacement of any fluid element.

If the displacement is vertical (e.g., from $A$ to $C$), then $\eta_*$ is zero, $\sin \phi$ is 1, and (7.6.8) reduces to (6.4.7). As long as $\partial\theta_*/\partial z_* > 0$ such displacements yield a positive restoring force. Indeed, all displacements will yield a *positive* restoring force unless the element is displaced so that its trajectory satisfies

$$0 < \tan \phi < \left(\frac{\partial z_*}{\partial y_*}\right)_{\theta_*}. \tag{7.6.9}$$

If fluid elements are displaced within the wedge defined by the horizontal geopotential surface and the surface of constant $\theta_*$, then $E_*$ will be negative, and the buoyancy force, rather than being restoring, will accelerate the fluid element further from its initial position. This is the essential mechanism of baroclinic instability, and it depends crucially on the slope of the potential-temperature surfaces, or, in an incompressible fluid, on the slope of the density surfaces. For trajectories within the wedge of instability, light fluid rises and heavy fluid sinks, releasing the potential energy of the basis state. Baroclinic instability is therefore a form of thermal convection.

To lowest order in Rossby number

$$-\frac{\partial\theta_*/\partial y_*}{\partial\theta_*/\partial z_*} = -\frac{D}{L}\varepsilon F\theta_s \frac{\partial\theta_0/\partial y}{\partial\theta_s/\partial z} \tag{7.6.10}$$

if (6.5.3) is used. Since the direction of the displacement is given equally well in terms of the velocity components, we have

$$\frac{\zeta_*}{\eta_*} = \frac{w_*}{v_*} = \frac{D}{L}\varepsilon\frac{w_1}{v_0} \tag{7.6.11}$$

if (6.3.12) is used. Thus for instability

$$\left(\frac{\partial z_*}{\partial y_*}\right)_{\theta_*}(\tan \phi)^{-1} = \left|-\frac{\partial\theta_0}{\partial y}\frac{v_0}{w_1}\right|S^{-1} > 1, \tag{7.6.12}$$

where by the scaling hypothesis the bracketed terms are $O(1)$. Thus (7.6.9) implies that baroclinic instability requires

$$S = \frac{L_D^2}{L^2} \leq O(1), \tag{7.6.13}$$

where $L$ is the horizontal scale of the motion and $L_D$ is the Rossby-deformation radius $N_s D/f_0$. Length scales of the perturbations favored for baroclinic instability therefore exceed the deformation radius, while by (7.5.24) and (7.5.25) much larger length scales will release the available energy very inefficiently. We can anticipate that scales of the order of $L_D$ are preferred for fluctuations produced by baroclinic instability. This, in turn, is a heuristic explanation of why the parameter setting $S = O(1)$ is so appropriate for synoptic-scale motions for both the atmosphere and the oceans.

Naturally, the displacements pictured above must be consistent with the potential vorticity dynamics. The argument to this point merely demonstrates that certain trajectories, if dynamically possible, will release the available potential energy of the mean current.

The results derived above may also be obtained directly from the vorticity equation (6.3.17). If (6.3.17) is linearized about the basic state given by the zonal flow $U_0$, we obtain

$$\left|\frac{\partial}{\partial t} + U_0 \frac{\partial}{\partial x}\right| \left[\frac{\partial^2 \phi}{\partial x^2} + \frac{\partial^2 \phi}{\partial y^2}\right] + \frac{\partial \phi}{\partial x}\left[\beta - \frac{\partial^2 U_0}{\partial y^2}\right] = \frac{1}{\rho_s}\frac{\partial}{\partial z}\rho_s w_1, \quad (7.6.14)$$

where, of course, $w_1$ is the vertical velocity of the perturbation. If (7.6.14) is multiplied by $-\rho_s \phi$ and integrated over the meridional plane, an equation for the kinetic energy

$$K(\phi) = \int_0^{z_T} \int_{-1}^1 dy\, dz \frac{\rho_s}{2}\left[\left(\frac{\overline{\partial \phi}}{\partial x}\right)^2 + \left(\frac{\overline{\partial \phi}}{\partial y}\right)^2\right] \quad (7.6.15)$$

results, i.e., after integration by parts,

$$\frac{\partial K}{\partial t} = + \int_0^{z_T} \int_{-1}^1 dy\, dz\, \rho_s \overline{\frac{\partial \phi}{\partial x}\frac{\partial \phi}{\partial y}}\frac{\partial U_0}{\partial y} + \int_0^{z_T} \int_{-1}^1 \rho_s \overline{\frac{\partial \phi}{\partial z} w_1}\, dy\, dz \quad (7.6.16)$$

if friction is ignored.

In the absence of horizontal shear the only source of kinetic energy for the perturbations is

$$\int_{-1}^1 \int_0^{z_T} \rho_s \overline{\frac{\partial \phi}{\partial z} w_1}\, dy\, dz = \int_{-1}^1 \int_0^{z_T} \rho_s \overline{\theta_0 w_1}\, dy\, dz, \quad (7.6.17)$$

where $\theta_0$ is the potential temperature of the perturbations. Thus in the absence of horizontal shear the perturbations may gain kinetic energy only if, when averaged over the mass of fluid, warm fluid (high $\theta_0$) rises and cold fluid (low $\theta_0$) sinks.

Using the relations

$$\left(\frac{\partial}{\partial t} + U_0 \frac{\partial}{\partial x}\right)\zeta = w_1,$$

$$\left(\frac{\partial}{\partial t} + U_0 \frac{\partial}{\partial x}\right)\eta = v_0, \quad (7.6.18)$$

it follows that

$$\theta_0 = -\eta \frac{\partial \Theta_0}{\partial y} - \zeta S. \quad (7.6.19)$$

Hence in the absence of horizontal shear

$$\frac{\partial K}{\partial t} = -\int_{-1}^{1}\int_{0}^{z_T} \rho_s \, dy \, dz \left\{ \overline{w_1 \zeta} S + \overline{w_1 \eta} \frac{\partial \Theta_0}{\partial y} \right\}$$

$$= -\int_{-1}^{1}\int_{0}^{z_T} S \frac{\rho_s}{2} \, dy \, dz \frac{\partial}{\partial t} \overline{\zeta^2} \times \left[ 1 + \frac{\overline{w_1 \eta}}{\overline{w_1 \zeta}} \frac{\partial \Theta_0}{\partial y} S^{-1} \right]. \qquad (7.6.20)$$

If, as in Figure 7.6.1, $\partial \Theta_0 / \partial y < 0$, then an increase of kinetic energy in an unstable disturbance, in which $(\partial / \partial t)\overline{\zeta^2} > 0$, requires that at least somewhere in the meridional plane

$$-\frac{\partial \Theta_0}{\partial y} S^{-1} \frac{\overline{w_1 \eta}}{\overline{w_1 \zeta}} > 1. \qquad (7.6.21)$$

If we note that

$$\frac{\overline{w_1 \eta}}{\overline{w_1 \zeta}} = O\left( \frac{v_0}{w_1} \right), \qquad (7.6.22)$$

then (7.6.21) is in fact the same condition as (7.6.12). The fact underscored by considering (7.6.16) is that in the absence of horizontal shear the slanted thermal convection described here is the only energy source for instability. Although the horizontal gradients of $\theta_*$ imply a vertical shear of the basic current, there is no conversion term related to a vertical Reynolds stress, $\overline{w_* u_*}$, acting on this shear. This is because $w_*$ is $O(\varepsilon)$ for quasigeostrophic flow and consequently the vertical Reynolds stresses are too feeble to convert the *kinetic* energy in the vertical shear. Thus the instability associated with the vertical shear of large-scale flows must be related to potential-energy conversion. By the arguments presented above this implies scales of motion of the order of the Rossby-deformation radius. In the following sections several examples of baroclinic instability will be discussed in detail.

## 7.7 Eady's Model

An elegantly simple model of baroclinic instability which shows the instability process in its purest form was introduced by Eady (1949). In this model the velocity $U_0$ is independent of $y$ and the horizontal potential-temperature gradient is constant, i.e.,

$$U_0 = z. \qquad (7.7.1a)$$

Thus,

$$\frac{\partial \Theta_0}{\partial y} = -1, \qquad (7.7.1b)$$

and so by (7.6.10)

$$\left( \frac{\partial z_*}{\partial y_*} \right)_{\theta_*} = \frac{D}{L} \varepsilon S^{-1}. \qquad (7.7.2)$$

Both $S$ and $\rho_s$ are taken as constants, and the effect of the earth's sphericity is purposely neglected by setting $\beta$ equal to zero. The flow is furthermore confined in the vertical direction by two rigid horizontal boundaries at $z = 0$ and $z = 1$. The effect of friction is also neglected. Note that for constant $S$, (7.7.2) implies that the potential-temperature surfaces have constant slope.

It follows from (7.7.1a) and (7.4.5) that for this flow the potential-vorticity gradient of the basic state identically vanishes, i.e.,

$$\frac{\partial \Pi_0}{\partial y} = 0, \tag{7.7.3}$$

so that (7.4.4) becomes

$$(z - c)\left\{ S^{-1}\frac{\partial^2 \Phi}{\partial z^2} + \frac{\partial^2 \Phi}{\partial y^2} - k^2 \Phi \right\} = 0. \tag{7.7.4}$$

On $z = 0$, (7.4.7) applies; in the absence of friction and bottom slope, it becomes

$$-c\frac{\partial \Phi}{\partial z} - \Phi = 0, \tag{7.7.5}$$

while at $z = 1$, (7.4.8) yields

$$(1 - c)\frac{\partial \Phi}{\partial z} - \Phi = 0. \tag{7.7.6}$$

Solutions of (7.7.4) which satisfy (7.4.6) may be sought in the form

$$\Phi(y, z) = A(z)\cos l_n y, \tag{7.7.7}$$

where

$$l_n = (n + \tfrac{1}{2})\pi, \qquad n = 0, 1, 2, \ldots, \tag{7.7.8}$$

and $n$ is any integer.

$A(z)$ then satisfies the ordinary differential equation

$$(z - c)\left[\frac{d^2 A}{dz^2} - \mu^2 A\right] = 0, \tag{7.7.9}$$

where

$$\mu^2 = (k^2 + l_n^2)S, \tag{7.7.10}$$

and the boundary conditions

$$+c\frac{dA}{dz} + A = 0, \qquad z = 0, \tag{7.7.11a}$$

$$(c - 1)\frac{dA}{dz} + A = 0, \qquad z = 1. \tag{7.7.11b}$$

Consider first the nonsingular solutions of (7.7:9), i.e., those solutions for which

$$\frac{d^2 A}{dz^2} - \mu^2 A = 0 \tag{7.7.12}$$

and which must be satisfied for complex $c$. The general solution of (7.7.12) may be written as

$$A(z) = a \cosh \mu z + b \sinh \mu z, \tag{7.7.13}$$

where $a$ and $b$ are arbitrary constants. The application of (7.7.11a,b) leads to two linear algebraic equations for $a$ and $b$, i.e.,

$$a + \mu c b = 0, \tag{7.7.14a}$$

$$a[(c - 1)\mu \sinh \mu + \cosh \mu] + b[(c - 1)\mu \cosh \mu + \sinh \mu] = 0. \tag{7.7.14b}$$

Nontrivial solutions for $a$ and $b$ can be found only if the determinant of the coefficients of $a$ and $b$ in (7.7.14a,b) vanishes. This condition yields a quadratic equation for $c$ as a condition for a solution of (7.7.12):

$$c^2 - c + \mu^{-1} \coth \mu - \mu^{-2} = 0 \tag{7.7.15}$$

or

$$c = \tfrac{1}{2} \pm \{\tfrac{1}{4} + \mu^{-2} - \mu^{-1} \coth \mu\}^{1/2}. \tag{7.7.16}$$

The identity

$$\coth \mu = \frac{1}{2}\left[\tanh \frac{\mu}{2} + \coth \frac{\mu}{2}\right] \tag{7.7.17}$$

allows (7.7.16) to be rewritten

$$c = \frac{1}{2} \pm \frac{1}{\mu}\left[\left(\frac{\mu}{2} - \coth \frac{\mu}{2}\right)\left(\frac{\mu}{2} - \tanh \frac{\mu}{2}\right)\right]^{1/2}. \tag{7.7.18}$$

Now, for all $\mu$, $\mu/2 \geq \tanh \mu/2$, hence when $\mu/2$ exceeds $\coth \mu/2$ the radicand, in (7.7.18) is positive and both roots for $c$ are real. On the other hand, for those values of $\mu/2$ such that $\mu/2$ is less than $\coth \mu/2$, the radicand is negative and $c$ is complex. The critical value of $\mu$ is therefore given by

$$\frac{\mu_c}{2} = \coth \frac{\mu_c}{2}, \tag{7.7.19}$$

whose numerical value is

$$\mu_c = 2.3994. \tag{7.7.20}$$

For $\mu > \mu_c$ the solutions for each $k$ and $n$ consist of two neutral waves, each with a real phase speed $c$, given by the roots of (7.7.18). The roots are shown in Figure 7.7.1(b). For very large $\mu$ (high wave number) one value of $c$ approaches zero, the velocity at the lower boundary, while the other root

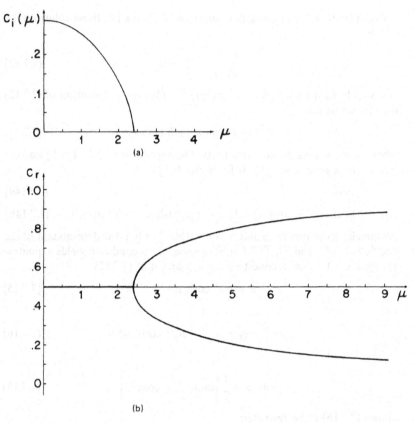

**Figure 7.7.1**  (a) The imaginary part of $c$, $c_i$, as a function of $\mu = (k^2 + l_n^2)^{1/2}S^{1/2}$. (b) The real part of $c$, $c_r$, as a function of $\mu$; note the coalescence at the critical wave number $\mu_c$.

approaches one, the velocity at the upper boundary. For large $\mu$ the first solution has the form

$$A(z) \sim e^{-\mu z}, \qquad c \to 0, \tag{7.7.21}$$

while the second mode, with $c \to 1$, has the form

$$A(z) \sim e^{\mu(z-1)}, \qquad c \to 1. \tag{7.7.22}$$

These two neutral waves are each trapped near either the lower and upper boundaries. In each case the opposite boundary is so far away that $\Phi$ is essentially zero there, and (7.4.22) shows that since $\partial \Pi_0 / \partial y$ is zero, $c_i$ must vanish if $\Phi$ is zero at one of the boundaries.

On the other hand, for $\mu < \mu_c$ (7.7.18) yields two roots for $c$ which are *complex conjugates*. In the absence of friction the coefficients of (7.4.4) and the accompanying boundary condition are real, aside from factors involving

c. Hence, in the general inviscid problem, if $\Phi$ is a solution of (7.4.4) with the accompanying eigenvalue $c$, $\Phi^*$ will also be a solution for the same $k$ with $c$ replaced by its complex conjugate $c^*$, an example of which is presented by (7.7.18). As $\mu$ approaches $\mu_c$ from above, the two real phase speeds coalesce. For $\mu < \mu_c$ the real part of the phase speed, $c_r$, is given by the mean velocity of the basic current, i.e., $c_r = 0.5$. Figure 7.7.1(a) shows $c_i(\mu)$ for the unstable mode $(c_i > 0)$; the other root $(c_i < 0)$ is obtained by reflection around the $\mu$-axis. It follows from (7.7.10) that for instability to occur for some real value of $k$, $S$ must satisfy the condition

$$S < \frac{\mu_c^2}{l_n^2} = 4 \frac{\mu_c^2}{\pi^2(2n+1)^2} = \frac{2.333}{(2n+1)^2}. \tag{7.7.23}$$

Clearly this condition for instability is most easily satisfied for the $n = 0$ mode, i.e., the least wiggly mode in $y$ is the most unstable. Aside from numerical factors, (7.7.23) was anticipated by (7.6.13), the discussion of the physical basis for baroclinic instability. The growth rate of the unstable mode is

$$kc_i = \frac{k}{\mu}\left[\left(\coth\frac{\mu}{2} - \frac{\mu}{2}\right)\left(\frac{\mu}{2} - \tanh\frac{\mu}{2}\right)\right]^{1/2}. \tag{7.7.24}$$

Since $\mu$ is a function of $k$ for a given $l_n$ and $S$, Equation (7.7.24) with (7.7.10) determines the behavior of the growth rate as a function of wave number. Figure 7.7.2(a) shows the growth rate for the mode $n = 0$ (the most unstable mode) and $S = 0.25$. For this value of $S$, $L = 2L_D$, so that the lateral extent of the region from $y = -1$ to $y = +1$ corresponds to a dimensional width of $4L_D$. Although $c_i(\mu)$ attains its greatest values at very long waves, the growth rate $kc_i$ vanishes as $k \to 0$ in accordance with (7.5.24). The growth rate vanishes for large $k$, for which $\mu > \mu_c$. At the intermediate value (for this $n$ and $S$)

$$k_m = 3.1277, \tag{7.7.25}$$

the growth rate achieves its maximum. This wave is the most unstable wave, and if the initial disturbance is composed of a synthesis of waves of all wave numbers, each wave possessing initially about the same amplitude, we can expect that the wave of wave number $k_m$ will emerge most rapidly and dominate the structure of the disturbance field. It is plausible, therefore, to associate the most unstable wave with the disturbance most favored by the instability mechanism to be observed in finite amplitude. The wave number $k_m$ corresponds to a *dimensional* wavelength, which for $S = 0.25$ is

$$\lambda_* = \frac{2\pi}{k_m}L = \frac{4\pi}{k_m}L_D = (4.018)L_D. \tag{7.7.26}$$

If $L_D$ is $10^3$ km, as in the atmosphere, (7.7.26) predicts that wavelengths of the order of 4,000 km will dominate the spectrum of atmospheric fluctuations. The quarter wavelength (i.e., half the distance from crest to trough) predicted by (7.7.26) is $O(L_D)$ and therefore $O(1,000$ km$)$, and is in such

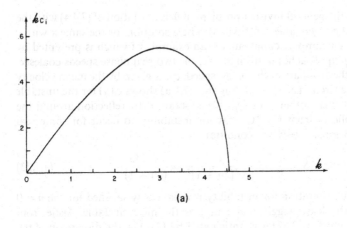

(a)

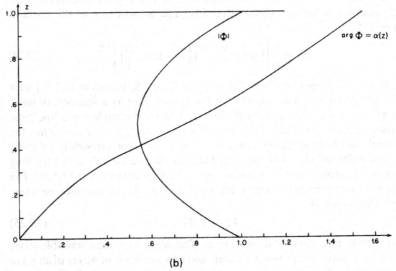

(b)

**Figure 7.7.2**   (a) The growth rate, $kc_i$, for the most unstable Eady mode, i.e., the mode proportional to $\cos \frac{1}{2}\pi y$, with $S = 0.25$. (b) The amplitude $|\Phi|$ and the phase $\alpha(z)$ as a function of height for the most unstable wave. Note that the increase of $\alpha$ with $z$ implies the tilting rearward of the wave with height, which in turn implies the release of potential energy by the disturbance.

excellent agreement with the observed scale of synoptic atmospheric disturbances that the mechanism of baroclinic instability becomes at once the most plausible explanation for the existence of the observed transient long waves in the atmosphere.

Now that $c$ is known, either (7.7.14a) or (7.7.14b) may be used to determine $b$ in terms of $a$, and therefore to yield the structure of the unstable wave, i.e., aside from a constant factor,

$$A(z) = \cosh \mu z - \frac{\sinh \mu z}{\mu c}. \qquad (7.7.27)$$

The solution (7.7.27) may be multiplied by an arbitrary constant. Linear theory is incapable of determining the amplitude of the disturbance. For $\mu < \mu_c$, this in turn yields

$$\phi = \text{Re } e^{kc_i t} e^{ik(x - 0.5t)}$$

$$\times \left[ \left( \cosh \mu z - \frac{0.5}{\mu |c|^2} \sinh \mu z \right) + \frac{ic_i \sinh \mu z}{\mu |c|^2} \right] \cos l_n y, \qquad (7.7.28a)$$

or

$$\phi = e^{kc_i t} |\Phi(z)| \cos l_n y \cos\{kx + \alpha(z) - 0.5kt\}, \qquad (7.7.28b)$$

where the distribution of wave amplitude with height is given by

$$|\Phi(z)| = \left[ \left( \cosh \mu z - \frac{0.5 \sinh \mu z}{\mu |c|^2} \right)^2 + \frac{c_i^2 \sinh^2 \mu z}{\mu^2 |c|^4} \right]^{1/2}, \qquad (7.7.29)$$

while the phase angle $\alpha(z)$ is given by

$$\alpha(z) = \tan^{-1} \left| \frac{c_i \sinh \mu z}{\mu |c|^2 \cosh \mu z - 0.5 \sinh \mu z} \right|. \qquad (7.7.30)$$

For $c_i \neq 0$ the phase of the wave will vary with height, and the lines of constant phase at each *instant* satisfy

$$x = \frac{-\alpha(z)}{k} + \text{constant}. \qquad (7.7.31)$$

Figure 7.7.2(b) shows $|\Phi|$ and $\alpha(z)$ for the most unstable wave. Note that since $\alpha$ is an increasing function of $z$, lines of constant phase must tilt upward and westward (against the current) with height. This is precisely the condition of (7.3.17) for the baroclinic conversion of energy from the mean field to the fluctuations. From Figure 7.7.2(b) we see that the wave at $z = 0$ leads the wave field at $z = 1$ by very nearly $\pi/2$ radians. That is, the crests and troughs of the isobar pattern at the upper level *lag* the surface wave by nearly 90°. The amplitude, $|\Phi|$, is very nearly symmetric about the midpoint $z = 0.5$. $|\Phi|$ attains its minimum value at this point, where $z = c_r$, called the *steering level*, i.e., where $U_0(z) = c_r$.

The northward heat flux at any $y$ and $z$ is given by

$$\rho_s \overline{\frac{\partial \phi}{\partial x} \frac{\partial \phi}{\partial z}}.$$

With (7.7.28b) this may be written

$$\rho_s \overline{\frac{\partial \phi}{\partial x} \frac{\partial \phi}{\partial z}} = \frac{|\Phi(z)|^2}{2} k \frac{\partial \alpha}{\partial z} \cos^2 l_n y \, e^{2kc_i t}. \qquad (7.7.32)$$

Since $\partial\alpha/\partial z$ is everywhere positive, the heat flux is everywhere down the basic temperature gradient, i.e., northward. The horizontal Reynolds stress vanishes identically, i.e.,

$$\rho_s \overline{\frac{\partial\phi}{\partial x}\frac{\partial\phi}{\partial y}} = 0, \tag{7.7.33}$$

since $u_0$ and $v_0$ are 90° out of phase, as they must be for *all* $U_0$ independent of $y$. Since $\partial\Pi_0/\partial y$ is identically zero, the potential-vorticity flux also vanishes, i.e.,

$$\overline{v_0 q} = 0, \tag{7.7.34}$$

from which, with (7.7.33), (7.7.34), and (7.3.21), the heat flux given by (7.7.32) must be independent of height. In fact, using the formulae for $|\Phi|$ and $\alpha(z)$ it can be shown that for the Eady wave (7.7.27)

$$\rho_s \overline{v_0\theta_0} = \rho_s \overline{\frac{\partial\phi}{\partial x}\frac{\partial\phi}{\partial z}} = \frac{kc_i}{2|c|^2} e^{2kc_i t} \cos^2 l_n y > 0. \tag{7.7.35}$$

For each $k$ and $n$, i.e., for each horizontal planform of the disturbance, the normal-mode analysis yields *two* modes in $z$ with real or imaginary $c$, depending on the magnitude of $\mu$. In principle, however, for each $k$ and $n$ an *initial* disturbance with an arbitrary vertical structure can be specified. If the normal modes were complete, this arbitrary disturbance could be represented as a sum of the normal modes. The two modes obtained from (7.7.12), (7.7.14), and (7.7.16) are obviously not complete, and it is natural to ask if other free solutions have been arbitrarily excluded. The answer is yes. In proceeding from (7.7.9) to (7.7.12) only the nonsingular solutions to (7.7.9) were obtained, i.e., only continuous solutions possessing continuous derivatives of any order. Consider, however, the case where $c$ is real. Then from (7.7.9) it follows that $A(z)$ must satisfy (7.7.12) for *all* $z$ except perhaps the single point $z_c$ where $z = c$. Thus if $c$ is real and lies anywhere in the range $(0, 1)$ the appropriate equation for $A(z)$ is, in fact,

$$\frac{d^2 A}{dz^2} - \mu^2 A = B\delta(z - z_c), \qquad z_c = c, \tag{7.7.36}$$

where $B$ is any constant, and $\delta$ is the Dirac delta function. This follows from the general fact that if a function $q(z)$ satisfies

$$(z - z_c)q(z) = 0, \tag{7.7.37}$$

then

$$q(z) = B\delta(z - z_c). \tag{7.7.38}$$

The meaning of (7.7.38) is understood from the integral limit

$$\lim_{z_0 \to 0} \int_{z_c - z_0}^{z_c + z_0} q(z)\, dz = B \tag{7.7.39}$$

while $q(z)$ is zero for all $z \neq z_c$. From (7.7.38), $(z - z_c)q(z)$ is clearly zero for $z \neq z_c$, while

$$\lim_{z_0 \to 0} \int_{z_c - z_0}^{z_c + z_0} (z - z_c)q(z) \, dz = 0 \cdot B = 0. \tag{7.7.40}$$

The solutions found previously corresponded to the choice $B = 0$. These are the normal modes, solutions which yield the unstable Eady waves. The solutions of (7.7.36) with $B \neq 0$ are singular only in the sense that the perturbation potential vorticity has a delta-function singularity at $z = z_c$.[†] There are an infinite number of such solutions, each corresponding to a real value of $c$ in the range of the basic velocity. These solutions must be added to the Eady modes to completely represent an arbitrary initial disturbance. Since $z_c$ is continuously distributed between (0, 1), these singular solutions represent a *continuous* spectrum of eigenvalues $c$ over the interval (0, 1). Although this continuous spectrum is required for the completeness of the total solution, the fact that the continuous spectrum corresponds only to stable waves allows us to focus entirely on the previously obtained normal modes insofar as the question of stability is concerned, and no further discussion of the nature of the continuous spectrum will be presented here.

Certain features of Eady's model are highly unrealistic for application to either the atmosphere or the ocean. Most important is the absence of an interior potential-vorticity gradient, in particular the absence of the planetary-vorticity gradient. Since $\partial \Pi_0 / \partial y$ is, in Eady's model, identically zero, (7.4.22) shows that the presence of a rigid boundary at $z = z_T$ is essential for the Eady instability. As $z_T \to \infty$, (7.4.22) and (7.7.23) both demonstrate that $c_i$ must approach 0. Although the Eady model demonstrates the essential character of the instability, the detailed structure of the unstable wave, which naturally must reflect the constraints of the potential-vorticity dynamics, may be expected to alter considerably in the presence of an ambient potential-vorticity gradient.

---

[†] The solution of (7.7.36) such that $A$ is continuous at $z = z_c$ can be shown to be, for any $z_c$ in the interval (0, 1),

$$A(z) = \frac{-B[(z_c - 1)\mu \sinh \mu + \cosh \mu]}{(z_c - c_1)(z_c - c_2)\mu^3 \sinh \mu} U(z_<)V(z_>),$$

where

$$U(z) = \mu z_c \cosh \mu z - \sinh \mu z,$$

$$V(z) = \frac{(z_c - 1)\mu \cosh \mu + \sinh \mu}{(z_c - 1)\mu \sinh \mu + \cosh \mu} \cosh \mu z - \sinh \mu z,$$

and

$$z_< = z, \quad z_> = z_c \quad \text{when } z < z_c,$$

$$z_< = z_c, \quad z_> = z \quad \text{when } z > z_c,$$

while $c_1$ and $c_2$ are the two roots of (7.7.18).

## 7.8 Charney's Model and Critical Layers

A more realistic model of baroclinic instability was formulated by Charney (1947). Although the model retains several of the simplifying features of Eady's model (e.g., no dependence of $U_0$ on $y$, the neglect of viscosity and constant vertical shear) it introduces certain additional important and realistic dynamical elements. In particular, the $\beta$-effect is included so that in the interior of the fluid an ambient potential vorticity gradient influences the motion of fluid elements. Furthermore, with application to the atmosphere in mind, Charney retained a finite value of the nondimensional density scale height

$$H^{-1} = \frac{1}{\rho_s}\frac{\partial \rho_s}{\partial z}. \tag{7.8.1}$$

For the sake of simplicity, both $H$ and $S$ (i.e., $N_s^2$) are taken to be constant. The zonal velocity in the basic state is

$$U_0 = z, \tag{7.8.2}$$

which implies that the scaling velocity $U$, is, in fact, equal to

$$U = D\frac{\partial U_*}{\partial z_*},$$

where $\partial U_*/\partial z_*$ is the dimensional vertical shear of the zonal flow while $D$ is the characteristic vertical scale of the motion. With (7.8.2a) the normal mode equation (7.4.4) becomes

$$(z - c)\left(\frac{d^2 A}{dz^2} - \frac{1}{H}\frac{dA}{dz} - \mu^2 A\right) + \left(\beta S + \frac{1}{H}\right)A = 0, \tag{7.8.3}$$

where solutions of the form

$$\Phi(y, z) = A(z)\cos(n + \tfrac{1}{2})\pi y, \qquad n = 0, 1, 2, \ldots, \tag{7.8.4}$$

are sought in order to satisfy (7.4.6). The potential vorticity gradient of the basic state is

$$\frac{\partial \Pi_0}{\partial y} = \beta + \frac{1}{SH}, \tag{7.8.5}$$

while

$$\mu^2 = (k^2 + (n + \tfrac{1}{2})^2 \pi^2)S. \tag{7.8.6}$$

Note that in terms of dimensional quantities

$$\beta S + \frac{1}{H} = \frac{\beta_0}{\partial U_*/\partial z_*}\frac{N^2}{f^2}D + \frac{D}{H_*}, \tag{7.8.7}$$

where $H_*$ is the dimensional density scale height, while

$$\mu^2 = (k_*^2 + l_*^2)L_D^2, \tag{7.8.8}$$

and $k_*$ is the dimensional zonal wave number and $l_*$ is the dimensional meridional wave number, $(n + \frac{1}{2})\pi/L$. Thus $\mu$ is simply the total wave number measured in units of the internal Rossby deformation radius $L_D = N_s D/f_0$. The scaling relations discussed above all involve the vertical scale of the motion, $D$. In Charney's model the upper boundary is taken at $z = \infty$, so neither the basic state velocity or static stability fields can specify the vertical scale. In fact, one of the novel features of Charney's model is the way in which the unstable perturbation field selects its own vertical scale. In Eady's model, on the other hand, the vertical scale of the unstable disturbances was the full finite depth of the fluid layer. That, of course, had to be the case since in Eady's model $\partial \Pi_0/\partial y$ is zero and the perturbation field must interact with the temperature gradients on *both* boundaries to satisfy (7.4.22). The presence of a nonzero basic potential vorticity gradient has an immense influence on the perturbation dynamics. As (7.4.22) demonstrates now that $\partial \Pi_0/\partial y$ is nonzero, instability is permitted in the absence of a boundary at $z = z_T$, i.e., if $z_T \to \infty$.

The presence of a horizontal temperature gradient at the lower boundary now can allow instability as long as $\partial \Pi_0/\partial y \, (\partial U_0/\partial z - S \, \partial \eta_B/\partial y) > 0$. The existence of a nonzero $\partial \Pi_0/\partial y$ has an important effect on selecting the vertical scale of the unstable disturbances and in this regard the discussions of Held (1978) and Branscome (1980) are particularly illuminating.

First, note that the factor $(\beta S + 1/H)$ whose nonzero value distinguishes the normal mode equation in Charney's model from that of Eady's can be written

$$\beta S + \frac{1}{H} = D\left(\frac{1}{h_*} + \frac{1}{H_*}\right), \tag{7.8.9}$$

where

$$h_* = \frac{\dfrac{f_0^2}{N_s^2} \dfrac{\partial U_*}{\partial z_*}}{\beta_0}. \tag{7.8.10}$$

The vertical length scale $h_*$ depends on the shear and $\beta_0$ and is the scale over which advection of planetary vorticity balances the stretching of planetary vorticity by the vertical velocity field, itself related to the advection of potential temperature. The dominant term in (7.8.9) depends on which of $h_*$ and $H_*$ is the *smaller*. Since all other terms in (7.8.3) are O(1) we anticipate that the vertical scale of the motion will be the *smaller* of the $h_*$ and $H_*$. Thus if the density scale height is large enough, then

$$D = h_* = \frac{\dfrac{f_0^2}{N_s^2} \dfrac{\partial U_*}{\partial z_*}}{\beta_0} < H_*, \tag{7.8.11a}$$

while if the inequality in (7.8.11a) is reversed

$$D = H_* < h_*. \tag{7.8.11b}$$

For weak shear, or strong $\beta_0$, the perturbation motion will shrink in vertical scale, $h_*$, and be concentrated at the lower boundary where the destabilizing effect of the horizontal temperature gradient can be felt.

For $\mu = O(1)$, the horizontal scale of the motion is simply the deformation radius, $L_D$. If (7.8.11a) is satisfied, i.e., if $D = h_* < H_*$, then

$$L_D = \frac{\dfrac{f_0}{N_s}\dfrac{\partial U_*}{\partial z_*}}{\beta_0},$$ (7.8.12)

which should be the order of magnitude of the length scale of the unstable disturbance (if any exist), while if $H_* < h_*$

$$L_D = \frac{N_s H_*}{f_0}.$$ (7.8.13)

Whereas in Eady's model the scales were independent of the motion field in Charney's model, the horizontal scale of the most unstable disturbance (say) can be expected to depend on the vertical shear of the basic flow.

Using the same scaling relations the characteristic growth rate will be the order of the horizontal wave number multiplied by $c_i$. Since $c$ is, by the semi-circle theorem, expected to be of the order of the basic state velocity, it follows that we can anticipate that

$$\sigma_* = k_* c_{i*} = O\left(\frac{U}{L_D}\right) = \frac{D}{L_D}\frac{\partial U_*}{\partial z_*}.$$ (7.8.14)

Since $L_D = N_s D/f_0$, the vertical scale drops out of the estimate in (7.8.14) so that

$$\sigma_* = O\left(\frac{f_0}{N_s}\frac{\partial U_*}{\partial z_*}\right)$$ (7.8.15)

in *both limits*; (7.8.11a) and (7.8.11b). In fact, (7.8.15) is also exactly the scale for the growth rate in Eady's model. Thus on the basis of these scaling arguments, we can anticipate important changes in the vertical and horizontal scales of the unstable disturbances due to the presence of the basic state potential vorticity gradient but, aside from $O(1)$ numerical factors, relatively little effect on the growth rate of the disturbance.

The second significant difference between Charney's model and Eady's is the fact that for $\partial \Pi_0/\partial y \neq 0$, (7.8.3) is singular at the point $z_c$, where

$$U_0(z_c) = c,$$ (7.8.16)

which for (7.8.2) is simply

$$z_c = c.$$ (7.8.17)

If $c_i$ is positive, i.e., for the growing disturbance, the singularity of the differential equation occurs only for complex values of $z$, and therefore the physical interval of the problem, $z > 0$ and real, is free of the singularity.

Nevertheless the singularity, considered in the complex $z$-plane, has important implications for the physical and mathematical structure of the disturbance. In particular, for small $c_i$ or for neutral waves, the singularity will have tremendous impact if $c_r$ lies within the range of $U_0$. The point $z_c$ defined by (7.8.16) is the critical point of (7.4.4), and the line $z = z_c$ is termed the critical level and its immediate neighborhood is called the critical layer. When $c$ is real, the critical point lies in the range of $z$ if $c$ lies in the range of $U_0$, while if $c_i$ is small and $c_r$ lies in the range of $U_0$ (as it does, recall, in Eady's model), the critical layer is apparent in the structure of the disturbance.

This fact may be seen in several ways. The general solution of (7.4.4), when $U_0$ is a function of $z$ only, may be obtained in terms of a generalized power-series expansion about the point $z = z_c$ by the method of Frobenius.* The general solution can be written as the sum of the two independent solutions

$$A_1(z) = (z - z_c)[1 + a_1(z - z_c) + a_2(z - z_c)^2 + \cdots] \qquad (7.8.18)$$

and

$$A_2(z) = A_1(z)\left[\frac{(\partial\Pi_0/\partial y)(z_c)}{(\partial U_0/\partial z)(z_c)}\right] S \log(z - z_c)$$
$$\qquad (7.8.19)$$
$$- [1 + b_1(z - z_c) + b_2(z - z_c)^2 + \cdots],$$

where the $a_n$'s and $b_n$'s depend on the detailed distribution of $U_0$, $\rho_s$, and $S$ about the point $z_c$. The crucial thing to note, however, is that $A_2(z)$ will contain a logarithmic singularity at the point $z = z_c$ unless $\partial\Pi_0/\partial y$ vanishes there. If, as in Charney's model, $(\partial\Pi_0/\partial y)(z_c)$ is nonzero, the log term in (7.8.19) is an essential part of the solution. The logarithm is, of course, a multivalued function and

$$\log(z - z_c) = \log|z - z_c| + i\alpha, \qquad (7.8.20a)$$

where we have written

$$z - z_c = |z - z_c|e^{i\alpha}. \qquad (7.8.20b)$$

Consider the case where $c_i$ is slightly greater than zero. Then as shown in Figure 7.8.1, the point $z = z_c$ will lie slightly above the real axis in the complex $z$-plane. To render the solution (7.8.19) physically meaningful, a branch of the logarithm must be chosen. The branch cut required to render the logarithm single valued is chosen so as not to impede, for $c_i > 0$, the real line, which is the physical domain. Now consider a point on the real line slightly to the left (i.e., below) the critical point. At this point $\alpha$ is very nearly $-\pi$ and abruptly changes to zero as the point under consideration moves across the critical layer to $z > z_c$. The physical significance of this rapid phase change across the critical layer is clear from (7.7.32), where the northward flux of heat is seen to be directly proportional to the rapidity with which $\alpha$ changes with $z$.

---

* See for example Hildebrand (1963).

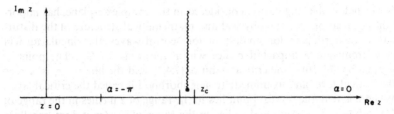

**Figure 7.8.1**   The singularity of the stability equation in the complex $z$-plane is at the point $z_c$, which lies slightly above the real axis if $c_i > 0$.

In the vicinity of the critical layer the general solution to (7.8.3) can be written as a linear combination of (7.8.18) and (7.8.19), i.e.,

$$A(z) = A_1(z) + RA_2(z),  \tag{7.8.21}$$

where $R$ is an arbitrary, complex constant. Since

$$\overline{v_0\theta_0}^x = \frac{ik}{4}e^{2kc_it}\left[A\frac{\partial A^*}{\partial z} - A^*\frac{\partial A}{\partial z}\right]\cos^2(n + \tfrac{1}{2})\pi y,  \tag{7.8.22}$$

a little algebra shows that as $z - z_c \to 0$,

$$\overline{v_0\theta_0}^x = -\frac{\alpha Sk}{2}\frac{|R|^2}{\partial U_0/\partial z}\frac{\partial\Pi_0}{\partial y}\cos^2(n + \tfrac{1}{2})\pi ye^{2kc_it},  \tag{7.8.23}$$

if (7.8.20) is used.

Thus, for all cases in which $c_r$ lies in the range of $U_0$, the rapid change in $\alpha$ in the critical layer region implies a heat flux and therefore a release of potential energy. The wave–mean flow interaction theory of Section 6.14 further implies that the rapid variation in $z$ of the heat flux will give rise locally to potential vorticity fluxes and alterations of the mean flow. Note that the heat flux in (7.8.23) is proportional to $\partial\Pi_0/\partial y$ at the critical layer. This implies the rather unexpected consequence that the presence of the $\beta$-effect can be expected to introduce the possibility of new, unstable modes in the presence of shear, although alone it provides a stabilizing restoring mechanism as manifested by the Rossby wave. It is clear that if $c$ is real (neutral waves) it must, in general, lie outside the range of $U_0$, else the complex quantity introduced into the solution by $\log(z - z_c)$ cannot be balanced by any other term in the boundary condition at $z = 0$.

The mathematical considerations have been alternatively described in helpful physical terms by Bretherton (1966). If (7.3.21) is integrated over the meridional plane, we obtain

$$\int_0^{z_T} dz \int_{-1}^1 dy\, \rho_s\overline{v_0 q} = \int_{-1}^1\left[\rho_s\frac{\overline{v_0\theta_0}}{S}\right]_{z=z_T} dy - \int_{-1}^1\left[\rho_s\frac{\overline{v_0\theta_0}}{S}\right]_{z=0} dy,  \tag{7.8.24}$$

so that the meridional flux of potential vorticity must be balanced by a heat flux at $z = 0$ and/or $z = z_T$. Since, by (7.3.26)

$$\overline{v_0 q} = -\left(\frac{\partial}{\partial t} \frac{\overline{\eta^2}}{2}\right) \frac{\partial \Pi_0}{\partial y}, \qquad (7.8.25a)$$

this flux of potential vorticity is negative for an unstable wave if $\partial \Pi_0 /\partial y > 0$ and therefore must be balanced by a positive heat flux at the lower boundary or a negative heat flux at the upper boundary. Since, from (7.3.28) and (7.3.29),

$$(\overline{v_0 \theta_0})_{z=z_T} = \left(\frac{\partial U_0}{\partial z}\right)_{z=z_T} \frac{\partial \overline{\eta^2}/2}{\partial t}, \qquad (7.8.25b)$$

$$(\overline{v_0 \theta_0})_{z=0} = \left(\frac{\partial U_0}{\partial z} - S\frac{\partial \eta_B}{\partial y}\right)_{z=0} \frac{\partial \overline{\eta^2}/2}{\partial t}, \qquad (7.8.25c)$$

the potential vorticity flux must be balanced by a positive heat flux at the lower boundary if $\partial U_0 /\partial z > 0$. This, in fact, is the content of (7.4.22). Furthermore, from (7.4.26)

$$\frac{\partial}{\partial t} \frac{\overline{\eta^2}}{2} = \frac{kc_i |\Phi|^2}{2|U_0 - c|^2} e^{2kc_i t}$$

$$= \frac{kc_i}{2}\left\{\frac{|\Phi|^2 e^{2kc_i t}}{(U_0 - c_r)^2 + c_i^2}\right\}, \qquad (7.8.26)$$

which is positive for $kc_i > 0$. Now consider disturbances for which $c_i$ is very small. In the limit $c_i \to 0$, $(\partial/\partial t)\overline{\eta^2}$ will vanish for all $z$ except at the critical level where $U_0 - c_r$ vanishes. At the critical level the dispersion of fluid elements $(\partial/\partial t)\overline{\eta^2}$ increases indefinitely, producing a flux of potential vorticity at this level alone. If (7.8.26) is integrated in a small neighborhood across the critical level, then as $c_i \to 0$,

$$\int_{z_{c-}}^{z_{c+}} \frac{\partial}{\partial t} \frac{\overline{\eta^2}}{2}\, dz = \lim_{c_i \to 0} \int_{z_{c-}}^{z_{c+}} \frac{kc_i}{2} \frac{|\Phi|^2 e^{2kc_i t}}{(U_0 - c_r)^2 + c_i^2}\, dz$$

$$= \mathrm{Re}\lim_{c_i \to 0} \int_{z_{c-}}^{z_{c+}} \frac{k|\Phi|^2 e^{2kc_i t}}{2i(U_0 - c)}\, dz \qquad (7.8.27)$$

$$= \mathrm{Re}\lim_{c_i \to 0} \int_{z_{c-}}^{z_{c+}} \frac{k|\Phi|^2 \left(\frac{\partial U_0}{\partial z}(z_c)\right)^{-1}}{2i(z - z_c)}\, dz.$$

Thus by the residue calculus, the indented integration around the singularity at $z = z_c$ as shown in Figure 7.8.2 yields, as $c_i \to 0$,

$$\int_{z_{c-}}^{z_{c+}} \frac{\partial}{\partial t}\left(\frac{\overline{\eta^2}}{2}\right) dz = \frac{\pi k}{2} \frac{|\Phi(z_c)|^2}{(dU_0/dz)(z_c)} + \mathrm{O}(c_i \log c_i) \qquad (7.8.28)$$

Thus, the integrated potential-vorticity flux, for small $c_i$, is due entirely to the

**Figure 7.8.2** The path of integration for (7.8.27) passes below $z_c$ if $c_i > 0$, and as $c_i \to 0$ is equivalent to the indented contour shown.

flux at the critical layer and is, from (7.8.24),

$$\int_0^{z_T} dz \int_{-1}^{1} dy\, \rho_s \overline{v_0 q} = -\frac{\pi k}{2} \int_{-1}^{1} dy \frac{|\Phi(z_c)|^2}{(\partial U_0/\partial z)_{z=z_c}} \frac{\partial \Pi_0}{\partial y}. \qquad (7.8.29)$$

However, as $c_i \to 0$ the heat flux at both $z = 0$ and $z = z_T$ vanishes, which violates (7.8.25a). Thus, unless $\partial \Pi_0/\partial y$ vanishes at the critical level, no stable normal mode is possible for which $c$ lies within the range of $U_0$. The existence in the limit $c_i \to 0$ of a nonzero potential vorticity flux in a narrow region around $z_c$ implies, by (7.3.21), a local jump in the heat flux independent of the size of $c_i$ which, uncompensated elsewhere, inevitably leads to the release of potential energy. As we noted before, the presence of a nonzero potential vorticity gradient in the fluid interior will destabilize waves with critical layers in the range of $U_0$.

Now let us turn our attention to the specific problem posed by (7.8.3). In most cases of interest, the vertical scale of the disturbance is not larger than the density scale height and it is convenient to choose

$$D = h_* = \frac{\dfrac{f_0^2}{N^2}\dfrac{\partial U_*}{\partial z_*}}{\beta_0}, \qquad (7.8.30)$$

so that (7.8.3) becomes

$$(z - c)\left(\frac{\partial^2 A}{\partial z^2} - \delta\frac{\partial A}{\partial z} - \mu^2 A\right) + (\delta + 1)A = 0, \qquad (7.8.31)$$

where

$$\delta = \frac{h_*}{H_*}. \qquad (7.8.32)$$

The boundary condition at $z = 0$ in the absence of friction and topographic slope is, from (7.4.7), in the present case

$$c\frac{dA}{dz} + A = 0, \qquad z = 0, \qquad (7.8.33)$$

and it will prove sufficient to insist that $\rho_s|A|^2$ remain finite as $z \to \infty$ for the remaining boundary condition.

The equation for $A$ can be reduced to a standard type by the transformation

$$A = (z - c)e^{\nu z}F(z), \qquad (7.8.34)$$

where

$$\nu = \frac{\delta}{2} - \left(\mu^2 + \frac{\delta^2}{4}\right)^{1/2}. \qquad (7.8.35)$$

The linear factor in (7.8.34) is suggested by (7.8.18) while the exponential factor reflects the fact that for large $z$ (7.8.3) reduces to the condition that the perturbation potential vorticity must vanish. If (7.8.34) is substituted into (7.8.31) it follows that $F$ satisfies

$$\xi \frac{d^2 F}{d\xi^2} + (2 - \xi)\frac{dF}{d\xi} - (1 - r)F = 0, \tag{7.8.36}$$

where $\xi$ is a new variable defined by

$$\xi = (z - c)(\delta^2 + 4\mu^2)^{1/2}, \tag{7.8.37}$$

and the parameter $r$ is given by

$$r = \frac{\delta + 1}{(\delta^2 + 4\mu^2)^{1/2}}. \tag{7.8.38}$$

Note that as the density scale height goes to infinity the vertical scale factor $v \to -\mu$, i.e., the vertical scale is linearly proportional to the horizontal scale.

At the lower boundary (7.8.33) applies and when written in terms of $F$ and $\xi$, becomes

$$\xi_0^2 \left[ \frac{dF}{d\xi} - \frac{(\alpha_1 - \frac{1}{2})}{2\alpha_1} F \right] = 0, \quad \text{on} \quad \xi = \xi_0 = -c(\delta^2 + 4\mu^2)^{1/2}. \tag{7.8.39}$$

Note that the position of the lower boundary in the $\xi$ variable depends on the eigenvalue $c$. The parameter

$$\alpha_1 = \frac{(\delta^2 + 4\mu^2)^{1/2}}{2\delta}. \tag{7.8.40}$$

Equation (7.8.36) is the confluent hypergeometric equation whose general solution may be written as (Abramowitz and Stegun 1964)

$$F(\xi) = c_1 M(a, 2, \xi) + c_2 U(a, 2, \xi), \tag{7.8.41}$$

where $c_1$ and $c_2$ are arbitrary constants,

$$a = 1 - r, \tag{7.8.42}$$

and $M$ and $U$ are defined as follows:

$$M(a, 2, \xi) = 1 + \frac{a\xi}{2} + \frac{a(a + 1)\xi^2}{2 \times 3 \times 2!} + \cdots + \frac{(a)_n \xi^n}{(2)_n n!} + \cdots$$

$$= \sum_{m=0}^{\infty} \frac{(a)_m \xi^m}{(2)_m m!}, \tag{7.8.43}$$

where for any number $b$

$$(b)_n = b(b + 1)(b + 2) \cdots (b + n - 1); \quad (b)_0 = 1. \tag{7.8.44}$$

When *r is not a positive integer* the second solution $U(\xi)$ may be written as

$$\Gamma(a)U(a, 2, \xi) = \frac{1}{\xi}$$

$$+ \sum_{m=0}^{\infty} \frac{\Gamma(a+m)[\log \xi + \psi(a+m) - \psi(1+m) - \psi(2+m)]}{\Gamma(a-1)m!\,(m+1)!} \xi^m,$$

(7.8.45)

where $\Gamma(x)$ is the gamma function defined as

$$\Gamma(x) = \int_0^{\infty} t^{x-1}e^{-t}\, dt,$$

(7.8.46)

which for integer $x$ reduces to the factorial function, i.e.,

$$\Gamma(n) = (n-1)!.$$

The function $\psi(x)$ in (7.8.45) is the logarithmic derivative of $\Gamma(x)$, i.e.,

$$\psi(x) = \frac{d}{dx} \log \Gamma(x).$$

(7.8.47)

For large $\xi$, $M$ and $U$ have the asymptotic behavior

$$M(a, 2, \xi) \sim \frac{2\xi^{a-2}}{\Gamma(a)} e^{\xi}$$

(7.8.48a)

$$U(a, 2, \xi) \sim \xi^{-a}.$$

(7.8.48b)

Thus for large $\xi$ or $z$ the function $M(a, 2, \xi)$ would, by (7.8.34), yield

$$A(z) \sim (z-c)^{-r} \exp\left\{\left[\frac{\delta}{2} + \left(\frac{\delta^2}{4} + \mu^2\right)^{1/2}\right](z-c)\right\}$$

(7.8.49)

and must be rejected. Thus, for $r$ not an integer, $c_1$ must be chosen zero in (7.8.41) and $F(\xi)$ has the form of $U(a, 2, \xi)$. On the other hand, when $r$ is a positive integer, the series for $M(a, 2, \xi)$ terminates and becomes a polynomial of degree $r-1$, so $M(a, 2, \xi)$ is an acceptable solution for all $\xi$. The second solution must be redefined for integral $r$. It can be shown to lead to an exponential increase for $A$ at large $z$ and must therefore be rejected; it is not given here. Thus, $F(\xi)$ is given by

$$F(\xi) = \begin{cases} c_1 M(a, 2, \xi), & r = n, \\ c_2 U(a, 2, \xi), & r \neq n, \end{cases}$$

where $n$ is any integer greater than zero.

Consider first the solution for integral $r$. The simplest solution corresponds to $r = 1$, i.e., $a = 0$, so that

$$F(\xi) = M(0, 2, z) = 1,$$

(7.8.50)

so that $A(z)$ has the simple form

$$A(z) = (z-c)e^{vz},$$

(7.8.51)

where $v$ is given by (7.8.35). The condition $r = 1$ specifies a particular relation between $\mu$ and $\delta$ according to (7.8.38). If (7.8.8), (7.8.10), and (7.8.32) are used, the condition $r = 1$ may be written in terms of dimensional quantities as

$$\frac{1}{\beta_0 H_*} \frac{\partial U_*}{\partial z_*} \frac{f_0^2}{N_s^2} = \left\{ \left[ 1 + 4(k_*^2 + l_*^2) \frac{N_s^2 H_*^2}{f_0^2} \right]^{1/2} - 1 \right\}^{-1}. \qquad (7.8.52)$$

Thus for *large* wave number, the vertical shear required by the condition $r = 1$ actually tends to zero while for *small* wave number (i.e., for wavelengths large compared to the Rossby deformation radius based on $H_*$) the required shear increases linearly with overall wavelength. The vertical scale of the solution for $r = 1$ is given by the inverse of $v$, which in terms of $r$ is given by the relation

$$v = \frac{\delta(r - 1) - 1}{2r}. \qquad (7.8.53)$$

Hence for $r = 1$, $v = -\frac{1}{2}$ so the eigenfunction decays away from the lower boundary with e-folding scale $2 h_*$. Note that this is true for all values of $\delta = h_*/H_*$.

Larger integer values of $r$ yield higher-order polynomial solutions for $A$. Each polynomial has $(r - 1)$ nodal points. For example, for $r = 2$

$$\delta = \frac{1}{\beta_0 H_*} \frac{\partial U_*}{\partial z_*} \frac{f_0^2}{N_s^2} = \left\{ r \left( 1 + 4(k_*^2 + l_*^2) \frac{N_s^2 H_*^2}{f_0^2} \right) - 1 \right\}^{-1}. \qquad (7.8.54)$$

Thus for a given wave number the solutions at higher $r$ correspond to *lower* values of the vertical shear. The lines of constant $r$ in the shear wave-number plane are shown in Figure 7.8.3. For all $r > 1$, the intercept of the line of constant $r$ with the zero wave-number axis occurs at a finite value of $\delta$, i.e., as $k_*^2 + l_*^2 \to 0$, $\delta \to (r - 1)^{-1}$. Thus for a fixed value of the vertical shear there will be a finite number of wave numbers which yield polynomial solutions. Note that as $H_* \to \infty$, the number of such solutions also becomes infinite.

In order that these polynomial solutions may be allowable normal modes they must satisfy the lower boundary condition (7.8.39). For example, for $r = 1$, (7.8.39) becomes

$$\xi_0^2(\alpha_1 - \tfrac{1}{2}) = 0. \qquad (7.8.55)$$

Since $\alpha_1$ always exceeds $\frac{1}{2}$, this implies that $\xi_0$ must vanish or that for $r = 1$

$$c = 0$$

must be a double root of (7.8.55). Similarly, it is possible to show that for all integer $r$, $c = 0$ is a double root of (7.8.39). Thus for $r = 2$, for which (7.8.53) yields $F(\xi)$, the lower boundary condition becomes

$$\xi_0^2 \left[ \frac{\xi_0}{2} - \left( 1 + \frac{1}{2\alpha_1 - 1} \right) \right]. \qquad (7.8.56)$$

Thus two of the roots of (7.8.56) come from the double root at $\xi_0 = 0$. There

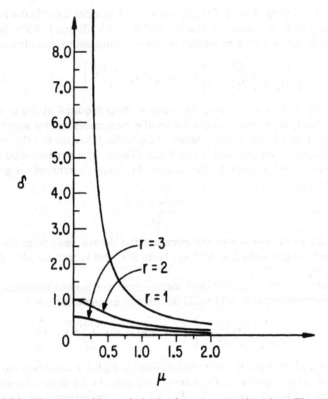

**Figure 7.8.3**  The curves of integral $r$ in the shear wave number plane. The ordinate

$$\delta = \frac{\dfrac{f_0^2}{N_s^2}\dfrac{\partial U_*}{\partial z_*}}{\beta_0 H_*}$$

while the abscissa is $N_s H_*/f_0 (k_*^2 + l_*^2)^{1/2}$, i.e., the wave number scaled by the Rossby deformation radius based on the density scale height. On these curves, $c_i$ is zero.

is a third root, corresponding to

$$\xi_0 = -c(\delta^2 + 4\mu^2)^{1/2} = 2\left(1 + \frac{\alpha_1/2}{\alpha_1 - \frac{1}{2}}\right). \tag{7.8.57}$$

Again, since $\alpha_1 > \frac{1}{2}$ this must correspond to a wave with $c < 0$, i.e., a retrograde wave whose phase speed lies outside the range of $U_0$. In general for all $r > 1$ there is a double root at $c = 0$ and $r - 1$ retrograde waves with $c < 0$.

The presence of the double root, $c = 0$, at curves of integer $r$ suggests that these are curves of marginal stability, i.e., thresholds of instability corresponding to a criticality condition like the one found in the Eady problem.

Indeed, it was originally thought that the curve $r = 1$ represented a boundary between stable and unstable waves so that (7.8.51) might suggest a minimum critical shear required for instability at each wave number. Obviously, the presence of the curves of integral $r$ for $r > 1$ at *lower* values of the shear denies the validity of this explanation. As Burger (1962) showed, the flow is unstable in the *entire* shear wave-number plane *except* on the lines of integral $r$ where $c$ is real. Rather than separating stable from unstable *regions*, the curves of integral $r$ separate parameter domains in which *different* vertical modes are unstable. That is, $c_i$ is nonzero on both sides of each integral $r$ curve. This very subtle result can be demonstrated directly with an argument due originally to Miles (1964a).

Consider nonintegral values of $r$, so that $U(a, z, \xi)$ is the appropriate solution for $F(\xi)$, but choose $r$ to be near an integer, $n + 1$, so that

$$(1 - r) = a = -n + O(\varepsilon) \tag{7.8.58}$$

where it is convenient to define $\varepsilon$ by the relation

$$\varepsilon = -\frac{1}{\pi} \tan \pi a. \tag{7.8.59}$$

The parameter $\varepsilon$ may be positive or negative. Now consider values of $c$ (and hence $\xi_0$) near zero, the value achieved on the integral curve. Now $\Gamma(x)$ has poles at the negative integers, so that the term $\psi(a)$ in (7.8.45) becomes singular as $\varepsilon \to 0$. In particular, the identity

$$\Gamma(x)\Gamma(1 - x) = \frac{\pi}{\sin \pi x} \tag{7.8.60}$$

allows us to write

$$\psi(x) = \psi(1 - x) - \frac{\pi}{\tan \pi x}. \tag{7.8.61}$$

Near $\xi = 0$, the leading term of (7.8.45) as $\varepsilon \to 0$ derives from the term $\xi^{-1}$ and the first term in the sum:

$$F(\xi) \sim \frac{1}{\xi} + \frac{\Gamma(a)}{\Gamma(a-1)} \psi(a) \sim \frac{1}{\xi} + \frac{a-1}{\varepsilon}$$
$$\sim \frac{1}{\xi} - \frac{n+1}{\varepsilon}, \tag{7.8.62}$$

while similarly, in the joint limit $\xi \to 0$, $\varepsilon \to 0$,

$$\frac{dF}{d\xi} \sim -\frac{1}{\xi^2} + \frac{1}{2}\frac{a(a-1)}{\varepsilon} = -\frac{1}{\xi^2} + \frac{n(n+1)}{2\varepsilon}, \tag{7.8.63}$$

where the $m = 1$ term in (7.8.45) must be retained to obtain (7.8.63). When (7.8.62) and (7.8.63) are substituted into (7.8.39) and only the lowest-order terms retained, we find that

$$\xi_0^2 = c^2 4\alpha_1^2 \delta^2 = \frac{2\varepsilon}{(n+1)(n+1-1/2\alpha_1)}. \tag{7.8.64}$$

It follows from (7.8.59) that immediately to the right (short-wave) side of the integral curves, $\varepsilon$ is negative, so that $\xi_0$ and hence $c$ are purely imaginary, e.g.,

$$c_i = (2\alpha_1 \delta)^{-1} \frac{(-2\varepsilon)^{1/2}}{[(n+1)(n+1-1/2\alpha_1)]^{1/2}}. \tag{7.8.65}$$

On the other hand, to the left of each of the integral curves this approximation yields a purely real $c$, e.g.,

$$\xi_0 = \pm \frac{(2\varepsilon)^{1/2}}{[(n+1)(n+1-1/2\alpha_1)]^{1/2}}. \tag{7.8.66}$$

To complete the argument for $\varepsilon > 0$, note that if the negative root for $\xi_0$ in (7.8.66) is chosen (i.e., $c_r > 0$), the $\log \xi$ term in (7.8.45) will provide a small complex term in (7.8.39), since

$$\log \xi_0 = \log |\xi_0| - i\pi$$

if $\xi_0$ is negative. When this higher-order correction involving the log term is retained (7.8.64) becomes, when $\varepsilon > 0$,

$$\xi_0^2 = \frac{2\varepsilon}{(n+1)(n+1-1/2\alpha_1)}[1 + i\varepsilon\pi]. \tag{7.8.67}$$

Thus, for $\varepsilon > 0$, the root for which $\xi_0$ is negative and real, in the first approximation, becomes

$$\xi_0 = -\frac{(2\varepsilon)^{1/2}}{\sqrt{(n+1)(n+1-1/2\alpha_1)}}\left[1 + i\varepsilon\frac{\pi}{2}\right] = -c/(2\alpha_1 \delta). \tag{7.8.68}$$

Hence, $c_r$ and $c_i$ are both slightly positive. The critical layer is now within the domain of the flow, and the disturbance is weakly growing, $c_i = O(\varepsilon^{3/2})$. Note that this instability has come about because $c$, in the first approximation, is real and lies in the range of $U_0$, so that a critical layer now exists. The considerations subsequent to (7.8.29) then apply, yielding a slightly unstable wave. Since $c_i$ in both (7.8.65) ($\varepsilon < 0$) and (7.8.68) ($\varepsilon > 0$) are decreasing functions of $n$, we can conclude that the dominant instability is associated with the $r = 1$ ($n = 0$) mode. In particular, the parameter region to the right of the $r = 1$ curve in Figure 7.8.3 presents the stronger instability, $c_i \sim \varepsilon^{1/2}$, as opposed to the long-wave side of the curve, where $c_i \sim \varepsilon^{3/2}$. The normal mode on the $r - 1$ curve and its continuation into the unstable domain at shorter wavelengths (or higher shear) is called the Charney mode. The sequence of modes that are encountered as $r$ is increased, i.e., at longer wavelengths are called the Green modes since their existence was first demonstrated by Green (1960).

Typically the Charney mode is the mode with the largest growth rate. However, the growth rates of the Green modes, which are numerically smaller, are typically of the same order as the growth rate of the Charney mode. Kuo (1973) has carried out detailed calculations for the normal mode problem in the limit $\delta = h_*/H_* \to 0$, i.e., for the case where the density scale height becomes large. In that limit (7.8.39) becomes

$$\xi_0^2 \left[ \frac{dF}{d\xi} - \frac{1}{2} F \right] = 0 \quad \text{on} \quad \xi = \xi_0, \tag{7.8.69}$$

so that the only parameter in the problem for $F$ is $r$ and hence the roots $\xi_0$ are functions of $r$ alone. Figure 7.8.4 shows Kuo's results. The notation

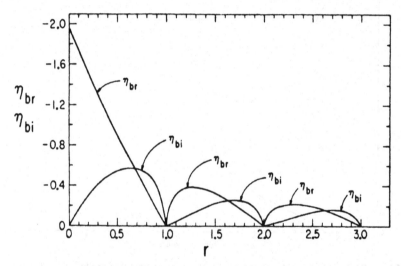

**Figure 7.8.4**  The real and imaginary parts of $c$ as calculated by Kuo (1973) for the case $\delta \to 0$. In this figure $\eta_{br} = \text{Re } \xi_0$, $\eta_{bi} = \text{Im } \xi_0$, where $\xi_0 = -2\mu c$ where $\mu$ is the nondimensional wave number.

$$\eta_{br} = \text{Re } \xi_0, \qquad \eta_{\beta i} = \text{Im } \xi_0$$

is used. As $\delta \to 0$

$$-\xi_0 = 2\mu c \tag{7.8.70}$$

the frequency and growth rate are proportional to $-\eta_{br}$ and $-\eta_{bi}$ respectively. Note in Figure 7.8.4 that the maximum growth rate in fact occurs in the parameter regime of the Charney mode and that the transition from one unstable domain to another occurs at integer values of $r$. The Green modes correspond to lower growth rates.

Figure 7.8.5 shows Kuo's calculation of the structure of the most unstable

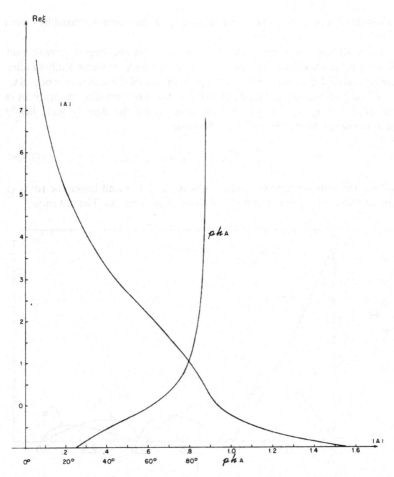

**Figure 7.8.5**   The amplitude and phase of the most unstable mode, i.e., at $r = 0.5$. (Courtesy H. L. Kuo (1973).)

mode. The wave's amplitude is intensified near the lower surface. Note that in distinction to the neutral Charney mode for which $A$ vanishes on the surface $z = 0$ the most unstable mode has its *maximum at the surface*. The phase of the unstable mode increases with height as it must in order to release the available potential energy. Note, however, that the region in which the phase changes with height is concentrated near the lower surface. Over most of the fluid the phase is nearly constant. Consequently the heat flux in the unstable wave will be similarly confined to a region of $O(h_*)$ near the lower boundary.

The Charney model, in distinction to the Eady model has no short wave cutoff. In the Eady model the vertical scale of the unstable disturbance had to be the full depth of the fluid layer. This fixed the deformation radius at $N_s D/f_0$ where $D$ is the layer depth. When the wavelength becomes short compared with this scale, the considerations of Section 7.6 apply. The length scale becomes small enough with respect to the deformation radius so that particle trajectories are forced to leave the wedge of instability. In the Charney mode, on the other hand, *as the wavelength decreases, the vertical scale shrinks* proportionally so that the angle of the trajectories in the meridional plane keeps the fluid motion within the wedge of instability. Thus the presence of the potential vorticity gradient allows the Charney mode to readjust its vertical scale so as to allow for energy releasing motions at arbitrarily small horizontal scales. Nevertheless, the energy conversion mechanism in Charney's model is identical to that in Eady's, i.e., the release of available potential energy by slanted convection. However, the introduction of the basic potential vorticity gradient has altered the modal structure. In addition, it has had the familiar effect of retarding the speed of propagation of the mode. The phase speeds of the unstable waves are now very near the minimum speed of the zonal flow rather than the *mean* speed as in Eady's model.

The Charney mode becomes stable at longer wavelengths as the $r = 1$ curve is met. However, the presence of the nonzero potential vorticity gradient introduces a whole new set of modes, the Green modes, not present in Eady's model. Although these new modes are more slowly growing than the Charney mode, these higher modes are of interest because of their vertical structure. Since they occur at smaller wave number, their vertical decay scale will tend to be smaller since $v$ goes to zero as the wave number goes to zero. Thus these modes, which are oscillating in $z$, are capable of penetrating to much greater heights in the atmosphere and usually have their maximum amplitudes well above the surface in distinction to the Charney mode. For this reason, the Green modes may play a role in linking the dynamics of the troposphere and stratosphere in the atmosphere.

It is somewhat unfortunate that the introduction of even a slightly more realistic model than Eady's introduces such elements of difficulty and subtlety in the analysis, for there are further steps towards realism that are desirable, such as the consideration of friction, horizontal shear, more realistic vertical shear, and nonlinearity. To discuss these several additional features of considerable physical interest, a model considerably simpler than Charney's, which yet retains the influence of the potential-vorticity gradient, is required. This goal is accomplished by the consideration of layer models for the instability which, *a priori*, place a limit on the complexity of the vertical structure of the motion field. The layer models, therefore, filter out modes connected with higher vertical structure, generally related to weaker instability, and lead to considerable simplification and the potential for the tractable consideration of more complex physical effects.

## 7.9 Instability in the Two-Layer Model: Formulation

The two-layer system, described in (6.16), provides a useful and simple model for flow instability, which by restricting the allowable vertical scale of disturbances filters out disturbances with small vertical scales.

The equations of motion for the two-layer model, in the absence of friction and bottom topography, are (6.16.34a,b), which may concisely be rewritten

$$\left[\frac{\partial}{\partial t} + \frac{\partial \psi_n}{\partial x}\frac{\partial}{\partial y} - \frac{\partial \psi_n}{\partial y}\frac{\partial}{\partial x}\right][\nabla^2 \psi_n - F_n(-1)^n(\psi_2 - \psi_1) + \beta y] = 0 \tag{7.9.1}$$
$$n = 1, 2,$$

where

$$\nabla^2 = \frac{\partial^2}{\partial x^2} + \frac{\partial^2}{\partial y^2}. \tag{7.9.2}$$

The stream function in each of the two layers is independent of depth. The key parameters are

$$\beta = \beta_0 \frac{L^2}{U} \tag{7.9.3}$$

$$F_n = \frac{f_0^2 L^2}{g[(\rho_2 - \rho_1)/\rho_0]D_n}. $$

The subscript refers to the layer; the upper layer corresponds to $n = 1$, and the lower to $n = 2$.

Consider the basic flow,

$$\psi_n = \Psi_n(y), \tag{7.9.4}$$

corresponding to the purely zonal flow

$$U_n(y) = -\frac{\partial \Psi_n}{\partial y}. \tag{7.9.5}$$

If $U_1(y) \neq U_2(y)$ a slope to the interface between the two layers will exist, for by (6.16.36)

$$\frac{\partial h_2}{\partial y} = \frac{\varepsilon D_2}{D} F_2[U_1 - U_2]. \tag{7.9.6}$$

The sloping interface is a source of available potential energy for a disturbance, while the horizontal shear of $U_n(y)$ is a source of kinetic energy. Let $\phi_n(x, y, t)$ be the disturbance stream function, so that

$$\psi_n = \Psi_n(y) + \phi_n(x, y, t) \tag{7.9.7}$$

which, when substituted into (7.9.1), yields

$$\left[\frac{\partial}{\partial t} + U_n\frac{\partial}{\partial x}\right]q_n + \frac{\partial \phi_n}{\partial x}\frac{\partial \Pi_n}{\partial y} + \left[\frac{\partial \phi_n}{\partial x}\frac{\partial q_n}{\partial y} - \frac{\partial \phi_n}{\partial y}\frac{\partial q_n}{\partial x}\right] = 0. \tag{7.9.8}$$

Here the perturbation potential vorticity $q_n$ is

$$q_n = \nabla^2 \phi_n - F_n(-1)^n(\phi_2 - \phi_1), \tag{7.9.9}$$

while the potential-vorticity gradient of the basic state is

$$\frac{\partial \Pi_n}{\partial y} = \beta - \frac{\partial^2 U_n}{\partial y^2} - F_n(-1)^n(U_1 - U_2). \tag{7.9.10}$$

The linear stability problem is obtained by neglecting terms of $O(\phi_n^2)$ in

(7.9.8), which yields

$$\left[\frac{\partial}{\partial t} + U_n \frac{\partial}{\partial x}\right] q_n + \frac{\partial \phi_n}{\partial x} \frac{\partial \Pi_n}{\partial y} = 0. \tag{7.9.11a}$$

As in the continuously stratified model, the lateral boundary conditions for the linear perturbations are

$$\frac{\partial \phi_n}{\partial x} = 0, \qquad y = \pm 1. \tag{7.9.11b}$$

The energy equation for the disturbances is obtained by multiplying (7.9.11a) by $-(D_n/D)\phi_n$, integrating over $y$, and summing over both layers:

$$\frac{\partial}{\partial t} \int_{-1}^{1} dy \left[ \frac{d_1}{2} \left\{ \overline{\left(\frac{\partial \phi_1}{\partial x}\right)^2} + \overline{\left(\frac{\partial \phi_1}{\partial y}\right)^2} \right\} \right.$$
$$\left. + \frac{d_2}{2} \left\{ \overline{\left(\frac{\partial \phi_2}{\partial x}\right)^2} + \overline{\left(\frac{\partial \phi_2}{\partial y^2}\right)^2} \right\} + \frac{\overline{(\phi_1 - \phi_2)^2}}{2} F_0 \right]$$
$$= \int_{-1}^{1} dy \left[ d_1 \overline{\frac{\partial \phi_1}{\partial x} \frac{\partial \phi_1}{\partial y}} \frac{\partial U_1}{\partial y} + d_2 \overline{\frac{\partial \phi_2}{\partial x} \frac{\partial \phi_2}{\partial y}} \frac{\partial U_2}{\partial y} \right.$$
$$\left. + F_0(U_1 - U_2) \overline{\frac{\partial \phi_2}{\partial x} \phi_1} \right], \tag{7.9.12}$$

where

$$d_n = \frac{D_n}{D},$$
$$F_0 = \frac{f_0^2 L^2}{g[(\rho_2 - \rho_1)/\rho_0]D} \tag{7.9.13}$$

while the overbar implies an average in $x$ over the period of the spatial perturbation field.

The interpretation of (7.9.12) is similar to that of (7.3.11), namely, that the time rate of change of the sum of the kinetic energies of each layer (weighted by their relative thickness) plus the change of the perturbation available potential energy, $(F_0/2)(\phi_1 - \phi_2)^2$, is produced by Reynolds stress acting in each of the layers on the horizontal shear of the basic current plus the conversion of available potential energy. This last term may be rewritten in a more transparent form, using (6.16.36):

$$F_0(U_1 - U_2) \overline{\frac{\partial \phi_2}{\partial x} \phi_1} = \left[ \varepsilon^{-1} \frac{\partial \bar{h}_2}{\partial y} \right] \overline{[v_2(\phi_1 - \phi_2)]}$$
$$= \varepsilon^{-2} \left[ \frac{\partial \bar{h}_2}{\partial y} \right] \overline{\left[ \frac{v_2 h_2'}{F_0} \right]} = \varepsilon^{-2} \frac{\partial \bar{h}_2}{\partial y} \overline{\left[ \frac{v_1 h_2'}{F_0} \right]}, \tag{7.9.14}$$

where $\bar{h}_2$ and $h_2'$ are the interface heights of the basic and perturbation fields respectively. Thus the potential-energy conversion is simply the meridional flux of the interface height multiplied by the slope of the interface of the basic state. The interface height is therefore dynamically analogous to the temperature field in the continuous model. It is left as an exercise for the reader to show that in analogy with (7.2.13), in each layer

$$\frac{\partial \bar{u}_n}{\partial t} = \overline{v}_n^{(1)} - \frac{\partial}{\partial y}\overline{v_n u_n}, \tag{7.9.15a}$$

where the superscript $^{(1)}$ reminds us that the $x$-averaged $v$ is nongeostrophic. Similarly, the $x$-average of (6.16.20a) yields

$$F_0 \frac{\partial}{\partial t}\overline{\eta}_2 = \overline{w}_2^{(1)} - F_0 \frac{\partial}{\partial y}\overline{v_2 \eta_2}, \tag{7.9.15b}$$

where from (6.16.16)

$$\overline{v_2 \eta_2} = \overline{\frac{\partial \phi_2}{\partial x}(\phi_2 - \phi_1)} = \overline{\frac{\partial \phi_1}{\partial x}(\phi_2 - \phi_1)} = \overline{v_1 \eta_2}. \tag{7.9.16}$$

Here $\eta_2$ is related to $h_2$ by (6.16.13).

Since from the vertical integral in each layer of the continuity equation

$$d_1 \frac{\partial \overline{v}_1^{(1)}}{\partial y} = +\overline{w}_2^{(1)}, \tag{7.9.17a}$$

$$d_2 \frac{\partial \overline{v}_2^{(1)}}{\partial y} = -\overline{w}_2^{(1)}, \tag{7.9.17b}$$

where $\overline{w}_2^{(1)}$ is the $x$-averaged vertical velocity at the interface, it follows that the $x$-averaged potential vorticity satisfies

$$\frac{\partial}{\partial t}\left[-\frac{\partial \bar{u}_1}{\partial y} + F_1 \bar{\eta}_2\right] = +\frac{\partial^2}{\partial y^2}\overline{u_1 v_1} - F_1 \frac{\partial}{\partial y}\overline{v_1 \eta_2},$$

$$\frac{\partial}{\partial t}\left[-\frac{\partial \bar{u}_2}{\partial y} - F_2 \bar{\eta}_2\right] = +\frac{\partial^2}{\partial y^2}\overline{u_2 v_2} + F_2 \frac{\partial}{\partial y}\overline{v_2 \eta_2}, \tag{7.9.18}$$

which is the two-layer analogue of (7.2.22). The flux of perturbation potential vorticity in each layer is

$$\overline{v_n q_n} = \overline{v_n\left(\frac{\partial^2 \phi_n}{\partial x^2} + \frac{\partial^2 \phi_n}{\partial y^2} - F_n(-1)^n(\phi_2 - \phi_1)\right)}; \tag{7.9.19}$$

hence

$$\overline{v_1 q_1} = -\left[\frac{\partial}{\partial y}\overline{u_1 v_1} - F_1 \overline{v_1 \eta_2}\right],$$

$$\overline{v_2 q_2} = -\left[\frac{\partial}{\partial y}\overline{u_2 v_2} + F_2 \overline{v_2 \eta_2}\right]. \tag{7.9.20}$$

Thus (7.9.18) is equivalent to the statement that the mean potential-vorticity change is due entirely to the perturbation flux of potential vorticity, i.e.,

$$\frac{\partial}{\partial t}\overline{q}_n = -\frac{\partial}{\partial y}\overline{v_n q_n}. \tag{7.9.21}$$

A particularly important relation for the change of the $x$-averaged zonal flow follows from summing over both layers, i.e.,

$$\frac{\partial}{\partial t}\left[\sum_{n=1}^{2} d_n \bar{u}_n\right] = -\frac{\partial}{\partial y}\sum_{n=1}^{2}\overline{u_n v_n}\, d_n, \tag{7.9.22}$$

since

$$\sum_{n=1}^{2} d_n \overline{v_n}^{(1)} = 0$$

from conservation of mass. If (7.9.20) is used, we obtain

$$\boxed{\frac{\partial}{\partial t}\left[\sum_{n=1}^{2} d_n \overline{u}_n\right] = + \sum_{n=1}^{2} d_n \overline{v_n q_n}} \ . \tag{7.9.23}$$

Hence at any latitude the change of the vertically summed zonal flow is given entirely in terms of the rectified (i.e., $x$-averaged) flux of perturbation potential vorticity. For linear disturbances, for which (7.9.11a) applies, it follows that

$$\frac{\partial}{\partial t}\left[\sum_{n=1}^{2} d_n \bar{u}_n\right] = -\frac{\partial}{\partial t}\sum_{n=1}^{2}\frac{d_n \overline{q_n^2}}{\partial \Pi_n/\partial y} \ . \tag{7.9.24}$$

Thus, as Held (1975) pointed out, if the basic potential-vorticity gradient has the same sign in both layers at a given value of $y$, an unstable disturbance (for which the positive definite quantity $\overline{q_n^2}$ must be increasing with $t$) must lead to a change in the vertically averaged zonal flow *opposite* to the sign of the basic potential-vorticity gradient.

## 7.10 Normal Modes in the Two-Layer Model: Necessary Conditions for Instability

Normal-mode solutions to (7.9.11a) may be sought in the form

$$\phi_n = \text{Re } \Phi_n(y)e^{ik(x - ct)}, \tag{7.10.1}$$

which upon substitution in (7.9.11) yields two coupled *ordinary* differential equations for the $\Phi_n$, i.e.,

$$(U_1 - c)\left[\frac{d^2\Phi_1}{dy^2} - k^2\Phi_1 - F_1(\Phi_1 - \Phi_2)\right] + \Phi_1\frac{\partial \Pi_1}{\partial y} = 0, \tag{7.10.2a}$$

$$(U_2 - c)\left[\frac{d^2\Phi_2}{dy^2} - k^2\Phi_2 - F_2(\Phi_2 - \Phi_1)\right] + \Phi_2\frac{\partial \Pi_2}{\partial y} = 0, \tag{7.10.2b}$$

where from (7.9.10)

$$\frac{\partial \Pi_1}{\partial y} = \beta - \frac{\partial^2 U_1}{\partial y^2} + F_1(U_1 - U_2), \tag{7.10.3a}$$

$$\frac{\partial \Pi_2}{\partial y} = \beta - \frac{\partial^2 U_2}{\partial y^2} + F_2(U_2 - U_1). \tag{7.10.3b}$$

The function $\Phi_n$ must satisfy the boundary condition (7.9.11b), which for nonzero $k$ is

$$\Phi_n = 0, \quad y = \pm 1, \quad n = 1, 2. \tag{7.10.3c}$$

The principal mathematical simplification derived from the two-layer formulation is the appearance of the coupled ordinary differential equations (7.10.2a,b) in lieu of the partial differential equation (7.4.4) which obtains for the continuously stratified model. Unstable modes correspond to those eigensolutions $\Phi_n$, $n = 1, 2$, whose corresponding eigenvalue $c$ has a positive imaginary part. Since the coefficients of (7.10.2a,b) are real, a solution $\Phi_n$ with eigenvalue $c$ implies the existence of the complex conjugate solution $\Phi_n^*$ with eigenvalue $c^*$. Hence instability is ensured in the inviscid problem if modes with $c_i \neq 0$ are possible.

Necessary conditions for instability may be derived by multiplication of (7.10.2a,b) by $[\Phi_1/(U_1 - c)]d_1$ and $[\Phi_2/(U_2 - c)]d_2$ respectively. If the resulting equations are integrated from $-1$ to $+1$ in $y$ and the result summed, we obtain

$$\sum_{n=1}^{2} \int_{-1}^{1} dy \left| \left| \frac{\partial \Phi_n}{\partial y} \right|^2 + k^2 |\Phi_n|^2 \right| d_n + F_0 \int_{-1}^{1} dy |\Phi_1 - \Phi_2|^2$$

$$= \sum_{n=1}^{2} \int_{-1}^{1} dy\, d_n \frac{|\Phi_n|^2}{U_n - c} \frac{\partial \Pi_n}{\partial y}. \tag{7.10.4}$$

Both the real and the imaginary part of (7.10.4) must vanish separately. The imaginary part of (7.10.4) yields

$$\boxed{c_i \sum_{n=1}^{2} d_n \int_{-1}^{1} dy \frac{|\Phi_n|^2}{|U_n - c|^2} \frac{\partial \Pi_n}{\partial y} = 0}. \tag{7.10.5}$$

If $c_i$ is not zero, i.e., if the mode is to be unstable, the potential-vorticity gradient of the basic state must be somewhere positive and somewhere negative, although it is *not* necessary that $\partial \Pi_n/\partial y$ vanish for any $y$ in the interval. Instead, (7.10.5) may be satisfied if the potential-vorticity gradient is positive in one layer and negative in the other. An important distinction between the layer model and the continuous model is immediately apparent. From (7.10.3a,b) it is clear that if $\beta$ is sufficiently large that both $\partial \Pi_1/\partial y$ and $\partial \Pi_2/\partial y$ are positive, the flow must be stable. In Charney's model of baroclinic instability there is no critical value of the vertical shear required for instability. The two-layer model, by filtering out disturbances of small vertical scale, must introduce a minimum critical shear for instability. This can be simply interpreted as a criterion that if modes of a *specified* vertical scale are to be unstable, a certain critical shear is required.

The real part of (7.10.4) yields, with (7.10.5),

$$\sum_{n=1}^{2} d_n \int_{-1}^{1} dy \left[ U_n \frac{\partial \Pi_n}{\partial y} \frac{|\Phi_n|^2}{|U_n - c|^2} \right]$$

$$= \sum_{n=1}^{2} d_n \int_{-1}^{1} dy \left[ \left| \frac{\partial \Phi_n}{\partial y} \right|^2 + k^2 |\Phi_n|^2 \right] + F_0 \int_{-1}^{1} dy \, |\Phi_1 - \Phi_2|^2 \qquad (7.10.6)$$

$$> 0,$$

so that the product $U_n \, \partial \Pi_n / \partial y$ must be somewhere positive for instability to occur. Again using the identity (7.5.11) applied to each of the $\Phi_n$, it follows that

$$\sum_{n=1}^{2} d_n \int_{-1}^{1} dy \, U_n \frac{\partial \Pi_n}{\partial y} \frac{|\Phi_n|^2}{|U_n - c|^2} \geq \sum_{n=1}^{2} d_n \int_{-1}^{1} dy \, |\Phi_n|^2 \left( k^2 + \frac{\pi^2}{4} \right). \qquad (7.10.7)$$

On the other hand

$$\sum_{n=1}^{2} d_n \int_{-1}^{1} dy \, U_n \frac{\partial \Pi_n}{\partial y} \frac{|\Phi_n|^2}{|U_n - c|^2} \leq \sum d_n \int_{-1}^{1} dy \frac{|\Phi_n|^2}{c_i^2} \left[ U_n \frac{\partial \Pi_n}{\partial y} \right]_{max} \qquad (7.10.8)$$

where the subscript max refers to the maximum value of the product of $U_n \, \partial \Pi_n / \partial y$ in the range $-1 \leq y \leq 1$. Combining (7.10.7) and (7.10.8) yields, as a bound on the growth rate,

$$\boxed{ k^2 c_i^2 \leq \left[ U_n \frac{\partial \Pi_n}{\partial y} \right]_{max} \frac{k^2}{\pi^2/4 + k^2}. } \qquad (7.10.9)$$

Whereas (7.5.25) (to which (7.10.9) is the two-layer analogy) is valid only when the surface velocity is independent of $y$, (7.10.9) is valid generally.

It is left as an exercise for the reader to show that the introduction of the function $\mathscr{S}_n = \Phi_n (U_n - c)^{-1}$ allows the steps leading to (7.5.14) and (7.5.18) to be repeated for the layer model. Thus, both (7.5.14) and (7.5.18) are equally valid for the two layer model, where $U_{min}$ and $U_{max}$ are the minimum and maximum values of $U_n$ in the interval $|y| \leq 1$.

The perturbation energy equation (7.9.12) may be written in terms of $\Phi_n$ since, for example,

$$\overline{\left( \frac{\partial \phi_n}{\partial x} \right)^2} = -\frac{k^2}{4} \overline{\{ \Phi_n e^{ik(x-ct)} - \Phi_n^* e^{-ik(x-c^*t)} \}^2}$$

$$= \frac{k^2}{2} |\Phi_n|^2 e^{2kc_i t}, \qquad (7.10.10)$$

while

$$\overline{\frac{\partial \phi_n}{\partial x}\frac{\partial \phi_n}{\partial y}} = \frac{ik}{4}\overline{\{\Phi_n e^{ik(x-ct)} - \Phi_n^* e^{-ik(x-c^*t)}\}\left\{\frac{d\Phi_n}{dy}e^{ik(x-ct)} + \frac{d\Phi_n^*}{dy}e^{-ik(x-c^*t)}\right\}}$$

$$= \frac{ik}{4}\left\{\Phi_n \frac{d\Phi_n^*}{dy} - \Phi_n^* \frac{d\Phi_n}{dy}\right\}e^{2kc_i t}, \tag{7.10.11}$$

and similarly

$$\overline{\phi_1 \frac{\partial \phi_2}{\partial x}} = -\frac{ik}{4}\{\Phi_1 \Phi_2^* - \Phi_2 \Phi_1^*\}e^{2kc_i t}.$$

Thus (7.9.12) becomes

$$\frac{kc_i}{2}\left[\sum_{n=1}^{2}d_n\int_{-1}^{1}dy\left\{|\Phi_n|^2 k^2 + \left|\frac{d\Phi_n}{dy}\right|^2\right\} + F_0\int_{-1}^{1}dy|\Phi_1 - \Phi_2|^2\right]$$

$$= \int_{-1}^{1}dy\sum_{n=1}^{2}d_n\left\{\left(\frac{ik}{4}\right)\left[\Phi_n\frac{d\Phi_n^*}{dy} - \Phi_n^*\frac{d\Phi_n}{dy}\right]\frac{dU_n}{dy}\right\} \tag{7.10.12}$$

$$- \int_{-1}^{1}dy\,F_0(U_2 - U_1)\frac{ik}{4}\{\Phi_1\Phi_2^* - \Phi_2\Phi_1^*\}.$$

Since

$$2|\Phi_2||\Phi_1| \le \frac{d_1|\Phi_1|^2 + d_2|\Phi_2|^2}{(d_1 d_2)^{1/2}},$$

$$2k\left|\frac{d\Phi_n}{dy}\right||\Phi_n| \le |\Phi_n|^2 k^2 + \left|\frac{d\Phi_n}{dy}\right|^2, \tag{7.10.13}$$

and

$$\left|\Phi_n\frac{d\Phi_n^*}{dy} - \Phi_n^*\frac{d\Phi_n}{dy}\right| \le 2|\Phi_n|\left|\frac{d\Phi_n}{dy}\right|,$$

$$|\Phi_1\Phi_2^* - \Phi_2\Phi_1^*| \le 2|\Phi_2||\Phi_1|, \tag{7.10.14}$$

the energy equation (7.10.12) implies

$$kc_i \le \frac{1}{2}\left[\left|\frac{dU_n}{dy}\right|_{max} + \frac{(F_1 F_2)^{1/2}k|U_1 - U_2|}{k^2 + \pi^2/4}\right], \tag{7.10.15}$$

so that the growth rate is again bounded in terms of the vertical and horizontal shear of the basic state.

## 7.11 Baroclinic Instability in the Two-Layer Model: Phillips' Model

Consider the basic state where $U_1$ and $U_2$ are independent of $y$ but differ in magnitude. This simple and illuminating model was first studied by Phillips (1954). The interface between the two layers will slope, and disturbances feeding on this available potential energy are possible. It follows from (7.9.12) that since $U_1$ and $U_2$ are independent of $y$, the available potential energy associated with the interface slope is the *only* source of energy for the perturbations. The fact that $U_1$ and $U_2$ are independent of $y$ renders the coefficients of (7.10.2a,b) constant, so that solutions which satisfy (7.10.3) may be sought in the form

$$\Phi_n = A_n \cos l_j y, \qquad (7.11.1)$$

where

$$l_j = (j + \tfrac{1}{2})\pi, \qquad j = 0, 1, 2, \ldots, \qquad (7.11.2)$$

and the $A_n$'s are *constants* and represent the wave amplitude in each layer. Substitution of (7.11.1) into (7.10.2) yields two coupled *algebraic* equations for $A_1$ and $A_2$, i.e., since now

$$\frac{\partial \Pi_n}{\partial y} = \beta - (-1)^n F_n (U_1 - U_2), \qquad (7.11.3)$$

it follows that $A_1$ and $A_2$ satisfy

$$A_1[(c - U_1)[K^2 + F_1] + \beta + F_1(U_1 - U_2)]$$
$$- A_2(c - U_1)F_1 = 0, \qquad (7.11.4a)$$

$$A_2[(c - U_2)[K^2 + F_2] + \beta - F_2(U_1 - U_2)]$$
$$- A_1(c - U_2)F_2 = 0, \qquad (7.11.4b)$$

where $K$ is the total wave number,

$$K^2 = k^2 + l_j^2. \qquad (7.11.5)$$

Nontrivial solutions for $A_1$ and $A_2$ are possible only if the determinant of the coefficients of $A_1$ and $A_2$ in (7.11.4) vanishes. This condition leads directly to a quadratic equation for $c$, whose solutions are

$$c = U_2 + \frac{U_s K^2 (K^2 + 2F_2) - \beta(2K^2 + F_1 + F_2)}{2K^2(K^2 + F_1 + F_2)}$$
$$\pm \frac{[\beta^2(F_1 + F_2)^2 + 2\beta U_s K^4(F_1 - F_2) - K^4 U_s^2(4F_1 F_2 - K^4)]^{1/2}}{2K^2(K^2 + F_1 + F_2)},$$
$$(7.11.6)$$

where

$$U_s \equiv U_1 - U_2. \tag{7.11.7}$$

In the absence of shear (i.e., when $U_s$ is zero), (7.11.6) reduces to the two solutions

$$c_1 = U_2 - \frac{\beta}{K^2},$$

$$c_2 = U_2 - \frac{\beta}{K^2 + F_1 + F_2}, \tag{7.11.8}$$

which, except for the Doppler shift by the constant current $U_2 = U_1$, is identical to the dispersion relation for the barotropic and baroclinic Rossby waves derived in Section 6.17. In the presence of shear the nature of the roots alters significantly. Consider first the case where $\beta$ is zero, i.e., where the effect of the earth's sphericity is ignored altogether. Then (7.11.6) becomes

$$c = \frac{U_1(K^2 + 2F_2) + U_2(K^2 + 2F_1)}{2(K^2 + F_1 + F_2)} \pm \frac{[-U_s^2(4F_1F_2 - K^4)]^{1/2}}{2(K^2 + F_1 + F_2)}. \tag{7.11.9}$$

Thus instability occurs in this case for all $U_s^2$ if

$$K^2 < 2(F_1 F_2)^{1/2}. \tag{7.11.10}$$

The smallest value $K^2$ attains occurs when $k$ is zero, so that for instability to occur for some $k > 0$,

$$(F_1 F_2)^{-1/2} < \frac{2}{l_j^2} = \frac{8}{\pi^2(2n+1)^2}, \tag{7.11.11}$$

which is qualitatively similar to the condition (7.7.23) for instability in Eady's model. That is, in the two-layer model as well as Eady's model, sufficiently short waves (compared with the deformation radius) will not grow. However, as long as

$$L > \pi f_0^{-1} \left( g \frac{\Delta \rho}{\rho_0} (D_1 D_2)^{1/2} \right)^{1/2}, \tag{7.11.12}$$

the two-layer model with uniform vertical shear in the absence of $\beta$ will be unstable, in good agreement with the continuous model.

Next consider the effect of $\beta$ in the case where the two layers have equal mass, i.e., where $D_1 = D_2$ and consequently $F_1 = F_2 = \hat{F}$. In this case (7.11.6) becomes

$$c = \frac{U_1 + U_2}{2} - \frac{\beta(K^2 + \hat{F})}{K^2(K^2 + 2\hat{F})}$$

$$\pm \frac{[4\beta^2\hat{F}^2 - K^4 U_s^2(4\hat{F}^2 - K^4)]^{1/2}}{2K^2(K^2 + 2\hat{F})}. \tag{7.11.13}$$

Unless $U_s^2$ is sufficiently great, the radicand in (7.11.13) will be positive for all $K^2$ and the flow will be stable. In order for the wave with total wave number $K$ to be unstable it is therefore necessary *and* sufficient that

$$U_s^2 > U_c^2 = \frac{4\beta^2 \hat{F}^2}{K^4(4\hat{F}^2 - K^4)} \qquad (7.11.14)$$

and

$$K^2 < 2\hat{F}. \qquad (7.11.15)$$

Thus instability in the two-layer model with $\beta > 0$ requires first that $K^2 < 2\hat{F}$ and, in addition, that $U_s$ exceed a critical shear which depends on $\beta$. When, as in the present case, $F_1 = F_2$, the criterion (7.11.14) is independent of the sign of $U_s$, i.e., independent of whether the *thermal* wind is from the east or west. Figure 7.11.1(a) shows $U_c$, the critical value of $U_s$ required for instability, as a function of $K^2$. Only the positive branch, $U_c > 0$, is shown, for the case $U_s < 0$ is obtained by reflection around the $K^2$ axis. The *minimum* critical shear occurs at

$$K^2 = 2^{1/2}\hat{F}, \qquad (7.11.16)$$

corresponding to the minimum critical shear

$$\boxed{(U_c)_{\min} = \pm \frac{\beta}{\hat{F}}}. \qquad (7.11.17)$$

Inspection of (7.11.3) shows that this minimum critical shear is precisely the shear required to satisfy the necessary condition for instability, (7.10.5). Thus by direct calculation the necessary condition (7.10.5) is also sufficient as long as (7.11.16) is possible, i.e., as long as

$$\frac{\pi^2}{4} < 2^{1/2}\hat{F}. \qquad (7.11.18)$$

When $U_s$ exceeds $(U_c)_{\min}$, the potential-vorticity gradient will be positive in one layer and negative in the other. The minimum value corresponds to the particular shear for which $\partial\Pi_n/\partial y$ is zero in either layer 1 or layer 2. If, for any $K$, $U_s$ slightly exceeds $U_c$—i.e., if

$$U_s = U_c(K) + \Delta, \qquad \Delta \ll U_c \qquad (7.11.19)$$

—then from (7.11.13),

$$c_i = \pm \frac{2^{1/2}\beta\hat{F}}{K^2(K^2 + 2\hat{F})}\left(\frac{\Delta}{U_c}\right)^{1/2} + O(\Delta). \qquad (7.11.20)$$

If (7.11.20) is compared with (7.8.53), it is clear that the transition to instability which occurs as the two-layer stability threshold is crossed is analogous to the *strong* instability in the continuous model in the absence of a critical layer. In the continuous model, there are an infinite number of such transitions, each corresponding to an integer-$r$ curve in Figure 7.8.3 and each

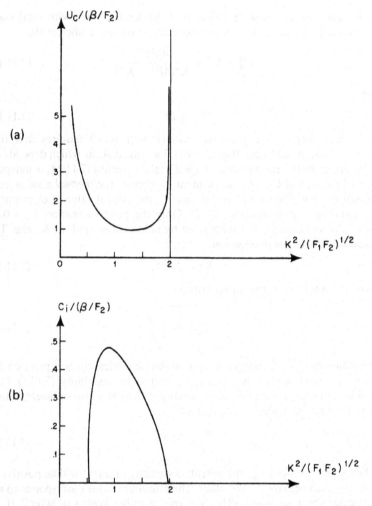

**Figure 7.11.1** (a) The critical shear $U_c$ as a function of wave number in the case $F_1 = F_2$. For $U_s > U_c$ and $K^2 < 2(F_1 F_2)^{1/2}$, $c$ is complex. Elsewhere $c$ is strictly real. (b) The imaginary part of $c$ as a function of wave number for $U_s = 2\beta/F_2$, i.e., for a shear which is twice the minimum critical shear.

corresponding to an increasingly more complex vertical structure. The two-layer model resolves only the *strong* instability corresponding to the *lowest* mode, $r = 1$, in the continuous model. This view of the relationship between the two-layer and the continuous model is strengthened by a comparison of Figure 7.11.1(b) with Figure 7.8.4. In Figure 7.11.1(b) $c_i$ is shown for a value of $U_s$ equal to twice the minimum critical shear. Only a single band of wave numbers corresponding to the lowest mode is unstable, in contrast to the multibanded structure of the continuous model. The two-layer model is inadequate insofar as it is unable to describe the weaker instability modes, but appears adequate for the most unstable mode.

Precisely at the minimum critical shear, where $U_s = \beta/\hat{F}$, all wave numbers are stable, and the phase speed is real and given by

$$c_r = U_2 + \frac{\beta(K^4 - 2\hat{F}^2)}{2K^2\hat{F}(K^2 + 2\hat{F})}(1 \pm 1). \qquad (7.11.21)$$

For one mode $c_r$ is equal to $U_2$ for all $K$; for the other mode

$$c_r = U_2 + \frac{\beta(K^4 - 2\hat{F}^2)}{K^2\hat{F}(K^2 + 2\hat{F})}. \qquad (7.11.22)$$

At the wave number given by (7.11.16) the two roots for $c$ coalesce. At this point the flow is marginally stable and $c_r$ is equal to $U_2$. (If $U_s < 0$, then $c_r \to U_1$ at $K^2 = 2^{1/2}F$.) The coalescence at the critical curve occurs for all $U_s > U_c$ and is the signature of the stability threshold. Figure 7.11.2 shows $c_r$ as a function of $K^2$ for $U_s = 2\beta/\hat{F_c}$ and displays two coalescence points corresponding to the two critical wave numbers for marginal stability at this shear. In the interval between these wave numbers, which delimit the unstable band of wave numbers, $c_r$ has a single value. Note that, once again, the most unstable mode corresponds to the gravest cross-stream mode, i.e., $j = 0$.

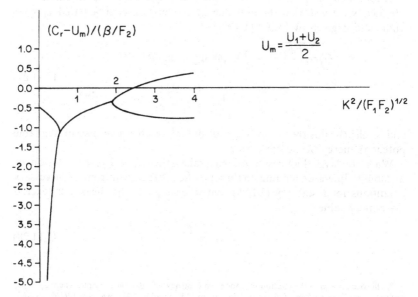

**Figure 7.11.2**   The real part of $c$ as a function of wave number when $U_s = 2\beta/F_2$.

Once $c$ is determined, the vertical structure of the wave is given by the ratio $A_2/A_1$, which may be obtained from either (7.11.4a) or (7.11.4b):

$$\frac{A_2}{A_1} = \frac{K^2 + F_1}{F_1} + \frac{\beta + F_1 U_s}{F_1(c - U_1)} = \left[\frac{K^2 + F_2}{F_2} + \frac{\beta - F_2 U_s}{F_2(c - U_2)}\right]^{-1}. \quad (7.11.23)$$

For example, for $U_s > 0$, at the minimum critical shear where $\beta = F_1 U_s = F_2 U_s$ and $K^2 = 2^{1/2}\hat{F}$, we have

$$\frac{A_2}{A_1} = \sqrt{2} - 1 \approx 0.414, \quad (7.11.24)$$

while if $U_s < 0$ the ratio is reversed at the minimum critical shear. When $U_s$ exceeds $U_c$, so that $c_i > 0$,

$$\frac{A_2}{A_1} = \frac{K^2 + F_1}{F_1} + \frac{\beta + F_1 U_s}{F_1|U_1 - c|^2}(c_r - U_1 - ic_i). \quad (7.11.25)$$

If $U_s > 0$, then $\beta + F_1 U_s > 0$ and the phase of $A_2$ with respect to $A_1$ is negative if $c_i > 0$. As in the continuous model, the phase of the unstable wave increases with height (for $U_s > 0$), and therefore, for $c_i > 0$, the wave in the upper layer *lags* the wave in the lower layer by the angle

$$\alpha = \tan^{-1}\left[c_i \frac{(\beta + F_1 U_s)/(F_1|U_1 - c|^2)}{(K^2 + F_1)/F_1 + (c_r - U_1)(\beta + F_1 U_s)/(F_1|U_1 - c|^2)}\right] \quad (7.11.26)$$

It follows directly from (7.10.11) that the Reynolds stress in each layer is identically zero, so that the only energy conversion, as anticipated, is baroclinic and is given by (7.10.11), i.e.,

$$\overline{U_s\phi_1\frac{\partial\phi_2}{\partial x}} = -\frac{ik}{4}U_s\{\Phi_1\Phi_2^* - \Phi_2\Phi_1^*\}e^{2kc_it}$$

$$= \frac{kc_i}{2}U_s|A_1|^2\frac{\beta + F_1 U_s}{|U_1 - c|^2}\cos^2 l_j y\, e^{2kc_it} \quad (7.11.27)$$

and is clearly positive when $kc_i > 0$ and represents a release of available potential energy of the basic flow.

When the layer thicknesses are unequal, so that $F_1 \neq F_2$, the criterion for instability differs, depending on the sign of $U_s$.* It is clear from the necessary conditions for instability (7.10.5) that when $U_s > 0$, the shear must exceed the critical value

$$U_{c+} = \frac{\beta}{F_2} \quad (7.11.28)$$

---

* This occurs also in the continuous model. In Charney's model, for example, where $z_T \to \infty$, instability can occur when $\partial U/\partial z < 0$ *only* if from (7.8.5) and (7.4.22)) we have $|\partial U/\partial z| > \beta hS$, which is analogous to (7.11.17).

for instability, while if $U_s < 0$, $-U_s$ must exceed

$$U_{c-} = \frac{\beta}{F_1}. \qquad (7.11.29)$$

Figure 7.11.3 shows the critical shear as a function of wave number for the case $D_1/D_2 = F_2/F_1 = 0.2$. It is clear from the figure that for both positive and negative $U_s$ the minimum critical shear is in fact the shear which just makes the basic potential-vorticity gradient zero in one of the two layers.

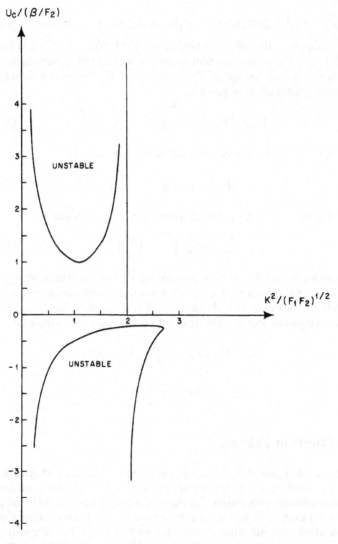

**Figure 7.11.3**  The curves of marginal stability when $D_1/D_2 = 0.2$.

Since $D_1 < D_2$, this occurs at a lower critical shear (by a factor $D_1/D_2$) for the case $U_s < 0$, because

$$\frac{U_{c+}}{U_{c-}} = \frac{F_1}{F_2} = \frac{D_2}{D_1}. \tag{7.11.30}$$

In this example, for which $F_1 > F_2$, an inspection of (7.11.6) and Figure 7.11.3 shows that instability is allowed at higher wave number for $U_s < 0$ than for $U_s > 0$. The curve for the critical value of $U_s$ is given by the condition

$$\beta^2(F_1 + F_2)^2 + 2\beta U_s K^4(F_1 - F_2) - K^4 U_s^2(4F_1 F_2 - K^4) = 0. \tag{7.11.31}$$

If the minimum value of the critical shear, (7.11.28) or (7.11.29), is inserted into (7.11.31), the total wave number corresponding to the marginally stable mode can be easily calculated. Thus, for $U_s = \beta/F_2$, the wave number at the minimum critical shear is given by

$$K_+^2 = [F_2(F_1 + F_2)]^{1/2}, \qquad U_s = \frac{\beta}{F_2}, \tag{7.11.32}$$

while if $U_s = -\beta/F_1$, the corresponding wave number is

$$K_-^2 = [F_1(F_1 + F_2)]^{1/2}, \qquad U_s = \frac{-\beta}{F_1}, \tag{7.11.33}$$

both of which reduce to (7.11.16) when $F_1 = F_2 = \hat{F}$. Since

$$\frac{K_+^2}{K_-^2} = \left(\frac{F_2}{F_1}\right)^{1/2} = \left(\frac{D_1}{D_2}\right)^{1/2}, \tag{7.11.34}$$

the marginally unstable wave will be longer when $U_s > 0$ than when $U_s < 0$ if $D_1 < D_2$. Since both $U_s$ and $K^2$ are known at the minimum critical shears where the radicand in (7.11.6) vanishes, it is a simple matter to calculate $c_r$ for the marginally stable wave. It is left to the reader to verify that

$$c_r = U_2 \quad \text{if } U_s = \frac{\beta}{F_2}, \quad K^2 = K_+^2,$$

$$c_r = U_1 \quad \text{if } U_s = \frac{-\beta}{F_1}, \quad K^2 = K_-^2. \tag{7.11.35}$$

## 7.12  Effects of Friction

In the examples considered thus far, the stability threshold, whether a critical wave number or a critical shear, was a manifestation of inviscid potential-vorticity constraints. On the stable side of the threshold the phase speed $c$ is purely real, and upon crossing the threshold two complex conjugate solutions appear whose imaginary parts are equal in magnitude but opposite in sign. Below the stability threshold in the inviscid problem, waves neither grow nor decay. The presence of friction generally alters this behavior, for if the wave is to simply survive it must extract sufficient energy to maintain itself against dissipation, while if it is to grow, it must drain from the basic flow an amount of energy in excess of its dissipative loss. In this section we will consider the nature of the instability in the presence of frictional Ekman layers. To keep the discussion as simple as possible, we will examine the baroclinic stability properties in the two-layer model with equal ambient depths of both layers and ignore the effect of the earth's sphericity. To keep as much symmetry as possible in the problem so as to facilitate discussion, it is assumed that both the upper and lower surfaces are rigid, so that, using (6.16.29) and (6.16.31), the equations for the two-layer model are, in the absence of $\beta$,

$$\left[\frac{\partial}{\partial t} + \frac{\partial \psi_1}{\partial x}\frac{\partial}{\partial y} - \frac{\partial \psi_1}{\partial y}\frac{\partial}{\partial x}\right][\nabla^2\psi_1 - \hat{F}(\psi_1 - \psi_2)] = -\frac{r}{2}\nabla^2\psi_1,$$

$$\left[\frac{\partial}{\partial t} + \frac{\partial \psi_2}{\partial x}\frac{\partial}{\partial y} - \frac{\partial \psi_2}{\partial y}\frac{\partial}{\partial x}\right][\nabla^2\psi_2 - \hat{F}(\psi_2 - \psi_1)] = -\frac{r}{2}\nabla^2\psi_2,$$

(7.12.1)

where

$$\hat{F} = F_1 = F_2,$$

$$r = \frac{(2A_V f_0)^{1/2}}{U}\frac{L}{D_2} = \frac{(2A_V f_0)^{1/2}}{U}\frac{L}{D_1}.$$

(7.12.2)

The basic state is the simple shear flow in which $U_1$ and $U_2$ are independent of $y$. The linear stability problem for the perturbations $\phi_n(x, y, t)$ then becomes

$$\left[\frac{\partial}{\partial t} + U_1\frac{\partial}{\partial x}\right][\nabla^2\phi_1 - \hat{F}(\phi_1 - \phi_2)]$$

$$+ \hat{F}(U_1 - U_2)\frac{\partial \phi_1}{\partial x} = -\frac{r}{2}\nabla^2\phi_1$$

(7.12.3a)

$$\left[\frac{\partial}{\partial t} + U_2\frac{\partial}{\partial x}\right][\nabla^2\phi_2 - \hat{F}(\phi_2 - \phi_1)]$$

$$- \hat{F}(U_1 - U_2)\frac{\partial \phi_2}{\partial x} = -\frac{r}{2}\nabla^2\phi_2.$$

(7.12.3b)

Normal-mode solutions may be sought in the form

$$\phi_n = \text{Re } A_n e^{ik(x - ct)}\cos l_j y,$$

(7.12.4)

where

$$l_j = (j + \tfrac{1}{2})\pi, \qquad j = 0, 1, 2, \ldots,$$

which satisfies the condition that $\phi$ vanishes at $y = \pm 1$.

If (7.12.4) is substituted into (7.12.3a,b) we obtain two, homogeneous linear equations for $A_1$ and $A_2$, i.e.,

$$A_1\left[(c - U_1)[K^2 + \hat{F}] + \frac{ir}{2}\frac{K^2}{k} + \hat{F}U_s\right] - A_2(c - U_1)\hat{F} = 0,$$

$$A_2\left[(c - U_2)[K^2 + \hat{F}] + \frac{ir}{2}\frac{K^2}{k} - \hat{F}U_s\right] - A_1(c - U_2)\hat{F} = 0,$$

(7.12.5)

where, as before

$$K^2 = k^2 + l_j^2,$$

$$U_s = U_1 - U_2.$$

(7.12.6)

Note that the coefficients of (7.12.5) are *complex* due to the presence of friction, and therefore it is no longer true that $c$ and its complex conjugate will both be eigenvalues of the normal-mode problem. Furthermore, (7.12.5) may be obtained directly from (7.11.4a,b) by the substitution $\beta \to irK^2/2k$ so that the solution for $c$ can be obtained from (7.11.6) by the same substitution. Finally, it is also evident that now the problem for $c$ depends on both $k$ and $K$, since the vorticity damping produced by the Ekman layer is directly proportional to the vorticity (i.e., $O(K^2)$), while the remaining inviscid terms are proportional to the advection of potential vorticity and hence proportional to $kK^2$ or $k\hat{F}$. Thus, the Ekman-layer frictional effect becomes dominant, as is apparent in (7.12.5), for *large* zonal wavelengths.

If $A_1$ and $A_2$ are to differ from zero, the determinant of their coefficients must vanish. This yields a quadratic equation for $c$ whose solutions are

$$c = \frac{U_1 + U_2}{2} - \frac{irK^2/k}{2K^2(K^2 + 2\hat{F})}(K^2 + \hat{F})$$

$$\pm \frac{i[K^4 U_s^2(4\hat{F}^2 - K^4) + r^2 K^4 \hat{F}^2/k^2]^{1/2}}{2K^2(K^2 + 2\hat{F})}.$$

(7.12.7)

In the absence of shear $c_i$ is negative for both roots; the decay rates are

$$kc_i = \begin{cases} -\dfrac{r}{2}, \\[2mm] -\dfrac{r}{2}\dfrac{K^2}{K^2 + 2F}, \end{cases}$$

(7.12.8)

corresponding to the decay of barotropic and baroclinic initial disturbances. For $U_s^2 > 0$ and $K^4 < 4F^2$ the radicand in (7.12.7) increases in magnitude until the square-root term balances the negative imaginary terms preceding it. This occurs for the critical value of $U_s$ given by

$$U_s = U_c(K, k) = \frac{r(K/k)}{(2\hat{F} - K^2)^{1/2}}$$

$$= \frac{rK}{[(K^2 - l_j^2)(2\hat{F} - K^2)]^{1/2}}.$$

(7.12.9)

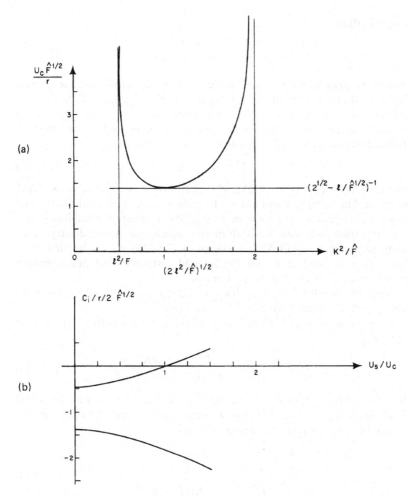

**Figure 7.12.1** (a) The curve of marginal stability. For $U_s < U_c$ all waves are *damped* and decay. (b) The imaginary part of $c$ as a function of $U_s/U_c$ at $K = K_m$ as defined by (7.12.10). Note that $U_c$ is a transition point for only one of the two roots.

The critical curve for a fixed $l_j^2$ is shown in Figure 7.12.1. The short-wave cutoff is again given by $K^2 = 2\hat{F}$ as in the inviscid theory. The *long* waves, $k \to 0$, are *stabilized* by the frictional effects of the Ekman layers. $U_c$, considered as a function of $K$ for a given modal number $l_j$, achieves a minimum at

$$K^2 = K_m^2 = (2\hat{F}l_j^2)^{1/2}, \qquad (7.12.10)$$

which is the geometric mean between the short-wave cutoff $K^2 = 2\hat{F}$ and the long-wave cutoff $K^2 = l_j^2 (k \to 0)$. The *minimum* critical shear corresponding to (7.12.10) is

$$(U_c)_{min} = \frac{r\hat{F}^{-1/2}}{2^{1/2} - l_j/\hat{F}^{1/2}}. \tag{7.12.11}$$

Again, the gravest cross-stream mode, $j = 0$, is the most unstable. Figure 7.12.1(b) shows $c_i$ as a function of $U_s$ at $K = K_m$. For $U_s < U_c$ both roots have $c_i < 0$, i.e., both normal modes decay on a time scale $O(r^{-1})$. As $U_s \to U_c$, $c_i$ approaches zero for only one mode, while for the other mode it remains negative and in fact becomes larger in magnitude. For

$$U_s = U_c + \Delta, \tag{7.12.12}$$

one root has a $c_i$ which is $O(\Delta)$, while the second root remains $O(1)$ and negative. The stability threshold is a transition curve for only one of the two modes. This contrasts with the inviscid problem, where the stability boundary is a transition point for both modes, which there suddenly acquire an imaginary part $c_i = \pm O(\Delta^{1/2})$ as shown by (7.11.20). $c_i$ as a function of $U_s$ has a branch point at $U_c$ for the inviscid instability, but has a smooth transition at $U_c$ if viscous effects dominate.

Note that for all $K^2 < 2\hat{F}$, each mode has $c_r = (U_1 + U_2)/2$, i.e., the wave moves with the mean velocity of the two layers.

Since $c$ is known, (7.12.5) may be used to obtain the vertical structure of the wave, i.e.,

$$\frac{A_2}{A_1} = \frac{K^2 + \hat{F}}{\hat{F}} + \frac{U_s}{c - U_1} + \frac{irK^2/2k}{(c - U_1)\hat{F}}. \tag{7.12.13}$$

It is important to note that the marginally stable wave, for which $c$ is real (and equal to $(U_1 + U_2)/2$), has a nonzero phase shift between the two layers, i.e., at the curve of marginal stability

$$\frac{A_2}{A_1} = \frac{K^2}{\hat{F}} - 1 - i\frac{rK^2}{k\hat{F}U_s},$$

$$= \frac{K^2}{\hat{F}} - 1 - i\frac{K(2\hat{F} - K^2)^{1/2}}{\hat{F}}. \tag{7.12.14}$$

By direct calculation, at the stability boundary

$$\frac{|A_2|}{|A_1|} = 1; \tag{7.12.15}$$

thus

$$\frac{A_2}{A_1} = e^{-i\alpha}, \tag{7.12.16}$$

where

$$\alpha = \tan^{-1}\left[\frac{K(2\hat{F} - K^2)^{1/2}}{K^2 - \hat{F}}\right]. \tag{7.12.17}$$

No phase shift occurs in the inviscid problem unless $c_i \neq 0$, i.e., unless the wave is growing. In the presence of friction the marginal wave, neither growing nor decaying, must have a phase shift in order to feed on the energy of the basic flow and maintain itself against dissipation.

The conversion of energy, by the marginal wave, from the zonal flow is given by (7.9.12), i.e., using $F_0 = \hat{F}/2$,

$$\frac{\hat{F}}{2} U_s \phi_1 \overline{\frac{\partial \phi_2}{\partial x}} = -\frac{\hat{F}}{2} U_s \frac{ik}{4} [A_1 A_2^* - A_1^* A_2] \cos^2 l_j y, \qquad (7.12.18)$$

which, with (7.12.14), yields

$$\frac{\hat{F}}{2} U_s \phi_1 \overline{\frac{\partial \phi_2}{\partial x}} = r \frac{K^2}{4} |A_1|^2 \cos^2 l_j y. \qquad (7.12.19)$$

The rate of energy dissipation in the lower Ekman layer is, for the marginal wave,

$$\frac{r}{2} d_2 \overline{(\nabla \phi_2)^2} = \frac{r}{4} \overline{(\nabla \phi_2)^2} = \frac{rk^2}{8} |A_2|^2 \cos^2 l_j y, \qquad (7.12.20)$$

while the rate of energy dissipation in the upper Ekman layer is similarly

$$\frac{r}{2} d_1 \overline{(\nabla \phi_1)^2} = \frac{rk^2}{8} |A_1|^2 \cos^2 l_j y. \qquad (7.12.21)$$

The total dissipation is the sum of (7.12.20) and (7.12.21), which by (7.12.15) is precisely equal to the rate of energy flow from the basic state to the wave given by (7.12.19). When $U_s$ exceeds $U_c$, the rate of energy conversion exceeds the dissipation. In the inviscid model there is no energy transfer until the stability threshold is crossed.

## 7.13 Baroclinic Instability of Nonzonal Flows

The physical basis for baroclinic instability, described in Section 7.6, emphasizes the release of available potential energy by motions in the direction of the basic horizontal temperature gradient. Because of the thermal wind relation, this implies that the energy-releasing portion of the fluid-element trajectory is perpendicular to the direction of the basic current. As the calculations of the preceding sections show, at a given total wave number $K$, the growth rate is maximized when $k/l_j$ is as large as possible. For then, geostrophically, the trajectories are as energy releasing as possible. This accounts for the preferential instability of the gravest cross-stream mode. However, the maximization of $v$ with respect to $u$ in the perturbation field also enhances the stabilizing effect of $\beta$, whose action on the vorticity field is also proportional only to $v$. If the $x$-component, $k$, of the wave vector were zero, the effect of $\beta$ would vanish for that wave, but then the fluid trajectories would be perpendicular to the horizontal temperature gradient of a zonal flow and be incapable of releasing energy. If, however, the basic current is oriented at some angle to a latitude circle, strictly zonal perturbation motions, which do not feel the $\beta$-effect, will still have a component of flow in the direction of the basic temperature gradient and be capable of releasing available potential energy. It can be anticipated, therefore, that such flows will not possess a minimum critical shear for instability except insofar as frictional dissipation is important.

If the basic flow is not strictly in the $x$-direction, the geostrophic pressure field associated with it will not be a solution of the *unforced* potential-vorticity equation, e.g., (7.9.1). Nonzonal basic flows imply the existence of an external forcing field. As a simple but important example, consider the two-layer model for oceanic flow in the presence of a wind-stress curl. If the derivation of (6.16.30) is modified to include the wind-stress curl as given by (4.11.10) and (6.16.20b), the two-layer equations become, in the absence of bottom friction,

$$
\left[ \frac{\partial}{\partial t} + \frac{\partial \psi_1}{\partial x} \frac{\partial}{\partial y} - \frac{\partial \psi_1}{\partial y} \frac{\partial}{\partial x} \right] [\nabla^2 \psi_1 - F_1(\psi_1 - \psi_2) + \beta y]
$$

$$
= \left[ \frac{\tau_0}{\rho_0 f_0 U D_1 \varepsilon} \right] \mathbf{k} \cdot \text{curl } \tau, \tag{7.13.1a}
$$

$$
\left[ \frac{\partial}{\partial t} + \frac{\partial \psi_2}{\partial x} \frac{\partial}{\partial y} - \frac{\partial \psi_2}{\partial y} \frac{\partial}{\partial x} \right] [\nabla^2 \psi_2 - F_2(\psi_2 - \psi_1) + \beta y] = 0, \tag{7.13.1b}
$$

which is the two-layer generalization of (4.11.11). Consider now the basic state

$$
\begin{aligned}
\psi_1 &= \Psi_1 = -U_1 y + V_1 x, \\
\psi_2 &= \Psi_2 = 0,
\end{aligned} \tag{7.13.2}
$$

where $U_1$ and $V_1$ are constants. The flow is directed across latitude circles with a velocity $V_1$. If (7.13.2) is inserted into (7.13.1a,b), we find that the second equation is trivially satisfied, while the first requires that

$$
\beta V_1 = \left[ \frac{\tau_0}{\rho_0 f_0 U D_1 \varepsilon} \right] \text{curl } \tau, \tag{7.13.3}
$$

which is the Sverdrup relation. The existence of $V_1 \neq 0$ requires a wind-stress curl, in this case one which is constant, although here that is merely a choice for the purposes of simplicity. To examine the stability of the flow represented by (7.13.2), we write, as before,

$$
\begin{aligned}
\psi_1 &= \Psi_1 + \phi_1(x, y, t), \\
\psi_2 &= \Psi_2 + \phi_2(x, y, t),
\end{aligned} \tag{7.13.4}
$$

which upon insertion into (7.13.1a,b) yields, after linearization,

$$\left[\frac{\partial}{\partial t} + U_1 \frac{\partial}{\partial x} + V_1 \frac{\partial}{\partial y}\right][\nabla^2\phi_1 - F_1(\phi_1 - \phi_2)]$$

$$+ \frac{\partial\phi_1}{\partial x}[\beta + F_1 U_1] + \frac{\partial\phi_1}{\partial y} F_1 V_1 = 0, \qquad (7.13.5a)$$

$$\left[\frac{\partial}{\partial t}\right][\nabla^2\phi_2 - F_2(\phi_2 - \phi_1)] + \frac{\partial\phi_2}{\partial x}[\beta - F_2 U_1] - \frac{\partial\phi_2}{\partial y} F_2 V_1 = 0. \qquad (7.13.5b)$$

It is extremely important to note that the wind-stress curl does *not* appear in the perturbation equations; rather it is balanced by the basic flow as in (7.13.3). This is a general feature of stability theory as discussed in Section 7.1. That is, although forcing may be present to produce the basic state, the only manifestation of that forcing for the stability problem is in the structure of the basic flow. The forcing field itself does not appear in the stability problem. Or, from another viewpoint, the stability of basic flows which are not solutions of the *unforced* equations of motion may be consistently considered in the context of the *unforced* perturbation equations, without the need to consider explicitly the forces required to produce the basic state.

Let us now examine the stability problem posed by (7.13.5a,b) for the case where the flow is unbounded in $x$ and $y$. The motive for this idealization is connected with the earlier results that baroclinic instability is favored for scales of the order of the deformation radius, which, for the oceanic case which we are considering, is far smaller than the extent of the oceanic basin. Indeed, the idealization that $U_1$ and $V_1$ are constants is appropriate when considering perturbation length scales small compared to the oceanic general circulation scale. Thus, in the present example we are considering an idealized model for the stability of the baroclinic mid-ocean flow distant from oceanic boundaries and their narrow, swift currents. The fluid is driven by a large-scale wind-stress curl which locally forces a basic flow at some angle to a latitude circle. This problem of the stability of the mid-ocean thermocline is of considerable oceanographic relevance and has been studied in different contexts by several investigators (e.g., Gill, Green, and Simmons (1974) and Robinson and McWilliams (1974)).

If plane-wave solutions of the form

$$\phi_n = A_n \exp i[kx + ly - \sigma t], \qquad n = 1, 2, \qquad (7.13.6)$$

are substituted into (7.13.5a,b), we obtain

$$A_1[(\sigma - U_1 k - V_1 l)(K^2 + F_1) + \beta k + F_1(U_1 k + V_1 l)]$$

$$- A_2(\sigma - U_1 k - V_1 l)F_1 = 0, \qquad (7.13.7a)$$

$$A_2[\sigma(K^2 + F_2) + \beta k - F_2(U_1 k + V_1 l)] - A_1 \sigma F_2 = 0, \qquad (7.13.7b)$$

where

$$K^2 = k^2 + l^2.$$

The following definitions are useful:

$$\tilde{\beta} = \frac{\beta k}{K} = \beta \cos \theta,$$

$$\tilde{U}_1 = \frac{U_1 k + V_1 l}{K} = (U_1^2 + V_1^2)^{1/2} \cos(\alpha - \theta), \qquad (7.13.8)$$

$$c = \frac{\sigma}{K},$$

where $\theta$ is the angle the wave vector **K** makes with the $x$-axis, while $\alpha$ is the angle of the basic shear flow with respect to the $x$-axis as shown in Figure 7.13.1. Without loss of generality, $k$ and hence $\tilde{\beta}$ may be taken as positive.

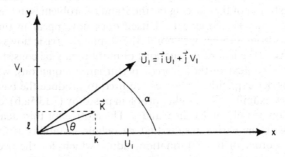

**Figure 7.13.1**   The orientation of the basic flow **U** is at an angle $\alpha$ with respect to the $x$-axis (a line of constant latitude), while the wave vector **K** is at an angle $\theta$ with respect to the $x$-axis.

The quantities $\tilde{U}_1$ and $\tilde{\beta}$ are the projection of the basic shear flow in the direction of the wave vector and the projection of the planetary-vorticity gradient on the path of the perturbation motion, respectively. The phase speed $c$ as defined by (7.13.8) may be complex. If so, the associated growth rate is $Kc_i$, while the speed of propagation of the wave crests in the direction of **K** is given by $c_r$. The definitions (7.13.8) allow (7.13.7a,b) to be rewritten, after division by $K$, as

$$A_1[(c - \tilde{U}_1)[K^2 + F_1] + \tilde{\beta} + F_1\tilde{U}_1] - A_2(c - \tilde{U}_1)F_1 = 0, \qquad (7.13.9a)$$

$$A_2[c[K^2 + F_2] + \tilde{\beta} - F_2\tilde{U}_1] - A_1(c)F_2 = 0, \qquad (7.13.9b)$$

which are precisely the equations (7.11.4a,b) for the stability of *zonal* flow with the correspondence

$$\beta \leftrightarrow \tilde{\beta}, \qquad (7.13.10a)$$

$$U_s = U_1 - U_2 \leftrightarrow \tilde{U}_1, \qquad (7.13.10b)$$

$$U_2 = 0. \qquad (7.13.10c)$$

The solution for $c$ follows, then, directly from (7.11.6):

$$c = \frac{\tilde{U}_1 K^2(K^2 + 2F_2) - \tilde{\beta}(2K^2 + F_1 + F_2)}{2K^2(K^2 + F_1 + F_2)}$$

$$\pm \frac{[\tilde{\beta}^2(F_1 + F_2)^2 + 2\tilde{\beta}\tilde{U}_1 K^4(F_1 - F_2) - K^4\tilde{U}_1^2(4F_1 F_2 - K^4)]^{1/2}}{2K^2(K^2 + F_1 + F_2)}.$$

$$(7.13.11)$$

The stability criteria deduced in Section 7.11 in terms of $\beta$ and $U_s$ may now be applied directly in the present case with the correspondence relations (7.13.10). The crucial difference is that both $\tilde{U}$ and $\tilde{\beta}$ are functions of the *orientation* of the wave vector with respect to the $x$-axis and the direction of the basic shear flow. In particular, the minimum critical shear required for instability is now either

$$\tilde{U}_{c+} = \frac{\tilde{\beta}}{F_2}, \qquad \tilde{U}_1 > 0, \qquad (7.13.12a)$$

or

$$\tilde{U}_{c-} = -\frac{\tilde{\beta}}{F_1}, \qquad \tilde{U}_1 < 0. \qquad (7.13.12b)$$

However, for $k = 0$, $\tilde{\beta}$ will vanish, so that *any* shear will be unstable to such a disturbance as long as $\tilde{U}_1 \neq 0$ for $k = 0$. This requires only that the shear flow have some nonzero component in the $y$-direction. That is, for $k = 0$,

$$c_i = \frac{V_1}{2} \frac{[4F_1 F_2 - l^4]^{1/2}}{l^2 + F_1 + F_2}. \qquad (7.13.13)$$

Thus, as long as $l^2 < 2(F_1 F_2)^{1/2}$, the basic flow will be unstable to the mode with $k = 0$, and therefore *no* minimum shear is required for instability. Of course, if the flow is nearly zonal, i.e., if $V_1$ is very small, the corresponding growth rates will be small. In fact, for a nonzonal flow the optimum direction for the wave vector (i.e., the orientation corresponding to the maximum growth rate) will not be strictly northward. Some intermediate position is favored between $k = 0$, which eliminates the $\beta$-effect, and $\theta = \alpha$, which maximizes the projection of the motion on the basic temperature gradient. For the purposes of illustration, consider the case where $F_1 = F_2$. Then if the radicand in (7.13.11) is negative, a little algebra yields

$$\frac{Kc_i}{\beta/\hat{F}^{1/2}} = \frac{[u_s^2 \cos^2(\alpha - \theta) a^4(4 - a^4) - 4 \cos^2 \theta]^{1/2}}{2a(2 + a^2)}, \qquad (7.13.14)$$

where

$$a^2 = \frac{K^2}{\hat{F}},$$

$$u_s = \frac{(U_1^2 + V_1^2)^{1/2}}{\beta/\hat{F}}, \qquad (7.13.15)$$

so that $u_s$ is a measure of the supercriticality of the flow, were it to be considered as zonal. That is, if $u_s < 1$, the flow would be stable if it were strictly zonal. For a given $a^2$ the maxima of $Kc_i$ correspond to the extrema of the function

$$I(\alpha, \theta) = W(a)\cos^2(\theta - \alpha) - \cos^2 \theta, \tag{7.13.16}$$

where

$$W(a) = \frac{u_s^2}{4} a^4(4 - a^4) \geq 0. \tag{7.13.17}$$

For fixed $\alpha$, $I$ and therefore $Kc_i$ have their maxima where

$$\frac{\partial}{\partial \theta} I(\alpha, \theta_m) = 0, \tag{7.13.18}$$

or where

$$\tan 2\theta_m = \left[ \frac{W(a) \sin 2\alpha}{W(a) \cos 2\alpha - 1} \right]. \tag{7.13.19}$$

For large $W$ (i.e., where $u_s$ considerably exceeds the critical value for the instability of zonal flow),

$$\theta_m \to \alpha, \tag{7.13.20}$$

i.e., the $\beta$-effect is relatively feeble at these high shears, and the most unstable wave is oriented parallel to the basic shear in order to maximize the release of energy. On the other hand, as $W \to 0$ (i.e., for very weak shears),

$$\theta_m \to \frac{\pi}{2} \tag{7.13.21}$$

to maintain the instability in the presence of $\beta$. In this limit the perturbation velocities are very nearly zonal. Figure 7.13.2 shows the growth rate as a function of the orientation of the wave vector for a basic flow with $\alpha = 45°$. The solid curve shows $Kc_i$ when $W(a) = 2.25$. This corresponds, at the critical wave number $K^2 = 2^{1/2}\hat{F}$, (7.11.16), to a shear flow $u_s$ which is 1.5 times the critical value for zonal-flow instability, i.e., this flow would be unstable even were it zonal. The flow is unstable to a broad range of wave-vector orientations and obtains its maximum value at $\theta = 57°$, i.e., somewhat closer to the direction of the basic shear than northward. The most unstable wave here makes a 12° angle with the flow axis and a 33° angle with respect to north. On the other hand, the dashed curve in Figure 7.13.2 shows the growth rate for $W(a) = 0.25$ for the same $\alpha$. At $a^2 = 2^{1/2}$, this corresponds to a shear that is only *half* the critical value required for instability if the flow were zonal. Yet as the figure demonstrates, the nonzonal flow is unstable. However, now the most unstable disturbance has a wave vector oriented more nearly to the north: $\theta_m$ is now 83°. The maximum growth rates are reduced by a factor of about four, but are the same order of magnitude as

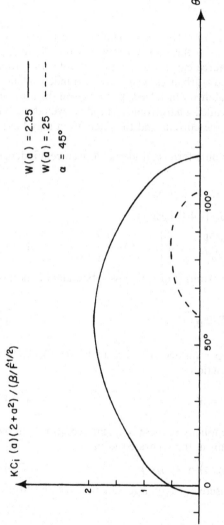

**Figure 7.13.2**    The growth rate as a function of the orientation of the wave vector for the case where the basic flow is directed N-NE (i.e., $\alpha = 45°$). The solid curve shows the growth rate for a basic shear one and a half times the value required for instability were the flow strictly zonal. The dashed curve shows the growth rate for the case where the basic shear would be only *half* the minimum critical shear were the flow strictly zonal.

before. Thus, except insofar as friction may damp the weakly unstable waves, baroclinic basic flows which are not strictly zonal will almost always be baroclinically unstable.

## 7.14 Barotropic Instability

The fluctuation-energy equation (7.3.11) (or (7.9.12) for the two-layer model) reveals that the horizontal shear of the basic current is a possible energy source for fluctuations. This source can exist in a current with no vertical shear and may be released by disturbances which are also independent of height. For this reason, an instability which feeds primarily on the horizontal shear of the basic current is called a barotropic instability, even though in more complex situations both the current and the fluctuation may have a baroclinic structure.

To study the instability in its pure form, consider a basic current independent of $z$, i.e.,

$$U_0 = U_0(y).$$

Then the normal-mode equation (7.4.4) becomes

$$(U_0 - c)\left[\frac{1}{\rho_s}\frac{\partial}{\partial z}\frac{\rho_s}{S}\frac{\partial \Phi}{\partial z} + \frac{\partial^2 \Phi}{\partial y^2} - k^2\Phi\right] + \left[\beta - \frac{d^2 U_0}{dy^2}\right]\Phi = 0. \quad (7.14.1)$$

In the absence of friction and bottom slope, the lower boundary condition (7.4.7) becomes

$$\frac{\partial \Phi}{\partial z} = 0, \quad z = 0. \quad (7.14.2)$$

For simplicity, imagine the upper surface at $z = z_T$ to be rigid as well, in which case, in the absence of friction,

$$\frac{\partial \Phi}{\partial z} = 0, \quad z = z_T. \quad (7.14.3)$$

Then, as in Section 6.12, the simplicity of these boundary conditions allows solutions of (7.14.1) to be sought in the separated form

$$\Phi(y, z) = A(y)\chi(z), \quad (7.14.4)$$

where $\chi$ is any one of the discrete set of eigenfunctions of the Sturm–Liouville problem (6.12.7), i.e.,

$$\frac{1}{\rho_s}\frac{d}{dz}\frac{\rho_s}{S}\frac{d\chi}{dz} = -\lambda\chi, \quad (7.14.5a)$$

with

$$\frac{\partial \chi}{\partial z} = 0, \quad z = 0, z_T, \quad (7.14.5b)$$

and where $\lambda$ is the associated eigenvalue. The function $A(y)$ satisfies the barotropic instability equation,

$$(U_0(y) - c)\left[\frac{d^2 A}{dy^2} - \mu^2 A\right] + \left[\beta - \frac{d^2 U_0}{dy^2}\right] A = 0, \tag{7.14.6}$$

where

$$\mu^2 = k^2 + \lambda. \tag{7.14.7}$$

Note that $\lambda = 0$ is always an eigenvalue of (7.14.5a,b) with an accompanying purely barotropic mode. The stability problem posed by (7.14.6) for each $\lambda$ is precisely identical to the stability of the barotropic flow to a *barotropic* disturbance with x-wave-number $\mu$. Thus the dynamics of an arbitrary baroclinic perturbation of the *barotropic* basic current can be described entirely in terms of the equivalent barotropic mode.

The lateral boundary conditions (7.4.6) become

$$A(y) = 0, \qquad y = \pm 1. \tag{7.14.8}$$

The necessary condition for instability, (7.4.22), can easily be simplified to yield Kuo's theorem (1949)

$$c_i \int_{-1}^{1} dy \frac{|A|^2}{|U_0 - c|^2}\left[\beta - \frac{d^2 U_0}{dy^2}\right] = 0. \tag{7.14.9}$$

Thus for inviscid instability of a barotropic current to occur, the northward gradient of the *absolute* vorticity must vanish somewhere. In distinction to baroclinic instability, no boundary term appears in the condition. Thus a sufficiently large value of $\beta$ can always stabilize a current with respect to barotropic instability. Since

$$\beta = \frac{\beta_0 L^2}{U}, \tag{7.14.10}$$

where $L$ is the characteristic scale of the motion, sufficiently broad currents with weak horizontal shear will be stable to barotropic instability. In previous sections we saw that baroclinic instability is favored when the horizontal scale of the motion is large compared to a Rossby-deformation radius. It is possible, therefore, to presume that baroclinic instability tends to be the favored mode for broad currents with a large $L$, while very narrow currents (thinner than a deformation radius) would tend to be barotropically unstable.

The mathematical structure of (7.14.1) is similar to the structure of the purely baroclinic problem which occurs when $U_0$ is a function of $z$ alone. It follows from this that the general solution of (7.14.6) is of the form given by (7.8.18) and (7.8.19), i.e., that $A(y)$ is a linear combination of $A_1(y)$ and $A_2(y)$,

where

$$A_1(y) = (y - y_c)[1 + a_1(y - y_c) + a_2(y - y_c)^2 + \cdots], \qquad (7.14.11a)$$

$$A_2(y) = A_1(y) \frac{\left(\beta - \dfrac{d^2 U_0}{dy^2}\right)_{y_c}}{\left(\dfrac{dU_0}{dy}\right)_{y_c}} \log(y - y_c)$$

$$- [1 + b_1(y - y_c) + b_2(y - y_c)^2 + \cdots], \qquad (7.14.11b)$$

where $y_c$ is the point where $U_0 - c$ vanishes. The presence of the logarithm in $A_2(y)$, coupled with the simplicity of the boundary conditions (7.14.8), has important implications. Consider, for example, a marginally stable wave, i.e., one parametrically adjacent to an unstable mode. If $c$ lies within the range of $U_0$, then the presence of the logarithm will alone introduce an imaginary contribution to the solution which cannot be balanced at the boundary, and hence (7.14.8) will be violated. This will occur unless the vorticity gradient *vanishes* at $y_c$. This may be observed alternately as follows. The Reynolds stress $-\overline{u_0 v_0}$ may be written as

$$-\overline{u_0 v_0} = \chi^2(z) \frac{ik}{4} \left[ A \frac{dA^*}{dy} - A^* \frac{dA}{dy} \right] e^{2kc_i t} \qquad (7.14.12)$$

and must vanish at $y = \pm 1$. If (7.14.6) is divided by $U_0 - c$, the result, and its complex conjugate, yield

$$\frac{d^2 A}{dy^2} - \mu^2 A + \left[ \beta - \frac{d^2 U_0}{dy^2} \right] \frac{A}{U_0 - c} = 0, \qquad (7.14.13a)$$

$$\frac{d^2 A^*}{dy} - \mu^2 A^* + \left[ \beta - \frac{d^2 U_0}{dy^2} \right] \frac{A^*}{U_0 - c^*} = 0. \qquad (7.14.13b)$$

If (7.14.13a) is multiplied by $A^*$ and subtracted from the product of (7.14.13b) with $A$, we obtain

$$\frac{d}{dy} \left[ A \frac{dA^*}{dy} - A^* \frac{dA}{dy} \right] = 2ic_i |A|^2 \frac{\beta - d^2 U_0/dy^2}{|U_0 - c|^2}. \qquad (7.14.14)$$

Thus, (7.14.12) implies that

$$\boxed{-\frac{d}{dy} \overline{u_0 v_0} = -\frac{kc_i}{2} |A|^2 \chi^2 \frac{\beta - d^2 U_0/dy^2}{|U_0 - c|^2} e^{2kc_i t},} \qquad (7.14.15)$$

where

$$|U_0 - c|^2 = (U_0 - c_r)^2 + c_i^2. \qquad (7.14.16)$$

The integral of (7.14.15) over $y$ from $-1$ to $+1$ yields (7.14.9), since $v_0$ vanishes at $y = \pm 1$. Now consider the limit of (7.14.15) as $c_i \to 0$, i.e., for a marginally stable wave. In this limit of vanishing $c_i$, the Reynolds stress becomes independent of $y$ for all $y$ except for possible discontinuous changes at those critical points where $U_0 - c_r$ vanishes, $c_i \to 0$. The jump in $-\overline{u_0 v_0}$ across at these critical points may be evaluated in the limit $c_i \to 0$ as in (7.8.16), so that the jump at $y = y_c$ may be written, if $(dU_0/dy)(y_c) > 0$,

$$\Delta(-\overline{u_0 v_0})_{y_c} = -\frac{k\pi}{2} \frac{|A(y_c)|^2}{(dU_0/dy)(y_c)} \left( \beta - \frac{d^2 U_0}{dy^2}(y_c) \right). \qquad (7.14.17)$$

If $(dU_0/dy)(y_c) < 0$, (7.14.17) applies with the opposite sign. Note the identity of the coefficient in (7.14.17) to the coefficient of the log term in (7.14.11b). The similarity is not fortuitous, since the discontinuous jump in the Reynolds stress as $c_i \to 0$ is due precisely to the discontinuous phase change with $y$ of $A$ across the critical point produced by the logarithm. As we have seen, the Reynolds stress, aside from these jumps, is constant in $y$ as $c_i \to 0$, and in fact vanishes on $y = \pm 1$. Hence, if $U_0$ is such that $U_0 - c$ vanishes for more than one value of $y$, either the *sum* of those jumps must add to zero or each jump itself must vanish. If $U_0(y)$ is monotonic in $y$, or if $U_0(y)$ is symmetric about the midpoint, the former alternative is not possible. For each jump in Reynolds stress to vanish we must have, in the limit $c_i \to 0$,

$$\beta - \frac{d^2 U_0}{dy^2}(y_c) = 0. \qquad (7.14.18)$$

Thus, marginally stable waves, adjacent to unstable waves, possess a phase speed $c$ given by

$$c = U_0(y_c), \qquad (7.14.19)$$

where $y_c$ is determined by (7.14.18). This assumes, of course, that $c$ lies within the range of $U_0$. Neutral waves, even if marginally stable, may *a priori* exist outside the range of $U_0$ or on the boundary of the range of $U_0$.

As an example, consider the problem studied by Kuo (1949, 1973) for which

$$U_0 = \frac{1 + \cos \pi y}{2} = \cos^2 \frac{\pi y}{2}. \qquad (7.14.20)$$

The basic flow is a symmetric jet, whose maximum nondimensional velocity of 1 is achieved at $y = 0$, while $U_0$ vanishes at $y = \pm 1$. The condition that the absolute vorticity gradient vanishes at some point in the interval requires that

$$b \equiv \frac{\beta}{\pi^2} = -\frac{\cos \pi y_c}{2} \qquad (7.14.21)$$

for some $y_c$ in the interval $(-1, 1)$, or since $U_0$ is symmetric, (7.14.21) must have a solution in the half interval $(-1, 0)$. When $\beta$ is zero this occurs at $y_c = -\frac{1}{2}$. As $\beta$ increases from zero, $y_c$ moves from $y = -0.5$ towards $y = -1$. When $b$ exceeds 0.5, $y_c$ lies outside the interval, the absolute vorticity gradient is always positive, and the flow is stable. Hence instability requires $b < \frac{1}{2}$, for which

$$\cos \pi y_c = -2b. \tag{7.14.22}$$

A neutral solution of (7.14.6) can be found with

$$c = U_0(y_c) = \frac{1}{2} + \frac{\cos \pi y_c}{2} = \frac{1}{2} - b. \tag{7.14.23}$$

With this value of $c$, it follows that

$$\frac{\beta - d^2 U_0/dy^2}{U_0 - c} = \pi^2, \tag{7.14.24}$$

so that (7.14.6) becomes

$$\frac{d^2 A}{dy^2} + (\pi^2 - \mu^2)A = 0. \tag{7.14.25}$$

For $\mu^2 \geq 0$, the only nontrivial solution of (7.14.25) is

$$A = \cos \tfrac{1}{2}\pi y, \tag{7.14.26}$$

so that

$$\boxed{\mu^2 = \mu_0^2 = \frac{3\pi^2}{4}} \,. \tag{7.14.27}$$

Consider now a wave with a slightly different wavelength, i.e., with

$$\mu^2 = \mu_0^2 + \Delta\mu^2, \tag{7.14.28}$$

and let

$$A = A_0(y) + \Delta A(y), \tag{7.14.29}$$

where

$$A_0(y) = \cos \tfrac{1}{2}\pi y. \tag{7.14.30}$$

In general, the phase speed $c$ will also be altered:

$$c = c_0 + \Delta c, \tag{7.14.31}$$

where $c_0$ is $\frac{1}{2} - b$. If (7.14.28), (7.14.29), and (7.14.31) are substituted into (7.14.6) and the fact that $(A_0, \mu_0^2, c_0)$ is a solution of (7.14.6) is used, the remaining terms, to lowest order, are

$$\left[\frac{d^2}{dy^2} + \frac{\pi^2}{4}\right]\Delta A = A_0 \,\Delta\mu^2 + \frac{\Delta c}{U_0 - c}\left[\frac{d^2 A_0}{dy^2} - \mu_0^2 A_0\right]$$

$$= A_0 \,\Delta\mu^2 - \frac{\Delta c \,\pi^2 A_0}{U_0 - c_0} \tag{7.14.32}$$

where (7.14.25) has been used. If (7.14.32) is multiplied by $A_0$ (given by

(7.14.30)) and integrated from $y = -1$ to $y = 0$, we obtain after repeated integration by parts

$$0 = \Delta\mu^2 \int_{-1}^{0} A_0^2 \, dy - \Delta c \, \pi^2 \int_{-1}^{0} \frac{A_0^2}{U_0 - c_0} \, dy. \tag{7.14.33}$$

The first integral in (7.14.33) is real and positive and is easily evaluated:

$$\int_{-1}^{0} A_0^2 \, dy = \tfrac{1}{2}. \tag{7.14.34}$$

The evaluation of the second integral is more subtle, since $U_0 - c_0$ vanishes in the interval at $y = y_c$. If $c_0$ is considered the limiting value of a phase speed with a *positive* imaginary part, then the singularity of the integrand in the integral

$$I = \int_{-1}^{0} \frac{A_0^2}{U_0 - c_0} \, dy \tag{7.14.35}$$

lies slightly above the real $y$-axis, as shown in Figure 7.14.1, and the indented contour for the integration *under* the singularity, as shown, is appropriate.

**Figure 7.14.1** The path of integration for the evaluation of the integrals in (7.14.35) is indented below the singularity at $y = y_c$ in the limit $c_i \to 0_+$ as shown.

The contribution from integrating on the semicircle around the pole at $y = y_c$ is

$$\pi i \, \text{Res}(y_c) = \frac{A_0^2(y_c)\pi i}{(dU_0/dy)(y_c)}; \tag{7.14.36}$$

hence $I$ may be written as

$$I = \mathcal{A} + i\mathcal{B}, \tag{7.14.37}$$

where

$$\mathcal{B} = \frac{\pi A_0^2(y_c)}{(dU_0/dy)(y_c)} > 0 \tag{7.14.38}$$

and

$$\mathcal{A} = \lim_{\delta \to 0} \left[ \int_{-1}^{y_c-\delta} + \int_{y_c+\delta}^{0} \right] \frac{A_0^2}{U_0 - c_0} \, dy. \tag{7.14.39}$$

$\mathcal{A}$ is the Cauchy principal value of the integral $I$ and is, of course, strictly real. It now follows from (7.14.33) that

$$\Delta c = \left(\frac{\Delta\mu^2}{2}\right) \frac{\mathcal{A} - i\mathcal{B}}{\mathcal{A}^2 + \mathcal{B}^2}. \tag{7.14.40}$$

Since $\mathscr{B}$ is positive, the imaginary part of $\Delta c$ will be positive if $\Delta\mu^2$ is negative. Since

$$k^2 = \mu^2 - \lambda = \mu_0^2 - \lambda + \Delta\mu^2, \qquad (7.14.41)$$

this implies that waves with slightly longer wavelengths than the marginally stable solution $(A_0, \mu_0)$ will be unstable.

Kuo (1973) has carried out detailed numerical calculations for the stability of this "cosine jet," and his results are shown in Figure 7.14.2, where isolines of the growth rate $kc_i$ are shown for the barotropic mode corresponding to $\lambda = 0$. Since $c$ is a function of $k$ only through $\mu$,

$$kc_i(\mu) = (\mu^2 - \lambda)^{1/2}c_i(\mu), \qquad (7.14.42)$$

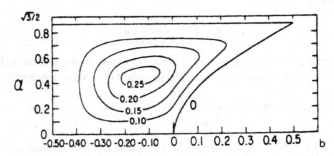

**Figure 7.14.2** Contours of constant $\alpha c_i$ as a function of $\alpha$ and $b$ for the cosine jet as calculated by Kuo (1973).

and therefore for any $\mu$, the *maximum* growth rate always occurs for zero $\lambda$, that is, the barotropic current is most unstable to *barotropic* perturbations. In Figure 7.14.2 the short-wave cutoff is given by $\alpha \equiv \mu/\pi = \sqrt{3/2}$ in agreement with (7.14.27), while the longer waves are unstable. The stability boundary in Figure 7.14.2 which slopes upwards and to the right and intersects the shortwave cutoff at $b = 0.5$, $\mu^2 = 3\pi^2/4$ is determined by the marginally stable solution

$$A = \cos^{2r}\left(\frac{\pi y}{2}\right),$$
$$c = 0, \qquad (7.14.43)$$

where

$$4r = 1 + (9 - 16b)^{1/2} \qquad (7.14.44)$$

and

$$\mu^2 = \pi^2(1 - r^2). \qquad (7.14.45)$$

Note that $\frac{1}{2} \le r \le 1$ for $b > 0$. The lower limit corresponds to the short-wave cutoff, while the upper limit is attained at $b = 0$, for which $\mu$ vanishes.

Negative values of $b$ in Figure 7.14.2 correspond (with a positive $\beta_0$) to a basic zonal current flowing from east to west. The solution (7.14.43) is only possible for $b > 0$, since for $b < 0$, either $r > 1$, for which $\mu^2 < 0$, or the negative branch of the radical in (7.14.44) is taken, in which case $A$ as given by (7.14.43) would be singular at $y = -1$. This analytical asymmetry between eastward and westward flowing jets is evident in the growth-rate diagram, where it is clear that for a given $\mu$, a westward-flowing jet is more unstable than its eastward image.

For the purely barotropic basic current, the energy equation (7.3.11) for the perturbations simplifies to

$$\frac{\partial E(\phi)}{\partial t} = -\int_{-1}^{1} dy \int_{0}^{z_T} dz \, \overline{u_0 v_0} \frac{dU_0}{dy}. \tag{7.14.46}$$

Since

$$-\overline{u_0 v_0} = v_0^2 \overline{\frac{\phi_y}{\phi_x}} = -v_0^2 \overline{\left(\frac{\partial x}{\partial y}\right)_\phi}, \tag{7.14.47}$$

it follows that for $E(\phi)$ to increase the phase of $\phi$ must tilt westward with increasing latitude in regions where $dU_0/dy > 0$, as shown in Figure 7.3.1(a), and tilt eastward with increasing $y$ where $dU_0/dy < 0$. This tilt to the wave field provides a rectified momentum flux from the center of the jet to its wings, tending to smooth the basic current profile, i.e., to reduce the shear. For the barotropic mode ($\lambda = 0$), no vertical velocity and hence no mean meridional circulation will be driven by the perturbations; hence the equation for the change of the mean flow, (7.2.13), becomes simply

$$\frac{\partial \bar{u}_0}{\partial t} = -\frac{\partial}{\partial y} \overline{v_0 u_0}, \tag{7.14.48}$$

which, with (7.14.15), may be written

$$\frac{\partial \bar{u}_0}{\partial t} = -\frac{kc_i}{2} \frac{|A|^2 \chi^2}{|U_0 - c|^2} \left[\beta - \frac{d^2 U_0}{dy^2}\right] e^{2kc_i t}. \tag{7.14.49}$$

Now (7.14.9) implies that the average of $\bar{u}_0$ over the interval in $y$ is unchanged by the perturbations and that the $x$-averaged zonal momentum is merely redistributed with latitude. The expression (7.14.49) shows that for $kc_i > 0$, $\bar{u}_0$ will decrease where $\beta - d^2 U_0/dy^2 > 0$ and increase where the absolute vorticity gradient is negative. At the center of any eastward-flowing jet, for example, $d^2 U_0/dy^2 < 0$, since this, with the vanishing of $dU_0/dy$, actually defines the point of maximum velocity. Hence at the center of the jet $\bar{u}_0$ must be *diminished* by the barotropic instability. South of the latitude $y_c$ where the absolute vorticity gradient changes sign, we have $\partial \bar{u}_0/\partial t > 0$ if $kc_i > 0$, and therefore the slower part of the jet is accelerated by the transfer of eastward momentum from the core of the jet. Barotropic instability tends, therefore, to *broaden* and weaken a narrow jet.

## 7.15  Instability of Currents with Horizontal and Vertical Shear

In previous sections the stability of currents with horizontal *or* vertical shear was examined to illustrate the fundamental nature of barotropic and baroclinic instability respectively. Yet all real flows have, to some degree, both horizontal and vertical shear, and an accurate study of their instability must allow for both. There is, though, a far deeper reason than accuracy for the examination of the stability of flows with both types of shears. When only one form of shear is present, only one source of energy for the disturbance is accessible, and the corresponding energy transfer mechanism in the fluctuation-energy equation must be positive for instability. Thus, a current with only vertical shear will be unstable to fluctuations which, on average, must transfer heat down the basic temperature gradient, which in turn tends to weaken the zonally averaged, meridional gradient. Similarly, a current with only horizontal shear will be unstable to perturbations which, on average, must transfer zonal momentum down the basic momentum (velocity) gradient, smoothing the x-averaged velocity field. In each of these cases, the perturbation fields act simply like large-scale mixing agents. When both vertical and horizontal shears are simultaneously present, so that both energy transformation mechanisms in (7.3.11) may act, their individual signs are unknown *a priori*. Only the fact that their sum must be positive for instability is required. There are, indeed, three possibilities. A disturbance may gain energy from both the basic temperature and momentum gradients, so that the Reynolds-stress conversion of mechanical energy and the eddy heat-flux conversion of available potential energy are both positive. It is also possible that the disturbance may be baroclinically unstable, releasing the available potential energy of the basic state while simultaneously generating Reynolds stresses which feed perturbation energy *into* the basic flow. This would imply that the rectified fluctuation momentum flux is counter to the basic momentum gradient, i.e., that, for example, the core of a baroclinic jet would tend to be accelerated by the fluctuations while its flanks are retarded by the disturbances. Naturally, for an unstable disturbance the extraction of one form of energy must exceed the return of energy to the zonally averaged flow. The third possibility is the mirror image of the second, i.e., that kinetic energy is released by the perturbation Reynolds stresses in a barotropic instability while the fluctuation heat flux enhances the basic temperature gradient. The ability of the energy transfers to have either sign is a reminder that the dynamics of these perturbations depends in a crucial way on the delicate phase relations in the disturbance, that is, on the tilt of the wave phase with height and latitude, and therefore that a conception of these fluctuations as simple analogies to molecular diffusion in the large is an excessive oversimplification. Naturally, from (7.3.26) it follows that the potential-vorticity flux for an unstable disturbance must be down the gradient of the basic potential-vorticity gradient.

These considerations have long been recognized as being particularly pertinent to the problem of the atmospheric general circulation. Figure 7.15.1 shows a cross section of the zonal winds as a function of latitude and height (the winds are time and longitude averages for each $y$ and $z$). In mid-latitude the winds are everywhere westerly.* Frictional dissipation at the earth's surface continuously tends to retard the westerlies, while small-scale turbulent mixing can be expected to transfer eastward momentum from the free atmosphere to be dissipated at the surface. Yet the surface westerlies and the net westerly flow with height are permanent, observable features of the atmospheric general circulation. What is responsible for the continuous supply of eastward momentum to the atmosphere in mid-latitudes? The equation for the mean zonal velocity (7.2.13) shows that $\bar{u}_0$ can be forced by the Coriolis acceleration of the mean meridional velocity and the convergence of the Reynolds stress. However, integration over the total depth yields, with (7.2.26),

$$\frac{\partial}{\partial t} \int_0^\infty \rho_s \bar{u}_0 \, dz = -\frac{E_v^{1/2}}{2\varepsilon} \rho_s(0)\bar{u}_0(y, 0) - \int_0^\infty \rho_s \frac{\partial}{\partial y} \overline{v_0 u_0} \, dz \quad (7.15.1)$$

if it is assumed that no contribution is obtained to the frictionally driven Ekman flow at $z = \infty$. The first term on the right-hand side of (7.15.1) will degrade the mean westerlies if, as observed, $\bar{u}_0(y, 0) > 0$. This result, at least in sign, is in fact independent of the detailed formulation of the frictional interaction between the atmosphere and the earth's surface. Consequently to maintain $\bar{u}_0$ against dissipation in the region of the mid-latitude westerlies, there must be a *convergence* of the Reynolds stress, $\overline{v_0 u_0}$. Reasoning essentially along these lines, Jeffreys (1933) first argued that ". . . cyclones are an essential part of the general circulation which could not exist without them." It was shown by Starr (1953) and his coworkers, rather dramatically, that indeed the cyclone fluctuations in mid-latitudes along with the forced, standing waves were responsible for the maintenance, in the mean, of the jet stream, i.e., that fluctuation momentum flows *up* the momentum gradient to energize the zonal flow in the face of frictional dissipation. This fact, coupled with the observation that the cyclone wave disturbances in mid-latitudes release available potential energy and transport heat poleward, focuses attention on the particular cycle of energy flow in which available potential energy is released by baroclinic instability in a current with vertical shear, whose horizontal shear yields Reynolds stresses which transfer fluctuation kinetic energy into mean zonal kinetic energy by the convergence of the Reynolds stress in regions of large westerly flow. Thus, the problem of the instability of currents with horizontal and vertical shear is fundamental to the conceptual picture of the dynamics of the atmospheric general circulation and the existence of swift currents like the jet stream. The degree to which such processes are at work in the oceanic circulation is yet unclear.

---

* Recall, in meteorological terminology this means *from* the west.

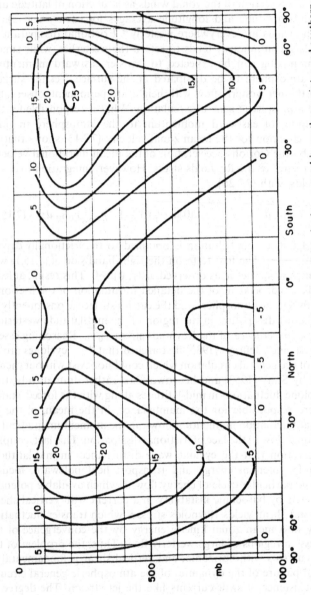

**Figure 7.15.1** A cross section of the observed zonal winds, time- and longitude-averaged in northern winter and southern summer conditions. (Reprinted from Lorenz (1967).)

Measurements (Schmitz 1977) indicate the existence of strong eddy fluxes of momentum in regions near the Gulf Stream, but the role of the fluxes in the momentum balance of the Stream remains to be fully revealed.

The joint instability problem is technically complex, for when $U_0$ is a function of $y$ and $z$, the normal-mode equation (7.4.4) is nonseparable. Progress can be more easily made in the context of the two-layer model (7.10.2a,b). Consider, for example, the basic state given by the broad zonal flow with no absolute vorticity extremum,

$$U_1 = U_s(1 - ay^2), \qquad 0 < a < 1,$$
$$U_2 = 0 \tag{7.15.2}$$

in the two-layer model of equal depths, i.e., $D_1 = D_2$. When $a$ is zero the problem reduces to the purely baroclinic problem examined in Section 7.11. The potential-vorticity gradients in each layer are (7.10.3a,b):

$$\frac{\partial \Pi_1}{\partial y} = \beta + \hat{F}U_s(1 - ay^2) + 2aU_s,$$

$$\frac{\partial \Pi_2}{\partial y} = \beta - \hat{F}U_s(1 - ay^2), \tag{7.15.3}$$

where

$$F_1 = F_2 = \hat{F}. \tag{7.15.4}$$

In the upper layer the potential-vorticity gradient is always positive, while in the lower layer $U_2 = 0$. The necessary conditions for instability, (7.10.5) and (7.10.6), imply that whatever instability occurs cannot be limited to a single layer. If there is to be an instability, it must be due to a baroclinic process which depends on the interaction between the two layers. The absence of an extremum in the absolute vorticity of the upper layer further suggests that barotropic instability is unlikely. The necessary condition for instability requires that $\partial \Pi_2 / \partial y$ be zero somewhere. This means that for instability it is necessary that

$$U_s > \frac{\beta}{\hat{F}}, \tag{7.15.5}$$

as in the case of baroclinic instability in Phillips' model.

If (7.15.5) is satisfied, $\partial \Pi_2 / \partial y$ will be negative for

$$|y| \leq y_0 = a^{-1/2} \left[ 1 - \frac{\beta}{\hat{F}U_s} \right]^{1/2}. \tag{7.15.6}$$

For $|y| > y_0$ both $\partial \Pi_1 / \partial y$ and $\partial \Pi_2 / \partial y$ are positive. From (7.9.24) it therefore follows that the vertically " integrated" zonal flow, $d_1 \bar{u}_1 + d_2 \bar{u}_2$, must be decreasing with time for $|y| > y_0$. Since

$$\frac{\partial}{\partial t} \int_{-1}^{1} (d_1 \bar{u}_1 + d_2 \bar{u}_2) \, dy = 0, \tag{7.15.7}$$

it follows that within the latitude band $-y_0 \leq y \leq y_0$, the zonal flow must be accelerating. Thus the baroclinic instability of the flow (7.15.2) must produce a convergence of zonal momentum flux near the maximum of the broad jet, tending to sharpen the jet and increase its kinetic energy.

Detailed calculations of the stability properties of (7.15.2) were carried out by the author (Pedlosky 1964). Note that the presence of the constant curvature of the basic velocity augments the $\beta$-effect in the upper layer. This tends to retard the phase speed even more than in the no-shear case, so that $c$ remains well below the range of $U_1$ and the normal-mode equations in consequence are nonsingular. Taylor-series representations of $\Phi_1$ and $\Phi_2$ may then be readily calculated and the stability characteristics determined by straightforward but tedious calculation of the condition that $\Phi_1$ and $\Phi_2$ vanish on $|y| = 1$. Figure 7.15.2 shows the curves of marginal stability for

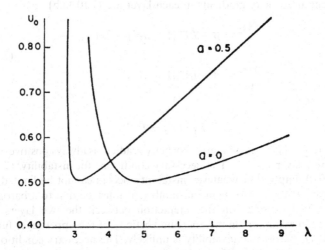

**Figure 7.15.2**  The curve of marginal instability as a function of zonal wavelength, $\lambda = 2\pi/k_0$. The profile of basic state velocity is $U_1 = U_s(1 - ay^2)$, $U_2 = 0$. The curve $a = 0$ corresponds to no horizontal shear.

the cases $a = 0$ (no horizontal shear) and $a = 0.5$ as a function of zonal wavelength. Note that the minimum critical shear required for instability is the *same* in both cases. The critical value of the vertical shear required for instability is just that value of the vertical shear which makes the basic potential-vorticity gradient in the lower layer somewhere negative so as to satisfy the necessary condition for instability. Thus direct calculation shows that with or without horizontal shear the necessary condition for instability is also sufficient. Note that the longer waves are stabilized and the shorter waves are destabilized by the effect of the horizontal shear. Figure 7.15.3(a,b) shows $c_r$ and $c_i$ in the two cases. The presence of the horizontal shear reduces $c_i$, and the curvature $2U_s a$ acts in the $\beta$ sense to reduce $c_r$ from the $a = 0$ case. Figure 7.15.4 shows the distribution of the Reynolds stress in the

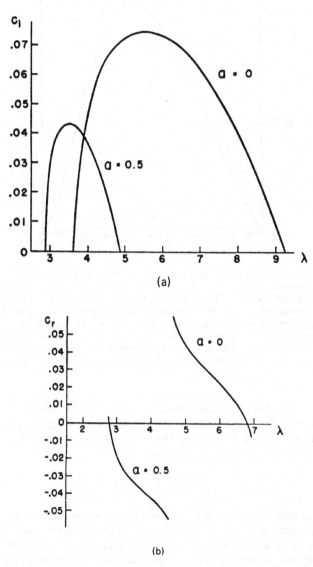

**Figure 7.15.3** (a) $c_i$ as a function of zonal wave length. $a = 0$ corresponds to the zero-horizontal-shear case of Section 7.11, while $a = 0.5$ is the case of a parabolic velocity profile in the thermal wind. $U_1 = U_s(1 - ay^2)$, $U_2 = 0$, $U_s = 0.6$, $\beta = 1.5$, $\hat{F} = 3$. (b) $c_r$ as a function of zonal wavelength as in (a).

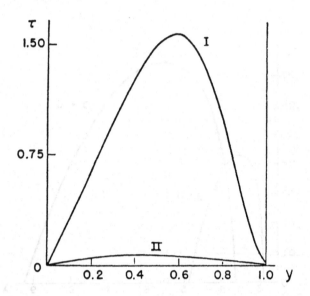

**Figure 7.15.4**   The Reynolds stress in the growing wave, $a = 0.5$, $\tau_n = -\overline{u_n^{(0)} v_n^{(0)}}$. Curve I corresponds to $n = 1$, i.e., the stress in the upper layer, while curve II corresponds to the stress in the lower layer, for $-1 \le y \le 0$ $\tau(y) = \tau(-y)$. Note that $\partial \tau_n / \partial y$ vanishes where the flow becomes locally stable at $y = y_0 \approx 0.5773$. For $|y| > y_0$ the mean zonal momentum is diminished by the transfer due to the Reynolds stresses of westerly momentum to the region $|y| < y_0$, thereby tending to sharpen the basic jet profile. Since the amplitude of the linear wave is arbitrary, the absolute value of $\tau_n$ is also arbitrary and depends on (amplitude)$^2$ of the perturbation field.

region $y > 0$. The Reynolds stress is antisymmetric about $y = 0$. For $U_s = 0.6$, $\beta = 1.5$, $\hat{F} = 3$, and $a = 0.5$ (the values for which the calculations were made), $y_0$ from (7.15.6) has the value 0.5773. Note that for $y$ greater than this value, according to the figure,

$$-\frac{\partial}{\partial y} \overline{v_n u_n} < 0, \qquad |y| > y_0,$$

so that in both layers the zonal momentum is decreasing. For $|y| < y_0$ the Reynolds-stress gradient changes sign, so that in the central portion of the jet there is an acceleration of the westerly flow due to the fluctuations. The calculations therefore verify the expectation that the kinetic energy of the zonal flow is fed by the Reynolds stresses of disturbances produced by baroclinic instability, i.e., release of basic available potential energy. Thus these growing waves provide the proper energy and momentum transformations required in a theory of the atmospheric general circulation and play the same role as the observed cyclone waves.

More complex basic states in which the basic current has an extremum of the potential vorticity *within* a single layer may possess a variety of unstable modes. Brown (1969) has examined numerically the stability of the flow

$$U_1 = U_s(1 + \cos \pi y),$$
$$U_2 = 0, \qquad |y| \le 1, \tag{7.15.8}$$

in which the velocity profile of the upper layer is identical, aside from a constant factor, to the profile examined for barotropic instability in Section 7.14. The potential vorticity gradients are

$$\frac{\partial \Pi_1}{\partial y} = \beta + U_s \pi^2 \cos \pi y + \hat{F} U_s (1 + \cos \pi y),$$
$$\frac{\partial \Pi_2}{\partial y} = \beta - \hat{F} U_s (1 + \cos \pi y). \tag{7.15.9}$$

The potential vorticity gradient of the upper layer vanishes at the point $y_c$ defined by

$$\cos \pi y_c = -\frac{\beta + \hat{F} U_s}{U_s \pi^2 + \hat{F} U_s}. \tag{7.15.10}$$

Thus for $y_c$ to be within the interval $(-1, 1)$, $\beta$ must not exceed $U_s \pi^2$. This criterion is precisely the condition required for barotropic instability. Detailed calculations were made by Brown (1969) for the parameter setting

$$L = 2,000 \text{ km},$$
$$U_{s*} = 30 \text{ m/s}.$$

Brown found that for long zonal wavelengths, greater than 7,000 km, the unstable wave released *both* kinetic and potential energy, thereby tending to smooth the horizontal gradients of both temperature and zonal velocity. The *most* unstable wave, however, occurred at shorter wavelengths (at about 4,500 kilometers) and, as in the case of the parabolic jet, transferred kinetic energy from the fluctuations to the basic zonal flow.

In currents where both barotropic and baroclinic instability are possible *a priori*, the energy-transfer characteristics of the most unstable wave can be determined only by detailed calculation, and depend on the detailed distribution of zonal velocity and potential vorticity of the basic state. Changing the parameters of the jet flow can profoundly affect whether basically baroclinic or barotropic mechanisms are favored for instability and the wavelength of maximum instability.

## 7.16 Nonlinear Theory of Baroclinic Instability

The energy equation for the perturbation fields shows that unstable waves must possess rectified fluxes of heat and/or momentum. At the same time the equations for the mean flow, particularly for the zonally averaged potential vorticity, show that the convergence of these perturbation fluxes will alter the zonal velocity, the mean meridional temperature gradient, and the meridional gradient of potential vorticity. Yet linear theory, applied to infinitesimal perturbations, neglects these changes, and so within the framework of linear theory the developing wave senses a basic state, identical to the initial state in its reservoir of energy, which remains formally unchanging with time. As a consequence of the linearization and the formal constancy of the mean state, the growth rate for the wave is also constant, leading inevitably to exponential growth for the perturbation. No matter how small the initial amplitude of the disturbance is, eventually this exponential growth will yield a perturbation amplitude so great that nonlinear effects can no longer be ignored. Linear theory, for a growing disturbance, is therefore not valid *uniformly* in time and only accurately describes the *initial* evolution and structure of the disturbance. In particular, the supercriticality of the current (i.e., its degree of instability as measured by the level of the shear, say, above its critical value) will alter with time as the basic flow is changed and consequently will affect the further development of the wave.

## 7.16a  Derivation of the Amplitude Equation

There are certain fundamental questions that only a nonlinear theory can answer. Do, for example, the effects of nonlinearity lead to an eventual halt in the growth of the wave? If so, how? If an amplitude limit exists, how is it approached? Does the wave maintain this maximum amplitude? In the presence of dissipation, we can imagine a new state in which a finite amplitude wave of steady amplitude continues to extract energy at a rate which balances the dissipation of wave energy and is equal to the external energy input to the basic flow. This would yield a picture amounting to a finite amplitude version of the energy balance which exists for the marginally stable wave described in Section 7.12. In most models of atmospheric or oceanic interest, however, dissipation is relatively weak. In the limit of very small dissipation, the steady wave state referred to earlier may itself be unstable. Moreover, in the complete *absence* of dissipation, the physical processes are essentially *reversible* with time and the evolution of the wave amplitude and phase can be expected to continue to remain unsteady. When dissipation is small but nonzero, the conjunction of near reversibility of the dynamics with a continuing dissipation of wave energy produces a wide range of possible behavior *a priori*.

We are also interested in understanding how the basic flow alters with time. The results of Section 6.14 suggest that both the growth of the wave amplitude and the dissipation of wave energy will lead to changes in the mean flow via divergences of the potential vorticity fluxes of the wave.

To calculate the effects of the wave on the mean flow, one might start by taking the growing unstable wave to calculate the eddy fluxes. However, due to the quadratic nonlinearity of the equations of motion, a wave growing like $\exp(kc_i t)$ will yield a correction to the zonal flow which grows like $\exp(2kc_i t)$, and indeed the $n$th term in any regular perturbation expansion will contain terms growing like $\exp(nkc_i t)$, so that the higher-order corrections grow very rapidly with time and quickly invalidate such representations of the effects of nonlinearity. In fact, they fail precisely when their predictions would be most interesting, i.e., when the alterations of the mean flow become large enough to alter the further evolution of the growing wave.

Progress can be made, however, when the initial supercriticality is small. For then, as the wave grows, only small changes in the basic flow are required to significantly affect the dynamics of the wave field and its ability to extract energy from the mean state. This, in turn, implies that the effects of nonlinearity will become significant for waves whose amplitudes, while finite, remain sufficiently small so that perturbation methods can be used. Furthermore, small supercriticality implies slow growth so that a separation of time scales exists between the evolution time of the wave amplitude and the advective time of its phase. A fluid particle passes through many wave crests before the amplitude of the wave has changed significantly. This separation of physical time scales allows an essential mathematical separation between the problem of determining the gross wave structure and the problem of determining the evolution of the wave amplitude.

Consider as an example the nonlinear extension of the problem discussed in Section 7.12, i.e., the instability of an initially purely baroclinic flow in the absence of $\beta$, but with the effects of dissipation considered. This will allow us to discuss the nature of the wave evolution as it depends on the degree of irreversibility in the system.

For a wave disturbance with a total wave number $K$ defined by (7.12.4), the condition for instability is given by (6.12.9) which states that for a given value of $\hat{F}$ (or $F_1$ and $F_2$), $K$ and $k$ (the zonal wave number), a critical value of the shear, $U_s$, is required for instability. In the present case, it is useful to rewrite this criterion and ask which value of $\hat{F}$ is required for instability for fixed $k$, and $U_s$. This is simply

$$\hat{F} = F_c(K, k, U_s, r) = \frac{K^2}{2} + \frac{r^2 K^2}{2k^2 U_s^2}, \tag{7.16.1}$$

where

$$K^2 = k^2 + l_j^2.$$

Figure 7.16.1 shows the critical curve as a function of $k$ for a fixed value of $l_j$ and $U_s$. The minimum occurs at $F = (1 + r/U_s l_j)^2$ (where $l_j$ is the meridional wave number of the mode) at a value of $k = (rl_j/U_s)^{1/2}$. The dotted curve is the criteria that would be obtained in the absence of friction, $F_c = K^2/2$, whose minimum occurs at $k = 0$. The short wave cutoff is given by the inviscid theory while the long waves, as in Section 7.12, are stabilized by Ekman friction.

Suppose we now consider placing a wavelike disturbance with a specified wave number $k$ on the basic state which has velocities $U_n$ such that $U_1 = -U_2$ so that, by (7.12.7) the real part of $c$ will be zero. Therefore a growing wave will be nonpropagating. Further, for a given $k$ and $U_s$ let us choose a value of $\hat{F}$ which only slightly exceeds the critical value at the wave number, i.e., we

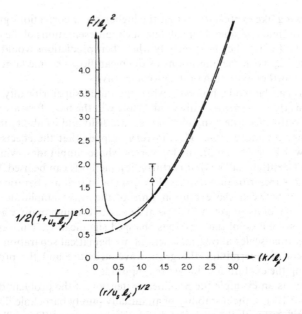

**Figure 7.16.1** The marginal stability diagram for baroclinic instability in the two-layer system with Ekman friction ($r \neq 0$) but $\beta = 0$. The critical parameter is $\hat{F} = f_0^2 L^2/(g \, \Delta\rho/\rho \, D_1)$. The figure shows the stabilization at small wave number $k$ by the effect of friction where $l_j$ is the cross-stream wave number. The supercriticality, $\Delta$, measures the distance above the critical curve, at a fixed $k$, at which the nonlinear analysis is carried out for a wave disturbance of that wave number.

suppose that

$$\hat{F} = F_c + \Delta = \frac{K^2}{2} + \frac{r^2 K^2}{2U_s^2 k^2} + \Delta, \tag{7.16.2}$$

and moreover, we require that

$$\Delta \ll F_c. \tag{7.16.3}$$

Now the nature of the dynamics of the wave depends very much on the degree to which dissipation is important. It will be considerably more interesting to examine the dynamics of a wave near the inviscid branch, as shown in Figure 7.16.1, i.e., for a wave number such that $r^2/U_s^2 k^2 \ll 1$. As we shall see, the qualitative nature of the behavior on the viscous branch can be recovered as a limit of the theory developed below.

From (7.16.2) it follows that for a given $\Delta \ll 1$, the effects of friction become important in determining the critical stability curve when $rK/kU_s \sim \Delta^{1/2}$. For $\Delta \ll 1$ and $rK/kU_s \sim \Delta^{1/2}$, (7.12.7) yields

$$kc_i = -\frac{3}{8}r \pm \frac{1}{2}\left[\frac{9}{16}r^2 + \Delta\frac{k^2 U_s^2}{K^2}\right]^{1/2}. \tag{7.16.4}$$

Thus as $\Delta$ passes through zero and becomes positive, one root for $kc_i$ becomes positive while the second root remains negative. If $\Delta < 0$, both roots are negative and the wave is damped. As long as $\Delta \ll 1$ and $rK/kU_s$ is $O(\Delta^{1/2})$, both roots are $O(\Delta^{1/2})$. Thus the appropriate time scale for the evolution of the disturbance in this parameter limit is $O(\Delta^{-1/2})$. This suggests rescaling the time variable to describe the amplitude evolution, i.e., it is natural to introduce

$$T = \sigma t \qquad (7.16.5)$$

as the new time variable, where $\sigma$ is $O(\Delta^{1/2})$ and whose precise value will be determined subsequently. Since the unstable wave is not propagating, the disturbance can be expected to be a function only of $T$ and not $t$. Had $U_1 + U_2 \neq 0$, it would have been necessary to consider the disturbance as a function of the "fast" time $t$ and the "slow" evolution time $T$, as in Section 3.26. In the present case, this is not necessary.

The disturbance stream function $\phi_n(x, y, T)$ will have a characteristic amplitude, $a_0$, with respect to the basic flow. It should depend on the supercriticality, $\Delta$, of the basic state, i.e., we expect $a_0$ to go to zero as $\Delta \to 0$. The potential vorticity flux of the growing wave is expected to be $O(a_0^2)$ and this must give rise to an alteration of the basic state of the same order as the supercriticality in order to be able to affect a stabilization of the initial basic state. This suggests that $a_0$ is $O(\Delta^{1/2})$.

Thus each of the critical parameters in the problem can be ordered with respect to $a_0$, the measure of nonlinearity, i.e.,

$$r = O(a_0),$$
$$\Delta = O(a_0^2), \qquad (7.16.6)$$
$$\sigma = O(a_0).$$

For small $\Delta$ (or small $a_0$) the nonlinear problem for the perturbation, derived from (7.12.1) is simply

$$\left(\sigma \frac{\partial}{\partial T} + \frac{U_s}{2} \frac{\partial}{\partial x}\right)\left(\nabla^2 \phi_1 - \left(\frac{K^2}{2} + \delta\right)(\phi_1 - \phi_2)\right) + \frac{\partial \phi_1}{\partial x}\left(\frac{K^2}{2} + \delta\right)U_s$$
$$= -J\left[\phi_1, \nabla^2\phi_1 - \left(\frac{K^2}{2} + \delta\right)(\phi_1 - \phi_2)\right] - \frac{r}{2}\nabla^2\phi_1, \qquad (7.16.7a)$$

$$\left(\sigma \frac{\partial}{\partial T} - \frac{U_s}{2} \frac{\partial}{\partial x}\right)\left(\nabla^2 \phi_2 - \left(\frac{K^2}{2} + \delta\right)(\phi_2 - \phi_1)\right) - \frac{\partial \phi_2}{\partial x}\left(\frac{K^2}{2} + \delta\right)U_s$$
$$= -J\left[\phi_2, \nabla^2\phi_2 - \left(\frac{K^2}{2} + \delta\right)(\phi_2 - \phi_1)\right] - \frac{r}{2}\nabla^2\phi_2. \qquad (7.16.7b)$$

In deriving (6.16.7a,b) we have used the fact that

$$\frac{\partial}{\partial t} = \sigma \frac{\partial}{\partial T},$$

$$U_1 = -U_2 \equiv \frac{U_s}{2},$$

and have defined

$$\delta = \Delta + \frac{r^2}{2} U_s^2 \frac{K^2}{k^2}, \qquad (7.16.8)$$

so that $\delta$ is the difference between $F$ and the *inviscid* instability threshold.

If the flow is bounded on $y = 0$ and $y = 1$ by rigid walls, then for the meridional velocity to vanish, there we require

$$\frac{\partial \phi_n}{\partial x} = 0, \qquad y = 0, 1. \qquad (7.16.9)$$

A portion of the evolving perturbation field must also represent the wave-induced change in the zonal flow. At $y = 0$ and 1, where $\partial \phi_n / \partial x$ vanishes, (7.2.13) implies that

$$\overline{v_n^{(1)}} = \frac{\partial}{\partial t} \overline{u_n^{(0)}} = -\frac{\partial}{\partial t} \frac{\overline{\partial \phi_n}}{\partial y}, \qquad (7.16.10)$$

where an overbar is an $x$-average and the superscripts refer to the order of the variable in the Rossby number expansion of the original equations. Since $v_n^{(1)}$ represents the *ageostrophic* meridional velocity, its $x$-average need not vanish.

Now consider the region near $y = 1$ as illustrated in Figure 7.16.2. In response to an $x$-independent flow $\bar{u}_1^{(0)}$ near $y = 1$, there will be a flux of fluid in the upper Ekman layer impinging on the wall at $y = 1$. This flux can be derived from the vertical integral of (4.5.32) and is equal to $E_V^{1/2} \bar{u}_0 / 2$ [which is the dimensionless version of (4.3.29)]. This mass flux must be returned via a

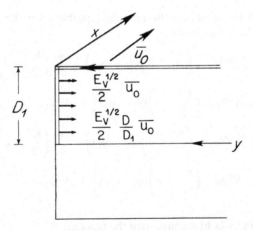

**Figure 7.16.2**   A schematic representation of the flow in the Ekman layer induced by a mean flow $\bar{u}_0$ near $y = 1$. The Ekman flux returns into the geostrophic interior and forces a nonzero ageostrophic velocity at $y = 1$.

thin frictional boundary layer on $y = 1$ whose details are unimportant. However, since the horizontal flow in the geostrophic region is independent of the meridional velocity emanating into the geostrophic region must be

$$\bar{v}_n = \overline{\varepsilon v_n^{(1)}}, \qquad \overline{v_n^{(1)}} = -\frac{E_V^{1/2}}{2\varepsilon}\frac{D}{D_1}\bar{u}_0 \quad \text{on} \quad y = 1, \qquad (7.16.11)$$

which, with (7.12.2) and (7.16.10) yields, as a boundary condition for the $x$-independent portion of the perturbation,

$$\left(\frac{\partial}{\partial t} + \frac{r}{2}\right)\frac{\partial\bar{\phi}_n}{\partial y} = 0, \qquad y = 0, 1, \qquad (7.16.12)$$

where we have noted that the same condition must hold on $y = 0$ as well. Since the correction to the zonal flow is zero at the instant the perturbation is placed on the basic state (7.16.12) implies that

$$\frac{\partial}{\partial y}\bar{\phi}_n = 0, \qquad y = 0, 1. \qquad (7.16.13)$$

Now $\phi_n$ will be a function of all the parameters of the problem each of which can be related by (7.16.6) to the amplitude parameter $a_0$. For small $a_0$, the solution for the perturbation fields can be found in the form of the asymptotic series

$$\phi_n(x, y, T, a_0) = a_0(\phi_n^{(1)} + a_0\phi_n^{(2)} + a_0^2\phi_n^{(3)} + \cdots), \qquad (7.16.14)$$

where the various $\phi_n^{(1)}$ are independent of $a_0$.

When (7.16.14) is substituted into (7.16.7a,b) and like orders of $a_0$ are equated, a sequence of linear problems emerge, the first of which is the $O(a_0)$ problem:

$$\frac{U_s}{2}\frac{\partial}{\partial x}\left[\nabla^2\phi_1^{(1)} + \frac{K^2}{2}(\phi_1^{(1)} + \phi_2^{(1)})\right] = 0, \qquad (7.16.15a)$$

$$\frac{U_s}{2}\frac{\partial}{\partial x}\left[\nabla^2\phi_2^{(1)} + \frac{K^2}{2}(\phi_2^{(1)} + \phi_1^{(1)})\right] = 0, \qquad (7.16.15b)$$

which is recognized as simply the problem for the marginal ($\Delta = 0$) inviscid ($r = 0$) wave for which $\partial\phi_n/\partial T = 0$.

A solution of the form

$$\phi_n^{(1)} = \text{Re } A_n(T)e^{ikx}\sin m\pi y = A_n(T)\frac{e^{ikx}\sin m\pi y}{2} + * \qquad (7.16.16)$$

may be sought where $m$ is any integer. The symbol Re reminds us that only the real part of the following expression is taken and the symbol $*$ refers to the complex conjugate of the preceding term. The form (7.16.16) which represents a single plane wave will be a solution of (7.16.15a,b) only if

$$K^2 = k^2 + m^2\pi^2, \qquad (7.16.17a)$$

and

$$A_1 = A_2 \equiv A. \tag{7.16.17b}$$

Note that the lowest-order horizontal and vertical structure of the wave are determined at this order without regard to the supercriticality, the dissipation, the temporal behavior, or the nonlinearity of the wave field. This results from the separation of scales due to (7.16.6) and the condition $a_0 \ll 1$. The vertical structure of wave, to this order is given by (7.16.17b), i.e., the wave to lowest order is barotropic and this follows directly from the linear theory (7.12.14) when $K^2 = 2\hat{F}$ and $c_i = 0$. Thus the wave (7.16.16) has the form of the marginal, inviscid wave of linear theory. However, in distinction to that theory $A$ is not constant but is evolving on the slower growth time.

The next order problem at $O(a_0^2)$ yields the following inhomogeneous problem for $\phi_n^{(2)}$. Using the fact that $\phi_1^{(1)} = \phi_2^{(1)}$, (7.16.7a,b) yields

$$\frac{U_s}{2} \frac{\partial}{\partial x} \left\{ \nabla^2 \phi_1^{(2)} + \frac{K^2}{2} (\phi_1^{(2)} + \phi_2^{(2)}) \right\} = -\frac{\sigma}{a_0} \left\{ \frac{\partial}{\partial T} + \frac{r}{2\sigma} \right\} \nabla^2 \phi_1^{(1)}$$
$$- J\{\phi_1^{(1)}, \nabla^2 \phi_1^{(1)}\}, \tag{7.16.18a}$$

$$\frac{U_s}{2} \frac{\partial}{\partial x} \left\{ \nabla^2 \phi_2^{(2)} + \frac{K^2}{2} (\phi_2^{(2)} + \phi_1^{(2)}) \right\} = -\frac{\sigma}{a_0} \left\{ \frac{\partial}{\partial T} + \frac{r}{2\sigma} \right\} \nabla^2 \phi_2^{(1)}$$
$$- J\{\phi_2^{(1)}, \nabla^2 \phi_2^{(1)}\}. \tag{7.16.18b}$$

Note that the ordering relations $\sigma = O(r) = O(a_0)$ have been used. A great simplification occurs when we realize that the solution (7.16.16) has the important property that the vorticity in the lowest-order solution is simply a constant multiple of the stream function, e.g., $\nabla^2 \phi_n^{(1)} = -K^2 \phi_n^{(1)}$. It follows immediately that

$$J(\phi_n^{(1)}; \nabla^2 \phi_n^{(1)}) = 0 \tag{7.16.19}$$

so that (7.16.18a,b) contains no nonlinear forcing term.

The remaining inhomogeneous terms on the right-hand side of (7.16.18a,b) each have the form $e^{ikx} \sin m\pi y$ which is the form of the solutions of (7.16.18a,b) whose homogeneous part is identical to (7.16.15a,b). This would appear to force a resonance in $\phi_n^{(2)}$, but that is illusory. This can be seen by attempting to find a particular solution of (7.16.18a,b) in the form

$$\phi_n^{(2)} = \text{Re } A_n^{(2)} e^{ikx} \sin m\pi y$$
$$= A_n^{(2)} \frac{e^{ikx}}{2} \sin m\pi y + *. \tag{7.16.20}$$

If (7.16.19) and (7.16.20) are inserted into (7.16.18a,b) we easily obtain

$$-K^2 \frac{U_s}{8} ik[A_1^{(2)} - A_2^{(2)}] = \frac{\sigma}{a_0} \frac{K^2}{2} \left[ \frac{dA}{dT} + \frac{r}{2\sigma} A \right], \tag{7.16.21a}$$

$$K^2 \frac{U_s}{8} ik[A_2^{(2)} - A_1^{(2)}] = \frac{\sigma}{a_0} \frac{K^2}{2} \left[ \frac{dA}{dT} + \frac{r}{2\sigma} A \right]. \tag{7.16.21b}$$

Hence the two equations are *redundant* and no resonance occurs. Solving for $A_2^{(2)}$ yields

$$A_2^{(2)} = A_1^{(2)} - \frac{4i}{kU_s}\frac{\sigma}{a_0}\left[\frac{dA}{dT} + \frac{r}{2\sigma}A\right].$$  (7.16.22)

If (7.16.22) is compared with (7.16.17b) it becomes clear that the first term on the right-hand side of (7.16.22) represents an $O(a_0^2)$ perturbation field with the same horizontal and *vertical* structure as the primary wave. If we take the viewpoint that the wave (7.16.16) represents completely the wave with the structure of the marginal wave then $A_1^{(2)}$ may be set equal to zero without loss of generality. Then to $O(a_0^2)$ the wave has the form

$$\phi_1 = a_0 A \frac{e^{ikx}}{2}\sin m\pi y + *,$$  (7.16.23a)

$$\phi_2 = a_0\left[A - a_0\frac{4i}{kU_s}\left(\frac{\sigma}{a_0}\right)\left[\frac{dA}{dT} + \frac{r}{2\sigma}A\right]\right]\frac{e^{ikx}}{2}\sin m\pi y + *.$$  (7.16.23b)

As long as $dA/dT + r/2\sigma A$ is nonzero, there will be a phase difference between the wave in upper and lower layers. If $A^{-1}[dA/dT + r/2\sigma A]$ is greater than zero, the wave in the upper layer will lag behind the wave in the lower layer. If $\sigma A^{-1} dA/dT = kc_i$, as in linear theory, the resulting phase shift would be precisely the phase shift predicted by linear theory (i.e., (7.12.5)). To this order, the analysis has only reproduced the relationships of linear theory. The crucial advance of the theory is that these delicate and important phase relations are given in terms of the general temporal behavior of $A$, which in finite amplitude can be expected to significantly depart from exponential growth. The phase relation is instead given in terms of the as yet unknown behavior in time of the wave amplitude, $A$.

To the particular solution (7.16.23a,b) we now add the homogeneous solution of (7.16.18a,b)

$$\phi_n^{(2)} = \Phi_n^{(2)}(y, T).$$  (7.16.24)

This solution represents an $O(a_0^2)$ correction to the zonal flow and trivially satisfies the homogeneous part of (7.16.18a,b) since it is independent of $x$. Solutions of this type could be added at each stage of the expansion but as we shall see this is the first order where such additions are required to balance the zonal-flow alterations forced by the nonlinear wave fluxes of the wave perturbation. For the sake of clarity, the total perturbation stream function up to and including $O(a_0^2)$ is rewritten below

$$\phi_1 = a_0 A \frac{e^{ikx}}{2}\sin m\pi y + * + a_0^2\Phi_1^{(2)}(y, T),$$  (7.16.25a)

$$\phi_2 = a_0 A \frac{e^{ikx}}{2}\sin m\pi y + * + a_0^2[X_2^{(2)}(x, y, T) + \Phi_2^{(2)}(y, T)],$$  (7.16.25b)

where

$$X_2^{(2)} = -\frac{4i}{kU_s}\frac{r}{a_0}\left\{\frac{dA}{dT} + \frac{r}{2\sigma}A\right\}\frac{e^{ikx}\sin m\pi y}{2} + *. \qquad (7.16.26)$$

Note that

$$\nabla^2 X_2^{(2)} = -K^2 X_2^{(2)}. \qquad (7.16.27)$$

To this order, the supercriticality $\Delta$ (or $\delta$) has not entered the problem and clearly the solutions must be considered further. This is, in addition, necessary in order to determine $A(T)$ and the alteration of the zonal flow $\Phi_n^{(1)}(y, T)$.

Collecting terms of $O(a_0^3)$ yields the final problem, i.e.,

$$\frac{U_s}{2}\frac{\partial}{\partial x}\left[\nabla^2\phi_1^{(3)} + \frac{K^2}{2}(\phi_1^{(3)} + \phi_2^{(3)})\right]$$

$$= -\frac{\sigma}{a_0}\frac{\partial}{\partial T}\left[\nabla^2\phi_1^{(2)} - \frac{K^2}{2}(\phi_1^{(2)} - \phi_2^{(2)})\right] - \frac{U_s}{2}\frac{\delta}{a_0^2}\frac{\partial}{\partial x}(\phi_1^{(1)} + \phi_2^{(1)})$$

$$- \frac{r}{a_0}\nabla^2\phi_1^{(1)} - J\left[\phi_1^{(1)}, \nabla^2\phi_1^{(2)} + \frac{K^2}{2}(\phi_2^{(2)} - \phi_1^{(2)})\right] \qquad (7.16.28a)$$

$$- J\left[\phi_1^{(2)}, \nabla^2\phi_1^{(1)} + \frac{K^2}{2}(\phi_2^{(1)} - \phi_1^{(1)})\right],$$

and

$$-\frac{U_s}{2}\frac{\partial}{\partial x}\left[\nabla^2\phi_2^{(3)} + \frac{K^2}{2}(\phi_1^{(3)} + \phi_2^{(3)})\right]$$

$$= -\frac{\sigma}{a_0}\frac{\partial}{\partial T}\left[\nabla^2\phi_2^{(2)} - \frac{K^2}{2}(\phi_2^{(2)} - \phi_1^{(2)})\right] + \frac{U_s}{2}\frac{\delta}{a_0^2}\frac{\partial}{\partial x}(\phi_1^{(1)} + \phi_2^{(1)})$$

$$- \frac{r}{a_0}\nabla^2\phi_2^{(2)} - J\left[\phi_2^{(1)}, \nabla^2\phi_2^{(2)} + \frac{K^2}{2}(\phi_1^{(2)} - \phi_2^{(2)})\right] \qquad (7.16.28b)$$

$$- J\left[\phi_2^{(2)}, \nabla^2\phi_2^{(1)} + \frac{K^2}{2}(\phi_1^{(1)} - \phi_2^{(1)})\right].$$

The Jacobian terms in each of (7.16.28a,b) may be easily evaluated especially if (7.16.27) is noted and this allows us to rewrite (7.26.28a,b) as

$$\frac{U_s}{2}\frac{\partial}{\partial x}\left[\nabla^2\phi_1^{(3)} + \frac{K^2}{2}(\phi_1^{(3)} + \phi_2^{(3)})\right]$$

$$= -\frac{\sigma}{a_0}\frac{\partial}{\partial T}\left[\frac{\partial^2\Phi_1^{(2)}}{\partial y^2} - \frac{K^2}{2}(\Phi_1^{(2)} - \Phi_2^{(2)})\right] - \frac{r}{2a_0}\frac{\partial^2\Phi_1^{(2)}}{\partial y^2}$$

$$+ \frac{\sigma}{a_0}\frac{K^2}{2}\frac{m\pi}{U_s}\left[\frac{d}{dT}|A|^2 + \frac{r}{\sigma}|A|^2\right]\sin 2m\pi y \qquad (7.16.29a)$$

$$- \frac{\partial}{\partial T}X_2^{(2)}\frac{K^2}{2}\frac{\sigma}{a_0}$$

$$- ik \frac{U_s}{2} \frac{\delta}{a_0^2} A e^{ikx} \sin m\pi y$$

$$- \left[ ik \frac{A}{2} e^{ikx} \sin m\pi y + * \right] \left[ \frac{\partial^3 \Phi_1^{(2)}}{\partial y^3} + \frac{K^2}{2} \frac{\partial}{\partial y} (\Phi_1^{(2)} + \Phi_2^{(2)}) \right]$$

$$- \frac{U_s}{2} \frac{\partial}{\partial x} \left[ \nabla^2 \phi_2^{(3)} + \frac{K^2}{2} (\phi_1^{(3)} + \phi_2^{(3)}) \right]$$

$$= - \frac{\sigma}{a_0} \frac{\partial}{\partial T} \left[ \frac{\partial^2 \Phi_2^{(2)}}{\partial y^2} - \frac{K^2}{2} (\Phi_2^{(2)} - \Phi_1^{(2)}) \right] - \frac{r}{2a_0} \frac{\partial^2 \Phi_2^{(2)}}{\partial y^2}$$

$$- \frac{\sigma}{a_0} \frac{K^2}{2} \frac{m\pi}{U_s} \left[ \frac{d}{dT} |A|^2 + \frac{r}{\sigma} |A|^2 \right] \sin 2m\pi y$$

$$\text{(7.16.29b)}$$

$$+ \frac{\partial}{\partial T} X_2^{(2)} \frac{3K^2}{2} \frac{\sigma}{a_0}$$

$$+ ik \frac{U_s}{2} \frac{\delta}{a^2} A e^{ikx} \sin m\pi y + \frac{rK^2}{a_0} X_2^{(2)}$$

$$- \left[ ik \frac{A}{2} e^{ikx} \sin m\pi y + * \right] \left[ \frac{\partial^3 \Phi_2^{(2)}}{\partial y^3} + \frac{K^2}{2} \frac{\partial}{\partial y} (\Phi_1^{(2)} + \Phi_2^{(2)}) \right] + *,$$

where $|A|^2 = AA^*$.

The first terms on the right-hand sides of (7.16.29a,b) are independent of $x$ and were they to be different from zero they would force a response in $\phi_n^{(3)}$ that would be linearly growing in $x$. Since the $x$ interval is infinite,† such solutions for $\phi_n^{(3)}$ would become large enough to invalidate the ordering implied by (7.16.14). Hence to keep our expansion valid, i.e., to obtain a legitimate asymptotic solution, we require that these terms vanish, i.e., that

$$\frac{\partial}{\partial T} \left\{ \frac{\partial^2 \Phi_n^{(2)}}{\partial y^2} + \frac{K^2}{2} (-1)^n (\Phi_1^{(2)} - \Phi_2^{(2)}) \right\} + \frac{r}{2\sigma} \frac{\partial^2 \Phi_n^{(2)}}{\partial y^2}$$

$$\text{(7.16.30)}$$

$$= (-1)^{n+1} \frac{K^2}{2} \frac{m\pi}{U_s} \left[ \frac{d}{dT} |A|^2 + \frac{r}{\sigma} |A|^2 \right] \sin 2m\pi y.$$

This is nothing more than the zonal average of the potential vorticity equation. The left-hand side represents the rate of change of the *correction* to the zonally averaged potential vorticity modified by a loss term due to Ekman friction. The terms on the right-hand side of (7.16.30) represent the eddy flux terms (7.9.18) evaluated for the case at hand. From the ideas developed in Section 6.14, we anticipate that this eddy flux of potential vorticity can be due only to either secular changes in the wave amplitude or to dissipation acting on the wave field, [see (6.14.33)]. The form of the evaluated flux term on the

† The same considerations also apply if the $x$-interval is finite, but where the solutions must be periodic and single valued in $x$.

right-hand side of (7.16.30) justifies that anticipation. Were $d|A|^2/dT$ to vanish, and if $r$ were zero, there would be no Eliassen–Palm flux divergence and no alteration of the $x$-averaged zonal flow.

In the present case, the Reynolds stress in the baroclinic wave is identically zero and the potential vorticity in each layer is changed only by the advection of the interface height (i.e., only by the heat flux). The wave can only alter the zonal flow by inducing a mean meridional circulation. Due to the symmetry of the problem in the vertical, this meridional circulation must be equal and opposite in the two layers to satisfy mass conservation. It may also be demonstrated directly from (7.16.30) and (7.16.13) that

$$\Phi_1^{(2)} = -\Phi_2^{(2)} \equiv M\Phi(y, T), \tag{7.16.31}$$

where $M$ is a constant to be determined shortly. The symmetry condition (7.16.31) thus preserves the symmetry of the shear flow of the basic state while $\Phi$ satisfies

$$\frac{\partial}{\partial T}\left[\frac{\partial^2 \Phi}{\partial y^2} - K^2\Phi\right] + \frac{r}{2\sigma}\frac{\partial^2 \Phi}{\partial y^2}$$

$$= \frac{K^2}{2M}\frac{m\pi}{U_s}\left[\frac{d}{dT}|A|^2 + \frac{r}{\sigma}|A|^2\right]\sin 2m\pi y, \tag{7.16.32a}$$

while

$$\frac{\partial \Phi}{\partial y} = 0, \qquad y = 0, 1. \tag{7.16.32b}$$

It is convenient to choose

$$M = \frac{K^2 m\pi}{2U_s} \tag{7.16.33a}$$

and to define

$$\gamma = \frac{r}{2\sigma}, \tag{7.16.33b}$$

so that (6.16.32a) takes the final form

$$\frac{\partial}{\partial T}\left[\frac{\partial^2 \Phi}{\partial y^2} - K^2\Phi\right] + \gamma\frac{\partial^2 \Phi}{\partial y^2} = \left[\frac{d}{dT}|A|^2 + 2\gamma|A|^2\right]\sin 2m\pi y, \quad (7.16.34)$$

which determines the mean flow correction entirely in terms of the evolving wave amplitude.

To determine $A(T)$ we must return to (7.16.29a,b). A portion of the right-hand side of each equation will be proportional to $e^{ikx}\sin m\pi y$ which is the form of the homogeneous solution of the equation and unless suitably restricted, this forcing will also produce a resonant behavior for $\phi_n^{(3)}$. To discover what further condition must be placed on the right-hand sides of (7.16.29a,b), which is equivalent to determining an equation for the amplitude $A$, it is only necessary to multiply each of (7.16.29a,b) by $k/2\pi\, e^{-ikx}\sin m\pi y$

and then integrate in $y$ from 0 to 1 and in $k$ over a wavelength, i.e., from $x = 0$ to $x = 2\pi/k$. The result yields

$$\frac{U_s ik}{2}\frac{k^2}{2}\left[\int_0^1 (\phi_2^{(3)} - \phi_1^{(3)})\sin m\pi y \, dy\right]$$

$$= -\frac{K^2}{8}\frac{\sigma}{a_0}\frac{\partial B}{\partial T} - ik\frac{U_s}{4}\frac{\delta}{a_0^2}A - ik\frac{A}{2}M\int_0^1 \sin^2 m\pi y\frac{\partial^3 \Phi}{\partial y^3}\,dy, \tag{7.16.35a}$$

and

$$\frac{U_s ik}{2}\frac{k^2}{2}\left[\int_0^1 (\phi_2^{(3)} - \phi_1^{(3)})\sin m\pi y \, dy\right]$$

$$= \frac{3}{8}K^2\frac{\sigma}{a}\frac{\partial B}{\partial T} + \frac{K^2 r}{8a_0}B + \frac{ikU_s\delta}{4a_0^2}A + ik\frac{A}{2}M\int_0^1 \sin^2 m\pi y\frac{\partial^3 \Phi}{\partial y^3}\,dy, \tag{7.16.35b}$$

where

$$B = -\frac{4i}{kU_s}\frac{\sigma}{a_0}\left[\frac{dA}{dT} + \frac{r}{2\sigma}A\right], \tag{7.16.36}$$

and where (7.16.31) has been used. Inspection of (7.16.35a,b) shows that the left-hand side of each equation is identical. This implies that the right-hand sides must also be equal. This condition results in the differential equation for $A$, i.e.,

$$\frac{d^2 A}{dT^2} + \frac{3}{4}\frac{r}{\sigma}\frac{dA}{dT} - \frac{k^2 U_s^2}{4k^2\sigma^2}\left[\delta - \frac{r^2 k^2}{2k^2 U_s^2}\right]A$$

$$- \frac{a_0^2}{\sigma^2}\frac{k^2 U_s M}{2K^2}A\int_0^1 \sin^2 m\pi y\frac{\partial^3 \Phi}{\partial y^3}\,dy = 0, \tag{7.16.37}$$

or, using (7.16.8) and (7.16.33b), the amplitude equation may be rewritten

$$\frac{d^2 A}{dT^2} + \frac{3}{2}\gamma\frac{dA}{dT} - \frac{k^2 U_s^2}{4K^2\sigma^2}\Delta A - \frac{a_0^2 k^2 U_s^2 M}{\sigma^2 2K^2}A\int_0^1 \sin^2 m\pi y\frac{\partial^3 \Phi}{\partial y^3}\,dy = 0. \tag{7.16.38}$$

The first three terms in (7.16.38) are linear and reflect the structure of the linear instability problem for $A$ near the marginal curve. If the final term in (7.16.38) is ignored (i.e., if $a_0^2 \to 0$) solutions for $A(T)$ are of the form $A = A_0 \exp \omega T$ where

$$\omega = -\frac{3}{4}\gamma \pm \frac{1}{2}\left[\frac{9}{4}\gamma^2 + \frac{\Delta k^2 U_s^2}{K^2\sigma^2}\right]^{1/2},$$

in complete agreement with (7.16.4), the linear result, if we note that $kc_i = \omega\sigma$ and use is made of (7.16.33b).

The final term in (7.16.38) represents the interaction of the growing wave with the alteration of the basic zonal flow. According to (7.16.38) this interaction depends only on the horizontal curvature of the zonal velocity profile induced by the potential vorticity flux of the growing wave. Since only the interaction of the wave with the evolving zonal flow enters the nonlinear stability problem, the effect of the nonlinearity can be thought of as a process by which the potential vorticity fluxes continuously produce a new velocity profile for the stability analysis. Thus, were $\Phi(y, T)$ known and independent of $T$ a new *linear* instability analysis would give rise to a new effective growth rate. Since $K^2 \simeq 2F$, one might plausibly argue that small changes in the shear alone would be ineffective in altering $kc_i$ [see (7.12.7)]. However, the addition of a curvature to the zonal velocity field adds an $O(a_0^2)$ correction to the potential vorticity gradient of the zonal flow and, like $\beta$, will alter the growth rate. In fact, were $\Phi$ constant in time, solutions of the form $A = A_0 \exp \omega T$ to (7.16.38) would be possible in which the new growth rate would depend on the slight curvature in the zonal velocity field. Thus (7.16.38) represents a sequence of linear "snapshot" stability analyses of a slowly evolving zonal flow. Of course, the problem is fundamentally nonlinear since $\Phi$ is not set externally but responds, *via* (7.16.34) to the evolving wave field.

To put the problem in final form we first choose

$$\sigma = \frac{kU_s}{2K} \Delta^{1/2} \tag{7.16.39}$$

to render the coefficient of the term linear in $A$ equal to unity. This is equivalent to choosing the time scale to be exactly equal to the inviscid $e$-folding time of linear theory.

Second, we choose

$$a_0 = \frac{U_s}{Km\pi} \Delta^{1/2} \tag{7.16.40}$$

to make the nonlinear term in the amplitude equation the same order as the linear term. This yields an estimate for the nondimensional amplitude of the problem in terms of $\Delta$ and the remaining parameters and allows the amplitude equation to be written in the relatively parameter-free form,

$$\frac{d^2 A}{dT^2} + \frac{3}{2}\gamma \frac{dA}{dT} - A + A \int_0^1 \sin 2\, m\pi y \frac{\partial^2 \Phi}{\partial y^2} \, dy = 0 \tag{7.16.41}$$

after an integration by parts of the final term. This equation coupled with (7.16.34) and (7.16.32b) determines both the wave amplitude and the correction to the mean flow. The latter, in terms of original parameters, is given in terms of the independent $O(a_0^2)$ correction to the stream function

$$a_0^2 \Phi_1^{(2)} = -a_0^2 \Phi_2^{(2)} = \frac{\Delta U_s}{2m\pi} \Phi(y, T). \tag{7.16.42}$$

As written, the behavior of the system depends on only two parameters, $\gamma$ and $K^2 = 2F_c$ and it is of particular importance to investigate the dependence of $A$ and $\Phi$ on $\gamma$, which is a measure of the dissipation and hence irreversibility in the system.

## 7.16b  The Inviscid System, $\gamma = 0$

When $\gamma = 0$, there will be a potential vorticity flux only while the wave is growing and this is reflected by the fact that (7.16.34) reduces to

$$\frac{\partial}{\partial T}\left[\frac{\partial^2 \Phi}{\partial y^2} - K^2 \Phi\right] = \frac{d|A|^2}{dT} \sin 2m\pi y, \tag{7.16.43}$$

which can be integrated in time immediately to yield

$$\frac{\partial^2 \Phi}{\partial y^2} - K^2 \Phi = (|A|^2 - |A_0|^2)\sin 2m\pi y, \tag{7.16.44}$$

where $A_0$ is the value of $A$ at $T = 0$ at which time the correction to the zonal flow is taken, by definition, to be zero. This equation may be solved for $\Phi$ with the use of (7.16.32b) to yield

$$\phi = -\frac{(|A|^2 - |A_0|^2)}{4m^2\pi^2 + K^2}\left[\sin 2m\pi y - \frac{2m\pi}{K}\frac{\sinh K(y - \frac{1}{2})}{\cosh K/2}\right]. \tag{7.16.45}$$

The hyperbolic functions enters as homogeneous solutions of (7.16.44) required to satisfy $\partial\Phi/\partial y = 0$, on $y = 0, 1$. The correction to the zonal vertical shear is

$$a_0^2(\bar{u}_1^{(2)} - \bar{u}_2^{(2)}) = -\frac{\Delta U_s}{2m\pi}\frac{\partial \Phi}{\partial y} \tag{7.16.46}$$

$$= \frac{\Delta U_s}{4m^2\pi^2 + K^2}[|A|^2 - |A_0|^2]\left\{\cos 2m\pi y - \frac{\cosh k(y - \frac{1}{2})}{\cosh k/2}\right\}.$$

The spatial form of the correction to the vertical shear when $|A|^2 > |A_0|^2$ was first calculated by Phillips (1954) and is shown in Figure 7.16.3 for the case $m = 1$, $k = \pi$, i.e., for a "square" wave whose $x$ and $y$ wave numbers are equal. Over most of the $y$-interval, and especially where the wave field has its maximum amplitude, the vertical shear is reduced by the Coriolis force acting on the meridional circulation as the wave amplitude increases from its initial value. Since it is the *rate of change* of the zonal flow which is equal to the Coriolis force acting on the meridional acceleration, it follows that the meridional acceleration itself is proportional to $d|A|^2/dT$ and this is consistent with (6.14.25) which relates the meridional circulation to the potential vorticity flux. By the thermal wind relation a reduction of the vertical shear implies a reduction of the slope of the $x$-average interface slope which occurs as the available potential energy is sapped by the growing wave. Note that as the wave grows, i.e., as long as $|A|^2 > |A_0|^2$, the shear will be reduced, however, if $|A|^2$ ever falls below its initial value, the shear of the basic current is actually increased.

However, as we have already noted, it is not simply the shear which

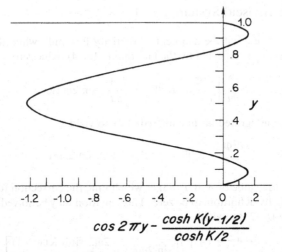

$$\cos 2\pi y - \frac{\cosh K(y-1/2)}{\cosh K/2}$$

**Figure 7.16.3**   The form (for $|A|^2 > |A_0|^2$) of the correction to the shear of the mean zonal flow. Note that it is negative over most of the channel. This is drawn for the case $m = 1, k = 2^{1/2}\pi$.

determines the nonlinear behavior of the equilibration process, but the curvature of the zonal velocity profile. Using (7.16.45) the required integral in (7.16.41) can be evaluated directly so that for $\gamma = 0$, (7.16.41) becomes

$$\frac{d^2A}{dT^2} - A + NA[|A|^2 - |A_0|^2] = 0, \tag{7.16.47}$$

where

$$N = \frac{4m^2\pi^2}{4m^2\pi^2 + K^2}\left[\frac{1}{2} - \frac{2K \tanh K/2}{K^2 + 4m^2\pi^2}\right].$$

Figure 7.16.4 shows $N$ as a function of $K$ for the $m = 1$ mode and it is seen that $N$ is *positive*. Now at the start of the growth of the unstable wave $A \sim A_0$ and the nonlinear term in (7.16.47) will be small. Hence, initially the unstable mode will grow as in linear theory, i.e., like $\exp(T)$. However when $e^T - 1$ is O(1), the nonlinear terms become important. Then as $A$ continues to grow, the *effective* growth rate

$$\left(\frac{1}{A}\frac{d^2A}{dT^2}\right)^{1/2} = [1 - N(|A|^2 - |A_0|^2)]^{1/2}$$

is reduced and eventually the growth will be halted. These indications of the nonlinear stabilization and ultimate equilibration of the wave are only part of the story of the amplitude dynamics. The amplitude equation (7.16.47) has the form of a mass-spring oscillator whose spring force is *repulsive* for $A \sim A_0$ but *restoring* for $|A|^2 > |A_0|^2 + N^{-1}$. Exploiting this analogy with the

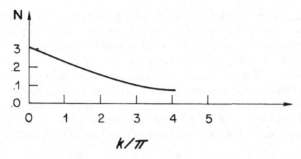

**Figure 7.16.4**   The nonlinear equilibration constant $N$ as a function of $k$ for $m = 1$.

oscillator, it is evident that when $A$ reaches its maximum value, it will not remain fixed at that point but will instead start to decrease until a minimum value, consistent with the repulsion at small amplitudes is achieved before starting the cycle anew.

If we let

$$A = R(T)e^{i\theta(T)},\qquad(7.16.48)$$

where $R$ is the (real) modulus of $A$ and (7.16.48) is substituted into (7.16.47), we obtain from the real and imaginary parts of the resulting equation two equations, i.e., from the imaginary part

$$\frac{d}{dT}\left(R^2\frac{d\theta}{dT}\right) = 0,\qquad(7.16.49)$$

while from the real part

$$\frac{d^2R}{dT^2} - R\left(\frac{d\theta}{dT}\right)^2 - R + RN[R^2 - R_0^2] = 0.\qquad(7.16.50)$$

The first of these implies that

$$\mathcal{L} = R^2\frac{d\theta}{dT}\qquad(7.16.51)$$

is a constant; hence (7.16.50) becomes

$$\frac{d^2R}{dT^2} - \frac{\mathcal{L}^2}{R^3} - R + NR(R^2 - R_0^2) = 0.\qquad(7.16.52)$$

Still guided by the analogy with the oscillator problem, an "energy" equation, or first integral of (7.16.52) may be obtained by multiplication of (7.16.52) by $dR/dT$. That step immediately demonstrates that

$$E = \frac{1}{2}\left(\frac{dR}{dT}\right)^2 + V(R)\qquad(7.16.53)$$

is constant, where

$$V(R) = -\frac{R^2}{2}(1 + NR_0^2) + \frac{NR^4}{4} + \frac{\mathcal{L}^2}{2R^2}.\qquad(7.16.54)$$

The constraint (7.16.53) is precisely analogous to the energy equation of a particle whose position coordinate is $R(T)$, subject to the force potential $V(R)$ and endowed with a total energy $E$ which is fixed by the values of $R$ and $dR/dT$ at $T = 0$. A case of particular interest occurs when at the initial instant $A^{-1} dA/dT = 1$. This is equivalent to making the *initial* vertical structure of the wave identical to the structure of the unstable wave of linear theory and then following its evolution. Since $d\theta/dT$ is equal to the imaginary part of $A^{-1} dA/dT$, it follows that at $T = 0$, $d\theta/dT$, and hence $\mathscr{L}$ vanish. Thus $d\theta/dT$ remains zero and $A$ may be considered real and equal to $R$. In this case

$$V(R) = -\frac{R^2}{2}(1 + NR_0^2) + N\frac{R^4}{4}, \qquad (7.16.55)$$

which is shown in Figure 7.16.5. Much of the qualitative value of the amplitude

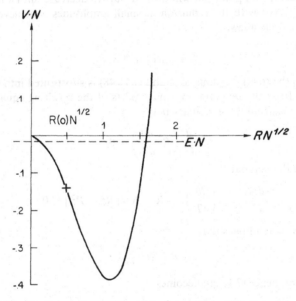

**Figure 7.16.5**   The "potential" $V(R)$ which governs the oscillation of the amplitude $R$ of the baroclinic wave. In the example shown $R_0 N^{1/2} = 0.5$. The "energy" level $E$ corresponding to an initially linear, growing baroclinic wave is also shown. The intersections of $E$ and $V(R)$ determine the maximum and minimum values of $R$ achieved in the amplitude oscillation.

evolution can be deduced directly from this figure. Observe that the initial amplitude $R_0$ always lies "uphill" from the "equilibrium" point, $(R_0^2 + N^{-1})^{1/2}$ where $dV/dR$ vanishes. At this point the effective growth rate, (7.16.47), vanishes. As in the case of a frictionless partial sliding down the potential "hill," inertia will carry the amplitude past the equilibrium point. $R$ will continue to grow until the point where $V(R) = E$, i.e., where $dR/dT = 0$. In the present case where $dR^{(0)}/dT = R_0$

$$E = \frac{1}{2}\left[\frac{dR}{dT}(0)\right]^2 + V(R_0) = -\frac{N}{4}R_0^4, \qquad (7.16.56)$$

and is negative as shown in the figure. The point $R_{max}$, where $V(R)$ equals $E$ is given by

$$R_{max}^2 = R_0^2 + N^{-1} + [1 + 2NR_0^2]^{1/2}N^{-1}. \qquad (7.16.57)$$

[Remember that the actual amplitude is scaled by $a_0$, (7.16.40), so that the maximum depends on the supercriticality, $\Delta$, of the basic state.]

The minimum value $R$ attains during the cycle is given by the smaller root of $V - E = 0$. Figure 7.16.5 shows that this second extremum is less than $R_0$ and is given by

$$R_{min}^2 = R_0^2 + N^{-1} - [1 + 2NR_0^2]^{1/2}N^{-1}. \qquad (7.16.58)$$

Initially the wave in the upper layer lags the lower layer in $x$ and available potential energy is released as in linear theory. The wave begins to grow and as its amplitude departs from its initial value, the zonal flow and the zonally averaged potential vorticity gradient alter due to the potential energy decreases in the mean state and the mean potential vorticity gradients alter until a point is reached where the instantaneous flow profile is stable according to linear theory. This occurs where $R = (R_0^2 + N^{-1})^{1/2}$. That is, at that moment the mean zonal flow (as opposed to the basic flow) has a structure which if considered in isolation from the wave field would be just marginally stable to infinitesimal disturbances. Of course, this "stabilization" has occurred because of the presence of the growing finite amplitude wave and illustrates the inappropriateness of using an instantaneous "snapshot" of an observed zonal flow to decide whether it will support a wave field by instability processes. In fact, at this "equilibrium" point $A^{-1}\,dA/dT$ is still positive and so there still is a phase shift between the waves in the two layers. In the presence of the available potential energy of the mean state, energy extraction continues, although more slowly, in a mean flow which is stable according to linear theory. The wave grows, slowly reducing the phase shift between the two layers until $R_{max}$ is reached. At this point the phase shift vanishes, the growth halts and further energy extraction is impossible. The wave then *reverses* its phase shift $(dR/dT < 0)$ and reduces its amplitude returning available potential energy *to* the mean flow. It again overshoots both the equilibrium point and its initial value until at the minimum point it ceases to return energy to the mean field. The mean flow at this point is even more unstable than the initial basic flow and the wave begins to grow again on this unstable current and the oscillation is repeated. It is important to recognize the reversible nature of this oscillation, which because $\gamma = 0$, keeps a perfect recollection of $R_0$ through the cycle. i.e., the oscillation depends explicitly on the initial value of the amplitude.

The first integral (7.16.53) may be rewritten

$$\left(\frac{dR}{dT}\right)^2 = \frac{N}{2}\left[(R_{max}^2 - R^2)(R^2 - R_{min}^2)\right], \tag{7.16.59}$$

which suggests the introduction of the variables

$$\xi = \frac{R}{R_{max}}, \tag{7.16.60a}$$

$$\theta = \left(\frac{N}{2}\right)^{1/2} R_{max} T, \tag{7.16.60b}$$

in terms of which (7.16.59) becomes

$$\left(\frac{d\xi}{d\theta}\right)^2 = (1 - \xi^2)(\xi^2 - k'^2), \tag{7.16.61}$$

where

$$\kappa'^2 = \frac{R_{min}^2}{R_{max}^2}. \tag{7.16.62}$$

Equivalently (7.16.61) may be rewritten in turn as

$$d\theta = \frac{d\xi}{\sqrt{(1 - \xi^2)(\xi^2 - \kappa'^2)}}. \tag{7.16.63}$$

During the oscillation $\xi$ increases from $\kappa'$ and grows to unity. The period of the oscillation is therefore twice the time interval required for $\xi$ to go from $\kappa'$ to 1. Hence the period of the oscillation is given by

$$T_p = \frac{2^{3/2} K(\kappa)}{N^{1/2} R_{max}}, \tag{7.16.64}$$

where $K(\kappa)$ is the complete elliptic integral of the first kind, i.e.,

$$K(\kappa) = \int_0^1 \frac{d\mu}{\sqrt{(1 - \mu^2)(1 - k^2\mu^2)}}, \tag{7.16.65}$$

where $\kappa^2 = 1 - \kappa'^2$. As $R_{max}$ increases $T_p$ decreases (although $K(\kappa)$ increases logarithmically as $\kappa \to 1$).

The amplitude itself follows from the solution of (7.16.63) and is obtained in terms of the dnoidal elliptic function, i.e.,

$$R = R_{max} \, \text{dn}[\theta - \theta_0, \kappa], \tag{7.16.66}$$

where $\theta_0$ is chosen such that at $T = 0$, $R = R_0$. The structure of $R(T)$ depends on the size of $\kappa$, i.e., on the ratio $R_{min}/R_{max}$. The period in $\theta$ of the oscillation is $2K(\kappa)$ and the behavior of $R$ is shown as a function of $T$ for several values of $\kappa$ in Figure 7.16.6. For $\kappa$ approaching unity, i.e., for $R_0^2 \ll N^{-1}$ [or in terms of the original variables, when the initial wave amplitude is small with respect to $U_s \Delta^{1/2}/(Km\pi N)$] the wave amplitude is small over most of its period, rising

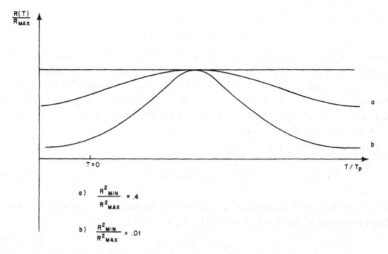

**Figure 7.16.6** The form of the amplitude oscillation over a single period: (a) $R_{min}^2/R_{max}^2 = 0.4$, (b) $R_{min}^2/R_{max}^2 = 0.01$. Note that the form is a function of amplitude.

rapidly to its maximum value and diminishing swiftly thereafter. The oscillation, although periodic, consists in the main of long intervals of undisturbed flow punctuated by intensified wave activity.

If initial data is chosen so that $E$ is positive reference to Figure 7.16.5 shows that the oscillation will pass through zero and oscillate symmetrically about zero. As we shall see shortly this second class of oscillation becomes particularly important when $\gamma \neq 0$.

## 7.16c The Effects of Friction, $\gamma \neq 0$

The inviscid system is reversible in time and consequently the amplitude oscillates periodically in time in a form and at a rate which depends on the initial value of the amplitude. On the other hand, when $\gamma \neq 0$, steady solutions for $A$ are possible. To examine this case we again write

$$A = Re^{i\theta}, \tag{7.16.67}$$

which when inserted into (7.16.41), and real and imaginary parts of the resulting equation are set equal to zero, yields

$$\frac{d}{dT}(R^2\dot{\theta}) = -\frac{3\gamma}{2}(R^2\dot{\theta}), \tag{7.16.68}$$

where $\dot{\theta} \equiv d\theta/dT$, and

$$\frac{d^2R}{dT^2} + \frac{3}{2}\gamma\frac{dR}{dT} - R + R\int_0^1 \sin m\pi y\frac{\partial\Phi}{\partial y}\,dy = 0. \tag{7.16.69}$$

Thus

$$R^2\dot{\theta} = \mathscr{L} \exp(-\tfrac{3}{2}\gamma T),$$

where $\mathscr{L}$ is the constant introduced previously when $\gamma = 0$. We see that in the presence of friction $d\theta/dT$ must always eventually vanish. This implies that we might as well consider $A$ to be a real variable. Note that if $A$ (or $R$) goes through zero the phase $\theta$ may change by $\pi$ without violating the condition $R^2\dot{\theta} = 0$. Thus in the present case it is simplest to return to (7.16.34) and (7.16.41) and consider $A$ real, e.g., $|A|^2 = A^2$.

For steady solutions (7.16.34) yields

$$\frac{\partial^2 \Phi}{\partial y^2} = 2A^2 \sin 2m\pi y, \qquad (7.16.70)$$

which, we note, is independent of $\gamma$ (however, its validity depends on $\gamma \neq 0$). Then the steady state form of (7.16.41) implies that

$$A(1 - A^2) = 0. \qquad (7.16.71)$$

Three solutions are possible. The first, obviously is $A = 0$. However, the results of linear theory tell us that if $A$ is *slightly* different from zero the amplitude will grow. That is, the solution $A = 0$ is linearly unstable and we do not anticipate solutions of (7.16.41) starting at some neighboring point to be able to reach and settle down at the unstable equilibrium point.

The other two solutions correspond to $A = \pm 1$, i.e., solutions whose amplitudes [scaled by (7.16.40)] are unity and differ only by a phase of $180°$. For some $\gamma$ and $m$ these solutions can be shown to be stable to *small* disturbances of $A$ from $A = \pm 1$. However, the relevance of the *linear* stability of these steady solutions is doubtful since *initially* the wave amplitude differs by an $O(1)$ amount from these possible, steady finite amplitude waves.

However, for large enough $\gamma$, (7.16.41) suggests that in order to achieve a balance in the equation, that $R$ should, in fact, be a function of

$$\tau = \frac{T}{\gamma} = \frac{2\sigma^2 t}{r} = \frac{tk^2 U_s^2 \Delta}{(2r)}, \qquad (7.16.72)$$

i.e., that now that natural time scale becomes $O(r/\Delta)$ rather than $O(\Delta^{-1/2})$. This is a longer time than $\Delta^{-1/2}$ as long as $r/\Delta^{1/2} > 1$. Thus this limit corresponds to the case where the Ekman spin downtime ($r^{-1}$ in dimensional units) is *short* compared with the inviscid $e$-folding time of an unstable wave. In terms of $\tau$, (7.16.34) and (7.16.41) then become

$$\frac{1}{\gamma^2}\frac{d^2A}{d\tau^2} + \frac{3}{2}\frac{dA}{d\tau} - A + A\int_0^1 \sin 2m\pi y \frac{\partial^2 \Phi}{\partial y^2}\, dy = 0, \qquad (7.16.73)$$

$$\frac{1}{\gamma^2}\frac{\partial}{\partial \tau}\left[\frac{\partial^2 \Phi}{\partial y^2} - k^2 \Phi\right] + \frac{\partial^2 \Phi}{\partial y^2} = \left[2A^2 + \frac{1}{\gamma^2}\frac{dA^2}{d\tau}\right]\sin 2m\pi y. \qquad (7.16.74)$$

For large $\gamma^2$, (7.16.74) yields the steady state relation (7.16.70) as an approximation to the relation between the wave amplitude and the zonal flow correction while (7.16.73) becomes the *first-order* equation

$$\frac{3}{2}\frac{dA}{d\tau} - A(1 - A^2) = 0 \tag{7.16.75}$$

after (7.16.70) is used. The reduction of order in time is a singular perturbation of the original system and it is left for the reader to show that the consequent loss of order is repaired by considering boundary layer in *time*, near $\tau = 0$ which serves to adjust a mismatch between $dA/d\tau$ as determined by (7.16.75) and arbitrary initial values of $dA/d\tau$ or $dA/dT$.

The reduction in order can be understood from the nature of $kc_i$ near the neutral curve. According to (7.16.4), when $r \gg \Delta^{1/2}$ (i.e., $\gamma \gg 1$), one root of (7.16.4) becomes

$$\frac{kc_i}{\sigma} \sim \frac{4}{3}\frac{\Delta}{r}\frac{k^2 U_s^2}{k^2 \sigma} = \frac{2}{3}\gamma^{-1}, \tag{7.16.76a}$$

in agreement with the linear form of (7.16.75), while the second root is the root

$$\frac{kc_i}{\sigma} \sim -\frac{3}{4}\frac{r}{\sigma} \tag{7.16.76b}$$

representing the rapidly decaying mode which carries the information concerning the initial value of $dA/dT$. This reduction of order for high $\gamma$ renders this problem qualitatively similar to the one which would result if the initial analysis were done near the viscous, long-wave branch where $r \sim kU_s$.

The solution to (7.16.75) is given by

$$A^2 = \frac{A_0^2 \exp(4\tau/3)}{1 + A_0^2(\exp(4\tau/3) - 1)}, \tag{7.16.77}$$

where $A_0$ is the initial value of $A$. Figure 7.16.7 shows the evolution of the wave amplitude. For large $\gamma$ the solution grows monotonically and then asymptotically approaches the equilibrium steady solution at $A^2 = 1$. Thus for sufficiently large $\gamma$ the steady solutions are stable and represent the end state for any initial condition on $A$. The final state is one in which a finite amplitude wave is continuously extracting energy from the mean flow and dissipating that energy via friction in the Ekman layers. Note that there is an implicit energy source continuously adding the required energy to the mean flow but as explained in Section 7.13 [see also the discussion in Section 7.17] also to balance the drain of energy by the wave. In this final state the wave amplitude is independent of initial conditions and completely steady.

The behavior of the wave amplitude in the inviscid limit and the strongly viscous limit are so different that it is natural to wonder about the amplitude behavior in the intermediate case, especially for moderately small $\gamma$. We can anticipate that for some sufficiently large $\gamma$, the steady solutions would occur. For small enough $\gamma$, one would expect oscillatory behavior, at least initially. In analogy with a damped oscillator, one might expect that oscillatory behavior would gradually give way to a final steady state. In fact, this will occur

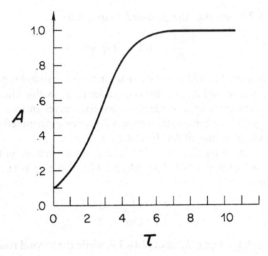

**Figure 7.16.7**   The amplitude evolution for large $\gamma$ as a function of $\tau = \gamma^{-1}T$. The evolution is nonoscillatory and tends to the steady state amplitude, $A^2 = 1$, independent of the initial amplitude.

for moderately small $\gamma$. However, the analogy with a damped oscillator with a steady forcing term is faulty in an important way, as the following argument will show. For, in fact, periodic solutions for $\gamma \neq 0$ are possible.

If $A$ is real and (7.16.41) is multiplied by $dA/dT$, we obtain

$$\frac{d}{dT}\left[\frac{1}{2}\dot{A}^2 - \frac{A^2}{2} + \frac{A^2}{2}\int_0^1 \sin 2m\pi y \frac{\partial^2 \Phi}{\partial y^2}\, dy\right]$$

$$= -\frac{3}{2}\gamma \dot{A}^2 + \frac{A^2}{2}\int_0^1 \frac{\partial}{\partial T}\frac{\partial^2 \Phi}{\partial y^2}\sin 2m\pi y\, dy, \qquad (7.16.78)$$

where $\dot{A} \equiv dA/dT$.

One condition for the existence of a periodic solution is that the integral of the right-hand side of (7.16.78) vanish when integrated over a period of the oscillation since the left-hand side clearly does. That is, for a periodic solution to exist, it is necessary that

$$\frac{3}{2}\gamma \overline{\dot{A}^2} = \overline{\frac{A^2}{2}\int_0^1 \frac{\partial}{\partial T}\frac{\partial^2 \Phi}{\partial y^2}\sin 2m\pi y\, dy}. \qquad (7.16.79)$$

This simply states that over a cycle the net dissipation must be balanced by a release of energy associated with the evolving mean field. If $\Phi(y, T)$ were in phase with $A$, i.e., if we could write $\Phi = Q(A)P(y)$ where $Q$ and $P$ are arbitrary functions of their arguments, it follows immediately that the right-hand side of (7.16.79) vanishes when integrated over a cycle. Hence the existence of a periodic solution depends on the lag between $\Phi$ and $A$.

To determine whether (7.16.79) can be nonzero write

$$\Phi = A^2 h(y) + V(y, T), \tag{7.16.80}$$

where $h(y)$ is chosen to satisfy

$$\frac{d^2 h}{dy^2} - k^2 h = \sin 2m\pi y, \tag{7.16.81}$$

then $V$ satisfies

$$\frac{\partial}{\partial T}\left(\frac{\partial^2 V}{\partial y^2} - k^2 V\right) + \gamma \frac{\partial^2 V}{\partial y^2} = \gamma A^2\left[2 \sin 2m\pi y - \frac{d^2 h}{dy^2}\right] \tag{7.16.82}$$

$$\equiv \gamma A^2 j(y),$$

while (7.16.79) becomes

$$\frac{3}{2}\gamma \overline{A^2} = \int_0^1 A^2 \frac{\partial}{\partial T}\frac{\partial^2 V}{\partial y^2} \sin 2m\pi y \, dy. \tag{7.16.83}$$

On the other hand (7.16.82) may be solved for $\partial V/\partial T$ in terms of $V$ and $A^2$ to yield

$$\frac{\partial V}{\partial T} = \int_0^1 G(y, y')\left[\gamma A^2 j(y') - \gamma \frac{\partial^2 V}{\partial y'^2}\right] dy', \tag{7.16.84}$$

where $G(y, y')$ is given by

$$G(y, y') = \frac{\sinh K(y - 1)\sinh Ky'}{K \sinh K}, \qquad y > y',$$

$$= \frac{\sinh K(y' - 1)\sinh Ky}{K \sinh K}, \qquad y > y'.$$

Thus

$$A^2 \frac{\partial}{\partial T}\frac{\partial^2 V}{\partial y^2} = \int_0^1 \frac{\partial^2 G}{\partial y^2}\left[\gamma \overline{A^4} j(y') - \gamma A^2 \frac{\partial^2 V}{\partial y'^2}\right] dy^1. \tag{7.16.85}$$

For all $\gamma$ an average of (7.16.82) over a period of the hypothetical cycle yields

$$\frac{\overline{\partial^2 V}}{\partial y^2} = \gamma \overline{A^2} j(y). \tag{7.16.86}$$

Now for *small* $\gamma$, (7.16.82) implies that to $O(\gamma)$ $\partial V/\partial T$ will be zero. Hence $V$ will be equal to its average value and thus

$$\int_0^1 A^2 \frac{\partial}{\partial T}\frac{\partial^2 V}{\partial y^2} \sin 2m\pi y = \gamma[\overline{A^4} - (\overline{A^2})^2]D, \tag{7.16.87}$$

where $D$ is

$$D = \int_0^1 dy \sin m\pi y \int_0^1 dy' \frac{\partial^2 G}{\partial y^2}(y, y')j(y'), \qquad (7.16.88)$$

so that (7.16.79) becomes

$$\overline{A^2} = \frac{D}{3}[\overline{A^4} - (\overline{A^2})^2]. \qquad (7.16.89)$$

$\overline{A^4}$ is always greater than $(\overline{A^2})^2$ for any oscillation since $\overline{A^4} - (\overline{A^2})^2 = \overline{(A^2 - \overline{A^2})^2} > 0$, thus a periodic solution is possible if $D$ is greater than zero. Furthermore, for small enough $\gamma$, we expect that the form of the oscillation must correspond to one of the class of inviscid oscillations described earlier.

Numerical calculations (Pedlosky and Frenzen, 1980) bear out these expectations but also provide some additional unexpected results. Figure 7.16.8(a) shows the result of a calculation for $K = 2^{1/2}\pi$ (i.e., $k = \pi$) and $\gamma = 0.12$. The solution for $R$ (or $A$) becomes periodic after a short adjustment period. The period of the oscillation is about 24 $e$-folding times and has, approximately, the form of an $E > 0$ inviscid oscillation. It is interesting to note that for these parameter settings the steady wave solution is *stable* to infinitesimal disturbances. However, since the initial data lies far from the steady solution

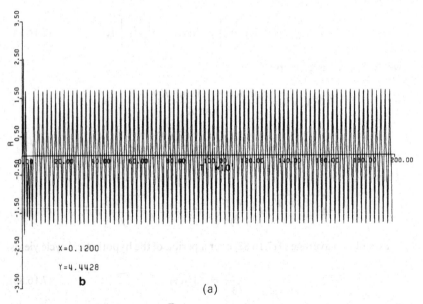

**Figure 7.16.8**  (a) For $K = \sqrt{2}\pi$, $\gamma = 0.12$, the amplitude evolution becomes periodic. (b) For $K = \sqrt{2}\pi$, $\gamma = 0.19$, the amplitude eventually attains a steady value, in this case $A = -1$, after a lengthy chaotic interval. (c) For $k = 2.45\pi$, $\gamma = 0.21$, the amplitude oscillates in an irregular, aperiodic manner.

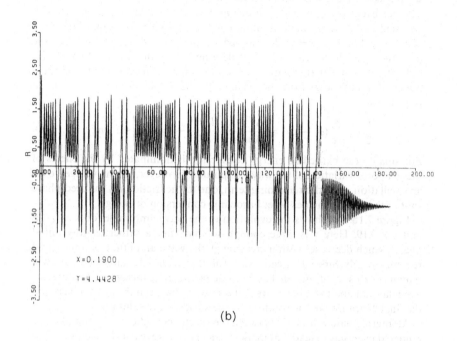

(b)

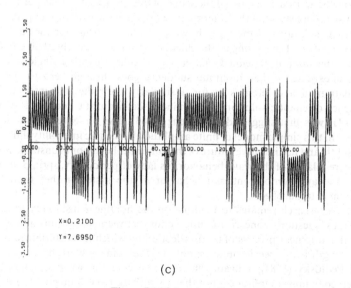

(c)

**Figure 7.16.8** (*continued*)

we may conclude that the steady solutions are not stable to finite disturbances. The persistent oscillation shown in Figure 7.16.8(a) is a limit cycle, i.e., it is an oscillation, independent of initial conditions. One may think of the *form* of the oscillation as being determined by the inviscid dynamics, while the *amplitude* (and hence the period) is determined by the frictional constraint (7.16.79) or (7.16.89). It is important to note that during the oscillation there is a net drain of energy from the mean flow. The baroclinic energy conversion is given by

$$\frac{\hat{F}}{2} U_s \overline{\frac{\partial \phi_2}{\partial x}(\phi_1 - \phi_2)} = \frac{\Delta^{3/2} U_s^2}{8km^2\pi^2}\left[\frac{d|A|^2}{dT} + 2\gamma|A|^2\right]\sin 2m\pi y. \quad (7.16.90)$$

For small $\gamma$ the baroclinic energy conversion is dominated by the effects of the temporal variation of the wave amplitude. However, when averaged over an oscillation cycle the remaining term, due to the frictionally induced phase shift in the wave provides a net baroclinic conversion of energy.

Figure 7.16.8(b) shows the evolution of the wave amplitude for $K = \sqrt{2}\pi$ and $\gamma = 0.19$. Here $\gamma$ is large enough so that after a rather lengthy initial phase (which decreases with increasing $\gamma$), the wave amplitude becomes (and remains) steady. Since the steady wave solution is linearly stable, the wave will remain in this state once it has been achieved. It is important to note the irregular and aperiodic behavior of the solution between the initial phase and the final "capture" of the solution by the steady wave solution.

At intermediate values of $\gamma$ between the strictly periodic solutions and those rendered eventually steady, numerical calculations reveal a rich and sensitive parameter dependence beyond the scope of this book to describe. Perhaps the most striking feature of this intermediate region is the existence of persistent, aperiodic solutions of the type shown in Figure 7.16.8(c) for the case $K = 2.45\pi$, $\gamma = 0.21$. For as long as the calculation was run (nearly 2,000 $e$-folding times), the solution remained *chaotic*, e.g., oscillating with a single sign and then unexpectedly the amplitude suddenly goes through zero to start an oscillation of a different sign for a similarly unpredictable interval. The presence of chaotic solutions to nonlinear, deterministic equations such as (7.16.41) was first suggested by Lorenz (1963). The presence of chaotic behavior in this simple model of finite amplitude baroclinic instability illustrates the subtle and nonintuitive nature of the nonlinear behavior of unstable waves. The irregularity of the wave behavior, its lack of periodicity, springs entirely from the autonomous internal behavior of the system rather than from random external effects.

Much research remains to be done on the nonlinear instability problem. Obvious questions concern the interaction between several unstable waves and the nonlinear process of mode selection by which it is determined which wavelength is favored in finite amplitude. Preliminary work by Hart (1981) and Pedlosky (1981) indicate that in some case the wave which will be observed in finite amplitude is not the one with the maximum growth rate but rather the wave which, at the same parameter setting, has the capacity for the largest amplitude.

## 7.17  Instability of Nonparallel Flow

The instability theory described in the previous sections of this chapter has been limited to the study of the instabilities of parallel flows, i.e., where for the basic state, at each $z$, every streamline is a straight line parallel to its neighboring streamline. The reason for this restriction is frankly the relative simplicity of the resulting problem compared to the more general problem of the instability of nonparallel flows. Progress in the latter case has been slight and compared to our understanding of the instability of parallel flows the instability of curvy, nonparallel flows is small. Yet, as we have seen in Section 3.26, nonparallel flows, such as the Rossby wave studied in that section, may become unstable in ways which are novel in the context of the traditional theory pertaining to parallel flows.

In this section a generalization applicable to nonparallel flows of the necessary conditions for instability developed in Section 7.3 will be given by methods similar to those used for parallel flows. The ideas presented here follow those of Arnold (1965) and Blumen (1969) although the development is somewhat different.

The quasigeostrophic potential vorticity equation (6.5.18) may be written in the form

$$\frac{\partial \Pi}{\partial t} + J(\psi, \Pi) = D(\psi) + G, \tag{7.17.1}$$

where

$$J(\psi, \Pi) \equiv \frac{\partial \psi}{\partial x} \frac{\partial \Pi}{\partial y} - \frac{\partial \psi}{\partial y} \frac{\partial \Pi}{\partial x}. \tag{7.17.2}$$

The right-hand side of (7.17.1) is a minor generalization of (6.5.8) in that the source of potential vorticity is no longer limited to heating but may be due to frictional forces, etc. The overall source term is split into two parts. One part, labeled $G$ is independent of $\psi$ and can be thought of as an external, fixed source of potential vorticity, i.e., the external driving mechanism for the motion. The term $D(\psi)$ is a source or sink term (depending on sign) and depends explicitly on $\psi$ as would for example a traditional parametrization of turbulent diffusion of potential vorticity of the form $D(\psi) = \kappa \nabla^2 \Pi$.

Suppose that the basic flow, i.e., the flow in the absence of perturbations

$$\psi = \Psi \tag{7.17.3}$$

is steady, i.e., that $\Psi$, although a function of $x$, $y$, and $z$ is independent of time. Then $\Psi$ satisfies, by definition,

$$J(\Psi, \Pi_0) = D(\Psi) + G. \tag{7.17.4}$$

Now let $\psi$ be perturbed so that

$$\psi = \Psi + \phi, \tag{7.17.5a}$$

$$\Pi = \Pi_0 + q, \tag{7.17.5b}$$

where

$$\Pi_0 = \frac{\partial^2 \Psi}{\partial x^2} + \frac{\partial^2 \Psi}{\partial y^2} + \frac{1}{\rho_s} \frac{\partial}{\partial z} \frac{\rho_s}{S} \frac{\partial \Psi}{\partial z} + \beta y, \tag{7.17.6a}$$

$$q = \frac{\partial^2 \phi}{\partial x^2} + \frac{\partial^2 \phi}{\partial y^2} + \frac{1}{\rho_s} \frac{\partial}{\partial z} \frac{\rho_s}{S} \frac{\partial \phi}{\partial z}. \tag{7.17.6b}$$

If (7.17.5a,b) is substituted into (7.17.1) and (7.17.4) is subtracted from the result of that operation, the equation for the perturbation field is obtained, i.e.,

$$\frac{\partial q}{\partial t} + J(\Psi, q) + J(\phi, \Pi_0) + J(\phi, q) = D'(\Psi, \phi), \tag{7.17.7}$$

where

$$D'(\Psi, \phi) \equiv D(\Psi + \phi) - D(\phi). \tag{7.17.8}$$

Note that if $D(\psi)$ is linear, $D'$ will depend only on $\phi$ and not on $\Psi$. In any case, note that each term in (7.17.7) vanishes as $\phi \to 0$. In particular, it is important to note that the external driving, $G$, does *not* appear in the finite amplitude problem for $\phi$ even when dissipation is present. The energy source implicit in the presence of $G$ is instead manifested by the structure of the basic flow, $\Psi$, which appears in the perturbation equation.

For the remainder of this section only linear perturbations not subject to dissipation will be studied, in which case (7.17.7) bcomes

$$\frac{\partial q}{\partial t} + J(\Psi, q) + J(\phi, \Pi_0) = 0. \tag{7.17.9}$$

The lower boundary, at $z = 0$ will be taken to be flat although the consideration of topography can easily be included. At that boundary the equation for the basic flow becomes from (6.6.10)

$$J\left(\Psi, \frac{\partial \Psi}{\partial z}\right) = \hat{D}(\Psi) + \hat{G}, \qquad z = 0, \tag{7.17.10}$$

with an obvious generalization of the right-hand side of (6.6.10) in analogy with (7.17.14). The perturbation field, when linearized satisfies, for non-dissipative flows

$$\frac{\partial}{\partial t} \frac{\partial \phi}{\partial z} + J\left(\Psi, \frac{\partial \phi}{\partial z}\right) + J\left(\phi, \frac{\partial \Psi}{\partial z}\right) = 0, \qquad z = 0, \tag{7.17.11}$$

while, as in Section 7.2, the upper boundary condition at $z = z_T$ is either identical to (7.17.11) or is equivalent to (7.4.18).

The problem specified by (7.17.9) and (7.17.11) is simplified considerably in the special but important case where

$$\Pi_0 = \Omega(\Psi, z), \tag{7.17.12a}$$

$$\frac{\partial \Psi}{\partial z} = \Theta(\Psi), \qquad z = 0, z_T, \tag{7.17.12b}$$

where $\Omega$ and $\Theta$ are arbitrary functions. That is, although in principle $\Psi$ and $\Pi_0$ are general functions of $x$, $y$, and $z$, we are restricting our attention to those cases where, at every $z$, $\Pi_0$ is constant on the streamline of the basic flow. Further, on the horizontal boundaries the basic state potential temperature is constant on the streamlines of the basic flow. There are a number of different situations in which the relations (7.17.12a,b) can occur. The most obvious one is the case where the basic flow is itself nondissipative and not forced, i.e., where *each* term on the right-hand sides of (7.17.4) and (7.17.10) is identically zero. Then both $J(\Psi, \Pi_0)$ and $J(\Psi, \partial\Psi/\partial z)$ must vanish and this yields (7.17.12a,b) directly. Another possibility is that while the individual source and sink terms are not individually zero, they cancel identically, i.e., that the left- and right-hand sides of the basic state equations are separately satisfied. This can happen when the structure of the basic flow possesses certain strong symmetries such that the Jacobian terms vanish identically leaving the determination of the basic flow to a balance between source and sink terms. A zonal basic flow in which $\Psi$ is independent of $x$ obviously is such an example as well as a flow with circular streamlines in which $\Psi$ is independent of azimuth angle. A third possibility is that the source and sink terms are individually *small but nonzero*. Then (7.17.12a,b) is satisfied to some lowest-order approximation and the determination of the basic flow involves a balance between the difference of source and sink terms and the relatively weak self-advection of potential vorticity by the small departures of the basic flow from the lowest-order flow which satisfies (7.17.12a,b). Thus it is not *necessary* that the basic flow is a completely free flow unaffected by dissipation for (7.17.12a,b) to apply.

Thus (7.17.9) becomes in such cases

$$\frac{\partial q}{\partial t} + J(\Psi, q) + J(\phi, \Psi)\frac{\partial\Omega}{\partial\Psi} = 0, \qquad (7.17.13)$$

while, similarly

$$\frac{\partial}{\partial t}\frac{\partial\phi}{\partial z} + J\left(\Psi, \frac{\partial\phi}{\partial z}\right) + J(\phi, \Psi)\frac{\partial\theta}{\partial\Psi} = 0, \qquad z = 0, z_T. \qquad (7.17.14)$$

In (7.17.13) and (7.17.14) $\partial\Omega/\partial\Psi$ and $\partial\Theta/\partial\Psi$ indicate the derivative at a given $z$ of the potential vorticity and potential temperature with respect to $\Psi$. It is important to note that at fixed $z$ these derivatives are themselves functions of only $\Psi$. Multiplication of (7.17.13) by $\rho_s q/(\partial\Omega/\partial\Psi)$ yields

$$\frac{\partial}{\partial t}\left(\frac{\rho_s}{2}q^2 \middle/ \frac{\partial\Omega}{\partial\Psi}\right) + J\left(\Psi, \frac{\rho_s}{2}q^2 \middle/ \frac{\partial\Omega}{\partial\Psi}\right) + \rho_s q J(\phi, \Psi) = 0, \qquad (7.17.15)$$

while multiplication by $\rho_s\phi$ leads, after a little algebra, to

$$\begin{aligned}
-\frac{\partial\rho_s E}{\partial t} + \nabla\cdot\left[\rho_s\phi\nabla_H\frac{\partial\phi}{\partial t} + \mathbf{k}\frac{\rho_s}{S}\phi\frac{\partial}{\partial t}\frac{\partial\phi}{\partial z}\right] \\
+ J\left(\frac{\phi^2}{2}, \rho_s\Psi\frac{\partial\Omega}{\partial\Psi}\right) + \rho_s\phi J(\Psi, q) = 0,
\end{aligned} \qquad (7.17.16)$$

where

$$E = \frac{(\nabla\phi)^2}{2} + \frac{\left(\dfrac{\partial\phi}{\partial z}\right)^2}{2S}$$

is the total perturbation energy. In (7.17.16)

$$\nabla_H \equiv \hat{\mathbf{i}}\frac{\partial}{\partial x} + \hat{\mathbf{j}}\frac{\partial}{\partial y}$$

where $\hat{\mathbf{i}}$ and $\hat{\mathbf{j}}$ are unit vectors in the $x$- and $y$-directions, while $\hat{\mathbf{k}}$ is the unit vector in the $z$-direction.

The final term in (7.17.16) may be rewritten, since

$$\phi J(\Psi, q) = J(\Psi, \phi q) - q J(\Psi, \phi). \tag{7.17.17}$$

If (7.17.16) is then subtracted from (7.17.15) we obtain

$$\frac{\partial}{\partial t}\left[\frac{\rho_s}{2}q^2 \bigg/ \frac{\partial\Omega}{\partial\Psi} + \rho_s E\right] + J\left(\Psi, \frac{\rho_s}{2}q^2 \bigg/ \frac{\partial\Omega}{\partial\Psi}\right)$$

$$+ J\left(\Psi, \frac{\rho_s}{2}\phi^2 \frac{\partial\Omega}{\partial\Psi}\right) - J(\Psi, \rho_s q\phi) \tag{7.17.18}$$

$$- \nabla\cdot\left[\rho_s\phi\nabla_H\frac{\partial\phi}{\partial t} + \hat{\mathbf{k}}\frac{\rho_s}{S}\phi\frac{\partial}{\partial t}\frac{\partial\phi}{\partial z}\right] = 0.$$

Now consider an integration of (7.17.18) over the entire volume, $V$, of the flow. If the lateral boundaries of the domain are streamlines of the *basic* flow, the integral of the three Jacobian terms in (7.17.18) will vanish. Furthermore, the term $\nabla\cdot(\phi\nabla_H \partial\phi/\partial t)$ will vanish when integrated of the boundary of the domain is a streamline of the perturbation flow. It then follows that

$$\frac{\partial}{\partial t}\iiint_V \rho_s\left[E + \frac{q^2}{2}\bigg/\frac{\partial\Omega}{\partial\Psi}\right]dx\,dy\,dz - \iint_A\left[\frac{\rho_s}{S}\phi\frac{\partial^2\phi}{\partial t\,\partial z}\right]dx\,dy\bigg|_{z=0}^{z=z_T} = 0, \tag{7.17.19}$$

where the surface term uses the notation

$$f(z)\bigg|_{z=0}^{z=z_T} = f(z_T) - f(0)$$

and where $A$ is the area of the domain defined by its horizontal boundary. The last term in (7.17.9) can be similarly rewritten if (7.17.14) is multiplied first by $(\partial\phi/\partial z)/(\partial\Theta/\partial\Psi)$ and then by $\phi$. After some algebra, it follows that

$$\iint_A\frac{\rho_s}{S}\phi\frac{\partial}{\partial t}\frac{\partial\phi}{\partial z}\,dx\,dy\bigg|_{z=0}^{z=z_T} = \frac{\partial}{\partial t}\iint_A\frac{\rho_s}{2S}\left(\frac{\partial\phi}{\partial z}\right)^2\bigg/\frac{\partial\theta}{\partial\Psi}\,dx\,dy\bigg|_{z=0}^{z=z_T}, \tag{7.17.20}$$

so that finally (7.17.19) becomes

$$\frac{\partial}{\partial t}\left[\iiint_V \rho_s\left(E + \frac{q^2}{2}\bigg/\frac{\partial\Omega}{\partial\Psi}\right)dx\,dy\,dz - \iint_A \frac{\rho_s}{2S}\left(\frac{\partial\phi}{\partial z}\right)^2\bigg/\frac{\partial\theta}{\partial\Psi}\,dx\,dy\,\bigg|_{z=0}^{z=z_T}\right] = 0.$$
(7.17.21)

If the upper boundary at $z = z_T$ is removed to infinity, we will assume that, as in Section 7.4, the perturbation field then vanishes so that the boundary term in (7.17.21) will make no contribution if $z_T \to \infty$.

The perturbation energy ($E$), the enstrophy ($q^2/2$), and the square of the surface potential temperature ($\partial\phi/\partial z)^2$ are each quadratic in the distrubance amplitude. If each of the coefficients of these terms in (7.17.21) is positive, the square bracket as a whole is positive definite in the disturbance amplitude and the bracket will be increasing with time in contradiction to the equality required by (7.17.21) should the disturbance amplitude increase with time. Thus a sufficient condition for stability, whose violation is a necessary condition for instability is that for all $x$, $y$, and $z$

$$\frac{\partial\Omega}{\partial\Psi} > 0,$$
(7.17.22a)

and

$$\frac{\partial\theta}{\partial\Psi} < 0, \qquad z = z_T,$$
(7.17.22b)

and

$$\frac{\partial\theta}{\partial\Psi} > 0, \qquad z = 0,$$
(7.17.22c)

for all $x$ and $y$.

If $z_T \to \infty$, the second condition is ignorable. These conditions are completely analogous to (7.3.31a,b,c) and, in fact, represent a generalization of them. In the special case where $\Psi$ is independent of $x$ and thus represents a parallel zonal flow,

$$\frac{\partial\Omega}{\partial\Psi} = \frac{\partial\Omega}{\partial y}\bigg/\frac{\partial\Psi}{\partial y} = -\frac{\partial\Pi_0/\partial y}{U_0},$$

while

$$\frac{\partial\theta}{\partial\Psi} = \frac{\partial\theta}{\partial y}\bigg/\frac{\partial\Psi}{\partial y} = -\frac{\partial U_0/\partial z}{U_0},$$
(7.17.23)

so that the conditions reduce to those already derived in Section 7.3. We note in passing that in the case of a zonal flow, a trivial Galilean transformation allows $U_0$ to take any fixed sign throughout the flow domain. It is a simple matter to show from this that the conditions derived from (7.3.37), in fact, can be derived directly from (7.3.31a,b,c) or their equivalents (7.17.22a,b,c).

It is interesting to observe that the condition (7.17.22a) is related to the uniqueness properties of solutions of the problem for $\Psi$ for a given $\Omega(\Psi)$, i.e.,

$$\frac{\partial^2 \Psi}{\partial x^2} + \frac{\partial^2 \Psi}{\partial y^2} + \frac{1}{\rho_s} \frac{\partial}{\partial z} \frac{\rho_s}{S} \frac{\partial \Psi}{\partial z} = \Omega(\Psi) - \beta y \qquad (7.17.24)$$

as consideration of the archetypal case $\Omega(\Psi) = \alpha \Psi$ illustrates. If $\alpha$ is positive, uniqueness occurs under the conditions of $\Psi$ vanishing on the boundaries of the domain, while $\alpha < 0$ allows free modes of unspecified amplitude. The relation between stability of a flow and the uniqueness properties of its $\Omega(\Psi)$ relation suggest that instability is connected to the ability of the basic flow to find more than one flow configuration consistent with conservation of potential vorticity and the boundary conditions.

A direct application of these ideas to flows discussed in earlier sections is possible as well. The free inertial mode described in Section 5.10, the so-called Fofonoff mode, corresponds to a barotropic flow for which $\partial \Psi / \partial z$ and $\partial \phi / \partial z$ are identically zero. The condition for stability then is that (7.17.22a) be satisfied. For the Fofonoff mode

$$\Omega(\Psi) = \frac{\Psi}{A^2}$$

in the notation of Section 5.10. Since $A^2 > 0$ it follows that the circulation discribed there *must be stable* to small disturbances. This is a remarkable result given the strong shears associated with boundary currents in that circulation.

The instability of the plane Rossby wave described in Section 3.26 can be analyzed similarly. The Rossby wave in that treatment was propagating, so that the basic state represented by the wave is unsteady. However, a dynamically irrelevant Galilean transformation will render the state steady if we limit ourselves to scales small with respect to the external deformation radius so that the term multiplied by $F$ may be neglected in (3.26.1). Then the stationary wave has

$$\Psi = -\hat{U}y + A \cos(kx + ly),$$

where by (3.18.9)

$$\hat{U} = \frac{\beta}{k^2 + l^2},$$

which we note is independent of the wave amplitude $A$. However

$$\Omega(\Psi) = -(k^2 + l^2)A \cos(kx + ly) - \beta y = -(k^2 + l^2)\Psi.$$

Hence the potential vorticity distribution satisfies the necessary condition for instability, $\partial \Omega / \partial \Psi < 0$, for any nonzero wave amplitude. As the detailed calculations of Section 3.26 show, it is always possible to find a disturbance field composed of other Rossby waves which complete a resonant triad that will destabilize the original basic flow. The inability of $\beta$ to stabilize the flow no matter how weak are the spatial gradients of the relative vorticity with respect to $\beta$ illustrates the important differences which arise when the flow is no longer a parallel flow.

# CHAPTER 8

# Ageostrophic Motion

## 8.1 Anisotropic Scales

The theory of quasigeostrophic motion developed in previous chapters required several conditions for its validity. Foremost, of course, was the condition that the time scale of the motion is large compared to $f^{-1}$. This is not always sufficient, as the discussion of the Kelvin wave in Section 3.9 indicated. When the wavelength of the Kelvin wave is long compared to the deformation radius, the frequency is small compared to $f$. Yet the equation of motion in the direction along the boundary does *not* reduce to geostrophic balance. A longshore* pressure gradient exists and is balanced by the acceleration of the longshore velocity, and the onshore flow is identically zero. This result is not inconsistent with the quasigeostrophic theory developed previously, for in the quasigeostrophic dynamics so far considered, we have assumed that there existed only a single horizontal scale for the motion, i.e., that the scaling of the motion is isotropic horizontally. The low-frequency Kelvin wave is not horizontally isotropic, since its offshore scale is the deformation radius, which for low frequencies is small compared with the wavelength. In this chapter we consider some of the refinements to quasigeostrophic theory that are required when the motion is strongly anisotropic.

In certain cases, such as the inertial boundary-current theory of Chapter 5, the disparity of the two horizontal scales affects primarily the vorticity

---

* Longshore = parallel to the boundary.

balance, while the momentum balance remains geostrophic to a tolerable degree. However, if the scales become too disparate and the pressure gradient in one direction becomes very feeble, the force balance in one direction will alter as the pressure gradient becomes, in *that* direction, too weak to balance the Coriolis force. Other forces then come into play, and these may be frictional or inertial, depending on the individual circumstances.

To illustrate these considerations in a definite way, consider the motion of a layer of homogeneous, incompressible fluid. Let us first consider the simplest circumstances, where friction may be ignored, and further consider only motions for which the sphericity of the earth is inconsequential, so that $f$ may be considered constant. This last assumption requires that the planetary-vorticity gradient be weak compared to other potential-vorticity gradients influencing the motion, and this is often the case for narrow fields of motion.

Let us orient the $x_*$-axis in the direction of rapid variations in the velocity and pressure fields, so that variations are slight in the $y_*$-direction. Let $L$ be the scale for the motion in the $y_*$-direction and $l$ be the scale in the $x_*$-direction, where by hypothesis

$$\frac{l}{L} \ll 1. \tag{8.1.1}$$

Similarly, let $U$ be the velocity *scale* in the $x_*$-direction and $V$ be the velocity scale in the $y_*$-direction. We introduce nondimensional variables as follows:

$$
\begin{aligned}
u_* &= Uu, & x_* &= lx, \\
v_* &= Vv, & y_* &= Ly, \\
w_* &= Ww, & z_* &= \delta Lz, \\
& & t_* &= t/\sigma.
\end{aligned}
\tag{8.1.2}
$$

In (8.1.2) the asterisked variables are dimensional. The vertical scale of the motion is $\delta L$, where $\delta$ is also small, and $\sigma^{-1}$, the time *scale*, satisfies by hypothesis

$$\frac{\sigma}{f} \ll 1. \tag{8.1.3}$$

The continuity equation for the homogeneous incompressible fluid is

$$\frac{V}{L}\frac{\partial v}{\partial y} + \frac{U}{l}\frac{\partial u}{\partial x} + \frac{W}{\delta L}\frac{\partial w}{\partial z} = 0. \tag{8.1.4}$$

To allow the *a priori* possibility that the mass flux can be balanced either in horizontal or in vertical planes, each of the terms in (8.1.4) must be of equal order. This leads to the familiar geometrical constraints on the relative velocity amplitudes,

$$W = \delta V, \tag{8.1.5a}$$

$$U = \frac{l}{L} V. \tag{8.1.5b}$$

The pressure scale is chosen in the expectation that the Coriolis acceleration due to the swift $O(V)$ flow will be of the same order as the pressure gradient in the $x$-direction, i.e.,

$$p_* = -\rho_0 g z_* + \rho_0 f V l p. \tag{8.1.6}$$

Note that this is equivalent to

$$p_* = -\rho_0 g z_* + \rho_0 f U L p. \tag{8.1.7}$$

In nondimensional units, the equations of motion in the $x$- and $y$-directions become

$$\left(\frac{\sigma L}{f l}\right)\frac{\partial v}{\partial t} + \frac{V}{f l}\left(u\frac{\partial v}{\partial x} + v\frac{\partial v}{\partial y} + w\frac{\partial v}{\partial z}\right) + u = -\frac{\partial p}{\partial y}, \tag{8.1.8a}$$

$$\left(\frac{\sigma l}{f L}\right)\frac{\partial u}{\partial t} + \frac{V l^2}{f l L^2}\left(u\frac{\partial u}{\partial x} + v\frac{\partial u}{\partial y} + w\frac{\partial u}{\partial z}\right) - v = -\frac{\partial p}{\partial x}. \tag{8.1.8b}$$

There are two separate measures of nonlinearity. In the $x$-momentum equation the advective Rossby number is (using (8.1.5b))

$$\varepsilon_L = \frac{V l^2}{f l L^2} = \frac{U}{f L} = \frac{U l}{f l L}, \tag{8.1.9a}$$

while in the $y$-momentum equation the corresponding Rossby number is

$$\varepsilon_l = \frac{V}{f l}. \tag{8.1.9b}$$

These two Rossby numbers are quite different in magnitude, since

$$\frac{\varepsilon_L}{\varepsilon_l} = \frac{l^2}{L^2} \ll 1. \tag{8.1.10}$$

Though $\varepsilon_L$ is small, $\varepsilon_l$ may be $O(1)$ if $l/L$ is sufficiently small. Similarly, the two Rossby numbers appropriate to the local time derivatives, from (8.1.8a,b), are clearly in the same ratio (i.e., $l^2/L^2$) as the advective Rossby numbers. The smallness of $\sigma/f$ is no guarantee that $\sigma L/f l$ will be small. If $\varepsilon_L$ and $(\sigma/f)(l/L)$ are small (i.e., if the time scales are long compared with $f^{-1}$) and if $l/L$ is small, then $v$ will be in geostrophic balance, i.e., to lowest order the *swift* component of the velocity will satisfy

$$v = \frac{\partial p}{\partial x}. \tag{8.1.11}$$

If either $V/fl$ or $\sigma L/fl$ is $O(1)$, then the velocity transverse to the speedy flow will *not* be in geostrophic balance. The quasigeostrophic boundary currents discussed in Chapters 3 and 5 require for their validity the smallness of $V/fl$, and the theories contain inaccuracies to that order. For a homogeneous fluid, (8.1.11) together with the hydrostatic approximation implies that $v$ is independent of $z$. This allows (8.1.8a) to be used to find $u$ in terms of $p$—i.e., with (8.1.11),

$$u = \frac{-\dfrac{\partial p}{\partial y} - \dfrac{\sigma}{f}\dfrac{L}{l}\dfrac{\partial^2 p}{\partial x \, \partial t} - \varepsilon_l \dfrac{\partial p}{\partial x}\dfrac{\partial^2 p}{\partial x \, \partial y}}{1 + \varepsilon_l \dfrac{\partial^2 p}{\partial x^2}}, \tag{8.1.12}$$

which may be interpreted as follows. When $l$ is $O(L)$ and all the time scales exceed $f^{-1}$, the term which predominantly balances the Coriolis acceleration due to $u$ is the pressure gradient in the $y$-direction. As $L$ increases for *fixed* $l$, this pressure gradient weakens, until $\varepsilon_l$ or $(\sigma/f)(L/l)$ is $O(1)$. Then the pressure gradient in the $y$-direction becomes as weak as the ageostrophic, inertial accelerations in that direction, and these latter accelerations enter the $O(1)$ momentum balance for *this* component of flow. For the linear Kelvin wave propagating in the $y$-direction, for example, $u$ is identically zero and a *low*-frequency balance can only occur between the first two terms in the numerator of (8.1.12) for a weak pressure gradient in the $y$-direction if $\sigma/f \sim l/L$. Since $\sigma$ is $O((gD)^{1/2}/L)$, where $L$ is the wavelength, this balance of terms will hold if $l$ is $O((gD)^{1/2}/f)$, which is simply the Rossby-deformation radius and, in fact, is the scale in $x$ of the Kelvin wave. Flows in which only one component of the velocity is geostrophic have been called *semigeostrophic* (Hoskins 1975). In the following sections we examine some dynamical phenomena that have this semigeostrophic property. There are two general points to bear in mind. First, the loss of geostrophy in one direction does not *necessarily* imply that the inertial accelerations are significant. For example, if the motion were completely steady and independent of $y$, then both the terms of order $(\sigma/f)(L/l)$ and $\varepsilon_l$ in the numerator of (8.1.12) would be identically zero. In such cases weak dissipative, frictional forces can enter as the ageostrophic agents. An example is described in Section 8.3. Secondly, the approach to the dynamical problem remains the same as in the quasigeostrophic context. That is, careful scaling will yield a suitably simplified dynamical system, and the geostrophy of one component of the flow ensures a degree of qualitative similarity to the simpler quasigeostrophic theory. For many processes it is often helpful to first assume that both $\varepsilon_L$ and $\varepsilon_l$ are small and construct the simple quasigeostrophic model, and then use its consequences as a guide for phenomena while $\varepsilon_L$ is small but $\varepsilon_l$ is $O(1)$.

## 8.2  Continental-Shelf Waves

The dispersion relation for quasigeostrophic topographic Rossby waves
in a homogeneous layer of fluid was derived in Section 3.15. In dimensional
form the frequency–wave-number relation may be written

$$\sigma_* = -\frac{f\mathbf{K}_* \cdot (\hat{z} \times \nabla_* H_0/D)}{K_*^2 + R^{-2}}. \tag{8.2.1}$$

The dimensional wave vector is $\mathbf{K}_*$, while $R$ is the deformation radius of the
(undisturbed) free surface $(gD)^{1/2}/f$. The thickness of the layer in the absence
of motion is $H_0$, and in quasigeostrophic theory we require that $H_0$ depart
only slightly from the constant value $D$ on the scale of the wavelength of the
Rossby wave. This latter condition is imposed in order that $\sigma_*/f$ may be
small. If $H_0$ varies by $O(1)$ over the wavelength, we may infer from (8.2.1)
that $\sigma_*$ will remain small only if the wave vector is nearly parallel to $\nabla_* H_0$.
In that situation the motion of the *fluid* is very nearly along the isolines of the
topography, and only a small amount of vortex stretching will occur, with a
correspondingly weak restoring mechanism for a displaced vortex tube. If
the wave vector is nearly parallel to $\nabla_* H_0$, then the length scale of the motion
in the direction along the isobaths is long compared with the scale of variation perpendicular to the isobaths. This is a condition on the length scales
for the existence of low-frequency oscillations when the layer thickness
varies strongly, and it falls precisely in the category discussed in Section 8.1.

The extension of the theory of topographic, low-frequency waves to the
case of strongly varying topography has direct application to the dynamics
of continental-shelf waves. These are waves propagating along continental
margins and exist in the region of the continental shelf. Direct observations
of the waves off the Oregon coast (Cutchin and Smith 1973) indicate the
existence of such coastally trapped waves with values of $\sigma_*/f$ of order 0.2.
The waves are apparently fundamentally barotropic in their dynamics and
exist over shelf topography, as shown in Figure 8.2.1. The depth of the fluid
layer is very small near the coast and slopes sharply downward, so that
within ten or twenty kilometers of the coast the depth has become of the
order of one hundred meters. At this point the shelf "breaks" and slopes
more steeply to depths of several thousand meters. The characteristic length
scale of the observed motions in the direction perpendicular to the coast is
tens of kilometers, and it is inferred from observation that the wavelength of
the oscillation along the coast is hundreds if not thousands of kilometers.
The strong slope of the shelf and continental slope invalidates the assumption required of quasigeostrophic theory that the depth change is only a
small departure from a constant value. Yet, as we shall see, the evident
applicability of the ideas of Section 8.1 leads to a theory for these waves
which bears a strong similarity to the quasigeostrophic theory.

Consider the motion over strongly varying topography which varies only
in x, bounded on one side by a straight coast as depicted in Figure 8.2.2. Let

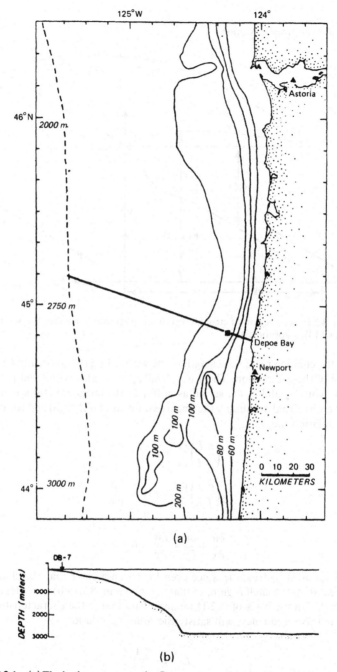

(a)

(b)

**Figure 8.2.1** (a) The bathymetry near the Oregon coast, showing the rapid increase of ocean depth with offshore distance. The dashed line indicates the base of the continental slope. The depth variation along the heavy solid line is shown in (b) (from Cutchin and Smith 1973).

629

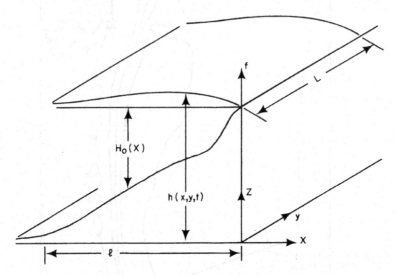

**Figure 8.2.2**   The region of strongly varying topography for which the wave equation (8.2.13) is appropriate.

$L$ be the characteristic wavelength of the wave along the coast, and $l$ its scale in the offshore direction. Then, if $x_*$ and $y_*$ are scaled with $l$ and $L$ respectively, while $u_*$ and $v_*$ are scaled as in (8.1.2), the linear equations of motion for the wave field are simply the linearized forms of (8.1.8a,b) and the continuity equation, i.e.,

$$\left(\frac{\sigma}{f}\frac{L}{l}\right)\frac{\partial v}{\partial t} + u = -\frac{\partial p}{\partial y}, \tag{8.2.2a}$$

$$\left(\frac{\sigma}{f}\frac{l}{L}\right)\frac{\partial u}{\partial t} - v = -\frac{\partial p}{\partial x}, \tag{8.2.2b}$$

and

$$\frac{\partial u}{\partial x} + \frac{\partial v}{\partial y} + \frac{\partial w}{\partial z} = 0.$$

The motion is hydrostatic, since even $l$ is much greater than the characteristic depth of the shelf region, so that $u$, $v$, and $p$ in (8.2.2a,b) are independent of depth. On the basis of (8.2.1) we anticipate that $\sigma$, the characteristic scale for the wave frequency, will satisfy the ordering relation

$$\frac{\sigma}{f} = \frac{l}{L} \ll 1. \tag{8.2.3}$$

Thus, to $O\left(\dfrac{l^2}{L^2}\right)$

$$v = \frac{\partial p}{\partial x}, \tag{8.2.4a}$$

$$\frac{\partial v}{\partial t} + u = -\frac{\partial p}{\partial y},$$ (8.2.4b)

which combine to yield the vorticity equation,

$$\frac{\partial}{\partial t}\left(\frac{\partial v}{\partial x}\right) = -\left(\frac{\partial u}{\partial x} + \frac{\partial v}{\partial y}\right)$$

$$= \frac{\partial w}{\partial z}.$$ (8.2.5)

The vorticity $\zeta$ is given entirely in terms of the longshore flow, since

$$\zeta = \frac{\partial v}{\partial x} - \frac{\partial u}{\partial y} = \frac{\partial v}{\partial x} + O\left(\frac{l^2}{L^2}\right).$$ (8.2.6)

Let $\eta_*$ be the displacement of the upper free surface. The obvious requirement that the longshore flow be geostrophic, coupled with the hydrostatic balance, implies that

$$\frac{g\eta_*}{l} = O(fV),$$ (8.2.7)

so that $\eta_*$ should be scaled as follows:

$$\eta_* = \frac{V}{fl}\frac{f^2 l^2}{gD} D\eta(x, y, t)$$

$$= D\varepsilon_l F\eta$$ (8.2.8)

where $D$ is a characteristic depth for the region and $F = f^2 l^2/gD$. Note that $F^{1/2}$ is the ratio of the offshore scale $l$ to the deformation radius. On the upper surface, in linear theory,

$$w_* = \frac{D}{L}Vw = \frac{\partial \eta_*}{\partial t_*} = \frac{\sigma}{f}\frac{V}{l}DF\frac{\partial \eta}{\partial t},$$ (8.2.9a)

or, with (8.2.3),

$$w = F\frac{\partial \eta}{\partial t},$$ (8.2.9b)

while at the lower boundary

$$w = \frac{u}{D}\frac{\partial H_0}{\partial x},$$ (8.2.10)

where $H_0$ is the *dimensional* thickness of the fluid region in the absence of motion. Integration of (8.2.5) over the fluid depth $H_0/D$ yields

$$\left(\frac{H_0}{D}\right)\frac{\partial \zeta}{\partial t} = F\frac{\partial \eta}{\partial t} + \frac{u}{D}\frac{\partial H_0}{\partial x}.$$ (8.2.11)

From (8.2.4a), (8.2.4b), and (8.2.6) and the condition of hydrostatic balance,

$$\zeta = \frac{\partial v}{\partial x} = \frac{\partial^2 p}{\partial x^2} \tag{8.2.12a}$$

$$u = -\frac{\partial p}{\partial y} - \frac{\partial^2 p}{\partial x \, \partial t} \tag{8.2.12b}$$

$$\eta = p. \tag{8.2.12c}$$

It follows that (8.2.11) may be written entirely in terms of $p$:

$$\frac{\partial}{\partial t}\left[\frac{\partial^2 p}{\partial x^2} + \frac{1}{H_0}\frac{\partial H_0}{\partial x}\frac{\partial p}{\partial x} - F\frac{D}{H_0}p\right] + \frac{1}{H_0}\frac{\partial H_0}{\partial x}\frac{\partial p}{\partial y} = 0. \tag{8.2.13}$$

In quasigeostrophic theory $H_0$ is replaced by $D$ wherever it appears in undifferentiated form. In the case at hand, the full variation of $H_0$ is retained, since it varies by $O(1)$ over the scale of the wave.

The wave equation (8.2.13) admits solutions of the form of traveling waves in the $y$-direction, i.e.,

$$p = \text{Re } \phi(x)e^{i(\alpha y - \omega t)}, \tag{8.2.14}$$

where $\alpha$ is the wave number in the $y$-direction and $\omega$ the corresponding nondimensional frequency. Substitution of (8.2.14) into (8.2.13) yields an equation for $\phi(x)$, i.e.,

$$\frac{d^2\phi}{dx^2} + \frac{1}{H_0}\frac{dH_0}{dx}\frac{d\phi}{dx} - \frac{\phi}{H_0}\left[FD + \frac{\alpha}{\omega}\frac{dH_0}{dx}\right] = 0. \tag{8.2.15}$$

Consider now the idealized shelf topography shown in Figure 8.2.3, for which

$$H_0 = \begin{cases} D & \text{if } x_* < -l \\ -\gamma D\dfrac{x_*}{l} = -\gamma Dx & \text{if } -l \le x_* \le 0 \end{cases} \tag{8.2.16}$$

For $x < -1$, where the bottom is flat, the general solution of (8.2.15) is

$$\phi(x) = Ae^{-F^{1/2}(\xi - 1)} + Be^{+F^{1/2}(\xi - 1)}, \tag{8.2.17}$$

where

$$\xi = -x.$$

In order for $\phi$ to remain bounded far from the coast (i.e., for large, positive $\xi$), the second solution in (8.2.17) must be rejected, so that

$$\phi = Ae^{-F^{1/2}(\xi - 1)}, \quad \xi > 1. \tag{8.2.18}$$

In this region the scale of the motion offshore is simply the deformation radius. The bottom is flat, so that conservation of potential wave vorticity is obtained, i.e.,

$$\frac{\partial}{\partial t}(\zeta - F\eta) = 0, \quad \xi > 1, \tag{8.2.19}$$

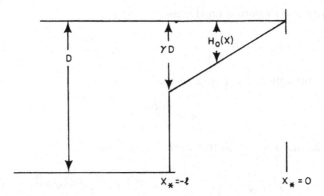

**Figure 8.2.3** An idealized depth profile. The depth of the deep ocean is $D$. The maximum shelf depth is $\gamma D$.

from which (8.2.18) follows directly, if (8.2.6) is duly noted. Over the sloping shelf, the amplitude of the motion is strongly affected by the topographic slope, and (8.2.15) becomes

$$x\frac{d^2\phi}{dx^2} + \frac{d\phi}{dx} - \phi\left[\frac{\alpha}{\omega} - \frac{F}{\gamma}\right] = 0, \qquad (8.2.20)$$

or, in terms of $\xi$,

$$\xi\frac{d^2\phi}{d\xi^2} + \frac{d\phi}{d\xi} + \phi\left[\frac{\alpha}{\omega} - \frac{F}{\gamma}\right] = 0. \qquad (8.2.21)$$

The general solution of (8.2.21) can be written directly in terms of the Bessel functions of zero order, i.e.,

$$\phi = aJ_0(2\mu^{1/2}\xi^{1/2}) + bY_0(2\mu^{1/2}\xi^{1/2}), \qquad (8.2.22)$$

where

$$\mu \equiv \frac{\alpha}{\omega} - \frac{F}{\gamma}. \qquad (8.2.23)$$

The second solution, $Y_0(2\mu^{1/2}\xi^{1/2})$ is logarithmically singular at the coast and must be rejected in order that $\eta$ may remain finite at $\xi = x = 0$. Thus

$$\phi(x) = aJ_0(2\mu^{1/2}\xi^{1/2}), \qquad 0 \le \xi \le 1. \qquad (8.2.24)$$

At the point $\xi = 1$, the two solutions must be joined. The two conditions required for the matching of the solution are (i) continuity of pressure, and (ii) continuity of the onshore transport. The first condition requires that $\phi$ be continuous at $\xi = 1$, i.e.,

$$aJ_0(2\mu^{1/2}) = A. \qquad (8.2.25)$$

The onshore transport in nondimensional units is

$$uH_0 = -H_0 \left[ \frac{\partial p}{\partial y} + \frac{\partial^2 p}{\partial x \, \partial t} \right] = -H_0 \left[ \frac{\partial p}{\partial y} - \frac{\partial^2 p}{\partial \xi \, \partial t} \right], \qquad (8.2.26)$$

whose continuity at $\xi = 1$ requires that

$$\left[ \phi + \frac{\omega}{\alpha} \frac{d\phi}{d\xi} \right] H_0$$

be continuous at $\xi = 1$, so that in the present case

$$A \left[ 1 - \frac{\omega}{\alpha} F^{1/2} \right] = a\gamma \left[ J_0(2\sqrt{\mu}) + \mu^{1/2} \frac{\omega}{\alpha} J_0'(2\sqrt{\mu}) \right], \qquad (8.2.27)$$

where $J_0'$ is the derivative of the Bessel function $J_0$ with respect to its argument. The identity

$$J_0'(x) = -\frac{x}{2} [J_0(x) + J_2(x)] \qquad (8.2.28)$$

allows (8.2.27) to be rewritten, with the aid of (8.2.25), as

$$J_0(2\mu^{1/2}) \left| 1 - [F^{1/2} + F] \frac{\omega}{\alpha} \right| = -\gamma \left[ 1 - \frac{F}{\gamma} \frac{\omega}{\alpha} \right] J_2(2\mu^{1/2}). \qquad (8.2.29)$$

This, with the definition (8.2.23), is the dispersion relation for the shelf wave. Note that

$$\frac{F}{\gamma} = \frac{f_0^2 l^2}{g(\gamma D)}$$

is proportional to the ratio of $l$ to the deformation radius based on the depth $\gamma D$ at the shelf break. For a typical shelf with a width of 20 km at the 100-meter isobath, $F/\gamma$ is $O(2 \times 10^{-3})$. $F$ is, of course, even smaller, so that to an excellent approximation, for modes such that $\omega/\alpha$ is $O(1)$, the dispersion relation reduces to

$$\gamma = -\frac{J_0(2\mu^{1/2})}{J_2(2\mu^{1/2})}. \qquad (8.2.30)$$

For each value of $\gamma$ there are an infinite number of roots of (8.2.30). Define $K_n$ as the $n$th such value of $2\mu^{1/2}$ which satisfies (8.2.30). For small $\gamma$, then, roots tend to the zeros of $J_0(K)$, and for increasing $\gamma$ the corresponding roots increase. Figure 8.2.4 shows the solutions of (8.2.30) for the first three modes over the range of $\gamma$ of physical interest, i.e., $0 \leq \gamma \leq 1$. The frequency for a given mode is then given by (8.2.23), which for small $F$ simply yields

$$2 \left( \frac{\alpha}{\omega_n} \right)^{1/2} = K_n, \qquad (8.2.31)$$

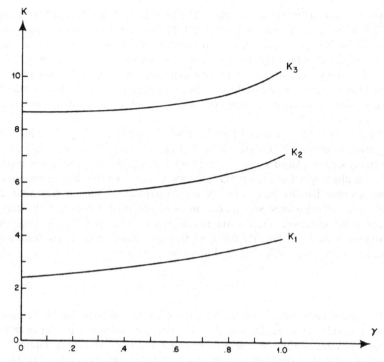

**Figure 8.2.4** The dependence of the first three eigenvalues of (8.2.30) on $\gamma$, where $K = 2\mu^{1/2}$.

or

$$\boxed{\omega_n = \frac{4\alpha}{K_n^2}}, \qquad n = 1, 2, 3, \ldots. \tag{8.2.32}$$

In dimensional units

$$\omega_{*n} = 4f \frac{(\alpha_* l)}{K_n^2}, \tag{8.2.33}$$

where $\alpha_*$ is the dimensional north–south wave number $\alpha/L$, and $l$ is the distance from the coast to the shelf edge. $K_n$, note, is only a weak function of depth at the shelf break.

The dynamical fields for each mode in the shelf region are given by the following expressions:

$$p = \eta = a \cos(\alpha y - \omega t + \theta) J_0(K_n \zeta^{1/2}), \tag{8.2.34a}$$

$$v = -\frac{aK_n}{2\zeta^{1/2}} \cos(\alpha y - \omega t + \theta) J_1(K_n \zeta^{1/2}), \tag{8.2.34b}$$

$$u = -a\alpha \sin(\alpha y - \omega t + \theta) J_2(K_n \zeta^{1/2}), \tag{8.2.34c}$$

where $\theta$ is an arbitrary phase angle. The amplitude of the motion beyond the shelf is, by (8.2.25) and (8.2.30), $O(\gamma)$. Hence, for moderately small $\gamma$ the motion is limited predominantly to the shelf, while for larger $\gamma$ the open sea also oscillates. This helps to explain why $K$ is such a weak function of $\gamma$. For small $\gamma$ the vortex-stretching restoring mechanism is weak but the mass of the system participating in the oscillation is correspondingly small, while for larger slopes the restoring mechanism is stronger but the system has greater effective inertia.

For small $\gamma$ the onshore flow very nearly vanishes at $x_* = -l$ and the motion is effectively isolated from the open ocean. A vertical wall could equally well be placed at $x = -1$ without affecting the structure of the mode.

The phase speed of the topographic wave is positive, so that, as for quasi-geostrophic Rossby waves, the phase propagates in such a way that an observer riding a crest sees higher ambient potential vorticity on his right. The continental-shelf waves are nondispersive, that is, for each mode the frequency depends linearly on $\alpha$, so that the phase speed is constant and equal to the group velocity. In dimensional units

$$C_{*y} = \frac{4fl}{K_n^2} = \frac{\partial \omega_*}{\partial \alpha_*}. \qquad (8.2.35)$$

This feature corresponds to the long-wave limit for Rossby waves. The shelf wave shares these qualitative features with the Rossby wave because the basic physics of the wave field is the same, i.e., the restoring mechanism is simply related to the vortex stretching in the planetary vorticity field. There are, however, important differences. Although the longshore velocity is geostrophic, the onshore velocity $u$ is not. Let us define a geostrophic velocity as

$$u_g = -\frac{\partial p}{\partial y} = a\alpha \sin(\alpha y - \omega t + \theta) J_0(K_n \xi^{1/2}). \qquad (8.2.36)$$

Then

$$\frac{u_g}{u} = -\frac{J_0(K_n \xi^{1/2})}{J_2(K_n \xi^{1/2})}. \qquad (8.2.37)$$

As the coast is approached, i.e., as $\xi \to 0$,

$$u = O(\xi) \qquad (8.2.38)$$

while $u_g$ goes to a constant value. At the shelf break $u_g/u = \gamma$ by (8.2.30). Hence, the geostrophic onshore velocity is too large near the coast and too small to be accurate at the shelf break. Thus, even though

$$\frac{\omega_*}{f} = O(\alpha_* l) \ll 1, \qquad (8.2.39)$$

the onshore component of the flow has $O(1)$ departures from geostrophic balance.

The reason for the neglect of the planetary-vorticity gradient can now be better appreciated. The contribution of the planetary-vorticity gradient to the vorticity balance of the wave on the shelf, compared to the vorticity produced by vortex-tube stretching of columns riding up the shelf, is given by the ratio

$$
\frac{v_* \beta_0}{u_* f H_0^{-1} \, \partial H_0 / \partial x} = O\left( \frac{\beta_0 L/l}{f \, \Delta H/Hl} \right)
$$

$$
= O \frac{\beta_0 L/f}{(\Delta H/H)}.
$$

$$(8.2.40)$$

If $\Delta H/H$ is $O(1)$, this ratio will be small for north–south length scales which are less than the distance from equator to pole. For if $\Delta H/H$ is $O(1)$, the change in the potential vorticity due to depth changes is $O(f/H)$, while that due to motion in the northward direction is only $O(\beta_0 L/H)$. Hence, to the very same degree that the $\beta$-plane approximation is valid, the $\beta$-effect should be ignored on the shelf if $\Delta H/H$ is $O(1)$. For motion off the shelf the $\beta$-effect should, of course, be retained, but the shelf wave has negligible energy there and the dynamics of this region is largely irrelevant for these trapped modes.

## 8.3 Slow Circulation of a Stratified, Dissipative Fluid

The narrow scale of the continental-shelf waves of the previous section is determined by the scale of the shelf itself. There are cases, however, where $l$, the narrow scale, is determined by dynamical processes intrinsic to the fluid dynamics rather than being imposed by external effects. This occurs frequently in the context of boundary-layer theory, where the short scale is the appropriate boundary-layer thickness. If the boundary-layer scale is thin enough, ageostrophic effects emerge. Furthermore, as the discussion of Chapters 4 and 5 demonstrated, a boundary-layer region may consist simultaneously of several different scales, i.e., there may be narrow boundary layers tucked within wider layers, and the degree of departure from geostrophy will vary from region to region.

To illustrate these considerations, consider the steady circulation driven by a wind stress as shown in Figure 8.3.1. A stratified fluid layer of uniform thickness $D$ is set in motion by a stress which is directed parallel to the $y$-axis. The fluid layer is semiinfinite and occupies the region $-\infty \le x \le 0$. The existence of a stress in the $y$-direction implies a mass flux in the upper Ekman layer in the $x$-direction, and the presence of the boundary, or coast, at $x = 0$ implies the necessity for vertical motion within the body of the fluid to allow flow into the Ekman layer at the junction of the layer and the coast. A simple model of this process for a homogeneous fluid was described in Section 5.12. The presence of stratification introduces new dynamical elements of interest. In particular, the intimate relationship between the driven

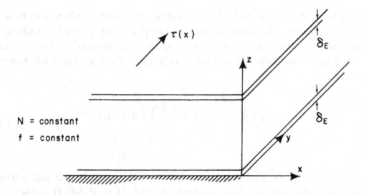

**Figure 8.3.1** The elements of the circulation problem. The stress $\tau(x)$ produces an Ekman flux in the $x$-direction in the upper Ekman layer. The Coriolis parameter $f$ and the Brunt–Väisälä frequency $N$ are constant.

vertical velocity, the consequent density anomaly, and the connected thermal-wind field profoundly alters the dynamics. In order to keep our discussion as simple as possible, let us restrict our attention, as in Section 5.12, to relative motions which are sufficiently weak that nonlinear effects may be ignored. In addition, we will assume that $f$ is constant. This, as we saw in Section 5.12, is appropriate for the discussion of the narrow boundary layer in which the vertical circulation exists and which is the focus of our attention here. We also assume that the fluid is a liquid whose density scale height is far greater than the depth of the fluid, so that, as shown in Section 6.8, the variation of the density can be ignored in the momentum equations except insofar as it provides a buoyancy force. Let $L$ be the characteristic scale for the spatial variation of the wind stress $\tau_*$, and use $L$ to nondimensionalize horizontal lengths. The wind stress is written as

$$\tau_* = \mathbf{j}\tau_0 \tau(x),    (8.3.1)$$

where $\mathbf{j}$ is a unit vector in the $y$-direction and $\tau_0$ is the characteristic magnitude of $\tau_*$. The scale for the horizontal velocity is chosen such that the velocity in the upper Ekman layer is O(1). By the results of Section 4.10, this suggests a scaling velocity

$$U = \frac{2\tau_0}{\rho f D E_V^{1/2}}    (8.3.2)$$

where

$$E_V = \frac{2A_V}{f D^2}    (8.3.3)$$

and $A_V$ is the coefficient of turbulent mixing of momentum in the $z$-direction.

If $\tau$ is independent of $y$, it seems sensible to suppose that the fluid circulation which it sets up is also independent of $y$. If *all* variables are assumed independent of $y$, the linearized horizontal momentum equations are, from (4.5.6a,b),

$$-v = -\frac{\partial p}{\partial x} + \frac{E_V}{2}\mathscr{L}u, \tag{8.3.4a}$$

$$u = \frac{E_V}{2}\mathscr{L}v, \tag{8.3.4b}$$

where the operator $\mathscr{L}$ is

$$\mathscr{L} = \frac{\partial^2}{\partial z^2} + \frac{1}{\lambda}\frac{\partial^2}{\partial x^2}, \tag{8.3.5a}$$

$$\lambda = \frac{A_V}{A_H}\frac{L^2}{D^2}, \tag{8.3.5b}$$

and where $A_H$ is the coefficient of lateral turbulent momentum mixing. The density field is partitioned as in (6.8.1), and if $D/L$ is sufficiently small, the hydrostatic approximation will remain valid throughout, i.e., the vertical equation of motion is simply

$$\frac{\partial p}{\partial z} = -\rho, \tag{8.3.5c}$$

while the continuity equation is

$$\frac{\partial u}{\partial x} + \frac{\partial w}{\partial z} = 0. \tag{8.3.5d}$$

The equation for $\rho$ is (6.8.2), whose linearized form is simply

$$w\left[\frac{1}{\rho_s}\frac{\partial \rho_s}{\partial z}\right] = -H_*\frac{L}{U}. \tag{8.3.6}$$

The transfer of heat, which $H_*$ represents, is modeled as follows. In analogy with the simple model for turbulent mixing of momentum, we write

$$H_* = -\left[\frac{K_V}{D^2}\frac{\partial^2\rho}{\partial z^2} + \frac{K_H}{L^2}\frac{\partial^2\rho}{\partial x^2}\right](\varepsilon F_s), \tag{8.3.7}$$

which, when combined with (8.3.6), yields

$$-wS = \frac{E_V}{2\sigma_V}\frac{\partial^2\rho}{\partial z^2} + \frac{E_V}{2\sigma_H\lambda}\frac{\partial^2\rho}{\partial x^2}, \tag{8.3.8}$$

where, as before,

$$S = \left[-\frac{1}{\rho_s}\frac{\partial \rho_s}{\partial z}g\right]\frac{D^2}{f^2L^2} = \frac{N_s^2 D^2}{f^2L^2}, \tag{8.3.9}$$

and where

$$\sigma_V = \frac{A_V}{K_V},$$

$$\sigma_H = \frac{A_H}{K_H}$$

(8.3.10)

are the ratios of the momentum and thermal transfer coefficients.

At the upper surface of the fluid, an Ekman layer exists to absorb the applied wind stress. This yields a flux of mass in the $x$-direction, which, from the discussion of Section 4.10, is

$$M_E^{(x)} = \frac{E_V^{1/2}}{2} \tau(x),$$

(8.3.11)

whose divergence yields the boundary condition for the fluid beneath the Ekman layer, i.e.,

$$w = \frac{E_V^{1/2}}{2} \frac{\partial \tau}{\partial x}, \qquad z = 1.$$

(8.3.12)

Similarly, the existence of the Ekman layer on the horizontal, rigid surface at the bottom of the fluid layer implies that for the fluid just above the lower Ekman layer,

$$w = \frac{E_V^{1/2}}{2} \frac{\partial v}{\partial x}, \qquad z = 0.$$

(8.3.13)

Thermal conditions must be specified on the boundaries of the fluid, and the following are chosen:

$$\frac{\partial \rho}{\partial z} = 0, \qquad z = 0,$$

(8.3.14a)

$$\rho = \rho_T(x), \qquad z = 1,$$

(8.3.14b)

$$\frac{\partial \rho}{\partial x} = 0, \qquad x = 0.$$

(8.3.14c)

The first of these implies that no vertical flux of heat through the lower boundary is produced by the density anomaly due to the motion. The density is specified at the upper surface, and the coast, at $x = 0$, is considered insulated against lateral heat flux. At the coast both $u$ and $v$ must vanish to satisfy the conditions of no normal flow and no slip. The hydrostatic approximation, however, simplifies the dynamics so completely that the no-slip condition on $w$ cannot be satisfied. In order to do so, an extraordinarily thin layer in which the dynamics is nonhydrostatic and whose width is

$$l_B = \left(\frac{\delta}{\lambda}\right)^{1/2} \frac{E_V^{1/2}}{(\sigma_H S)^{1/4}}$$

(8.3.15)

is required. It can be shown that this layer (whose relative width tends to zero as $\delta \to 0$) carries a negligible fraction of the vertical mass flux and in all other major dynamical respects is utterly inconsequential. To avoid this unnecessary further complexity, the no-slip condition for $w$ can be consistently relaxed.

The problem posed above illustrates two important points. When the pressure field is independent of one horizontal coordinate, the horizontal velocity in the direction in which $p$ varies must be ageostrophic. In the present case, $u$ is driven by the wind stress and is a crucial feature of the circulation, yet *nowhere* can it be in geostrophic balance. Second, the geostrophic flow, in this case in the $y$-direction, will be by far the most intense component of flow, so that to lowest order the flow field is primarily a rectilinear flow in geostrophic balance. The determination of this flow requires, however, the detailed consideration of the relatively weak ageostrophic motion and the dissipation associated with it. The problem formulated here is one example of the circulation problem for rectilinear currents which, in principle, must be solved before the stability considerations of Chapter 7 can be applied. This process, as noted there, is often short-circuited by the arbitrary specification of the geostrophic current structure, while here we illustrate the nature of the required calculation.

Certain further parametric restrictions are required to render the solution of the problem tractable even in this simplified model. Not surprisingly, we are interested primarily in cases where

$$E_V \ll 1, \qquad (8.3.16)$$

i.e., where the scales associated with friction, such as $\delta_E$, are small compared with $L$ and $D$. $L$ is the scale associated with the spatial variation of $\tau_*$, and this scale is typically large compared with the internal deformation radius $L_D$, so that

$$S^{1/2} = \frac{L_D}{L} = \frac{N_s D}{fL} \ll 1. \qquad (8.3.17)$$

The ratios $\sigma_H$ and $\sigma_V$ are difficult to determine, and we will choose the simplest reasonable setting,

$$\sigma_H = \sigma_V = \sigma = O(1). \qquad (8.3.18)$$

Two further restrictions,

$$1 \ll \lambda, \qquad (8.3.19a)$$

$$E_V^{1/2} \ll \lambda \sigma S \ll 1, \qquad (8.3.19b)$$

will ease the subsequent analysis. The first of these may be interpreted as follows. In regions where vertical mixing is important, horizontal mixing will be equally important on scales

$$L_H = \left(\frac{A_H}{A_V}\right)^{1/2} D. \qquad (8.3.20)$$

The parameter $\lambda$, as given in (8.3.5b), measures the ratio $(L/L_H)^2$, and (8.3.19a) simply states that the density anomalies created by the motion mix over a scale which is small compared to $L$. The implications of (8.13.19b) will become clearer as the discussion progresses.

In the interior of the fluid, where the chosen scales aptly describe the motion, it follows from (8.3.4a,b,c) that

$$v_I = \frac{\partial p_I}{\partial x}, \tag{8.3.21a}$$

$$u_I = O(E_V)v, \tag{8.3.21b}$$

$$\rho_I = -\frac{\partial p_I}{\partial z}, \tag{8.3.21c}$$

where the subscript $I$ reminds us that this simplified set of equations is valid in the *interior*. The interior vertical velocity, from (8.3.8), is $O(E_V/\sigma S)\rho_I$. Since $\sigma S$ is small, and $\rho_I$ and $v_I$ are of the same order by the thermal-wind equation, it follows that $w_I$ exceeds $u_I$ by $O(\rho S)^{-1}$, and thus conservation of mass implies that to lowest order

$$\frac{\partial w_I}{\partial z} = 0, \tag{8.3.22}$$

which, with (8.3.12), yields

$$w_I = \frac{E_V^{1/2}}{2} \frac{\partial \tau}{\partial x} \tag{8.3.23}$$

throughout the interior. The interior density field then satisfies

$$\mathscr{L}\rho_I = -\frac{\sigma S}{E_V^{1/2}} \frac{\partial \tau}{\partial x}, \tag{8.3.24}$$

and since $\lambda \gg 1$, this simply becomes

$$\frac{\partial^2 \rho_I}{\partial z^2} = -\frac{\sigma S}{E_V^{1/2}} \frac{\partial \tau}{\partial x}, \tag{8.3.25}$$

whose solution, satisfying (8.3.14a,b), is

$$\rho_I = \frac{1}{2} \frac{\sigma S}{E_V^{1/2}} \frac{\partial \tau}{\partial x} (1 - z^2) + \rho_T(x). \tag{8.3.26}$$

A positive wind-stress curl and the associated interior vertical motion, in the presence of the basic stratification, will induce a positive density anomaly throughout the fluid interior, to which must be added the density produced by the surface heating.

The thermal-wind relation

$$\frac{\partial v_I}{\partial z} = -\frac{\partial \rho_I}{\partial x} \tag{8.3.27}$$

allows $v_I$ to be determined as

$$v_I = -\frac{\sigma S}{2E_V^{1/2}} \frac{\partial^2 \tau}{\partial x^2}\left(z - \frac{z^3}{3}\right) - z\frac{\partial \rho_T}{\partial x} + V_0(x), \qquad (8.3.28)$$

where $V_0(x)$ is an arbitrary barotropic flow, obviously unconstrained by the thermal-wind relationship. However, (8.3.13) implies, with (8.3.23) and (8.3.28), that on $z = 0$

$$w = \frac{E_V^{1/2}}{2}\frac{\partial \tau}{\partial x} = \frac{E_V^{1/2}}{2}\frac{dV_0}{dx}, \qquad (8.3.29)$$

or

$$V_0 = \tau(x) + c, \qquad (8.3.30)$$

where $c$ is an arbitrary constant. Now the total onshore mass flux in the upper Ekman layer is, from (8.3.11), $(E_V^{1/2}/2)\tau(x)$, while the total offshore mass flux in the lower Ekman layer is, from (4.3.31), simply

$$\frac{E_V^{1/2}}{2}V_0 = \frac{E_V^{1/2}}{2}(\tau + c). \qquad (8.3.31)$$

There is a negligible mass flux in the interior, where $u$ is very small, and in order that the total onshore flow may balance the offshore flux, $c$ must vanish, or

$$V_0 = \tau(x). \qquad (8.3.32)$$

Thus the interior flow is completely determined by the properties of the local values of the stress and surface density. This applies only to the region far removed from the boundary at $x = 0$. It is reminiscent of the dynamics of a barotropic fluid and may be understood in this fashion: Although the motion is not barotropic, and a thermally driven current exists, the scale of the motion so much exceeds the deformation radius that the dynamical *constraints* are barotropic, i.e., (8.3.22) is, in fact, a consequence of the Taylor–Proudman theorem, which applies in the interior when $S \ll 1$.

The interior fields do not satisfy the boundary conditions on $x = 0$, i.e., the no-slip condition on $v$ or the insulating condition for $\rho$. More dramatically, as Figure 8.3.2 emphasizes, the circuit of the mass flux is not closed in the interior unless, fortuitously, $\tau(x)$ vanishes at the coast. If $\tau(x)$ vanishes as $x \to -\infty$, for example, the total vertical flux of fluid in the interior is

$$\frac{E_V^{1/2}}{2}\int_{-\infty}^0 \frac{\partial\tau(x)}{\partial x}\,dx = \frac{E_V^{1/2}}{2}\tau(0) \qquad (8.3.33)$$

so that, as shown in the figure, a vertical mass flux equal to $-E_V^{1/2}\tau(0)/2$ must arise in a boundary layer at $x = 0$. The boundary layer must be sufficiently narrow that the dynamics of the layer can significantly depart from the interior dynamics. The Taylor–Proudman constraint must be broken, so that $w$ is no longer directly tied to the local value of $\tau$ by (8.3.23).

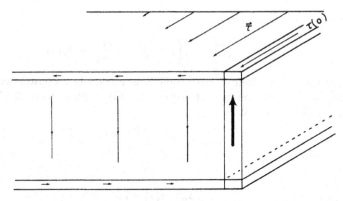

**Figure 8.3.2** A schematic view of the stress-driven circulation. The downwelling in the interior must be compensated by intense upwelling at the boundary.

This requires that (8.3.22) no longer apply, which in turn suggests that *locally* the horizontal scale of the motion be such that $S$, based on *that* scale, is O(1). That is, for both terms in the continuity equation to be of the same order, the horizontal scale of the layer returning the vertical mass flux must be of the order of the deformation radius. We may either, now, start anew and choose a new length scale to nondimensionalize the equations or, equivalently and more economically, investigate this narrow upwelling region by introducing a new boundary-layer variable

$$\xi = -\frac{x}{(\sigma_H S)^{1/2}},$$ (8.3.34)

i.e.,

$$\xi = -\left(\frac{x_*}{L_D}\right)\frac{1}{(\sigma_H)^{1/2}},$$

so that $\xi$ is the $x$-coordinate scaled with the deformation radius. The factor of $\sigma_H$ in (8.3.34) is introduced for convenience in the analysis. To examine the dynamics on this scale only requires that each variable in this boundary layer region be considered a function of $z$ and $\xi$. The magnitude of each field is then determined easily. Since the width of the region is $(\sigma S)^{1/2}$ and the total vertical mass flux over this region must be $O(E_V^{1/2})$, the vertical velocity in this region must be $O(E_V^{1/2}/(\sigma S)^{1/2})$. The continuity equation then implies that $u$ is $O(E_V^{1/2})$ in this region, which, with (8.3.4b), implies that $v$ is $O(\lambda \sigma S/E_V^{1/2})$. The flow in the $y$-direction remains in geostrophic balance, which determines the appropriate scale for $p$ as $\lambda(\sigma S)^{3/2}/E_V^{1/2}$, and the hydrostatic relation fixes this latter magnitude as the scale for $\rho$ as well in this upwelling layer. Hence, in this region

$$w = \left(\frac{E_V}{\sigma S}\right)^{1/2}\tilde{w}(\xi, z),$$ (8.3.35a)

$$u = E_V^{1/2}\tilde{u}(\xi, z),$$ (8.3.35b)

$$v = \frac{\lambda\sigma S}{E_V^{1/2}}\,\tilde{v}(\xi, z),$$ (8.3.35c)

$$p = \frac{\lambda(\sigma S)^{3/2}}{E_V^{1/2}}\,\tilde{p}(\xi, z),$$ (8.3.35d)

$$\rho = \frac{\lambda(\sigma S)^{3/2}}{E_V^{1/2}}\,\tilde{\rho}(\xi, z),$$ (8.3.35e)

where the tilde ˜ labels the variables as appropriate to the layer whose width is the deformation radius. This boundary layer is called the *hydrostatic layer* to distinguish it from the narrow layer described in (8.3.15), where w is as large as in (8.3.35a), but which carries only a negligible fraction of the mass flux, i.e., a relative flux of order

$$\frac{l_B}{(\sigma S)^{1/2}} = \left[\frac{\delta E_V}{\lambda(\sigma S)^{1/2}}\right]^{1/2} \ll 1.$$ (8.3.36)

Substituting (8.3.35) into (8.3.4a,b), (8.3.5c,d), and (8.3.8) yields, using (8.3.34),

$$\tilde{v} = -\frac{\partial\tilde{p}}{\partial\xi} + O(E_V/\lambda\sigma S),$$ (8.3.37a)

$$\tilde{u} = \frac{1}{2}\frac{\partial^2\tilde{v}}{\partial\xi^2} + O(\lambda\sigma S),$$ (8.3.37b)

$$-\tilde{w} = \frac{1}{2}\frac{\partial^2\tilde{\rho}}{\partial\xi^2} + O(\lambda\sigma S),$$ (8.3.37c)

$$-\frac{\partial\tilde{u}}{\partial\xi} + \frac{\partial\tilde{w}}{\partial z} = 0,$$ (8.3.37d)

$$\tilde{\rho} = -\frac{\partial\tilde{p}}{\partial z}.$$ (8.3.37e)

The condition $\lambda\sigma S < 1$ guarantees that the hydrostatic layer is sufficiently narrow so that horizontal diffusion of density and momentum dominates vertical mixing. Note, too, that $\tilde{u}$ is produced entirely by the nongeostrophic mixing of $\tilde{v}$ momentum. A single equation for $\tilde{v}$ may easily be derived by substitution of (8.3.37b) and (8.3.37c) into (8.3.37d) with the subsequent use of the thermal-wind relation

$$\frac{\partial\tilde{v}}{\partial z} = \frac{\partial\tilde{p}}{\partial\xi}$$ (8.3.38)

to yield

$$\frac{\partial}{\partial\xi}\left[\frac{\partial^2\tilde{v}}{\partial\xi^2} + \frac{\partial^2\tilde{v}}{\partial z^2}\right] = 0.$$ (8.3.39)

It is convenient to split each field into its interior representation plus a boundary-layer correction field which must vanish as the boundary-layer region merges with the region exterior to it. For $\tilde{v}$, for example, we write

$$\tilde{v} = v_I(x, z) + \frac{\lambda \sigma S}{E_V^{1/2}} \tilde{v}_c(\xi, z). \tag{8.3.40}$$

Since the dynamical equations are linear and the interior fields separately satisfy the same dynamical equations (although the detail balances differ), the correction field $\tilde{v}_c$ satisfies the same *equations* as $\tilde{v}$. The boundary conditions, however, differ; in particular $\tilde{v}_c$ must vanish as $\xi \to \infty$. This simplification is, of course, the very reason the fields are split as in (8.3.40). Thus

$$\frac{\partial^2 \tilde{v}_c}{\partial \xi^2} + \frac{\partial^2 \tilde{v}_c}{\partial z^2} = 0. \tag{8.3.41}$$

On $z = 1$, the *correction* field sees no applied stress, since $\tau$ is absorbed by the interior portion of each field. Hence, on $z = 1$,

$$\frac{\partial \tilde{v}_c}{\partial z} = 0, \qquad z = 1, \tag{8.3.42}$$

while at $z = 0$, (8.3.13) yields, after use is made of (8.3.35),

$$\tilde{w}_c = -\frac{\lambda \sigma S}{E_V^{1/2}} \frac{\partial \tilde{v}_c}{\partial \xi}, \qquad z = 0. \tag{8.3.43}$$

We now apply the condition (8.3.19b) that $\lambda \sigma S \gg E_V^{1/2}$ and observe that its consequence is that to lowest order

$$\tilde{v}_c = 0, \qquad z = 0. \tag{8.3.44}$$

The stratification is sufficiently large that the vertical velocity pumped out of the lower Ekman layer by a $v = O(\lambda \sigma S/E_V^{1/2})$, i.e., an Ekman $w$ of $O(\lambda(\sigma S)^{1/2})$, is far too large to rise against the ambient vertical density gradient, which has limited $w$ to $O(E_V^{1/2}/(\sigma S)^{1/2})$. The ratio of these vertical velocities is

$$\frac{\text{Ekman pumping}}{\text{Boundary-layer vertical velocity}} = O\left(\frac{\lambda \sigma S}{E_V^{1/2}}\right). \tag{8.3.45}$$

For $\lambda \sigma S/E_V^{1/2} \gg 1$ an apparent mismatch exists between what will be pumped out of the Ekman layer and what can be accepted into the stratified fluid above the Ekman layer.* To resolve this mismatch the velocity in the $y$-direction must adjust itself so that to $O(\sigma S/E_V^{1/2})$ it vanishes at $z = 0$, so that no $w$ greater than $O(E_V^{1/2}/(\sigma S)^{1/2})$ is pumped out of the lower Ekman layer. The presence of such substantial stratification obliges the horizontal

---

* The validity of (8.3.13) applied to the intersecting of the hydrostatic layer wth the Ekman layer requires that in this intersecting region $E_V \, \partial^2/\partial z^2 \gg E_V/\lambda \, \partial^2/\partial x^2$. Since the scale in $z$ is $O(E_V^{1/2})$ and the scale in $x$ is $O((\sigma S)^{1/2})$, this requires only that $\lambda \sigma S \gg E_V$.

velocity to satisfy the no-slip condition at the lower boundary *without* the aid of the Ekman layer. This constraint is produced by the presence of stratification and is satisfied with the aid of the vertical shear of $\tilde{v}_c$, which is, in turn, only possible in the presence of stratification. The solution of (8.3.41) which vanishes at infinity and which satisfies (8.3.42) and (8.3.44) is

$$\tilde{v}_c = \sum_{n=0}^{\infty} V_n e^{-\mu_n \xi} \sin \mu_n z, \qquad \mu_n = (n + \tfrac{1}{2})\pi, \qquad (8.3.46a)$$

in terms of which

$$\tilde{p}_c = -\sum_{n=0}^{\infty} V_n e^{-\mu_n \xi} \cos \mu_n z, \qquad (8.3.46b)$$

$$\tilde{u}_c = \frac{1}{2} \sum_{n=0}^{\infty} V_n \mu_n^2 e^{-\mu_n \xi} \sin \mu_n z, \qquad (8.3.46c)$$

$$\tilde{w}_c = \frac{1}{2} \sum_{n=0}^{\infty} V_n \mu_n^2 e^{-\mu_n \xi} \cos \mu_n z. \qquad (8.3.46d)$$

The total vertical mass flux in this hydrostatic layer is

$$E_V^{1/2} \int_0^{\infty} \tilde{w}_c \, d\xi = \frac{E_V^{1/2}}{2} \sum_{n=0}^{\infty} V_n \mu_n \cos \mu_n z \qquad (8.3.47a)$$

$$= \frac{E_V^{1/2}}{2} \frac{\partial \tilde{p}_c}{\partial \xi}(0, z). \qquad (8.3.47b)$$

The constants $V_n$ must be determined by matching to the boundary. Now, in the hydrostatic layer, $\tilde{v}_c$ far exceeds $v_I$ in magnitude. This is a consequence of the strong density gradients produced by the vertical motion. However, $v$ must satisfy the no-slip condition on $\xi = x = 0$. If the hydrostatic layer fields were the sole correction fields to the interior fields, then $\tilde{v}_c$ would have to vanish on $\xi = 0$; this would imply that each $V_n$ is zero, and the layer would be annihilated. This perplexity is resolved by recalling that in the interior, *vertical* mixing balances vertical motion, while in the hydrostatic layer, the density anomaly produced by vertical motion is balanced by only lateral mixing. A buffer region between the interior and hydrostatic layer is required wherein vertical and horizontal diffusion are equally important. As (8.3.20) shows, this region has a horizontal scale $L_H$, so that, based on this scale, $\lambda$ is one. In the language of boundary-layer theory, this means that between the hydrostatic layer and the interior an additional boundary layer is required whose horizontal scale is such that the dynamic fields in this region are functions of $z$ and the boundary-layer variable

$$\eta = -x\lambda^{1/2} = -\frac{x_*}{L_H}. \qquad (8.3.48)$$

In this region, between the hydrostatic layer and the interior, the dynamical fields may be written

$$\bar{v} = v_I(x, z) + \frac{\lambda \sigma S}{E_V^{1/2}} \bar{v}_c(\eta, z), \tag{8.3.49a}$$

$$\bar{\rho} = \rho_I(x, z) + \frac{\lambda^{1/2} \sigma S}{E_V^{1/2}} \bar{\rho}_c(\eta, z), \tag{8.3.49b}$$

$$\bar{p} = p_I + \lambda^{1/2} \frac{\sigma S}{E_V^{1/2}} \bar{p}_c, \tag{8.3.49c}$$

$$u = u_I + (\lambda \sigma S) E_V^{1/2} \bar{u}_c, \tag{8.3.49d}$$

$$w = w_I + (\lambda \sigma S) \lambda^{1/2} E_V^{1/2} \bar{w}_c, \tag{8.3.49e}$$

where the notation $\bar{\phantom{x}}_c$ refers to the correction fields that must be added to the interior fields to represent $v$, $\rho$, $p$, etc. in this region. Of course, in this region, where $\eta$ is $O(1)$,

$$\zeta = -\frac{x}{(\sigma S)^{1/2}} = \frac{\eta}{(\lambda \sigma S)^{1/2}} \tag{8.3.50}$$

is very large, so the hydrostatic-layer correction fields are negligibly small. The condition $\lambda \sigma S \ll 1$ implies the diffusion layer is far broader than the hydrostatic layer. The amplitude of the correction for $v$ is determined by the anticipation that as $x \to 0$, the *total* field for $v$ will be, smack up against $\lambda = 0$,

$$v = v_I(0, z) + \frac{\lambda \sigma S}{E_V^{1/2}} \bar{v}_c(0, z) + \frac{\lambda \sigma S}{E_V^{1/2}} \tilde{v}_c(0, z) \tag{8.3.51}$$

so that $\bar{v}_c$ may be used *with* $\tilde{v}_c$ to match the no-slip condition on $v$. The size of the other fields are then determined by the equations of motion when written in terms of $\eta$ and $z$. It follows by substitution of (8.3.49) into the basic equations that

$$\bar{v}_c = -\frac{\partial \bar{p}_c}{\partial \eta} + O(E_V^2), \tag{8.3.52a}$$

$$\bar{\rho}_c = -\frac{\partial \bar{p}_c}{\partial z}, \tag{8.3.52b}$$

$$-\frac{\partial \bar{u}_c}{\partial \eta} + \frac{\partial \bar{w}_c}{\partial z} = 0, \tag{8.3.52c}$$

$$\bar{u}_c = \frac{1}{2} \left( \frac{\partial^2 \bar{v}_c}{\partial \eta^2} + \frac{\partial^2 \bar{v}_c}{\partial z^2} \right), \tag{8.3.52d}$$

$$-2\lambda \sigma S \bar{w}_c = \frac{\partial^2 \bar{\rho}_c}{\partial \eta^2} + \frac{\partial^2 \bar{\rho}_c}{\partial z^2}. \tag{8.3.52e}$$

The condition $\lambda \sigma S \ll 1$ implies that to lowest order

$$\frac{\partial^2 \bar{\rho}_c}{\partial \eta^2} + \frac{\partial^2 \bar{\rho}_c}{\partial z^2} = 0. \tag{8.3.53}$$

The solutions of (8.3.53) which vanish for large $\eta$ and satisfy the *homogeneous* conditions*

$$\frac{\partial \bar{\rho}_c}{\partial z} = 0, \qquad z = 0,$$

$$\bar{\rho}_c = 0, \qquad z = 1, \tag{8.3.54}$$

are

$$\bar{\rho}_c = \sum_{n=0}^{\infty} R_n e^{-\mu_n \eta} \cos \mu_n z, \qquad \mu_n = (n + \tfrac{1}{2})\pi, \tag{8.3.55}$$

in terms of which

$$\bar{v}_c = - \sum_{n=0}^{\infty} R_n e^{-\mu_n \eta} \sin \mu_n z. \tag{8.3.56}$$

The latter expression is obtained from the thermal-wind equation supplemented by the condition, valid here as in the hydrostatic layer, that $\bar{v}_c$ must vanish at $z = 0$ to ensure a proper match between the weak vertical velocity and the Ekman pumping out of the lower Ekman layer.

The representations for the total fields for $v$ and $\rho$ consist of the interior fields plus the correction in the diffusion layer plus the correction fields required in the hydrostatic layer, i.e.,

$$v = v_I(x, z) + \frac{\lambda \sigma S}{E_V^{1/2}} \sum_{n=0}^{\infty} [V_n e^{-\mu_n \xi} - R_n e^{-\mu_n \eta}] \sin \mu_n z$$

$$\rho = \rho_I(x, z) + \frac{\lambda \sigma S}{E_V^{1/2}} \sum_{n=0}^{\infty} \left[ -V_n(\sigma S)^{1/2} e^{-\mu_n \xi} + \frac{R_n}{\lambda^{1/2}} e^{-\mu_n \eta} \right] \cos \mu_n z. \tag{8.3.57}$$

To lowest order (i.e., to $O(\lambda \sigma S / E_V^{1/2})$) the condition of no slip on $v$ is satisfied by

$$R_n = V_n. \tag{8.3.58}$$

Since

$$\frac{\partial \rho}{\partial x} = \frac{\partial \rho_I}{\partial x} - \frac{\lambda \sigma S}{E_V^{1/2}} \sum_{n=0}^{\infty} [V_n e^{-\mu_n \xi} - R_n e^{-\xi_n \eta}] \cos \mu_n z, \tag{8.3.59}$$

the condition that $\partial \rho / \partial x$ vanishes on $x = \eta = \xi = 0$ is satisfied, to the same order, by the condition (8.3.58). This follows directly from the fact that $v$ is

---

* The interior density field, of course, satisfies the inhomogeneous boundary conditions (8.3.14a,b).

geostrophic; hence if $v$ vanishes on $x = 0$ for all $z$, then $\partial v / \partial z$ and hence $\partial \rho / \partial x$ will also vanish as a consequence of the thermal-wind relation.

The constants $V_n$ are determined as follows. The vertical mass flux in the diffusion layer is $O(\lambda \sigma S) E_V^{1/2}$ and hence inconsequential. The vertical mass flux in the interior is $O(E_V^{1/2})$ and must be balanced by the mass flux in the hydrostatic layer. The vertical mass flux in the interior is furthermore independent of $z$, so the same must be true for the hydrostatic layer. The hydrostatic layer is fed by the mass flux in the lower Ekman layer as shown in Figure 8.3.3. The magnitude of this flux is $-(E_V^{1/2}/2)\tau(0)$ and is

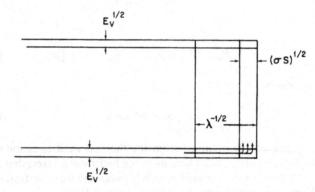

**Figure 8.3.3** The boundary-layer regions and a sketch of the mass flow in the lower corner.

equal to the mass flux exiting at the top into the upper Ekman layer to flow offshore, if, as in Figure 8.3.2, $\tau(0)$ is negative. Thus for all $z$, from (8.3.47)

$$E_V^{1/2} \int_0^\infty \tilde{w}_c \, d\xi = \frac{E_V^{1/2}}{2} \sum_{n=0}^\infty V_n \mu_n \cos \mu_n z = -\frac{E_V^{1/2}}{2} \tau(0), \qquad (8.3.60)$$

or using the orthogonality relations of the cos $\mu_n z$ on the interval $(0, 1)$, the Fourier series in (8.3.60) may be inverted to obtain $V_n$ as

$$V_n = -2\tau(0) \frac{(-1)^n}{\mu_n^2}. \qquad (8.3.61)$$

Thus

$$v = v_I(x, z) - \frac{2\lambda \sigma S}{E_V^{1/2}} \tau(0) \sum_{n=0}^\infty \{e^{-\mu_n \xi} - e^{-\mu_n \eta}\} \frac{(-1)^n}{\mu_n^2} \sin \mu_n z, \qquad (8.3.62)$$

while, ignoring the small correction in the diffusion layer,

$$w = w_I(x, z) - \frac{E_V^{1/2}}{(\sigma S)^{1/2}} \tau(0) \sum_{n=0}^\infty (-1)^n e^{-\mu_n \xi} \cos \mu_n z. \qquad (8.3.63)$$

Repeated use of the identity

$$\sum_{n=0}^{\infty} (-1)^n e^{-n\theta} = \frac{1}{1 + e^{-\theta}}, \qquad (8.3.64)$$

whose validity requires only that Re $\theta > 0$, allows the series in (8.3.63) to be summed to yield, with (8.3.23),

$$w = \frac{E_V^{1/2}}{2} \frac{\partial \tau}{\partial x} - \frac{E_V^{1/2}}{(\sigma S)^{1/2}} \frac{\tau(0)}{2} \frac{\cos \pi z/2 \cosh \pi \xi/2}{\sinh^2 \pi \xi/2 + \cos^2 \pi z/2}. \qquad (8.3.65)$$

If $\tau(0)$ is negative the surface stress forces the flow in the upper Ekman layer offshore. This offshore mass flux is fed by strong vertical motion within a deformation radius of the boundary. The cross-stream profile of $w$, which depends only weakly on the turbulent mixing coefficients through the ratio $\sigma_H = (A_H/K_H)^{1/2}$, is shown in Figure 8.3.4. As the upper surface is approached, the profile of $w$ becomes sharper and narrower until it enters, as in

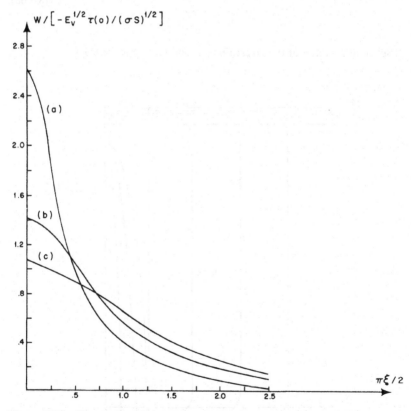

**Figure 8.3.4**  The profile of the upwelling, vertical velocity (a) at $z = 0.75$, (b) at $z = 0.5$, (c) at $z = 0.25$.

the homogeneous counterpart of Section 5.11, an extremely narrow corner region whose width is $O((A_H/fL^2)^{1/2})$ in our nondimensional units. The ageostrophic onshore flow may be calculated directly from (8.3.65), and the continuity equation and is given by

$$u = -E_V^{1/2}\frac{\tau(0)}{4}\frac{\sin \pi z \sinh \pi \xi/2}{\cos^2 \pi z/2 + \sinh^2 \pi \xi/L}. \qquad (8.3.66)$$

The circulation in the vertical plane may be represented equally well in terms of the stream function

$$\chi = \frac{E_V^{1/2}}{2}\left[\tau(x) + \tau(0)\frac{2}{\pi}\tan^{-1}\left(\frac{\sinh \pi \xi/2}{\cos \pi z/2}\right) - \tau(0)\right], \qquad (8.3.67)$$

from which

$$w = \frac{\partial \chi}{\partial x},$$

$$(8.3.68)$$

$$u = -\frac{\partial \chi}{\partial z}.$$

A schematic sketch of the streamlines is shown in Figure 8.3.5.

$$\odot \ \vec{\tau}$$

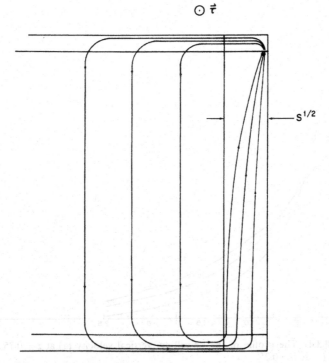

**Figure 8.3.5**   A sketch of the streamlines of the steady circulation in the $x, z$ plane.

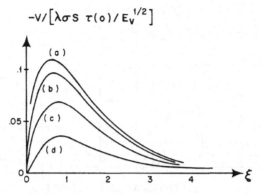

**Figure 8.3.6** The profile of the longshore velocity: (a) $z = 1.0$, (b) $z = 0.75$, (c) $z = 0.5$, (d) $z = 0.25$, for $\sigma = 1$, $\lambda S = 0.5$ (Allen 1973).

The thermal anomaly produced by the vertical motion produces, in turn, a strong $(O(\lambda \sigma S/E_V^{1/2}))$ current along the coast in the direction of the stress. Figure 8.3.6 shows the profile of this current as calculated by Allen (1973) for the case $\lambda S = 0.5$. The current along the boundary is in geostrophic balance. Its lateral extent is determined by $L_H$ and so depends, unwholesomely, on the ratios of $A_V$ to $A_H$, neither of which is easily prescribed. It is important to keep in mind that this strong, rectilinear, geostrophic current is produced by the weak ageostrophic circulation in the vertical plane and this circulation picks out its own narrow scale, in this case the deformation radius, for which the ageostrophic flow dynamics is consistent with the maintenance of the geostrophic flow. The wider diffusion layer does *not* yield any important ageostrophic flow in the x-direction. The flow remains fundamentally in geostrophic balance until the narrow upwelling layer is entered, in which $v$ remains geostrophic but in which a significant ageostrophic u-field is produced.

## 8.4 The Theory of Frontogenesis

Zones of sharp horizontal temperature change, called fronts are conspicuous features on almost every surface weather map. The spatial scale of variation of a typical front is so much less than the scale associated with the large-scale waves in the general circulation that they are usually represented on weather maps as though they were lateral discontinuities in temperature. The transition in temperature is, of course, continuous, as shown in Figure 8.4.1, but the scale of the front ($\leq 100$ km) and the size of the change in temperature across the front ($\sim 15°C$) ensure that the thermal wind associated with the front must correspond to a substantial cross-front Rossby number.

The frontal region slopes with height in the atmosphere, the cold, dense air lying beneath the warmer, lighter air. The slope of the zone ranges from 1/300 to 1/50, in contrast to the slope of a typical surface of potential

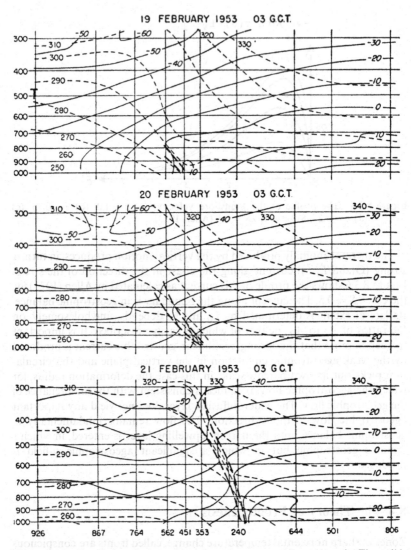

**Figure 8.4.1** Successive cross sections through a zone of frontogenesis. The solid lines are lines of constant temperature. Dashed lines are isolines of potential temperature (Petterssen 1956).

temperature in the atmosphere, which is 1/1000. Typical shears of the horizontal wind are O(30 m/s) in 100 km, so that

$$\varepsilon_l \approx \frac{30 \text{ m/s}}{10^{-4} \text{s}^{-1} \times 10^5 \text{ m}} = 3, \qquad (8.4.1)$$

and therefore ageostrophic effects must be important in the dynamics of fronts. The existence of such sharp zones embedded within the more smoothly varying atmospheric circulation naturally poses a problem which has long intrigued atmospheric scientists. In addition we must note the intense practical importance of being able to predict the strength, location, and movement of fronts for the purpose of weather prediction. The observation (Voorhis 1969) of zones of a similar nature in the ocean suggests that the mechanism for the formation of fronts is fundamentally independent of dynamical features which are special to either the atmosphere or ocean.

Early attempts (Bergeron 1928, Stone 1966) to explain the formation of fronts focused on a class of horizontal velocity fields called *deformation fields*, the streamlines for which are shown in Figure 8.4.2. Such deformation

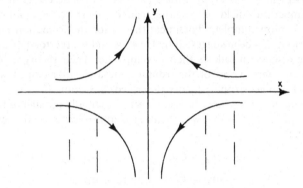

**Figure 8.4.2**    The streamlines (solid curve) of a confluent deformation field will tend to pack together isotherms (dashed lines) which lie athwart the axis of confluence (*x*-axis).

fields, which are *local* features of large-scale horizontal wave motions, will clearly tend to concentrate a large-scale preexisting temperature gradient, squeezing the isotherms together. A simple case is shown in the figure. Let the (dimensional) deformation field be described by

$$u_* = -\alpha x_* ,$$

$$v_* = \alpha y_* ,$$

$$(8.4.2)$$

where $\alpha$ is a constant with the dimensions (time)$^{-1}$. If the potential-temperature or density field is initially oriented so that isotherms are parallel to the $y$-axis, then at the ground, where $w$ vanishes, the potential temperature, if conserved, must satisfy

$$\frac{\partial \theta_*}{\partial t_*} = -u_* \frac{\partial \theta_*}{\partial x_*} = \alpha x_* \frac{\partial \theta_*}{\partial x_*} , \qquad (8.4.3)$$

since $\theta_*$ is independent of $y$. The solution of (8.4.3) is easily verified to be

$$\theta_* = \theta_0(x_* e^{\alpha t_*}), \qquad (8.4.4)$$

where $\theta_0(x_*)$ is the surface distribution of potential temperature at the initial instant. The temperature gradient at the surface is

$$\frac{\partial \theta_*}{\partial x_*} = e^{\alpha t_*} \theta_0'(x_* e^{\alpha t_*}), \qquad (8.4.5)$$

where $\theta_0'$ is the derivative of $\theta_0$ with respect to its argument. The *surface* temperature gradient, therefore, increases exponentially in time in regions of confluence in the deformation field. Naturally, as time goes by and the temperature field changes as a result of this confluence, the changing thermal wind in the $y$-direction will produce, by Coriolis accelerations, flow in the $x$-direction to alter the initial deformation velocity field. Thus the solution (8.4.4) will be valid only initially, and even then describes the structure of the temperature field only at the ground, where vertical motion is prohibited. A more complete dynamical theory for the formation process, or frontogenesis, has been suggested by Hoskins and Bretherton (1972), and the following discussion follows their development. Their theory explicitly recognizes the narrowness of the frontal zone and is developed within the context of semigeostrophic theory described in Section 8.1.

Consider the case where the front forms essentially parallel to the $y$-axis. Then, as in (8.1.), the following scaling is appropriate for the length and velocity fields:

$$\begin{aligned}
u_* &= Uu, & x_* &= lx, \\
v_* &= Vv, & y_* &= Ly, \\
w_* &= U\frac{D}{l}w, & z_* &= D, \\
& & t_* &= \frac{t}{\sigma}.
\end{aligned} \qquad (8.4.6)$$

The existence of fronts in the ocean, and their relatively shallow extent in the atmosphere, suggests that a theory in which the density scale height is much greater than the motion scale $D$ will be adequate to illuminate the basic physical mechanism. The density can therefore be scaled as

$$\rho_* = \bar{\rho} + \bar{\rho}\frac{D}{H}\rho(x, y, z, t), \qquad (8.4.7a)$$

where $\bar{\rho}$ is a constant and $H$ is a characteristic value for the density scale height, $((-1/\rho_*)\partial\rho_*/\partial z_*)^{-1}$. The pressure is scaled as in (8.1.6), i.e.,

$$p_* = -\bar{\rho}gz_* + \bar{\rho}fULp(x, y, z, t). \qquad (8.4.7b)$$

If $\alpha$ typifies the rate of confluence of the initial deformation field, it is appropriate to choose the scaling parameters such that

$$\frac{U}{l} = \frac{V}{L} = \sigma = \alpha. \qquad (8.4.8)$$

The deformation radius is

$$L_D = \frac{D}{f}\left(\frac{-g}{\rho_*}\frac{\partial\rho_*}{\partial z_*}\right)^{1/2}. \tag{8.4.9}$$

We choose $L_D$ as an appropriate scale against which to measure the frontal scale. Thus, we choose

$$l = L_D = \frac{D}{f}\left(\frac{g}{H}\right)^{1/2}. \tag{8.4.10}$$

This is the scale the front will achieve in a completely quasigeostrophic theory, i.e., it is the natural scale for the deformation of the density surfaces. We will be particularly interested in examining the scale of the solution of the semigeostrophic problem in contrast. The length scale $L$ is considered large compared with $l$, and except for this constraint, its precise magnitude is unimportant. Thus we have set the stage for the consideration of the dynamics of a flow which is originally mildly narrow in the $x$-direction, (possessing the deformation radius as its scale there), but extensive in the $y$-direction. The fundamental problem is whether the kinematic confluence described earlier will naturally sharpen a mildly varying temperature field to frontal dimensions.

If the scaled variables are substituted into the equations of motion and terms of $O((l/L)^2)$ are neglected, we obtain the semigeostrophic set

$$v = \frac{\partial p}{\partial x}, \tag{8.4.11a}$$

$$\varepsilon_l\left[\frac{\partial v}{\partial t} + u\frac{\partial v}{\partial x} + v\frac{\partial v}{\partial y} + w\frac{\partial v}{\partial z}\right] + u = -\frac{\partial p}{\partial y}, \tag{8.4.11b}$$

$$\rho = -\varepsilon_l\frac{\partial p}{\partial z}, \tag{8.4.11c}$$

$$\left[\frac{\partial \rho}{\partial t} + u\frac{\partial \rho}{\partial x} + v\frac{\partial \rho}{\partial y} + w\frac{\partial \rho}{\partial z}\right] = 0, \tag{8.4.11d}$$

$$\frac{\partial u}{\partial x} + \frac{\partial v}{\partial y} + \frac{\partial w}{\partial z} = 0, \tag{8.4.11e}$$

where

$$\varepsilon_l = \frac{V}{fl} = \frac{UL}{fl\,l} = \frac{\alpha L}{f\,l}$$

may be considered an $O(1)$ parameter. For atmospheric motions with $D/H \ll 1$, the density $\rho$ may be replaced in the dynamical equations by *minus* the potential temperature as explained in Section 6.8.

Let $\omega$ be the vertical component of the absolute vorticity, i.e.,

$$\omega = 1 + \varepsilon_l\frac{\partial v}{\partial x}. \tag{8.4.12}$$

Then the vorticity equation, derived by taking an $x$-derivative of (8.4.11b), is

$$\frac{d\omega}{dt} = \frac{\partial\omega}{\partial t} + u\frac{\partial\omega}{\partial x} + v\frac{\partial\omega}{\partial y} + w\frac{\partial\omega}{\partial z}$$

$$= -\omega\left(\frac{\partial u}{\partial x} + \frac{\partial v}{\partial y}\right) - \varepsilon_l\frac{\partial w}{\partial x}\frac{\partial v}{\partial z}. \tag{8.4.13}$$

The first term on the right-hand side is clearly the convergence of preexisting absolute-vorticity filaments, while the second term represents the production of a vertical component of vorticity by the tilting of the horizontal vorticity, $\partial v/\partial z$, by the vertical velocity. This term, $O(\varepsilon_l)$, is absent in quasigeostrophic theory for which $\varepsilon_l$ is small. It is left as an exercise for the reader to show that the statement of conservation of potential vorticity here takes the form

$$\frac{d}{dt}\Pi = 0, \tag{8.4.14a}$$

where

$$\Pi = -\omega\frac{\partial\rho}{\partial z} + \varepsilon_l\frac{\partial v}{\partial z}\frac{\partial\rho}{\partial x}, \tag{8.4.14b}$$

so that with the thermal-wind equation

$$\varepsilon_l\frac{\partial v}{\partial z} = -\frac{\partial\rho}{\partial x},$$

it follows that

$$\Pi = -\omega\frac{\partial\rho}{\partial z} - \varepsilon_l^2\left(\frac{\partial v}{\partial z}\right)^2. \tag{8.4.14c}$$

Consider now the deformation field imposed, say, by some larger-scale motion system,

$$w = 0, \tag{8.4.15a}$$

$$u = -x, \tag{8.4.15b}$$

$$v = y, \tag{8.4.15c}$$

which will be a solution of (8.4.11a,b,c) if

$$\rho = \rho_0(z), \tag{8.4.15d}$$

$$p = xy - \frac{\varepsilon_l y^2}{2} + p_0(z), \tag{8.4.15e}$$

where

$$\varepsilon_l\frac{\partial p_0}{\partial z} = -\rho_0. \tag{8.4.16}$$

Note that for this field $\omega$ is unity (i.e., there is no relative vorticity) and $\Pi$ is simply $-\partial\rho_0/\partial z$. At this point we add to this field of motion a density or temperature field with initially weak horizontal gradients, whose isolines are parallel to the $y$-axis. Thus now the *total* density field is

$$\rho = \rho'(x, z, t). \tag{8.4.17a}$$

The associated velocity field will also change, and provisionally we assume that the changes in $u$, $v$, $w$, and $p$, denoted by primes, are *independent* of $y$, an assumption whose validity must later be verified. Thus

$$w = w'(x, z, t), \tag{8.4.17b}$$

$$u = -x + u'(x, z, t), \tag{8.4.17c}$$

$$v = y + v'(x, z, t), \tag{8.4.17d}$$

$$p = xy - \frac{\varepsilon_l y^2}{2} + p'(x, z, t), \tag{8.4.17e}$$

which, when substituted into (8.4.11a,b,c,d,e), yields the equations of motion for the prime variables which represent the response of the fluid to the deformation field acting on the horizontal density gradient, i.e.,

$$v' = \frac{\partial p'}{\partial x}, \tag{8.4.18a}$$

$$\rho' = -\varepsilon_l \frac{\partial p'}{\partial z}, \tag{8.4.18b}$$

$$\varepsilon_l \left[ \frac{\partial v'}{\partial t} + (u' - x) \frac{\partial v'}{\partial x} + v' + w' \frac{\partial v'}{\partial z} \right] + u' = 0, \tag{8.4.18c}$$

$$\frac{\partial \rho'}{\partial t} + (u' - x) \frac{\partial \rho'}{\partial x} + w' \frac{\partial \rho'}{\partial z} = 0, \tag{8.4.18d}$$

$$\frac{\partial u'}{\partial x} + \frac{\partial w'}{\partial z} = 0. \tag{8.4.18e}$$

We first of all note that solutions which are independent of $y$ are consistent with (8.4.18), i.e., only the confluent part $u$ of the original deformation field leads to spatially varying coefficients in (8.4.18). Furthermore, since $\partial p'/\partial y$ is zero, $u'$ is entirely ageostrophic. The advection of the density field in (8.4.18d) by the geostrophic $u$ due to the deformation field would, *alone*, produce a surface distribution of $\rho$ similar to (8.4.4). The ageostrophic $u'$ is produced by the acceleration of $v'$ (which is geostrophic), and this will alter the frontogenetic process at the surface. This ageostrophic motion in the vertical plane may be described by the stream function $\chi(x, z)$, since from (8.4.18e)

$$w' = -\frac{\partial \chi}{\partial x}, \tag{8.4.19a}$$

$$u' = \frac{\partial \chi}{\partial z}. \tag{8.4.19b}$$

An equation for $\chi$ may be derived by differentiating (8.4.18c) with respect to $z$ and (8.4.18d) with respect to $x$ and then using the thermal-wind equation and (8.4.19a,b) to obtain

$$
\frac{\partial}{\partial z}\left[\omega\frac{\partial\chi}{\partial z}\right] - \frac{\partial}{\partial z}\left[\frac{\partial\chi}{\partial x}\varepsilon_l v_z'\right] - \frac{\partial}{\partial x}\left[\frac{\partial\chi}{\partial z}\varepsilon_l\frac{\partial v'}{\partial z}\right]
$$
$$
+ \frac{\partial}{\partial x}\left[\frac{\partial\chi}{\partial x}\left(-\frac{\partial\rho'}{\partial z}\right)\right] = 2\frac{\partial\rho'}{\partial x}, \tag{8.4.20}
$$

where

$$
\omega = 1 + \varepsilon_l\frac{\partial v'}{\partial x}. \tag{8.4.21}
$$

Thus, if at any instant $\rho'$ and $v'$ are known, (8.4.20) can be solved for $\chi$, in terms of which $u'$ and $w'$ are known. Then, (8.4.18c,d) can be stepped forward in time to calculate the new field of $\rho'$ and $v'$, and (8.4.20) could be used again to calculate $\chi$. The process could be continued indefinitely to yield the solution in this manner, although it is not the method of solution to be described. Note, though, that the structure of the problem for $\chi$ depends on the nature of $v'$ and $\rho'$, and these fields evolve with time. Hence, the circulation in vertical planes across the frontal region alters in a fundamentally nonlinear fashion as the geostrophic, hydrostatic fields, $v'$ and $\rho'$, change. *Initially*, when $\omega \sim 1$ and $v'$ is small, the problem for $\chi$ is

$$
\frac{\partial^2\chi}{\partial z^2} + \frac{\partial^2\chi}{\partial x^2}\left(-\frac{\partial\rho_0}{\partial z}\right) = 2\frac{\partial\rho'}{\partial x}. \tag{8.4.22}
$$

If $\partial\rho_0/\partial z$ is $-1$, i.e., if the initial stratification is uniform, (8.4.22) is simply an inhomogeneous form of Laplace's equation. The form of (8.4.20) will change as $v'$ and $\rho'$ change, but qualitatively (8.4.20) will remain the same. This notion is made more precise by considering the *characteristic coordinates* of (8.4.20). A partial differential equation of the form

$$
A\frac{\partial^2\chi}{\partial z^2} + 2B\frac{\partial^2\chi}{\partial x\,\partial z} + C\frac{\partial^2\chi}{\partial x^2} + G\left(\chi, \frac{\partial\chi}{\partial z}, \frac{\partial\chi}{\partial x}, x, z\right) = 0, \tag{8.4.23}
$$

where $A$, $B$, and $C$ are independent of the second derivatives of $\chi$ with respect to $x$ and $z$, possesses an intrinsic coordinate frame whose characteristic curves in $x$, $z$ space are determined from the differential relations (e.g., Sommerfeld 1949, Chapter II)

$$
\frac{dx}{dz} = +\frac{B}{A} \pm \frac{1}{A}[B^2 - AC]^{1/2}. \tag{8.4.24}
$$

In the present case we have for the characteristic curves,

$$
\frac{dx}{dz} = -\frac{\varepsilon_l}{\omega}\frac{\partial v'}{\partial z} \pm \left[\omega\frac{\partial\rho'}{\partial z} + \varepsilon_l^2\left(\frac{\partial v}{\partial z}\right)^2\right]^{1/2}, \tag{8.4.25a}
$$

or

$$\frac{dx}{dz} = -\frac{\varepsilon_l}{\omega}\frac{\partial v'}{\partial z} \pm [-\Pi]^{1/2}. \qquad (8.4.25b)$$

At $t = 0$, $\Pi$ is positive for every fluid element, and since $\Pi$ is conserved for each fluid element, $\Pi$ must remain everywhere positive. Thus the characteristic curves are everywhere complex, which implies that (8.4.20) remains, for all time, an *elliptic* partial differential equation, of which type Laplace's equation is the archetype. This qualitative result ensures that, as in the case of Laplace's equation, all singularities and all maxima and minima must occur on the boundaries of the region. In particular, we therefore can anticipate that the maximum temperature gradient will occur on the horizontal boundary.

Consider the case where

$$\frac{\partial \rho_0}{\partial z} = -1.$$

Then $\Pi$ is initially and subsequently equal to unity throughout the domain of the flow. In this case (8.4.25b) can be integrated directly to obtain the characteristic curves

$$x = iz - \varepsilon_l v' + a_1, \qquad (8.4.26a)$$

$$x = -iz - \varepsilon_l v' + a_2, \qquad (8.4.26b)$$

where $a_1$ and $a_2$ are arbitrary constants. This, in turn, suggests the consideration of new real coordinates obtained by addition and subtraction of (8.14.26a,b), viz.,

$$\xi = x + \varepsilon_l v', \qquad (8.4.27a)$$

$$Z = z. \qquad (8.4.27b)$$

This transformation was suggested by Hoskins and Bretherton (1972), who noted that for arbitrary initial $\Pi$, $\xi$ satisfied the simple relation

$$\frac{d\xi}{dt} = u + \varepsilon_l \frac{dv'}{dt}$$

$$= -x + u' + \varepsilon_l \frac{dv'}{dt} \qquad (8.4.28)$$

$$= -x - \varepsilon_l v'$$

$$= -\xi.$$

Thus following each fluid element,

$$\xi = \xi_0 e^{-t}, \qquad (8.4.29)$$

where $\xi_0$ is the value of $x$ the element had at $t = 0$. $\xi$ is therefore the $x$-position of the fluid element would have if it had moved with the geostrophic, confluent velocity. Consider any dependent variable $Q$ as a function of $x$, $z$, and $t$ through $\xi$, $Z$, and $T$, where $T = t$. A simple application of the chain rule for differentiation yields

$$\frac{\partial Q}{\partial x} = \omega \frac{\partial Q}{\partial \xi},$$

(8.4.30a)

$$\frac{\partial Q}{\partial z} = \frac{\partial Q}{\partial Z} + \varepsilon_l \frac{\partial v'}{\partial z} \frac{\partial Q}{\partial \xi},$$

(8.4.30b)

$$\frac{\partial Q}{\partial t} = \frac{\partial Q}{\partial T} + \varepsilon_l \frac{\partial v'}{\partial t} \frac{\partial Q}{\partial \xi},$$

(8.4.30c)

and, in particular,

$$\frac{\partial v'}{\partial z} = \omega \frac{\partial v'}{\partial Z},$$

(8.4.31a)

$$\omega = \left(1 - \varepsilon_l \frac{\partial v'}{\partial \xi}\right)^{-1},$$

(8.4.31b)

$$\Pi = -\omega \frac{\partial \rho'}{\partial Z}.$$

(8.4.31c)

Since

$$\frac{\partial \rho'}{\partial x} = \omega \frac{\partial \rho'}{\partial \xi},$$

(8.4.32a)

it follows that

$$\frac{\partial \rho'}{\partial \xi} = -\varepsilon_l \frac{\partial v'}{\partial Z},$$

(8.4.32b)

so that the thermal-wind equation retains its form in this new coordinate frame. This follows from a more basic fact. If we introduce $\phi'$, defined by

$$p' = \phi' - \frac{\varepsilon_l}{2} v'^2,$$

(8.4.33a)

as suggested by (8.4.17e), then it follows that

$$v' = +\frac{\partial \phi'}{\partial \xi}$$

(8.4.33b)

$$\rho' = -\varepsilon_l \frac{\partial \phi'}{\partial Z}.$$

(8.4.33c)

Thus, in terms of $\phi'$, the geostrophic relation in $\xi$, $Z$ coordinates is retained if $\phi'$ replaces $p'$.

The true motive for introducing the $\xi$, $Z$ coordinate system can now be appreciated in terms of the elegant simplicity it brings to the equations of motion. It follows from (8.4.30a,b,c) that

$$\frac{dQ}{dt} = \frac{\partial Q}{\partial T} + \frac{\partial Q}{\partial \xi}\frac{d\xi}{dt} + \frac{\partial Q}{\partial Z}\frac{dZ}{dt}$$

$$= \frac{\partial Q}{\partial T} - \xi\frac{\partial Q}{\partial \xi} + w'\frac{\partial Q}{\partial Z}.$$

(8.4.34)

The density equation (8.4.18d) becomes

$$\frac{\partial \rho'}{\partial T} - \xi\frac{\partial \rho'}{\partial \xi} + w'\frac{\partial \rho'}{\partial Z} = 0.$$

(8.4.35)

Since

$$w' = -\frac{\partial \chi}{\partial x} = -\omega\frac{\partial \chi}{\partial \xi}$$

(8.4.36a)

and

$$\omega\frac{\partial \rho'}{\partial Z} = -1,$$

(8.4.36b)

it follows that

$$\frac{\partial \rho'}{\partial T} - \xi\frac{\partial \rho'}{\partial \xi} + \frac{\partial \chi}{\partial \xi} = 0.$$

(8.4.37)

Thus in $\xi$ and $Z$, the equation for $\rho'$ is identical with the equation in $x$ and $z$ that would be obtained if $\rho'$ were horizontally advected by only the geostrophic deformation field and if $w'$ were given by $\partial\chi/\partial\xi$ and acted only on the initial vertical temperature gradient. Thus (8.4.37) has the *form* of the quasi-geostrophic approximation to the density equation. The effect of finite $\varepsilon_l$ is realized only in the coordinate transformation from $\xi$ to $x$. The equation for $v'$ in $\xi$, $Z$ coordinates is

$$\varepsilon_l\left[\frac{\partial v'}{\partial T} - \xi\frac{\partial v'}{\partial \xi} + v'\right] + \frac{\partial \chi}{\partial z} = 0.$$

(8.4.38)

The use of (8.4.32b), (8.4.37), and (8.4.38) yields

$$\boxed{\frac{\partial^2\chi}{\partial Z^2} + \frac{\partial^2\chi}{\partial \xi^2} = 2\frac{\partial \rho'}{\partial \xi}}$$

(8.4.39)

which is identical in *form* with (8.4.22). Hence, in these intrinsic coordinates, the problem for $\chi$ and hence $u'$ and $w'$ remains Laplace's equation for all $t$. The potential vorticity is

$$\Pi = -\omega\frac{\partial \rho'}{\partial Z} = -\frac{\partial \rho'/\partial Z}{1 - \varepsilon_l(\partial v'/\partial \xi)},$$

(8.4.40)

or noting (8.4.33b,c) and the fact that $\Pi$ is unity,

$$\varepsilon_l \left[ \frac{\partial^2 \phi'}{\partial \xi^2} + \frac{\partial^2 \phi'}{\partial Z^2} \right] = 1. \tag{8.4.41}$$

An equation for the density may be immediately obtained by differentiating (8.4.41) with respect to $Z$ to yield

$$\boxed{\frac{\partial^2 \rho'}{\partial \xi^2} + \frac{\partial^2 \rho'}{\partial Z^2} = 0}. \tag{8.4.42}$$

At the lower boundary, $Z = 0$, the appropriate boundary condition is simply the vanishing of $w$. We also apply this condition at an upper boundary, i.e., at $Z = 1$. For the *atmosphere* such a rigid lid is certainly artificial; yet if, as we can anticipate, the frontal development is intensified at the boundary, the presence of the second boundary will be physically irrelevant. With $w$ equal to zero at both boundaries, $\chi$ must also be constant on each boundary; hence on $Z = 0$ and $Z = 1$

$$\frac{\partial \chi}{\partial x} = \omega \frac{\partial \chi}{\partial \xi} = 0. \tag{8.4.43a}$$

The equation for $\rho'$ on each boundary, (8.4.37), therefore reduces to

$$\frac{\partial \rho'}{\partial T} - \xi \frac{\partial \rho'}{\partial \xi} = 0, \qquad z = 0, 1. \tag{8.4.43b}$$

In the $\xi$, $Z$ coordinate frame the density (or temperature) on the boundary responds as though (8.4.3) were valid, i.e., the boundary temperature in $\xi$ contracts in response to the geostrophic confluence of the deformation velocity. The solution of (8.4.43b) is

$$\rho' = \rho_0(\xi e^T, 0), \qquad Z = 0, \tag{8.4.44a}$$

$$\rho' = \rho_0(\xi e^T, 1), \qquad Z = 1, \tag{8.4.44b}$$

where $\rho_0(\xi, Z)$ is the initial density distribution. At $T = 0$, we have assumed (since $\Pi$ is unity) that $\partial \rho'/\partial Z$ is constant; hence

$$\rho' = 1 + \tilde{\rho}_0(\xi e^T), \qquad Z = 0, \tag{8.4.45a}$$

$$\rho' = \tilde{\rho}_0(\xi e^T), \qquad Z = 1, \tag{8.4.45b}$$

corresponding to the known *initial* density field

$$\rho'(\xi, Z, 0) = (1 - Z) + \tilde{\rho}_0(\xi). \tag{8.4.46}$$

Thus (8.4.42) must be solved, subject to (8.4.45a,b). Note that time enters the problem for $\rho'$ only as a parameter and enters only to define the characteristic scale in $\xi$ of the surface density field. This point can be made explicit by introducing the "length"

$$L_0(T) = e^{-T},$$

in terms of which

$$\rho' = 1 + \tilde{\rho}_0\left(\frac{\xi}{L_0}\right), \qquad Z = 0, \tag{8.4.47a}$$

$$\rho' = \tilde{\rho}_0\left(\frac{\xi}{L_0}\right), \qquad Z = 1. \tag{8.4.47b}$$

Let

$$\rho' = (1 - Z) + \tilde{\rho}_0\left(\frac{\xi}{L_0}\right) - \tilde{\theta}(\xi, Z, T); \tag{8.4.48}$$

then $\tilde{\theta}$ satisfies

$$\frac{\partial^2 \tilde{\theta}}{\partial \xi^2} + \frac{\partial^2 \tilde{\theta}}{\partial Z^2} = \frac{\partial^2 \tilde{\rho}_0}{\partial \xi^2}, \tag{8.4.49}$$

$$\tilde{\theta}(\xi, 0) = \tilde{\theta}(\xi, 1) = 0. \tag{8.4.50}$$

This problem may be conveniently solved with the aid of the Fourier transform

$$\Theta(k, Z) = \frac{1}{(2\pi)^{1/2}} \int_{-\infty}^{\infty} \tilde{\theta}(\xi, Z)e^{-ik\xi} \, d\xi,$$

$$\tilde{\theta}(\xi, Z) = \frac{1}{(2\pi)^{1/2}} \int_{-\infty}^{\infty} \Theta(k, Z)e^{ik\xi} \, dk, \tag{8.4.51}$$

which, when applied to (8.4.49) and (8.4.50), yields

$$\frac{d^2\Theta}{dZ^2} - k^2\Theta = \frac{1}{\sqrt{2\pi}} \int_{-\infty}^{\infty} \frac{\partial^2 \tilde{\rho}_0}{\partial \xi^2} e^{-ik\xi} \, d\xi$$

$$= \frac{L_0^{-1}}{\sqrt{2\pi}} \int_{-\infty}^{\infty} \frac{\partial^2 \tilde{\rho}_0}{\partial \eta^2} e^{-i(kL_0)\eta} \, d\eta, \tag{8.4.52}$$

where

$$\eta \equiv \frac{\xi}{L_0}, \tag{8.4.53a}$$

and

$$\Theta(k, 0) = \Theta(k, 1) = 0. \tag{8.4.53b}$$

Define

$$F(k) = \frac{1}{(2\pi)^{1/2}} \int_{-\infty}^{\infty} \frac{\partial \tilde{\rho}_0}{\partial \xi} e^{-ik\xi} \, d\xi, \tag{8.4.54}$$

i.e., $F(k)$ is the Fourier transform of the *initial* horizontal density gradient. Thus (8.4.52) becomes

$$\frac{d^2\Theta}{dz^2} - k^2\Theta = +ikF(kL_0)$$

whose solution, satisfying (8.4.53b), is

$$\Theta = -\frac{i}{k} F(kL_0) \left[ 1 - \frac{\cosh k(Z - \frac{1}{2})}{\cosh k/2} \right], \qquad (8.4.55)$$

which yields

$$\vartheta(\xi, Z) = \frac{1}{(2\pi)^{1/2}} \int_{-\infty}^{\infty} (-i) \frac{F(kL_0)}{k} \left[ 1 - \frac{\cosh k(Z - \frac{1}{2})}{\cosh k/2} \right] e^{+ik\xi} \, dk. \quad (8.4.56a)$$

The initial density gradient may be written as the sum of an even and an odd function of $\xi$.* Thus, $F(kL_0)$ may similarly be written as $F_e(kL_0) - iF_o(kL_0)$, where $F_e$ and $F_o$ are even and odd functions of $kL_0$ respectively. In terms of $F_e$ and $F_o$ (which are real functions of $kL_0$), (8.4.56a) becomes

$$\vartheta(\xi, Z) = \left( \frac{2}{\pi} \right)^{1/2} \int_0^{\infty} \left[ 1 - \frac{\cosh k(Z - \frac{1}{2})}{\cosh k/2} \right]$$
$$\times \left[ -\frac{F_o(kL_0)}{k} \cos k\xi + \frac{F_e(kL_0)}{k} \sin k\xi \right] dk \qquad (8.4.56b)$$

From the thermal-wind equation and (8.4.40) it is possible to determine $v'$ now that $\rho'$ is known:

$$\varepsilon_l v'(\xi, Z) = -\left( \frac{2}{\pi} \right)^{1/2} \int_0^{\infty} \frac{\sinh k(Z - \frac{1}{2})}{\cosh k/2}$$
$$\times [F_o(kL_0) \sin k\xi + F_e(kL_0) \cos k\xi] \, dk. \qquad (8.4.57)$$

The problem for the vertical circulation (8.4.39) may now be solved if use is made of (8.4.48) and (8.4.56). It is left to the reader to show that

$$\chi(\xi, Z) = \left( \frac{2}{\pi} \right)^{1/2} \int_0^{\infty} \left[ (Z - \frac{1}{2}) \frac{\sinh k(Z - \frac{1}{2})}{\cosh k/2} \right.$$
$$\left. - \frac{1}{2} \tanh k/2 \frac{\cosh k(Z - \frac{1}{2})}{\cosh k/2} \right] \qquad (8.4.58)$$
$$\times [F_o(kL_0) \sin k\xi + F_e(kL_0) \cos k\xi] \, dk.$$

For each value of $T$, and hence of $L_0$, the quantities $\rho'$, $v'$, and $\chi$ are known as functions of $\xi$ and $Z$. The final step in the solution is the transformation from $\xi$ to $x$ via the relation (8.4.27a), i.e.,

$$x = \xi - \varepsilon_l v'(\xi, Z, T), \qquad (8.4.59)$$

which may be accomplished by drawing lines of constant $\xi$ in $x, z$ space and attributing the value of (say) $\rho'$ at the point $\xi, Z$ to its equivalent point in $x, z$.

Hoskins (1971) has carried out the detailed calculations outlined above in the meteorological context wherein $\rho'$ is simply replaced by $-\theta'$ (within the

---

* Recall that at $T = 0$, $\xi = x$.

large-scale height approximation). He chose as an initial distribution of potential temperature, in dimensional units,

$$\theta'_* = CZ_* + \left(2\frac{\Delta\theta_*}{\pi}\right)\tan^{-1}(x_*/L), \qquad (8.4.60)$$

with

$$C = 3 \text{ K km}^{-1},$$
$$2\,\Delta\theta_* = 24 \text{ K}, \qquad (8.4.61)$$
$$D = 8 \text{ km}.$$

These choices correspond to a deformation radius $L_D$ of O(800 km). The length $L$ is arbitrary provided it is large compared to $L_D$. The form of (8.4.60) implies that initially the total horizontal temperature drop from the cold air at $-\infty$ to the warm air at $x = +\infty$ is $2\,\Delta\theta_*$. As time goes on the scale of the density field at the surface, were *only* the deformation field acting on $\theta_*$, would simply be

$$L_{0*} = Le^{-\alpha t*}. \qquad (8.4.62)$$

Figure 8.4.3 shows Hoskins's calculation of the $\theta_*$ and $v'_*$ fields when $L_{0*}$ is 148 km. Isolines of $\theta_*$ are drawn for every 2.4 K, while $v'_*$ contours every 4 m s$^{-1}$ are shown in the region $Z_* \leq D/2$. Note the 200-km length scale for

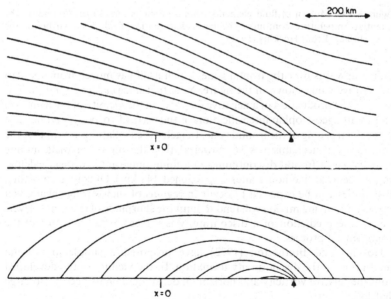

**Figure 8.4.3**   Lines of constant potential temperature (upper panel) (contours every 2.4 K). Lines of constant $v$ (lower panel) (contours every 4 m s$^{-1}$). Reprinted from Hoskins (1971).

scale. It is evident that sharp gradients in both $v'_*$ and $\theta'_*$ have developed near the surface. The frontal zone slopes upward with a slope of $\sim 1/100$, and the scale of the frontal region is considerably less than the deformation radius. Values of maximum $v'_*$ are $O(38 \text{ m s}^{-1})$ at the surface. With $\alpha = 10^{-5} \text{ s}^{-1}$, the time required to reduce $L_0$ from 200 km to 148 km is 8.4 hours. Although $\alpha/f < 1$, the calculated value of $\zeta_* /f$, which is a measure of $\varepsilon_l$, is found to triple from 1.3 to 3.9.

Figure 8.4.4 shows Hoskins's calculation of the vertical circulation in the frontal zone. The circulation is thermodynamically *direct*, with warm air

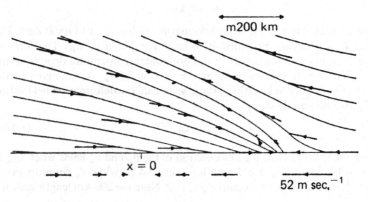

**Figure 8.4.4**  Motion of fluid elements from a previous time superimposed on the potential-temperature contours. The basic deformation field is shown below the lower surface. From Hoskins (1971).

gliding upwards over the frontal zone, which itself has moved from $x = 0$ to the positive value shown in the figure. Note that since $\partial v'_* /\partial z_* \sim \partial \theta'_* /\partial x_*$ is positive, the production of a negative $\partial w'_* /\partial x_*$ in the vicinity of the front will lead to an ageostrophic intensification of the vertical vorticity, and hence a further sharpening of the frontal zone. In fact, the calculations reveal that the frontogenetic mechanism is so powerful that the model predicts infinite vorticity, i.e., a frontal discontinuity in a finite time (in the present calculation it occurred five hours after $L_{0*}$ reached 148 km). Of course, in reality, instabilities can be expected to occur in regions of such strong shears, and the accompanying mixing will tend to limit the sharpness of the zone. Nevertheless, the prediction of a discontinuity is a measure of the vigor of the frontogenetic process.

Hoskins and Bretherton (1972) have proposed a simple and useful description of the discontinuity formation. In the vicinity of the developed front the relative vorticity and horizontal density gradients become so large that for

$$\Pi = -\frac{\partial \rho}{\partial z} - \varepsilon_l \left( \frac{\partial v}{\partial x} \frac{\partial \rho}{\partial z} - \frac{\partial v}{\partial z} \frac{\partial \rho}{\partial x} \right) \tag{8.4.63}$$

to remain bounded (in the case presented here $\Pi$ is unity everywhere) it must follow that near the front to lowest order

$$\frac{\partial v}{\partial x}\frac{\partial \rho}{\partial z} - \frac{\partial v}{\partial z}\frac{\partial \rho}{\partial x} \approx 0. \tag{8.4.64}$$

Thus in the frontal zone

$$v = A(\rho), \tag{8.4.65}$$

where $A(\rho)$ is an arbitrary function of $\rho$. Hence surfaces of constant $v$ and $\rho$ coincide. Further, since with (8.4.65) and the thermal-wind relation

$$\left(\frac{\partial z}{\partial x}\right)_v = -\frac{\partial v/\partial x}{\partial v/\partial z} = A'(\rho)\varepsilon_l, \tag{8.4.66}$$

it follows that a surface of constant $\rho$ (and hence constant $v$) must have constant *slope* in the $x$, $z$ plane, as shown in Figure 8.4.5. Since $\rho$ is conserved, it follows that the slope of each $\rho$ and $v$ surface remains constant with time. From (8.4.64)

$$\frac{\partial v}{\partial x} = -\varepsilon_l \frac{(\partial v/\partial z)^2}{\partial \rho/\partial z} \tag{8.4.67}$$

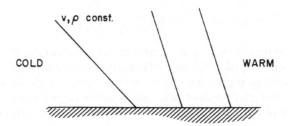

**Figure 8.4.5** The slope of the surfaces of constant $v$ and $\rho$ in the frontal zone.

hence it follows that $\partial v/\partial x$ must be *positive* (cyclonic) in frontal regions, since $\partial \rho/\partial z$ is always negative (statically stable). Consider, now, the motion of the $v$ and $\rho$ planes. If their slopes remain individually constant, the velocity in the $x$-direction of a $v$-plane must be a function only of $v$, i.e.,

$$\left(\frac{\partial x}{\partial t}\right)_v = G(v), \tag{8.4.68}$$

or

$$\frac{\partial v}{\partial t} + G(v)\frac{\partial v}{\partial x} = 0. \tag{8.4.69}$$

Now, for $\partial v/\partial x \gg 1$ it follows from (8.4.11b) that at $z = 0$, where $w$ vanishes,

$$G(v) = u, \tag{8.4.70}$$

and in particular

$$\boxed{\frac{\partial u}{\partial x} = G'(v)\frac{\partial v}{\partial x}}$$  (8.4.71)

Since $\partial v/\partial x > 0$, it follows that $G'(v)$ must be negative in regions of confluence. The crucial point (8.4.71) demonstrates is that the local rate of confluence, $\partial u/\partial x$, increases with the local value of the relative vorticity. As the latter increases by the frontogenetic dynamics, the drawing together of the isopycnals accelerates, leading to the final collapse to a discontinuity.

If (8.4.69) is differentiated with respect to $x$, we obtain, on $z = 0$,

$$\frac{d\zeta}{dt} = \frac{\partial \zeta}{\partial t} + u\frac{\partial \zeta}{\partial x} = -\frac{\partial u}{\partial x}\frac{\partial v}{\partial x}$$  (8.4.72)

$$= -G'(v)\zeta^2,$$

where $\zeta = \partial v/\partial x$. Following a fluid element, for which $G'(v)$ is *constant*, (8.4.72) may be easily integrated to yield

$$\zeta = [G'(v)(t - \tau)]^{-1},$$  (8.4.73)

where $\tau$ is defined in terms of the vorticity at $t = 0$:

$$\tau^{-1} = -G'(v)\left(\frac{\partial v}{\partial x}\right)_{t=0} = \left(\frac{\partial u}{\partial x}\right)_{t=0}$$  (8.4.74)

Hence $\tau^{-1}$ is the time scale associated with the initial deformation rate, and when $t$ becomes equal to $\tau$, $\zeta$ becomes infinite—i.e., a discontinuity in $v$ and hence $\rho$ forms on the surface $z = 0$. Thus, the time scale for complete collapse is given by $\alpha$ for the deformation-field model, although as we have seen, sharp zones of steep gradients in $v$ and $\rho$ develop considerably sooner.

## 8.5 Equatorial Waves

Near the equator the normal component of the earth's rotation vanishes and the geostrophic balance can no longer be expected. It is beyond the scope of this text to discuss at length the general nature of equatorial dynamics. However, one feature of equatorial dynamics common to both the atmosphere and the oceans is the phenemenon of the equatorially trapped wave.

Some of the extraordinary behavior of equatorial waves can be anticipated if reference is made to the results of Section 3.10, where the linear theory for Poincaré, Kelvin, and Rossby waves is discussed. The (dimensional) frequency of a Poincaré wave is given by

$$\sigma_P^2 = f^2 + gD_0(k^2 + l^2),$$  (8.5.1)

where $k$ and $l$ are the $x$ and $y$ wave numbers respectively, while the frequency of the Rossby mode may be written

$$\sigma_R = -\frac{\beta_0 k}{k^2 + l^2 + f^2/c_0^2}, \tag{8.5.2}$$

where $\beta_0$ is the planetary-vorticity gradient and $c_0$ is defined as

$$c_0 = (gD_0)^{1/2}, \tag{8.5.3}$$

where $D_0$ is the undisturbed depth. The minimum frequency of the Poincaré wave is $f$ while the maximum frequency (numerically) of the Rossby wave is $+\beta_0 c_0 /2f$. In mid-latitude regions these frequency limits are well separated, and indeed, this fact formed the heuristic basis in Chapter 3 for the development of the quasigeostrophic theory. In very low latitudes, where $f$ is small, we must expect that the dynamical regimes of gravity and Rossby waves to merge and their dynamics to become interrelated. Consider, too, a Poincaré wave of *fixed* $\sigma$ which exceeds $f$ in regions near the equator, where $f$ is small, and imagine how the spatial nature of the oscillation must change at higher latitudes, where $f$ exceeds $\sigma$. Since $k$ is real (for reasons of periodicity in $x$), $l^2$ must become negative for large enough $y$ where $f > \sigma$, implying an *evanescent* character of the wave at large $y$. This is a hint that the equatorial region may act as a waveguide, trapping oscillations in a zone around the equator. Naturally, this argument is only suggestive, for (8.5.1) is strictly valid only in mid-latitudes and for nearly constant $f$. Nevertheless, as we shall see, this heuristic argument satisfactorily anticipates the result of the detailed analysis. This argument may also be expected to apply to *baroclinic* modes if $D_0$ is replaced by the equivalent depth for the baroclinic mode (6.12.22).

To examine these ideas more precisely, consider the linearized equations of motion applied to the wave dynamics in a latitude band about the equator where the Coriolis parameter $f$ may be written

$$f = \beta_0 y_*, \tag{8.5.4}$$

where $y_*$ is the (dimensional) distance northward from the equator and where now

$$\beta_0 = \frac{2\Omega}{r_0}, \tag{8.5.5}$$

$r_0$ being the radius of the earth. Nondimensional variables are introduced by the relations

$$(u_*, v_*) = U(u, v), \qquad (x_*, y_*) = L(x, y),$$

$$w_* = \frac{D}{L} U w, \qquad z_* = Dz,$$

$$t_* = \frac{t}{\beta_0 L}, \tag{8.5.6}$$

$$p_* = p_s(z) + \rho_s(z)\beta_0 L^2 U p(x, y, z, t),$$

$$\rho_* = \rho_s(z) + \rho_s(z)\frac{\beta_0 L^2 U}{gD} \rho(x, y, z, t).$$

This scaling is fundamentally the same as that used for synoptic-scale motions in mid-latitudes with the major exception that $\beta_0 L$, rather than $f$, typifies the Coriolis parameter used to scale the pressure and density. Note that the scaling for the pressure anticipates that the Coriolis force and the pressure-gradient force are of the same *order* and not that geostrophic balance will be achieved. If (8.5.6) are substituted into the equations of motion and only linear terms are retained, the resulting equatorial $\beta$-plane equations are

$$\frac{\partial u}{\partial t} - yv = -\frac{\partial p}{\partial x}, \tag{8.5.7a}$$

$$\frac{\partial v}{\partial t} + yu = -\frac{\partial p}{\partial y}, \tag{8.5.7b}$$

$$\rho = -\frac{1}{\rho_s}\frac{\partial}{\partial z}(\rho_s p), \tag{8.5.7c}$$

$$\frac{\partial u}{\partial x} + \frac{\partial v}{\partial y} + \frac{1}{\rho_s}\frac{\partial}{\partial z}(\rho_s w) = 0. \tag{8.5.7d}$$

As in Chapter 6, the closure of this system requires the statement of energy conservation. For nondissipative waves in the atmosphere, this reduces to conservation of potential temperature, while for the ocean the statement of density conservation suffices. We will examine the former explicitly; as in Chapter 6, the oceanic case may be directly obtained from the atmospheric counterpart by merely letting the density scale height become large in comparison with the vertical scale of the wave motion and then identifying the potential-temperature fluctuation with the negative of the density fluctuation. That is, for the atmosphere we write the potential temperature as

$$\theta_* = \theta_s(z) + \theta_s(z)\frac{\beta_0 L^2 U}{gD}\theta(x, y, z, t), \tag{8.5.8}$$

where $\theta$ is related to $\rho$ and $p$ by the linearized form of (6.5.1), which with (8.5.7c) may be written as

$$\theta = \frac{\partial p}{\partial z} - p\theta_s^{-1}\frac{d\theta_s}{dz}. \tag{8.5.9}$$

As earlier, we restrict attention to cases where

$$\frac{1}{\theta_s}\frac{d\theta_s}{dz} \ll 1, \tag{8.5.10}$$

so that (8.5.7c) may be replaced by*

$$\theta = \frac{\partial p}{\partial z}, \tag{8.5.11}$$

---

* Note that the neglect of the second term in (8.5.9) will be valid unless $\partial p/\partial z$ is identically zero, that is, when the motion is strictly barotropic.

while the linearized form of the statement of potential-temperature conservation becomes

$$\frac{\partial \theta}{\partial t} + w \left[ \frac{gD}{\beta_0^2 L^4 \theta_s} \frac{d\theta_s}{dz} \right] = 0. \tag{8.5.12}$$

Let $N_0$ be a characteristic value of the Brunt–Väisälä frequency $((g/D\theta_s) \times (d\theta_s/dz))^{1/2}$, so that

$$N^2 = N_0^2 s(z), \tag{8.5.13}$$

where $s(z)$ is an $O(1)$ function which characterizes the distribution of $N^2$ with $z$. It is apparent from (8.5.12) that a natural characteristic length scale exists, i.e.,

$$L_e = \left( \frac{N_0 D}{\beta_0} \right)^{1/2} \tag{8.5.14}$$

which is the *equatorial*, Rossby internal-deformation radius. This interpretation follows from the defining relation of the deformation radius in mid-latitude for $L_D$ as

$$L_D = \frac{N_0 D}{f}. \tag{8.5.15}$$

If $f$ is written as $\beta_0 L_D$ for the equatorial problem, (8.5.14) follows immediately.

In the absence of imposed lateral boundaries, the only length scales in the wave problem are the zonal wavelength and $L_e$. We *choose* $L_e$ as the characteristic horizontal scale, $L$, so that (8.5.12) becomes simply

$$\frac{\partial \theta}{\partial t} + ws(z) = 0. \tag{8.5.16}$$

Following Moore and Philander (1977), we seek separable solutions in the form

$$\begin{Bmatrix} u \\ v \\ p \end{Bmatrix} = \begin{Bmatrix} U(x, y, t) \\ V(x, y, t) \\ P(x, y, t) \end{Bmatrix} G(z), \tag{8.5.17}$$

$$w = \left[ s^{-1}(z) \frac{dG}{dz} \right] W(x, y, t),$$

and

$$\theta(x, y, z, t) = P(x, y, t) \frac{dG}{dz}. \tag{8.5.18}$$

If (8.5.17) is substituted into (8.5.7a,b), (8.5.11), and (8.5.16), we obtain

$$\frac{\partial U}{\partial t} - yV = -\frac{\partial P}{\partial x}, \tag{8.5.19a}$$

$$\frac{\partial V}{\partial t} + yU = -\frac{\partial P}{\partial y}, \tag{8.5.19b}$$

$$\frac{\partial P}{\partial t} + W = 0, \tag{8.5.19c}$$

while (8.5.7d) becomes

$$\frac{\partial U}{\partial x} + \frac{\partial V}{\partial y} + \frac{W}{G\rho_s}\frac{d}{dz}\left[\frac{\rho_s(z)}{s(z)}\frac{dG}{dz}\right] = 0. \tag{8.5.19d}$$

In order for the separation implied by (8.5.17) and (8.5.18) to succeed, we infer from (8.5.19d) that $G(z)$ must satisfy the vertical-structure equation

$$\frac{1}{\rho_s}\frac{d}{dz}\left[\frac{\rho_s}{s}\frac{dG}{dz}\right] = -m^2 G, \tag{8.5.20}$$

where $m^2$ is the separation constant. Note that (8.5.20) is identical to the vertical-structure equation (6.12.7) that arose for mid-latitude motions. With (8.5.20), (8.5.19c) and (8.5.19d) may be combined to yield

$$m^2\frac{\partial P}{\partial t} + \frac{\partial U}{\partial x} + \frac{\partial V}{\partial y} = 0, \tag{8.5.21}$$

which, with (8.5.19a,b), forms a closed system for $U$, $V$, and $P$. The same equations apply directly to the oceanic case for $U$, $V$, and $P$. The only alteration is that for the oceanic case $\rho_s$ may be taken as a constant in (8.5.20). $P$ may be eliminated from (8.5.19a,b) with the aid of (8.5.21) to yield

$$L_1(U) = \frac{\partial^2 U}{\partial t^2} - \frac{1}{m^2}\frac{\partial^2 U}{\partial x^2} = y\frac{\partial V}{\partial t} + \frac{1}{m^2}\frac{\partial^2 V}{\partial y\, \partial x} \tag{8.5.22a}$$

and

$$L_2(V) = \frac{\partial^2 V}{\partial t^2} - \frac{1}{m^2}\frac{\partial^2 V}{\partial y^2} = -y\frac{\partial U}{\partial t} + \frac{1}{m^2}\frac{\partial^2 U}{\partial y\, \partial x}. \tag{8.5.22b}$$

It is convenient to eliminate $U$ and achieve a statement of the wave problem entirely in terms of $V$. This is accomplished by operating on (8.5.22b) with the operator $L_1$ and then using (8.5.22a,b) to eliminate $U$ to obtain

$$\frac{\partial}{\partial t}\left[\left(\frac{\partial^2}{\partial x^2} + \frac{\partial^2}{\partial y^2}\right)V - m^2 y^2 V - m^2\frac{\partial^2 V}{\partial t^2}\right] + \frac{\partial V}{\partial x} = 0. \tag{8.5.23}$$

This formulation will be apt for motions for which $V$ is not trivially zero. Our earlier experience with the Kelvin wave of Section 3.9 alerts us to the possibility that nontrivial modes may exist for which $V$ is identically zero. To examine this possibility, let $V$ be zero, in which case (8.5.22a,b) reduces to

$$\frac{\partial^2 U}{\partial t^2} - \frac{1}{m^2}\frac{\partial^2 U}{\partial x^2} = 0, \tag{8.5.24a}$$

$$\frac{\partial^2 U}{\partial y\,\partial x} - m^2 y\frac{\partial U}{\partial t} = 0. \tag{8.5.24b}$$

The first equation implies that $U$ is a linear combination of

$$U_+ = U_+\left(x + \frac{t}{m}, y\right) \tag{8.5.25a}$$

and

$$U_- = U_-\left(x - \frac{t}{m}, y\right), \tag{8.5.25b}$$

where $U_+$ and $U_-$ are arbitrary functions. The first, $U_+$, represents a nondispersive westward-moving wave; the second, $U_-$, a nondispersive eastward-moving wave. Substitution of (8.5.25a,b) into (8.5.24b) determines the meridional structure of each wave, i.e.,

$$\frac{\partial}{\partial y}\left(\frac{\partial U_+}{\partial x}\right) - my\left(\frac{\partial U_+}{\partial x}\right) = 0, \tag{8.5.26a}$$

$$\frac{\partial}{\partial y}\left(\frac{\partial U_-}{\partial x}\right) + my\left(\frac{\partial U_-}{\partial x}\right) = 0, \tag{8.5.26b}$$

so that

$$U_\pm = A_\pm\left(x \pm \frac{t}{m}\right)e^{\pm my^2/2}. \tag{8.5.27}$$

Without loss of generality we may consider $m > 0$, for negative $m$ merely interchanges the two Kelvin waves. It is clear from (8.5.27) that only $U_-$ remains bounded for large $y$, and $U_+$ must consequently be rejected as an acceptable solution. Thus the eastward-propagating Kelvin wave, with $V = 0$, of the form

$$U = A\left(x - \frac{t}{m}\right)e^{-my^2/2}, \tag{8.5.28}$$

is a possible equatorially trapped wave. This is consistent with the result of Chapter 3, in which it was shown that for $f > 0$ (i.e., in the northern hemisphere) the Kelvin wave propagates so that the disturbance is a maximum to the right of an observer looking in the direction of propagation. The associated pressure field is given by

$$P = \frac{1}{m}A\left(x - \frac{t}{m}\right)e^{-my^2/2}, \tag{8.5.29}$$

from which it follows that for all $y$, including the equator, the zonal velocity in the *Kelvin* wave is in *geostrophic* balance, i.e.,

$$yU = -\frac{\partial P}{\partial y}.$$ (8.5.30)

The speed of propagation of the Kelvin wave is $1/m$, or in dimensional units

$$\boxed{c_* = \frac{\beta_0 L_e^2}{m} = \frac{N_0 D}{m}},$$ (8.5.31)

which is independent of the earth's rotation. We momentarily defer the details of the question of the calculation of $m$. For *free* waves, however, $m^2$ is obtained as an eigenvalue of (8.5.20) after suitable boundary conditions are specified. The natural "width" of the equatorially trapped Kelvin mode is the $e$-folding scale

$$y_* = L_e\left(\frac{2}{m}\right)^{1/2}.$$ (8.5.32)

Wave modes with $V \neq 0$ are generally dispersive, i.e., the phase speed depends on the zonal wavelength, and so solutions of (8.5.23) are sought in the form

$$V = \text{Re } e^{i(kx - \sigma t)}\psi(y),$$ (8.5.33)

where, without loss of generality, we take $\sigma > 0$. The meridional structure equation for $\psi(y)$ is, then, from (8.5.23),

$$\frac{d^2\psi}{dy^2} + \psi\left[m^2(\sigma^2 - y^2) - \frac{k}{\sigma} - k^2\right] = 0,$$ (8.5.34)

which the reader may profitably compare with (3.10.5). The meridional-structure equation is reduced to standard form by the introduction of the variable

$$\eta = m^{1/2}y,$$ (8.5.35)

in terms of which (8.5.34) becomes

$$\frac{d^2\psi}{d\eta^2} + \psi\left[\left(m^2\sigma^2 - \frac{k}{\sigma} - k^2\right)m^{-1} - \eta^2\right] = 0.$$ (8.5.36)

The only solutions of (8.5.36) which are bounded for large $\eta$ (i.e., large $y$) are of the form

$$\psi(y) = \psi_j(y) = \frac{e^{-\eta^2/2}H_j(\eta)}{(2^j j!\, \pi^{1/2})^{1/2}},$$ (8.5.37)

where $j$ is any nonnegative integer, including zero, and where $H_j(\eta)$ is the

Hermite polynomial defined by

$$H_j(\eta) = (-1)^j e^{\eta^2} \frac{d^j}{d\eta^j} e^{-\eta^2}. \qquad (8.5.38)$$

Thus the first six polynomials are

$$H_0 = 1, \qquad\qquad\qquad H_1 = 2\eta$$
$$H_2 = 4\eta^2 - 2, \qquad\qquad H_3 = 8\eta^3 - 12\eta, \qquad (8.5.39)$$
$$H_4 = 16\eta^4 - 48\eta^2 + 12, \qquad H_5 = 32\eta^5 - 160\eta^3 + 120\eta.$$

The $\psi_j(\eta)$ with $j$ even (odd) are even (odd) functions of $\eta$ in the interval $(-\infty, \infty)$ with $j$ nodal points. The functions are oscillatory for $|\eta| < (2j + 1)^{1/2}$ and evanescent beyond these latitudes.

The $\psi_j$'s each satisfy the differential equation

$$\frac{d^2\psi_j}{d\eta^2} + \psi_j((2j + 1) - \eta^2) = 0, \qquad (8.5.40)$$

from which it follows that $\sigma$, $k$, and $m$ must satisfy the dispersion relation

$$\boxed{m^2\sigma^2 - \frac{k}{\sigma} - k^2 = (2j + 1)m}, \qquad j = 0, 1, 2 \ldots, \qquad (8.5.41)$$

while the $\psi_j(\eta)$ satisfy the orthogonality relations

$$\int_{-\infty}^{\infty} \psi_j(\eta)\psi_p(\eta) \, d\eta = \delta_{jp}. \qquad (8.5.42)$$

Note that if we set $j = -1$, (8.5.41) possesses as a solution

$$\frac{\sigma}{k} = \frac{1}{m}, \qquad (8.5.43)$$

which is simply the dispersion relation for the *Kelvin wave*. Hence the dispersion relation *includes* the Kelvin wave in this expanded interpretation. $V$ is, of course, zero for the Kelvin wave, but $U$ may be written in terms of $\psi_0(\eta)$, for the simple harmonic Kelvin wave, as

$$U(x, t) = \text{Re } e^{ik(x - t/m)}\psi_0(\eta). \qquad (8.5.44)$$

In general the dispersion relation is most easily written for $k$ as a function of $m$ and $\sigma$, i.e., for each $j$,

$$\boxed{k = -\frac{1}{2\sigma} \pm \tfrac{1}{2}[(\sigma^{-1} - 2m\sigma)^2 - 8mj]^{1/2}}. \qquad (8.5.45)$$

The case where $j$ is zero presents some subtle features. When $j$ vanishes, there are two apparent roots of (8.5.45), namely

$$k = -m\sigma \qquad (8.5.46a)$$

and

$$k = -\frac{1}{\sigma} + m\sigma. \tag{8.5.46b}$$

The first root corresponds to the frequency relation for a *westward*-propagating Kelvin mode. Examination of (8.5.22b) then shows that the corresponding $U$-field becomes unbounded for large $y$ and hence (8.5.46a) must be rejected. The second solution is the Yanai wave, or the "mixed" Rossby–gravity mode. For large $\sigma$, (8.5.46b) is

$$\sigma \sim \frac{k}{m} \tag{8.5.47a}$$

and so is asymptotic to the Kelvin mode, while for small $\sigma$

$$\sigma \sim -\frac{1}{k}, \tag{8.5.47b}$$

which is the familiar, high-zonal-wave-number limit for the Rossby wave.

For $j \geq 1$, both roots of (8.5.45) lead to acceptable wave fields, and the full dispersion relation is shown in Figure 8.5.1, where $\omega = \sigma m^{1/2}$ is plotted as a function of $K = km^{-1/2}$.

For all $j \geq 1$, the modes split into two classes: first, a relatively high-frequency inertia–gravity set analogous to the Poincaré modes of Chapter 3, and second, a low-frequency class of Rossby modes. The Kelvin mode and the Yanai, or "mixed," mode are also shown. It is important to note that while the pure inertia–gravity modes and the Rossby modes possess, at a given frequency, waves with both positive and negative group velocities in the $x$-direction, the Yanai and Kelvin modes possess only *eastward* group velocities for all $k$.

If the dispersion relation is differentiated with respect to $k$, we find that the group speed in the $x$-direction is

$$\frac{\partial \sigma}{\partial k} = \frac{1 + 2k\sigma}{2\sigma^2 m^2 + k/\sigma}, \tag{8.5.48}$$

so that for each mode the frequency extrema occur on the locus $2k\sigma = 2K\omega = -1$ in the wave-number–frequency diagram. The corresponding frequency, i.e., the points where $\partial \sigma / \partial k$ vanish, are then determined by substituting $k = -1/2\sigma$ into (8.5.41) to obtain

$$\sigma_{\min} = \frac{((j+1)/2)^{1/2} + (j/2)^{1/2}}{m^{1/2}}, \tag{8.5.49}$$

which is the minimum frequency of the $j$th inertia–gravity wave, and

$$\sigma_{\max} = \frac{((j+1)/2)^{1/2} - (j/2)^{1/2}}{m^{1/2}}, \tag{8.5.50}$$

which is the maximum frequency for $j$th Rossby mode. For each $j$ the Poincaré modes have an eastward (westward) group speed for $k$ greater (less) than $-(2\sigma_{\min})^{-1}$, while for the $j$th Rossby mode the group speed is eastward (westward) for $k$ less (greater) than $-(2\sigma_{\max})^{-1}$.

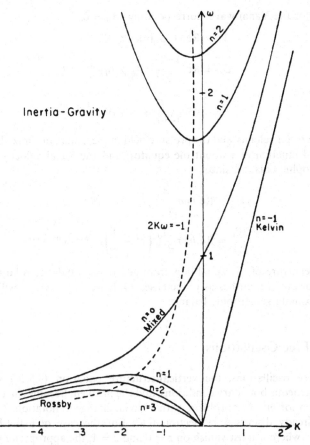

**Figure 8.5.1** The dispersion relation for equatorial waves [from Cane and Sarachik (1976)].

Once $V$ is known for each mode, the remaining fields can be calculated from (8.5.22a) and (8.5.21). For all $j \geq 1$ (i.e., for the Poincaré and Rossby modes) we have

$$V_j = \text{Re } A_j \psi_j(\eta) e^{i(kx - \sigma t)}, \qquad j \geq 1, \tag{8.5.51a}$$

$$U_j = \text{Re } \frac{iA_j}{m^{1/2}} \left[ \left( \frac{j}{2} \right)^{1/2} \frac{\psi_{j-1}(\eta)}{(\sigma + k/m)} \right.$$

$$\left. + \left( \frac{j+1}{2} \right)^{1/2} \frac{\psi_{j+1}(\eta)}{(\sigma - k/m)} \right] e^{i(kx - \sigma t)}, \tag{8.5.51b}$$

$$P_j = \text{Re } \frac{iA_j}{m^{3/2}} \left[ -\left( \frac{j}{2} \right)^{1/2} \frac{\psi_{j-1}(\eta)}{(\sigma + k/m)} \right.$$

$$\left. + \left( \frac{j+1}{2} \right)^{1/2} \frac{\psi_{j+1}(\eta)}{(\sigma - k/m)} \right] e^{i(kx - \sigma t)}, \tag{8.5.51c}$$

while for the Yanai wave, corresponding to $j = 0$,

$$V_0 = \text{Re } A_0 \psi_0(\eta)e^{i(kx - \sigma t)}, \tag{8.5.52a}$$

$$U_0 = \text{Re } \frac{im^{1/2}\sigma}{2^{1/2}} A_0 \psi_1(\eta)e^{i(kx - \sigma t)}, \tag{8.5.52b}$$

$$P_0 = \text{Re } \frac{i\sigma}{(2m)^{1/2}} A_0 \psi_1(\eta)e^{i(kx - \sigma t)}. \tag{8.5.52c}$$

The zonal velocity and the pressure field in the Yanai, or "mixed," mode is an odd function of $y$ about the equator, and the zonal velocity is not in geostrophic balance, since

$$yU_0 = \text{Re } im\sigma \frac{A_0}{\pi^{1/4}} y^2 e^{-my^2/2} e^{i(kx - \sigma t)}, \tag{8.5.53a}$$

$$-\frac{\partial P_0}{\partial y} = \text{Re } im\sigma \frac{A_0}{\pi^{1/4}} \left(y^2 - \frac{1}{m}\right) e^{-my^2/2} e^{i(kx - \sigma t)}. \tag{8.5.53b}$$

The departure of the zonal flow from geostrophic balance is largest at the equator and diminishes as $y^2$ increases. Only for $|y| \gg m^{-1/2}$ will $U_0$ be in approximate geostrophic balance.

## 8.5a  Free Oscillations

For free oscillations, the vertical-structure equation (8.5.20) will have homogeneous boundary conditions, and the problem becomes an eigenvalue problem for $m^2$. The problem, with a trivial change in notation, is the same one already discussed in Section 6.12. For the case of a flat-bottomed ocean,* where $w$ must vanish on $z = 0$ and $z = 1$, the appropriate boundary condition for (8.5.20) is, using (8.5.16) and (8.5.18),

$$\frac{dG}{dz} = 0, \qquad z = 0, 1. \tag{8.5.54}$$

The eigenvalue problem (8.5.20), (8.5.54) defines an eigenvalue set

$$m^2 = m_i^2, \qquad i = 0, 1, 2, 3, \tag{8.5.55}$$

where the $i = 0$ mode corresponds to $m = 0$, i.e., the barotropic mode. When $m$ is zero, it is clear from (8.5.51), (8.5.52), and (8.5.53) that the resulting oscillations are not equatorially trapped. In fact though, for the barotropic mode the approximation that the vertical scale of the motion is small compared to $g/N_0^2$ (the scale height defined by the static stability) is no longer valid, and more detailed calculations in that case imply, as we would expect, that the appropriate meridional scale is the *external* deformation radius associated with the motion of the upper surface, which in equatorial

---

* If topography is considered, the separation (8.5.17) is no longer possible, as shown in Section 6.15.

regions is

$$\mathcal{R}_{eq} = \frac{(gD)^{1/4}}{\beta_0^{1/2}} \gg L_e. \tag{8.5.56}$$

Characteristic values for $\mathcal{R}_{eq}$ are O(3,000 km), so that although the consideration of the dynamics of the free-surface motion (neglected in (8.5.54)) does produce equatorial trapping, it is a very mild form of trapping. The scale $\mathcal{R}_{eq}$ is sufficiently large that the barotropic mode can probably not be considered as isolated from mid-latitude dynamics. For the baroclinic modes, $L_e$ as given by (8.5.14) is the gross, appropriate scale, corresponding to eigenvalues $m_i^2 \neq 0$. For each mode the characteristic scale depends on $m_i$ through (8.5.35); hence the true meridional scale is, for each mode,

$$\frac{L_e}{(m_i)^{1/2}} = \left(\frac{N_0 D}{\beta_0 m_i}\right)^{1/2} \equiv \frac{(gh_i)^{1/4}}{\beta_0^{1/2}}, \tag{8.5.57}$$

where $h_i$ is the so-called *equivalent depth*, i.e.,

$$h_i = \frac{N_0^2 D^2}{gm_i^2}. \tag{8.5.58}$$

That is, $h_i$ is the depth of a homogeneous ocean whose external deformation radius is equal to the *internal* deformation radius of the $i$th baroclinic mode. The equivalent depth is *not* the characteristic vertical scale for the motion of the $i$th mode. The characteristic vertical scale is

$$\frac{D}{m_i} = \frac{(gh_i)^{1/2}}{N_0} = (Hh_i)^{1/2}, \tag{8.5.59}$$

where

$$H = \left(\frac{N_0^2}{g}\right)^{-1} \tag{8.5.60}$$

is a characteristic value of the density scale height. Moore and Philander (1977) report the values of $h_i$ in Table 8.5.1 for the first 5 baroclinic modes for a typical equatorial distribution of $N^2$. Note that the quoted values of $(gh_i)^{1/2}$ yield the Kelvin wave speeds for the $i$th mode, while the last two columns of the table display the characteristic meridional trapping scale and periods of the first five internal Kelvin modes.

Table 8.5.1

| $i$ | $h_i$ (cm) | $(gh_i)^{1/2}$ (cm/s) | $L_e/m_i^{1/2}$ (km) | $T_i = m_i^{1/2}/(\beta L_e)$ (days) |
|---|---|---|---|---|
| 1 | 60 | 240 | 325 | 1.5 |
| 2 | 20 | 140 | 247 | 2.0 |
| 3 | 8 | 88 | 197 | 2.6 |
| 4 | 4 | 63 | 165 | 3.1 |
| 5 | 2 | 44 | 139 | 3.6 |

The process of reflection of equatorially trapped baroclinic modes from an oceanic western boundary is a fascinating one, especially in view of the explanation offered in Section 5.8 for mid-latitude westward intensification of the oceanic circulation in terms of reflected Rossby waves. Consider now the equatorial version of this problem. Let the incoming wave be a low-frequency equatorial planetary wave with unit amplitude and meridional mode number $J$. For the group velocity to be westward, the appropriate root of (8.5.45), for a given $\sigma$, is

$$k_+ = -\frac{1}{2\sigma} + \tfrac{1}{2}[(\sigma^{-1} - 2m_i\sigma)^2 - 8m_i J]^{1/2}, \qquad (8.5.61)$$

i.e., a *long* planetary wave. The corresponding $U$ and $V$ for the incoming wave are, from (8.5.51a,b),

$$V_{\text{incoming}} = \text{Re } e^{i(k_+ x - \sigma t)}\psi_J(\eta), \qquad (8.5.62a)$$

$$U_{\text{incoming}} = \text{Re } \frac{i}{m_i^{1/2}}\left[+(J/2)^{1/2}\frac{\psi_{J-1}(\eta)}{(\sigma + k_+/m_i)}\right.$$

$$\left. +\left(\frac{J+1}{2}\right)^{1/2}\frac{\psi_{J+1}(\eta)}{(\sigma - k_+/m_i)}\right] \qquad (8.5.62b)$$

where

$$\eta = (m_i)^{1/2}y. \qquad (8.5.62c)$$

This solution, representing an incoming wave, must be supplemented by plane-wave solutions at the *same* frequency $\sigma$ representing waves with group velocity to the *east*. Reference to Figure 8.5.2 shows that this set of reflected waves will, in general, consist of a *finite* set of Rossby waves with $j \leq J$ *plus* a Yanai and Kelvin wave. That is, the reflected wave field will have the general representation, from (8.5.51a,b), (8.5.52a,b), and (8.5.44),

$$V_{\text{reflected}} = \text{Re }\left[\sum_{j=1}^{J} A_j e^{i(k-x-\sigma t)}\psi_j(\eta) + A_0 e^{i[(m\sigma-\sigma-1)x-\sigma t]}\psi_0(\eta)\right], \qquad (8.5.63a)$$

$$U_{\text{reflected}} = \text{Re }\left[\sum_{j=1}^{J}\frac{A_j}{m_i^{1/2}}\left(+\frac{(j/2)^{1/2}}{\sigma + k_-/m_i}\psi_{j-1}(\eta)\right.\right.$$

$$\left.\left. +\frac{((j+1)/2)^{1/2}}{\sigma - k_-/m_i}\psi_{j+1}(\eta)\right)e^{i(kx-\sigma t)}\right] \qquad (8.5.63b)$$

$$+ \text{Re }\left[\frac{im_i^{1/2}\sigma}{2^{1/2}} A_0\psi_1(\eta)e^{i[(m\sigma-\sigma-1)-\sigma t]}\right.$$

$$\left. + A_{\text{Kel}}\psi_0(\eta)e^{-i\sigma(t-m_i x)}\right],$$

where $A_0$ and $A_{\text{Kel}}$ are the amplitudes of the reflected Yanai and Kelvin waves respectively. The $A_j$ for $j \geq 1$ are the amplitudes of the reflected Rossby waves. The reflected wave number for each short, planetary wave is

$$k_- = -\frac{1}{2\sigma} - \tfrac{1}{2}[(\sigma^{-1} - 2m\sigma)^2 - 8mj]. \qquad (8.5.64)$$

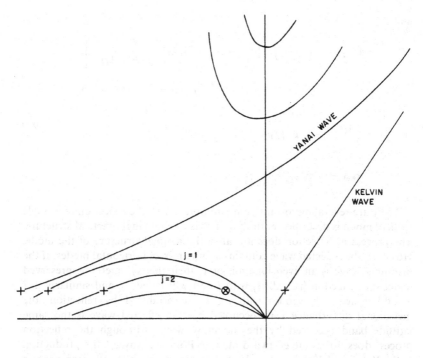

**Figure 8.5.2** A schematic of the dispersion diagram. The position of the Rossby wave incident on a western meridional boundary is marked by $\otimes$. The positions in the diagram of the reflected Yanai, Rossby, and Kelvin wave are marked by $+$.

Note that $k_-$ is a function of $j$, i.e., $k_- = k_-(j)$; however, we have not expressed that dependence explicitly until now, for notational simplicity. It is important to bear in mind that each reflected planetary wave will have a different zonal wave number and that the wave number $k_-$ *increases* in magnitude with *decreasing* $j$. The lowest-$j$ waves are therefore shorter in the $x$-direction *and* confined to a *narrower* region around the equator.

At $x = 0$,

$$U_{\text{incoming}} + U_{\text{reflected}} = 0, \tag{8.5.65}$$

which, with (8.5.62b) and (8.5.63b), implies that (Moore and Philander 1977)

$$A_J = -\frac{\sigma - k_-/m_i}{\sigma - k_+/m_i},$$

$$A_{J-1} = 0,$$

$$A_{J-2} = -\left(\frac{J}{J-1}\right)^{1/2}[\sigma - k_-(J-2)]$$

$$\times \left\{\frac{A_J}{\sigma + k_-(J)/m_i} + \frac{1}{\sigma + k_+(J)/m_i}\right\}, \tag{8.5.66a}$$

and

$$A_{j-2} = -\left(\frac{j}{j-1}\right)^{1/2} [\sigma - k_-(j-2)]A_j \left[\frac{\sigma - k_-(j-2)}{\sigma + k_-(j)}\right],$$

$$\text{for } j = J - 1, J - 2, \dots, 3,$$

$$A_0 = \frac{2^{1,2}A_2}{[\sigma + k_-(2)/m_i]\sigma}, \qquad (8.5.66b)$$

$$A_{\text{Kel}} = -\frac{iA_1}{2^{1/2}(\sigma + k_-(1))}.$$

There are several important points to note: First, the reflection of a single *vertical* mode is described entirely in terms of that single vertical structure. The process of reflection does not alter the *vertical* structure of the mode. However, the reflected wave contains a finite mix of horizontal modes. If the incoming wave is an even function of $y$, then this symmetry is preserved under the reflection (i.e., all $A_j$ with $j$ odd will be zero), and similarly if $J$ is odd. Since only modes with $j \leq J$ are produced by reflection, the meridional disturbance zone occupied by the reflected wave is the same latitude band occupied by the incoming wave. Although the reflection process does not excite eastward-moving Poincaré waves,* the production of the Yanai and Kelvin waves is an essential part of the reflection process at a western boundary. As Cane and Sarachik (1977) have noted, the ability of the relatively rapidly moving (see Table 8.5.1) Kelvin wave to carry energy away from the western boundary suggests that less incoming energy can be trapped in equatorial regions for the formation of western boundary currents than is the case for midlatitudes as described in Section 5.8.

The reflection process at an oceanic eastern boundary is severely complicated by the absence of Yanai and Kelvin modes capable of transmitting energy to the west. Moore (1968) has shown that in this case the reflected wave field is represented by a sum of $\psi_j$-modes with $j > J$. The most important qualitative result of Moore's work is his demonstration, complex beyond the scope of our discussion, that a portion of the incoming energy at an eastern boundary escapes poleward as a Kelvin wave, traveling away from the equator along the meridional boundaries. This is not possible at the western boundary, where the Kelvin wave must travel equatorward along the boundary.

For the atmosphere, there are no baroclinic modes for the free-oscillation problem, as discussed in Section 6.12. As in the case of the ocean, a free barotropic mode exists. If the approximation $(D/\theta_*) \, \partial\theta_*/\partial z_* \ll 1$ is not made, calculations show (e.g., Holton 1975) that the characteristic horizon-

---

* Unless, of course, the *incoming* wave is a Poincaré wave, in which case the reflected wave is a similar finite sum of Poincaré, Yanai, and Kelvin waves, with $k_-$ and $k_+$ interchanged.

tal scale is the *external* Rossby-deformation radius, which, for an isothermal atmosphere, is

$$\mathcal{R}_{eq} = (g\gamma H)^{1/4}/\beta_0^{1/2}$$

where $H$ is the density scale height and $\gamma$ is the ratio of the specific heats $c_p$ and $c_v$ for air. This barotropic scale is sufficiently large (i.e., of the order of 3,000 km) that the validity of the characterization of this mode as equatorially trapped is problematic.

## 8.5b Forced Oscillations

The problem of forced, equatorially trapped waves is significant for both the atmosphere and the oceans. In the former case, the periodic release of heat in strata of the atmosphere with absorbing constituents such as ozone and water vapor, and nonadiabatic heat release by the condensation of moisture in the troposphere, have been suggested as sources of forcing for equatorial waves. In the ocean, the time-dependent wind stress acting over the depth of the surface mixed layer provides an unsteady forcing for equatorial motions. To model these mechanisms we return to (8.5.7a,b) and (8.5.12) and rewrite them to include inhomogeneous terms representing applied forcing, i.e.,

$$\frac{\partial u}{\partial t} - yv = -\frac{\partial p}{\partial x} + X(x, y, z, t), \qquad (8.5.67a)$$

$$\frac{\partial v}{\partial t} + yu = -\frac{\partial p}{\partial y} + Y(x, y, z, t) \qquad (8.5.67b)$$

$$\frac{\partial}{\partial t}\frac{\partial p}{\partial z} + ws(z) = \frac{\partial Q}{\partial z}(x, y, z, t), \qquad (8.5.67c)$$

where $X$, $Y$ are imposed sources of momentum in the meridional and zonal directions and $\partial Q/\partial z$ is an imposed source of heat. These sources are assumed to be known, although in fact they may well depend on $u$, $v$, and $p$. In deriving (8.5.67c) we have used (8.5.11), which, with (8.5.7c), is considered unaltered by the forcing. It is convenient to define the operator

$$\frac{1}{\rho_s}\frac{\partial}{\partial z}\frac{\rho_s}{s(z)}\frac{\partial}{\partial z} \equiv \mathcal{D}^2. \qquad (8.5.68)$$

The continuity equation (8.5.7d) may be used with the momentum equations and (8.5.67c) to eliminate $p$ and to yield

$$\frac{\partial^2 u}{\partial x^2} + \frac{\partial^2}{\partial t^2}[\mathcal{D}^2 u] = -\frac{\partial^2 v}{\partial y\,\partial x} + y\frac{\partial}{\partial t}\mathcal{D}^2 v + \frac{\partial}{\partial t}\mathcal{D}^2 X - \frac{\partial}{\partial x}\mathcal{D}^2 Q, \quad (8.5.69a)$$

$$\frac{\partial^2 v}{\partial y^2} + \frac{\partial^2}{\partial t^2}[\mathcal{D}^2 v] = -\frac{\partial^2 u}{\partial y\,\partial x} - y\frac{\partial}{\partial t}\mathcal{D}^2 u + \frac{\partial}{\partial t}\mathcal{D}^2 Y - \frac{\partial}{\partial y}\mathcal{D}^2 Q. \quad (8.5.69b)$$

For free oscillations, where $X$, $Y$, and $Q$ are zero, (8.5.69a,b) reduce to (8.5.22a,b) if the separation (8.5.20) is assumed.

If $u$ is eliminated between (8.5.69a,b), a single equation for $v$ results:

$$\frac{\partial^4}{\partial t^4}\mathscr{D}^4 v + \frac{\partial^2}{\partial t^2}\nabla_1^2\mathscr{D}^2 v + \frac{\partial^2}{\partial x\,\partial t}\mathscr{D}^2 v + y^2\frac{\partial^2}{\partial t^2}\mathscr{D}^4 v = J, \qquad (8.5.70)$$

where

$$\nabla_1^2 = \frac{\partial^2}{\partial x^2} + \frac{\partial^2}{\partial y^2}, \qquad (8.5.71a)$$

$$J = \left[\frac{\partial^2}{\partial t^2} + \frac{\partial^2}{\partial x^2}\mathscr{D}^2\right]\left[\frac{\partial}{\partial t}\mathscr{D}^2 Y - \frac{\partial}{\partial y}\mathscr{D}^2 Q\right]$$

$$- \left[y\mathscr{D}^2 + \frac{\partial^2}{\partial x\,\partial y}\right]\left[\frac{\partial}{\partial t}\mathscr{D}^2 X - \frac{\partial}{\partial x}\mathscr{D}^2 Q\right]. \qquad (8.5.71b)$$

Suppose now that $J$ is periodic in $x$ and $t$, i.e., let

$$J = \text{Re } J_0(y, z)e^{i(kx - \sigma t)}. \qquad (8.5.72)$$

Clearly, any forcing can be resolved as a linear combination of such forcings, each of which may be dealt with separately as a consequence of the linearity of (8.5.70). If solutions for $v$ are then sought in the form

$$v = \text{Re } V(y, z)e^{i(kx - \sigma t)}, \qquad (8.5.73)$$

then $V(y, z)$ must satisfy

$$\left[(\sigma^2 - y^2)\mathscr{D}^4 V - \left[\frac{d^2}{dy^2} - k^2 - \frac{k}{\sigma}\right]\mathscr{D}^2 V\right] = \frac{J_0(y, z)}{\sigma^2}. \qquad (8.5.74)$$

Separable solutions to the homogeneous portion of (8.5.74) will be possible only if the operator multiplying $\mathscr{D}^2 V$ is equal to a constant multiple of $\sigma^2 - y^2$, i.e., only if

$$V = A_j(z)V_j(y), \qquad (8.5.75)$$

where

$$\frac{d^2}{dy^2}V_j + [\lambda_j^2(\sigma^2 - y^2) - \frac{k}{\sigma} - k^2]V_j = 0, \qquad (8.5.76)$$

with $\lambda_j^2$ the aforementioned constant. The important point is that (8.5.76) is identical to (8.5.34) with $\lambda_j^2$ replacing $m^2$. Solutions of (8.5.76) which are equatorially trapped are therefore

$$V_j(y) = \psi_j(\lambda_j^{1/2}y), \qquad (8.5.77)$$

where $\lambda_j$ must satisfy the eigenvalue relation (8.5.41):

$$\sigma^2\lambda_j^2 - 2(j + \tfrac{1}{2})\lambda_j - \frac{k}{\sigma} - k^2 = 0, \qquad (8.5.78a)$$

or

$$\lambda_j = \lambda_{j\pm} = \frac{j+\frac{1}{2}}{\sigma^2}\left[1 \pm \left\{1 + \frac{k\sigma(1 + k\sigma)}{(j+\frac{1}{2})^2}\right\}^{1/2}\right]. \tag{8.5.78b}$$

For free oscillations $m^2$ in (8.5.36) is determined as an eigenvalue of the homogeneous vertical-structure equation, and $\sigma^2$ is one of the infinite set of eigenvalues of (8.5.36) for each corresponding $m_i$. For oscillations at a fixed frequency, $\lambda_j$ is an eigenvalue of (8.5.76), which is identical to (8.5.36) but now for a *fixed* $\sigma$. The orthogonality relation for the $\psi_j$'s may be immediately derived from (8.5.76) as

$$\int_{-\infty}^{\infty} (\sigma^2 - y^2)V_j(\lambda_j y)V_l(\lambda_l y)\, dy = 0, \qquad \lambda_l \neq \lambda_j, \tag{8.5.79}$$

and it is important to note that (8.5.79) applies for the two differing $\lambda_j$'s corresponding to the same index $j$ as given by (8.5.78b). The $V_j(y)$ form a complete set in the interval $(-\infty, \infty)$ for those functions which go to zero (i.e., trapped) for large $|y|$. Considering only such motions and such forcing, $V$ may be written

$$V = \sum_{\lambda_j} A_j(z)V_j(y). \tag{8.5.80}$$

If (8.5.80) is inserted in (8.5.74) and the orthogonality relationship (8.5.79) is used, we obtain

$$\mathcal{D}^2 B_j + \lambda_j^2 B_j = M_j(z), \tag{8.5.81}$$

where

$$B_j = \mathcal{D}^2 A_j,$$
$$M_j = \frac{\int_{-\infty}^{\infty} V_j(y)J_0(y, z)\, dy}{\int_{-\infty}^{\infty} (\sigma^2 - y^2)V_j^2\, dy}, \tag{8.5.82}$$

which may be solved for $B_j$ and then $A_j$. The effect of the forcing now appears only in the vertical-structure equation (8.5.81), which is now inhomogeneous. It is clear from the form of (8.5.81) that for each lateral mode the vertical scale of the forced response is given by $\lambda_j^{-1}$ in regions external to the strata where $M_j(z)$ is nonzero. That is, except in the vicinity of the forcing, the vertical scale of the response for each $j$ is a function of $\sigma$ and $k$ through (8.5.78b)—i.e., the vertical scale, rather than being intrinsic to the fluid, is a sensitive function of the forcing parameters. In *dimensional units* the vertical scale is given by

$$\lambda_{j*}^{-1} = \frac{D}{\lambda_j}; \tag{8.5.83}$$

restoring dimensional units with the aid of the fundamental scales of (8.5.6) and (8.5.14), we have

$$\lambda_{j*} = (j+\tfrac{1}{2})\frac{N_0\beta_0}{\sigma_*^2}\left[1 \pm \left\{1 + \frac{(k_*\sigma_*/\beta_0)(1 + k_*\sigma_*/\beta_0)}{(j+\frac{1}{2})^2}\right\}^{1/2}\right]. \tag{8.5.84}$$

It is evident from (8.5.84) that all the $\lambda_j$ are real. However, in order that the $V_j$ may go to zero for large $|y|$ it is also necessary that we accept as permissible only those roots whose $\lambda_j$ are *positive*. If $k_* \sigma_* > 0$, corresponding to eastward phase speed for the forcing, the negative root in (8.5.78b) will yield a $\lambda_j < 0$ and must be rejected. On the other hand, if $-1 < k_* \sigma_* / \beta_0 < 0$, both roots for $\lambda_j$ are positive. This distinction corresponds to the existence of the free Rossby normal mode in the restricted region $k_* \sigma_* < 0$ and inside the boundary delineated by the Yanai mode. Indeed, the negative root corresponds to a forced Rossby wave, while the positive root corresponds to a forced inertia–gravity wave. For very low frequencies both roots for $\lambda_j$ become large, corresponding to waves with short vertical scales and severe meridional trapping. Wunsch (1977), for example, has suggested that the observations of a complex multicellular structure found by Luyten and Swallow (1976) in the current structure of the Indian Ocean is a manifestation of this very phenomenon as a response to periodic forcing by the monsoon winds with an annual period. In the atmosphere, the forced Yanai and the Kelvin waves have both been inferred from observation (Holton 1975), although the energy source for the waves still remains unclear.

# Selected Bibliography

## Section 1.2

Defant, Albert, 1961. *Physical Oceanography*, Vol. 1. Pergamon Press, 728 pp.

Fuglister, F. C. 1963. *Gulf Stream '60.* Progress in Oceanography I, Pergamon Press, 265–383.

Kochanski, A. 1955. Cross sections of the mean zonal flow and temperature along 80° W. *J. Meteorol.* **12**, 95–106.

Lorenz, E. N. 1967. *The Nature and Theory of the General Circulation of the Atmosphere.* World Meteorological Organization, #218, Geneva, Switzerland.

Palmén, E. and Newton, C. W. 1969. *Atmospheric Circulation Systems.* Academic Press, 603 pp.

Pickard George L. 1975. *Descriptive Physical Oceanography.* Pergamon Press, 214 pp.

Sverdrup, H. U., Johnson, M. W., and Fleming, R. H. 1942. *The Oceans.* Prentice-Hall, 1087 pp.

## Section 1.3

*U.S. Standard Atmosphere.* 1962. N.A.S.A., U.S. Government Printing Office, Washington, D.C.

## Section 1.4

Batchelor, G. K. 1967. *An Introduction to Fluid Dynamics*. Cambridge University Press, 615 pp. Chapters 1, 2, 3.

Bryan, Kirk and Cox, Michael D. 1972. An approximate equation of state for numerical models of ocean circulation. *J. Physical Ocean.* **2**, 510–514.

Holton, J. R. 1972. *An Introduction to Dynamic Meteorology*. Academic Press, 319 pp.


## Section 2.5

Ertel, H. 1972. Ein neuer hydrodynamischer Wirbesatz. *Meteorolol. Z.* **59**, 277–281.


## Section 2.7

Taylor, G. I. 1923. Experiments on the motion of solid bodies in rotating fluids. *Proc. Roy. Soc. A* **104**, 213–218.


## Section 3.1

Rossby, C. G., et al. 1939. Relation between variations in the intensity of the zonal circulation of the atmosphere and the displacements of the semi-permanent centers of action. *J. Marine Res.* **2**, 38–55.

Stommel. H. 1948. The westward intensification of wind-driven ocean currents. *Trans. Amer. Geoph. Union* **29**, 202–206.


## Section 3.22

Longuet-Higgins, M. S. 1964. On group velocity and energy flux in planetary wave motions. *Deep-Sea Research* **11**, 35–42.


## Section 3.24

Jeffreys, H. A. and Jeffreys, B. S. 1962. *Methods of Mathematical Physics*. Cambridge University Press.

Rossby, C. G. 1945. On the propagation of frequencies and energy in certain types of oceanic and atmospheric waves. *J. Meteor.* **2**, 187–204.

# Section 3.25

Flierl, G. R. 1977. Simple applications of McWilliams's "A note on a consistent quasi-geostrophic model in a multiply connected domain." *Dynamics of Atmospheres and Oceans* 1, 443–454.

Greenspan, H. P. 1968. *The Theory of Rotating Fluids.* Cambridge University Press, 327 pp.

Longuet-Higgins, M. S. 1964. Planetary waves on a rotating sphere. *Proc. Royal Soc. A* 279, 446–473.

Pedlosky, J. 1965. A study of the time dependent ocean circulation. *J. Atmos. Sci.* 22, 267–272.

# Section 3.26

Gill, A. E. 1974. The stability of planetary waves on an infinite beta-plane. *Geophysical Fluid Dynamics* 6, 29–47.

Longuet-Higgins, M. S. and Gill, A. E. 1967. Resonant interactions between planetary waves. *Proc. Roy. Soc. A* 299, 120–140.

Lorenz, E. N. 1972. Barotropic instability of Rossby wave motion. *J. Atmos. Sci.* 29, 258–269.

# Section 3.27

Fjortoft, R. 1953. On the changes in the spectral distribution of kinetic energy for a two-dimensional, non-divergent flow. *Tellus* 5, 225–237.

# Section 3.28

Batchelor, G. K. 1953. *Theory of Homogeneous Turbulence.* Cambridge University Press, 197 pp.

Batchelor, G. K. 1969. Computation of the energy spectrum in homogeneous two-dimensional turbulence. *The Physics of Fluids*, Supplement II, pp. 233–239.

Charney, J. G. 1971. Geostrophic turbulence. *J. Atmos. Sci.* 28, 1087–1095.

Rhines, P. B. 1975. Waves and turbulence on a beta-plane. *J. Fluid Mech.* 69, 417–443.

Rhines, P. B. 1977. The dynamics of unsteady ocean currents. In *The Sea*, Vol. VI, Wiley, pp. 189–318.

# Section 4.2

Schlichting, H. 1968. *Boundary Layer Theory.* McGraw-Hill, 745 pp., Chapter XIX.

Sutton, O. G. 1949. *Atmospheric Turbulence.* Methuen and Co., 107 pp.

Taylor, G. I. 1915. Eddy motion in the atmosphere. *Philosophical Trans. Royal Soc. A* CCXV, 1–26.

## Section 4.3

Ekman, V. W. 1905. On the influence of the earth's rotation on ocean currents. *Arkiv. Matem., Astr. Fysik*, Stockholm **2** (11).

## Section 4.4

Cole, J. D. 1968. *Perturbation Methods In Applied Mathematics*. Blaisdell, 260 pp.

Van Dyke, M. 1964. *Perturbation Methods in Fluid Mechanics*. Academic Press, 229 pp.

## Section 4.5

Cole, J. D. 1968. *Perturbation Methods In Applied Mathematics*. Blaisdell, 260 pp.

## Section 4.7

Greenspan, H. P. and Howard, L. N. 1963. On a time dependent motion of a rotating fluid. *J. Fluid Mech.* **22**, 449–462.

## Section 4.11

Charney, J. G. and Eliassen, A. 1949. A numerical method for predicting the perturbations of the middle-latitude westerlies. *Tellus* **1**, 38–54.

## Section 4.13

Stewartson, K. 1957. On almost rigid rotations. *J. Fluid Mech.* **3**, 17–26.

## Section 4.14

Basdevant, C., Legras, B., and Sadourny, R. 1981. A study of barotropic model flows: Intermittency, waves, and predictability. *J. Atmos. Sci.* **38**, 2305–2326.

Batchelor, G. K. 1969. Computation of the energy spectrum in homogeneous two-dimensional turbulence. *The Physics of Fluids*, Supplement II: pp. 233–239.

Kraichnan, R. H. 1967. Inertial ranges in two-dimensional turbulence. *The Physics of Fluids*, Supplement II: pp. 233–239.

Kraichnan, R. H. 1971. Inertial-range transfer in two- and three-dimensional turbulence. *J. Fluid Mech.* **47**, 525–535.

Lilly, D. K. 1969. Number simulation of two-dimensional turbulence. *The Physics of Fluids*, Supplement II: pp. 240–249.

# Section 5.1

Defant, Albert. 1961. *Physical Oceanography*, Vol. 1. Pergamon Press, 728 p.

Stommel, H. 1960. *The Gulf Stream*, University of California Press.

Stommel, H. and Yoshida, K. 1972. *Kuroshio: Its Physical Aspects*. University of Tokyo Press.


# Section 5.3

Leetmaa, A., Niiler, P., and Stommel, H. 1977. Does the Sverdrup relation account for the mid-Atlantic circulation? *J. Marine Res.* **35**, 1–10.

Sverdrup, H. U. 1947. Wind-driven currents in a baroclinic ocean; with application to the equatorial currents of the eastern Pacific. *Proc. Nat. Acad. Sci.* **33**, 318–326.

Welander, P. 1959. On the vertically integrated mass transport in the oceans. In *The Atmosphere and the Sea in Motion*. Ed., B. Bolin. Rockefeller Institute Press. 75–101.


# Section 5.4

Munk, W. H. 1950. On the wind-driven ocean circulation. *J. Meteor.* **7**, 79–93.

Munk, W. H. and Carrier, G. F. 1950. The wind-driven circulation in ocean basins of various shapes. *Tellus* **2**, 158–167.

Pedlosky, J. and Greenspan, H. P. 1967. A simple laboratory model for the oceanic circulation. *J. Fluid Mech.* **27**, 291–304.


# Section 5.5

Stommel, H. 1948. The westward intensification of wind-driven ocean currents. *Trans. Amer. Geophys. Union* **99**, 202–206.


# Section 5.6

Charney, J. G. 1955. The Gulf Stream as an inertial boundary layer. *Proc. Nat. Acad. Sci.* **41**, 731–740.

Greenspan, H. P. 1962. A criterion for the existence of inertial boundary layers in oceanic circulation. *Proc. Nat. Acad. Sci.* **48**, 2034–2039.


# Section 5.7

Moore, D. W. 1963. Rossby waves in ocean circulation. *Deep-Sea Res.* **10**, 735–748.

## Section 5.8

Pedlosky, J. 1965. A note on the western intensification of the oceanic circulation. *J. Marine Res.* **23**, 207–209.

## Section 5.10

Fofonoff, N. P. 1954. Steady flow in a frictionless homogeneous ocean. *J. Marine Res.* **13**, 254–262.

## Section 5.11

Beardsley, R. C. and Robbins, K. 1975. The "sliced cylinder" laboratory model of the wind-driven ocean circulation. Part 1. Steady forcing and topographic Rossby wave instability. *J. Fluid. Mech.* **69**, 27–40.

Bryan, K. 1963. A numerical investigation of a non-linear model of a wind-driven ocean. *J. Atmos. Sci.* **20**, 594–606.

Veronis, G. 1966. Wind-driven ocean circulation—Part 2. Numerical solutions of the non-linear problem. *Deep-Sea Res.* **13**, 31–55.

## Section 5.12

Pedlosky, J. 1968. An overlooked aspect of the wind-driven oceanic circulation. *J. Fluid Mech.* **32**, 809–821.

## Section 5.13

Schulman, E. E. 1975. A study of topographic effects. In *Numerical Models of Ocean Circulation*. Nat. Acad. Sci. 147–165.

## Section 6.2

Batchelor, G. K. 1967. An introduction to fluid dynamics, Cambridge University Press (Appendix 2).

Phillips, N. A. 1963. Geostrophic motion. *Reviews of Geophysics* **1**, 123–176.

## Section 6.3

Burger, A. 1958. Scale considerations of planetary motions of the atmosphere. *Tellus* **10**, 195–205.

Charney, J. G. 1947. *On the Scale of Atmospheric Motions.* Geofys. Publikasjoner, Norske Videnskaps-Akad Oslo 17.

Charney, J. G. and Drazin, P. G. 1961. Propagation of planetary scale disturbances from the lower into the upper atmosphere. *J. Geophys. Res.* **66**, 83–109.


## Section 6.10

Lorenz, E. 1955. Available potential energy and the maintenance of the general circulation. *Tellus* **7**, 157–167.


## Section 6.12

Chapman, S. and Lindzen, R. S. 1970. *Atmospheric Tides.* Gordon and Breach, 200 pp. Chapter 3.

Kundu, P. K., Allen, J. S., and Smith, R. L. 1975. Modal decomposition of the velocity field near the Oregon coast. *J. Phys. Oceanog.* **5**, 683–704.


## Section 6.13

Holton, J. R. 1975. The dynamic meteorology of the stratosphere and mesosphere. *Amer. Meteor. Soc.*, 216 pp.

Smagorinsky, J. 1953. The dynamical influences of large scale heat sources and sinks on the quasi-stationary mean motions of the atmosphere. *Quart. J. Roy. Meteor. Soc.* **79**, 342–366.


## Section 6.14

Andrews, D. G. and McIntyre, M. E. 1976. Planetary waves in horizontal and vertical shear: The generalized Eliassen–Palm relation and the mean zonal acceleration. *J. Atmos. Sci.* **33**, 2031–2048.

Benney, D. J. and Bergeron, R. F. 1969. A new class of nonlinear waves in parallel flows. *Studies Appl. Math.* **48**, 181–204.

Charney, J. G. and Drazin, P. G. 1961. Propagation of planetary scale disturbances from the lower into the upper atmosphere. *J. Geophys. Res.* **66**, 83–109.

Edmond, H. J. Jr., Hoskins, B. J., and McIntyre, M. E. 1980. Eliassen–Palm cross sections for the troposphere. *J. Atmos. Sci.* **37**, 2600–2616.

Eliassen, A. and Palm, E. 1961. On the transfer of energy in stationary mountain waves. *Geofys. Publ.* **22**, 1–23.

Smagorinsky, J. 1953. The dynamical influences of large scale heat sources and sinks on the quasi-stationary mean motions of the atmosphere. *Quart. J. Roy. Meteor. Soc.* **79**, 342–366.

## Section 6.15

Rhines, P. 1970. Edge-, bottom-, and Rossby waves in a rotating stratified fluid. *Geophys. Fluid Dyn.* **1**, 273–302.

## Section 6.16

Phillips, N. A. 1951. A simple three-dimensional model for the study of large-scale extratropical flow patterns. *J. Meteor.* **8**, 381–394.

## Section 6.19

Sverdrup, H. U. 1947. Wind-driven currents in a baroclinic ocean; with application to the equatorial currents of the eastern Pacific. *Proc. Nat. Acad. Sci.* **33**, 318–326.

## Section 6.20

Needler, G. T. 1985. The absolute velocity as a function of conserved measurable quantities. *Prog. Oceanog.* **14**, 421–429.

## Section 6.21

Blandford, R. 1965. Notes on the theory of the thermocline. *J. Mar. Res.* **23**, 18–29.

Bryan, K. and Cox, M. D. 1968. A non-linear model of an ocean driven by wind and differential heating. Parts I and II. *J. Atmos. Sci.* **25**, 945–978.

Carslaw, H. S. and Jaeger, J. C. 1959. *Conduction of Heat in Solids*, Oxford Press, 510 pp. 388.

Needler, G. T. 1967. A model for thermohaline circulation in an ocean of finite depth. *J. Marine Res.* **25**, 329–342.

Robinson, A. R. and Stommel, H. 1959. The oceanic thermocline and the associated thermohaline circulation. *Tellus* **11**, 295–308.

Robinson, A. R. and Welander, P. 1963. Thermal circulation on a rotating sphere; with application to the oceanic thermocline. *J. Marine Res.* **21**, 25–38.

Welander, P. 1971a. Some exact solutions to the equations describing an ideal-fluid thermocline. *J. Mar. Res.* **29**, 60–68.

Welander, P. 1971b. The thermocline problem. *Philos. Trans. Royal Soc. Lond.* A **270**, 69–73.

## Section 6.22

Luyten, J. R., Pedlosky, J., and Stommel, H. 1983. The ventilated thermocline. *J. Phys. Oceanogr.* **13**, 292–309.

Pedlosky, J., 1983. Eastern boundary ventilation and the structure of the thermocline. *J. Phys. Oceanogr.* **13**, 2038–2044.

Rhines, P. B. and Young, W. R. 1982. Homogenization of potential vorticity in planetary gyres. *J. Fluid Mech.* **122**, 347–367.


## Section 6.23

Pedlosky, J. and Young, W. R. 1983. Ventilation, potential vorticity homogenization and the structure of the ocean circulation. *J. Phys. Oceanogr.* **13**, 2020–2037.

Rhines, P. B. and Young, W. R. 1982a. A theory of the wind-driven circulation. I. Mid-ocean gyres. *J. Mar. Res.* **40** (Suppl.), 559–596.

Rhines, P. B. and Young, W. R. 1982b. Homogenization of potential vorticity in planetary gyres. *J. Fluid Mech.* **122**, 347–367.

Young, W. R. and Rhines, P. B. 1982. A theory of the wind-driven circulation. II. Circulation models with western boundary layers. *J. Mar. Res.* **40** (Suppl.), 849–972.


## Section 6.24

Pedlosky, J. 1984. The equations for geostrophic motion in the ocean. *J. Phys. Oceanogr.* **14**, 448–456.


## Section 7.1

Charney, J. G. 1947. The dynamics of long waves in a baroclinic westerly current. *J. Meteor.* **4**, 135–163.

Eady, E. T. 1949. Long waves and cyclone waves. *Tellus* **1**, 33–52.


## Section 7.3

Charney, J. G. and Stern, M. 1962. On the stability of internal baroclinic jets in a rotating atmosphere. *J. Atmos. Sci.* **19**, 159–172.


## Section 7.4

Charney, J. G. and Pedlosky, J. 1963. On the trapping of unstable planetary waves in the atmosphere. *J. Geophys. Res.* **68**, 6441–6442.

Pedlosky, J. 1964. The stability of currents in the atmosphere and the oceans. Part I. *J. Atmos. Sci.* **27**, 201–219.


## Section 7.5

Howard, L. N. 1961. Note on a paper of John Miles. *J. Fluid Mech.* **10**, 509–512.

Pedlosky, J. 1964. The stability of currents in the atmosphere and the oceans. Part I. *J. Atmos. Sci.* **21**, 201–219.


## Section 7.6

Charney, J. G. 1947. The dynamics of long waves in a baroclinic westerly current. *J. Meteor.* **4**, 135–163.

Eady, E. T. 1949. Long waves and cyclone waves. *Tellus* **1**, 33–52.

Pedlosky, J. 1971. Geophysical fluid dynamics. In *Mathematical Problems in the Geophysical Sciences*. Ed., W. H. Reid. Amer. Math. Soc. 1–60.


## Section 7.7

Eady, E. T. 1949. Long waves and cyclone waves. *Tellus* **1**, 33–52.

Pedlosky, J. 1964. An initial value problem in the theory of baroclinic instability. *Tellus* **XVI**, 12–17.


## Section 7.8

Abramowitz, M. and Stegun, I. A. 1964. *Handbook of Mathematical Functions*. National Bureau of Standards. Chapter 13.

Branscome, L. E. 1983. The Charney stability problem: approximate solutions and modal structures. *J. Atmos. Sci.* **40**, 1393–1409.

Bretherton, F. P. 1966. Critical layer instability in baroclinic flows. *Quart. J. Roy. Meteor. Soc.* **92**, 325–334.

Burger, A. P. 1962. On the non-existence of critical wave lengths in a continuous baroclinic stability problem. *J. Atmos. Sci.* **19**, 31–38.

Charney, J. G. 1947. The dynamics of long waves in a baroclinic westerly current. *J. Meteor.* **4**, 135–163.

Garcia, R. V. and Norscini, R. 1970. A contribution to the baroclinic instability problem. *Tellus* **22**, 239–250.

Gill, A. E., Green, J. S. A. and Simmons, A. J. 1974. Energy partition in the large-scale ocean circulation and the production of mid-ocean eddies. *Deep-Sea Res.* **21**, 497–528.

Green, J. S. A. 1960. A problem in baroclinic instability. *Quart. J. Roy. Meteor. Soc.* **86**, 237–251.

Held, I. M. 1978. The vertical scale of an unstable baroclinic wave and its importance for eddy heat flux parameterization. *J. Atmos. Sci.* **35**, 572–576.

Hildebrand, F. B. 1963. *Advanced Calculus for Applications.* Prentice-Hall, 646 pp. Chapter 4.

Kuo, H. L. 1952. Three dimensional disturbances in a baroclinic zonal current. *J. Meteor.* **9**, 260–278.

Kuo, H. L. 1973. Dynamics of quasi-geostrophic flows and instability theory. In *Advances in Applied Mechanics* **13**. 247–330.

Lin, C. C. 1955. *The Theory of Hydrodynamic Instability.* Cambridge Univ. Press, 155 pp. Chapter 8.

Miles, J. W. 1964a. A note on Charney's model of zonal-wind instability. *J. Atmos. Sci.* **21**, 451–452.

Miles, J. W. 1964b. Baroclinic instability of the zonal wind. *Rev. of Geophys.* **2**, 155–176.

Miles, J. W. 1964c. Baroclinic instability of the zonal wind. Parts I, II, *J. Atmos. Sci.* **21**, 550–556, 603–609.

Phillips, N. A. (1963) Geostrophic Motion. *Rev. of Geophysics* **1**, 123–176.

## Section 7.9

Held, I. M. 1975. Momentum transport by quasi-geostrophic eddies. *J. Atmos. Sci.* **32**, 1494–1497.

Phillips, N. A. 1954. Energy transformations and meridional circulations associated with simple baroclinic waves in a two-level, quasi-geostrophic model. *Tellus* **6**, 273–286.

## Section 7.10

Pedlosky, J. 1963. Baroclinic instability in two-layer systems. *Tellus* **15**, 20–25.

Pedlosky, J. 1964. The stability of currents in the atmosphere and the oceans. Part I. *J. Atmos. Sci.* **21**, 201–219.

## Section 7.11

Phillips, N. A. 1954. Energy transformations and meridional circulations associated with simple baroclinic waves in a two-level, quasi-geostrophic model. *Tellus* **6**, 273–286.

## Section 7.12

Barcilon, V. 1964. Role of Ekman layers in the stability of the symmetric regime in a rotating annulus. *J. Atmos. Sci.* **21**, 291–299.

## Section 7.13

Gill, A. E., Green, J. S. A., and Simmons, A. J. 1974. Energy partition in the large-scale ocean circulation and the production of mid-ocean eddies. *Deep-Sea Res.* **21**, 497–528.

Robinson, A. R. and McWilliams, J. C. 1974. The baroclinic instability of the open ocean. *J. Phys. Oceanog.* **4**, 281–294.

## Section 7.14

Dickinson, R. E. and Clare, F. J. 1973. Numerical study of the unstable modes of a hyperbolic-tangent barotropic shear flow. *J. Atmos. Sci.* **30**, 1035–1049.

Howard, L. N. and Drazin, P. G. 1964. On instability of parallel flow of inviscid fluid in a rotating system with variable Coriolis parameter. *J. Math. Phys.* **43**, 83–99.

Kuo, H. L. 1949. Dynamic instability of two-dimensional nondivergent flow in a barotropic atmosphere. *J. Meteor.* **6**, 105–122.

Kuo, H. L. 1973. Dynamics of quasi-geostrophic flows and instability theory. In *Advances in Applied Mechanics* **13**. 247–330.

## Section 7.15

Brown. J. A., Jr. 1969. A numerical investigation of hydrodynamic instability and energy conversions in the quasigeostrophic atmosphere. Parts I, II. *J. Atmos. Sci.* **26**, 352–365, 366–375.

Charney, J. G. 1951. On baroclinic instability and the maintenance of the kinetic energy of the westerlies. In *Proc. 9th Gen. Assembly, UGGI (Assoc. Meteor.) Brussels.* 47–63.

Green, J. S. A. 1970. Transfer properties of the large-scale eddies and the general circulation of the atmosphere. *Quart. J. Roy. Meteor. Soc.* **96**, 157–185.

Jeffreys, H. 1933. The function of cyclones in the general circulation. In *Procès-Verbaux de l'Association de Météorologie, UGGI (Lisbon)*, Part II. 219–230. Reprinted in *Theory of Thermal Convection.* Ed., B. Saltzman. Dover, 1962.

Lorenz, E. N. 1967. *The Nature and Theory of the General Circulation of the Atmosphere.* World Meteorological Organization #218, Geneva, Switzerland.

Pedlosky, J. 1964. The stability of currents in the atmosphere and the ocean: Part II. *J. Atmos. Sci.* **21**, 342–353.

Schmitz, W. J., Jr. 1977. On the circulation in the western North Atlantic. *J. Marine Res.* **35**, 21–28.

Starr, V. P. 1953. Note concerning the nature of the large scale eddies in the atmosphere. *Tellus* **5**, 494–498.

## Section 7.16

Drazin, P. G. 1970. Non-linear baroclinic instability of a continuous zonal flow. *Quart. J. Roy. Meteor. Soc.* **96**, 667–676.

Hart, J. E. 1973. On the behavior of large-amplitude baroclinic waves. *J. Atmos. Sci.* **30**, 1017–1034.

Hart, J. E. 1981. Wave number selection in nonlinear baroclinic instability. *J. Atmos. Sci.* **38**, 400–408.

Lorenz, E. N. 1963a. The mechanics of vacillation. *J. Atmos. Sci.* **20**, 448–464.

Lorenz, E. N. 1963b. Deterministic nonperiod flow. *J. Atmos. Sci.* **12**, 130–141.

Pedlosky, J. 1970. Finite amplitude baroclinic waves. *J. Atmos. Sci.* **27**, 15–30.

Pedlosky, J. 1971. Finite amplitude baroclinic waves with small dissipation. *J. Atmos. Sci.* **28**, 587–597.

Pedlosky, J. 1972a. Limit cycles and unstable baroclinic waves, *J. Atmos. Sci.* **29**, 53–63.

Pedlosky, J. 1972b. Finite amplitude baroclinic wave packets. *J. Atmos. Sci.* **29**, 680–686.

Pedlosky, J. 1981. The nonlinear dynamics of baroclinic wave ensembles. *J. Fluid Mech.* **102**, 169–209.

Pedlosky, J. and Frenzen, C. 1980. Chaotic and period behavior of finite-amplitude baroclinic waves. *J. Atmos. Sci.* **37**, 1177–1196.

Phillips, N. A. 1954. Energy transformations and meridional circulations associated with simple baroclinic waves in a two-level, quasi-geostrophic model. *Tellus* **6**, 273–286.

Smith, R. K. and Reilly, J. M. 1977. On a theory of Amplitude Vacillation in Baroclinic Waves: Some Numerical Solutions. *J. Atmos. Sci.* **34**, 1256–1260.

## Section 7.17

Arnold, V. I. 1965. Conditions for nonlinear stability of stationary plane curvilinear flows of an ideal fluid. *Dokl. Akad. Nauk SSSR* **162**, 975–978.

Blumen, W. 1968. On the stability of quasigeostrophic flow. *J. Atoms. Sci.* **25**, 929–931.

## Section 8.1

Hoskins, B. J. 1975. The geostrophic momentum approximation and the semi-geostrophic approximation. *J. Atmos. Sci.* **32**, 233–242.

## Section 8.2

Cutchin, D. L. and Smith, R. L. 1973. Continental shelf waves: low frequency variations in sea-level and currents over the Oregon continental shelf. *J. Phys. Oceanog.* **3**, 73–82.

Leblond, P. H. and Mysak, L. A. 1977. Trapped coastal waves and their role in shelf dynamics. In *The Sea*. Wiley-Interscience. Vol. 6, Chapter 10, 459–495.

Mysak, L. A. 1967. On the theory of continental shelf waves. *J. Mar. Res.* **25**, 205–227.

Robinson, A. R. 1964. Continental shelf waves and the response of sea level to weather systems. *J. Geophys. Res.* **69**, 367–368.


## Section 8.3

Allen, J. S. 1973. Upwelling and coastal jets in a continuously stratified ocean. *J. Phys. Oceanog.* **3**, 245–257.

Barcilon, V. and Pedlosky, J. 1966. Linear theory of rotating stratified fluid motions. *J. Fluid Mech.* **29**, 1–16.

Barcilon, V. and Pedlosky, J. 1966. A unified linear theory of homogeneous and stratified rotating fluids. *J. Fluid Mech.* **9**, 609–621.

Pedlosky, J. 1974. On coastal jets and upwelling in bounded basins. *J. Phys. Oceanog.* **4**, 3–18.


## Section 8.4

Bergeron, T. 1928. Über die dreidimensional verknüpfend Wetteranalyse I. *Geofys. Publikasjoner* **5**, 1–11.

Hoskins, B. J. 1971. Atmospheric frontogenesis models: some solutions. *Quart. J. Roy. Meteor. Soc.* **97**, 139–153.

Hoskins, B. J. and Bretherton, F. P. 1972. Atmospheric frontogenesis models: Mathematical formulation and solution. *J. Atmos. Sci.* **29**, 11–37.

Petterssen, S. 1956. *Weather Analysis and Forecasting. Vol. 1. Motion and Motion Systems.* McGraw Hill, 422 pp. Chapter 11.

Sommerfeld, A. 1949. *Partial Differential Equations in Physics.* Academic Press, 329 pp.

Stone, P. H. 1966. Frontogenesis by horizontal wind deformation fields. *J. Atmos. Sci.* **23**, 455–465.

Voorhis, A. D. 1969. The horizontal extent and persistence of thermal fronts in the Sargasso Sea. *Deep-Sea Res.* **16** (supplement), 331–337.

Williams, R. T. 1967. Atmospheric frontogenesis: a numerical experiment. *J. Atmos. Sci.* **24**, 627–641.


## Section 8.5

Cane, M. A. and Sarachik, E. S. 1976. Forced baroclinic ocean motions: I. The linear equatorial unbounded case. *J. Mar. Res.* **34**, 629–665.

Cane, M. A. and Sarachik, E. S. 1977. Forced baroclinic ocean motions: II. The linear equatorial bounded case. *J. Mar. Res.* **35**, 395–432.

Holton, J. R. 1975. *The Dynamic Meteorology of the Stratosphere and Mesosphere.* Amer. Meteor. Soc., 216 pp.

Luyten, J. R. and Swallow, J. C. 1976. Equatorial undercurrents. *Deep-Sea Res.* **23**, 1005–1007.

Moore, D. W. 1968. Planetary gravity waves in an equatorial ocean. Ph.D. Thesis. Harvard University, Cambridge, Massachusetts.

Moore, D. W. and Philander, S. G. H. 1977. Modeling of the tropical oceanic circulation. In *The Sea.* Eds., E. D. Goldberg et al. Wiley-Interscience. Vol. 6, Chapter 8.

Wallace, J. M. 1971. General circulation of the tropical lower stratosphere. *Rev. Geophys. Space Phys.* **11**, 191–222.

Wunsch, C. 1977. Response of an equatorial ocean to a periodic monsoon. *J. Phys. Oceanog.* **7**, 497–511.

# Index

9 780387 963877

9 780387 963877